LONDON MATHEMATICAL SOCIETY LECTURE NOTE SERIES

Managing Editor: Professor J.W.S. Cassels, Department of Pure Mathematics and Mathematical Statistics, University of Cambridge, 16 Mill Lane, Cambridge CB2 1SB, England

The titles below are available from booksellers, or, in case of difficulty, from Cambridge University Press.

46	*p*-adic Analysis: a short course on recent work, N. KOBLITZ
59	Applicable differential geometry, M. CRAMPIN & F.A.E. PIRANI
66	Several complex variables and complex manifolds II, M.J. FIELD
86	Topological topics, I.M. JAMES (ed)
87	Surveys in set theory, A.R.D. MATHIAS (ed)
88	FPF ring theory, C. FAITH & S. PAGE
89	An F-space sampler, N.J. KALTON, N.T. PECK & J.W. ROBERTS
90	Polytopes and symmetry, S.A. ROBERTSON
92	Representation of rings over skew fields, A.H. SCHOFIELD
93	Aspects of topology, I.M. JAMES & E.H. KRONHEIMER (eds)
96	Diophantine equations over function fields, R.C. MASON
97	Varieties of constructive mathematics, D.S. BRIDGES & F. RICHMAN
98	Localization in Noetherian rings, A.V. JATEGAONKAR
99	Methods of differential geometry in algebraic topology, M. KAROUBI & C. LERUSTE
100	Stopping time techniques for analysts and probabilists, L. EGGHE
104	Elliptic structures on 3-manifolds, C.B. THOMAS
105	A local spectral theory for closed operators, I. ERDELYI & WANG SHENGWANG
107	Compactification of Siegel moduli schemes, C.-L. CHAI
109	Diophantine analysis, J. LOXTON & A. VAN DER POORTEN (eds)
113	Lectures on the asymptotic theory of ideals, D. REES
114	Lectures on Bochner-Riesz means, K.M. DAVIS & Y.-C. CHANG
116	Representations of algebras, P.J. WEBB (ed)
119	Triangulated categories in the representation theory of finite-dimensional algebras, D. HAPPEL
121	Proceedings of *Groups - St Andrews 1985*, E. ROBERTSON & C. CAMPBELL (eds)
128	Descriptive set theory and the structure of sets of uniqueness, A.S. KECHRIS & A. LOUVEAU
130	Model theory and modules, M. PREST
131	Algebraic, extremal & metric combinatorics, M.-M. DEZA, P. FRANKL & I.G. ROSENBERG (eds)
132	Whitehead groups of finite groups, ROBERT OLIVER
133	Linear algebraic monoids, MOHAN S. PUTCHA
134	Number theory and dynamical systems, M. DODSON & J. VICKERS (eds)
137	Analysis at Urbana, I, E. BERKSON, T. PECK, & J. UHL (eds)
138	Analysis at Urbana, II, E. BERKSON, T. PECK, & J. UHL (eds)
139	Advances in homotopy theory, S. SALAMON, B. STEER & W. SUTHERLAND (eds)
140	Geometric aspects of Banach spaces, E.M. PEINADOR & A. RODES (eds)
141	Surveys in combinatorics 1989, J. SIEMONS (ed)
144	Introduction to uniform spaces, I.M. JAMES
146	Cohen-Macaulay modules over Cohen-Macaulay rings, Y. YOSHINO
148	Helices and vector bundles, A.N. RUDAKOV *et al*
149	Solitons, nonlinear evolution equations and inverse scattering, M. ABLOWITZ & P. CLARKSON
150	Geometry of low-dimensional manifolds 1, S. DONALDSON & C.B. THOMAS (eds)
151	Geometry of low-dimensional manifolds 2, S. DONALDSON & C.B. THOMAS (eds)
152	Oligomorphic permutation groups, P. CAMERON
153	L-functions and arithmetic, J. COATES & M.J. TAYLOR (eds)
155	Classification theories of polarized varieties, TAKAO FUJITA
156	Twistors in mathematics and physics, T.N. BAILEY & R.J. BASTON (eds)
158	Geometry of Banach spaces, P.F.X. MÜLLER & W. SCHACHERMAYER (eds)
159	Groups St Andrews 1989 volume 1, C.M. CAMPBELL & E.F. ROBERTSON (eds)
160	Groups St Andrews 1989 volume 2, C.M. CAMPBELL & E.F. ROBERTSON (eds)
161	Lectures on block theory, BURKHARD KÜLSHAMMER
162	Harmonic analysis and representation theory, A. FIGA-TALAMANCA & C. NEBBIA
163	Topics in varieties of group representations, S.M. VOVSI
164	Quasi-symmetric designs, M.S. SHRIKANDE & S.S. SANE
166	Surveys in combinatorics, 1991, A.D. KEEDWELL (ed)
168	Representations of algebras, H. TACHIKAWA & S. BRENNER (eds)
169	Boolean function complexity, M.S. PATERSON (ed)
170	Manifolds with singularities and the Adams-Novikov spectral sequence, B. BOTVINNIK
171	Squares, A.R. RAJWADE
172	Algebraic varieties, GEORGE R. KEMPF
173	Discrete groups and geometry, W.J. HARVEY & C. MACLACHLAN (eds)
174	Lectures on mechanics, J.E. MARSDEN
175	Adams memorial symposium on algebraic topology 1, N. RAY & G. WALKER (eds)
176	Adams memorial symposium on algebraic topology 2, N. RAY & G. WALKER (eds)
177	Applications of categories in computer science, M. FOURMAN, P. JOHNSTONE & A. PITTS (eds)
178	Lower K- and L-theory, A. RANICKI
179	Complex projective geometry, G. ELLINGSRUD *et al*
180	Lectures on ergodic theory and Pesin theory on compact manifolds, M. POLLICOTT
181	Geometric group theory I, G.A. NIBLO & M.A. ROLLER (eds)

London Mathematical Society Lecture Note Series. 254

Galois Representations in Arithmetic Algebraic Geometry

Edited by

A. J. Scholl
University of Durham

R. L. Taylor
Harvard University

CAMBRIDGE
UNIVERSITY PRESS

CAMBRIDGE UNIVERSITY PRESS
Cambridge, New York, Melbourne, Madrid, Cape Town, Singapore, São Paulo

Cambridge University Press
The Edinburgh Building, Cambridge CB2 8RU, UK

Published in the United States of America by Cambridge University Press, New York

www.cambridge.org
Information on this title: www.cambridge.org/9780521644198

© Cambridge University Press 1998

First published 1998

A catalogue record for this publication is available from the British Library

ISBN 978-0-521-64419-8 paperback

Transferred to digital printing 2007

CONTENTS

PREFACE

This volume grew out of the London Mathematical Society symposium on "Galois representations in arithmetic algebraic geometry" held in Durham from the 9th to the 18th of July 1996. We understood our title rather loosely and the symposium considered many recent developments on the interface between algebraic number theory and arithmetic algebraic geometry. There were six expository courses on

1. Galois module structure
2. Shimura varieties in mixed characteristic
3. p-adic comparison theorems
4. the work of Kato on the Birch-Swinnerton-Dyer conjecture
5. polylogarithms
6. rigid analysis and modular forms

We are very grateful to the organisers of each of theses courses (Chinburg, Oort, Fontaine, Kato, Goncharov and Coleman) as well as all the other lecturers who worked hard to make these courses highly successful. In addition to the short courses there were 14 research seminars. We would also like to thank these lecturers and particularly those who have contributed to this volume. The symposium received generous financial support from the EPSRC and from the EU (through the network on "automorphic forms and arithmetic algebraic geometry"). We were particularly pleased that this enabled a large number of young European researchers to attend. Finally we would like to thank Steve Wilson and the department of mathematics at Durham University for their help with the organisation of the meeting.

This volume contains both expository and research articles. We are particularly grateful to the authors (Erez, Mazur, Moonen and Schneider) who have put a lot of time into preparing what we feel will be a useful collection of expositions. (Schneider's introductory lectures on rigid analysis were particularly well received at the symposium and we hope that we did right in pressuring him to write them up despite their relatively elementary nature.) The rest of the articles are research papers mostly based either on one of the short courses or on one of the individual lectures at the symposium. We would like to thank all the authors for allowing us to publish their work here.

Tony Scholl
Richard Taylor

PARTICIPANTS

B. Agboola (Princeton)
F. Andreatta (Utrecht)
P. Balister (London)
P.R. Bending (Oxford)
P. Berthelot (Rennes)
A. Besser (UCLA)
B. Birch (Oxford)
D. Blasius (UCLA)
S. Bloch (Chicago)
G. Boeckle (Strasbourg)
C. Breuil (Ecole Polytechnique)
K. Buecker (Grenoble)
O. Bultel (Oxford)
D. Burns (KCL)
C. Bushnell (KCL)
K. Buzzard (Berkeley)
N. Byott (Exeter)
J.W.S. Cassels (Cambridge)
R. Chapman (Exeter)
I. Chen (Oxford)
F. Cherbonnier (Orsay)
T. Chinburg (Pennsylvania)
J. Coates (Cambridge)
R. Coleman (Berkeley)
P. Colmez (ENS Paris)
D. Delbourgo (Cambridge)
R. de Jeu (Durham)
J. de Jong (Harvard)
F. Diamond (Cambridge)
M.E.T. Dickinson (Oxford)
T. Dokshitzer (Utrecht)
B. Edixhoven (Rennes)
B. Erez (Bordeaux)
G. Faltings (Max-Planck Institute)
I. Fesenko (Nottingham)
L. Figueiredo (Rio de Janeiro)
M. Flach (Caltech)
J. Fontaine (Orsay)
A. Frohlich (Cambridge)
A. Goncharov (MIT)
R. Greenberg (Washington)
B. Gross (Harvard)
M. Harris (Paris 7)
S. Howson (Cambridge)
C. Huyghe (Rennes)
L. Illusie (Orsay)
F. Jarvis (Oxford)
K. Kato (Tokyo)

G. Kings (Munster)
V. Kolyvagin (Johns Hopkins)
H. Knosper (Munster)
S. Kudla (Maryland)
L. Lafforgue (Orsay)
E. Landvogt (Munster)
S. Lichtenbaum (Brown)
A. Macintyre (Oxford)
J. Manorhamayum (Cambridge)
D. Mauger (Paris-Nord)
B. Mazur (Harvard)
L. Merel (Paris)
J. Merriman (Kent)
W. Messing (Minnesota)
B.J.J. Moonen (Munster)
H. Nakamura (IAS)
J. Nekovar (Cambridge)
R. Noot (Rennes)
L. Nyssen (Strasbourg)
Y. Ochi (Cambridge)
R. Odoni (Glasgow)
F. Oort (Utrecht)
G. Pappas (Princeton)
A. Plater (Bordeaux 1)
F. Pop (Heidelberg)
M. Rapoport (Wuppertal)
K. Ribet (Berkeley)
K. Rubin (Ohio State)
N. Schlappacher (Strasbourg)
A. Schmidt (Heidelberg)
P. Schneider (Munster)
A.J. Scholl (Durham)
J-P. Serre (Paris)
N. Shepherd-Barron (Cambridge)
V. Snaith (McMaster)
R. Taylor (Oxford)
M.J. Taylor (UMIST)
J. Tilouine (Paris-Nord)
T. Tsuji (Tokyo)
A. Vasiu (ETH Zurich)
M. Volkov (Orsay)
R. Weissauer (Mannheim)
A. Werner (Munster)
J. Wilson (Oxford)
S.M.J. Wilson (Durham)
J.P. Wintenberger (Strasbourg)
S. Wortmann (Cologne)
M.A. Young (Durham)

Lecture programme

	9:30–10:30	11:15-12:15	2:30–3:30	4:00–5:00	5:15–6:15
Tuesday 9th July	A1	C1	D1	I1	I2
Wednesday 10th July	B1	C2	E1	I3	I4
Thursday 11th July	A2	C3	E2	I5	I6
Friday 12th July	B2	C4			D2
Saturday 13th July	I7	F1	D3	I8	
Monday 15th July	A3	F2	B3	I9	
Tuesday 16th July	B4	F3	D4	I10	I11
Wednesday 17th July	A4	E3	I12	I13	I14
Thursday 18th July	I15	F4	I16	I17	I18

A: GALOIS MODULE STRUCTURE
1) Ted Chinburg (Pennsylvania): *Geometric group actions and Galois structure.*
2) George Pappas (Princeton): *The generalised Frohlich conjecture in any dimension.*
3) Martin Taylor (UMIST, Manchester): *Hermitian Euler characteristics and ϵ-constants.*
4) David Burns (Kings, London): *Motivic Galois structure invariants.*

B: SHIMURA VARIETIES IN MIXED CHARACTERISTIC
1) Bas Edixhoven (Rennes): *Introduction. Moduli schemes of abelian varieties.*
2) Ben Moonen (Münster): *Models of Shimura varieties.*
3) Johan de Jong (Harvard; Princeton University): *Integral crystalline cohomology.*
4) Adrian Vasiu (ETH Zurich; Berkeley): *Points of integral canonical models of Shimura varieties of preabelian type.*

C: P-ADIC COMPARISON THEOREMS
1) Pierre Colmez (Paris VI): *Fontaine's rings and the conjectures.*
2) Jean-Marc Fontaine (Orsay): *Abelian varieties.*
3) Bill Messing (Minnesota): *Syntomic and crystalline cohomology.*
4) Takeshi Tsuji (Kyoto): *p-adic vanishing cycles.*

D: THE WORK OF KATO ON THE BIRCH-SWINNERTON-DYER CONJECTURE
1) Tony Scholl (Durham): *Kato's Euler system.*
2/3) Kazuya Kato (Tokyo)
4) Karl Rubin (Ohio State): *Euler systems and Selmer groups.*

E: POLYLOGARITHMS

1) Sasha Goncharov (MIT): *Geometry of polylogarithms and regulators.*
2) Sasha Goncharov (MIT): *L-functions of elliptic curves at $s = 2$.*
3) Sasha Goncharov (MIT): *Multiple polylogarithms and motivic Galois groups.*

F: RIGID ANALYSIS AND MODULAR FORMS

1/2) Peter Schneider (Münster): *Rigid Analysis.*
3) Robert Coleman (Berkeley): *Serre's p-adic Fredholm theory in families.*
4) Robert Coleman (Berkeley): *The curve of q-expansions.*

I: INDIVIDUAL LECTURES

1) Florian Pop (Heidelberg): *An introduction to anabelian geometry I.*
2) Jean-Pierre Serre (Paris): *Euler-Poincaré characteristics of profinite groups.*
3) Barry Mazur (Harvard): *Open questions about rational points on curves and varieties.*
4) Fred Diamond (Cambridge): *The Taylor-Wiles construction and multiplicity one.*
5) Hiroaki Nakamura (IAS): *An introduction to anabelian geometry II.*
6) Dick Gross (Harvard): *Modular forms for groups over **Q** whose real points are compact.*
7) Pierre Colmez (Paris): *Iwasawa theory of de Rham representations.*
8) Gerd Faltings (Bonn): *Fundamental groups of algebraic curves.*
9) Gerd Faltings (MPI): *Almost étale extensions.*
10) Steve Kudla (Maryland): *Height pairings and derivatives of Eisenstein series.*
11) Michael Rapoport (Wuppertal): *Special cycles on Siegel threefolds.*
12) Victor Kolyvagin (Johns Hopkins): *On the arithmetic of cyclotomic fields.*
13) Michael Harris (Paris): *p-adic uniformization and local Galois correspondences.*
14) Ken Ribet (Berkeley): *Torsion points on $X_0(37)$.*
15) Takeshi Tsuji (Kyoto): *C_{st}.*
16) Rainer Weissauer (Mannheim): *Siegel modular forms mod p.*
17) Loic Merel (Berkeley): *Arithmetic of elliptic curves and diophantine equations.*
18) Johan de Jong (Harvard/Princeton): *Alterations.*

The Eigencurve

R. COLEMAN AND B. MAZUR

in memory of Bernard Dwork

Let p be a prime number and $\mathbf{C}_p$ the completion of an algebraic closure of the field $\mathbf{Q}_p$ of p-adic numbers. Let N be an integer relatively prime to p. To describe our main object, we assume that $p > 2$, that the group of units in the ring $\mathbf{Z}/N\mathbf{Z}$ is of order prime to p, and we restrict our attention (at least in this introduction) to classical modular cuspidal eigenforms $f = \sum_{n=1}^{\infty} a_n q^n$ on $\Gamma_0(pN)$ of weight $k \geq 2$, with Fourier coefficients in $\mathbf{C}_p$, and normalized so that $a_1 = 1$. By "eigenform" let us mean eigenform for the Hecke operators T_ℓ for primes ℓ not dividing pN, for the Atkin-Lehner operators U_q for primes q dividing pN and for the diamond operators $\langle d \rangle$ for integers d prime to Np. By the **slope** of such an eigenform we mean the non-negative rational number $\sigma = \operatorname{ord}_p(a_p)$, where a_p is (both) the p-th Fourier coefficient of f and the U_p-eigenvalue of f. Assume that the newform associated to f is either a newform for $\Gamma_0(N)$ or $\Gamma_0(pN)$ (the latter case can occur only if $\sigma = (k-2)/2$).

By the work of Hida (cf. [H-ET] for an exposition of this theory, and for further bibliography given there) one knows that any such eigenform f of weight $k \geq 2$ and slope 0 is a member of a p-adic analytic family f_κ of (overconvergent) p-adic modular eigenforms of slope 0 parameterized by their p-adic weights κ (and, such that $f_k = f$) for κ ranging through a small p-adic neighborhood of k in (p-adic) weight space. Let us call this result *p-adic analytic variation of slope 0 eigenforms*. But Hida's results are, in fact, much more precise: If $\Lambda_N := \mathbf{Z}_p[[(\mathbf{Z}/N\mathbf{Z})^* \times \mathbf{Z}_p^*]]$, Hida constructs a finite flat Λ_N-algebra (let us call it $\mathbf{T}_{p,N}^o$) which is universal, in a certain sense, for slope 0 (overconvergent) eigenforms of tame level N, and such that the associated rigid analytic space to $\mathbf{T}_{p,N}^o$ (let us call it $C_{p,N}^o$) is the rigid-analytic space parameterizing p-adic analytic families of slope 0 eigenforms. If $\mathcal{W}_N$ is weight space, i.e., the rigid analytic space associated to Λ_N, then the p-adic families f_κ alluded to above are obtained from the finite flat projection $C_{p,N}^o \to \mathcal{W}_N$, Hida having proved that this mapping is

Reprinted from 'Galois Representations in Arithmetic Algebraic Geometry', edited by A. J. Scholl & R. L. Taylor. ©Cambridge University Press 1998

étale at any classical modular point of weight ≥ 2 in $C^o_{p,N}$.

The natural question arising from this work of Hida for eigenforms of slope 0 is to find the appropriate generalization of that theory valid for arbitrary finite slope eigenforms. The slope 0 theory has a significant simplifying advantage over the general theory in that there is available a clean idempotent operator (call it e^o) projecting to the slope 0 part of the theory. This idempotent e^o, a p-adic idempotent reminding one a bit of the "holomorphic projector" in classical analysis, is a bounded operator on the Banach spaces involved, and in fact it has no denominators and therefore acts as an idempotent on all aspects of the theory (e.g., parabolic cohomology, as well as spaces of modular forms); moreover the image of this idempotent is of finite type over whatever is the natural base ring.

In [C-BMF] a satisfactory analogue of Hida's p-adic analytic variation theorem for slope 0 eigenforms was established for finite slope classical eigenforms (at least for those satisfying a mild condition; cf. Cor. B5.7.1 of [C-BMF]).

The ultimate aim of the theory developed in this article is to provide a more global counterpart to the work done in [C-BMF] and construct a rigid analytic curve $C_{p,N}$ (analogous to $C^o_{p,N}$) which parameterizes *all* finite slope overconvergent p-adic eigenforms of tame level N, and to study its detailed geometry: in particular, we wish to understand the nature of the projection of $C_{p,N}$ to weight space $\mathcal{W}_N$.

For reasons of space, and time, we do this only in the case of $p > 2$, and tame level $N = 1$ in the present article. We do, however, treat noncuspidal eigenforms as well as eigenforms on $\Gamma_1(p^m)$, as opposed to the more restricted class of eigenforms delineated at the beginning of this introduction. For more precise, yet still introductory, statements concerning our main results, the reader might turn to section 1.3 (and in particular to the Theorems A,B,C,D,E,F,G formulated there). For the rest of this introduction, we suppose that $p > 2$, and discuss the case of tame level $N = 1$.

We construct a rigid analytic curve $C_p = C_{p,N=1}$ over $\mathbf{Q}_p$ whose $\mathbf{C}_p$-valued points parameterize *all* finite slope overconvergent p-adic eigenforms of tame level $N = 1$ with Fourier coefficients in $\mathbf{C}_p$. We call C_p the *(p-adic) eigencurve (of tame level $N = 1$)*. Hida's rigid space $C^o_p = C^o_{p,N=1}$ which parameterizes slope 0 eigenforms of tame level 1 occurs as a component part (cf. section 1.2) of our eigencurve C_p, but in contrast to Hida's theory, the natural projection of C_p to weight space is *not* of finite degree. The eigencurve C_p has a natural embedding

$$c \mapsto (\rho_c, 1/u_c)$$

into the rigid analytic space $X_p \times \mathbf{A}^1$ where X_p is the rigid analytic space attached to the universal deformation ring R_p of certain Galois (pseudo-) representations and $\mathbf{A}^1$ is the affine line. For a discussion of pseudo-representations, see Chapter 5 below. The particular residual pseudo-representations for which R_p is the universal deformation ring we call p-**modular** residual representations (of the Galois group $G_{\mathbf{Q},\{p,\infty\}}$). These are the residual pseudo-representations coming from the classical modular eigenforms of finite slope and level a power of p (See section 5.1 for their definition.); there are only finitely many such residual pseudo-representations, and R_p is a complete semi-local noetherian ring.

If $c \in C_p$ corresponds to the overconvergent eigenform f_c, the first coordinate of c with respect to this embedding is the Galois pseudo-representation $r_c \in X_p(\mathbf{C}_p)$ attached to the eigenform f_c and the second coordinate $(1/u_c)$ is the inverse of $u_c :=$ the U_p-eigenvalue of the eigenform f_c. We define the eigencurve $C_p \subset X_p \times \mathbf{A}^1$ to be the rigid analytic subspace cut out by an ideal generated by certain specific rigid analytic functions, these functions being *Fredholm series over R_p* (cf. 1.2) obtained by pullback to $X_p \times \mathbf{A}^1$ of the characteristic series of certain completely continuous systems of operators (cf. 4.3). In general, a Fredholm series over a complete local noetherian ring R is an entire power series (with constant term 1) in one variable T over R, and we call a *Fredholm variety* over R (see section 1.2 below) a rigid analytic subspace of $X_p \times \mathbf{A}^1$ which is cut out by an ideal generated by a collection of Fredholm series over R (here T is the variable parameterizing the affine line $\mathbf{A}^1$).

Let us refer to the nilreduced rigid analytic space subjacent to the eigencurve as the **reduced eigencurve** C_p^{red}. For a brief discussion of the process of passing to the nilreduction of a rigid analytic space, see section 1.2 below. We define irreducible components there and prove under sufficiently general hypotheses that every point in such a space lies on one.

We show that each irreducible component of the reduced eigencurve C_p^{red} is isomorphic (at least outside of a discrete set of points) to some reduced irreducible Fredholm hypersurface over the Iwasawa ring Λ (i.e., it is a Fredholm variety defined by a single (irreducible) Fredholm series with coefficients in the Iwasawa algebra Λ) via an isomorphism that preserves projection to weight space. Each irreducible component of a Fredholm hypersurface over Λ is again a Fredholm hypersurface over Λ, and it is seen (Cor. 1.3.12 below) to either be isomorphic to the open complement of a finite set of points in the rigid analytic space attached to a finite flat Λ-algebra, or else to be of infinite degree. When (and if) the first case holds, we would be in a situation very analogous to Hida's theory for slope 0 eigenforms. But so far we *have not yet been able to determine* for any positive slope irreducible component of C_p^{red} which of these two alternatives hold!

It follows from our results that the natural projection of each irreducible component of the reduced eigencurve to weight space is componentwise almost surjective in the sense that its image avoids at most a finite number of $\mathbf{C}_p$-valued points in the component of weight space within which it lands; in particular, each irreducible component of C_p^{red} has eigenforms of all (but a finite number of) weights. For a more precise statement, see Theorem B of 1.5.

It also follows from our results that any convergent eigenform of finite slope and integral weight (of tame level 1) is overconvergent if and only if its q-expansion is approximable p-adically by the q-expansions of classical eigenforms (of tame level 1). (Note that the q-expansions of Serre's p-adic modular forms are the limits of the q-expansions of classical modular (but not necessarily eigen) forms.) For a more precise statement, see Theorem G of 1.5.

In Chapter 7, we construct a second rigid space D, by means of the Banach module theory of [C-BMF], which is evidently a curve. We prove the above assertions (in particular, that it is a curve) about C_p^{red} by proving them about D and showing that

$$D \cong C_p^{\mathrm{red}}.$$

From the its construction, one sees that the projection of D to weight space is *locally in-the-domain finite flat*, meaning that D is covered by admissible affinoid domains $\mathcal{U}$ such that the restriction of projection to weight space to $\mathcal{U}$ is a finite flat mapping of $\mathcal{U}$ onto its image in $\mathcal{W}$.

The image of a rigid analytic morphism can be quite intricate, and, in particular, the projection $C_p^{\mathrm{red}} \to X_p$ has an infinite number of double points (e.g., those coming from the classical modular eigenforms of level 1). The image of C_p^{red} in X_p contains the "infinite ferns" studied in the articles [GM-FM], [GM 3] and [M-IF].

We may pull back the universal pseudo-representation on X_p to C_p^{red}, via the natural projection, to obtain a rigid analytically varying pseudo-representation on C_p^{red} which is realizable, at least on the complement of a certain discrete set of points on C_p^{red} as an $\mathcal{O}_{C_p^{\mathrm{red}}}$-linear continuous Gal $(\bar{\mathbf{Q}}/\mathbf{Q})$-representation unramified outside p with the property that the restriction of this representation to a $\mathbf{C}_p$-valued point $c \in C_p$ yields the p-adic Galois representation attached to the eigenform f_c.

Open questions. Is C_p reduced? (I.e., does $C_p = D$?) Is C_p smooth? Does C_p^{red} have a finite or an infinite number of components? Do any of these components have infinite genus? Is every component of C_p^{red} of slope > 0 of *infinite* degree over weight space, or are there components of positive slope that are of finite degree over weight space? Do there exist p-adic analytic families of overconvergent eigenforms of finite slope parameterized by a punctured disc, and converging, at the puncture, to an overconvergent eigenform of infinite slope? (If so, one would want to *complete* the eigencurve a bit by including these missing points of infinite slope.) Having constructed the (global) eigencurve allows us to ask more global questions along the lines of conjectures made by one of the authors of the present article, with Fernando Gouvêa, and bears on recent work of Daqing Wan [Wa]: Are the ramification points of the natural projection from C_p to weight space infinite in number? We would like to know where those ramification points are; specifically given a point $c \in C_p$ of weight $\kappa \in \mathcal{W}$ let us say that an affinoid subdomain $\mathcal{U} \subset \mathcal{W}$ containing κ is a *weight-parameter space* for c if there is an affinoid neighborhood $\mathcal{V} \subset C_p$ of c whose natural projection to $\mathcal{U}$ is finite étale. Can one find, given any point c corresponding to a classical eigenform of weight k of slope strictly less than $k - 1$, a weight-parameter space for c of radius greater than the inverse of a linear function of the slope of c? One can prove, using Wan's results [Wa], that there are such weight-parameter spaces of radius greater than the inverse of a quadratic function of the slope if the slope is strictly less than $k - 1$ and not equal to $(k - 1)/2$.

We mentioned above that, excluding a discrete set of points (call this set $\Delta \subset C_p^{\text{red}}$), the reduced eigencurve parametrizes a rigid-analytically varying family of Galois representations, i.e. we have a continuous representation

$$\rho : \text{Gal}\,(\bar{\mathbf{Q}}/\mathbf{Q}) \to \text{GL}_2(\mathcal{O}_{C_p^{\text{red}} - \Delta})$$

Can this family of Galois representations ρ be extended over the excluded set Δ? Is there a continuous representation of $\text{Gal}\,(\bar{\mathbf{Q}}/\mathbf{Q})$ into the group of units in a rigid Azumaya algebra of rank 4 over $\mathcal{O}_{C_p}$ which, in the appropriate sense, extends ρ? Can one construct ρ (or at least ρ restricted to an appropriate $C_p^{\text{red}} - \Delta$) from the cohomology of modular curves in the following sense: Let $\mathcal{H}$ be the polynomial algebra over Λ in the countable set of generators denoted T_ℓ for prime numbers $\ell \neq p$ and an *extra* generator denoted U_p. For a more complete discussion of this Λ-algebra $\mathcal{H}$ see Chapter 6. There is a natural homomorphism of $\mathcal{H}$ to the ring $\mathcal{O}_{C_p}$ of rigid analytic functions on the eigencurve; in particular, in the discussion below we view $\mathcal{O}_{C_p}$ as $\mathcal{H}$-algebra. Define the discrete abelian group

$$\mathcal{M} := \varinjlim H^1(X_1(p^n); \mathbf{Q}_p/\mathbf{Z}_p),$$

the direct limit of étale cohomology groups of the modular curves $X_1(p^n)$, this limit being compiled via the mappings on cohomology induced from the natural projections $X_1(p^{n+1}) \to X_1(p^n)$ for all $n \geq 1$ (in the present brief description we omit saying *which* natural projections these are). The Galois group $G_{\mathbf{Q},\{p,\infty\}}$ acts on this direct limit in the natural way, as does the algebra $\mathcal{H}$ (each of the generators acting as the corresponding Hecke or Atkin-Lehner operator) and the actions of $\mathcal{H}$ and $G_{\mathbf{Q},\{p,\infty\}}$ commute, allowing us to view $\mathcal{M}$ as an $\mathcal{H}[[G_{\mathbf{Q},\{p,\infty\}}]]$-module. Let $\mathcal{M}^* := \mathrm{Hom}(\mathcal{M}; \mathbf{Q}_p/\mathbf{Z}_p)$ be the pontrjagin dual of $\mathcal{M}$, viewed as compact $\mathcal{H}[[G_{\mathbf{Q},\{p,\infty\}}]]$-module. There are various (possibly equivalent) ways of *spreading the cohomology $\mathcal{H}$-module $\mathcal{M}^*$ over the eigencurve*. The simplest way to do this is to form $V :=$ the completed tensor product of the $\mathcal{H}$-module $\mathcal{M}^*$ with the $\mathcal{H}$-algebra $\mathcal{O}_{C_p}$. Viewing V as quasi-coherent sheaf over the eigencurve, the $G_{\mathbf{Q},\{p,\infty\}}$-action on $\mathcal{M}^*$, which commutes with the action of $\mathcal{H}$, induces a $\mathcal{O}_{C_p}$-linear action on V. Is it the case that, over C_p, or possibly just over some large portion of C_p, the quasi-coherent sheaf V is locally free of rank 2, and the $G_{\mathbf{Q},\{p,\infty\}}$-representation on it is equivalent to the family of representations ρ discussed above? On the converse side it would be good to find an intrinsic operation which cuts out of $\mathcal{M}^*$ precisely those Galois representations which are attached to overconvergent eigenforms. All these questions have analogues when the cohomology group

$$\mathcal{M} := \varinjlim H^1(X_1(p^n); \mathbf{Q}_p/\mathbf{Z}_p)$$

is replaced by (direct limits of) parabolic cohomology of higher weights, and one might ask for a theory, following Hida's work, which deals with all weights. Relevant to this are the results of Hida [H-HA] and Gouvêa [G-ApM].

Does Glenn Stevens' construction of the p-adic L-function of a (p-adic) eigenform on $\Gamma_0(pN)$ (cf. [St]) work well over the eigencurve, and, in particular, does it give an L-function (with the coefficients of its Taylor expansion rigid analytic functions on C_p) which interpolates all the classical p-adic L-functions? We hope to show in a later publication, that at least a weak version of this is true. The locus of zeroes of this L-function, once constructed, would itself be a rigid-analytic curve (call it $\mathcal{L}_p$) admitting a natural rigid analytic projection to C_p. This deserves study. particularly intriguing is the nature of this projection in the neighborhood of a double-zero of the L-function.

It would be interesting to construct a local version of the eigencurve using the theory of crystalline Galois representations, in the following sense. Let X now refer to the rigid-analytic universal deformation space of a fixed absolutely irreducible representation of the Galois group of $\mathbf{Q}_p$ into $GL_2(\mathbf{F}_p)$. It is known that X is a smooth 5-dimensional rigid-analytic ball. By a crystalline point in (the 6-dimensional) rigid-analytic space $X \times \mathbf{A}^1 - \{0\}$, let us mean a pair (x, u) where $x \in X$ classifies a crystalline representation, and $u \in \mathbf{A}^1 - \{0\}$ is one of its the Frobenius eigenvalues in the Fontaine-Dieudonné module attached to the crystalline representation x. (For a general overview of crystalline, semi-stable, and potentially semi-stable representations, see [F].) Is it the case that the rigid-analytic closure of the set of crystalline points in $X \times \mathbf{A}^1 - \{0\}$ is a *3-dimensional rigid-analytic subvariety* ? Is it a rigid analytic subspace of affine three-space?

There are also certain foundational questions which deserve much better understanding than we have, at present. For example, one still lacks a satisfactory conceptual definition of what it means for a modular form (of general p-adic weight) to be *overconvergent*. This notion is clear for integral weight, but at present, for general weight, we have only an ad hoc procedure based on the availability of families of Eisenstein series of general weight (See Chapter 2 below: you multiply your modular form by such an Eisenstein series to get the product to be of weight 0 and then ask that this be overconvergent as a rigid analytic function on a suitable affinoid in the modular curve). As a consequence of this awkward strategy, treatment of any serious property about overconvergent modular forms depends upon our detailed understanding of the family of Eisenstein series. It would be good to

a. have a more direct definition of overconvergent modular forms and of families of overconvergent modular forms, and at the same time

b. have a closer understanding of the properties of families of Eisenstein series (including knowledge of their zero-free regions).

An improvement in our state of knowledge of **b** would help in our understanding of the detailed geometry of the eigencurve.

It would be important, as well, to have some detailed computations of specific affinoid subdomains of C_p. In this regard, see forthcoming work of Matthew Emerton who (augmenting earlier calculations of Coleman and Teitelbaum) gives a complete description of the geometry of that part of the eigencurve for $p = 2$ and tame level 1 having minimal slope for their weight (cf. [Em]). See also [CTS] where some results on the low slope part of the 3-adic eigencurve are proven. We might mention here that in these examples $(p = 2, 3)$, the slope tends to zero as one approaches the "boundary" of weight space. Are there components of the p-adic eigencurve (for some p) where this phenomenon does not occur?

To be sure, a satisfactory general theory of the eigencurve must deal with *all* tame levels N (and as a prelude for this, one necessary task, which we carry out, is to set up the deformation theory of pseudo-representations with fixed tame level). In the present paper, although we make the construction of the eigencurve only for $N = 1$, in some sections (where it is easy to do so) we work with more general level N in preparation for the general theory. Our conventions concerning level will be signaled at the start of each section.

Is there an a priori deformation-theoretical approach to the eigencurve, and to the Galois representations that the eigencurve parameterizes? (We explain more precisely what we mean by this at the end of Chapter 1.) Our lack of such an approach accounts for some of the difficulty we have in analyzing local properties of the eigencurve. Is there a natural formal scheme over $\mathbf{Z}_p$ whose associated rigid analytic space is the eigencurve? One is, in any event, guaranteed (by the preprint [LvP]) that any connected one-dimensional, separated, rigid analytic space over $\mathbf{Q}_p$ is the generic fiber of some formal scheme which is flat over $\mathbf{Z}_p$. It might be interesting to study irreducible components of the closed fiber of a formal scheme whose generic fiber is (a piece of) the eigencurve. In this connection, J. Teitelbaum has a computer program that produces such irreducible components in a range of slopes (for $p = 2, 3$).

In a later publication we hope to present more foundational material regarding the connection between Katz modular function and convergent eigenforms (including the proof of the compatibility of the action of the diamond operators, i.e., the proof of Prop. 3.4.2 below which we omitted from this article). We are deeply indebted to Kevin Buzzard for his extremely helpful comments throughout the preparation of this article. We also wish to thank Brian Conrad for his helpful suggestions on Chapter 1 and Matthew Emerton for his close reading of an early draft.

Table of Contents.

1. Rigid analytic varieties **10**

1.1 Rigid analytic spaces attached to complete local noetherian rings.

1.2 Irreducible components and component parts.

1.3 Fredholm varieties.

1.4 Weight space.

1.5 The eigencurve as the Fredholm closure of the classical modular locus. (Statement of the main theorems.)

2. Modular forms **35**

2.1 Affinoid sub-domains of modular curves.

2.2 Eisenstein series.

2.3 Katz p-adic modular functions.

2.4 Convergent modular forms and Katz modular functions.

3. Hecke algebras **47**

3.1 Hecke eigenvectors and generalized eigenvectors.

3.2 Action on $M_k(Np^m, v; K)$.

3.3 Action on Katz Modular Functions.

3.4 Action on $M^\dagger(N)$.

3.5 Action on weight κ forms.

3.6 Remarks about cusp forms and Eisenstein series.

4. Fredholm determinants **55**

4.1 Completely continuous operators and Fredholm determinants.

4.2. Factoring characteristic series.

4.3. Analytic variation of the Fredholm determinant.

4.4. The Spectral Curves.

5. Galois representations and pseudo-representations **68**

5.1. Deforming representations and pseudo-representations.

5.2. Pseudo-representations attached to Katz modular functions.

6. The eigencurve **83**

6.1 The definition of the eigencurve.

6.2. The points of the eigencurve are overconvergent eigenforms.

6.3. The projection of the eigencurve to the spectral curves.

6.4 The Eisenstein curve.

7. The eigencurve constructed as finite cover of spectral curves **91**

Chapter 1. Rigid analytic varieties.

1.1 Rigid analytic spaces attached to complete local noetherian rings.

Let p be a prime number, and $\mathbf{C}_p$ the completion of a fixed algebraic closure of $\mathbf{Q}_p$. Let v be the continuous valuation on $\mathbf{C}_p$ such that $v(p) = 1$. Define the absolute value $|\ |$ by $|a| = p^{-v(a)}$, for $a \in \mathbf{C}_p$. By the ring of integers $\mathcal{O}_K$ in a complete subfield K of $\mathbf{C}_p$ we mean the set of elements of absolute value at most 1. For a rigid analytic variety Y (see §9.3 of [BGR]), $A(Y)$ will denote the ring of rigid functions on Y, for $a \in A(Y)$, $|a|$ will denote the spectral semi-norm of a (which may be infinite valued) and $A^0(Y)$ will denote the sub-$\mathcal{O}_K$-algebra of rigid functions with spectral semi-norm at most 1. For an affinoid algebra B, we let $\mathrm{Max}(B)$ denote the corresponding rigid variety over K.

Now let $R = \mathcal{O}_K$ be the ring of integers in a fixed finite extension K of $\mathbf{Q}_p$ in $\mathbf{C}_p$. Let k denote the residue field of R. Let A be a complete local noetherian R-algebra with maximal ideal m_A and residue field $A/m_A = k$. Consider the functor $A \longrightarrow X_A$ which attaches to each such complete noetherian local ring, its associated rigid analytic space over K. We refer to section 7 of [de J] for the construction, and for its basic properties (cf. loc. cit. Definition 7.1.3, 7.1.4a, and 7.1.5). Briefly, the construction may be given as follows: For each real number $r > 0$, let A_r be the p-adic completion of the quotient by its p-power torsion, of the ring

$$A[\{y_{r,(a)}\}]$$

where (a) ranges over unordered tuples of elements in m_A subject to the relations

$$p^{[k(a)r]}y_{r,(a)} = \prod(a),$$

where $k(a)$ is the length of (a) and the product is over the entries of (a). (Note: $A_r = 0$ if r is sufficiently large.) If $r \geq s > 0$ there is a natural A-homomorphism from A_s to A_r determined by $y_{r,(a)} \mapsto p^{[k(a)r]-[k(a)s]}y_{s,(a)}$. If r is rational, this ring is finitely generated over A. Indeed, if $\{t_1, \ldots, t_n\}$ generate m_A and $r = g/h$ with g, h positive integers, then A_r is generated by $y_{(a)}$ where the entries of (a) are in $\{t_1, \ldots, t_n\}$ and the length of (a) is at most h, because in this case, $m_A^h \subseteq p^g A_r$. It follows that the admissible formal schemes, in the sense of [BL-FRI&II], $\mathrm{Spf}A_r$, where r is rational, glue together into an admissible formal scheme over R and we can make a rigid space X_A out of this formal scheme as in [BL-FRI&II] covered by the affinoids $X_r := \mathrm{Max}(A_r \otimes K)$. In a word, X_A is the union of the affinoids X_r's for all positive rational r. In particular, the $\mathbf{C}_p$-valued points of X_A are given by (continuous) homomorphisms of the R-algebra A into $\mathbf{C}_p$ (or, equivalently, into $\mathcal{O}_{\mathbf{C}_p}$):

$$X_A(\mathbf{C}_p) = \mathrm{Hom}_R(A, \mathbf{C}_p),$$

and we have a natural ring homomorphism $A \to A^0(X_A)$. To a complete Noetherian semi-local ring, we attach the union of the rigid spaces attached to each of its localizations at maximal ideals. An important special case for us is $A = \mathbf{Z}_p[[T]]$ (here $R = \mathbf{Z}_p$, and $K = \mathbf{Q}_p$). For $r \leq 1$, A_r is the ring of rigid functions defined over $\mathbf{Q}_p$ and bounded by 1 on the disk $B[0, p^{-r}] = \{x \in \mathbf{C}_p \mid |x| \leq p^{-r}\}$ in $\mathbf{C}_p$, and X_A is the open unit disk, $B(0, 1)$, viewed as rigid analytic space over $\mathbf{Q}_p$. More generally, if one takes A to be the ring of rigid analytic functions bounded by 1 on a residue class (formal fiber) in an affinoid (see [Bo]), X_A will be canonically isomorphic to that residue class. All of the rigid spaces we shall be considering in this article are separated and have a countable admissible covering by affinoid subsets. Following the terminology of [JP] one says that a rigid space X over a complete non-archimedean valued field K is **paracompact** if it admits an admissible covering by affinoid subsets, any one of which meets only finitely many others. If a rigid space X over a complete non-archimedean valued field K admits an admissible covering by a countable increasing sequence of affinoid subdomains $X_0 \subset X_1 \subset \cdots \subset X_i \subset \cdots \subset X$, let us call X **nested**. For any complete noetherian ring R, its associated rigid space X_R described above is **nested**, as is $X_R \times \mathbf{A}^1$, the product of X_R with the affine line, which will be introduced in the next section. According to [LvP], any connected, separated, one-dimensional rigid space X over a complete non-archimedean valued field K is paracompact (this is no longer the case in higher dimensions), and is the generic fiber of a formal scheme which is flat over $\mathrm{Spf}(\mathcal{O}_K)$, where $\mathcal{O}_K$ is the ring of integers of K. As a general reference for the relationship between formal schemes and rigid analytic spaces, the reader might consult [L], [BL-FRI&II], [JP], [LvP], and pp. 429-436 of [Ra].

1.2. Irreducible components and component parts.

If $\iota\colon S \to Z$ is a morphism of rigid spaces, ι is said to be a **closed immersion** and Z is said to be a **Zariski-closed analytic subvariety** (via ι) if there exists an admissible affinoid covering $\{U_i\}_{i\in I}$ of Z such that for all $i \in I$ the induced map $\iota\colon \iota^{-1}(U_i) \to U_i$ is a closed immersion of affinoid varieties [BGR §9.5.3]. It follows that for any open affinoid subvariety U of X, $\iota\colon \iota^{-1}(U) \to U$ is a closed immersion of affinoid varieties [BGR Prop. 9.5.3/2]. If S is a Zariski-closed analytic subvariety of Z, we shall refer to S, for short, as **Zariski-closed** and note that (since affinoids are noetherian) the category of Zariski-closed subspaces of a given rigid space Z enjoys the local descending chain condition: any infinite descending chain of Zariski-closed subspaces of Z stabilizes finitely when restricted to any given affinoid in Z. If $\iota\colon S \to X$ is a closed analytic subvariety of X we let the **interior** of S be the rigid subspace which is the union of affinoids of the form $\iota^{-1}(U)$ where U is an open affinoid subvariety of X and $\iota\colon \iota^{-1}(U) \to U$ is an isomorphism. The **rigid Zariski-closure** of a morphism $\iota\colon S \to Z$ is the minimal element in the category of closed immersions $W \to Z$ through which ι factors. The fact that the rigid Zariski-closure of S exists follows from the fact that if $S \to Z$ factors through $W \to Z$ and $W' \to Z$ it factors through $W \times_Z W' \to Z$, and from the local descending chain condition described above. If $S \subseteq X_A \times_K \mathbf{A}_K^1(\mathbf{C}_p)$, consider the rigid space over $\mathbf{C}_p$ which is the disjoint union of the points S (we denote this space by the same letter S). We define the the rigid Zariski closure of the set of points S to be the rigid Zariski closure of the morphism $i : S \to X$.

If X is a rigid analytic space over K, by the **nilreduction** of X, denoted X^{red} (cf. page 389 of [BGR]) we mean the unique rigid analytic subspace of X over K (constructed in loc. cit.) whose admissible affinoid subdomains $\mathcal{U}^{\mathrm{red}} = \mathrm{Max}(\mathcal{O}_{\mathcal{U}^{\mathrm{red}}})$ are in one:one correspondence with the admissible affinoid subdomains $\mathcal{U} = \mathrm{Max}(\mathcal{O}_{\mathcal{U}})$ of X, where $\mathcal{O}_{\mathcal{U}^{\mathrm{red}}}$ is the quotient of $\mathcal{O}_{\mathcal{U}}$ by its nilradical. The natural closed immersion $X^{\mathrm{red}} \subset X$ induces a one:one correspondence on sets of $\mathbf{C}_p$-valued points. One says that X is nilreduced (or reduced) if $X^{\mathrm{red}} = X$.

Definition. Suppose Z is a rigid space. Then by a **component part** of Z we mean a (non-empty) rigid subspace of Z which is the rigid Zariski-closure in Z of an admissible open of Z. We will say that Z is **irreducible** if it has only one component part (i.e., itself). An **irreducible component** of Z is a component part which is itself irreducible.

Notes. It may be worth explicitly pointing out that the definition of irreducible above ignores nilpotent elements, and in particular (as in standard algebraic geometric usage) our definition allows the possibility that irreducible components be *nonreduced*; moreover, embedded components are simply not counted. As we'll see, in Corollary 1.2.1/1, if Z is an affinoid, Z possesses only a finite number of distinct irreducible component parts, and any component part of Z may be viewed at least set-theoretically as a (finite, of course) union of irreducible component parts. Let $\Phi(Z)$ denote the set of irreducible component parts of Z. One should note that $\Phi(-)$ is *not* in general functorial in Z (e.g., consider the case of $Z =$ a single point mapping to the intersection of two irreducible components of a rigid analytic space Z'). Nevertheless, rigid analytic morphisms $\iota : Z \to Z'$ which embed Z as an open subdomain in Z' induce natural mappings $\Phi(\iota) : \Phi(Z) \to \Phi(Z')$ and we shall make use of this in what follows. Brian Conrad and Ofer Gabber have independently proposed another definition of irreducible components of X when X is reduced. One first passes to the normalization of X (which one has to prove is a well defined rigid space), takes connected components and then defines the irreducible components of X to be the images of these components. Using this definition one can show that irreducible components have the right localization properties. Conrad has established that, in this case, these irreducible components are the same as ours.

Lemma 1.2.1. *Suppose X is an affinoid. Then X is irreducible if and only if the map $A(X) \to A(X)_x$ is an injection for all closed points x of X. (Here $A(X)_x$ denotes the localization of $A(X)$ at x.)*

Proof. Suppose $0 \neq f \in A(X)$ such that $f \mapsto 0$ in $A(X)_x$ for some $x \in MaxA(X)$. Then there exists an admissible open neighborhood U of x in X such that $f \mapsto 0$ in $A(U)$. It follows that f maps to 0 in the ring of functions on the Zariski-closure Z of U and hence $Z \neq X$. Now suppose that $A(X)$ injects into $A(X)_x$ for all x and Z is a component part of X equal to the Zariski-closure of an admissible open U of X. Then by Proposition 9.5.3/1 of [BGR] Z is an affinoid of the form $MaxA(X)/I$ for some ideal I of $A(X)$. It follows that I maps to zero in $A(X)_x$ for any closed point of U and so $I = 0$ and $Z = X$. ∎

Corollary 1.2.1/1. *(i) If X is an affinoid, the irreducible components of X are affinoids $Z_I := MaxA(X)/I$ where I is a maximal ideal among the set S of ideals which map to zero in $A(X)_x$ for some closed point x. (ii) (In particular:) If $A(X)$ is an integral domain X is irreducible. (iii) The number of irreducible components X is equal the number of minimal prime ideals of $A(X)$.*

Proof. (i) It follows from the previous lemma that if I satisfies the above hypotheses then Z_I is irreducible. Conversely, suppose Z_I is an irreducible component part of X for some ideal I of $A(X)$ and U is an admissible open of X contained in Z_I. Then I maps to zero in $A(X)_x$ for all closed points x of U so $I \in S$. If $I \subset J$ where $J \in S$, then Z_J would be a component part of Z_I equal to Z_I if and only if $I = J$. (ii) follows from (i). (iii) It is clear that if $I \in S$ then I is contained in some minimal prime ideal of $A(X)$. Now suppose P is a minimal prime ideal of $A(X)$. Let I_P be the set of $f \in A(X)$ such that $f \mapsto 0$ in $A(X)_x$ for some closed point x of Z_P. Then I_P is the ideal of all $f \in A(X)$ such that there exists a $t \notin P$ such that $tf = 0$. The ideal I_P is contained in P and contains a power of P. It follows that I_P is not contained in any other minimal prime ideal. From its definition we see that I_P is a maximal element of S contained in P. It follows that I_P is a maximal element of S and the map $P \mapsto Z_{I_P}$ is a one to one correspondence between the minimal prime ideals of $A(X)$ and the irreducible components of X. ∎

We say an affinoid X is **equidimensional** if the Krull dimension of $\mathcal{O}_{X,x}$, where x is a closed point of X is independent of x. If this happens, this number equals the Krull dimension of $A(X)$.

Lemma 1.2.2. *Suppose X is a reduced equidimensional affinoid. Then the singular locus of X has codimension at least 1 in X.*

Proof. The proof of this follows the lines of the classical proof in algebraic geometry (see [Ha] Chapter 1, section 5). First, we can express $A(X)$ as the quotient of $K\langle T_1, \ldots, T_n \rangle$ by an ideal generated by elements $f_1, \ldots, f_m$ for some m and n. If X has dimension r, the singular locus of X is the set of points where the rank of the corresponding Jacobian matrix has rank $< n - r$ so is a Zariski closed subaffinoid. To show it has codimension at least 1 one show that the non-singular locus on each irreducible component Z of X is non-empty. By rigid Noether normalization, Corollary 6.1.3/2 of [BGR], we can find a find a finite map ϕ from Z onto $\mathbf{B}^r$. Then the singular locus of Z is contained in the set of ramification points of ϕ. Let F be the fraction field of $A := A(\mathbf{B}^r)$. Then since $A(Z)$ is an integral domain and $Z \to \mathbf{B}^r$ is finite and surjective, $A(Z)$ injects into $A(Z) \otimes_A F$ which is a finite extension of F. It follows that Z_U is flat over the complement U of the zeroes of a non-zero element in $A(\mathbf{B}^r)$ and hence the discimininant ideal D in $A(U)$ of Z_U/U is defined and non-zero. Moreover, the ramification points of ϕ in Z_U lie over the zeroes of D. Thus the non-sinular locus of Z maps onto a non-empty open of $\mathbf{B}^r$. ∎

Proposition 1.2.3. *Suppose X is a reduced affinoid of equidimension n. If I is an ideal of $A(X)$ and $V = \mathrm{Max}\{A(X)/I\}$, then V is a component part of X if and only if V is reduced and the Krull dimension of $\mathcal{O}_{V,v}$ equals n for all closed points of V.*

Proof. If V is a component part, $I = \bigcap_{J \in S} J$ where S is a set of minimal prime ideals. It follows that V is reduced and the Krull dimension of $\mathcal{O}_{V,v}$

equals n for all closed points v of V. Suppose V satisfies the latter hypotheses. Let Y be the maximal component part of X contained in V. Suppose $Y \neq V$. Let J be the ideal corresponding to Y. Since V is reduced it follows that there exist an $f \in J$ and a $v_0 \in V$ such that f doesn't vanish at v_0. We can assume $f(v_0) = 1$. Let B be the subdomain of V defined by

$$\{v \in V : |f(v) - 1| \leq |p|\}.$$

Then $B \cap Y = \emptyset$. Now If v is point of V which is smooth in X then V contains an affinoid neighborhood of v in X and hence V contains the component part of X which is the Zariski closure of this neighborhood. Hence B is contained in the singular locus of X. Since this is a subaffinoid of codimension at least one, it follows that the Krull dimension of V at v_0 which equals the Krull dimension of B at v_0 is at most dimension $n - 1$. ∎

Proposition 1.2.4. *Suppose X is a reduced affinoid and Y is an affinoid subdomain. Then if Z is a component part of X, $Z \cap Y$ is a component part of Y.*

Proof. This follows in the equidimensional case, from the fact that Y is reduced by Corollary 7.3.2/8 of [BGR] and Prop. 1.2.3. The general case of the lemma follows from the equidimensional case and the fact that X is the union of equidimensional component parts. Indeed, if Z is a component part of X and V is the equidimensional part of X of dimension n, the part of Z of equidimension n is a component part of V and we can apply the equidimensional part of the proposition to V, $Y \cap V$ and $Z \cap V$. ∎

Ofer Gabber, using the normalization definition of irreducible component mentioned above has proved that arbitrary rigid spaces can be covered by irreducible components. However, the following result is sufficient for our purposes:

Proposition 1.2.5. *Suppose X is a reduced rigid analytic space which has an admissible open cover by of affinoids $\{X_i : i \in \Omega\}$, where Ω is some index set such that $X_{ij} := X_i \cap X_j$ is an affinoid for each i and j. Then every closed point of X lies on an irreducible component of X.*

Proof. This proposition is true if X itself is a reduced affinoid (e.g., as can be seen from Cor. 1.2.1/1). In particular (fixing on some index $0 \in \Omega$) Proposition 1.2.5 is true for the affinoid subdomain $X_0 \subset X$. It will then follow from

Lemma 1.2.6. *Let A_0 be an irreducible component part of X_0. Then A_0 is contained in a (unique) irreducible component of X.*

Proof. We will define, inductively, a collection of component parts, $A_{n,i}$ of the affinoid subdomain X_i for pairs of integers $i \geq 0$ and $n \geq 0$. We start the induction with $A_{0,0} := A_0$ which is an irreducible component part of X_0 and $A_{0,i} = \emptyset$ for $i \neq 0$. Suppose, for a given integer $n \geq 0$ we have defined a component part $A_{n,i}$ of X_i for all $i \in \mathbf{Z}$. We proceed to define the component parts $A_{n+1,i}$ of X_i. Put $Y_{n,j,k} := X_{j,k} \cap A_{n,k}$, which is a component part of $X_{j,k}$ by Proposition 1.2.4. Let $A_{n+1,j}$ be the union over k of the Zariski closures in X_j of $Y_{n,j,k}$ (there are only a finite number of these, as each of them is a component part of X_j). For each j these $A_{n,j}$ stabilize for large n since they are all component parts and $A_{n+1,i} \supseteq A_{n,i}$. Let Z be the union of the $A_{n,i}$ over all n and i. Then (using Prop. 1.2.4) Z is a rigid subspace of X and a component part by definition. We claim that Z is irreducible. Suppose U is any non-empty affinoid subdomain contained in Z. We will show that if $\bar{U}$ is the rigid Zariski closure of U in X, then $\bar{U} = Z$. We may suppose that $U \subset A_{n,i}$ for some n and i. Let V be the rigid Zariski closure in X_i of U. Then $V \subseteq A_{n,i}$. It follows by construction that if $n > 0$ there exists a j so that $A_{n-1,j} \cap V$ contains a non-empty affinoid subdomain of X_i. Since this is also contained in Z we can repeat this process inductively and conclude that $\bar{U}$ contains a non-empty affinoid subdomain of $A_{0,j}$ for some $j \geq 0$. Therefore $\bar{U}$ contains $A_{0,0} = A_0$. In particular, $\bar{U}$ contains the union of the $A_{0,j}$ for all $j \geq 0$. Now, working up the inductive definition of $A_{n,i}$ over all n, one sees that $Y_{n,j,k} \subset \bar{U}$ for all n, j, k and consequently, $\bar{U} = Z$. ∎

1.3. Fredholm varieties.

If A is a topological ring, recall that $A\langle T \rangle \subset A[[T]]$ is the set of power series whose coefficients tend to 0 in A. Let $R = \mathcal{O}_K \subset K$ where K is a complete subfield of $\mathbf{C}_p$, as in the previous section. For any integer m, the closed disk of radius p^m about the origin; that is, the rigid analytic subspace of $\mathbf{A}^1_K$ whose $\mathbf{C}_p$-valued points consist of $\{x \in \mathbf{C}_p \mid |x| \leq p^m\}$ will be denoted, as usual, $B[0, p^m]_K = \text{Max}(\mathcal{O}_K \langle p^m T \rangle)$. We have $\mathbf{A}^1_K = \cup_{m=1}^{\infty} B[0, p^m]_K$. If X is a rigid space over K and $F(T)$ is a power series with coefficients in $A(X)$ we will say $F(T)$ is **entire over** X if it is the series associated to a rigid and analytic function on $X \times_K \mathbf{A}^1_K$. If X is a reduced affinoid, an equivalent formulation is: $F(T)$ is entire over X if and only if the power series $F(T)$ lies in (the intersection of) the subrings $A(X)\langle p^m T \rangle \subset A(X)[[T]]$ for all natural numbers m. This is so because $X \times_K \mathbf{A}^1_K = \cup_{m=1}^{\infty} X \times_K B[0, p^m]_K$. The set of entire power series over X, then, forms an $A(X)$-subalgebra of the ring of power series $A(X)[[T]]$, denoted $A(X)\{\{T\}\} \subset A[[T]]$.

If A is a local ring with maximal ideal m_A, and $P(T) = \sum_{n=1}^{\infty} a_n T^n$ is a power series with coefficients $a_n \in A$, we will say that $P(T)$ is **entire over the ring** A if the coefficient a_n is in $m_A^{c_n}$, where c_n are integers such than the sequence of real numbers $\{c_n/n\}$ tends to ∞. The set of entire power series (in T) over a local ring A forms an A-subalgebra of the ring of power series $A[[T]]$ which we denote $A\{\{T\}\} \subset A[[T]]$. For example, if A is a field, then $A\{\{T\}\} = A[T]$, the ring of polynomials in T over A. If A is a complete noetherian local R-algebra with finite residue field, where R is the ring of integers in a finite extension K of $\mathbf{Q}_p$, and if X_A is the rigid space over K associated to A, then we have a natural A-algebra homomorphism

$$A\{\{T\}\} \to A(X_A)\{\{T\}\}.$$

Definition. If A is a local ring, a **Fredholm series** over A is an entire power series over A with constant term 1. If X is a rigid space, a **Fredholm series** over X is an entire power series over $A(X)$ with constant term 1. By the **Fredholm function** associated to a Fredholm series $P(T)$ over $A(X)$ we mean the rigid analytic function on $X \times \mathbf{A}^1$ determined by $P(T)$ (i.e., the image of $P(T)$ under the natural homomorphism $A\{\{T\}\} \to A(X_A)\{\{T\}\}$). If A is a complete noetherian local R-algebra with finite residue field, where R is the ring of integers in a finite extension K of $\mathbf{Q}_p$, and if X_A is the rigid space over K associated to A, an A-**Fredholm hypersurface** in $X_A \times_K \mathbf{A}_K^1$ is the rigid analytic subvariety cut out by a Fredholm series over A. An A-**Fredholm subvariety** in $X_A \times \mathbf{A}^1$ is a rigid analytic subvariety of $X_A \times \mathbf{A}^1$ cut out by a non-empty (possibly infinite) collection of A-Fredholm series. If $\mathcal{S}$ is a subset of the $\mathbf{C}_p$-valued points of $(X_A \times \mathbf{A}^1)(\mathbf{C}_p)$, by the **Fredholm closure** of $\mathcal{S}$ we mean the smallest Fredholm subvariety of $X_A \times \mathbf{A}^1$ whose $\mathbf{C}_p$-valued points contain $\mathcal{S}$.

Note. We extend the above notions to complete semi-local noetherian rings A in the evident manner. For example, if A is a complete semi-local noetherian ring, a power series $P(T) \in A[[T]]$ is a **Fredholm series** over A if the image of $P(T)$ in $A_m[[T]]$ is a Fredholm series, where m ranges through the maximal ideals of A, and A_m is the completion of A at m. Fredholm subvarieties are contained in the complement of $X_A \times_K (0, 1)$ in $X_A \times_K \mathbf{A}_K^1$. Inversion of the second coordinate in $X_A \times \mathbf{A}^1$ therefore sends Fredholm subvarieties to closed, rigid analytic subvarieties of $X_A \times_K B[0, 1] \backslash \{0\}$, where $B[0, 1] \backslash \{0\}$ is the punctured closed unit disk. The rigid Zariski-closure of a set of $\mathbf{C}_p$-valued points $\mathcal{S} \subseteq X_A \times_K \mathbf{A}_K^1(\mathbf{C}_p)$ is a reduced rigid analytic space and therefore the rigid Zariski-closure of $\mathcal{S}$ factors through the nilreduction of the the Fredholm closure of $\mathcal{S}$. We do not, at present, know whether it is generally true that the rigid Zariski-closure of $\mathcal{S}$ is *equal to* this nilreduction.

Lemma 1.3.2. *Suppose X is a reduced affinoid.*

    a. *The only invertible Fredholm series $H(T)$ over X is the constant series $H(T) \equiv 1$.*

b. For any positive integer n, put $Y_n = X \times_K B[0, p^n] \subset Y = X \times_K \mathbf{A}_K^1$. Let $G(T), F(T)$ be two Fredholm series over X with the property that for each n, if $G_n(T), F_n(T)$ denote the restrictions of these Fredholm series to Y_n, we have that $G_n(T)$ divides $F_n(T)$ in the ring of rigid analytic functions on Y_n. Then there is a Fredholm series $H(T)$ over X such that

$$F(T) = G(T)H(T).$$

Proof. **a.** We may assume $X \neq \emptyset$. Let $H(T) = 1 + a_1 T + \cdots + a_n T + \cdots$ be an entire series over A and suppose there exists a $k > 0$ such that $a_k \neq 0$. Then since X is reduced there exists a closed point x of $X(L)$ for some finite extension L of K such that $a_k(x) \neq 0$. Thus we can assume that $\{x\} = X$ and so $A(X) = L$ in which case the lemma is obvious.

b. For each $n \geq 1$ let $H_n(T)$ denote the rigid analytic function on Y_n such that $F_n(T) = G_n(T)H_n(T)$, and note that $H_n(T)$ is uniquely determined by this property since $G_n(T)$, having constant term 1, cannot be a zero divisor. This system of functions $\{H_n\}_n$ is compatible under restriction and patches together to give us a Fredholm series $H(T)$ with the required properties. ∎

Lemma 1.3.3. *Suppose X is a reduced affinoid over K. For any positive integer n, put $Y_n = X \times_K B[0, p^n]$, as before. Let $Z \to X \times_K \mathbf{A}_K^1$ be a closed immersion such that for each n its intersection with Y_n, $Z_n := Z \times_K Y_n$, is the affinoid cut out by some rigid analytic function $g_n(T) \in A(X)\langle p^n T \rangle$ on Y_n with $g_n(0) = 1$. Then there is a Fredholm series $G(T)$ over X such that for $n \geq 1$,*

$$\lim_{n \leq n' \to \infty} g_{n'}(T)|_{Y_n} = G(T)|_{Y_n},$$

where Z is the subvariety cut out by $G(T)$.

Proof. Our hypotheses imply that we may write

$$g_{m+1}(T)|_{Y_m} = g_m(T)h_m(T)$$

where $h_m(T) \in A(Y_m)^*$ and $h_m(0) = 1$. Write

$$h_m(T) = \sum_{k=0}^{\infty} a_{m,k} T^k$$

where $a_{m,k} \in A(X)$. The fact that $h_m(T)$ is a unit on Y_m and $h(0) = 1$ implies

$$|a_{m,k}| < p^{-mk} \quad \text{for} \quad k > 0.$$

It follows, since X is reduced and so the spectral norm determines the topology on $A(X)$, that the product

$$\prod_{m \geq n} h_m(T)$$

converges to a unit on Y_n for all $n \geq 1$. This concludes the proof. ∎

Definition. We say that an affinoid algebra A over K is **relatively factorial** if every $F(T) \in A\langle T \rangle$ with constant term 1 (i.e., $F(0) = 1$) can be factored into a (finite) product of primes (i.e., elements which generate prime ideals).

We call an affinoid X over K relatively factorial if the affinoid algebra $A(X)$ is. It follows that such a factorization is unique up to units. (The point is that no element in $A\langle T \rangle$ with constant term 1 is a zero divisor.) The Tate algebra $K\langle T_1, \ldots, T_n \rangle$ is an example of a relatively factorial (in fact, factorial) algebra by Theorem 5.2.6/1 of [BGR].

Proposition 1.3.4. *Suppose X is a relatively factorial affinoid over K such that $A(X)$ is an integral domain and $F(T)$ is a Fredholm series over $A(X)$ which cuts out the (Fredholm) hypersurface Z. Then every component part W of Z is a Fredholm hypersurface whose defining equation is a divisor of $F(T)$ in the ring $A(X)\{\{T\}\}$.*

Proof. Let $V = X \times_K \mathbf{A}_K^1$ and, as before, for positive integers n let $Y_n = X \times_K B[0, p^n]$ and $Z_n = Z \times_V Y_n$. Suppose U is an admissible open in Z whose rigid Zariski-closure is W. Let C_n be the Zariski-closure of $U \times_V Y_n$ in Y_n.

Because X is relatively factorial, we can write

$$F(T)|_{Y_n} = \prod_{\mathcal{Z} \in \Phi(Z_n)} s_{\mathcal{Z}}(T)$$

where the product is taken over all the irreducible component parts $\mathcal{Z}$ of Z_n, and $s_{\mathcal{Z}}(T)$ is a rigid analytic function on Y_n (with the property that $s_{\mathcal{Z}}(0) = 1$) defining the irreducible component part $\mathcal{Z}$. Because $U \times_V Y_n$ is an admissible open of $Z_n = Z \times_V Y_n$ and $A(X)$ is an integral domain, there exists a collection $\{V_{n,m}\}$ of irreducible reduced affinoid subdomains of Y_n such that $U \times_V V_{n,m}$ is the affinoid cut out by $F(T)$ restricted to $V_{n,m}$ and the collection $\{U \times_V V_{n,m}\}$ is an admissible open cover of $U \times_V Y_n$ (for m ranging through an appropriate index set, call it $\mathcal{M}$).

Put

$$h_{n,m}(T) = \prod_{\mathcal{Z}} s_{\mathcal{Z}}(T),$$

where the product is taken over those irreducible component parts $\mathcal{Z} \in \Phi(Z_n)$ which are in the image of the natural mapping

$$\Phi(U \times_V V_{n,m}) \to \Phi(Z_n),$$

there being such a natural mapping (as discussed in the note in section 1.2) since $U \times_V V_{n,m}$ is a subdomain of Z_n. Then $U \times_V V_{n,m}$ is the affinoid in $V_{n,m}$ cut out by the restriction of $h_{n,m}(T)$ to $V_{n,m}$. An easy argument gives that the map $A(Y_n)/h_{n,m}(T)A(Y_n)$ to $A(V_{n,m})/h_{n,m}(T)A(V_{n,m})$ is an injection from which it follows that the Zariski-closure of $U \times_V V_{n,m}$ in Y_n is the affinoid cut out by $h_{n,m}(T)$. Therefore C_n is the affinoid in Y_n cut out by

$$k_n(T) := \prod_{\mathcal{Z}} s_{\mathcal{Z}}(T),$$

where the product is now taken over those irreducible component parts $\mathcal{Z} \in \Phi(Z_n)$ which are in the union of the images of the natural mappings $\Phi(U \times_V V_{n,m}) \to \Phi(Z_n)$, for m ranging through the full index set $\mathcal{M}$. So, for $n' \geq n$, the restriction of $k_{n'}(T)$ to Y_n may be expressed as follows,

$$k_{n'}(T)_{|Y_n} = \prod_{\mathcal{Z}} s_{\mathcal{Z}}(T),$$

where $\mathcal{Z}$ runs through those elements of $\Phi(Z_n)$ which have the property that they are contained in an element of $\Phi(Z_{n'})$ which is in the image of

$$\Phi(U \times_V V_{n',m'}) \to \Phi(Z_{n'})$$

for some m'. Since these collections of irreducible component parts of Z_n are increasing as n' grows and contained the finite set $\Phi(Z_n)$ they stabilize for large n'. Putting $g_n(T) := k_{n'}(T)_{|Y_n}$ for n' large enough so that the resulting function $g_n(T)$ is unchanged if we augment n' any further, we see that the rigid analytic function $g_n(T)$ defines the subspace $W \times_V Y_n \subset Z_n$ and (of course) divides $F_n(T) := F(T)_{|Y_n}$. By Lemma 1.3.3, there is a Fredholm series $G(T)$ over X which defines W. By Lemma 1.3.2 (b) we have that $G(T)$ divides $F(T)$, i.e., we have a factorization $F(T) = G(T)H(T)$ where $H(T)$ is a (well-defined) Fredholm series over X. ∎

We note that it follows from the proof of this proposition that $\cup_n C_n = W$ and

$$W \times_V Y_n = \bigcup_{n' \geq n} C_{n'} \times_V Y_n.$$

Corollary 1.3.5. *With notation as above, if Z is the rigid Zariski-closure in V of the affinoid $B := B_{\mathcal{Z}}$ cut out by $s_{\mathcal{Z}}(T)$ in Y_n, then Z is irreducible.*

Proof. Suppose Z is not irreducible. Then it follows from the proposition that $F(T) = G(T)H(T)$ for two non-trivial entire series $G(T)$ and $H(T)$

such that $s_Z(T)$ divides the restriction of one and only one of $G(T)$ or $H(T)$ to Y_n. It follows that the rigid Zariski-closure of B is contained in the component part of Z corresponding to $G(T)$ or $H(T)$ which is a contradiction. ∎

Proposition 1.3.6. *Suppose X is a relatively factorial affinoid over K such that $A(X)$ is an integral domain and $F(T)$ is an entire series over X such that $F(0) = 1$ and the hypersurface Z cut out by $F(T)$ is irreducible. Then there exists a prime element $G(T)$ in $A(X)\{\{T\}\}$ such that $G(T)^m = F(T)$ and $G(0) = 1$ for some positive integer m.*

Proof. Set $V = X \times \mathbf{A}^1$. Let W_n be the induced reduced affinoid corresponding to $Z \times_V Y_n$. If

$$\prod_{k=1}^{r_n} \pi_{n,k}(T)^{m_{n,k}}$$

is the prime factorization of $F(T)|_{Y_n}$ $((\pi_{n,k}) \neq (\pi_{n,j})$ if $k \neq j)$ then W_n is the affinoid determined by

$$\prod_{k=1}^{r_n} \pi_{n,k}(T).$$

Now, if $\ell \leq n$, the affinoid determined by the restriction of $\pi_{n,k}(T)$ to Y_ℓ is a product of the $\pi_{\ell,j}(T)$ to some exponents but, by Corollary 7.3.2/10 of [BGR], it is reduced since the fiber product of the affinoid B determined by $\pi_{n,k}(T)$ in Y_n and Y_ℓ is an affinoid subdomain of B. Hence all these exponents are at most 1. Since (by the dimension argument used above) every $\pi_{\ell,j}(T)$ divides the restriction of one and only one of the $\pi_{n,k}(T)$ we see that the fiber product of W_n and Y_ℓ equals W_ℓ. It follows that $W := \cup_n W_n$ is a Zariski closed subspace of V that satisfies the conditions of Lemma 1.3.3 and hence is the hypersurface cut out by a Fredholm series $G(T)$. Moreover, for each n we know there exists an element $H_n(T) \in A(Y_n)$ such that

$$G(T)H_n(T) = F(T).$$

By Lemma 1.3.2.b, these $H_n(T)$ glue together into a Fredholm series $H(T)$ such that

$$F(T) = G(T)H(T).$$

Let U be the closed subvariety of Z determined by $H(T)$. On Y_n it is the affinoid determined by

$$\prod_{k=1}^{r_n} \pi_{n,k}(T)^{m_{n,k}-1}.$$

Suppose there exists n and k such that $m_{n,k} - 1 \neq 0$, then since the rigid Zariski-closure in U of the affinoid cut out by $\pi_{n,k}$ in Y_n is an irreducible component of W, and since (we claim) W is irreducible, it must be equal to W and hence $G(T)$ divides $H(T)$. Continuing in this way we conclude $F(T) = G(T)^m$ where m is equal to any of the non-zero $m_{n,k}$.

Proof of claim. Suppose that W is not irreducible. Then we see that $G(T)$ has a non-trivial factorization into Fredholm series.

$$G(T) = R(T)S(T).$$

It follows that there exist integers n and k, j such that $1 \leq k, j \leq n$, $\pi_{n,k}(T)$ divides $R(T)|_{Y_n}$, and $\pi_{n,j}(T)$ does not. In particular, it follows that the rigid Zariski-closure of the affinoid in Y_n determined by $\pi_{n,k}(T)^{m_{n,k}}$ is an irreducible component of the hypersurface cut out by $R(T)^{m_{n,k}}$ but the rigid Zariski-closure of the affinoid determined by $\pi_{n,j}(T)^{m_{n,j}}$ is not. Hence the rigid Zariski-closure of the first affinoid is not Z, a contradiction. Since W is both reduced and irreducible, it follows that $G(T)$ is prime. ∎

Theorem 1.3.7. *Suppose X is a relatively factorial affinoid over K such that $A(X)$ is an integral domain. Then, if $F(T)$ is an entire series over X such that $F(0) = 1$, there exist distinct prime entire series $P_i(T)$ and positive integers n_i such that $P_i(0) = 1$ and*

$$F(T) = P_1(T)^{n_1} \cdot P_2(T)^{n_2} \cdots P_k(T)^{n_k} \cdots$$

Moreover, this factorization is unique up to order.

Proof. Let notation be as in Proposition 1.3.6. For each n and $1 \leq k \leq n$, we know from Corollary 1.3.5 that the rigid Zariski-closure in Z of the affinoid $B_{n,k}$ in Y_n cut out by $\pi_{n,k}(T)$ is an irreducible component of Z, $Z_{n,k}$, which after Propositions 1.3.4 and 1.3.6 is cut out by the $m_{n,k}$-th power of a prime entire series $P_{n,k}(T)$ such that $P_{n,k}(0) = 1$ for some positive integer $m_{n,k}$. To hide these unsightly double indices, let $P_\gamma(T)$ (for $\gamma = 1, 2, 3, \ldots$) be the list of these $P_{n,k}$'s, putting the indices (n, k) in lexographic order; and correspondingly, denote $m_{n,k}$ by m_γ. Then the product

$$\prod_{\gamma \geq 1} P_\gamma^{m_\gamma}(T)$$

converges by the argument in Lemma 1.3.3 since $P_\gamma^{m_\gamma}(T)$ is a unit on Y_n for large γ and $P_\gamma^{m_\gamma}(T)(0) = 1$. The product converges to $F(T)$ by Lemma 1.3.2. This yields the existence of the factorization. Now suppose $Q(T)$ and $H(T)$ are entire series with constant term 1 in $A(X)\{\{T\}\}$ such that $Q(T)$ is prime and $Q(T)H(T) = F(T)$. Then for every γ, it follows that $P_\gamma(T)$ divides $H(T)$ or $Q(T)$. One of them, say $P_\kappa(T)$, must divide $Q(T)$ since $Q(T)$ is not a unit and $A(X)\{\{T\}\}$ is an integral domain. Hence

$Q(T) = P_\kappa(T)$ since both series are prime and there are no non-trivial units with constant term 1 in $A(X)\{\{T\}\}$ by Lemma 1.3.2. From this, we may deduce that the factorization of $F(T)$ is unique. ∎

In the following, if $U \subset X$ is an affinoid subdomain, and $Z \to X$ a rigid space (over X) we let $Z_U = Z \times_X U$ denote the fiber product.

Lemma 1.3.8. *Suppose X is a relatively factorial affinoid such that $A(X)$ is an integral domain. Let $F(T) = G(T)H(T)$ be an equation in Fredholm series over X such that $G(T) \neq 1$. Suppose $Z(G) \subset Y = X \times \mathbf{A}^1$, the hypersurface cut out by $G(T)$ is a non-empty component part of $Z(F) \subset Y$, the hypersurface cut out by $F(T)$. Then if U is a non-empty affinoid subdomain of X, $Z(G)_U$ is a non-empty component part of $Z(F)_U$.*

Proof. The rigid space $Z(G)_U$ is non-empty, for if it were empty then all the coefficients of G would have to vanish on $U \subset X$ but then Lemma 1.3.1 implies $G(T) = 1$. The only way for $Z(G)_U$ *not to be* a component part of $Z(F)_U$ is if there were an irreducible component $V \subset Z(G) \subset Y$ such that the following holds. If $Y_n = X \times B[0, p^n]$, $V_n = V \times_Y Y_n$. But from Lemma 1.3.1, it follows that $H(T)$ vanishes identically when restricted to V_n^{red} for all n contradicting the assertion that $Z(G)$ is a component part of $Z(F)$. ∎

Lemma 1.3.9. *Suppose X is an affinoid over K such that $A(X)$ is an integral domain, L is a finite Galois extension of K with Galois group Γ and $P(T)$ is a prime entire series over $X_L := X \times L$ such that $P(0) = 1$. Then if $P_1(T), P_2(T), \ldots, P_r(T)$ is the orbit of $P(T)$ under the action of Γ,*

$$Q(T) := \prod_i P_i(T)$$

is a prime entire series over X.

Proof. Suppose there exist entire series $F(T)$ and $G(T)$ over X such that $Q(T)$ divides $F(T)G(T)$ in $A(X)\{\{T\}\}$. Then in $A(X_L)\{\{T\}\}$, $P(T)$ divides $F(T)$ or $G(T)$. Suppose

$$P(T)H(T) = F(T).$$

It follows that $P^\sigma(T)H^\sigma(T) = F(T)$ for all $\sigma \in \Gamma$. Since $P^\sigma(T)$ doesn't divide $P(T)$ unless it equals $P(T)$ by Lemma 1.3.2, we conclude that there exists an entire series $R(T)$ over X_L such that

$$Q(T)R(T) = F(T).$$

Since Q and F have coefficients in $A(X)$ and

$$A(X_L)\{\{T\}\}$$

is an integral domain, it follows that $R(T)$ is an entire series over X. ∎

Lemma 1.3.10. *Suppose that A is a complete local R algebra as in §1.1 and $A = A^0(X_A)$. Suppose $F(T) \in A\{\{T\}\}$ and we have a factorization*

$$F(T) = G(T)H(T)$$

with $G(T)$, $H(T) \in A(X_A)\{\{T\}\}$ and $G(0) = H(0) = 1$. Then $G(T)$, $H(T) \in A\{\{T\}\}$.

Proof. Suppose $G(T) = \sum_{n \geq 0} a_n T^n$ and there exists an n such that $a_n \notin A$. Then, since $A = A^0(X_A)$, there exists an $x \in X_A(\bar{K})$ such that $|a_n(x)| > 1$. Denoting $G_x(T) := \sum_{n \geq 0} a_n(x) T^n$ we see that $G_x(T) \in \bar{K}\{\{T\}\}$, has a zero in $\bar{K}$ of absolute value strictly less than 1. The same must be true about $F_x(T)$ which is patently false. Hence, $G(T) \in A\{\{T\}\}$ and we reach the same conclusion for $H(T)$. ∎

The proof of the next theorem follows the general format of the proof of Theorem 1.3.7.

Theorem 1.3.11. *Let $A = \mathcal{O}[[T_1, \ldots, T_m]]$ be a power series ring in m variables over $\mathcal{O}$, the ring of integers in a field K which is finite over $\mathbf{Q}_p$. Denote its associated rigid space over K by X_A. Let $F(T)$ be a non-trivial Fredholm series over A. Then there exist unique distinct prime Fredholm series over A, $P_i(T)$, and positive integers n_i such that*

$$F(T) = P_1(T)^{n_1} \cdots P_k(T)^{n_k} \cdots.$$

Proof. Recall the notation introduced in the construction of the rigid analytic space X_A in section 1.1. In particular, for each rational number, $1 \geq r > 0$, we have an affinoid subdomain of $X_r \subset X_A$. We may identify X_A with the open unit n dimensional polydisk $v(t_i) > 0$ and X_r with the affinoid subpolydisk defined over $\mathbf{Q}_p$, $v(t_i) \geq r$. Let $B_m := A(X_{1/m}) \otimes_{\mathbf{Q}_p} K$ where $K = \mathbf{Q}_p(\mu_m, p^{1/m})$. Then $B_m \cong K\langle X_1, \ldots, X_n \rangle$ and so is relatively factorial. It follows from Theorem 1.3.7, Lemma 1.3.9 and Lemma 1.3.3 that there exists a countable set I_n such that

$$F(T)|_{X_{1/n}} = \prod_{i \in I_n} P_{n,i}(T)^{m_{n,i}},$$

where $P_{n,i}(T) \in A(X_{1/n})\{\{T\}\}$ are prime entire series on $X_{1/n}$ and the $m_{n,i}$ are positive integers. Now if $n \geq \ell$, arguing as in the proof of Proposition 1.3.6 we see that for each $i \in I_\ell$, there exists a unique $j(n, \ell, i) \in I_n$ such that $P_{\ell,i}(T)$ divides $P_{n,j(n,\ell,i)}(T)|_{X_{1/\ell}}$ and it divides it exactly once. In fact, $m_{\ell,i} = m_{n,j(n,\ell,i)}$. Let $S(n, \ell, j)$ be the subset of I_ℓ such that $P_{n,j}(T)|_{X_{1/\ell}}$

is divisible by $P_{\ell,i}(T)$. Because of Lemma 1.3.8, this set is non-empty and because of Lemma 1.3.3, for $j \in I_n$,

$$P_{n,j}(T)|_{X_{1/\ell}} = \prod_{i \in S(n,\ell,j)} P_{\ell,i}(T).$$

Now let

$$Q_{\ell,i}(T) = \lim_{n \to \infty} P_{n,j(n,\ell,i)}(T).$$

This equals

$$\prod_{k \in T(\ell,i)} P_{\ell,k}(T),$$

where $T(\ell,i) = \cup_n S(n,\ell,j(n,\ell,i))$. If J_ℓ is a subset of I_ℓ such that

$$\bigcup_{i \in J_\ell} T(\ell,i) = I_\ell \quad \text{and} \quad T(\ell,i) \cap T(\ell,j) = \emptyset$$

if $i, j \in J_\ell$, $i \neq j$, then

$$F(T)|_{X_{1/\ell}} = \prod_{i \in J_\ell} Q_{\ell,i}(T)^{m_{\ell,i}}.$$

Now it follows from the definitions that

$$Q_{n,j(n,\ell,i)}(T)|_{X_{1/\ell}} = Q_{\ell,i}(T),$$

and hence for each $i \in I_\ell$ there exists a unique entire series $Q_i(T)$ over X_A such that $Q_i(T)|_{X_{1/\ell}} = Q_{\ell,i}(T)$ and so

$$F(T) = \prod_{i \in J_\ell} Q_i(T)^{m_{\ell,i}}.$$

It follows from Lemma 1.3.10 that $Q_i(T) \in A\{\{T\}\}$ and it follows by an argument similar to that used in Theorem 1.3.7 that $Q_i(T)$ is prime and that the factorization is unique. ∎

Comments about degree and the geometry of Fredholm hypersurfaces. Suppose, for concreteness, that $A = \mathcal{O}[[T_1, \ldots, T_m]]$ is a power series ring in m variables over $\mathcal{O}$, the ring of integers in a field K which is finite over $\mathbf{Q}_p$. Let $X = X_A$ be the open unit ball of dimension m over K. Any Fredholm series over A (which is different from the constant function 1) is either a polynomial of (finite) degree $d > 0$ over A,

$P(T) = 1 + a_1T + a_2T^2 + \cdots + a_dT^d$, with a_d nonzero, or else it is an entire power series $P(T) = 1 + a_1T + a_2T^2 + \cdots$, where there are arbitrarily large integers n for which the corresponding coefficient a_n doesn't vanish. Consider the Fredholm hypersurface $Z \subset X \times \mathbf{A}^1$ whose defining equation is given by $P(T) = 0$. We view Z as rigid space over K. In the first case, i.e., when $P(T)$ is a polynomial of degree d, consider the monic polynomial

$$P^*(U) := U^d P(1/U) = U^d + a_1U^{d-1} + a_2U^{d-2} + \cdots a_d.$$

Form the finite flat A-algebra $E := A[U]/P^*(U)$, and consider the quotient algebra $E \to \bar{E} := E/(U) = A/(a_d)$. Both E and $\bar{E}$ are complete semi-local noetherian $\mathcal{O}$-algebras. Let X_E and $X_{\bar{E}}$ denote their respective rigid analytic spaces over K. We have a canonical immersion $X_{\bar{E}} \hookrightarrow X_E$, and the rigid space Z is the complement of $X_{\bar{E}}$ in X_E. The natural projection $X_E \to X$ is finite flat of degree d, and therefore the projection $Z \to X$ is a mapping of generic degree d (with a decrease in the degree of fibers occurring over the zero-locus of the function a_d on X). If the image of a_d is a unit in $A \otimes_{\mathcal{O}} K$, then $X_{\bar{E}}$ is empty, and $Z = X_E$.

In the case where $P(T)$ has an infinite number of nonvanishing coefficients we shall say that the Fredholm hypersurface Z is of **infinite degree** over X; this is a natural terminology, for, in this case, there are infinitely many points in the fiber of Z above each of the points x lying outside a proper subspace of X.

This discussion, together with Theorem 1.3.11 above, immediately gives:

Corollary 1.3.12. *If Z is the Fredholm hypersurface defined by a Fredholm series over A, then each irreducible component Z_o of Z is either*

1. *a rigid space of the form $X_E \setminus X_{\bar{E}}$ where $X_E \to X$ is the rigid space associated to a finite flat A-algebra, and where $X_{\bar{E}} \subset X_E$ is the rigid space associated to a closed subscheme of Spec E which projects isomorphically onto a proper closed subscheme of Spec A.*

or else is

2. *of infinite degree over X.*

Corollary 1.3.13. *When A is isomorphic to $\mathcal{O}[[T]]$, any irreducible component Z_o of a Fredholm hypersurface defined by a Fredholm series over A has the property that its image in X omits at most a finite number of points.*

Note that if A is the completion at any maximal ideal of the Iwasawa algebra Λ_N of the next section, where $\phi(N)$ is prime to p, then A satisfies the hypotheses of Cor. 1.3.13.

1.4 Weight space.

Let p be a prime number, and fix $\mathbf{C}_p$ as in 1.1. Put $\mathbf{q} = p$ if $p > 2$ and $\mathbf{q} = 4$ if $p = 2$. Let N be a positive integer prime to p. Put $\mathbf{Z}_{p,N}^* := \varprojlim (\mathbf{Z}/Np^n\mathbf{Z})^*$, and

$$\Lambda_N = \varprojlim \mathbf{Z}_p[(\mathbf{Z}/Np^n\mathbf{Z})^*] = \mathbf{Z}_p[[(\mathbf{Z}/N\mathbf{Z})^* \times \mathbf{Z}_p^*]],$$

the projective limits being taken as n tends to infinity. The ring Λ_N (sometimes referred to as the *Iwasawa algebra*) is the completed topological group algebra with $\mathbf{Z}_p$ coefficients of the profinite group $\mathbf{Z}_{p,N}^*$. The group structure of the finite group $(\mathbf{Z}/Np^n\mathbf{Z})^*$ induces, in the standard way, a commutative (associative) Hopf algebra structure on the group ring $\mathbf{Z}_p[(\mathbf{Z}/Np^n\mathbf{Z})^*]$ (which one can view as the affine Hopf Algebra associated to the Cartier dual of the constant group scheme $(\mathbf{Z}/Np^n\mathbf{Z})^*$ over Spec $\mathbf{Z}_p$). Passing to the limit as n tends to ∞, we get a (completed) associative and commutative Hopf algebra structure

$$\Lambda_N \to \Lambda_N \hat{\otimes}_{\mathbf{Z}_p} \Lambda_N.$$

We have a canonical factorization,

$$\mathbf{Z}_{p,N}^* = (\mathbf{Z}/N\mathbf{q}\mathbf{Z})^* \times (1 + \mathbf{q}\mathbf{Z}_p),$$

the first factor being identified with the torsion subgroup of $\mathbf{Z}_{p,N}^*$ and the second factor being the wild part of $\mathbf{Z}_{p,N}^*$; i.e., the multiplicative group of elements congruent to 1 mod $\mathbf{q}N$. If $a \in \mathbf{Z}_{p,N}^*$, we let $[a]$ denote the corresponding element in Λ_N and $\langle\langle a \rangle\rangle$ *to avoid confusion with the diamond operators* the projection of a to $1 + \mathbf{q}\mathbf{Z}_p$ with respect to the above factorization. We have a corresponding factorization of the Iwasawa algebra,

$$\Lambda_N \cong \mathbf{Z}_p[(\mathbf{Z}/N\mathbf{q}\mathbf{Z})^*] \otimes \mathbf{Z}_p[[(1 + \mathbf{q}\mathbf{Z}_p)]],$$

where the second factor admits a (not very natural) isomorphism with a power series ring $\mathbf{Z}_p[[t]]$ (in one variable t over $\mathbf{Z}_p$ by sending $[1 + \mathbf{q}] \in \mathbf{Z}_p[[(1 + \mathbf{q}N\mathbf{Z}_p)]]$ to $1 + t \in \mathbf{Z}_p[[t]]$. If the finite group $(\mathbf{Z}/N\mathbf{q}\mathbf{Z})^*$ is of order prime to p (equivalently, if $p > 2$ and no prime divisor of N is congruent to 1 mod p) then Λ_N is a regular (noetherian) semi-local ring of Krull dimension two, and, conversely, if Λ_N is regular, then N has the property just stated.

Denote by $\mathcal{W} = \mathcal{W}_N$ the rigid analytic space over $\mathbf{Q}_p$, X_{Λ_N} associated to the formal Spf $(\mathbf{Z}_p)$-scheme Spf (Λ_N) as in §1.1. The completed Hopf algebra structure on Λ_N endows $\mathcal{W}_N$ with the structure of (one-parameter) commutative rigid analytic group where multiplication corresponds to the usual multiplication of characters. We refer to $\mathcal{W}_N$, with its structure as

rigid analytic group over $\mathbf{Q}_p$, as **weight space of tame level** N. The
$\mathbf{C}_p$-valued points of $\mathcal{W}_N$ are the continuous $\mathbf{C}_p$-valued characters on $\mathbf{Z}_{p,N}^*$.
If N divides M, the natural projection $\mathbf{Z}_{p,M} \to \mathbf{Z}_{p,N}$ induces an injective
homomorphism $\mathcal{W}_N \hookrightarrow \mathcal{W}_M$ which allows us to identify the rigid analytic
group $\mathcal{W}_N$ with an open and closed rigid analytic subgroup of $\mathcal{W}_M$. In
particular, for any N we identify the connected component of the identity in
the rigid analytic group $\mathcal{W}_N$ with the connected component of the identity,
call it $\mathcal{B}$, in $\mathcal{W}_1$, which can be identified, by restriction, with the continuous
characters on $1+\mathbf{q}\mathbf{Z}_p$. The group of connected components, i.e., the quotient
group $\mathcal{D}_N := \mathcal{W}_N/\mathcal{B}$ may be canonically identified with the $\mathbf{C}_p^*$-dual of the
finite group $(\mathbf{Z}/N\mathbf{q}\mathbf{Z})^*$ and since we have a natural restriction map from
$\mathcal{W}_N$ to $\mathcal{B}$ we may regard $\mathcal{W}_N$ as $\mathcal{D}_N \times \mathcal{B}$. The rigid analytic group $\mathcal{B}$ is
isomorphic to the open disk of radius 1 about the identity element in $\mathbf{C}_p^*$
with its induced multiplicative group structure, and the $\mathbf{C}_p$-valued points of
$\mathcal{B}$ are identified with continuous homomorphisms $\mathbf{Z}_p[[(1 + \mathbf{q}N\mathbf{Z}_p)]] \to \mathbf{C}_p$.
Setting $\Lambda^{(0)} = \mathbf{Z}_p[[(1 + \mathbf{q}\mathbf{Z}_p)]]$, we have that $\Lambda^{(0)}$, the ring of rigid analytic
functions over $\mathbf{Q}_p$ and bounded by 1 on $\mathcal{B}$, is the *classical* Iwasawa algebra,
and the standard completed Hopf algebra structure on $\Lambda^{(0)}$ induces the rigid
analytic group structure on $\mathcal{B}$. If $\kappa \in \mathcal{W}_N$, we let $\langle \kappa \rangle \in \mathcal{B} \subset \mathcal{W}_N$ be the
character $a \mapsto \kappa(\langle\langle a \rangle\rangle)$. It is trivial on $(\mathbf{Z}/N\mathbf{q}\mathbf{Z})^*$.

Definition. If $p > 2$ and $\{s \in \mathbf{C}_p : v(s) > -1 + 1/(p-1)\}$, let $\eta_s : \Lambda_N \to \mathbf{C}_p$
denote the continuous character $a \mapsto \langle\langle a \rangle\rangle^s$, and denote by $\mathcal{B}^*$ the space of
all these characters η_s; i.e.,

$$\mathcal{B}^* := \{\eta_s : a \mapsto \langle\langle a \rangle\rangle^s \mid s \in \mathbf{C}_p, v(s) > -1 + 1/(p-1)\}.$$

If $p = 2$, we let

$$\mathcal{B}^* := \{\eta_s : a \mapsto \langle\langle a \rangle\rangle^s \mid s \in \mathbf{C}_p, v(s) > -1\}.$$

By an **accessible weight-character** (of tame level N) we shall mean a
$\mathbf{C}_p$-valued point of $\mathcal{W}_N$ of the form $\chi\eta_k$ where $k \in \mathbf{Z}_p$ and χ is a character
of finite order. We sometimes refer to such a point κ by the coordinates
(χ, k).

Let $\mathbf{1}$ be the trivial character and $\tau \in \mathcal{W}$ be the the unique character
which is trivial on $(\mathbf{Z}/N\mathbf{Z})^* \times (1 + \mathbf{q}\mathbf{Z}_p)$ and the identity on $\mu(\mathbf{Z}_p)$. If k is
an integer, the k-th power map $(a \mapsto a^k)$ on $\mathbf{Z}_{p,N}^*$ composed with projection
to $\mathbf{C}_p^*$ is an accessible weight-character in the sense just defined, and has
coordinates (τ^k, k). We refer to this homomorphism as the character of
integral weight k (and trivial nebentypus). More generally, if $\epsilon : \mathbf{Z}_{p,N}^* \to$
$\mathbf{C}_p^*$ is any continuous character of finite order, the point $\epsilon \cdot \tau^k \cdot \eta_k \in \mathcal{W}_N(\mathbf{C}_p)$

will be said to be an **arithmetic** point of **weight k and nebentypus character ϵ.**

Most of the rigid analytic spaces we will deal with in this article will come along with a natural (rigid analytic) morphism to $\mathcal{W}_N$ (for some appropriate tame level N). We shall refer to this morphism by the phrase *projection to weight space.* All the rigid analytic morphisms we encounter will preserve projections to weight space.

1.5. The eigencurve as the Fredholm closure of the classical modular locus. (Statement of the main theorems)

For a prime number p and N (a tame level) an integer prime to p , let S denote the finite set of places of $\mathbf{Q}$ consisting of the infinite place, p, and the prime divisors of N. Let $G_{\mathbf{Q},S}$ be the Galois group of the maximal algebraic extension of $\mathbf{Q}$ in $\mathbf{C}$) which is unramified outside S. One would eventually wish to treat all primes p and all tame levels N but in the present article we prove our main theorems only for $p > 2$ and tame level $N = 1$. For the rest of this section, then, we let p be an odd prime number and $N = 1$.

In section 5.1 below we shall define a certain collection of residual $G_{\mathbf{Q},S}$-representations into finite fields of characteristic p which include, up to equivalence, (the semi-simplification of) all characteristic p residual Galois representations coming from classical modular forms of p-power level. These we call p-**modular** residual representations of $G_{\mathbf{Q},S}$ of tame level 1. We also recall in detail in 5.1 the notion of pseudo-representation, following [Wi], [T], [Ny], [Ro] and [H-NO]. Briefly, if D is a topological ring, to any continuous representation $\rho : G_{\mathbf{Q},S} \to \mathrm{GL}_2(D)$

$$g \mapsto \begin{pmatrix} a(g)\ b(g) \\ c(g)\ d(g) \end{pmatrix},$$

which sends complex conjugation in $G_{\mathbf{Q},S}$ to the diagonal matrix

$$\begin{pmatrix} 1 & 0 \\ 0 & -1 \end{pmatrix},$$

its associated 2-dimensional pseudo-representation r with values in D is taken to be the pair of (continuous) functions $\alpha, \delta : G \to D$ given by the formulas, $\alpha(g) = a(g);\ \delta(g) = d(g)$, noting that α and δ, so defined, are invariants of the equivalence class of ρ. It is convenient to define, as well, the function $\xi : G \times G \to D$ by the formula $\xi(g, h) = b(g)c(h)$, and although the function ξ is determined by either function α or δ because

$$\xi(g, h) = \alpha(gh) - \alpha(g)\alpha(h) = \delta(hg) - \delta(h)\delta(g),$$

we refer to the triple of functions $r := (\xi, \alpha, \delta)$, as *the* pseudo-representation associated to ρ. A (general) pseudo-representation is roughly a triple of functions satisfying the same relations as an arbitrary (ξ, α, δ) obtained from a representation in the above way. For any 2-dimensional pseudo-representation $r = (\xi, \alpha, \beta)$ on $G_{\mathbf{Q},S}$ with values in D and any $g \in G_{\mathbf{Q},S}$ we set

$$\text{trace}(r(g)) = \alpha(g) + \beta(g) \quad \text{and} \quad \det(r(g)) = \alpha(g)\beta(g) - \xi(g).$$

In 5.1 we shall consider the (complete noetherian, semi-local) ring R_p defined as the universal (p-complete, noetherian, semi-local) deformation ring for (degree two) modular pseudo-representations of $G_{\mathbf{Q},S}$ of tame level 1. Being universal, there is a pseudo-representation r^{univ} from $G_{\mathbf{Q},S}$ with values in R_p which specializes to any deformation of a p-modular pseudo-representation. Here let us just note that R_p includes among its factor rings the universal Galois deformation rings of all absolutely irreducible modular residual representations (i.e., residual representations over finite fields of characteristic p) attached to newforms of levels dividing powers of p (see 5.1).

As we show in section 5.1, R_p may viewed in a natural way as a Λ-algebra (with its weight Λ-algebra structure).

Let $X_{R_p} = X_p$ be the rigid analytic space over $\mathbf{Q}_p$ associated to the complete semi-local ring R_p regarded as a $\mathbf{Z}_p$-algebra, as in §1.1, endowed with the projection to weight space, $\pi : X_p \to \mathcal{W}$ coming from the weight Λ-algebra structure of R_p. Using [Wi], one can see that any two-dimensional continuous representation $\rho : G_{\mathbf{Q},S} \to \text{GL}_2(\mathbf{C}_p)$ of tame level 1 whose associated residual representation is p-modular corresponds to a (unique) point $x_\rho \in X_p(\mathbf{C}_p)$ (given by its associated pseudo-representation).

To a normalized (classical) modular eigenform f on $\Gamma_1(M)$ over $\mathbf{C}_p$, where M is a positive multiple of p, of weight k and character χ with q-expansion

$$f(q) = c_0 + q + c_2 q^2 + c_3 q^3 + \cdots$$

one can attach an odd degree 2 representation ρ_f over $\mathbf{C}_p$ (up to equivalence) whose associated pseudo-representation $r_f := (\xi, \alpha, \delta)$ satisfies

$$\text{trace}(r_f(Frob_l)) = c_l \quad \text{and} \quad \det(r_f(Frob_l)) = \chi(l)l^{k-1}$$

for primes l not dividing M induced from the universal pseudo-representation via a unique homomorphism $R_p \to \mathbf{C}_p$, i.e., a unique $\mathbf{C}_p$-valued point $x_f \in X_p(\mathbf{C}_p)$.

If, further, M is a power of p and f is of finite slope, i.e., if its U_p-eigenvalue $u := c_p$ is nonzero, denote by $\tilde{x}_f$ the point $(x_{\rho_f}, 1/u) \in (X_p \times \mathbf{A}^1)(\mathbf{C}_p)$.

Definition 1. The **classical modular locus of tame level 1**,

$$\mathcal{M} \subset (X_p \times \mathbf{A}^1)(\mathbf{C}_p)$$

is the set of points $\tilde{x}_f = (x_f, 1/u)$ where f runs through all (classical) modular eigenforms of finite slope, as above, on $\Gamma_1(p^m)$ for $m \geq 1$.

In this article we shall give two different constructions of rigid analytic curves which parameterize the collection of all overconvergent eigenforms of tame level $N = 1$ and of finite slope (i.e., having non-zero U_p-eigenvalue).

Definition 2. The **eigencurve** $C = C_p$ of tame level $N = 1$ is the Fredholm closure (see Section 1.2) in $X_p \times \mathbf{A}^1$ of the classical modular locus (of tame level 1).

Note: We have jumped the gun in our terminology for the above definition does not immediately allow us to see that C_p is a *curve*. In this paper we shall *not* define the eigencurve of general tame level, and only concern ourselves with tame level $N = 1$. Thus, in what follows, we will drop the tag "of tame level $N = 1$" and refer to the eigencurve of tame level 1 simply as **the eigencurve**. It follows from the above definition that C_p and its nilreduction C_p^{red} are *nested* in the sense defined in section 1.1, and therefore Proposition 1.2.5 implies that C_p^{red} is the union of its irreducible components.

Here is a statement of some of the main results of this paper. By the reduced eigencurve, we mean the nilreduction of the eigencurve C_p

Theorem A. *Let $C_o \subset X_p \times \mathbf{A}^1$ be an irreducible component part of the reduced eigencurve. Then there exists an element $\gamma_0 \in R_p^*$ such that the mapping*

$$X_p \times \mathbf{A}^1 \to \mathcal{W} \times \mathbf{A}^1$$

induced by the Λ-algebra homomorphism $\Lambda\{\{T\}\} \to R_p\{\{T\}\}$ $(T \mapsto \gamma_o \cdot T)$ when restricted to C_o is generically an isomorphism of C_o onto a Fredholm hypersurface in $\mathcal{W} \times \mathbf{A}^1$.

By "generically" we simply mean, since we are dealing with a rigid analytic *curve*, after the exclusion of a discrete set of points (i.e., a 0-dimensional Zariski-closed subspace).

Theorem A follows directly from Corollary 7.6.2, and its proof. ∎

Theorem B. *The natural projection of any irreducible component of the reduced eigencurve to weight space is* **component-wise almost surjective** *in the sense that given any irreducible component of the reduced eigencurve, the complement of its image in the unique irreducible component of weight space containing that image is (empty, or) consists of at most a finite number of weights.*

Proof. This is Corollary 7.4.2 below. The argument for it uses Theorem A together with the Corollary 1.3.13.

Theorem C. *The projection of the reduced eigencurve to weight space is* **locally in-the-domain finite flat** *in the sense that C_p^{red} is covered by admissible affinoid domains $\mathcal{U}$ such that the restriction of projection to weight space to $\mathcal{U}$ is a finite flat mapping of $\mathcal{U}$ onto its image in $\mathcal{W}$.*

Proof. This follows directly from the construction of D in Chapter 7 (Prop. 7.2.2) together with the fact that $D \cong C_p^{\text{red}}$ (Theorem. 7.5.1).

It follows from Theorem C that C_p is a *curve* in the sense that it is an equidimensional rigid analytic space of dimension 1.

Since the universal pseudo-representation ring R_p is a complete semi-local ring whose maximal ideals are in one-one correspondence with (certain) equivalence classes of semi-simple residual representations $G_{\mathbf{Q},S} \to \mathrm{GL}_2(\bar{\mathbf{F}}_p)$ (see section 5.1 below) we have:

Theorem D. *If two classical modular points of tame level N, $\tilde{x}_f, \tilde{x}_{f'} \in (X_p \times \mathbf{A}^1)(\mathbf{C}_p)$ lie on the same irreducible component of the reduced eigencurve, then $f \equiv f'$ modulo the maximal ideal of $\mathcal{O}_{\mathbf{C}_p}$ in the sense that their Fourier expansions are congruent modulo that maximal ideal (and their associated residual representations have equivalent semi-simplifications).*

In view of Theorem D, given $\bar{\rho} : G_{\mathbf{Q},S} \to \mathrm{GL}_2(\bar{\mathbf{F}}_p)$, a semi-simple representation, we will say that a component C_j of the eigencurve is of **type** $\bar{\rho}$ if one (equivalently: all) of its classical modular forms have the semi-simplifications of their associated residual representations equivalent to $\bar{\rho}$. Define $C_{\bar{\rho}}$ to be the rigid analytic subspace of C_p which is the union of all components of the eigencurve of type $\bar{\rho}$. We say a rigid subspace V of a rigid space U is **admissibly closed-open** if there exists a rigid subspace W of U disjoint from V such that $\{V, W\}$ is an admissible open cover of U.

Corollary 1.5.1. *The subspaces $C_{\bar{\rho}}$ are admissibly closed-open in C_p.*

Proof. Indeed, the proof we sketched for theorem D understates the case a bit: Since the ambient rigid-analytic space $X_p \times \mathbf{A}^1$ breaks up unto the disjoint union of rigid analytic spaces $X_{\bar{\rho}} \times \mathbf{A}^1$ where $X_{\bar{\rho}}$ is the universal deformation space of the pseudo-representation (of $G_{\mathbf{Q},S}$) associated to $\bar{\rho}$, for $\bar{\rho}$ running through the semi-simple residual' representations described above and since $C_{\bar{\rho}} = C_p \cap X_{\bar{\rho}} \times \mathbf{A}^1$, Corollary 1.5.1 follows.

The classical modular forms of finite slope correspond to (certain) points on the eigencurve. What interpretation is there for the remaining points? For this we have Theorem 6.2.1 below, a paraphrase of which is given by the following theorem.

Theorem E. *There is a natural one-one correspondence*

$$c \longleftrightarrow f_c$$

between the $\mathbf{C}_p$-valued points c on the eigencurve and overconvergent (see section 5.2) modular eigenforms f_c of finite slope of tame level 1. This one-one correspondence is characterized by the conditions that $\pi(c)$, the natural projection of c to X_p, when viewed as a pseudo-representation, is the pseudo-representation associated to the overconvergent modular eigenform f_c by the Gouvêa-Hida Theorem (see section 5.2), and the natural projection of c to $\mathbf{A}^1_{\mathbf{Q}_p}$ is the reciprocal of the U_p-eigenvalue of f_c.

We note that the notion of an overconvergent eigenform of integral weight was introduced in [K-PMF] and its generalization to more general arithmetic weight characters was made in [C-CO] and [C-COHL] and to non-arithmetic weight in [C-BMF] (see also [G-ApM]). We might note that the compatibility of these two definitions was not formally established previously but is now, as a consequence of Corollary 2.2.6.

Theorem F. *The reduced eigencurve is the rigid Zariski-closure of the classical modular locus.*

Proof. By Theorem E, the points of the eigencurve correspond to overconvergent eigenforms. By Theorem B, every irreducible component of the eigencurve has points of all but a finite number of possible weights in the component of weight space to which it projects. In particular, for each irreducible component C_o of the eigencurve, there is a positive integer k_o such that C_o has points of all integral weights $\geq k_o$. One immediately sees from this that every irreducible component C_o has points of positive integral weight k and of slope $< k - 1$ (and different from $\frac{k-1}{2}$). By Corollary B5.7.1 of [C-BMF] there is an affinoid disk in C_o whose projection to weight space is an isomorphism $C_o \to W_o \subset W$ onto an affinoid W_o in W, and such that for a topological dense set $\mathcal{D}_o$ in the set of arithmetic weights in $W_o(\mathbf{Q}_p)$ the points of C_o mapping to $\mathcal{D}_o$ are given by classical eigenforms. This proves Theorem F.

A similar argument to that used above also gives the following result (which is Proposition 7.6.5 below).

Theorem G. If $\{f_n\}$ is a sequence of normalized eigenforms of tame level 1 with Fourier coefficients in $\mathbf{C}_p$ such that f_n has weight $\kappa_n \in \mathcal{W}(\mathbf{C}_p)$, $v(a_p(f_n))$ is bounded independently of n, the sequence κ_n converges in $(\mathbf{Z}/(p-1)\mathbf{Z}) \times \mathbf{Z}_p$ to some weight $\kappa \in \mathcal{W}(\mathbf{C}_p)$ and the sequence $\{f_n(q) \in \mathbf{C}_p[[q]](1/q)\}$ converges coefficientwise to a series $f(q)$. Then, $f(q)$ is the q-expansion of an overconvergent modular eigenform f of tame level 1, weight κ, and finite slope. Let f be a Katz modular eigenfunction (see section 2.2 below) of tame level 1, of finite slope, with (normalized) Fourier expansion

$$\tilde{f} = q + a_2 q^2 + a_3 q^3 + \cdots$$

with coefficients $a_j \in \mathbf{C_p}$. Suppose that f is of accessible weight-character (cf. section 1.4) (χ, s) with $s \in \mathbf{Z}_p$. Then f is overconvergent if and only if f is the limit in the q-expansion topology of a sequence of classical, normalized, cuspidal-overconvergent (cf. section 3.6) modular eigenforms of tame level 1.

In section 5.2 we show that the eigencurve is universal for certain rigid analytic families of (overconvergent, finite slope) modular eigenforms. We lack, however, a completely satisfactory theory here, for we do not have an a priori concept of infinitesimal deformation of (overconvergent, finite slope) modular eigenforms. By "a priori" we mean prior to constructing the eigencurve: it would be good, for example, when given an overconvergent eigenform of finite slope f, to have a theory which produces the Zariski tangent space of the eigencurve at the point f, and produces it in a reasonably computable manner (perhaps as an appropriate cohomology group).

We will now summarize some of the key properties of the eigencurve in a purely rigid-geometric way without mentioning deformations. The p-adic eigencurve of tame level 1 is a rigid analytic curve C with the following properties: There is a rigid analytic morphism κ from C to $\mathcal{W}$ and rigid analytic functions $t_n, n \geq 1$ with the following properties: Suppose

$$f(q) = \sum_{n \geq 0} a_n q^n$$

is a normalized eigenform on $X_1(p^r)$, $r > 0$, of weight k, such that $a_p \neq 0$ whose character on $(\mathbf{Z}/p^r\mathbf{Z})^*$ is $\epsilon \chi \tau^{-k}$, where $\chi \in \mathcal{B}$. Then there exists a unique point x_f on C such that $\kappa(x_f) = \epsilon \chi \eta_k$ and

$$t_n(x_f) = a_n.$$

Call these points the classical points. On the other hand, if $x \in C(\mathbf{C}_p)$ such that $\kappa(x) = \epsilon \chi \eta_k$ where k is an integer, $\chi \in \mathcal{B}$ is a character of finite order and $v(t_p(x)) < k - 1$ then x is classical. Moreover, as we stated above, C is the rigid Zariski-closure of the classical locus. This is not, however, enough to characterize C.

Chapter 2. Modular Forms.

2.1 Affinoid sub-domains in modular curves.

If $M \geq 1$ is an integer, and S is a scheme, $\mu_{M/S}$ denotes the finite flat group scheme over S given by as the kernel of multiplication by M in the multiplicative group (over S). Let p be a prime number. Fix an integer N prime to p, and $m \geq 1$ and let us begin our discussion in level Np^m. Our running assumption in this section is that, unless otherwise explicitly mentioned, $Np^m \geq 5$. Let $Y_1(N \cdot p^m)$ be the (uncompactified) modular curve classifying isomorphism classes of pairs $(\mathcal{E}, \alpha)$ where $\mathcal{E}$ is an elliptic curve over a $\mathbf{Q}_p$-scheme S, $\alpha : \mu_{N \cdot p^m} \hookrightarrow \mathcal{E}$ is an injection (over the base S). This is equivalent to the $[\Gamma_1[(Np^m)]$ problem (over $\mathbf{Q}_p$) as discussed in [KM]. Since $Np^m \geq 5$, the above problem is rigid and representable in the terminology of 4.7 of [KM] (cf. pp. 327, 328 of [KM]) and we have a universal family

$$(u : \mathbf{E}_1(Np^m) \to Y_1(Np^m), \ \alpha).$$

Let ω be the invertible line bundle on $Y_1(Np^m)_{/\mathbf{Q}_p}$ given by

$$\omega \ = \ u_* \Omega^1_{\mathbf{E}_1(N \cdot p^m)/X_1(N \cdot p^m)}.$$

By the discussion of (10.13) (cf. the summarizing table (10.13.9.1)) of [KM], since $Np^m \geq 5$ we have an extension of the line bundle ω to a line bundle (which we denote by the same letter) on the compactification $X_1(Np^m)_{/\mathbf{Q}_p}$, and a canonical Kodaira-Spencer isomorphism

$$\omega^{\otimes 2} \cong \Omega^1_{X_1(Np^m)_{/\mathbf{Q}_p}}(\log \text{ cusps}),$$

of line bundles on $X_1(Np^m)$. Assuming now that $N \geq 5$, the reader is invited to consider, as well, a noncompact model (over $\mathbf{Z}_p$) of this moduli scheme given by the analysis in [KM] of the $[\Gamma_1[(Np^m)]$ problem (over $\mathbf{Z}_p$) whose closed fiber is the (geometrically irreducible) incomplete Igusa curve of level Np^m (cf. Chapter 12 of [KM]).

It is sometimes useful to keep in mind the entire tower of modular curves

$$(*) \ \ldots \to X_1(N \cdot p^{n+1}) \to X_1(N \cdot p^n) \to \ldots X_1(N \cdot p^{m+1}) \to X_1(N \cdot p^m)$$

as n tends to ∞ $(m \leq n)$. Consider, for example, a triple $(\mathcal{E}, \alpha_N, \alpha_\infty)$ where $\mathcal{E}$ is an elliptic curve over $\bar{\mathbf{Z}}_p$ with ordinary reduction mod p (i.e., with either good ordinary reduction, or multiplicative type reduction), where $\alpha : \mu_N \hookrightarrow \mathcal{E}$ is an injective homomorphism (over Spec $\bar{\mathbf{Z}}_p$) and where α_∞

is an embedding $\mathbf{G}_m[p^\infty] \hookrightarrow \mathcal{E}[p^\infty]$ of the p-divisible group attached to $\mathbf{G}_m$ (over Spec $\bar{\mathbf{Z}}_p$) into the p-divisible group attached to $\mathcal{E}$. Such a triple $(\mathcal{E}, \alpha_N, \alpha_\infty)$ (which we will sometimes refer to as a p-*trivialized* elliptic curve with tame $\Gamma_1(N)$-structure) gives rise to a cofinal system of Spec $\bar{\mathbf{Z}}_p$-valued points of

$$(**) \quad \ldots \to Y_1(N \cdot p^{n+1}) \to Y_1(N \cdot p^n) \to \ldots Y_1(N \cdot p^{m+1}) \to Y_1(N \cdot p^m),$$

as follows. For each n restrict α_∞ to get an embedding $\alpha_{p^n} : \mu_{p^n} \hookrightarrow \mathcal{E}$, and define $\alpha_{Np^n} : \mu_{Np^n} \hookrightarrow \mathcal{E}$ to be the unique embedding which restricts to α_N and α_{p^n} respectively on μ_N and μ_{p^n}. We then associate to the triple $(\mathcal{E}, \alpha_N, \alpha_\infty)$ the cofinal system $n \mapsto (\mathcal{E}, \alpha_{Np^n})$. Conversely, a cofinal system of Spec $\bar{\mathbf{Z}}_p$-valued points of $(**)$ comes from such a triple $(\mathcal{E}, \alpha_N, \alpha_\infty)$. Given such a triple, the embedding α_∞ induces an isomorphism $\hat{\mathbf{G}}_m \cong \hat{\mathcal{E}}$ of formal completions at the origin. The image of the standard differential dt/t on $\hat{\mathbf{G}}_m$ under this isomorphism then gives us a specific differential w generating the fiber of the line bundle ω over the point $(\mathcal{E}, \alpha_{Np^n})$ of $Y_1(N \cdot p^n)(\bar{\mathbf{Q}}_p)$, for any $n \geq 1$. (Although we will not use this point of view in the present paper, one way of thinking of this, is to invoke the projective limit $Y_1(N \cdot p^\infty)$ of the projective system $(*)$ of curves and to view the process of stipulating a generating section $w \in \omega$ over $Y_1(N \cdot p^\infty)$ as giving us a canonical trivialization of the line bundle ω over $Y_1(N \cdot p^\infty)$. See I.1 of [G] for slightly more discussion of this.)

Let us now return to the modular curve $X_1(N \cdot p^m)$, taken over the base $\mathbf{Q}_p$.

First, let $Z_1(N \cdot p^m)$ denote the inverse image under reduction to $\mathrm{Spec}(\mathbf{F}_p)$ of the complement of the supersingular points on the irreducible component containing the cusp ∞ in the above model of $X_1(N \cdot p^m)$ over $\mathrm{Spec}(\mathbf{Z}_p)$. We have that $Z_1(N \cdot p^m)$ is an affinoid subdomain of the rigid analytic space $X_1(N \cdot p^m)$ over $\mathbf{Q}_p$ (see [C-RLC]). In §B2 of [C-BMF] a system of affinoid neighborhoods $Z_1(N \cdot p^m)(v)$ (called there $X_1(N \cdot p^m)(v)$) of $Z_1(N \cdot p^m)$ in $X_1(N \cdot p^m)$ is described, where the parameter v (which we sometimes refer to as the *radius*) is a rational number allowed to vary in the range $0 \leq v < p^{(2-m)}/(p+1)$. Let I_m denote the set of such v. In particular, $Z_1(N \cdot p^m)(0) = Z_1(N \cdot p^m)$ and when $v > 0$ these neighborhoods are strict (cf. the definition of strict neighborhood below).

We shall briefly sketch the definition of $Z_1(N \cdot p)(v)$ here for $p \geq 5$ (for the full story, see [C-BMF], section B2). Let A be the level 1, characteristic p, Hasse invariant modular form. For a non-negative rational number $v \in \mathbf{Q}$, $Z_1(Np)(v)$ is the affinoid subdomain of $X_1(Np)$ whose points over $\mathbf{C}_p$ correspond to pairs $(\mathcal{E}, \alpha)$ over $S = Spec(R)$, where R is the ring of integers in $\mathbf{C}_p$, as above such that $v(A(\bar{\mathcal{E}}, \bar{\eta})) \leq v$ where η generates $\Omega^1_{\mathcal{E}/S}$ and $(\bar{\mathcal{E}}, \bar{\eta})$

is the reduction of $(\mathcal{E}, \eta)$ modulo p, and $\alpha(\mu_p)$ is the canonical subgroup (see Chapter 3 of [K-PMF] where the canonical subgroup is defined for all N and $p \geq 5$). We will also have occasion to consider the modular curves $X_1(Np^m, \ell)$, where ℓ is a prime not dividing Np, which classify triples $(\mathcal{E}, \alpha, C)$ where $\mathcal{E}$ and α are as above and C is a subgroup scheme of $\mathcal{E}$ of order ℓ. Then we have an affinoid subdomain $Z_1(Np^m, \ell)$ with affinoid neighborhoods $Z_1(Np^m, \ell)(v)$ for $v \in I_m$ defined in a similar way.

In what follows let us fix a rational number v and focus attention on the affinoid sub-domain $Z_1(N \cdot p^m)(v)$ of that width. We also reinstitute the hypothesis $N \cdot p^m \geq 5$.

Definition. Let $\mathcal{V}$ be a rigid analytic space. By an **overconvergent family** of rigid analytic functions on $Z_1(N \cdot p^m)$ parameterized by $\mathcal{V}$ we mean a rigid analytic function F on $Z_1(N \cdot p^m) \times \mathcal{V}$ with the property that there is an admissible open covering of $\mathcal{V}$ by affinoids V_j $(j \in J)$ such that there are positive numbers $v_j \in I_m$ $(j \in J)$ for which the restriction of F to V_j extends to a rigid analytic function on $Z_1(N \cdot p^m)(v_j) \times V_j$. More generally,

Definition. If $X \to Y$ is a morphism of rigid spaces over K, we say that X is **affinoid** over Y if for each affinoid subdomain Z in Y, X_Z is an affinoid. Suppose $W \to Y$ is a map of rigid spaces and X is a rigid subpace of W which is affinoid over Y, then we say that an admissible open subspace V of W is a **strict** neighborhood of X over Y in W if there is an admissible covering of Y by affinoids V_j such that for each j there exists an affinoid neighborhood U_j of X_{V_j} in V_{V_j} such that the morphism $\bar{X}_{V_j} \to \bar{U}_{V_j}$ factors through a subscheme of $\bar{U}_{V_j}$ finite over $\bar{Y}$. Finally, if X, W and Y are as above, we say that a rigid function f on X is **overconvergent** in W over Y if f extends to some strict neighborhood of X in W over Y. When Y is $Max(K)$, we just say f is overconvergent on X in W.

We denote the ring of these functions on X which are overconvergent in W over Y by $A^\dagger(X/Y, W)$ and the subring of those functions bounded by 1 we denote $A^\dagger(X/Y, W)^0$. We drop Y from the notation when it is a point and we drop W when it is understood from the context. In particular, the ring of overconvergent families of rigid analytic functions on $Z_1(Nq)$ parameterized by $\mathcal{V}$ is denoted $A^\dagger(Z_1(Nq)_{\mathcal{V}}/\mathcal{V})$. If $B_K[0,1]$ and $B_K(0,1)$ denote the affinoid and wide open unit disks over K, $A^\dagger(B_{\mathbf{Q}_p}[0,1]_{B_{\mathbf{Q}_p}(0,1)}/B_{\mathbf{Q}_p}(0,1))^0$ may be identified as the ring of power series in the parameter variable T over $\Lambda^{(0)}$ which are of the form $\sum_{n=0}^\infty \lambda_n T^n$ such that for some positive real number a, $\lambda_n \in I_1^{[an]}$ where I_1 is the maximal ideal of $\Lambda^{(0)}$. (For more discussion of this notion see §A5 of [C-BMF] as well as §1 of [C-CCS].) It is less straightforward to define the notion of overconvergent modular form or family of overconvergent modular forms. Cf. section 2.4 below for the general definition but, for now, we define over-convergence only for integral weights:

Let ω^k for $k \in \mathbf{Z}$ denote the k-th tensor power of ω, and consider the vector space of rigid analytic sections of ω^k over the affinoid $Z_1(N \cdot p^m)(v)$. More explicitly, letting K be a complete subfield of $\mathbf{C}_p$, and $Z_1(N \cdot p^m)(v)_{/K}$ the affinoid over K induced from $Z_1(N \cdot p^m)(v)$ by base change, define the K-vector space

$$M_k(Np^m, v; K) := \omega^k\big(Z_1(N \cdot p^m)(v)_{/K}\big),$$

which has a natural K-Banach space structure, as described in section B2 of [C-BMF].

If $v > 0$, an element of $M_k(Np^m, v; K)$ is called a **v-overconvergent modular form** (of weight k and level $N \cdot p^m$). By definition an **overconvergent modular form** (of tame level N with coefficients in K) is a v-overconvergent modular form of level $N \cdot p^m$ for some $v > 0$, and some $m > 0$, i.e., it is an element in the p-adic Frechet space given by the union of these p-adic Banach spaces. (Compare this with the definition of overconvergent function, and form in A5 and B4 of [C-BMF] as well as in [C-COHL]).

The elements of $M_k(Np^m, v; K)$ for $v = 0$, i.e., the space of rigid analytic sections of ω^k over $Z_1(N \cdot p^m)$ will be referred to simply as **convergent**; convergent modular forms on $X_1(Np^m)$ of weight k coincide with Katz p-adic modular functions of that tame level and weight-character (see section 2.4 below).

The standard automorphisms and correspondences on the curve $X_1(N \cdot p^m)$ (which we shall list below) induce natural continuous operators on the Banach space $M_k(Np^m, v; K)$. We explain this in detail in the next chapter. For now, we will discuss the diamond operators.

For any positive integer m, the group $(\mathbf{Z}/Np^m\mathbf{Z})^*$ acts on the modular curve $X_1(Np^m)$ over $\mathbf{Q}_p$ as follows. For $r \in (\mathbf{Z}/Np^m\mathbf{Z})^*$, the operator $\langle r \rangle$ sends the pair $(\mathcal{E}, \alpha)$ to $(\mathcal{E}, r \cdot \alpha)$. This gives us a compatible action of the profinite group

$$\mathbf{Z}_{p,N}^* = \varprojlim (\mathbf{Z}/Np^m\mathbf{Z})^*$$

on the tower (*) and more specifically, for every positive integer m, we have compatible actions of the finite quotient group $(\mathbf{Z}/Np^m\mathbf{Z})^*$ of $\mathbf{Z}_{p,N}^*$ on the pair

$$(**) \quad (X_1(Np^m), \ \omega_{X_1(Np^m)}).$$

Lemma 2.1.1. *The action described above induces a continuous homomorphism from the topological group* $\mathbf{Z}_{p,N}^*$ *to the group* $\mathrm{Aut}_{\mathrm{rig.an.}}(Z_1(Np^m)(v))$,

$\omega_{Z_1(Np^m)(v)}$) *of rigid analytic automorphisms of the pair consisting of the affinoid* $Z_1(Np^m)(v)$ *in* $X_1(Np^m)$ *and the restriction of the line bundle* ω *to this affinoid. This induces a natural action of the Iwasawa algebra* Λ_N *as an algebra of continuous operators on the Banach space* $M_k(Np^m, v; K)$.

2.2 Eisenstein series.

In this section we work in tame level $N = 1$. We recall and continue the development of the theory of the fundamental family of modular forms, i.e., the Eisenstein family. Some of this theory was discussed and/or proved on pages 446-448 of [C-BMF] as well as in [C-CCS]. See also section 7.1 of Hida's book [H-ET] for further proofs and discussion.

Let $\mathcal{W} = \mathcal{W}_N$ for $N = 1$ and $\mathcal{W}^+$ denote that part of weight space $\mathcal{W}$ consisting of characters κ such that $\kappa(-1) = 1$. We have the p-adic ζ-function $\zeta^*(\kappa)$ (notation as in section B1 of [C-BMF]) which is a rigid analytic function on $\mathcal{W}^+$ away from the point $\kappa = 1$ (at which it has a simple pole). If $\kappa = (\chi, s)$ is an accessible weight,

$$\zeta^*(\kappa) = L_p(\chi, 1 - s),$$

where $L_p(\chi, s)$ is the Kubota-Leopoldt p-adic L-function.

For $n \geq 1$ set

$$\sigma_\kappa^*(n) := \sum_{d|n,(d,p)=1} \kappa(d)d^{-1},$$

which we view as an Iwasawa function on $\mathcal{W}^+$. If $\kappa \neq 1$ and $\zeta^*(\kappa) \neq 0$, set

$$E_\kappa(q) = 1 + \frac{2}{\zeta^*(\kappa)} \cdot \sum_{n=1}^{\infty} \sigma_\kappa^*(n)q^n,$$

and for $\kappa = 1$, put $E_\kappa(q) = 1$. Compare the discussion on page 447 of loc. cit. Let us refer to this family of q-expansions $E_\kappa(q)$, parameterized by the rigid analytic subspace

$$\mathcal{W}_{\text{eis}} = \{\kappa \in \mathcal{W}^+ \text{ and } \zeta^*(\kappa) \neq 0\}$$

as the **basic Eisenstein family.** Note that $\mathcal{B} \subset \mathcal{W}_{eis}$. We also use the notation

$$\mathbf{E}^o(q) := 1 + \epsilon_1 q + \epsilon_2 q^2 + \cdots,$$

where the coefficients ϵ_n are the rigid analytic functions in $A(\mathcal{W}_{\text{eis}})$ with the property that $\epsilon_n(\kappa) = 2\sigma_\kappa^*(n)/\zeta^*(\kappa)$. The power series $\mathbf{E}^o(q) \in A(\mathcal{W}_{\text{eis}})[[q]]$ has the corresponding property that

$$\kappa(\mathbf{E}^o(q)) = E_\kappa(q).$$

One immediately checks that the power series $\mathbf{E}^o(q)$ is fixed under the action of the U-operator on $A(\mathcal{W}_{\text{eis}})[[q]]$ which takes power series $\sum_{n=0}^{\infty} a_n q^n$ to $\sum_{n=0}^{\infty} a_{pn} q^n$. Recall that if $\kappa = (\chi, k) \in \mathcal{W}^+$ is arithmetic and $k > 0$ and $\zeta^*(\kappa) \neq 0$, then $E_\kappa(q)$ is the q-expansion of a weight k classical (see Miyake [Mi] Chapter 7) (hence v-overconvergent, for any v) modular eigenform on $X_1(\text{l.c.m.}(\mathbf{q}, \text{cond}_\chi))$ with character $\chi \tau^{-k}$. Here cond_χ is the smallest power of p such that χ is trivial on $1 + \text{cond}_\chi \cdot \mathbf{Z}_p$. More generally, if $\kappa = (\chi, k)$ with k any integer we have that $\zeta^*(\kappa) E_\kappa(q)$ is the q-expansion of a weight k overconvergent modular eigenform on $X_1(\text{l.c.m.}(\mathbf{q}, \text{cond}_\chi))$ with character $\chi \tau^{-k}$ as we will show in Corollary 2.2.6 below when χ trivial on $\mu(\mathbf{Z}_p)$. (The general case follows from this and Theorem 2.1 of [C-CCS] applied to $\zeta^*(\kappa) E_\kappa(q)/E_{\langle \kappa \rangle}(q)$. (See also Proposition II.3.22 of [G-ApM].)

As discussed in [C-BMF, §B1] and elsewhere, if κ is trivial on the subgroup of roots of unity, $\mu(\mathbf{Q}_p)$, in $\mathbf{Q}_p$, then

$$|E_\kappa(q) - 1| < 1.$$

Definition. The **restricted Eisenstein family** $\mathbf{E}$ is the restriction of the basic Eisenstein family of tame level 1 to the subspace $\mathcal{B} \subset \mathcal{W}$ (cf. notation of section 1.4).

Remark 1. For any tame level N, as we saw in section 1.4, the rigid analytic group $\mathcal{W}_N$ is the disjoint union of a finite number of $\mathcal{B}$-cosets of the form $\chi \cdot \mathcal{B}$ where χ is a character of finite order in $\mathcal{D}_N$. This is important in that it will allow us to transform modular forms of arbitrary weights to functions on modular curves by division by Eisenstein series in the restricted Eisenstein family. But here we considered only tame level $N = 1$, and now restrict to the special case where $\chi = 1$ is the trivial character.

Some important facts about the restricted Eisenstein family are given below.

Proposition 2.2.1. *The q-expansion, $\mathbf{E}(q)$ of the restricted Eisenstein family has coefficients in $\Lambda^{(0)}$:*

$$\mathbf{E}(q) = 1 + \epsilon_1 q + \epsilon_2 q^2 + \cdots \in \Lambda^{(0)}[[q]],$$

and the coefficients ϵ_j for $j \geq 1$ lie in the maximal ideal of $\Lambda^{(0)}$.

Proof. This is well known. For example, it is a consequence of Iwasawa's theorem on p-adic L-functions [I] (cf. Theorem 16 of [S-FMZ]).

Theorem 2.2.2 (additivity). *Let $Z = Z_1(\mathbf{q})$. There exists a rigid analytic function on $Z_{\mathcal{B} \times \mathcal{B}}$ bounded by one and overconvergent over $\mathcal{B} \times \mathcal{B}$ (i.e. an element of $A^0(Z_{\mathcal{B} \times \mathcal{B}}/\mathcal{B} \times \mathcal{B}))^\dagger$) with q-expansion*

$$\frac{E_\alpha(q) E_\beta(q)}{E_{\alpha\beta}(q)}.$$

Proof. Let $E = E_{(\tau^0, 1)}$. This is the classical weight one Eisenstein series on $X_1(\mathbf{q})$ used in Chapter B of [C-BMF]. For $v(s) > 1 - 1/(p-1)$ we know, by Corollary B4.5.2 of [C-BMF], that $E_{(\tau^0, s)}(q)/E(q)^s$ is the q-expansion of a function on $Z_{\mathcal{B}^*}$ overconvergent over $\mathcal{B}^*$. It is without zeroes on $Z_1(\mathbf{q})$, since its q-expansion is congruent to 1, hence on $Z_1(\mathbf{q})(v)$ for some $v > 0$. It follows that if $v(s)$ and $v(t)$ are greater that $1 - 1/(p-1)$, $\alpha = (\tau^0, s)$ and $\beta = (\tau^0, t)$ then

$$\frac{(E_\alpha(q)/E(q)^s) \cdot (E_\beta(q)/E(q)^t)}{E_{\alpha\beta}(q)/E(q)^{s+t}}$$

is the q-expansion of a function on $Z_{\mathcal{B}^* \times \mathcal{B}^*}$ overconvergent over $\mathcal{B}^* \times \mathcal{B}^*$. Hence, we obtain the conclusion of the theorem with $\mathcal{B}$ replaced by $\mathcal{B}^*$. We also note that the q-expansion coefficients of this function lie in $\Lambda \hat{\otimes} \Lambda$. We want to apply a generalization of Theorem 2.1 of [C-CCS] (i.e., Theorem 2.2.4 below) which we must now prepare to prove. Let I_n be the maximal ideal of $\Lambda_n := \mathbf{Z}_p[[S_1, \ldots, S_n]]$. For $a \in \Lambda_n$ let $v_\Lambda(a) = \min\{m : a \in I_n^m\}$. Suppose $t = (t_1, \ldots, t_n)$ is an n-tuple of real numbers < 1. We define a norm $\| \ \|_t$ on Λ_n by setting for

$$\| \sum_M b_M S^M \|_t = \max_M \{|b_M| t^M\}.$$

Here M is a multi-index. The norm $\| \ \|_t$ is the spectral norm on the disk $\mathcal{B}[t]$, $|S_i| \leq t_i$, which is an affinoid when $t_i \in p^{\mathbf{Q}}$ for $1 \leq i \leq n$. Set $v_t(a) = -\log \|a\|_t$.

Lemma 2.2.3. *There exist positive constants $A(t)$ and $B(t)$ such that for all non-zero $a \in \Lambda_n$,*

$$A(t)v_\Lambda(a) \leq v_t(a) \leq B(t)v_\Lambda(a).$$

Proof. Let $a = p^{m_0} S^M$ where $M = (m_1, \ldots, m_n)$ and $m_0 + \sum M = m$. Then,

$$\|a\|_t = |p|^{m_0} \cdot t^M,$$

so

$$v_t(a) = m_0 \log p + \sum_i m_i(-\log t_i).$$

Thus the lemma follows if we take $B(t) = \max\{\log p, -\log t_1, \ldots, -\log t_n\}$ and $A(t) = \min\{\log p, -\log t_1, \ldots, -\log t_n\}$. ∎

As a corollary we get that a series over Λ_n which is convergent on $\mathcal{B}^n \times \mathbf{B}[0, 1]$ is overconvergent over the open unit n-polydisk $\mathcal{B}^n$ if and only if it is overconvergent over any single affinoid subpolydisk around the origin. We are ready to prove the generalization of Theorem 2.1 of [C-CCS] anticipated above. Let $\mathcal{B}_n$ be the n-dimensional unit polydisk. Then Λ_n may be regarded as the ring of rigid analytic functions on $\mathcal{B}_n$ defined over $\mathbf{Q}_p$ and bounded by 1.

Theorem 2.2.4 (q-expansion Principle). *Suppose, $t \in \mathbf{R}^n$ and $0 < t_i < 1$. Then, if $G \in A^\dagger(Z_{B[t]}/B[t])$ and $G(q) \in \Lambda_n[[q]]$, G uniquely analytically continues to an element of $A^\dagger(Z_{\mathcal{B}_n}/\mathcal{B}_n)^0$.*

Proof. The proof goes exactly like that of Theorem 2.1 in [C-CCS] and we employ the notation of that paper. In particular if Y is an affinoid, $A^0(Y)[X]^\dagger$ denotes the ring of power series with coefficients in $A(Y)$ which overconverge on $B[0,1] \times Y$ over Y and are bounded by 1 there. In particular, if Y is the origin consider as a rigid space over $\mathbf{Q}_p$, $A^0(Y)[X]^\dagger = \mathbf{Z}_p[X]^\dagger$. We apply Lemma 2.2 of [C-CCS] which asserts, that there exists a finite morphism f from $Z^\dagger$ onto $\mathbf{B}[0,1]^\dagger$ such that $f^{-1}(0) = \infty$ and $\bar{f}$ is separable. We base change f to a finite morphism from $(Z_{B[t]}/B[t])^\dagger$ to $(\mathbf{B}[0,1]_{B[t]}/B[t])^\dagger$ and let Tr_f be the corresponding trace map. We let X be the standard parameter on $\mathbf{A}^1$ and conclude that for $r \in A^\dagger(Z)^0$, $Tr_f(rG)$ is in both $\Lambda_n[[X]]$ and $A^\dagger(\mathbf{B}[0,1]_{B[t]}/B[t])$. It follows from this that if D generates the discriminant ideal in $\mathbf{Z}_p[X]^\dagger$ of $A^\dagger(Z)^0/\mathbf{Z}_p[X]^\dagger$, then $DG \in A^\dagger(Z_{\mathcal{B}_n}/\mathcal{B}_n)$. Since $\bar{f}$ is separable, $p \nmid D$. Then Lemma 2.3 of [C-CCS] generalizes immediately to

Lemma 2.2.5. *Let $t \in \mathbf{R}^n$, $0 < t_i < 1$. Suppose $a(X) \in A^0(B[t])[X]^\dagger$ and suppose that there exists a $D(X) \in \mathbf{Z}_p[X]^\dagger$ such that $p \nmid D(X)$ and $D(X)a(X) \in \Lambda_n[X]^\dagger$, then $a(X) \in \Lambda_n[X]^\dagger$.*

However, we feel that one of the assertions made in the proof of Lemma 2.3 of [C-CCS] requires more justification. Namely, we will now justify the convergence of the sum (in the notation of [C-CCS]) $S := \sum_n \lambda_n h_n(X)$. Recall, $h_n(X)$ is an element of $\mathbf{Z}_p\langle X \rangle$ satisfying

$$X^n - r_n(X) = D(X)h_n(X),$$

where $D(X) \in \mathbf{Z}_p[X]^\dagger$ and $r_n(X)$ is either 0 or a polynomial of degree strictly less than d over $\mathbf{Z}_p$. This is already enough to show that $h_n(X)$ is overconvergent but we need more. Let $\delta^{-1} > s > 1$ so that $D(X)$ converges on $B[0,s]$ and has no zeroes on the half open annulus $A(1,s]$. It follows that $h_n(X)$ converges on $B[0,s]$. Let $C = |1/D(X)|_{A[s,s]}$. It follows that for large n, $|h_n(X)|$ is at most Cs^n on the circle $A[s,s]$ and hence (by the maximum modulus principle) on the disk $B[0,s]$. It follows, since $|\lambda_n|_t \leq \delta^n$, that the sum S converges to an element of $A^0(B[t])[X]^\dagger$ as claimed. We conclude the proof of the Theorem 2.2.4 the same way as that of Theorem 2.1 of [C-CCS]: Suppose $b_1, \ldots, b_d$ is a basis for $A^0(Z^\dagger)$ over $\mathbf{Z}_p[X]^\dagger$. We may write,

$$G = a_1(X)b_1 + a_2(X)b_2 + \cdots + a_d(X)b_d$$

where $a_i(X) \in A^0(B[t])[X]^\dagger$. Then as $DG \in A^\dagger(Z_{\mathcal{B}_n}/\mathcal{B}_n)^0$, it follows, since $b_1, \ldots, b_n$ is also a basis for $A^\dagger(Z_{\mathcal{B}_n}/\mathcal{B}_n)^0$ over $\mathcal{B}_n$, that $D(X)a_i(X) \in \Lambda[X]^\dagger$. Thus we may apply the previous lemma and deduce Theorem 2.2.4. ∎

We now give a complete proof of a claim made in §B1 of [C-BMF].

Corollary 2.2.6. *If $\kappa \in \mathcal{B}$ is arithmetic of weight k and nebentypus ϵ, then $E_\kappa(q)$ is the q-expansion of an overconvergent modular form in $M_k(p^m, v, \mathbf{Q}_p(\mu_{p^m}))$ for some $v > 0$ where p^m is $LCM(\mathbf{q}, cond_\epsilon)$.*

Proof. This is automatically true for $k > 0$ since then E_κ is classical of the appropriate weight and level. Moreover, in this case, knowledge about the field of definition follows from the q-expansion principle. In general, we proceed by induction on $-k$. Suppose the corollary is true when $k > -n$. Let $\alpha \in \mathcal{B}$ be the arithmetic character with coordinates $(\tau^0, 1)$. Suppose $k = -n$. Then Theorem 2.2.4 implies $E_\kappa(q)$ equals

$$\frac{E_{\alpha\kappa}(q)}{E_\alpha(q)} F(q),$$

where F is an overconvergent function on $Z_1(N\mathbf{q})$. Since E_α reduces modulo p to the $(p-1)$-st root of the Hasse invariant on $Z_1(\mathbf{q})$, it has no zeroes there and hence no zeroes in a strict neighborhood of this affinoid. The corollary now follows as the induction hypothesis applies to $E_{\alpha\kappa}(q)$ which has weight $-n + 1$. ∎

Proposition 2.2.7. *Let ℓ be a prime number. The ratio of $\mathbf{E}(q)$ and $\mathbf{E}(q^\ell)$,*

$$\mathbf{E}_\ell(q) := \mathbf{E}(q)/\mathbf{E}(q^\ell),$$

is the q-expansion of a rigid analytic function on $Z_1(\mathbf{q}, \ell) \times \mathcal{B}$ if $\ell \neq p$, and on $Z_1(\mathbf{q}) \times \mathcal{B}$ if $\ell = p$, overconvergent over $\mathcal{B}$. The function $\mathbf{E}_\ell(q)$ has Fourier coefficients in $\Lambda^{(0)}$, its constant term being 1, and all other coefficients lying in the maximal ideal of $\Lambda^{(0)}$.

Proof. The proof of existence of the overconvergent rigid analytic function runs along the same lines as the proof of Corollary 2.1.1 of [C-CCS] which is this proposition with $\ell = p$, as did that of the additivity theorem, with Z replaced by $Z_1(\mathbf{q}, \ell)$ if $\ell \neq p$. The statement about the coefficients follows from Proposition 2.2.1. ∎

Remarks. For fuller control of the geometry of the eigencurve it would be useful to know explicit affinoid regions of the type $X \times Z_1(Np^m)(v)$ on which some, or all of the functions $\mathbf{E}_\ell(q)$ converge. (For a preliminary result along these lines, in the case of $p = 2$ see forthcoming work of M. Emerton.)

We also feel that it is generally a good idea for anyone who is just beginning the study of p-adic modular forms to spend time specifically concentrating on, and understanding the nature of, the Eisenstein family. It is at the root of much of the theory. For example:

1. (*It is the ur-example of a p-adic family.*) The Eisenstein family provides a very explicit example of a p-adically varying Fourier expansion (of eigenforms) parameterized by p-adic weight which interpolates to give classical modular eigenforms at positive integral weights. To our knowledge, Serre was the first person to signal the p-adic analytic variation of the Fourier coefficients of the Eisenstein family, and the fact that this could be used to provide an alternate construction of p-adic L-functions.

2. (*It provides a way of understanding Hida families.*) Given a classical parabolic modular eigenform f of tame level N and weight $k \geq 2$ which is (p-)ordinary (i.e., which has slope zero), the restricted Eisenstein family can be used in conjunction with Hida's projection operator to the subspace of ordinary modular forms, to enable one to see, a bit more concretely than is offered by the general theory, the analytically varying family of ordinary eigenforms of which f is a member. Briefly, just multiply f by $E_\kappa(q)$ where κ ranges through $\mathcal{B}$ and then project this family to a family of ordinary modular forms using Hida's projection operator. In the special case where the rank of the space of ordinary modular eigenforms of that tame level is 1 then things are particularly easy: the projection would then be family of ordinary modular eigenforms (for all Hecke operators) parameterized by an open subspace of weight space. But the general case is hardly more difficult.

3. (*Multiplying by Eisenstein series provides us with a convenient way of passing from weight 0 to general weight.* In this way we get a convenient definition of overconvergence for modular forms of arbitrary p-adic weight, and for families of modular forms by requiring the corresponding form of weight 0, or family of forms of weight 0, to be overconvergent. See, for example, the definitions given in B4 of [C-BMF], and repeated in section 2.4 below. Moreover, the multiplier function $\mathbf{E}_p$ enables one to compare the operation of the Atkin-Lehner operator U_p on weight 0 modular functions to the same operator on modular forms of general weight. See [C-BMF] and section 2.4. The fact that the Fourier coefficients of $\mathbf{E}_p$ are nice is one of the key ingredients for a good p-adic analytic dependence (on weight) of U_p-eigenvalues.

2.3. Katz p-adic Modular Functions.

For the beginning of this section fix any prime number p and an arbitrary tame level N. Denote the Iwasawa algebra Λ_N by Λ, for short. By a Λ-adic ring, we mean an algebra over Λ which is complete with respect to the Λ-adic topology, i.e., B is such a ring if and only if B is a Λ-algebra and $B \cong \varprojlim B/\mathrm{rad}(\Lambda)^n B$ where $\mathrm{rad}(\Lambda)$ is the radical of Λ.

Definition. Let B be a Λ-adic ring. By a **Katz p-adic modular function** over B (of tame level N) we mean a function F which assigns to any isomorphism class $(\mathcal{E}, \alpha_\infty, \nu)$ of trivialized elliptic curves with level N-structure (ν) over a Λ-adic B-algebra D an element $F(\mathcal{E}, \alpha_\infty, \nu) \in D$, and whose formation commutes with base change.

For a discussion of this notion and its variants (e.g., cuspidal Katz p-adic modular functions) see Chapter 4 of [K-PMF], section I.3.1 of [G-ApM], and all of [K-2], especially the appendix there. If F is a Katz modular function of tame level N over the Λ-adic ring B and $a \in \mathbf{Z}_{p,N}^*$, $a = (b,c)$ where $b \in (\mathbf{Z}/N\mathbf{Z})^*$ and $c \in \mathbf{Z}_p^*$, we set

$$F|\langle a \rangle (\mathcal{E}, \alpha_\infty, \nu) = F(\mathcal{E}, c^{-1}\alpha_\infty, b\nu),$$

for all isomorphism classes of triples $(\mathcal{E}, \alpha_\infty, \nu)$ of trivialized elliptic curves with level N structure over a B-algebra D. If $\psi : \mathbf{Z}_{p,N}^* \to B^*$ is a continuous homomorphism we will say that F has **weight-character** ψ if $F|\langle a \rangle = \psi_D(a)F$ for all $a \in \mathbf{Z}_{p,N}^*$ where ψ_D is the composition of ψ with the homomorphism $B^* \to D^*$, induced from the structure homomorphism of the B-algebra D.

For any p-adically complete $\mathbf{Z}_p$-algebra B, let $V_{n,m}(B)$ be as in [G, §I.3.1]. Let $Np^m \geq 5$. The ring $V_{n,m}(B)$ is the ring of functions on $\hat{Z}_1(Np^m) \times B/p^n B$ where $\hat{Z}_1(Np^m)$ is the underlying formal scheme over $\mathbf{Z}_p$ which is in the affinoid connected component $Z_1(Np^m)$ of ∞ in the ordinary locus of $X_1(Np^m)$. Then the ring of Katz holomorphic p-adic modular functions over B (of tame level N) is

$$\varprojlim_n \varinjlim_m V_{n,m}.$$

For this, see ibid. and [K-2].

Proposition 2.3.1. *The q-expansion $\mathbf{E}(q)$ is the q-expansion of a Katz p-adic modular function $\mathcal{E}$ over $\Lambda^{(0)}$ with the identity weight-character.*

Proof. Fix an integer $n \geq 0$. Let $R_n = \Lambda^{(0)}/([1 + \mathbf{q}]^{p^n} - (1 + \mathbf{q})^{2p^n})$. Then $\Lambda = \varprojlim R_n$. Let $\mathbf{E}_n(q)$ be the restriction of $\mathbf{E}(q)$ to R_n. Let $\chi(a) = \langle \langle a \rangle \rangle^2 \psi(a)$ where ψ is a character of finite order on $1 + \mathbf{q}\mathbf{Z}_p$. Then since $\chi(\mathbf{E}(q))$ is the q-expansion of a classical weight 2 modular form with character $\psi\tau^{-2}$, and

$$[1 + \mathbf{q}]^{p^n} - (1 + \mathbf{q})^{2p^n} = \prod_\psi \left([1 + \mathbf{q}] - (1 + \mathbf{q})^2 \psi(1 + \mathbf{q})\right),$$

where the product is over all characters ψ on $1 + \mathbf{q}\mathbf{Z}_p$ trivial on $1 + \mathbf{q}p^n\mathbf{Z}_p$, $\mathbf{E}_n(q)$ is the q-expansion of a Katz p-adic modular function over the normalization of R_n. It follows from the q-expansion principle that it is, in fact, the q-expansion of a Katz p-adic modular function over R_n and hence, passing to the inverse limit, $\mathbf{E}(q)$ is the q-expansion of a Katz p-adic modular function over $\Lambda^{(0)}$. ∎

2.4 Convergent modular forms and Katz modular functions.

Let p be a prime number, N a tame level, and $m \geq 1$, and assume, for this section, that $Np^m \geq 5$.

Definition. We say $F(q) = \sum_{n=0}^{\infty} a_n q^n$, $a_n \in K$, is the q-expansion of a **convergent** (resp. **overconvergent**) **form** of tame level N and weight-character $\kappa \in \mathcal{W}$ over K if $F(q)/E_{\langle \kappa \rangle}(q)$ is the q-expansion of a rigid (resp. overconvergent) function on $Z_1(Nq)$ in $X_1(Nq)$ of character $\kappa/\langle \kappa \rangle$ for the action of $(\mathbf{Z}/Nq\mathbf{Z})^*$. The K-vector space of these q-expansions with the property that the corresponding function converges on $Z_1(Nq)(v)$ will be denoted $M_\kappa(K, N, v)$.

If $\mathcal{U}$ is an admissible open subspace of $\mathcal{B}$ we also say that

$$F(q) = \sum_{n=0}^{\infty} a_n q^n,$$

$a_n \in A(\mathcal{U})$, is the **q-expansion of a family of convergent** (resp. **overconvergent**) **forms** over $\mathcal{U}$ of tame level N if $F(q)/\mathbf{E}(q)$ is the q-expansion of a rigid (resp. overconvergent) function on $\mathcal{U} \times Z_1(Nq)$. We say this family has type $\delta \in \mathcal{D}_N$ if this function has character δ for the action of $(\mathbf{Z}/Nq\mathbf{Z})^*$. We call the $A(\mathcal{U})$ module of these forms $M_\mathcal{U}^\dagger(N)$, and for $v > 0$, we call $M_\mathcal{U}^\dagger(N)(v)$ the submodule consisting of those families over $\mathcal{U}$ whose respective functions converge on $\mathcal{U} \times Z_1(Nq)(v)$.

Thus, in particular, it follows from the above definition that $\mathbf{E}(q^p)$ is the q-expansion of a family of overconvergent forms over $\mathcal{B}$ of tame level 1 and type 1. Note that when κ is arithmetic of weight k, $M_\kappa(K, N, v)$ is contained in $M_k(Np^m, v, K)$ for $p^m = LCM\{\mathbf{q}, \mathrm{cond}_\chi\}$ where χ is the wild part of κ for small enough $v > 0$ because of Corollary 2.2.6. Also, as an immediate consequence of Theorem 2.2.2 we get

Proposition 2.4.1. *If $F(q)$ and $G(q)$ are the q-expansions of overconvergent forms of weight-characters α and β, then $F(q)G(q)$ is the q-expansion of an overconvergent form of weight character $\alpha\beta$.*

Theorem 2.4.2. *Let F be a convergent family of modular forms over an affinoid subspace X of $\mathcal{W}$, with q-expansion coefficients in $A^0(X)$. Then the q-expansion of F is the q-expansion of a Katz modular function $\tilde{F}$ over $A^0(X)$.*

Proof. What we have is that $F(q)/\mathbf{E}(q)$ is the q-expansion of a rigid analytic function on $X \times Z_1(\mathbf{q})$. But

$$A^0(X \times Z_1(\mathbf{q})) = A^0(X) \hat{\otimes} A^0(Z_1(\mathbf{q}))$$

is just $V_{\infty,i}(A^0(X))$, where $i = 2$ if $p = 2$ and $i = 1$ otherwise, so $F(q)/\mathbf{E}(q)$ is the q-expansion of a Katz modular function over $A^0(X)$. The result now follows from 2.3.1 since $A^0(X)$ is naturally a Λ-adic ring. ∎

Chapter 3. Hecke Algebras

Fix a prime number p and a tame level N relatively prime to p. The precise conditions on these will be given at the head of each section below. Whenever we deal with a level written as Np^m we assume that $Np^m \geq 5$.

3.1 Hecke eigenvectors and generalized eigenvectors.

In this section we work with arbitrary p and level N. By $\mathcal{H} = \mathcal{H}_N$ let us mean the (commutative) polynomial algebra over the topological ring Λ_N in the infinitely many variables labelled T_ℓ for prime numbers ℓ not dividing $N \cdot p$ and U_q for primes q dividing $N \cdot p$. We view $\mathcal{H}$ as a topological Λ_N-algebra given its weak topology. More explicitly, if for any finite set of monomials $\mathcal{S}$ in the variables T_ℓ and U_q we let $\mathcal{H}_\mathcal{S}$ denote the free Λ_N-submodule of $\mathcal{H}$ of finite rank generated by the monomials in $\mathcal{S}$, given the product topology it inherits from the topology on Λ_N, a subset of the topological ring $\mathcal{H}$ is open if and only if its intersection with every $\mathcal{H}_\mathcal{S}$ is open. In particular, continuous ring homomorphisms from $\mathcal{H}$ have the property that their restriction to Λ_N are continuous. Purely formally, we may define the universal q-expansion

$$\sum_{n=1}^{\infty} \mathcal{T}_n \cdot q^n \ \in \ \mathcal{H}[[q]]$$

by the following rules: For prime numbers ℓ which do not divide Np, we put $\mathcal{T}_\ell = T_\ell$; for prime numbers ℓ which divide Np, we put $\mathcal{T}_\ell = U_\ell$; and we define $\mathcal{T}_n$ for any positive integer n by the finite recursive relations summarized by the equality of formal Dirichlet series

$$\sum_{n=1}^{\infty} \mathcal{T}_n \cdot n^s = \prod_\ell (1 - \mathcal{T}_\ell \cdot \ell^{-s} + [\ell] \cdot \ell^{-2s})^{-1},$$

where the product is taken over all prime numbers ℓ, and where $[\ell]$ is defined to be 0 if ℓ divides pN and to be the image of $\ell \in \varprojlim(\mathbf{Z}/Np^n\mathbf{Z})^* \subset \Lambda_N$ if ℓ doesn't divide pN.

If $\Psi : \mathcal{H} \to \mathbf{C}_p$ is a $\mathbf{C}_p$-valued character (i.e., a continuous ring homomorphism) by the **weight-character** of Ψ we mean the homomorphism

$w = w_\Psi : \Lambda_N \to \mathbf{C}_p$ which is the restriction of Ψ to Λ_N. It is sometimes useful to refer to the q-expansion

$$\sum_{n=1}^{\infty} \Psi(\mathcal{T}_n) \cdot q^n \in \mathbf{C}_p[[q]]$$

as the **Fourier expansion** of the character Ψ. For any such $\Psi : \mathcal{H} \to \mathbf{C}_p$, it is convenient to define the following ideal: First consider the ring $\mathcal{H} \otimes_{\Lambda_N} \mathbf{C}_p$ where the tensor product is made via the weight-character $w : \Lambda_N \to \mathbf{C}_p$. Define

$$m_\Psi \subset \mathcal{H} \otimes_{\Lambda_N} \mathbf{C}_p$$

to be the ideal generated by all elements of the form $\tau \otimes 1 \otimes \Psi(\tau)$ for $\tau \in \mathcal{H}$.

Definition 1. Given a continuous homomorphism $w : \Lambda_N \to \mathbf{C}_p$ and a power series

$$f(q) := \sum_{n=1}^{\infty} a_n \cdot q^n \in \mathbf{C}_p[[q]],$$

say that $f(q)$ is a **normalized Hecke eigenvector** of weight $\mathbf{w}$ if there is a character $\Psi : \mathcal{H} \to \mathbf{C}_p$ of weight-character w whose Fourier expansion is $f(q)$.

There are two other equivalent (and more standard) ways of formulating this notion of $f(q)$ being a (normalized) Hecke eigenvector of weight w. To give these definitions, let us fix a weight-character w, and define the **weight** w **action** of $\mathcal{H}$ on $\mathbf{C}_p[[q]]$ as follows: we let the coefficient ring Λ_N act on $\mathbf{C}_p[[q]]$ via scalars through the homomorphism w, and define the action of $\mathcal{T}_\ell$ by the standard rule: For primes ℓ,

$$f|_w \mathcal{T}_\ell = \sum_{n=1}^{\infty} b_n \cdot q^n \in \mathbf{C}_p[[q]],$$

where $b_n = a_{n/\ell} + \ell^{-1} w([\ell]) \cdot a_{n\ell}$, where we set $a_{n/l} = 0$ if $l \not\mid n$ and $w([\ell]) = 0$ if $\ell | Np$. The weight w action of $\mathcal{H}$ on $\mathbf{C}_p[[q]]$ extends naturally to an action of the tensor product $\mathcal{H} \otimes_{\Lambda_N} \mathbf{C}_p$ considered above (this tensor product being made via the weight-character $w : \Lambda_N \to \mathbf{C}_p$).

Definition 2. The power series $f(q) := \sum_{n=1}^{\infty} a_n \cdot q^n \in \mathbf{C}_p[[q]]$ is a **normalized Hecke eigenvector** of weight $\mathbf{w}$ if and only if the coefficient a_1, of q^1 in its Fourier expansion is equal to 1, and if there is a character $\Psi : \mathcal{H} \to \mathbf{C}_p$ of weight-character $\mathbf{w}$ such that $f(q)$ is an eigenvector for the weight $\mathbf{w}$ action of $\mathcal{T}_\ell$ on $\mathbf{C}_p[[q]]$ with eigenvalue $\Psi(\mathcal{T}_\ell)$, for all prime numbers ℓ. Equivalently we may ask that $a_1 = 1$, and that the power series

$f(q) \in \mathbf{C}_p[[q]]$ be annihilated by the ideal $m_\Psi \subset \mathcal{H} \otimes_{\Lambda_N} \mathbf{C}_p$ acting on $\mathbf{C}_p[[q]]$. If this is the case, then we refer to Ψ as the **Hecke eigenvalue character** of $f(q)$.

Definition 3. We will say that a nonzero power series $f(q) := \sum_{n=1}^\infty a_n \cdot q^n$ in $\mathbf{C}_p[[q]]$, is a **generalized Hecke eigenvector** or just a **generalized eigenvector** of weight w if there is a character Ψ as above, of weight-character w, and some positive integer ν such that $f(q) \in \mathbf{C}_p[[q]]$ is annihilated by the ideal $(m_\Psi)^\nu \subset \mathcal{H} \otimes_{\Lambda_N} \mathbf{C}_p$. If $f(q)$ is such a generalized eigenvector, we refer to Ψ as the **Hecke eigenvalue character** underlying $f(q)$.

3.2. Action on $\mathbf{M_k}(\mathbf{Np^m}, \mathbf{v}; \mathbf{K})$.

In this section we work with arbitrary p and square-free tame level N, and as usual, when a level Np^m is invoked, we assume that $Np^m \geq 5$. For prime numbers ℓ not dividing $N \cdot p$ we have the Hecke operators T_ℓ, and for primes q dividing $N \cdot p$, we have the Atkin-Lehner operators U_q. These operators come from correspondences, all defined, when $Np^m \geq 5$, on the modular curve $X_1(Np^m)$ (cf. [M-W]) all commute with each other, and also commute with the diamond correspondences $\langle d \rangle$ for $d \in (\mathbf{Z}/Np^m\mathbf{Z})^*$.

Lemma 3.2.1. *For $q \neq p$, the Hecke, Atkin-Lehner and diamond correspondences T_l, U_q and $\langle d \rangle$ induce rigid analytic correspondences on the affinoid $Z_1(Np^m)(v)$ and compatible correspondences on the restriction of ω to this affinoid. These operators induce continuous Λ_N-linear operators on the Banach space $M_k(Np^m, v; K)$ which commute with each other.*

Proof. The main point is that if E is an elliptic curve over a scheme S of characteristic p and $Ver: E^{(p)} \to E$ is Verschiebung, then if ν is a generator of $\Omega^1_{E/S}$,

$$Ver^*\nu = A(E,\nu)\nu^{(p)}.$$

This is equivalent to the dual formula of Katz on page 97 of [K-PMF]. From this one can deduce that if $h: F \to E$ is an étale isogeny, then

$$A(F, h^*\nu) = A(E,\nu). \tag{1}$$

This means one can define operators on $M_k(Np^m, v; K)$ in the standard way. Eg., suppose $m = 1$ and ℓ is a prime, such that $(\ell, Np) = 1$. Then if $(\mathcal{E}, \alpha, C)$ is a triple corresponding to an R-valued point on $Z_1(Np^m, l)(v)$, where R is the ring of integers in a finite extension of $\mathbf{Q}_p$ and $v \in I_1$, and η is a generator of $\Omega^1_{E/R}$, then

$$v(A(\overline{E}, \overline{\eta})) = v,$$

where $(\overline{E}, \bar{\eta})$ is the reduction of (E, η) modulo p. Then the image of $\alpha(\mu_p)$ in E/C is the canonical subgroup and (1) implies

$$v = v(A(\overline{E/C}, \bar{\xi}))$$

where ξ is any generator of $\Omega^1_{(E/C)/R}$. Thus the two maps from $X_1(Np, \ell)$ to $X_1(Np)$ take $Z_1(Np, \ell)(v)$ onto $Z_1(Np)(v)$ for $v \in I_1$. ∎

We also have a completely continuous operator corresponding to U_p acting on $M_k(Np^m, v; K)$ for $v \in I_m$ (see section B2 of [C-BMF]). **Note:** One must stipulate, specifically, whether one means to take these operators as the ones induced from *direct image* or the *inverse image* of the geometric automorphisms or correspondences listed. Our convention will be to take the direct image, i.e.,

$$\langle r \rangle = \langle r \rangle_*, \quad T_\ell = T_{\ell\,*}, \quad U_q = U_{q\,*}.$$

Had we made the other choice, i.e., had we taken the inverse image, we would have had

$$\langle r^{-1} \rangle^* = \langle r \rangle_*, \quad T_\ell^* = \langle \ell \rangle \cdot T_{\ell\,*},$$

etc.; cf. the discussion and formulas in [M-W] Ch. 2, section 5.

We view $\mathcal{H}$ as acting on the (Banach) Λ_N-modules $M_k(Np^m, v; K)$ in the evident manner. We will also have use for notation for the sub-Λ_N-algebra $\mathcal{H}' \subset \mathcal{H}$ generated by all the T_ℓ's, for prime numbers ℓ not dividing $N \cdot p$. Thus when $N = 1$, $\mathcal{H}$ is the polynomial algebra over $\mathcal{H}'$ generated by the single symbol U_p, i.e., by the Atkin-Lehner operator at p. If Φ is a $\mathbf{C}_p$-valued character on $\mathcal{H}$, and if a modular form $f \in M_k(Np^m, v; \mathbf{C}_p)$ is an eigenvector for the action of $\mathcal{H}$ with Hecke eigenvalue character equal to Φ then if $f(q) \in \mathbf{C}_p$ is the Fourier expansion of f and if the coefficient of q in $f(q)$ is 1, then $f(q)$ is a normalized eigenvector in the sense of §3.1, of weight w_Φ. We refer to f and $f(q)$ simply as **normalized (overconvergent, modular) eigenforms**; if $f(q)$ is a generalized eigenvector in the sense §3.1, we refer to f and $f(q)$ as **generalized (overconvergent, modular) eigenforms**.

3.3 Action on Katz Modular Functions.

Let N be any square-free tame level. By the **Katz-Hecke Algebra** $\mathbf{T}_p(N)$ we shall mean the completion of the Λ_N-algebra generated by the Hecke operators and diamond operators (as described in §2.3) over $\mathbf{Z}_p$ acting (faithfully) on the space $\mathbf{V}$ of all Katz p-adic modular functions of tame level N with respect to the compact open topology on the ring of $\mathbf{Z}_p$ endomorphisms of $\mathbf{V}$ (see section II.1 of [G-ApM]). As Hida shows, in Theorems 3.1 and 3.2 of [H-HA],

$$\mathbf{T}_p(N) \cong \varprojlim_\nu h_k(Np^\nu)$$

for any weight $k \geq 2$ where $h_k(Np^\nu)$ is the Hecke algebra acting on weight k modular forms of level Np^ν over $\mathbf{Z}_p$.

As we will show in section 5.2 (for p odd and tame level $N = 1$) there is a natural homomorphism $R_p \to \mathbf{T}_p(1) =: \mathbf{T}_p$, where R_p is the universal deformation ring of two dimensional p-adic pseudo-representations of the Galois group of $\mathbf{Q}$ of tame level N.

Proposition 3.3.1. *If $p \neq 2$, the ring $\mathbf{T}_p$ is a complete semi-local noetherian ring with finite residue fields.*

Remark. We will give a proof of Noetherian-ness, motivated by the proof of Corollary III.5.7 [G-ApM], in Corollary 5.2.3 below. We will prove the rest of the proposition now (for any tame level N).

Proof. . As remarked, for any fixed weight $k \geq 2$, $\mathbf{T}_{p,N}$ is the projective limit of the Hecke algebras $h_k(Np^\nu)$ acting (faithfully) on weight k modular forms of level Np^ν over $\mathbf{Z}_p$. This is enough to show it is complete, since $h_k(Np^\nu)$ is. By duality, any homomorphism of $h_k(Np^\nu)$ into a field of characteristic p comes from an eigenform of this level modulo p. Since there are only finitely many, say n, systems of eigenvalues of tame level N (cf. the Proposition 5.1.1), for large ν, there exist maximal ideals $m_1(\nu), \ldots, m_n(\nu)$ of $h_k(Np^\nu)$ such that $m_i(\nu+1)$ is the inverse image of $m_i(\nu)$ and

$$h_k(Np^\nu) = \bigoplus_i (h_k(Np^\nu))_{m_i(\nu)}.$$

Hence, $m_i := \varprojlim_\nu m_i(\nu)$ is a maximal ideal of $\mathbf{T}_{p,N}$, $(\mathbf{T}_{p,N})_{m_i}$ is a complete local ring and

$$\mathbf{T}_{p,N} = \bigoplus_i (\mathbf{T}_{p,N})_{m_i}. \quad \blacksquare$$

3.4. Action on $\mathbf{M}^\dagger(\mathbf{N})$.

In this section we work with arbitrary p, square-free tame level N, and we assume that $Np^m \geq 5$. Let $\mathbf{E}(q) \in \Lambda^{(0)}[[q]]$ be the q-expansion of the restricted Eisenstein family, so that

$$\kappa(\mathbf{E}(q)) = E_\kappa(q)$$

for all $\kappa \in \mathcal{B}$. By Proposition 2.2.7, for any prime number ℓ the ("multiplier") function $\mathbf{E}_\ell$ whose q-expansion is $\mathbf{E}(q)/\mathbf{E}(q^\ell)$ is an overconvergent function on $Z_1(\mathbf{q} \cdot \ell)$ if $\ell \neq p$, on $Z_1(p)$ if $\ell = p \geq 5$, and when $p = 2$ or $p = 3$ and $\ell = p$ we work, rather, on the affinoid $Z_1(5\mathbf{q})$, making use of the strategy explained in section 2.5.

We set $M^\dagger(N) = M_\mathcal{B}^\dagger(N)$ where the latter was defined in §2.4. This is an alteration of that given in section B.4 of [C-BMF] where the definition was the same except that $\mathcal{B}$ was replaced by $\mathcal{B}^*$. For prime numbers ℓ, we will define operators $\mathcal{T}_\ell$ copying Lemma B5.1 of [C-BMF] *mutatis mutandis*. If $F \in M^\dagger(N)$ and $\ell \in (\mathbf{Z}/N\mathbf{Z})^* \times \mathbf{Z}_p^*$ define

$$F|\langle \ell \rangle^* = \left([\langle\langle \ell \rangle\rangle] \cdot \mathbf{E} \right) \left(\frac{F}{\mathbf{E}} | \langle \ell \rangle \right).$$

For prime numbers ℓ let ψ_ℓ be the operator on $A(\mathcal{B})[[q]]$ given by

$$\psi_\ell \left(\sum_n a_n q^n \right) = \sum_n a_{n\ell} q^n.$$

Lemma 3.4.1. *For each prime number ℓ there is a unique continuous operator $\mathcal{T}(\ell)$ on $M^\dagger(N)$ such that*

$$F|\mathcal{T}(\ell)(q) = \psi_\ell(F(q))$$

when $\ell|Np$, and

$$(F|\mathcal{T}(\ell))(q) = \psi_\ell(F(q)) + \ell^{-1}(F|\langle \ell \rangle^*)(q^\ell)$$

when ℓ does not divide Np. In fact, for each affinoid X of $\mathcal{B}$, and each prime ℓ there is a $v_\ell(X) > 0$ such that the operator $T(\ell)$ preserves the subspace $M(N, X)(v)$ for all v in the range $0 < v \le v_\ell(X)$.

The proof of the existence of the operators goes exactly the same as that of Lemma B5.1 of *ibid.*, to obtain the assertion about $\mathcal{T}(\ell)$ and $v_\ell(X)$, one must use Proposition 2.2.7 and the property of the Hasse invariant used in the proof of Lemma 3.2.1. The nature of the $v_\ell(X)$ depends on the analytic properties of $\mathbf{E}_\ell(q)$.

Proposition 3.4.2. *The mapping $F \mapsto \tilde{F}$ from overconvergent modular forms to Katz modular functions respects the Hecke module structure.*

Proof. We omit the proof of this proposition in this article, noting firstly that the main thing to check here is the action of the diamond operators since the other operators can be checked by looking at q-expansions, and noting secondly that this check on the action of the diamond operators is *not* trivial: it will be done in a subsequent article.

If we could show that the $v_\ell(X)$ could be chosen independently of ℓ we would be in a position to put a natural topology on the algebra generated by the above operators (but see Chapter 7).

3.5. Action on weight κ forms.

We use the same formulas as in the lemma of the last section. The existence of the operators follows as a corollary of that lemma by specialization. It follows from Proposition 3.4.2 that action of each of the Katz Hecke operators corresponding to $\mathcal{T}(\ell)$ and $\langle d \rangle$ preserve $M_\kappa(K, N, v)$ (see section 2.4) for $v > 0$ and sufficiently small.

3.6. Remarks about cusp forms and Eisenstein series.

For this discussion, let $p > 2$ and let us assume the tame level N is equal to 1. Let f be a (p-adic) overconvergent modular eigenform (of tame level 1) of finite slope. Consider its Fourier expansion $\tilde{f} = \sum_{n=0}^\infty a_n q^n$ (with coefficients $a_n = a_n(f) \in \mathbf{C}_p$).

Definition. We say that f is a **cuspidal-overconvergent** eigenform if $a_0 = 0$; that is, if the constant term of its Fourier expansion vanishes.

We have included the hyphen between cuspidal and overconvergent because our condition only requires the parabolicity condition $a_0 = 0$ at the cusp ∞ and therefore, for example, even certain classical Eisenstein series whose Fourier expansion at the cusp ∞ have no constant term, but whose Fourier expansion at the cusp 0 have nonzero constant term, would be counted by the above definition as cuspidal-overconvergent, even though when considered as a classical eigenform they would not be cuspidal. This is specifically the case for the eigenform on $X_0(p)$ which was called in [GM-FM,CPS] the "evil twin" of the Eisenstein series of weight k and level 1.

Let us examine the properties of a **non**-cuspidal-overconvergent eigenform f of weight κ, of tame level 1, and of finite slope, whose Fourier expansion we may normalize to be in the form: $\tilde{f} = 1 + a_1 q + \cdots$. Let λ_ℓ denote the eigenvalue of the operator T_ℓ on f (for primes $\ell \neq p$) and λ_ℓ the eigenvalue of U_p, we have

$$\lambda_\ell = \lambda_\ell \cdot a_0(f) = a_0(f|T_\ell) = a_0(f)(1 + \ell^{-1}\kappa([\ell])) = (1 + \ell^{-1}\kappa([\ell])),$$

and similarly $\lambda_p = \lambda_p \cdot a_0(f) = a_0(f|U_p) = a_0(f) = 1$. It follows that the eigenvalues of the operators T_ℓ and U_p on such an f are determined solely by the weight κ, and the eigenvalues for the operators $\mathcal{T}_n$ for $n \geq 1$ are given by $\sigma_\kappa^*(n)$ as in section 2.2. These eigenvalues are equal to the corresponding eigenvalues of the operators in question on E_κ, the member of weight κ of the basic Eisenstein family, as discussed in section 2.2 (for the weights κ for which E_κ is defined; i.e., when $\zeta^*(\kappa) \neq 0$). In particular, all these eigenforms f are of slope 0. Now make a similar computation on the coefficient a_1, using the above evaluation of the eigenvalues, to give a formula for *all* of the coefficients $a_n(f)$ $(n \geq 2)$ as specific multiples of $a_1(f)$ where the multiples depend only up n and the weight κ. For example :

$$a_\ell(f) = a_1(f|T_\ell) = \lambda_\ell \cdot a_1(f) = (1 + \ell^{-1}\kappa([\ell])) \cdot a_1(f).$$

Proposition 3.6.1. *The only normalized (in the sense of the previous paragraph) non-cuspidal-overconvergent eigenforms of tame level 1 are Eisenstein series. Specifically, for any weight κ, if $\zeta^*(\kappa) \neq 0$ then the unique normalized non-cuspidal-overconvergent eigenform of weight κ of tame level 1 is E_κ, while if $\zeta^*(\kappa) = 0$ there are no normalized non-cuspidal-overconvergent eigenforms of weight κ of tame level 1.*

Proof: Step 1. We first show that if $\kappa \neq 0$, and if f and g are two normalized non-cuspidal-overconvergent eigenforms (of tame level 1) of weight κ, then $a_1(f) = a_1(g)$. Suppose that we are given two such eigenforms with $a_1(f) \neq a_1(g)$; we may assume with no loss of generality that $a_1(g) \neq 0$. We already know that these eigenforms have the same eigenvalues for all the T_ℓ's and for U_p. Now consider the modular form

$$f - \frac{a_1(f)}{a_1(g)} \cdot g,$$

which is an eigenform of weight κ and has Fourier expansion identically equal to the non-zero constant $1 - a_1(f)/a_1(g)$. But an overconvergent eigenform whose Fourier expansion is a constant is necessarily of weight 0 (for a discussion of this fact, see, e.g. sections 4.4, 4.5 of [K-PMF]).

Step 2. We next show that if $\zeta^*(\kappa) = 0$, there are no normalized non-cuspidal-overconvergent eigenforms (of tame level 1) of weight κ. For in this case, despite appearances, E_κ is cuspidal-overconvergent, and also $\kappa \neq 0$ (for at $\kappa = 0$, ζ^* has a pole and not a zero). If there were a normalized non-cuspidal-overconvergent eigenform f (of tame level 1) of weight κ, then by the discussion above, for an appropriate constant c, $f - c \cdot E_\kappa$ would have its Fourier expansion equal to the constant 1, which is impossible since $\kappa \neq 0$.

Step 3. It remains to deal with the case of weight $\kappa = 0$. Here, the Eisenstein series E_κ has its Fourier expansion equal to the constant 1, and if there were another normalized non-cuspidal-overconvergent eigenform f (of tame level 1) of weight 0, then $f - E_\kappa$ would have its Fourier expansion equal to a nonzero constant times $\sum_{n=1}^\infty \sigma_{-1}^*(n)q^n$. But this latter Fourier series is *not* the Fourier series of an overconvergent eigenform, by Lemma 4 of [CGJ], and therefore such an f does not exist.

Chapter 4. Fredholm determinants.

4.1. Completely continuous operators and Fredholm determinants.

Let p be any prime number, N a square-free tame level. The levels Np^m and qN which occur below are assumed to be ≥ 5. The results employed in this section are mainly due to Serre [S 1]. A special role is played in this theory by the Atkin-Lehner operator $U_p \in \mathcal{H}$ for the important reason that it acts *completely continuously* (cf. section A1 of [C-BMF] for the definition) on $M_k(Np^m, v; K)$ for $k \in \mathbf{Z}$ (cf. 3.11 and 3.12 of [K-PMF], B3 of [C-BMF], and Proposition II.3.15 of [G-ApM]), and we'll show it acts essentially completely continuously on the p-adic Fréchet space of overconvergent forms of any weight $\kappa \in \mathcal{W}$ over $\mathbf{C}_p$

$$M_\kappa := \varinjlim_{v \to 0^+} M_\kappa(\mathbf{C}_p, v).$$

Since the composition of a completely continuous operator and a continuous operator is again completely continuous, it follows that any operator in the ideal $\mathcal{U} \subset \mathcal{H}$ generated by U_p is completely continuous in its action on the Banach space $M_k(Np^m, v; K)$. Let $\tilde{U} \in \mathcal{U}$ be any element in this ideal, and denote by $\tilde{U}_\kappa$ the operator of M_κ that it induces. When $k \in \mathbf{Z}$, the p-adic Fredholm theory applies, giving us well-defined Fredholm determinants

$$P_{\tilde{U}}(Np^m, k, \chi, T) := \det(1 - T \cdot \tilde{U}_k | M_k(Np^m, v; K; \chi)).$$

Here, for χ a character on $(\mathbf{Z}/Np^m\mathbf{Z})^*$, let $M_k(Np^m, v; K; \chi)$ be the subspace of $M_k(Np^m, v; K)$ on which the diamond operators $\langle d \rangle$, $d \in (\mathbf{Z}/qN\mathbf{Z})^*$ act via the character χ. These Fredholm determinants are independent of the choice of radius v (provided that v is in the range stipulated above) as one can see using proposition 4.3.2 below (Cor. II.3.18 of Gouvêa [G-ApM] and section A3 of [C-BMF]).

As we'll see, in section 4.3, the operator $\tilde{U}_\kappa$ acting on M_κ has a Fredholm determinant $P_{\tilde{U}}(\kappa, T)$. The Fredholm determinants are power series in the variable T (indeed they are entire functions of T with coefficients in the ring of integers of K, or $\mathbf{C}_p$, with constant term 1, and hence are Fredholm functions in the sense of section 1.2). For the basics regarding these p-adic Fredholm determinants, see Chapter A of [C-BMF]. The key fact about them is given in Theorem 4.1.1 below. For any $\tilde{U}$ as above, and any element $\tilde{u} \in \mathbf{C}_p$ define

$$M_{\kappa, \{\tilde{U} - \tilde{u}\}} \subset M_\kappa$$

to be the $\mathbf{C}_p$ vector subspace consisting of vectors in the kernel of some positive power of the operator $\tilde{U} - \tilde{u} \cdot I$ (where I is the identity operator) acting on M_κ. We shall refer to an element of $M_{\kappa,\{\tilde{U}-\tilde{u}\}} \subset M_\kappa$ as a **generalized eigenform** for $\tilde{U}$ (with eigenvalue $\tilde{u}$). The subspace $M_{\kappa,\{\tilde{U}-\tilde{u}\}}$ is stable under the action of $\mathcal{H}$.

Theorem 4.1.1. *Suppose* $\kappa \in \mathcal{W}_N$. *Over* $\mathbf{C}_p$, *the power series* $P_{\tilde{U}}(\kappa,T)$ *may be written as a product*

$$P_{\tilde{U}}(\kappa,T) = \prod_{j=1}^{J}(1 - \tilde{u}_j \cdot T)^{e_j}$$

*for J either $+\infty$ or a positive finite integer. Here the $\tilde{u}_j$'s are distinct elements in $\mathbf{C}_p - \{0\}$; the e_j's are positive integers ($e_j = $ the **multiplicity** of $\tilde{u}_j$ in the product expansion above) and the ordering of the $\tilde{u}_j$ is such that the rational numbers $\sigma_j = \mathrm{ord}_p(\tilde{u}_j)$ are increasing (but not necessarily strictly, of course). If $J = +\infty$ then σ_j tends towards ∞. For any (nonzero) element $\tilde{u} \in \mathbf{C}_p - \{0\}$ the $\mathbf{C}_p$ vector subspace,*

$$M_{\kappa,\{\tilde{U}-\tilde{u}\}} \subset M_\kappa$$

of generalized eigenforms for $\tilde{U}$ with eigenvalue $\tilde{u}$ is of $\mathbf{C}_p$-dimension equal to the multiplicity of $\tilde{u}$ in the product expansion of the Fredholm determinant $P_{\tilde{U}}(k,\chi,T)$. That is, the dimension of $M_{\kappa,\{\tilde{U}-\tilde{u}\}}$ is zero if $\tilde{u}$ is not equal to one of the $\tilde{u}_j$'s and is equal to e_j if $\tilde{u} = \tilde{u}_j$.

Proof. cf. Propositions 11 and 12 of section 7 in [S-ECC]. ∎

Definition. Let M be a K-vector space and $L : M \to M$ a K-linear operator. A nonzero vector $m \in M$ will be said to have L-slope $\sigma \in \mathbf{Q}$ if there is a polynomial $Q(T) \in K[T]$ whose Newton polygon has a single side which is slope $-\sigma$ such that m is in the kernel of $Q(L)$.

For σ a positive rational number, and $\tilde{U}$ as above, denote by

$$M_{\kappa,\{\mathrm{slope}(\tilde{U})=\sigma\}} \subset M_\kappa$$

the $\mathbf{C}_p$-vector subspace of vectors of slope σ. The space $M_{\kappa,\{\mathrm{slope}(\tilde{U})=\sigma\}}$ is generated by the vector spaces $M_{\kappa,\{\tilde{U}-\tilde{u}_j\}}$ where $\tilde{u}_j$ ranges through the inverses of the zeroes of $P_{\tilde{U}}(\kappa,T)$ such that $\mathrm{ord}_p(\tilde{u}_j) = \sigma$. The subspace $M_{\kappa,\{\mathrm{slope}(\tilde{U})=\sigma\}}$ is stable under the action of $\mathcal{H}$.

Proposition 4.1.2. *Let $\alpha \in \mathcal{H}$ be of the form $1 + p \cdot \tau$ for $\tau \in \mathcal{H}$. Put $\tilde{U} = \alpha \cdot U_p$. The $\mathbf{C}_p$-vector subspace*

$$M_{\kappa, \{\text{slope}(\tilde{U}) = \sigma\}}$$

depends only on κ and σ. It is independent of the choice of α.

Remark. More generally, Proposition 4.1.2 holds when α is a unit in the p-completion, $\mathcal{H}_p := \varprojlim \mathcal{H}/p^n \mathcal{H}$ of the ring $\mathcal{H}$.

Proof. Let $\alpha, \alpha' \in \mathcal{H}$ be two elements of the form required in our proposition, and put $\tilde{U} = \alpha \cdot U_p$, and $\tilde{U}' = \alpha' \cdot U_p$. For any $\tilde{U}$-eigenvalue $\tilde{u}$ of slope σ we shall show that any element in $M_{\kappa, \{\tilde{U} - \tilde{u}\}}$ is generated by $\tilde{U}'$-generalized eigenforms with $\tilde{U}'$-slope equal to σ. This will establish the inclusion

$$M_{\kappa, \{\text{slope}(\tilde{U}) = \sigma\}} \subset M_{\kappa, \{\text{slope}(\tilde{U}') = \sigma\}}$$

and, by symmetry, the opposite inclusion as well. Fix, then, a $\tilde{u}$ as above. Note that the operators $\alpha, \alpha' \in \mathcal{H}$ induce nonsingular linear transformations of the $\mathbf{C}_p$-vector space

$$M := M_{\kappa, \{\tilde{U} - \tilde{u}\}},$$

and more stringently they induce automorphisms of an $\mathcal{O}_{\mathbf{C}_p}$-lattice in M. Using the same symbols $\alpha, \alpha', \tilde{U}$, etc. for the endomorphisms of M that they induce, let $\gamma = \alpha' \cdot \alpha^{-1} : M \to M$. The eigenvalues of γ are all units in $\mathcal{O}_{\mathbf{C}_p}$. Decompose the finite dimensional $\mathbf{C}_p$-vector space M as a direct sum of (generalized) eigenspaces for the operator γ. We work on each of these (generalized) eigenspaces in turn. Let $g \in \mathbf{C}_p$ be an eigenvalue of γ and consider now the $\mathbf{C}_p$-subspace $M_g \subset M$ consisting of generalized eigenforms for γ with eigenvalue g. We have the following formula involving (commuting) elements in the algebra $\text{End}_{\mathbf{C}_p}(M_g)$:

$$\gamma\tilde{U} - g\tilde{u} \cdot I = (\gamma - g \cdot I) \cdot \tilde{U} + g \cdot (\tilde{U} - \tilde{u} \cdot I).$$

Since $\gamma - g \cdot I$ and $\tilde{U} - \tilde{u} \cdot I$ are nilpotent operators on M_g (and since all the operators in the above formula commute with each other) it follows that $\gamma\tilde{U} - g\tilde{u} \cdot I = \tilde{U}' - g\tilde{u} \cdot I$ is nilpotent as well on M_g, concluding the proof of our Proposition.

Remark. We denote this $\mathbf{C}_p$-vector space $M_{\kappa, \{\text{slope}(\tilde{U}) = \sigma\}}$ simply as

$$M_{\kappa, \sigma}$$

to reflect the fact that it depends only on σ and not on the choice of α (provided that α is chosen as stipulated in the above proposition).

4.2. Factoring Characteristic Series.

Suppose A is a Banach algebra, $(M, |\ |)$ is an orthonormizable Banach A-module, and V is a completely continuous operator on M such that its norm $|V|$ is ≤ 1. In A2 of [C-BMF] the characteristic series of V operating on M, denoted $\det(1 - TV) \in A^o\{\{T\}\}$, was defined.

By Corollary A2.6.1 of [C-BMF], if A is semi-simple, $\det(1 - TV)$ only depends on (V and) the topology of M. Note that the definition of *semi-simple* used here is that of [C-BMF]. Specifically, a normed ring A is semi-simple if the intersection of its maximal ideals is the zero-ideal, and if, for every maximal ideal $m \subset A$, the residual norm on A/m is multiplicative. Note (cf. "Example (i)" on page 427 of [C-BMF]) that if A is a reduced affinoid algebra over a complete subfield of $\mathbf{C}_p$ and A is viewed as Banach algebra with norm equal to the supremum norm, then A is semi-simple. It follows from Corollary A2.6.1 and Lemma A1.4 of [C-BMF] that

Proposition 4.2.1. *If A is semi-simple and $\mathrm{Max}(A)$ is a rigid space which has an admissible covering by affinoid subdomains X_i such that there exists an orthonormizable norm on M_{X_i} equivalent to the induced norm then any completely continuous operator on M over A has a characteristic series in $A\{T\}$ which depends only on the topology on M.*

Theorem 4.2.2. *Suppose A is a semi-simple Banach algebra, $(M, \|\ \|)$ is an orthonormizable Banach module and N is a free submodule of M of finite rank such that there exists a continuous projector from M onto N. Then, locally, there exists a norm on M equivalent to $\|\ \|$ such that both N and M/N with their induced norms are orthonormizable.*

Proof. Suppose $\{e_i\}_{i \in I}$, for some index set I, is an orthonormal basis for M and $\{n_i\}_{i \in S}$ is a basis for N where S is a finite subset of I. Write

$$n_i = \sum_{j \in I} a_{i,j} e_j.$$

It follows that we may cover $Spec(A)$ with affine opens such that for each such open U, there exist $\{j_i\}_{i \in S}$ such that

$$\det\left(a_{i,j_k} \right)_{i,k \in S}$$

is invertible on U. We may suppose $U = Spec(A)$ and $j_i = i$. We may now change our basis for N so that $a_{ij} = \delta_{ij}$ for $j \in S$. Now $\{n_i, e_j\}_{i \in S, j \notin S}$ is a basis for M. We define a new norm $\|\ \|''$ on M by requiring this set to be an orthonormal basis. We claim that $\|\ \|''$ is equivalent to $\|\ \|$. Suppose $m \in M$ and

$$m = \sum_{i \in S} b_i n_i + \sum_{j \notin S} b_j e_j.$$

Then
$$||m|| \leq \max\{|b_i|||n_i|, |b_j|: i \in S, j \notin S\}$$
$$\leq R||m||'',$$

where $R = Max\{1, |n_i|: i \in S\}$. If we set

$$c_i = \begin{cases} b_i & \text{for } i \in S \\ b_i + \sum_{j \in S} b_j a_{ji} & \text{for } i \notin S \end{cases}$$

Then,

$$m = \sum_i c_i e_i$$

and

$$||m||'' \leq \max\{|c_i|||e_i||'', |c_j|: i \in S, j \notin S\}$$
$$\leq R||m||$$

because $||n_i|| = ||e_i||''$. We will henceforth suppose $n_i = e_i$ for $i \in S$. Suppose $\pi: M \to N$ is a continuous projector. Then $||\pi||$ is defined. Let $F = \text{Ker}\,\pi$ and $\pi' = 1 - \pi$. Clearly $\{e_i, \pi'(e_j): i \in S, j \notin S\}$ is a basis for M. Let $|| \; ||'$ be the norm on M determined by making this an orthonormal basis. We claim that $|| \; ||'$ is equivalent to $|| \; ||$. Let

$$m = \sum_{i \in S} b_i e_i + \sum_{i \notin S} b_i \pi'(e_i).$$

Then,

$$m = \sum_{i \in S} b_i e_i - \pi\Big(\sum_{i \notin S} b_i e_i\Big) + \sum_{i \notin S} b_i e_i.$$

It follows that

$$||m|| \leq Max\{1, ||\pi||\}||m||'.$$

Also if $m = \sum d_i e_i$, then $b_i = d_i$ if $i \notin S$ and

$$b_i = d_i - \sum_{j \notin S} d_j c_{ji},$$

if

$$\pi(e_j) = \sum_{k \in S} c_{jk} e_k.$$

We see that

$$||m||' \leq Max\{1, ||\pi||\}||m||.$$

The theorem follows. ∎

It follows from the above results and Corollary A2.6.2 of [C-BMF] that

Corollary 4.2.3. *Suppose A, M and N are as in the theorem and u is a completely continuous operator on M over A such that u preserves N. Then, if u_F is the induced operator on F, u_F has a characteristic series and*

$$\det(1 - Tu) = \det(1 - Tu|_N)\det(1 - Tu_F).$$

Remark. The proof of the last sentence of Theorem A4.5 in [C-BMF] is not complete, but that sentence follows from this corollary.

4.3. Analytic variation of the Fredholm determinant.

Fix a tame level N such that $(\phi(N\mathbf{q}), p) = 1$. The constructions in this section will depend upon this N (but we do not systematically indicate this dependence upon N in the notation below). Under our hypothesis, the ring Λ_N is a regular (complete local noetherian) ring. The object of this section is to state a result (Theorem 4.3.1) which shows that the Fredholm determinants $P_{\tilde{U}}(\kappa, T)$, for κ arithmetic, vary p-adic analytically in the variable κ. This is a mild modification of work done in [C-BMF] and [C-CCS] (cf. Theorem B and Appendix I in [C-BMF] and especially Theorem 5.1 of [C-CCS]). Although by the conventions adopted in the present section the prime number p is odd, we should note that a result similar to the one we will be stating below is valid for $p = 2$ as well.

Now, for arbitrary $\kappa \in \mathcal{W}$ we let $P_{\tilde{U}}(\kappa; T)$ denote to the Fredholm determinant of the completely continuous system of operators $\tilde{U}$ on the system of Banach modules $M_\kappa^\dagger(N, 1/i)$ for i large enough, where $M_\kappa^{N,\dagger}(v)$ is the Banach module over $\mathbf{C}_p$ consisting of v-overconvergent modular forms of tame level N.

Theorem 4.3.1. *For each $\tilde{U} \in \mathcal{U}$ there is an entire power series,*

$$P_{\tilde{U}}(T) = \sum a_n(\tilde{U}) \cdot T^n \quad \in \Lambda_N\{\{T\}\}$$

uniquely determined by the property that for every weight $\kappa \in \mathcal{W}_N$, the image of

$$P_{\tilde{U}}(T) \in \Lambda_N\{\{T\}\} \subset \Lambda_N[[T]]$$

under the homomorphism $\Lambda_N[[T]] \to \mathbf{C}_p[[T]]$ induced by $\kappa : \Lambda_N \to \mathbf{C}_p$ is equal to the Fredholm determinant $P_{\tilde{U}}(\kappa; T) \in \mathbf{C}_p[[T]]$.

Here $P_{\tilde{U}}(\kappa; T)$ refers to the Fredholm determinant of the completely continuous system of operators $\tilde{U}$ on the system of Banach modules $M_\kappa^\dagger(N, 1/i)$ for i large enough, where $M_\kappa^{N,\dagger}(v)$ is the Banach module over $\mathbf{C}_p$ consisting of v-overconvergent modular forms of tame level N.

Before we begin the proof we need to talk about completely continuous systems of operators. First suppose $(M, \| \; \|)$ is a Banach module over a

Banach algebra $(A, |\ |)$. By an **orthogonal** basis of M over A we mean a set $\{e_i\}_{i \in I}$ such that every element m of M is the limit of a unique series of the form

$$\sum_{i \in I} a_i e_i,$$

where $a_i \in A$ and $\lim_{i \to \infty} ||a_i e_i|| = 0$ such that $||m|| = \max_{i \in I}\{|a_i| \cdot ||m_i||\}$. If $|M^*| = |A^*|$, where $M^* = M - \{0\}$, an orthogonal basis can be converted to an orthonormal basis. The proof of Proposition 3.1 of [C-CCS] yields:

Proposition 4.3.2. *Suppose M and N are Banach A-modules, $f: M \to N$ is a continuous map of Banach A-modules, and $U_M: M \to M$ and $U_N: N \to N$ are completely continuous (A-linear) maps of Banach A-modules such that*

$$
\begin{array}{ccc}
M & \overset{U_M}{\longrightarrow} & M \\
\downarrow f & & \downarrow f \\
N & \overset{U_N}{\longrightarrow} & N
\end{array}
$$

commutes. Then if there exists an orthogonal basis B of M such that $f(B)$ is an orthogonal basis of N,

$$\det(1 - T U_M) = \det(1 - T U_N) \in A\{\{T\}\}.$$

Now suppose given a sequence of Banach A-modules M_n for $n \geq 1$ and continuous A-linear mappings for each $n \geq 1$, $f_n : M_n \to M_{n+1}$ with the property that there is an orthogonal basis B_n of M_n such that $B_{n+1} := f_n(B_n)$ is an orthogonal basis of M_{n+1}. Let $M = \varprojlim M_n$. Suppose further that $|M_n^*| = |A^*|$ for all $n \geq 1$, and that there exists an orthogonal basis of M_0 whose image in M_n is orthogonal for each n. Then if U is an operator on M which restricts to a completely continuous operator on M_n for each n, we set

$$\det(1 - TU) = \det(1 - TU|_{M_n}) \in A\{\{T\}\},$$

for any n. Here we have used Proposition 4.3.2 to guarantee that $\det(1 - TU|_{M_n})$ is independent of n. We say that U is a **completely continuous operator on the sequence** $\{M_n\}_{n \geq 1}$.

If the restriction of the operator U to M_n is of norm ≤ 1 for all $n \geq 1$, then $\det(1 - TU) \in A^o\{\{T\}\}$.

More generally, we must allow the Banach algebra A to vary as well. So, suppose we have:

i. a sequence of Banach algebras A_n for integers $n \geq 1$, and contractive (this means norm decreasing) ring homomorphisms $A_m \to A_n$ for $m > n \geq 1$,

ii. a function $n \mapsto s_n$ from the set of non-negative integers to non-negative integers, and for each $n \geq 1$, a sequence of Banach A_n-modules $\{M_{n,i}\}_{i \geq s_n}$ such that $|M_{n,i}^*| = |A^*|$ (for integers $i \geq s_n$),

iii. continuous A_n-module homomorphisms $f_{n,i} : M_{n,i} \to M_{n,i+1}$ for all $n \geq 1$ and $i \geq s_n$ such that there is an orthogonal basis $B_{n,i}$ of $M_{n,i}$ with the property that $f_{n,i}(B_{n,i}) \subset M_{n,i+1}$ is an orthogonal basis of $M'_{n,i+1}$, and for all $m \geq n \geq 1$ and i sufficiently large, $M_{m,i}$ is the completed tensor product of A_m and $M_{n,i}$.

Definition. We call the above structure a **system of Banach modules.** Given such a system of Banach modules $\mathcal{M}$, to give a **completely continuous system of operators** $\mathcal{V}$ on $\mathcal{M}$ one must give a completely continuous operator $V_{n,i}$ on $M_{n,i}$ over A_n, for each $n \geq 1$ and $i \geq s_n$, such that all the obvious diagrams commute and such that $V_{m,i} = V_{n,i} \otimes 1$ if $m \geq n \geq 1$ and $i \geq s_m$.

Given a system of Banach modules $\mathcal{M}$ as above, we will say that $\mathcal{M}$ is a system of Banach modules over the sequence of Banach algebras $\{A_n\}_n$. Form the projective limit,

$$\mathcal{A} := \lim_{\leftarrow} A_m,$$

of the sequence of Banach algebras and contractive mappings, $A_m \to A_n$, given in **i.** above, to obtain a topological ring which we will call the **limit ring** $\mathcal{A}$ of the system $\mathcal{M}$. Consider the topological ring $\mathcal{A}^o := \lim_{\leftarrow} A_m^o \subset \mathcal{A}$ (it is, in fact, a subring of $\mathcal{A}$ since projective limit over the ordered set of natural numbers is left-exact).

Proposition 4.3.3. *Let $\mathcal{V}$ be a completely continuous system of operators on the system of Banach modules $\mathcal{M}$. Let $\mathcal{A} = \lim_{\leftarrow} A_m$ be the limit ring of $\mathcal{M}$ and $\mathcal{A}^o \subset \mathcal{A}$ as above. There is a unique formal power series which we denote*

$$\det(1 - T\mathcal{V}) \in \mathcal{A}[[T]]$$

such that for each $n \geq 1$, the natural projection of $\det(1 - T\mathcal{V})$ to $A_n[[T]]$ is the entire power series

$$\det(1 - TV_{n,i}|M_{n,i})$$

for any $i \geq s_n$. If the operators $V_{n,i}$ of the system $\mathcal{V}$ all have operator norm ≤ 1, (the norm of an operator on a Banach module was defined in section A1 of [C-BFM]) then $\det(1 - T\mathcal{V}) \in \mathcal{A}^0[[T]]$.

Proof. This follows from Proposition 4.3.2 together with Theorem A2.1 and Lemma A2.5 of [C-BMF].

We call $\det(1 - T\mathcal{V}) \in \mathcal{A}[[T]]$ the **Fredholm determinant** of the completely continuous system of operators $\mathcal{V}$.

Proof of Theorem 4.3.1.

We now describe a specific system of completely continuous operators to which Proposition 4.3.2 applies.

The sequence of Banach algebras. To give the Banach algebras, let us refer to the terminology of section 1.1 where to any complete noetherian local ring R is attached a rigid space $X = X_R$ given as a union of a nested sequence of affinoid subdomains, denoted $X_r = X_{R,r} \subset X_R$ for positive rational numbers r, where $X_r \subset X_{r'}$ if $r \geq r'$. Let us use the same notation, when R is a complete noetherian semi-local ring, to obtain X_R as a union of a nested sequence of affinoid subdomains, $X_{R,r} \subset X_R$, by taking $X_r = X_{R,r}$ to be the disjoint union of the correspondingly denoted affinoids for each of the (finite) connected components of X_R.

Now put $R = \Lambda_N$ and, for every integer $n \geq 1$ define the Banach algebra A_n to be the affinoid algebra of $X_{\Lambda_N, 1/n} \subset X_{\Lambda_N}$.

Note that for each $n \geq 1$ there is a natural continuous ring homomorphism $\Lambda_N \to A_n^o \subset A_n$, and the mappings $A_n \to A_{n+1}$ are Λ_N-algebra homomorphisms, as well as contractive homomorphism of Banach algebras.

Proposition 4.3.4. *For the above sequence of Banach algebras, the natural continuous ring homomorphism*

$$\Lambda_N \to \mathcal{A}^0$$

is an isomorphism of topological rings. Any power series in $\Lambda_N[[T]]$ whose projection to $A_n[[T]]$ is an entire power series (over A_n) for all $n \geq 1$ is entire over Λ_N,

Proof. . The first sentence of the proposition is (essentially) Proposition 1.1 and Corollary 1.1.1 of [C-CCS]. For the convenience of the reader we sketch the outlines of its proof. Given our assumptions on the level, the ring Λ_N is a product of regular local rings of the form $\mathcal{O}[[S]]$; i.e., power series rings in one variable over complete discrete valuation rings $\mathcal{O}$ which are finite over $\mathbf{Z}_p$. It suffices to prove the analogous statement with $R = \mathcal{O}[[S]]$. Let, then, R be such a ring, let $|\ |$ be the multiplicative norm on the discrete valuation ring $\mathcal{O}$ whose value on any uniformizer is $1/p \in \mathbf{R}$. Let K denote the field of fractions of $\mathcal{O}$ which we can take to be a (finite degree) field extension of $\mathbf{Q}_p$ contained in $\mathbf{C}_p$. Denote the maximal ideal of R by $m_R = (\pi, S) \subset R$ (here π is a uniformizer of $\mathcal{O}$). As discussed (at least for the discrete valuation ring $\mathbf{Z}_p$) in section 1.1, $X = X_R$ is the open unit disk, and one sees that for $n \geq 1$, A_n is the ring of rigid functions defined over K and bounded by 1 on the disk $B[0, p^{-1/n}] = \{x \in \mathbf{C}_p \mid |x| \leq p^{-1/n}\}$ in $\mathbf{C}_p$.

For every real number t in the range $0 < t < 1$ define the t-**norm** $|\ |_t$ on a power series $\sum_{n=0}^{\infty} b_n S^n \in R$ by the formula

$$\left| \sum_{n=0}^{\infty} b_n S^n \right|_t = \mathrm{Max}_n \{ \ |b_n| t^n \ \}$$

which (if $t \in |\mathbf{C}_p|$) is the norm obtained by mapping an element of R into $A^o(B[0,t])$ and then taking the supremum norm of its image. One checks that for $f \in R$,

$$|f|_t \leq |f|_s \leq |f|_t^{\log_t(s)}.$$

Now fix $t = p^{-1/n}$ for an integer $n \geq 1$, and let $f \in (m_R)^{\nu}$ for some $\nu > 0$. Then

$$|f|_t \leq \mathrm{Max}\{ \ |\pi|^{\nu}, t^{\nu} \ \} = \mathrm{Max}\{ \ p^{-\nu}, p^{-\nu/n} \ \} = p^{-\nu/n}.$$

Moreover, if

$$\nu \leq \mathrm{Min}\{ \ \log_t(|f|_t), -\log_p(|f|_t) \ \} = \mathrm{Min}\{ \ -n\log_p(|f|_t), -\log_p(|f|_t) \ \},$$

we have

$$f \in (m_R)^{\nu}.$$

These two facts give us that all the norms $|\ |_t$ for $t = p^{-1/n}$ and $n \geq 1$ are equivalent and induce the m_R-adic topology on R. It follows that the ring homomorphisms $R \to A_n^o$ are closed, and $R \cong \varprojlim A_n^o$, the isomorphism being in the category of topological rings. If $f \in (m_R)^{\nu}$ for $\nu > 0$ and $f \notin (m_R)^{\nu - 1}$, we have the inequalities (recalling that $t = p^{-1/n}$):

$$-\log_p(|f|_t) \leq \nu \leq -n \cdot \log_p(|f|_t).$$

The second sentence of Proposition 4.3.4 then follows. The above inequalities reduce, in the case, of $n = 1$ (i.e., $t = 1/p$) to the equality $\nu = -\log_p(|f|_t)$.

The system of Banach modules.

Here we begin by defining, for integers $n, i \geq 1$, the $A_n = A(X_{\Lambda_N, 1/n})$-modules $M'_{n,i}$ as the ring of rigid analytic functions on the reduced affinoid

$$X_{\Lambda_N, 1/n} \times Z_1(N\mathbf{q})(v),$$

where the radius $v = 1/i$. Having stipulated this, we have already established a structure giving much of what is required in **ii.** and **iii.** to have a system of Banach modules.

Specifically, for each $n \geq 1$, we take $s'_n := 1$ and consider the sequence of Banach A_n-modules

$$\{M'_{n,i}\}_{i \geq s'_n},$$

for integers $i \geq s'_n = 1$. We have that $|M'^*_{n,i}| = |A^*|$. Denote by

$$f'_{n,i} : M'_{n,i} \to M'_{n,i+1}$$

(for all $n, i \geq 1$) the continuous A_n-module homomorphisms induced by the natural inclusion mappings of the corresponding affinoids. Moreover, for all $m \geq n \geq 1$ and $i \geq 1$ we have that $M'_{m,i}$ is the completed tensor product of A_m and $M'_{n,i}$. Note also that *each* Banach A_n-module $M'_{n,i}$ is orthonormizable, by Lemma A5.1 and the remark immediately following that lemma, in [C-BMF]. What remains to be done is to check that for each Banach A_n-module $M'_{n,i}$, we may find an orthogonal basis $B'_{n,i} \subset M'_{n,i}$ such that $f'_{n,i}(B'_{n,i}) \subset M'_{n,i+1}$ is an orthogonal basis of $M'_{n,i+1}$. This follows from Proposition 3.1 and Corollary 4.2.1 of [C-CCS]. Furthermore, by Proposition A5.2 of [C-BMF], combined with what has been said above, we see that the operator U (cf. section 3.4) on $M'_{n,i}$ is a completely continuous operator (on the Banach A_n-module $M'_{n,i}$).

Remark. By the discussion in section 2.4, one has that for each $n \geq 1$ there is an integer, call it σ_n, which is large enough so that if $i \geq \sigma_n$, the mapping

$$F \mapsto F \cdot \mathbf{E}(q)$$

defines an isomorphism of the Banach A_n-module, $M'_{n,i}$ of rigid analytic functions on $X_{\Lambda_N, 1/n} \times Z_1(N\mathbf{q})(v)$ onto the Banach A_n-module

$$M^\dagger(N)_{X_{\Lambda_N, 1/n}}(1/i)$$

of v-overconvergent modular forms over $X_{\Lambda_N, 1/n}$ of tame level N, where $v = 1/i$. If $i \geq \sigma_n$, put $M_{n,i} := M^\dagger(N)_{X_{\Lambda_N, 1/n}}(1/i)$, and let us concentrate on the system of Banach modules given by the $M_{n,i}$ ($i \geq \sigma_n$). We note that the Banach A_n-module $M_{n,i}$ inherits an orthogonal A_n-basis from $M'_{n,i}$ which is brought (under $f_{n,i}$) to an orthogonal A_n-basis of $M_{n,i+1}$.

The sequence of completely continuous operators.

We now return to $\tilde{U} = \alpha \cdot U_p$ given in the hypothesis of Theorem 4.3.1. This operator is the product of an element $\alpha \in \mathcal{H}$ which may be written as a polynomial in a finite number of Hecke operators $\mathcal{T}_{\ell_j}$ ($j = 1, \ldots, \mu$) and the operator U_p, the coefficients of this polynomial being elements of Λ_N. The operator U_p, when viewed as acting on overconvergent modular forms $F \in M_{n,i}$ (for i large enough) is given by the formula

$$F|U_p = \mathbf{E}U(F/\mathbf{E}).$$

Recalling Lemma 3.4.1 (and the notation $v_\ell(X)$ described in that Lemma) let us put

$$s_n := \text{Max}\{ \sigma_n, v_{U_p}(X_{\Lambda_N, 1/n}), v_{\ell_j}(X_{\Lambda_N, 1/n}); j = 1, \ldots, \mu \}.$$

Define (for $n \geq 1$ and $i \geq s_n$) the transformations

$$V_{n,i} : M_{n,i} \to M_{n,i}$$

to be the A_n-linear operator on $(1/i)$-overconvergent modular forms given by $\tilde{U} = \alpha \cdot U_p$. Since U_p is completely continuous on $M_{n,i}$, so are all the operators $V_{n,i}$. By the formulas given for the operators $\mathcal{T}_{\ell_j}$ and for U_p in section 3.4, we see that the operators $V_{n,i}$ are all of norm ≤ 1.

We have established

Lemma 4.3.5. *The above system* $\mathcal{V} := \{V_{n,i}\}_{n,i}$ *is a system of completely continuous operators of norm* ≤ 1 *on the system of Banach modules* $\mathcal{M} := \{M_{n,i}\}_{n,i}$.

It follows from Propositions 4.3.3 and 4.3.4, that the system of operators $\mathcal{V}$ has a Fredholm determinant,

$$P_{\mathcal{V}}(T) := \det(1 - T\mathcal{V}) \in \Lambda_N[[T]]$$

such that for each $n \geq 1$, the natural projection of $P_{\mathcal{V}}(T)$ to $A_n[[T]]$ is the entire power series $\det(1 - TV_{n,i}|M_{n,i})$ for any $i \geq s_n$. Moreover, by comparing the definitions one sees directly that for every weight $\kappa \in \mathcal{W}_N$, the image of $P_{\tilde{U}}(T)$ under the homomorphism $\Lambda_N[[T]] \to \mathbf{C}_p[[T]]$ induced by $\kappa : \Lambda_N \to \mathbf{C}_p$ is equal to the Fredholm determinant $P_{\tilde{U}}(\kappa; T) \in \mathbf{C}_p[[T]]$. This concludes the proof of Theorem 4.3.1.

Remark. In the special instance where the element $\tilde{U}$ is the Atkin-Lehner operator U_p itself, the coefficients of the Fredholm determinant may be given in closed form as in formulae of Appendix I of [C-BMF].

4.4 The Spectral Curves.

In this section we again suppose that $(\phi(N\mathbf{q}), p) = 1$ and $N\mathbf{q} \geq 5$.

Definition. For each $\tilde{U} \in \mathcal{U}$, define $Z_{\tilde{U}} \subset \mathcal{W}_N \times \mathbf{A}^1$ to be the rigid space cut out by the Fredholm determinant,

$$Z_{\tilde{U}} : P_{\tilde{U}}(T) = 0,$$

viewed as rigid analytic (Fredholm) closed hypersurface (over $\mathbf{Q}_p$) in the (smooth) rigid analytic surface $\mathcal{W}_N \times \mathbf{A}^1$ (cf. section 1.2; recall that $\mathcal{W}_N$ is weight space of tame level N and $\mathbf{A}^1$ is the affine line parameterized by the variable T).

From its definition we see that $Z_{\tilde{U}} \subset \mathcal{W}_N \times \mathbf{A}^1$ is a rigid analytic Zariski-closed subvariety which is (equi-dimensional) and of dimension equal to 1 (i.e., it is a rigid analytic curve). Since $Z_{\tilde{U}}$ is rigid-Zariski-closed in the nested (cf. 1.1) rigid analytic space $\mathcal{W}_N \times \mathbf{A}^1$ it follows that $Z_{\tilde{U}}$ is nested. We will refer to the curve $Z_{\tilde{U}}$ as the **Spectral curve** attached to $\tilde{U}$. It is easy to find examples where $Z_{\tilde{U}}$ is disconnected. Can $Z_{\tilde{U}}$ have an infinite number of connected components? As in Appendix I of [C-BMF] one sees that the natural projection mapping

$$\pi : Z_{\tilde{U}} \to \mathcal{W}_N$$

can be of infinite degree.

Theorem 4.4.1. *For any $\kappa \in \mathcal{W}_N(\mathbf{C}_p)$ and any non-zero element $\tilde{u} \in \mathbf{C}_p$, the $\mathbf{C}_p$-valued point $(\kappa, 1/\tilde{u}) \in \mathcal{W}_N \times \mathbf{A}^1$ lies on the curve $Z_{\tilde{U}}$ if and only if there exists an overconvergent eigenform (eigenform for all Hecke and Atkin-Lehner operators) f with Fourier coefficients in $\mathbf{C}_p$ with the following characteristics: The overconvergent form f is of tame level N, of weight-character κ, and has $\tilde{U}$-eigenvalue equal to $\tilde{u}$.*

Addendum. We also have that the $\mathbf{C}_p$-vector space of overconvergent (generalized) eigenforms with the above characteristics is of dimension equal to the ramification index of $Z_{\tilde{U}}$ over $\mathcal{W}_N$ at $(\kappa, 1/\tilde{u})$.

Proof. . This follows immediately from Theorems 4.1.1 and 4.3.1 (compare also: Lemma A2.5 of [C-BMF] and Proposition 12 of [S-ECC] or Theorem A4.5 of [C-BMF]).

Chapter 5. Galois representations and pseudo-representations.

5.1. Deforming representations and pseudo-representations.

Let $p > 2$ be a prime number, and let our tame level N be an arbitrary positive integer not divisible by p. Let S be the finite set of places of $\mathbf{Q}$ consisting of ∞ and the prime divisors of Np. Let $G_{\mathbf{Q},S}$ denote the Galois group of the maximal algebraic extension of $\mathbf{Q}$ in $\mathbf{C}$ which is unramified outside the set S. Let $\mathbf{c} \in G_{\mathbf{Q},S}$ denote complex conjugation.

We begin by describing the residual representations of interest to us. By the catch-phrase *residual representation* we shall mean a two-dimensional vector space $\bar{V}$ over a finite field $\mathbf{F}$ of characteristic p endowed with an *odd* $\mathbf{F}$-linear continuous $G_{\mathbf{Q},S}$-action. By *odd* we mean that the determinant of the operation $\mathbf{c}$ on $\bar{V}$ is $-1 \in \mathbf{F}$. In particular, we will be considering residual representations that come from modular eigenforms, by which we mean the following:

If m is an integer and f is a (classical, *cuspidal*) modular eigenform on $\Gamma_1(Np^m)$ with ring of Fourier coefficients $\mathcal{O}_f \subset \mathcal{O}_{\mathbf{C}_p}$ where $\mathcal{O}_f$ is a (finite) complete local ring extension of $\mathbf{Z}_p$ with field of fractions of $\mathcal{O}_f$ denoted by $\mathcal{K}_f$ and with residue field isomorphic to $\mathbf{F}_f$, let V_f denote the associated two dimensional (irreducible) continuous representation of $G_{\mathbf{Q},S}$ over $\mathcal{K}_f$ attached to f in the usual sense (cf. Section 3 of [D-D-T]). Let $\Omega_f \subset V_f$ be a $G_{\mathbf{Q},S}$-stable $\mathcal{O}_f$-lattice in the $\mathcal{K}_f$-vector space V_f. Let us allow ourselves the abuse of notation

$$\bar{V}_f = \Omega_f \otimes_{\mathcal{O}_f} \mathbf{F}_f,$$

but note that in the case where $\bar{V}_f$ is not absolutely irreducible, the isomorphism class of the $\mathbf{F}_f[[G_{\mathbf{Q},S}]]$-module $\bar{V}_f$ is not uniquely determined by f; it depends, as well, on the lattice $\Omega_f \subset V_f$ chosen.

Definition. Let us say that a $\mathbf{F}[[G_{\mathbf{Q},S}]]$-module $\bar{V}$ **comes from** a modular eigenform (of tame level N) if its extension of scalars to a finite field extension field $\mathbf{F} \subset \mathbf{F}'$ is isomorphic as module over the ring $\mathbf{F}'[[G_{\mathbf{Q},S}]]$ to a Galois representation $\bar{V}_f \otimes_{\mathbf{F}_f} \mathbf{F}'$, where f is a (classical) modular eigenform of tame level N, and $\mathbf{F}_f \to \mathbf{F}'$ is a homomorphism of fields.

By a p**-modular, tame level** N, **residual** representation (a p**-modular** representation, for short) let us mean a two-dimensional vector space $\bar{V}$ over a finite field $\mathbf{F}$ of characteristic p endowed with an $\mathbf{F}$-linear continuous $G_{\mathbf{Q},S}$-action and satisfying the following properties.

1. The $\mathbf{F}[[G_{\mathbf{Q},S}]]$-module $\bar{V}$ descends to no proper subfield of $\mathbf{F}$. That is, there is no proper subfield $\mathbf{F}' \subset \mathbf{F}$ and sub- $\mathbf{F}'[[G_{\mathbf{Q},S}]]$-module $\bar{V}' \subset \bar{V}$ whose change of scalars to $\mathbf{F}$ is isomorphic to $\bar{V}$. Equivalently (since the Schur

index of these representations is 1) the character of the $G_{\mathbf{Q},S}$-representation $\bar{V}$ maps surjectively to $\mathbf{F}$.

2. The $\mathbf{F}[[G_{\mathbf{Q},S}]]$-module $\bar{V}$ comes from some classical modular eigenform f on $\Gamma_1(Np^m)$ for some positive integer m. We call the pseudo-representation associated to the semi-simplification of a p-modular representation a p-**modular pseudo-representation**.

Proposition 5.1.1. *For given N and $p > 2$ there are only a finite number of isomorphism classes of p-modular, tame level N, residual representations.*

Proof.

Step 1. Let us first show that there are only a finite number of such representations which are absolutely irreducible. Denote by $\bar{\rho}$ an absolutely irreducible p-modular residual representation satisfying the further requirements in our proposition above. By basic representation-theoretic results over finite fields, one has that $\bar{\rho}$ is defined over the field generated over $\mathbf{F}_p$ by the traces of the $\bar{\rho}(g)$'s for $g \in G_{\mathbf{Q},S}$ (for example, see the Corollary of section 6 of [M-DTR] so we have from hypothesis 1. that $\mathbf{F}$ is generated by traces, as described above. Moreover, the isomorphism class (over $\mathbf{F}$) of the $\mathbf{F}$-representation $\bar{\rho}$ is determined by the character

$$g \mapsto \mathrm{Trace}_{\bar{F}}\left(\bar{\rho}(g)\right).$$

For this, cf. the Corollary of section 5 of [M3] . So we must show that there are only a finite number of character-functions occurring as characters of absolutely irreducible representations $\bar{\rho}$ satisfying our hypotheses. A quick way of doing this is to make use of the work surrounding Serre's Conjecture (see [S-RD2] and the discussion given in [Ri]). Specifically, note, that if p is odd, and $\bar{\rho}$ arises from a modular form on $\Gamma_1(Np^m)$ (with N prime to p) then it arises from a modular form on $\Gamma_1(N)$. This is a well-known fact (*Remarque* on p. 195 of [S-RD2]) and a proof is given in [Ri] (Theorem 2.1). Now it is also known (cf. Theorem 4.3 of [Ed]; see also the discussion in [Ri]) that any *modular* residual Galois representation $\bar{\rho}$ with values in a finite field of characteristic p and tame level N arises from an eigenform on $\Gamma_1(N)$ and having weight $k(\rho)$ defined by Serre ([S-RD2] section 2). Since one easily computes from the formulas given in section 2 of [S-RD2] that $k(\rho) \leq p^2$ since it follows that if $\bar{\rho}$ is a p-modular residual Galois representation of fixed tame level N, then $\bar{\rho}$ is associated to one of a finite number of modular eigenforms, from which the proposition follows. ∎

Step 2. We are left with the task of showing finiteness of the set of isomorphism classes of (absolutely) reducible p-modular, tame level N, residual representations ($p > 2$). Here we first show that the characters of such representations (with values in the algebraic closure $\bar{\mathbf{F}}_p$) are finite in number.

But any such character is the sum of two degree 1 characters $\psi_1 + \psi_2$ of $G_{\mathbf{Q},S}$ where S is the collection consisting of the prime at ∞ and the prime divisors of pN. Moreover the conductors of the characters ψ which can occur are bounded. It follows (by the Kronecker-Weber theorem if nothing else) that only a finite number of ψ's can occur. Fix two such characters ψ_1, ψ_2, let $K_{\{\psi_1,\psi_2\}}/\mathbf{Q}$ denote the splitting field of the two characters (meaning the fixed field in $\bar{\mathbf{Q}}$ of the subgroup of $G_{\mathbf{Q},S}$ given by the intersection of the kernels of the homomorphisms ψ_1, $\psi_2 : \bar{\mathbf{Q}} \to \bar{\mathbf{F}}_p$). Now consider the set of isomorphism classes of two-dimensional $\bar{\mathbf{F}}_p$-vector spaces $\bar{V}$ equipped with continuous representations $\bar{\rho} : G_{\mathbf{Q},S} \to \mathrm{Aut}_{\mathbf{F}}(\bar{V})$ with character equal to $\psi_1 + \psi_2$, and admitting a line (i.e., a one-dimensional $\bar{\mathbf{F}}_p$-vector subspace) stabilized by the action of $G_{\mathbf{Q},S}$ with character ψ_1. The corresponding quotient space (which is also a line) has $G_{\mathbf{Q},S}$ character equal to ψ_2. Consider such a $\{\bar{V}, \bar{\rho}\}$. The subgroup $H = \ker(\bar{\rho}) \subset G_{\mathbf{Q},S}$ has, as fixed field, an Abelian, exponent p, extension-field of $K_{\{\psi_1,\psi_2\}}$ which is unramified outside the set of primes of $K_{\{\psi_1,\psi_2\}}$ lying above S. Since the maximal Abelian, exponent p, extension-field of $K_{\{\psi_1,\psi_2\}}$ which is unramified outside primes lying above S is finite, we see that the collection of all $\{\bar{V}, \bar{\rho}\}$'s we are presently considering have the property that there is a fixed finite quotient group Φ of $G_{\mathbf{Q},S}$ through which the $\bar{\rho}$'s all factor. It follows that there are only a finite number of $\{\bar{V}, \bar{\rho}\}$'s up to isomorphism. ∎

We shall now study *continuous, two-dimensional, odd $G_{\mathbf{Q},S}$-representa-tions* over commutative topological rings A in which 2 is invertible; that is, continuous A-linear representations of $G_{\mathbf{Q},S}$ into the topological group of automorphisms of free modules M of rank two over such rings A, for which the complex conjugation $\mathbf{c}$ has determinant (over A) equal to $-1 \in A$. Since 2 is invertible in A, by the symmetrizing and anti-symmetrizing with respect to $\mathbf{c}$, we have a canonical decomposition of A-modules, $M = M^+ \oplus M^-$ where

$$M^{\pm} := \{\, m \in M \mid \mathbf{c}(m) = \pm m \,\}.$$

By the $\mathbf{c}$-**normal matrix description** for the G-action on M we mean the matrix which describes the G-action in terms of the above direct sum decomposition:

$$g \mapsto \begin{pmatrix} a(g) & b(g) \\ c(g) & d(g) \end{pmatrix},$$

where $a(g) \in \mathrm{Hom}_A(M^+, M^+)$, $b(g) \in \mathrm{Hom}_A(M^-, M^+)$, $c(g) \in \mathrm{Hom}_A(M^+, M^-)$, and $d(g) \in \mathrm{Hom}_A(M^-, M^-)$.

By a $\mathbf{c}$-**normal A-basis** for such an $A[G_{\mathbf{Q},S}]$-module M is meant a A-basis $\{x, y\}$ such that x is an A-generator of M^+ and y is an A-generator of M^-. Equivalently, it is a basis for which $\mathbf{c}(x) = x$ and $\mathbf{c}(y) = -y$.

Remark Let $A = D$, a complete noetherian semi-local (commutative) ring in which 2 is invertible. Then any (odd) two-dimensional representation-module M over D admits a c-normal D-basis . To see this, consider the quotient of D by its Jacobson radical $\text{rad}(D)$ which may be decomposed as a product of fields $D/\text{rad}(D) = \prod_{j=1}^{\nu} k_j$, and let $\epsilon_j : D \to k_j$ be the projection to the j-th factor. Now form the direct sum decomposition of D-modules, $M = M^+ \oplus M^-$ and the analogous direct sum decomposition of $D/\text{rad}(D)$-modules

$$M/\text{rad}(D) \cdot M = (M/\text{rad}(D) \cdot M)^+ \oplus (M/\text{rad}(D) \cdot M)^-.$$

By noting that the dimensions of the vector spaces $(M/\text{rad}(D) \cdot M)^+ \otimes_D k_j$ over k_j are all 1 one sees that $(M/\text{rad}(D) \cdot M)^\pm$ is free of rank one over $D/\text{rad}(D)$. Choosing generators $\bar{x}$ and $\bar{y}$ of the free rank one modules $(M/\text{rad}(D) \cdot M)^+$ and $(M/\text{rad}(D) \cdot M)^-$ respectively, and lifting these to elements x and y of the D- modules M^+ and M^-, we see by Nakayama's lemma that x and y are generators of M^+ and M^-, respectively, giving us a surjective D-module homomorphism

$$h : D \oplus D \to M$$

defined by $(a, b) \mapsto ax \oplus by$. Since M and $D \oplus D$ are both free of rank 2 over D the surjection h is immediately seen to be an isomorphism. That is, $\{x, y\}$ constitutes a c-normal D-basis of M. Now let us return to A, a commutative topological ring in which 2 is invertible and M a free rank two A-module endowed with an odd, continuous, A-linear representation of $G = G_{\mathbf{Q},S}$. Let $\rho : G \to \text{Aut}_A(M)$ denote this G-action. Following [Wi], [T] (cf. also [H-NO]), and a suggestion of Buzzard, define the (continuous) functions

$$\alpha = \alpha_\rho : G \to A; \quad \delta = \delta_\rho : G \to A$$

by the rules:

$$\alpha(g) := \frac{\text{trace}_A(\rho(g)) + \text{trace}_A(\rho(cg))}{2},$$

$$\delta(g) := \frac{\text{trace}_A(\rho(g)) - \text{trace}_A(\rho(cg))}{2}.$$

Define the continuous function $\xi = \xi_\rho : G \times G \to A$ to be

$$\xi(g, h) =:= \alpha(gh) - \alpha(g)\alpha(h).$$

If M admits a c-normal basis for the G-action on M and if we consider the matrix representation with respect to such a basis:

$$g \mapsto \begin{pmatrix} a(g) & b(g) \\ c(g) & d(g) \end{pmatrix},$$

we have: $\alpha(g) = a(g)$, $\delta(g) = d(g)$, and $\xi(g, h) = b(g)c(h)$.

Lemma. *The triple of continuous functions*

$$r := (\xi, \alpha, \delta)$$

satisfy the following relations (for all $g, h, k, \ell \in G = G_{\mathbf{Q},S}$):

$$\alpha(g \cdot h) = \alpha(g)\alpha(h) + \xi(g, h)$$
$$\delta(g \cdot h) = \delta(g)\delta(h) + \xi(h, g) \tag{1}$$

$$\xi(gh, k) = \alpha(g)\xi(h, k) + \delta(h)\xi(g, k)$$
$$\xi(g, hk) = \alpha(k)\xi(g, h) + \delta(h)\xi(g, k) \tag{2}$$

$$\xi(g, h)\xi(k, \ell) = \xi(g, \ell)\xi(k, h) \tag{3}$$

and

$$\alpha(1) = \delta(1) = 1; \quad \alpha(\mathbf{c}) = 1; \quad \delta(\mathbf{c}) = -1$$
$$\xi(g, h) = 0 \quad \text{if either } g \text{ or } h \text{ is 1 or } \mathbf{c}. \tag{4}$$

Proof. We leave this as an exercise to the reader, with the suggestion that each of the formula-checks can be made to follow from a single computation in an an appropriate universal situation. For example, one can check formula (1), the second equation being the only one that needs checking, by replacing G by the group $\mathcal{G}$ generated by elements denoted $\mathbf{c}, g, h$ with the single relation $\mathbf{c}^2 = 1$, and replacing A by the $\mathbf{Z}[1/2]$-algebra $\mathcal{A}$ which is obtained from the polynomial ring in twelve variables over $\mathbf{Z}[1/2]$ (four variables making up the matrix coefficients of a 2×2-matrix C, G, H for each of the three elements $\mathbf{c}, g, h$) by inverting the determinants of G, H, and insisting that the determinant of C be 1 and its trace be 0. To check formula (1) in the context of the lemma, it suffices to check the analogue of formula (1) for the representation $\mathbf{c} \mapsto C$, $g \mapsto G$, $h \mapsto H$ of $\mathcal{G}$ on the free module $\mathcal{M} := \mathcal{A} \oplus \mathcal{A}$ or, since $\mathcal{A}$ is an integral domain, it suffices to check these formulas after extension of scalars to the field of fractions $\mathcal{A}, \mathcal{K}$. But, since $\mathcal{K}$ is a field, $\mathcal{M} \otimes_{\mathcal{A}} \mathcal{K}$ admits a c-normal basis, with respect to which it is easy to compute everything.

Recall that a triple $r = (\xi, \alpha, \delta)$, where $\xi : G \times G \to A$ and α, $\delta : G \to A$ are continuous functions, is called a **(continuous) pseudo-representation** of $G = G_{\mathbf{Q},S}$ **with values in** A if formulas (1)-(4) are satisfied. From (1) we see that r is determined by the pair of functions (α, δ). If M is a free A-module of rank two equipped with a continuous, odd, A-linear G-action, and if $\rho : G \to \mathrm{GL}_2(A)$ is the corresponding representation, the A-valued pseudo-representation $r = (\xi_\rho, \alpha_\rho, \delta_\rho)$ as defined above will be referred to as the pseudo-representation *attached* to the G-module M, or to the representation ρ. If $r = (\xi, \alpha, \delta)$ is a pseudo-representation with values in A, we define

$$\text{Trace}(r) = \alpha + \delta : G \to A,$$

and, for $g \in G$, put

$$\text{Det}(r)(g) = \alpha(g) \cdot \delta(g) - \xi(g, g).$$

One checks, using (1)-(4), that $\text{Det}(r) : G \to A^*$ is a (continuous) homomorphism. If r is the pseudo-representation attached to ρ, we have that $\text{Trace}(r) = \text{Trace}(\rho)$ and $\text{Det}(r) = \text{Det}(\rho)$. It follows from a result of Carayol and Serre (cf. the Corollary of section 5 of [M-DTR]) that if $A = D$ is a a complete noetherian local ring, and if ρ is absolutely irreducible, then the pseudo-representation r determines ρ up to equivalence.

The theory we have discussed, requiring that M be free of rank two over A, generalizes immediately to the situation where M is locally free. But let us focus on this in the special case of continuous pseudo-representations of $G = G_{\mathbf{Q},S}$ with values in the ring of rigid analytic functions $A = A(Y)$, where Y is a rigid analytic space. If Y is a rigid analytic space, consider pairs (V, ρ) where V is now a rank two vector bundle over the rigid analytic Y and ρ is a continuous, odd, $\mathcal{O}_Y$-linear representation of $G = G_{\mathbf{Q},S}$ on V,

$$\rho : G \to \text{Aut}_{\mathcal{O}_Y}(V),$$

and where the action of complex conjugation, $\rho(c)$ induces a decomposition of V into the direct sum of two line bundles, i.e. invertible $\mathcal{O}_Y$-modules,

$$V = V^+ \oplus V^-,$$

where $\rho(c)$ fixes V^+, and acts as multiplication by -1 on V^-.

We leave to the reader the exercise of defining the pseudo-representation r with values in $A(Y)$ associated to such a representation ρ.

One of the aspects of the usefulness of pseudo-representations is that given any pseudo-representation $r = (\xi, \alpha, \delta)$ with values in a field K which is not of characteristic 2, a theorem of Wiles [Wi] guarantees that r is indeed the pseudo-representation attached to a representation $\rho : G_{\mathbf{Q},S} \to \text{GL}_2(K)$. Moreover, the representation ρ is irreducible if and only if there exists a pair of elements $(s_0, t_0) \in G \times G$ such that $\xi(s_0, t_0) \neq 0$. Compare Proposition 1.1 of [H-NO]). A simple version of this theorem over rigid analytic spaces can be proved using the same method, and we now prepare for its formulation.

Definition. *Let* Y *be a rigid analytic space over* $\mathbf{Q}_p$ *and* $r = (\xi, \alpha, \delta)$ *a continuous pseudo-representation of* $G_{\mathbf{Q},S}$ *into the ring* $A(Y)$ *of rigid analytic functions on* Y. *By the* **ideal of reducibility** *of the pseudo-representation* $I_r \subset \mathcal{O}_Y$ *let us mean the sheaf of* $\mathcal{O}_Y$-*ideals generated by all the functions* $\xi(s, t) \in A(Y)$ *for (all pairs)* $(s, t) \in G \times G$.

For each point $y \in Y(\mathbf{C}_p)$ evaluation of the pseudo-representation r at the point y gives a $\mathbf{C}_p$-valued pseudo-representation r_y to which (by the theorem of Wiles referred to above) one can attach an equivalence class of representations $\rho_y : G_{\mathbf{Q},S} \to \mathrm{GL}_2(\mathbf{C}_p)$. We have that ρ_y is irreducible if and only y is not contained in the support of the ideal I_r.

Theorem 5.1.2. *Let* Y *be a rigid analytic space over* $\mathbf{Q}_p$ *and* $r = (\xi, \alpha, \delta)$ *a continuous pseudo-representation of* $G_{\mathbf{Q},S}$ *with values in* $A(Y)$. *Suppose that there is a pair of elements* $s_0, t_0 \in G_{\mathbf{Q},S}$ *such that* $\xi(s_0, t_0) \in A(Y)$ *is a non-zero-divisor, and is a generator of the* $\mathcal{O}_Y$-*module* I_r. *Then there is a continuous representation*

$$\rho : G_{\mathbf{Q},S} \to \mathrm{GL}_2(A(Y))$$

whose associated pseudo-representation is r.

Proof. This is just an exercise in globalizing Wiles' theorem; one can use the argument of Proposition 1.1 in [H-NO], with hardly any change, to prove the above theorem. Explicitly, for $g \in G$, we repeat the formulas of [H-NO] (after adjusting for the changes in powers of 2). A representation ρ is given by

$$g \mapsto \begin{pmatrix} \alpha(g) & \frac{\xi(g,t_0)}{\xi(s_0,t_0)} \\ \xi(s_0, g) & \delta(g) \end{pmatrix}.$$

Remark. In particular, if Y is a smooth rigid analytic curve, and $r = (\xi, \alpha, \delta)$ a continuous pseudo-representation of $G_{\mathbf{Q},S}$ with values in $A(Y)$ then Y can be covered by open rigid analytic subspaces $\mathcal{U}$ over each of which the restriction of the pseudo-representation r satisfies the hypothesis of the above theorem, or else has the property that the function ξ is identically 0. If the open $\mathcal{U}$ falls into the first category, Theorem 5.1.2 applies to the restriction of r to $\mathcal{U}$ giving us a rigid analytic $G_{\mathbf{Q},S}$-representation over $\mathcal{U}$ whose corresponding pseudo-representation is r restricted to $\mathcal{U}$; if it fall in the second category, the corresponding functions α and δ restricted to $\mathcal{U}$ are continuous characters of G with values in $\mathcal{O}_{\mathcal{U}}^*$, and r restricted to $\mathcal{U}$ is the pseudo-representation associated to direct sum of the two degree one $G_{\mathbf{Q},S}$-representations given by the characters α and δ. Can one get a finer, more global theorem along the above lines by considering $\mathcal{O}_Y$-linear representations of $G_{\mathbf{Q},S}$ into the groups of units in Azumaya algebras (of rank 4) over $\mathcal{O}_Y$?

Deformations of pseudo-representations compile well forming universal deformation rings as we demonstrate below in the two dimensional case. For the general case, under the assumption that the pseudo-representation is irreducible, see [Ny] or [Ro].

Theorem 5.1.3. *Let k be a finite field of characteristic $p \neq 2$. Let $\bar{\psi}$ denote a pseudo-representation of $G = G_{\mathbf{Q},S}$ with values in k. Then there is a universal lifting of $\bar{\psi}$ to a pseudo-representation ψ^{univ} of G with values in a complete noetherian local ring $R^{univ}(\bar{\psi})$ with residue field k which is universal in the sense that given any pseudo-representation ψ with values in a complete noetherian local ring D with residue field k, such that ψ is a lifting of $\bar{\psi}$, there is a unique continuous homomorphism of local rings*

$$\pi_\psi : R^{univ}(\bar{\psi}) \to D$$

reducing to the identity on k, such that ψ is the composition of ψ^{univ} and π_ψ.

Proof. One can prove this easily using Schlessinger's criterion [Sch]; compare [Ny]. Briefly, fix a pseudo-representation $\bar{\psi} = (\bar{\alpha}, \bar{\delta}, \bar{\xi})$ of $G = G_{\mathbf{Q},S}$ into k. We consider deformations of $\bar{\psi}$. Namely, let $\hat{\mathcal{C}}$ be the category of complete noetherian local rings with residue field k. For $A \in \hat{\mathcal{C}}$, let $F(A)$ be the set of pseudo-representations of G into A lifting $\bar{\psi}$. Suppose we have a Cartesian diagram of elements of $\hat{\mathcal{C}}$:

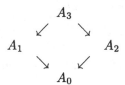

The natural map

$$b \colon F(A_3) \to F(A_1) \times_{F(A_0)} F(A_2)$$

is an isomorphism and we will write down its inverse. Suppose, $\psi_i = (\alpha_i, \delta_i, \xi_i)$ are pseudo-representations with values in A_i for $i = 1, 2$ and the images of these in A_0 are the same. Then

$$e(\psi_1, \psi_2) =: ((\alpha_1, \alpha_2), (\delta_1, \delta_2), (\xi_1, \xi_2))$$

is a pseudo-representation with values in A_3. Clearly, e is the inverse of b. Thus conditions H_1, H_2 and H_4 of [Sch] are automatic. The only thing we have to check is H_3. That is, we must show that for $A = k[\epsilon]$, the dual numbers over k, that $F(A) = F(k[\epsilon])$, which is a k-vector space, is finite dimensional, or equivalently, is finite.

Case 1. $\bar{\psi}$ is irreducible (which, by definition, means that there is a pair of elements s and t in G such that $\bar{\xi}(s,t) \neq 0$). Then as mentioned above (and compare any of our basic references; e.g., [H-NO,§1]) we know that for any $\psi \in F(k[\epsilon])$ there exists an irreducible representation $\rho \colon G \to GL_2(F(k[\epsilon]))$ such that $\rho(\mathbf{c}) = \begin{pmatrix} -1 & 0 \\ 0 & 1 \end{pmatrix}$ whose associated pseudo-representation equals ψ. Because any two representations that have the same character, take the value $\begin{pmatrix} -1 & 0 \\ 0 & 1 \end{pmatrix}$ on $\mathbf{c}$, and are irreducible, must be conjugate by a matrix of the form $\begin{pmatrix} l & 0 \\ 0 & k \end{pmatrix}$ where l and k are in $F(k[\epsilon])^*$, if we fix one representation, $\bar{\rho}$, into $GL_2(k)$ whose associated pseudo-representation is $\bar{\psi}$, then there is a unique strict equivalence class of liftings of it with pseudo-representation ψ. By [M-DGR], the vector space of strict equivalence classes of liftings of $\bar{\rho}$ is isomorphic to $H^1(G, Ad(\bar{\rho}))$ which is of finite dimension over k.

Case 2. $\bar{\psi}$ is reducible. That is, $\bar{\xi}(s,t) = 0$ for all s and t in G. It follows from (1) and (4) that, $a =: \bar{\alpha}$ and $d =: \bar{\delta}$ are characters on G with values in k^*. Let $\psi = (\alpha, \beta, \xi) \in F(k[\epsilon])$. Write $\alpha(s) = a(s) + \epsilon R(s)$, $\beta(s) = b(s) + \epsilon S(s)$ and $\xi(s,t) = \epsilon X(s,t)$. Then (3) is trivial and if $R' = R/a$ and $S' = S/d$, (1) implies that

$$R'(st) - \big(R'(s) + R'(t)\big) = X(s,t)/a(st) \tag{5}$$
$$S'(st) - \big(S'(s) + S'(t)\big) = X(s,t)/d(st). \tag{6}$$

Moreover, if $y(s,t) = X(s,t)/d(st)$ and $\chi(s) = a(s)/d(s)$, (2) implies:

$$\begin{aligned} y(st,r) &= \chi(s)y(t,r) + y(s,r) \\ y(s,ru) &= y(s,r)\chi(u) + y(s,u). \end{aligned} \tag{7}$$

In other words, y is a continuous bi-cocycle of G with values in k where the action of G on k is via the character χ. That is, $r \mapsto y(\cdot, r)$ is a one-cocycle h_r and if V is the k vector space of one-cocycles, the map $r \mapsto (h_{r-1})$ is a one-cocycle with values in V where G now acts via the character χ^{-1}. Since, G satisfies the finiteness condition Φ_p of [M-DGR], we see first that V is a finite dimensional and then that the space of these cocycles is finite dimensional. Finally, given y and hence ξ, we see that the difference of any two R''s satisfying (5) or of any two S''s satisfying (6) is a homomorphism from G into k. Since there are only finitely many such homomorphisms, the set of liftings of $\bar{\psi}$ to $k[\epsilon]$ is finite. ∎

Remark. In the case where the pseudo-representation $\bar{\psi}$ comes from a continuous, odd, two-dimensional representation $\bar{\rho}$ of $G = G_{\mathbf{Q},S}$ with values in the finite field k with the property that $\bar{\rho}$ is absolutely irreducible (or more generally, such that the only endomorphisms commuting with ρ are scalars) then there exists a universal deformation ring $R = R^{univ}(\bar{\rho})$ as in [M-DGR] and a universal deformation

$$\rho^{univ} : G_{\mathbf{Q},S} \to \mathbf{GL}_2(R)$$

of the residual representation $\bar{\rho}$. In this case, we have that the universal deformation ring $R^{univ}(\bar{\rho})$ is canonically isomorphic to the universal deformation ring $R^{univ}(\bar{\psi})$ of the pseudo-representation $\bar{\psi}$ attached to $\bar{\rho}$, and (making the canonical identification of these two rings) the pseudo-representation attached to ρ^{univ} is ψ^{univ}.

From now on, in this article, we suppose that $N = 1$.

By Proposition 5.1.1, there are only a finite number of non-isomorphic p-modular (modular, tame level 1, residual, characteristic p) representations. Let $\bar{\psi}_1, \bar{\psi}_2, \ldots, \bar{\psi}_m$ be the set of the distinct pseudo-representations corresponding to these p-modular (modular, tame level 1, residual, characteristic p) representations. So the set of $\bar{\psi}_j$'s is in natural one:one correspondence with the set of (isomorphism classes of) semi-simplifications of p-modular (modular, tame level 1, residual, characteristic p) representations. We will refer to any one of these pseudo-representations ψ_j $(j = 1, \ldots m)$ simply as a *p-modular pseudo-representation*. For each p-modular ψ, let R_ψ denote its universal deformation ring as constructed in Theorem 5.13. Put

$$R_p = \prod_{j=1}^{j=m} R_{\psi_j}.$$

We think of R_p as the universal p-adic deformation ring of residual representations which are p-modular, tame level 1, (pseudo-)representations, and will refer to R_p as *the* **universal deformation ring** (of tame level 1) for short. Of course, to be more precise we should perhaps qualify the adjective *universal* to emphasize that the only residual representations we consider are the p-modular ones, defined at the beginning of this section. The ring R_p is a semi-local complete noetherian ring, because of the proposition proved at the beginning of this section, and it comes along with a universal pseudo-representation,

$$r^{univ} = (\xi^{univ}, \alpha^{univ}, \delta^{univ})$$

with values in R_p.

For each p-modular pseudo-representation ψ, let X_ψ denote the rigid
analytic space over $\mathbf{Q}_p$ associated to R_ψ, and let $X_p = \coprod_{j=1}^m X_\psi$ be the
rigid analytic space over $\mathbf{Q}_p$ associated to R_p. We will refer to X_p as the
universal deformation (rigid analytic) space (of tame level 1). As
hinted above, there is a well-behaved part of R_p and a not-so-well-behaved
part. Specifically, when ψ is an irreducible p-modular pseudo-representation
(in the sense that its associated ξ-function does not vanish identically),
then ψ is the pseudo-representation of an irreducible (odd, degree two) G-
representation. Any such representation is absolutely irreducible (because
an irreducible, but absolutely reducible, degree two representation is seen
to have Abelian image, and therefore it stabilizes the two eigenvectors of
the image of $\mathbf{c}$). Let $\bar{V}$ denote a k-vector space of dimension two, with k-
linear G-action having ψ as corresponding pseudo-representation. Then $\bar{V}$,
being an absolutely irreducible residual G-representation, has an associated
universal deformation space (cf. [M-DGR]) which we denote $R_{\bar{V}}$ equipped
with a continuous G-action on a (free rank two) $R_{\bar{V}}$-module V (the universal
deformation of the G-module $\bar{V}$) which satisfies the universal lifting property
for deformations of the G-module $\bar{V}$ (or equivalently, the G-representation
to $\mathrm{Aut}_{R_{\bar{V}}} V$ is universal for strict equivalence classes of liftings of the G-
representation on $\bar{V}$). Passage to the corresponding pseudo-representation
of the G action on V induces a natural isomorphism (cf. [Ny]) $R_\psi \cong R_{\bar{V}}$.

Let ψ_j ($j = 1, \ldots \mu$) be the subset of the set of p-modular pseudo-
representations which are irreducible, and ψ_j ($j = \mu + 1, \ldots m$) the p-
modular pseudo-representations which are not. For each $j = 1, \ldots \mu$ fix
a degree two G-representation $\bar{V}_j$ with pseudo-representation equal to ψ_j.
Define

$$R_p^{irr} := \prod_{j=1}^\mu R_{\psi_j} \cong \prod_{j=1}^\mu R_{\bar{V}_j},$$

and

$$R_p^{red} := \prod_{j=\mu+1}^m R_{\psi_j}.$$

Then

$$R_p = R_p^{irr} \times R_p^{red},$$

and corresponding to this product decomposition, we have a decomposition
of associated rigid analytic spaces,

$$X_p = X_p^{irr} \bigcup X_p^{red}.$$

The essential difference between the two factors R_p^{irr} and R_p^{red} for us is that over the first factor we have a continuous representation

$$\rho^{univ-irr} : G \to \mathrm{GL}_2(R_p^{irr})$$

satisfying a universal property *vis a vis* irreducible p-modular residual representations (of tame level 1) whereas over the second factor we must make do with a continuous pseudo-representation (which nevertheless does indeed satisfy a corresponding universal property for pseudo-representations). The representation $\rho^{univ-irr}$ induces a rigid analytic, $\mathcal{O}_{X_p^{irr}}$-linear G-representation on the free $\mathcal{O}_{X_p^{irr}}$-bundle of rank two over X_p^{irr}. In contrast, we do *not* show the existence of an analogous rigid analytic, $\mathcal{O}_{X_p^{red}}$-linear G-representation over X_p^{red}. Nevertheless, Theorem 5.1.2 will provide partial results along these lines, applicable to open smooth neighborhoods of the modular curve.

The ring R_p has "a" natural Λ-algebra structure. This is defined by noting that the (determinant) mapping $\mu : G \to R_p$ of the universal pseudo-representation, given by the formula

$$\mu(g) := \alpha^{univ}(g)\delta^{univ}(g) - \xi^{univ}(g)$$

is, in fact, a determinant: it is a continuous homomorphism to R_p^* and factors through the Abelianization $\chi : G^{ab} \cong \mathbf{Z}_p^*$, where we have identified G^{ab} with $\mathbf{Z}_p^*$ via the cyclotomic character χ; that is, if ζ is a p-power root of unity, $g(\zeta) = \chi(g) \cdot \zeta$. This mapping $\mathbf{Z}_p^* \to R_p$ extends uniquely to a continuous ring homomorphism which we will denote here $\mu_{det} : \Lambda = \mathbf{Z}_p[[\mathbf{Z}_p^*]] \to R_p$ and which gives a Λ-algebra structure to R_p that might be referred to as the **determinant Λ-algebra structure** on R_p (for, tensoring R_p via this Λ-algebra structure with a character

$$\kappa : G \to \mathbf{Z}_p^*$$

gives the universal ring of pseudo-deformations of $\bar{\psi}$ possessing determinant character κ; for more discussion about this see section 24 of Chapter 5 of [M-DTR]). One should note that when one works with modular forms, as we are doing, it is more conventional to twist this Λ-algebra structure $\mu = \mu_{det}$ to get a Λ-algebra structure on R_p that gives directly the *weight* of the modular form, rather than the *determinant* of its (pseudo-)representation. The conversion formula is

$$\mu_{wt}([\gamma]) = \gamma \cdot \mu_{det}([\gamma]),$$

where for $\gamma \in \mathbf{Z}_p^*$, we denote its image in the completed group ring Λ by $[\gamma]$, and the multiplication-law in the right-hand side of the above formula is via the natural $\mathbf{Z}_p$-algebra structure of R_p. We will view R_p from now on with its *weight* Λ-algebra structure, i.e., the one given by μ_{wt} structure.

The rigid analytic space X_p is thereby endowed with a canonical projection to weight space $\mathcal{W}_1 = \mathcal{W}$.

5.2. Pseudo-representations attached to Katz modular functions.

Keep the notation of the previous section. Let $\mathbf{T}_p := \mathbf{T}_p(1)$ denote the completion of the algebra generated by Hecke operators acting on Katz modular functions of tame level 1 as described in section 3.3. Then as proven in §3 of [H2],

Theorem 5.2.1 (Gouvêa-Hida).   *There exists a continuous pseudo-representation π from $G_{\mathbf{Q},\{p\}}$ to $\mathbf{T}_p$ such that, for primes $l \nmid p$,*

$$\text{``trace''}(\pi(Frob_l)) = T(l) \quad \text{and} \quad \text{``det''}(\pi(Frob_l)) = \langle l \rangle^*/l.$$

In particular, as $\mathbf{T}_p$ is the inverse limit of Noetherian rings (see section 3.3), we get a natural (unique) ring homomorphism

$$R_p \to \mathbf{T}_p$$

bringing the universal pseudo-representation of tame level 1 to π.

Corollary 5.2.2. *Suppose m is a maximal ideal of $\mathbf{T}_p$ such that the corresponding residual representation $\bar{\rho}$ is absolutely irreducible (cf. section 5.1). Then, there exists a representation $\rho_m: G_{\mathbf{Q},S} \to GL_2((\mathbf{T}_p)_m)$ such that, for $l \nmid Np$,*

$$trace(\rho(Frob_l)) = T(l) \quad \text{and} \quad det(\rho(Frob_l)) = \langle l \rangle^*/l.$$

Corollary 5.2.3. *If $p > 2$, the ring $\mathbf{T}_p$ of Hecke operators on Katz modular functions of tame level 1 is noetherian.*

Proof. We know that $\mathbf{T}_p$ is a complete semi-local p-adic ring, so all we have to show is that the completion $(\mathbf{T}_p)_m$ is noetherian for any maximal ideal m. Let $\mathbf{T}_p'$ denote the subring of $\mathbf{T}_p$ which is the completion of the Λ-algebra generated by the Hecke operators $T(l)$ for $(l, p) = 1$. Then it follows that $\mathbf{T}_p' \cong \varprojlim h_k'(p^k)$ where $h_k'(p^k)$ is the subalgebra of $h_k(p^k)$ generated by the diamond operator and the Hecke operators $T(l)$ for $(l, p) = 1$. The pseudo-representation π above actually takes values in $\mathbf{T}_p'$. Let $m' = m \cap \mathbf{T}_p'$ and let $\bar{\pi}_m$ denote the reduction of $\pi\colon G_{\mathbf{Q},\{p\}} \to \mathbf{T}'_p$ modulo m'.* It follows that there exists a homomorphism of $R(\bar{\psi}_m)$ into $(\mathbf{T}_p)_m$ whose image contains the image of $T(l)$ and $\langle l \rangle^*$ for all $l \neq p$. Since $R(\bar{\psi}_m)$ is compact, using the Tchebotarev density theorem, we see that this homomorphism is surjective. Hence this complete local ring $(\mathbf{T}'_p)_{m'}$ is Noetherian. (When m corresponds to an absolutely irreducible representation this is the conclusion of Corollary III.5.7 of [G-ApMF].) We need the following general proposition to conclude the proof:

Proposition 5.2.4. *Suppose R_n are complete p-adic local rings and $R_{n+1} \to R_n$ are surjections. Then $R = \varprojlim_n R_n$ is a complete local ring.*

Let m be its maximal ideal and k its residue field. The ring R is Noetherian if and only if the dimensions of the Zariski tangent spaces $\dim_k mR_n/m^2 R_n$ are bounded independently of n.

Proof. The Noetherianness of R implies the uniform boundedness of the dimensions of the Zariski tangent spaces. If now $\dim_k mR_n/m^2 R_n$ is bounded independently of n, it follows that $d =: \dim_k m/m^2$ is finite. Suppose $t_1, \ldots, t_d \in m$ is a basis modulo m^2. Then there is a natural map from $W(k)$ to R and the map

$$W(k)[[T_1, \ldots, T_d]] \to R,$$

which sends T_i to t_i is a surjection. Since $W(k)[[T_1, \ldots, T_d]]$ is Noetherian we may conclude the proof. ∎

We can now complete the proof of the corollary. Let $\mathbf{T}_n = h_2(p^n)$ and $\mathbf{T}'_n = h'_2(p^n)$. Then $\mathbf{T}_n = \mathbf{T}^*_n[u_n]$, where $U = \varprojlim u_n$. Let $k = \mathbf{T}/m$ and $k' = \mathbf{T}'/m'$. Since, $(\mathbf{T}')_{m'}$ is Noetherian, there exists $a_1, \ldots, a_d$ in m' which reduce to a basis of $m'/(m')^2$ over k'. Let $g(x) \in k'[x]$ be the minimal monic polynomial for $U \bmod m$ over k', which exists since k is finite. Suppose f is its degree. Now suppose n is such that $k' \cong \mathbf{T}'_n/m'\mathbf{T}'_n := m'_n$ and $k \cong \mathbf{T}/m\mathbf{T} := m_n$. We identify these isomorphic fields. Let $\tilde{g}$ be a lifting of g to a monic polynomial in $\mathbf{T}'_n[x]$. It follows that $a_0 =: \tilde{g}(u_n) \in m'_n$. We can write any element α of $\mathbf{T}_n$ in the form

$$\alpha = \sum_j b_j(u_n)\tilde{g}(u_n)^j,$$

---

* We note that, $\mathbf{T}'_p$ is also a complete semi-local ring with finite residue fields. Moreover, the map $Max(\mathbf{T}_p) \to Max(\mathbf{T}'_p)$ is surjective by Atkin-Lehner theory and bijective away from the ordinary points.

82   R. Coleman & B. Mazur

where $b_j(x) \in \mathbf{T}'_n[x]$ and $\deg b_j(x) < f$. Then $\alpha \in m_n$ if and only if $b_0(x) \in m'_n \mathbf{T}'_n[x]$. This means that, in this case, we can write

$$b_0(x) = \sum_{k=1}^{d} c_k(x) a_k$$

where $c_k(x) \in \mathbf{T}'_n[x]$. We conclude,

$$\alpha \equiv b_1(u_n)a_0 + c_1(u_n)a_1 + \cdots c_d(u_n)a_d \bmod m T_n.$$

In particular,
$$\dim_k m_n/m_n^2 \leq \dim_{k'} m'_n/(m'_n)^2 + 1,$$

which concludes the proof. ∎

Corollary. $\dim_k m/m^2 \leq \dim_{k'} m'/(m')^2 + 1$.

Chapter 6. The Eigencurve.

6.1. The definition of the eigencurve.

Keep the notation of Chapter 5. We assume that $p > 2$, and recall that we are working with tame level $N = 1$. Let R_p be the universal deformation space for (p-modular) pseudo-representations of tame level 1, as before, and X_p is the associated rigid analytic (universal deformation) space of tame level 1. Recall that R_p has a natural $\mathcal{H}'$-algebra structure, and is the receiving ring for the universal pseudo-deformation r^{univ} of $G_{\mathbf{Q},S}$ with the notational conventions of Chapter 5. Let Y_p denote the rigid analytic variety given as the product

$$Y = Y_p = X_p \times \mathbf{A}^1,$$

where the parameter for $\mathbf{A}^1$ will be given by the variable T. We prepare to apply the constructions of section 1.2 to Y. In particular, putting $R = R_p$ for short, we have a natural homomorphism of the ring of entire power series $R\{\{T\}\}$ into the ring $A(Y)$ of rigid analytic functions on Y.

Let $\alpha \in \mathcal{H}'$ be any element such that the image, $\iota \cdot \alpha \in R$, is a unit in R (where $\iota : \mathcal{H}' \to R$ denotes the canonical homomorphism). Let $\tilde{U} \in \mathcal{U}$ be the product of α and U_p. Let $P_{\tilde{U}}(T) \in \Lambda\{\{T\}\}$ be the Fredholm determinant of the completely continuous system of operators associated to $\tilde{U}$ as discussed in Chapters 3 and 4 (and was also denoted $P_\alpha(T)$ in Chapter 5). Put

$$\phi_{\tilde{U}} := P_{\tilde{U}}(T/(\iota \cdot \alpha))$$

which, by Theorem 4.3.1 and our hypothesis about α, is a Fredholm series in the ring $R\{\{T\}\}$, and consequently determines a Fredholm function on Y_p, which we will denote by the same letter. Consider the ideal $\mathcal{I}$ in $A(Y)$ generated by the Fredholm functions $\phi_{\tilde{U}}$ for all α's which satisfy our hypothesis.

Definition 1. The **eigencurve of tame level** 1 is the closed, rigid analytic $\mathbf{Q}_p$-subvariety,

$$C = C_p \subset Y_p = X_p \times \mathbf{A}^1,$$

whose ideal of definition is $\mathcal{I} \in A(Y)$.

Remark. We shall be proving that C is a curve (cf. below) but this fact is not evident from the definition.

Equivalently, we can think of C as follows: For every $\alpha \in \mathcal{H}'$ such that $\iota \cdot \alpha$ is a unit in R we define the mapping $r_\alpha : X_p \times \mathbf{A}^1 \to \mathcal{W} \times \mathbf{A}^1$ by

$$r_\alpha(x,t) = \left(\pi(x), \frac{t}{\iota \cdot \alpha(x)} \right),$$

where $\pi : X_p \to \mathcal{W}$ is the canonical projection to weight space (see section 5.1). Then

$$C_p := \bigcap_\alpha r_\alpha^{-1}(Z_{\alpha \cdot U}),$$

where the spectral curve $Z_{\alpha \cdot U}$ is the locus of zeroes of the Fredholm function $P_{\alpha \cdot U}(T)$ on $\mathcal{W} \times \mathbf{A}^1$.

By its definition, the curve C_p comes along with a rigid analytic projection to the universal deformation space $C_p \to X_p$ and therefore by composition of this morphism with the canonical projection, $\pi : X \to \mathcal{W}$, we have a canonical projection, $C_p \to \mathcal{W}$, of C_p to weight space. There are also natural morphisms of C_p to $Z_{\tilde{U}}$ for every $\tilde{U} = \alpha \cdot U_p$ with $\iota \cdot \alpha \in R^*$, and these natural morphisms are compatible with projection to weight space. Explicitly, we have the ring homomorphism

$$\iota_{\tilde{U}} : \Lambda\{\{\tilde{T}\}\} \to R\{\{T\}\}$$

which extends the natural ring homomorphism $\iota : \Lambda \to R$ by sending the variable $\tilde{T}$ to $T/(\iota \cdot \alpha)$. The ring homomorphism $\iota_{\tilde{U}}$ brings the Fredholm determinant $P_{\tilde{U}}(\tilde{T}) \in \Lambda\{\{\tilde{T}\}\}$ to $\phi_{\tilde{U}} \in R\{\{T\}\}$. Since $P_{\tilde{U}}(\tilde{T})$ generates the defining ideal of the rigid analytic curve (the Fredholm hypersurface)

$$Z_{\tilde{U}} \subset \mathcal{W} \times \mathbf{A}^1$$

and since $\phi_{\tilde{U}}$ is in the defining ideal of $C_p \subset X_p \times \mathbf{A}^1$, $\iota_{\tilde{U}}$ induces a rigid analytic morphism, call it $\lambda_{\tilde{U}} : C_p \to Z_{\tilde{U}}$.

Given a $\mathbf{C}_p$-valued point $c \in C_p$ let us refer to its two coordinates $c = (r, 1/u) \in X_p \times \mathbf{A}^1$ as its associated **pseudo-representation** $r = r_c$ and inverse U_p-**eigenvalue** respectively, and we will refer to the quantity $u = u_c$ as the U_p-**eigenvalue** of c. The point c induces a ring homomorphism

$$\Psi_c : \mathcal{H} \to \mathbf{C}_p$$

as follows. The associated pseudo-representation $r = r_c$ induces a homomorphism $\Psi_r : \mathcal{H}' \to \mathbf{C}_p$ as in the previous discussion and Ψ_c is the unique (ring homomorphism) extension of Ψ_r to the polynomial ring $\mathcal{H} = \mathcal{H}'[U_p]$ which sends U_p to $u \in \mathbf{C}_p$. As in the discussion in Section 3.1 we may attach to the ring homomorphism Ψ_c a Fourier expansion $\mathcal{F}_c$ which we will refer to as the **Fourier expansion of the point** $c \in C_p$:

$$\mathcal{F}_c = \sum_{n=1}^{n=\infty} \Psi_c(\mathcal{T}_n) q^n \in \mathbf{C}_p[[q]],$$

where the conventions regarding the symbols $\mathcal{T}_n$ are as in Chapter 3.

The point $c \in C_p$ is determined by its Fourier expansion $\mathcal{F}_c$.

6.2. The points of the eigencurve are overconvergent eigenforms.

Recall that an eigenform f for (the full Λ-algebra of operators) $\mathcal{H}$, with Fourier expansion $\tilde{f} = \sum_{n=0}^{\infty} a_n q^n$, is called **normalized** if either $a_1 = 1$, or else we are in the specific case of weight 0 where it is possible for a_1 to vanish, in which case the standard recurrence relations that hold for the Fourier coefficients of a $\mathcal{H}$-eigenform force *all* the nonconstant coefficients of $\tilde{f}$ to vanish; in this idiosyncratic case, let us call f **normalized** if $\tilde{f}$ is the constant power series 1. For a positive integer n this normalized eigenvector has $\mathcal{T}_n$-eigenvalue $\sigma_{-1}^*(n)$.

Any eigenform for $\mathcal{H}$ is a multiple of a normalized eigenform.

Theorem 6.2.1. *There is a one-one correspondence between the set of normalized overconvergent modular eigenforms with Fourier coefficients in $\mathbf{C}_p$, of tame level 1, weight $w \in \mathcal{W}(\mathbf{C}_p)$, having nonzero U_p-eigenvalue, and the set of $\mathbf{C}_p$-valued points of weight w on the eigencurve $C = C_p$. If f is a normalized overconvergent modular eigenform (of tame level 1) with Fourier coefficients in $\mathbf{C}_p$ and with nonzero U_p-eigenvalue with Fourier expansion*

$$\tilde{f} = \sum_{n=0}^{n=\infty} a_n q^n \in \mathbf{C}_p[[q]],$$

with $a_1 = 1$, then the point $c \in C$ to which f corresponds has its Fourier expansion equal to $\tilde{f} - a_0$ (i.e., the Fourier expansion of c is that of f deprived of its constant term).

Proof. Starting with an overconvergent modular eigenform f with the characteristics formulated in the statement of the above theorem, one sees (by Theorem 4.4.1) that there is a point $z_{\tilde{U}} \in Z_{\tilde{U}}$ for every $\tilde{U} \in \mathcal{U}$ corresponding to f and therefore, one immediately sees from the construction of the rigid analytic space C_p that there is a ($\mathbf{C}_p$-valued) point $c \in C_p$ corresponding to f as asserted in the statement of our theorem. It is somewhat more delicate to go the other way.

For this, let us start with a $\mathbf{C}_p$-valued point c of the eigencurve $C = C_p$. We shall fix c in the discussion below, and define a list of objects dependent upon c. Note that we are working systematically, in the proof below, over $\mathbf{C}_p$. Denote the weight of c by w (i.e., w is the $\mathbf{C}_p$-valued point of $\mathcal{W}$ which is the image of c under the canonical projection $C_p \to \mathcal{W}$).

For $\tilde{U} = \alpha \cdot U_p$ with $\alpha \in \mathcal{H}'$ such that $\iota\alpha$ is in R^*, let $z_{\tilde{U}} := \lambda_{\tilde{U}}(c) \in Z_{\tilde{U}}$. The points $z_{\tilde{U}} = (w, 1/\tilde{u}) \in \mathcal{W} \times \mathbf{A}^1$ are, of course, of fixed weight w. Moreover, we have that $ord_p(\tilde{u})$ is equal to $\sigma :=$ the **slope** of c, i.e. $\sigma = ord_p(u)$ where u is the U_p-eigenvalue of c (Note that $ord_p(\tilde{u}) = ord_p(u)$ since $\iota\alpha \in R^*$). Applying Theorem 4.1.1 to the points $z_{\tilde{U}} \in Z_{\tilde{U}}$, we have that, for *every* $\tilde{U}$ as above, there is at least one overconvergent modular form of weight w which is an eigenform (not just a generalized eigenform) for $\tilde{U} = \alpha U_p$ with eigenvalue $\tilde{u}$. Moreover, the $\tilde{U}$-eigenvalue of each such eigenform is equal to $\tilde{u} =$ the $\tilde{U}$-eigenvalue of c. The $\mathbf{C}_p$-vector space $\mathcal{V}_{\tilde{U}}$ of these overconvergent $\tilde{U}$- eigenforms (of weight w, tame level 1, and $\tilde{U}$-eigenvalue equal to $\tilde{u}$) is finite dimensional, and is stabilized by the action of the commutative Λ-algebra $\mathcal{H}$. Therefore $\mathcal{V}_{\tilde{U}}$ contains eigenforms for the action of the entire algebra $\mathcal{H}$. It follows that $\mathcal{V}_{\tilde{U}}$ contains at least one normalized $\mathcal{H}$ -eigenform. Let $\mathcal{F}_{\tilde{U}}$ denote the set of all normalized eigenforms in $\mathcal{V}_{\tilde{U}}$. That is, $\mathcal{F}_{\tilde{U}}$ is the set of normalized overconvergent modular forms of tame level 1 which are $\mathcal{H}$ -eigenforms and such that their $\tilde{U}$-eigenvalue is equal to $\tilde{u}$. Then for every $\tilde{U} = \alpha \cdot U$, $\mathcal{F}_{\tilde{U}}$ is a finite non-empty set.

Lemma 6.2.2. *The intersection of the sets $\mathcal{F}_{\tilde{U}}$ (running over all $\tilde{U}$'s as above) is nonempty and, in fact, consists in precisely a single eigenform f.*

Remarks and Proof of Lemma 6.2.2. Assume, for a second, that there is an eigenform

$$f \in \bigcap_{\tilde{U}} \mathcal{F}_{\tilde{U}}.$$

It follows that f has the same Fourier expansion as c. To see this, first note that f has the same U_p-eigenvalue as c (because $f \in \mathcal{F}_{U_p}$). Moreover, for $\tau \in \mathcal{H}'$ and $\tilde{U} = (1 + p \cdot \tau) \cdot U_p$, since $f \in \mathcal{F}_{\tilde{U}}$, we have that f has the same $\tilde{U}$-eigenvalue as c. Since the operators U_p and $\tilde{U}$ commute, it follows that f also has the same τ-eigenvalue as c. This argument yields two things. First, we see that it suffices to show existence of f, for then its uniqueness follows. Second, we see that it suffices to show that

$$\bigcap_{\tilde{U}} \mathcal{F}_{\tilde{U}}$$

is nonempty where the intersection is taken over a special list of $\tilde{U}$'s. Specifically, let $\ell_1, \ell_2, \ldots$ be an enumeration of all the prime numbers different from p so that $U_p, T_{\ell_1}, T_{\ell_2} \ldots$ is an enumeration of the full list of Hecke and Atkin-Lehner operators.

Sublemma 6.2.3. *If for a fixed infinite sequence of positive integers $a_1, a_2, \ldots$ and for the list*

$$\tilde{U}_0 = U_p$$

$$\tilde{U}_1 = (1 + p^{a_1} T_{\ell_1})\tilde{U}_0$$

$$\tilde{U}_2 = (1 + p^{a_2} T_{\ell_2})\tilde{U}_1$$

$$\cdots$$

of operators, the intersection $\bigcap_{\tilde{U}} \mathcal{F}_{\tilde{U}}$ is nonempty as $\tilde{U}$ runs through $\tilde{U}_k$ for $k \geq 0$, then this intersection is nonempty as $\tilde{U}$ runs through all $(1+p\cdot\tau)\cdot U_p$.

In the proof of Lemma 6.2.2 that we give below, we shall assume given an enumeration of the primes (different from p) as above, and we will be generating inductively a particular sequence of positive integers $a_1, a_2, \ldots$ as required by sublemma 6.2.3 to prove lemma 6.2.2.

Step 1. Consider all the $\tilde{U}$'s of the form $\tilde{U} = \alpha \cdot U_p$ where $\alpha = 1 + p \cdot \tau$ for $\tau \in H'$. For these $\tilde{U}$'s, the eigenforms of fixed weight w and eigenvalue of fixed slope σ all lie in the vector space

$$M_{\kappa,\sigma}$$

defined at the end of section 4.1 (here, $\kappa = w$) which we denote simply M, for short; recall that M comes along with a natural $\mathcal{H}$-module structure. By Proposition 4.1.2, this (finite dimensional) M is generated by all over-convergent generalized $\tilde{U}$-eigenforms of weight w and $\tilde{U}$-slope equal to σ, where $\tilde{U}$ is any element of $\mathcal{U}$ as above. In particular, it is generated by all overconvergent generalized U_p-eigenforms of weight w and U_p- slope equal to σ. Since there are only a finite number of eigenvalues for the operator U_p acting on the (finite dimensional) space M we may find an integer a_1 large enough to *distinguish* U_p-eigenvalues in the sense that if λ and λ' are distinct U_p- eigenvalues in its action on M, then $ord_p(\lambda - \lambda') < a_1$.

Step 2. Put $\alpha_1 := 1 + p^{a_1} \cdot T_{\ell_1}$, and $\tilde{U}_1 = \alpha_1 \cdot U_p$. Here, ℓ_1 is the *first* prime given in our enumeration above, and a_1 is the integer obtained in Step 1. Let

$$f_1 \in \mathcal{F}_{\tilde{U}_1},$$

so that the $\tilde{U}_1$-eigenvalue of f_1 is the same as that of c. We claim that it then follows that *both* the U_p- and the $T_{\ell_1}\emptyset$eigenvalues of f_1 are the same as those of c. To see this, let us establish notation for the U_p- and the T_{ℓ_1}-eigenvalues of f_1 and of c by the formulas:

$$U_p \cdot f_1 = u_{p,(f)} \cdot f_1 \quad ; \quad U_p \cdot c = u_{p,(c)} \cdot c$$

$$T_{\ell_1} \cdot f_1 = t_{\ell_1,(f)} \cdot f_1 \quad ; \quad T_{\ell_1} \cdot c = t_{\ell_1,(c)} \cdot c.$$

Denoting by $\tilde{u}_1$ the common $\tilde{U}_1$-eigenvalue of f_1 and c, we have two expressions for $\tilde{u}_1$; namely:

$$\tilde{u}_1 = (1 + p^{a_1} \cdot t_{\ell_1,(f)}) \cdot u_{p,(f)} = (1 + p^{a_1} \cdot t_{\ell_1,(c)}) \cdot u_{p,(c)}.$$

Since all the elements of the above formula lie in the ring of integers of $\mathbf{C}_p$, it follows that $ord_p(u_{p,(f)} - u_{p,(c)}) \geq a_1$ which implies, by Step 1, that $u_{p,(f)} = u_{p,(c)}$, giving also that $t_{\ell_1,(f)} = t_{\ell_1,(c)}$. Since there only a finite number of eigenvalues for the operator $\tilde{U}_1$ acting on the (finite dimensional) space M we may find an integer $a_2 \geq a_1$ large enough so that if λ and λ' are distinct $\tilde{U}_1$- eigenvalues in the space M, then $ord_p(\lambda - \lambda') < a_2 - a_1$.

Step 3. We follow the pattern of Step 2. Put $\alpha_2 := 1 + p^{a_2} \cdot T_{\ell_2}$, and $\tilde{U}_2 = \alpha_2 \cdot \tilde{U}_1$. Let

$$f_2 \in \mathcal{F}_{\tilde{U}_2},$$

so that the $\tilde{U}_2$-eigenvalue of f_2 is the same as that of c. With the analogous notational conventions as in Step 2 (i.e., the eigenvalue of an operator on f_2 or on c is denoted by the symbol denoting the operator, only in lower case, and with an (f) or a (c) in the subscript) we have:

$$\tilde{u}_2 = (1 + p^{a_2} \cdot t_{\ell_2,(f)}) \cdot \tilde{u}_{1,(f)} = (1 + p^{a_2} \cdot t_{\ell_2,(c)}) \cdot \tilde{u}_{1,(c)}.$$

By an analogous argument to that of Step 2, since a_2 is large enough to distinguish $\tilde{U}_1$-eigenvalues, we get from the above equation first that the $\tilde{U}_1$-eigenvalue of f_2 and of c are equal, and then that the T_{ℓ_2}- eigenvalues of f_2 and of c are equal as well. Since the $\tilde{U}_1$-eigenvalue of f_2 and of c are equal, a direct application of the argument of Step 2 then gives that f_2 and c agree on their T_{ℓ_1}- and U_p-eigenvalues as well. Let $a_3 \geq a_2$ be an integer large enough, now, to distinguish $\tilde{U}_2$-eigenvalues in the action of $\tilde{U}_2$ on M.

Step 4. Proceeding by induction as in Steps 2 and 3 we may obtain a sequence of integers $a_1 \leq a_2 \leq \ldots a_k \ldots$ and corresponding operators $\tilde{U}_k$ for $k = 1, 2, \ldots$ and overconvergent modular forms $f_k \in \mathcal{F}_{\tilde{U}_k} \subset M$ which have the property that f_k and c agree on their U_p-eigenvalues as well as their T_{ℓ_j}-eigenvalues for $j = 1, \ldots, k$. Since the sets $\mathcal{F}_{\tilde{U}_k}$ are finite, it follows that there is at least one element $f \in \bigcap_{k=1}^{\infty} \mathcal{F}_{\tilde{U}_k} \subset M$ which, by the previous discussion, is then an eigenform for U_p and the Hecke and Atkin-Lehner operators T_ℓ for all prime numbers $\ell \neq p$. This f is then an overconvergent (normalized) modular eigenform of tame level 1 whose eigenvalues for the action of any element in $\mathcal{H}$ is the same as that of c. This establishes Theorem 6.2.1.

6.3. The projections of the eigencurve to the spectral curves.

But the above proof gives more, and to record what it gives, let us consider multiplicities of various spaces of generalized eigenforms. Define:

$\mu(c) :=$ the dimension over $\mathbf{C}_p$ of the space of all overconvergent generalized eigenforms (for $\mathcal{H}$) in the $\varkappa$-isotypic component corresponding to c in $M = M_k(\mathbf{C}_p;)_\sigma$.

Similarly, for $\tilde{U} \in \mathcal{U}$ of the appropriate form (i.e., $\tilde{U} = (1 + p \cdot \tau) \cdot U_p$ with $\tau \in \mathcal{H}$) let $z \in Z_{\tilde{U}}$ be the image of c under $\lambda_{\tilde{U}}$, and put

$\mu(z) :=$ the dimension over $\mathbf{C}_p$ of the space, call it $M_{\tilde{U}} \subset M$, of all overconvergent modular forms in $M = M_k(\mathbf{C}_p)_\sigma$ which have the same weight as c, and which are generalized $\tilde{U}$-eigenforms with the same $\tilde{U}$-eigenvalue as c.

We have the following geometric properties of the projections of the $Z_{\tilde{U}}$'s to weight space, and the morphisms $\lambda_{\tilde{U}} : C \to Z_{\tilde{U}}$ (using the Reisz Theory developed in A4 of [C-BMF]).

Proposition 6.3.1. *The structure of $Z_{\tilde{U}}$ over weight space.*

Let $\tilde{U} = (1 + p \cdot \tau) \cdot U_p$ with $\tau \in \mathcal{H}$. Then $Z_{\tilde{U}}$ is a Fredholm hypersurface in $\mathcal{W} \times \mathbf{A}^1$.

Proof. This follows from Theorems 4.3.1 and 4.4.1.

Proposition 6.3.2. *The structure of $\lambda_{\tilde{U}} : C \to Z_{\tilde{U}}$.*

    a. *The morphism $\lambda_{\tilde{U}} : C \to Z_{\tilde{U}}$ is surjective.*

    b. *If z is point in $Z_{\tilde{U}}(\mathbf{C}_p)$ then*

$$\mu(z) = \sum \mu(c)$$

where the sum ranges over the points c in $C(\mathbf{C}_p)$ which map to z.

Proof. Part **a** follows from the fact that every point z in $Z_{\tilde{U}}(C_p)$ corresponds to a finite dimensional space of eigenforms for $\tilde{U}$ with fixed weight and slope by theorems 4.1.1 and 4.3.1 which is stable under the action of $\mathcal{H}$. It follows that this space contains an eigenform and if c is the coresponding point of C, c maps to z.

Part **b** follows from the fact that the space of generalized eigenforms corresponding to a point $z \in Z_{\tilde{U}}(\mathbf{C}_p)$ is of dimension $\mu(z)$ and is the direct sum isotypic components for the action of $\mathcal{H}$. Moreover each such isotypic component corresponds to a point $c \in C(\mathbf{C}_p)$ which maps to z and has dimension $\mu(c)$. ∎

We now state some consequences of the above Propositions and of Theorem 6.2.1 (and its proof). We will have occasion to refer below to the sequence of operators $\tilde{U}_k$ for $k = 1, 2, \ldots$ given in that proof.

Corollary 6.3.3. *For $c \in C$ as in Theorem 6.2.1 with weight w, we have:*

    1. *If k is sufficiently large,*

$$\mathcal{F}_{\tilde{U}_k} = \{f_c\}.$$

2. Let $z_k \in Z_{\tilde{U}_k}$ denote the image of c under $\lambda_{\tilde{U}_k}$. If k is sufficiently large, then

$$\mu(c) = \mu(z_k).$$

Proof. The first statement was proved in the course of the proof of Theorem 6.2.1. But the second comes along with it for if k is sufficiently large to guarantee that $\mathcal{F}_{\tilde{U}_k} = \{f_c\}$, then any $\tilde{U}_k$-generalized eigenform in $M_{\tilde{U}_k}$ is an $\mathcal{H}$-generalized eigenform in the isotypic component corresponding to c. ∎

6.4. The Eisenstein curve.

Define the **Eisenstein pseudo-representation** $r_\kappa^{\mathrm{Eis}} = (\xi_\kappa^{\mathrm{Eis}}, \alpha_\kappa^{\mathrm{Eis}}, \delta_\kappa^{\mathrm{Eis}})$ of weight κ by the formulas $\xi_\kappa^{\mathrm{Eis}}(g, h) \equiv 0$, $\alpha_\kappa^{\mathrm{Eis}}(g) \equiv 1$, and let $\delta_\kappa^{\mathrm{Eis}}$ denote the continuous $\mathbf{C}_p^*$-valued character of $G_{\mathbf{Q},\{p,\infty\}}$ which is the determinant-character of representations attached to modular eigenforms of weight κ, as discussed at the end of section 5.1. Specifically, consider the composition of the homomorphisms

$$\eta_\kappa : G_{\mathbf{Q},\{p,\infty\}} \to \mathbf{Z}_p^* \to \Lambda^* \to \mathbf{C}_p^*,$$

where the first homomorphism is χ, the canonical homomorphism arising from action on p-power roots of 1 (so for $\zeta \in \bar{\mathbf{Q}}_p$ a p-power root of unity, $g(\zeta) = \zeta^{\chi(g)}$) the second is the natural injection, and the third is the restriction of κ to multiplicative groups of units. Calling χ again the $\mathbf{C}_p^*$-valued character induced by χ (i.e., the composition $\chi : G_{\mathbf{Q},\{p,\infty\}} \to \mathbf{Z}_p^* \subset \mathbf{C}_p^*$, we put

$$\delta_\kappa^{\mathrm{Eis}} := \chi^{-1} \cdot \eta_\kappa.$$

More systematically, one can simply define a single pseudo-representation $r^{\mathrm{Eis}} = (\xi^{\mathrm{Eis}}, \alpha^{\mathrm{Eis}}, \delta^{\mathrm{Eis}})$ of $G_{\mathbf{Q},\{p,\infty\}}$ with values in Λ by taking $\xi^{\mathrm{Eis}} \equiv 0$ and $\alpha^{\mathrm{Eis}} \equiv 0$ and δ^{Eis} the evident character $G_{\mathbf{Q},\{p,\infty\}} \to \Lambda^*$ so that $\kappa(r^{\mathrm{Eis}}) = r_\kappa^{\mathrm{Eis}}$ for all $\mathbf{C}_p$-valued weights κ. By universality for p-modular pseudo-representations, we then get that r^{Eis} is the pseudo-representation induced from a continuous ring-homomorphism $\pi^{\mathrm{Eis}} : R_p \to \Lambda$ (which sends the radical of R_p to the maximal ideal in Λ). By $X^{\mathrm{Eis}} \subset X_p$ let us mean the rigid analytic subspace of X_p associated to the closed subscheme $\mathrm{Spec}\,(\Lambda) \subset$ $\mathrm{Spec}\,(R_p)$ and define the **Eisenstein curve** C_p^{Eis} to be the rigid subspace of $X_p \times \mathbf{A}^1$ defined by:

$$C_p^{\mathrm{Eis}} := X^{\mathrm{Eis}} \times 1 \subset X_p \times \mathbf{A}^1,$$

that is, the U_p-eigenvalue ((in the sense discussed in section 6.1) of the points of C_p^{Eis} is taken to be identically equal to 1. By Propositions 3.6.1, and 6.2.1, we see that the set of $\mathbf{C_p}$-valued points of C_p^{Eis} is contained in the set of $\mathbf{C_p}$-valued points of the eigencurve C_p, and, moreover, any point of C_p which is not in C_p^{Eis} is represented by a normalized cuspidal-overconvergent eigenform (of tame level 1 and finite slope).

Chapter 7. The eigencurve as a finite cover of a spectral curve.

In this chapter, we will construct a curve which we will call D, which we will prove is isomorphic to the reduced rigid space associated to the eigencurve. We perform this construction by patching together generalizations of the pieces made in the proof of Theorem B5.7 of [C-BMF].

7.1. Local pieces.

We will need the following generalization of Theorem B5.7 of [C-BMF] (see also Theorem 1 of [CST]): We let p be an odd prime number, and work with tame level 1. We keep the notation of the previous chapters except for the following changes: For $\alpha \in \mathcal{H}$ let $P_\alpha(T)$ denote the Fredholm determinant of αU_p, Z_α the spectral curve of αU_p and λ_α the natural map from C_p to Z_α. (These were called $P_{\alpha U_p}(T)$, $Z_{\alpha U_p}$ and $\lambda_{\alpha U_p}$, respectively, in previous sections.) We view Z_α as a rigid analytic Fredholm curve over $\mathbf{Q}_p$. Let $\pi_\alpha : Z_\alpha \to \mathcal{W}$ be the projection of Z_α to weight space. Now fix $\alpha \in \mathcal{H}$. If V be an irreducible affinoid subdomain in Z_α, then by Lemma A5.6 of [C-BMF], its image $Y := \pi_\alpha(V)$ in $\mathcal{W}$ is an affinoid subdomain of $\mathcal{W}$. Denote by $Z_Y \subset Z_\alpha$ the inverse image of Y under the morphism π_α. Let $\mathcal{C}_\alpha$ be the collection of affinoid subdomains V in Z_α such that V is finite over its image $Y := \pi_\alpha(V)$ and which are admissibly closed-open in Z_Y. Let $\mathcal{C}_\alpha^i$ denote the set of irreducible members of $\mathcal{C}_\alpha$. We know using Proposition A5.8 of [C-BMF] that $\mathcal{C}_\alpha$ is an admissible covering of Z_α.*

Fixing $\alpha \in \mathcal{H}$, for each affinoid subdomain $Y \in \mathcal{W}$, there is a one:one correspondence

$$V \longleftrightarrow \{ P_\alpha(T) = Q_V(T)H(T) \},$$

where V runs through the subset of $\mathcal{C}_\alpha^i$ with image Y, and where the equations $P_\alpha(T) = Q_V(T)H(T)$ run through all factorizations of $P_\alpha(T)$ over $A := A(Y)$, with $Q_V(T)$ a polynomial in $A[T]$ whose constant term is one

---

* It was said in [C-BMF] that the proof of Proposition A5.8 given there only works over $\mathbf{C_p}$, but it actually works more generally. First the phrase "$x \in V$" should be replaced by "closed points x of V." Then the affinoid W found in the proof of Lemma A5.9 can be defined, in the same way, for arbitrary K.

and whose leading coefficient is a unit in A, and with $H(T)$ an entire function in $A\{\{T\}\}$ such that $(Q_V, H) = 1$. If V and $Q_V(T)$ are connected via
the above one:one correspondence, we have that

$$V = \text{Max } A(Y)\langle T\rangle/(Q_V^*(T)),$$

recalling the notation $Q_V^*(T) := T^d Q_V(T^{-1})$, where $d = \text{degree}(Q_V(T))$.
Let $M_Y^\dagger(v)$ be the Banach A-module of v-overconvergent modular forms of
tame level 1 over the affinoid domain Y, as in section 2.4, for sufficiently
small $v > 0$, and $M_Y^\dagger = \cup_v M_Y^\dagger(v)$ the A-module of overconvergent modular
forms of tame level 1 over the affinoid domain Y. For each $v > 0$ we obtain
a direct sum decomposition

$$M_Y^\dagger(v) = N(V; v) \bigoplus F(V; v)$$

of closed $A(Y)$-modules for which $Q_V^*(\tilde{U})$ annihilates $N(V; v)$ and is invertible on $F(V; v)$. Using the results of section 4.3 we see that $N(V; v)$
is independent of v for v close to 1. The submodule $N(V) \subset M_Y^\dagger$ is then
defined to be $N(V; v)$ for v sufficiently close to 1.

Remark. In the notation of section A4 of [C-BMF], we may write

$$N(V) = N_{\tilde{U}_Y}(Q_V),$$

where $\tilde{U}_Y$ is the restriction of $\tilde{U} = \alpha U$ to $M_Y^\dagger$.

By Theorem A4.5 of [C-BMF], $N(V)$ is locally free over $A = A(Y)$ of
constant rank $d_V := \deg Q_V(T)$. Define $\mathbf{T}(V)$, the **overconvergent finite
slope Hecke algebra over** V, to be the image of $\mathcal{H} \otimes A$ in $End_A(N(V))$.
The A-algebra $\mathbf{T}(V)$ is complete since $End_A(N(V))$ is finite over $A = A(Y)$.
We have that $\mathbf{T}(V)$ is the ring of rigid analytic functions on an affinoid $D(V)$
over $\mathbf{Q}_p$ finite flat over Y of degree d. The argument for this can be found
on pp. 437-438 of [C-BMF]. In fact, A is a PID and so $N(V)$ is free. In
summary, so far, we have associated to every element V of $\mathcal{C}_\alpha^i$ with image
$Y \subset \mathcal{W}$, an affinoid $D(V)$ which is finite and flat over Y.

$$V/Y \quad \longmapsto \quad D(V)/Y .$$

Since any member of $\mathcal{C}_\alpha$ is a finite disjoint union of elements in $\mathcal{C}_\alpha^i$, we may
extend the above association (by taking disjoint unions) to any member V
of $\mathcal{C}_\alpha$. Now let μ denote the projection of $\mathcal{W} \times \mathbf{A}^1$ to $\mathbf{A}^1$. We will apply
the q-expansion principle to show:

Theorem 7.1.1. *Suppose $V \in C_\alpha$ such that $Y := \pi_\alpha(V)$. Let $L \subset \mathbf{C}_p$ be a finite extension of $\mathbf{Q}_p$. For an L-valued point $x \in D(V)(L)$, let $\eta_x \colon \mathbf{T}_V \to L$ denote the corresponding homomorphism. Now suppose κ is a weight-character in $Y(L)$, and let $(D(V))_\kappa$ denote the fiber in $D(V)$ over κ. For each $x \in (D(V))_\kappa(L)$ the vector space W_x over L of weight-character κ overconvergent eigenforms F over L such that*

$$F|\mathcal{T}_n = \eta_x(\mathcal{T}_n)F$$

(recalling the notation of 3.1) is one dimensional. Moreover, the correspondence $x \mapsto W_x$ is a bijection between $(D(V))_\kappa(L)$ and the one dimensional spaces of eigenforms of weight-character κ whose inverse $(\alpha U)_\kappa$-eigenvalue lies in $\mu(V_\kappa(L))$.

The proof runs along the same lines as that of Theorem B5.7 of [C-BMF]. We construct a pairing between $\mathbf{T}(V)$ and $N(V)$ using q-expansions in the same way. Specifically, if $\tau \in \mathbf{T}(V)$ and $f \in N(V)$, define

$$\langle \tau, f \rangle := a_1(\tau \cdot f),$$

where a_1 denotes the coefficient of q^1 in the Fourier expansion. The main difference between the present setting and Theorem B5.7 of [C-BMF] is that the pairing here is not perfect, in general, because of the possible existence of a non-zero constant term in the q-expansions. Our pairing leads to an exact sequence of A-modules

$$0 \to \mathbf{T}(V) \to Hom_A(N(V), A) \xrightarrow{\rho} M \to 0, \qquad (7.1.1)$$

where M is the residue field of $\mathcal{B}$ at the trivial character $\mathbf{1}$ if $\mathbf{1} \in N(V)_1$ and $\rho(h)$ is then $h_1(\mathbf{1})$ in this case, otherwise $M = 0$. The proof of this requires our comparison of convergent modular forms and Katz modular functions, Theorem 2.4.2, and the injectivity of the q-expansion map on Katz modular functions by the second lemma of section XI of [K-LME]. Keeping this in mind, we can (almost) reformulate this result in terms of q-expansions. In particular, we obtain,

Corollary 7.1.2. *Let notation be as in the theorem and set*

$$F_x(q) = \sum_{n \geq 1} \eta_x(\mathcal{T}_n) q^n.$$

Then if κ is not trivial or if κ is trivial and η_x is not the homomorphism which takes $\mathcal{T}_n$ to $\sigma_{-1}^*(n)$, $F_x(q)$ is the q-expansion of a normalized overconvergent eigenform over L (minus its constant term) of weight-character κ.

Here $\sigma_{-1}^*(n)$ is, as in 2.2, the sum of the reciprocals of the squares of the positive divisors of n which are relatively prime to p. We have a natural finite map of $D(V)$ to V since the image of $Q_V^*(\alpha U)$, where $Q_V^*(T) = T^d Q_V(T^{-1})$, in $\mathbf{T}(V)$ is zero, by Theorem A4.5 of [BMF] (see also Corollary 4.2.3 above). This map is not necessarily an injection but it is when V is reduced (this follows from the Cayley-Hamilton theorem). Since both V and $D(V)$ have degree d over Y we deduce:

Proposition 7.1.3. *If $Q_V(T)$ is square free, then the map from $D(V)$ to V is generically an isomorphism and moreover $D(V)$ is reduced.*

The last assertion follows from the following lemma, taking X to be $D(V)$ and W to be $\pi_\alpha(V)$.

Lemma 7.1.4. *Suppose $X \to W$ is a morphism of affinoids of dimension one such that W is reduced and irreducible and $A(X)$ is finite free over $A(W)$. Then if X is generically reduced, it is reduced.*

Proof. Indeed, since X is generically reduced, of dimension one and finite over Y there exists an open Y in W such that X_Y is reduced (just take Y to be the complement of the image of the finitely many points where X is not reduced). Thus if $e \in A(X)$ and $e^2 = 0$, the image of e in $A(X_Y)$ must be zero. But, the other hypotheses imply that the map $A(X) \to A(X_Y)$ is an injection. Thus $e = 0$. ∎

7.2. Gluing.

Fix $\alpha \in \mathcal{H}$, and let $Z = Z_\alpha$. $\mathcal{C} := \mathcal{C}_\alpha$.

Lemma 7.2.1. *If V_1 and V_2 belong to $\mathcal{C}$ so does $V_1 \cap V_2$. Moreover, if $V_1 \subseteq V_2$, $D(V_1)$ is naturally an affinoid subdomain of $D(V_2)$.*

Proof. Let $M = M^\dagger$. Let Y_i be the image of V_i in W and $Y_{12} = Y_1 \cap Y_2$. Then, it is easy to see that $V_i \cap Z_{Y_{12}} \in \mathcal{C}$ so we may assume $V_i = V_i \cap Z_{Y_{12}}$. The hypotheses imply that there exists functions e_1 and e_2 on $Z_{Y_{12}}$ such that $e_i|_{V_j} = \delta_{i,j}$. It follows that V_1 is the disjoint union of two subdomains U_1 and U_2, disconnected from each other, where e_2 equals 1 and 0 respectively. Hence each of these must be finite over Y_1. But $U_1 = V_1 \cap V_2$. Now suppose $V_1 \subseteq V_2$. First as we argued above $V_3 := V_2 \cap Z_{Y_1} \in \mathcal{C}$. Moreover, it is not hard to see that $D(V_3)$ is naturally isomorphic to $D(V_2)_{Y_1}$ which is an affinoid subdomain of $D(V_2)$. Thus we may suppose $Y_1 = Y_2$. Let $Q_i(T)$ be the factor of $P_\alpha(T)$ over Y_1, such that $Q_i(0) = 1$, corresponding to V_i. Then, again arguing as above, $V_2 = V_1 \cup V_1'$ where V_1 and V_1' are disconnected and finite over Y_1. It follows that

$$Q_2(T) = Q_1(T)Q_1'(T),$$

where $Q_1(T)$ and $Q_2(T)$ are relatively prime. We may apply the theory of Riesz decomposition (as in section A4 of [C-BMF]) because αU is associated to a completely continuous system (cf. section 4.3) of operators on $M_{Y_1} = \cup_v M_{Y_1}^\dagger(v)$. We have

$$N_{Q_2} = N_{Q_1} \oplus N_{Q_1'}.$$

From this we see that

$$D(V_2) = D(V_1) \coprod D(V_1')$$

and so $D(V_1)$ is an affinoid subdomain of $D(V_2)$ as asserted. ∎

Construction of the curve D_α. Now we will glue all the $D(V)$'s, for $V \in \mathcal{C}_\alpha$, together to make a rigid analytic space D_α using Proposition 9.3.2/1 of [BGR]. For $U, V \in \mathcal{C}_\alpha$, let $D(U, V)$ denote the image of $D(U \cap V)$ in $D(U)$. We know this is an affinoid subdomain. Let $\phi_{U\,V}$ be the natural map from $D(U, V)$ to $D(V, U)$. Clearly, $\phi_{U\,V} \circ \phi_{V\,U} = id$, $D(U, U) = D(U)$ and $\phi_{U\,U} = id$. Also $\phi_{U\,V}$ induces an isomorphism

$$\phi_{U\,V\,W}: D(U, V) \cap D(U, W) \to D(V, U) \cap D(V, W)$$

such that

$$\phi_{U\,V\,W} = \phi_{W\,V\,U} \circ \phi_{U\,W\,V}$$

because $D(U, V) \cap D(U, W)$ is the image of $D(U \cap V \cap W)$ in $D(U)$. Thus the triple $(\{D(U)\}, \{D(U, V)\}, \{\phi_{U,V}\})_{U,V \in \mathcal{C}}$ satisfies all the hypotheses of the aforementioned proposition and the $D(U)$'s glue together into a rigid curve D_α Note that D_0 is empty. Let z_α denote the projection from D_α onto Z_α. By construction,

Proposition 7.2.2. *The morphism $z_\alpha: D_\alpha \to Z_\alpha$ is finite, and the projection to weight space $\pi : D_\alpha \to \mathcal{W}$ has the property that it is locally in-the-domain finite flat in the sense that D_α is covered by affinoid subdomains $D_\alpha(V) \subset D_\alpha$ which have the property that their images $Y := \pi(D_\alpha(V)) \subset \mathcal{W}$ are affinoid subdomains of $\mathcal{W}$ and moreover $D_\alpha(V)$ is finite flat over Y. The rigid analytic spaces D_α are curves; that is, they are equidimensional of dimension 1.*

7.3. The relationships among the curves D_α.

Consider the set $\mathcal{P}$ of all pairs $S = (X, N)$ such that:

  a. X is an affinoid in $\mathcal{W}$.

  b. N is a $\mathbf{T}_X$-submodule of $M_X^\dagger$ (here $\mathbf{T}_X = \mathcal{H} \otimes A(X)$),

  c. N is locally free of finite rank over $A(X)$ and there exists a $\mathbf{T}_X$-module projector from $M_X^\dagger$ onto N which comes from a continuous projector from $M_X(v)$ on N, where $v > 0$ is sufficiently close to 0 so that all the elements on N are the images of elements in $M_X(v)$. (Note that the module N, being of finite rank, is contained in the image of the Banach space $M_V^\dagger(v)$ for $v > 0$ and sufficiently small, and the results of section 4.2 apply.)

d. The operator $\mathcal{T}(p)$ is invertible on N. (In this way the elements of N correspond to families of forms of finite slope).

One source of such pairs is the following: Let notation be as above. Suppose $V \in \mathcal{C}_\alpha$ and $Q_V(T) \in A(\pi_\alpha(V))[T]$ the corresponding polynomial. Let N_V be the kernel of $Q_V^*(\alpha U_p)$ in $M_X^\dagger$. Then, using Theorem A4.5 of [C-BMF], we see that $S_V := (\pi_\alpha(V), N_V) \in \mathcal{P}$. Even though $M_X^\dagger$ is not naturally a Banach-module over $A(X)$, writing $M_X^\dagger = \cup_v M_X^\dagger(v)$ we may view $M_X^\dagger$ as being given by a system of Banach modules, as in the terminology of section 4.3, and using Proposition 4.3.3 an operator $\tilde{U}$ in the ideal $\mathcal{U}$ has a characteristic series $\det(1 - T\tilde{U}|M_X^\dagger)$ in $A(X)\{\{T\}\}$. It follows from Corollary 4.2.3 that if $\alpha \in \mathcal{H}$ and $\tilde{U} = \alpha U_p$,

$$\det(1 - T\tilde{U}|M_X^\dagger) = \det(1 - T\tilde{U}|N) \cdot \det(1 - T\tilde{U}|M_X^\dagger/N).$$

Suppose $S = (X, N)$ is a pair in $\mathcal{P}$. Let $\mathbf{T}(S)$ denote the ring generated over $A(X)$ by the image of $\mathbf{T}_X$ in $End_X(N)$. It is the algebra of rigid functions on a one dimensional affinoid $D(S)$. If $V \in \mathcal{C}_\alpha$, $D(S_V) = D(V)$ and $\mathbf{T}(S_V) = \mathbf{T}(V)$. We also see that if $S = (X, N)$ and $S' = (X', N')$ are elements of $\mathcal{P}$ such that $X' \subset X$ and $N' \subset N_{X'} := N \otimes_{A(X)} A(X')$ then we have an induced natural ring homomorphism from the image of $\mathbf{T}(S)$ to $\mathbf{T}(S')$ which in turn induces a natural map

$$D(S') \to D(S),$$

which we will refer to below as the "functorial map". We also set

$$\mathbf{T}_{p,OF} = \varprojlim \mathbf{T}(S)$$

(the "OF" stands for overconvergent of finite slope). Using Proposition 3.4.2 and Theorem 2.4.2 we obtain compatible maps of $\mathbf{T}_p$ into $\mathbf{T}(S)$ for $S \in \mathcal{P}$ and hence a map from $\mathbf{T}_p$ into $\mathbf{T}_{p,OF}$.

Construction of a natural mapping from $D(S)$ into D_α. Suppose $\alpha \in \mathcal{H}$, $S = (X, N) \in \mathcal{P}$ and α is invertible on N. Let $\tilde{U} = \alpha U$.

By Corollary A2.6.2 of [C-BMF] and Theorem 4.2.2 (since $\tilde{U}$ is stabilizes N and $A(X)$ is semi-simple) we may factor the Fredholm characteristic series $P_\alpha(T)$ over X,

$$P_\alpha(T) = Q_S(T)H_S(T)$$

where $Q_S(T) = \det(1 - T\tilde{U}|_N)$ is a polynomial over $A(X)$ whose degree is the rank of N and whose leading coefficient is a unit in $A(X)$, and where $H_S(T) \in A(X)\{\{T\}\}$, and in contrast to members of $\mathcal{C}_\alpha$, here $H_S(T)$ is not necessarily relatively prime to $Q_S(T)$.

Define

$$Z_S := Z_{\alpha,S} := Max\big(A(X)[T]/Q(T)\big)$$

which is an affinoid and note that we have natural mappings

$$D(S) \to Z_S \to Z := Z_\alpha.$$

Now because $\mathcal{C} := \mathcal{C}_\alpha$ is an admissible cover of Z, there exists a finite collection $Z_1, \ldots, Z_n$ of members of $\mathcal{C}$ which covers the image of Z_S, i.e., such that the collection $\{Z_S \times_Z Z_i\}_i$ is an admissible open cover of Z_S. Let $X_i = \pi_\alpha(Z_i)$ and $X'_i = X_i \cap X$. Corresponding to Z_i there is a Riesz theory idempotent e_i which acts on $M^\dagger_{X_i}$. The idempotent e_i also may be thought of as a rigid analytic function on Z_{X_i} which is 1 on Z_i and 0 on its complement. As $e_i \in \mathbf{T}_{X_i}$ (actually in the completion of the subring generated by $\tilde{U}$) e_i acts on N_{X_i}. Also, over $X_i \cap X_j$, e_i and e_j commute. Let $f_i = 1 - e_i$ and $E_i = e_i N_{X'_i}$ and $F_i = f_i N_{X'_i}$. Then

$$N_{X'_i} = E_i \oplus F_i.$$

Consider the pairs $S_i := (X'_i, E_i)$ and $\Sigma_i := (X'_i, F_i)$ in $\mathcal{P}$. Then, we have a factorization

$$Q(T) = \det(1 - T\tilde{U}|_{E_i}) \det(1 - T\tilde{U}|_{F_i}),$$

which, given our notational conventions, may be written

$$Q(T) = Q_{S_i}(T) \cdot Q_{\Sigma_i}(T),$$

the polynomial factors Q_{S_i} and Q_{Σ_i} being relatively prime because of the Riesz theory. Moreover, we have the affinoids

$$Z_{S_i} := Z_{\alpha,S_i} := Max\big(A(X'_i)[T]/Q_{S_i}(T)\big),$$

and

$$Z_{\Sigma_i} := Z_{\alpha,\Sigma_i} := Max\big(A(X'_i)[T]/Q_{\Sigma_i}(T)\big).$$

It is easy to see that $S_i := (X'_i, E_i)$ and

$$S_{ij} := (X_{ij} = X'_i \cap X'_j \,, \ E_{ij} := e_i e_j N_{X'_i \cap X'_j})$$

all lie in $\mathcal{P}$. We will have later use for the natural mapping

$$D(S_i) \to D(Z_i).$$

This map is obtained as follows. First, consider the pair $\mathcal{Z}_i := (X_i, e_i N_{X_i}) \in \mathcal{P}$. Since $X_i' \subset X_i$ and $e_i N_{X_i'} \subset e_i N_{X_i} \otimes_{A(X_i)} A(X_i')$, we obtain a functorial map (see above) $D(S_i) \to D(\mathcal{Z}_i)$. Next note that $D(\mathcal{Z}_i) = D(Z_i)$.

Lemma 7.3.2. *The natural mapping*

$$Z_{S_i} \to Z_S \times_Z Z_i$$

is an isomorphism.

Proof.

$$Z_S \times_Z Z_i = (Z_S)_{X_i} \times_Z Z_i$$

and

$$(Z_S)_{X_i} = Z_{S_i} \coprod Z_{\Sigma_i}$$

because $Q_{S_i}(T)$ and $Q_{\Sigma_i}(T)$ are relatively prime. As $Z_{\Sigma_i} \times_Z Z_i = \emptyset$ (again using the Riesz theory)

$$Z_S \times_Z Z_i = Z_{S_i} \times_Z Z_i.$$

The lemma then follows since the natural map from Z_{S_i} to Z is contained in Z_i.

Lemma 7.3.3. *The natural map*

$$D(S_i) \to D(S) \times_Z Z_i$$

is an isomorphism.

Proof. As in Lemma 7.3.2,

$$D(S) \times_Z Z_i = D(S_i) \times_Z Z_i$$

and because the map from $D(S_i)$ to Z factors through Z_{S_i} we see that this equals $D(S_i)$.

Lemma 7.3.4. *The natural map $D(S_{ij}) \to D(S_i) \times_{D(S)} D(S_j)$ is an isomorphism. Moreover, the restriction to $D(S_{ij})$ of either the map from $D(S_i)$ to $D \times_Z Z_i$ or $D(S_j)$ to $D \times_Z Z_j$ is the same map to $D \times_Z Z_{ij}$ (where $Z_{ij} = Z_i \times_Z Z_j$).*

Proof. This comes from

$$D(S_i) \times_{D(S)} D(S_j) = \left(D(S) \times_Z Z_i\right) \times_{D(S)} \left(D(S) \times_Z Z_j\right)$$
$$= D(S) \times_Z Z_{ij}$$

We saw above that $Z_{ij} \in \mathcal{C}$, but we can be more precise. Over $X_{ij} := X_i \cap X_j$, we have the decomposition

$$M = e_i e_j M \oplus e_i f_j M \oplus e_j f_i M \oplus f_i f_j M.$$

This corresponds to a factorization, over $X_i \cap X_j$,

$$P_\alpha(T) = A(T)B(T)C(T)D(T)$$

where $D(T) \in A(X_{ij})\{\{T\}\}$ and $A(T)$, $B(T)$, $C(T)$ are polynomials in $A(X_{ij})[T]$, and

$$Q_{Z_i}(T) = A(T)B(T)$$
$$Q_{Z_j}(T) = A(T)C(T).$$

We see that $Q_{Z_{ij}}(T) = A(T)$ and if $Z_{ij} \neq \emptyset$, the image of Z_{ij} in $\mathcal{B}$ equals X_{ij} and the projector corresponding to Z_{ij} is $e_i e_j$. By Lemma 7.3.3, we see that $D(S_{ij}) = D(S) \times_D D(Z_{ij})$ and the map from $D(S_{ij})$ factors through the projection to $D(Z_{ij})$. From this the lemma follows. ∎

The Lemmas 7.3.3 and 7.3.4 allow us to patch together mappings.

Proposition 7.3.5. *Let* $S \in \mathcal{P}$. *There is a unique map from* $D(S)$ *to* D_α *characterized by the following property: Suppose* $V \in \mathcal{C}_\alpha$, *with image* $Y = \pi_\alpha(V)$. *Let* e *be the corresponding idempotent acting on* $M_Y^\dagger$ *and* $S' = (Y \cap X, eN_Y) \in \mathcal{P}$. *Then the following diagram commutes:*

$$
\begin{array}{ccc}
D(S) & \to & D_\alpha \\
\uparrow & & \uparrow \\
D(S') & \to & D(V)
\end{array}
$$

where the left and the bottom arrows are the natural ones coming from the inclusions $S'(N) \subseteq N(S)_{Y \cap X}$ *and* $S'(N) \subseteq N(V)_{Y \cap X}$.

Proof. It follows from Lemma 7.3.3 that the $D(S_i)$ cover $D(S)$ and from Lemma 7.3.4 that all the morphisms $D(S_i) \to D(Z_i)$ (constructed above) glue together to give a morphism from $D(S)$ to D_α. To check that one gets the same map from a different covering of the image of Z_S by a collection of elements of $\mathcal{C}_\alpha$ it is enough to check this when the second covering is a refinement of the first in which case it follows easily from Lemmas 7.3.2-4. We may deduce from the above that we have a natural maps $\rho_{\beta,\alpha} : D_\beta \to D_\alpha$ for all $\beta \in \alpha\mathcal{H}$ as follows: Suppose $V_1, V_2 \in \mathcal{C}_\beta$ then, as β is invertible on $N(V_i)$, by the above we have morphisms $D(V_1) \to D_\alpha$ and $D(V_2) \to D_\alpha$. We must show that they agree on $D(V_1) \times_{D_\beta} D(V_2)$ which is $D(V_1 \times_{Z_\beta} V_2)$. Hence, it is enough to prove this when $V_2 \subseteq V_1$ (i.e. $V_1 \times_{Z_\beta} V_2 = V_2$). Suppose $Z_1, \ldots, Z_n$ cover the image of $Z_{S(V_1)}$. Then as $N(V_2)$ is a submodule of $N(V_1)_{\pi_\alpha(V_2)}$, $Z_1, \ldots, Z_n$ also cover the image of $Z_{S_{V_2}}$. We can therefore use $Z_1, \ldots, Z_n$ to make our maps. Let in the above notation, with $S = S_{V_j}$, $S_i(V_j) = S_i$, $X_i(V_j) = X_i'$. We must check that the following diagram commutes:

$$\begin{array}{ccc} D(S_i(V_1)) & \to & D(Z_i) \\ \uparrow & & \| \\ D(S_i(V_2)) & \to & D(Z_i), \end{array}$$

but this follows from

$$e_i(N(V_2)_{X_i(V_2)}) \subseteq e_i((N(V_1)_{X(V_2)})_{X_i(V_2)}) = (e_i(N(V_1)_{X_i(V_1)})_{X_i(V_2)}.$$

Lemma 7.3.6. *If* $\beta \in \alpha \mathbf{T}_{p,OF}$ *and* $\gamma \in \beta \mathbf{T}_{p,OF}$

$$\rho_{\beta,\alpha} \circ \rho_{\gamma,\beta} = \rho_{\gamma,\alpha}.$$

Proof. Let $U \in \mathcal{C}_\gamma$, $V \in \mathcal{C}_\beta$, $W \in \mathcal{C}_\alpha$, $X = \pi_\gamma(U)$, $Y = \pi_\beta(V)$ and $Z = \pi_\alpha(W)$. Let e_V and e_U be the idempotents corresponding to V and W (We suppress mentioning in the notation where they are defined here and in the following). Then if $S' = (X \cap Y, e_V N(U))$ and $S'' = (Y \cap Z, e_W N(V))$, using Proposition 7.3.5, we see that the following diagram commutes:

$$\begin{array}{ccccc} D(U) & \xrightarrow{\rho_{\gamma,\beta}} & D_\beta & \xrightarrow{\rho_{\beta,\alpha}} & D_\alpha \\ \uparrow & & \uparrow & & \\ D(S') & \longrightarrow & D(V) & & \uparrow \\ & & \uparrow & & \\ & & D(S'') & \longrightarrow & D(W). \end{array}$$

Now if $S''' = (X \cap Y \cap Z, e_W e_V N(U))$, it is easy to see that

$$\begin{array}{ccc} D(S') & \to & D(V) \\ \uparrow & & \uparrow \\ D(S''') & \to & D(S'') \end{array}$$

commutes, where the right and bottom arrows are the natural ones. Now using the previous corollary and the fact that the $D(S''')$'s constructed in this way cover $D(U)$ we see that the composition of $\rho_{\beta,\alpha}$ and $\rho_{\gamma,\beta}$ restricted to $D(U)$ is $\rho_{\gamma,\alpha}$. This completes the proof of the lemma. ∎

Since $\rho_{\alpha,\alpha}$ is the identity,

Corollary 7.3.7. *If* β *and* α *generate the same ideal in* $\mathbf{T}_{p,OF}$, *then* $\rho_{\beta,\alpha}$ *is an isomorphism.*

Let $D = D_1$. The above gives us maps from D to Z_α for all $\alpha \in \mathbf{T}_{p,OF}^*$. The curve D will be shown to be isomorphic to the reduced eigencurve.

If $\beta \in \mathcal{H}$ such that $\iota \cdot \beta$ is invertible, then the image of β in $\mathbf{T}_{p,OF}$ is invertible because the homomorphism from $\mathcal{H}$ into $\mathbf{T}_{p,OF}$ is the composition of the homomorphisms

$$\mathcal{H} \to R_p \to \mathbf{T}_p \to \mathbf{T}_{p,OF}$$

(see sections 3.3, 5.2 and 7.2).

Remark. If α and β are non-associate elements of $\mathbf{T}_{OF}$ then D_α and D_β may be very different. For example, if $\alpha = 0$, D_α will be empty although as we'll see $D := D_1$ is quite large. If α is an idempotent, D_α is an admissibly closed-open subspace of D.

7.4. D is reduced.

In this section, we will prove D is nested (cf. 1.1), reduced, and each irreducible component of D maps surjectively and generic isomorphically onto a Fredholm hypersurface (see Prop. 7.4.5 below). By construction we have a map $w \colon D \to \mathcal{W}$. By the results of §7.3, $D \cong D_\alpha$ for all $\alpha \in \mathbf{T}^*_{p,OF}$. This means that we have finite maps $z_\alpha \colon D \to Z_\alpha$, and again by construction the following diagrams commute:

where the mappings to weight space are the natural ones; i.e., the mapping $D \to \mathcal{W}$ is w and the mapping $Z_\alpha \to \mathcal{W}$ is π_α. Let $\sigma_\alpha \colon Z_\alpha \to \mathbf{A}^1$ be the natural projection to the affine line. If $\alpha \in \mathbf{T}^*_{p,OF}$, the map $s \colon P \mapsto -v(\sigma_\alpha(z_\alpha(P)))$ is independent of α and we call it ν. Indeed, if we think of a closed point on D as a one dimensional space of eigenforms, as we can by Theorem 7.1.1, s is just the slope of these forms. We know that every point on the nilreduction D^{red} of D lies on an irreducible component part by Proposition 1.2.5. Let $Z = Z_1$.

Lemma 7.4.1. *Suppose $\alpha \in \mathbf{T}^*_{p,OF}$. If Y is an irreducible component of D^{red} then the map $Y \to Z_\alpha^{red}$ factors through a surjective map onto an irreducible component part of Z^{red}.*

Proof. Let $V \in \mathcal{C}_\alpha$. Let $\mathcal{D}_1, \ldots, \mathcal{D}_n$ be the irreducible components in $D(V)^{red}$ in $Y_V = Y \cap D(V)$. These correspond, by Corollary 1.2.1, to minimal prime ideals $P_1, \ldots, P_n$ of $\mathbf{T}_V$. Since the map from $D(V)$ to V is surjective on closed points, it follows that the pullback Q_i of P_i to V is a minimal prime ideal of $A(V)$. and let $\mathcal{Z}_i = \mathrm{Max} A(V)/Q_i$ be the irreducible component of of V^{red} corresponding to Q_i. Since $D(V)$ to V is finite, it follows from the going-up theorem, Theorem 5.11 of [A-M], that the map from $\mathcal{D}_i$ to $\mathcal{Z}_i$ is surjective. Let $\mathcal{Z}_V$ be the component part of V^{red} which is the union of the irreducible components $\mathcal{Z}_i$ for $i = 1 \cdots n$. The natural map from Y_V to Z_V is finite and surjective. The $\mathcal{Z}_V$'s for $V \in \mathcal{C}_\alpha$ glue together into a component part $\mathcal{Z}$ of Z_α^{red} and the maps glue together into a finite surjective map from Y onto $\mathcal{Z}$. Since Z_α^{red} is Fredholm hypersurface it follows that $\mathcal{Z}$ is as well. Since Y is irreducible so is $\mathcal{Z}$. ∎

Write $\Lambda = \prod_{j=1}^{p-1} \Lambda^{(j)}$ where each of the $p-1$ factor rings $\Lambda^{(j)}$ are isomorphic to a power series ring in one variable over $\mathbf{Z}_p$, and where the product decomposition of rings corresponds to the decomposition of the rigid space $\mathcal{W}$ as a (disjoint) union of its $p-1$ irreducible components, $\mathcal{W} = \coprod_{j=1}^{p-1} \mathcal{W}^{(j)}$, where $\Lambda^{(j)}$ is the algebra over $\mathbf{Q}_p$ of rigid analytic functions bounded by 1 on the rigid analytic space $\mathcal{W}^{(j)}$.

Corollary 7.4.2. *Let* $w : D \to \mathcal{W}$ *be the projection mapping to weight space. If* $\mathcal{D}$ *is an irreducible component of* D^{red} *then* w *maps* $\mathcal{D}$ *into a unique irreducible component,* $\mathcal{W}^{(j_\mathcal{D})}$, *of* $\mathcal{W}$ *(where* $1 \leq j_\mathcal{D} \leq p-1$*) and the mapping*

$$w : \mathcal{D} \to \mathcal{W}^{(j_\mathcal{D})}$$

is almost surjective on $\mathbf{C}_p$-*valued points in the sense that the complement of* $\mathcal{D}(\mathbf{C}_p)$ *in* $\mathcal{W}^{(j_\mathcal{D})}(\mathbf{C}_p)$ *is finite.*

Proof. Let $\mathcal{D}$ be an irreducible component of D^{red}. Then $\mathcal{D}$ projects onto an irreducible component of Z_α^{red} by Lemma 7.4.1. To prove Corollary 7.4.2 it suffices, then, to prove for irreducible components of Z_α^{red}. By Theorem 4.3.1, Z_α is cut out by a Fredholm series $P_{\tilde{U}}(T)$ over Λ. Consider the corresponding factorization

$$P_{\tilde{U}}(T) = \prod_{j=1}^{p-1} P_{\tilde{U}}(T)^{(j)},$$

where $P_{\tilde{U}}(T)^{(j)} \in \Lambda^{(j)}\{\{T\}\}$ for $j = 1 \cdots p-1$. Since $\Lambda^{(j)}$ is isomorphic to a power series ring in one variable over $\mathbf{Z}_p$, Theorem 1.3.11 applies, allowing us to identify the irreducible components of Z_α^{red} (and also of Z_α) with factors of the Fredholm series $P_{\tilde{U}}(T)^{(j)}$ for $j = 1 \cdots p-1$. Corollary 1.3.13 then applies, which concludes the proof of our corollary. ∎

If W is an affinoid open in $\mathcal{W}$ and $t \in \mathbf{Q}$, the conditions

$$w(P) \in W$$
$$\nu(P) \leq t$$

determine an affinoid open $D(W,t)$ in D.

This follows because the conditions

$$\pi_\alpha(Q) \in W$$
$$v(\sigma_\alpha(Q)) \leq t$$

determine an affinoid open $Z_\alpha(W,t)$ in Z_α for any α of which $D(W,t)$ is the pullback so is an affinoid by the rigid Chevalley lemma Proposition 9.4.4/1 of [BGR].

Lemma 7.4.3. *For W an affinoid open in $\mathcal{W}$ and $t \in \mathbf{Q}$. There exists an $\alpha \in \mathbf{T}^*_{p,OF}$ such that $Z_\alpha(W,t)$ is reduced.*

Proof. We know $D(W,t)^{red}$ has finitely many irreducible components by Corollary 1.2.1/1. Using Proposition 1.2.5 we see that these are contained in a finite number of irreducible components of D^{red}, $\mathcal{D}_1, \ldots, \mathcal{D}_n$. These components $\mathcal{D}_\nu$ each map almost surjectively (in the sense described in Corollary 7.4.2) to some component $\mathcal{W}^{(j_\nu)}$ of $\mathcal{W}$. In particular, for each ν, there is a point δ_ν on $\mathcal{D}_\nu$ of some arithmetic weight k_ν and slope s_ν with $0 \leq s_\nu < (k_\nu - 1)/2$. Now choose $\alpha \in \mathbf{T}^*_{p,OF}$ as in Chapter 6 so as to distinguish all the classical forms of weight k_ν and slope s_ν for all $\nu = 1, \cdots, n$.

Claim. $Z_\alpha(W,t)$ *is reduced.*

Let $\mathcal{Z}_\nu$ be an irreducible component part of Z_α containing the image ζ_ν of δ_ν under $z_\alpha : D_\alpha \to Z_\alpha$. Then by the arguments in Chapter 6 (Props. 6.3.1- 6.3.3) $\mathcal{Z}_\nu$ is reduced (There is only one modular form corresponding to ζ_ν. This implies $\mathcal{Z}_\nu$ is generically reduced, but since it is an irreducible Fredholm hypersurface it must be reduced.) Now the claim and our lemma follow because $Z_\alpha(W,t)$ is the union over ν of $\mathcal{Z}_\nu \cap Z_\alpha(W,t)$. ∎

Lemma 7.4.4. $D(W,t)$ *is reduced.*

Proof. If we knew $Z_\alpha(W,t)$ was an element of $\mathcal{C}_\alpha$ we could apply Proposition 7.1.3 to conclude that $D(W,t)$ is reduced. But it may not be so we must work harder. We go to the proof of Proposition A5.8 of [BFM]. We must generalize it slightly and replace $\mathbf{B}^1_K$ with W. (We also use the proof of Lemma A5.9 in loc. cit. which works even when $d = 0$). Translating what was proved there, we know that for $u \in \mathbf{Q}$, $u > t$, and sufficiently close to t, there exists a finite covering $\{U_0, U_1, \ldots, U_d\}$ of W by affinoids in $\mathcal{W}$ such that the the affinoid V_i in Z_α, whose points in Z_α consist of all Q such that

$$\pi_\alpha(Q) \in U_i \quad \text{and} \quad v(\sigma_\alpha(Q)) \leq u$$

lie in $\mathcal{C}_\alpha$ and is finite over U_i. Moreover, $S_n := \bigcup_{i \geq n} U_i$ contains the affinoid W_n in W consisting of all $P \in W$ such that the fiber of π_α restricted to

$Z_\alpha(W, t)$ at P has degree at least n. We can also assume such that U_n is the rigid Zariski closure of $W_n \cap U_n$. Since U_n is the Zariski closure of $W_n \cap U_n$ and $Z_\alpha(W, t)$ is reduced, it follows that the Zariski closure of $(V_n)_{W_n \cap U_n}$ is V_n. But $(V_n)_{W_n \cap U_n} = (Z_\alpha(W, t))_{U_n}$ which is reduced by Corollary 7.3.3/10 of [BGR] since $Z_\alpha(W, t)$ is reduced. Thus V_n is generically reduced. Now we can apply Proposition 7.1.3 to conclude that $D(V_n)$ is reduced. Since the $D(V_n)$ cover $D(W, t)$ our lemma follows. ∎

Proposition 7.4.5. *The rigid analytic curve D is reduced, nested, and and each irreducible component of D maps surjectively and generic isomorphically onto a Fredholm hypersurface.*

Proof. The $D(W, t)$'s form an admissible cover of D, so D is reduced by Lemma 7.4.4. Now suppose $\mathcal{D}$ is an irreducible component of D. We may suppose $\mathcal{D}$ is the $\mathcal{D}_1$ mentioned above. We claim it is generically isomorphic to $\mathcal{Z}_1$. We know $D(W, t)$ is generically isomorphic to $Z_\alpha(W, t)$ (here α is as in the previous lemmas). It follows that $\mathcal{D}_1 \cap D(W, t)$ is generically isomorphic to $\mathcal{Z}_1 \cap Z_\alpha(W, t)$ since these affinoids are open and Zariski dense in their respective spaces. Since $\mathcal{Z}_1$ is nested, so is D. Our proposition follows.

7.5. Equality of D and C^{red}.

Set $C = C_p$. Recall C^{red} denotes the nilreduction of C. We maintain the notation of the previous section so in particular, we have morphisms $z_\alpha \colon D \to Z_\alpha$ (which is contained in $\mathcal{W} \times \mathbf{A}^1 - \{0\}$) for all $\alpha \in \mathbf{T}^*_{p,OF}$. We also have a morphism

$$\delta \colon D \to X_p \times (\mathbf{A}^1 - \{0\}),$$

given by $c \mapsto (\rho(c), 1/u(c))$ where $\rho(c)$ is the pseudo-representation and $u(c)$ is the U_p-eigenvalue attached to c, or more precisely: The morphism from D to X_p arises from the map $\mathbf{T}_p \to \mathbf{T}_{p,OF}$ described in section 7.3 and the pseudo-representation described in section 5.2. The map from D to $\mathbf{A}^1 - \{0\}$ is the composition of the morphism from D to the spectral curve $Z =: Z_1$ and the natural projection of $\sigma_1 \colon Z \to \mathbf{A}^1 - \{0\}$. We constructed the eigencurve C, in section 6.1, as follows: For every $\alpha \in \mathcal{H}'$ such that $\iota \cdot \alpha$ is a unit in R_p we have a map r_α from $Y := X_p \times (\mathbf{A}^1 - \{0\})$ to $\mathcal{W} \times (\mathbf{A}^1 - \{0\})$ such that

$$r_\alpha(x, t) = \big(w(x), 1/t\,(\iota \cdot \alpha(x))\big).$$

Then C is the rigid analytic subspace of Y which is the intersection of the subspaces

$$r_\alpha^{-1}(Z_\alpha) := Y \times_{\mathcal{W} \times (\mathbf{A}^1 - \{0\})} Z_\alpha,$$

where α ranges over the elements in $\mathcal{H}'$ as above. The key point is that

$$r_\alpha \circ \delta = z_\alpha.$$

Thus δ factors through C. It follows that we have a natural map $\omega \colon D \to C$ such that the diagrams

$$D \xrightarrow{\ \omega\ } C$$

$$\searrow \quad \swarrow$$

$$Z_\alpha$$

commute for all $\alpha \in \mathbf{T}^*_{OC}$.

Theorem 7.5.1. *The map ω factors through a map to C^{red} and induces an isomorphism (which we continue to denote by the same letter)*

$$\omega : D \cong C^{red}.$$

Proof. That ω factors through C^{red} follows from the fact that D is reduced. That ω is a bijection on closed points follows from Theorems 6.2.1 and 7.1.1. To prove it is an isomorphism of rigid spaces, we will first prove that it is a generic isomorphism and then prove it is a local isomorphism. For the first assertion we need the following lemma. For the second assertion we will show below that the morphism from D to $X_p \times \mathbf{A}^1$ is locally a closed immersion.

Lemma 7.5.2. *Suppose E_0 and E_1 are reduced rigid spaces (over K a complete subfield of $\mathbf{C}_p$), and $h_0 : E_0 \to E_1$ and $h_1 : E_1 \to E_0$ are rigid analytic mappings which induce mappings on $\bar{K}$-valued points which are two-sided inverses of one another; i.e., the compositions of $h_0 : E_0(\bar{K}) \to E_1(\bar{K})$ and $h_1 : E_1(\bar{K}) \to E_0(\bar{K})$ in either order yield the identity mapping on the relevant set of $\bar{K}$-valued points. Then $h_0 : E_0 \cong E_1$ and $h_1 : E_1 \cong E_0$ are rigid analytic isomorphisms between E_0 and E_1 and are two-sided inverses of one another as rigid analytic mappings.*

Proof. It will suffice to show that if E is a reduced rigid space over K, and $h \colon E \to E$ is a morphism of rigid spaces over K such that h is the identity on $\bar{K}$-valued points, then h is the identity. Let S be an admissible cover of E by affinoid subdomains. Let $V \in S$. Then there exists an admissible cover U_i of V by affinoid opens and $V_i \in S$ such that $h|_{U_i}$ factors through the inclusion $V_i \to E$. Now since, $h(V(\bar{K})) = V(\bar{K})$, $h(U_i(\bar{K})) \subseteq (V \cap V_i)(\bar{K})$ and since V is an affinoid subdomain of E, $V \cap V_i$ is an affine subdomain of V_i. It follows from Proposition 7.2.2/1 of [BGR] that $h|_{U_i}$ actually factors through $V \cap V_i$ and so $h|_V$ factors through the inclusion $V \to E$. Thus we may suppose E is an affinoid. Let $A = A(E)$. We need to show $h^* \colon A \to A$ is the identity. Suppose $a \in A$. Let m be a maximal ideal of A and (extending scalars, if necessary) suppose $c \in \overline{K}$ such that $a - c \in m$. Then $h^*a - c \in m$ and so $h^*a - a \in m$ for all maximal ideals m. Since A is reduced (and semi-simple), it follows that $h^*a = a$. ∎

Now we apply this to our eigencurve. Let $\alpha \in \mathcal{H}$ whose image in $\mathbf{T}_{p,OF}$ is invertible and suppose V is a reduced component part of Z_α. Let D_V be the fiber above V in D and C_V the fiber above V in C^{red}. (Thus, by definition, C_V is reduced.) Then we have a commutative diagram

$$D_V \xrightarrow{\omega} C_V$$
$$\searrow \quad \swarrow$$
$$V$$

and we know ω is one:one on closed points and the restriction of z_α to D_V is generically an isomorphism onto V using the fact that D is reduced and Lemma 7.4.1. Let S be the set of points of V where z_α is not an isomorphism, $V' = V - S$, $D' = D_V - z_\alpha^{-1}(S)$. $C' = C_V - \lambda_\alpha^{-1}(S)$, ω' the restriction of ω to D' etc. Let $\mu: V' \to D'$ be the inverse of $z_\alpha: D' \to V'$. Then

$$(\mu \circ \lambda') \circ \omega' = id.$$

Applying Lemma 7.5.2 with $E_0 = D'$, $E_1 = C'$ and $h_0 = \omega'$, $h_1 = \mu \circ \lambda'$ that $\omega|_{D_V}: D_V \to C_V$ is generically an isomorphism. Since every component part of D composed of finitely many irreducible components is generically isomorphic to a reduced component part of Z_α for some $\alpha \in \mathbf{T}_{p,OF}^*$ we may conclude that ω is a generic isomorphism. Now we will show ω is a local isomorphism. Let V be a open in $\mathcal{C} := \mathcal{C}_1$ and Y its image in $\mathcal{W}$. Let $\mathcal{X}$ be an affinoid open in X_p and U a affinoid in $\mathbf{A}^1$ such that the image of $\mathcal{X}$ in $\mathcal{W}$ is Y and such that the image of $D(V)$ in $X_p \times \mathbf{A}^1$ is contained in $\mathcal{X} \times U$ (more precisely, such that the morphism $D(V) \to X_p \times \mathbf{A}^1$ factors through $\mathcal{X} \times U$ for after we get this second property, we can get the first by taking fiber products). Since $A(D(V))$ is generated over $A(Y)$ by the image of the Hecke operators which can be identified as functions on $X_p \times \mathbf{A}^1$ and $A(Y)$ maps into $A(\mathcal{X} \times U)$ the map from $D(V)$ to $\mathcal{X} \times U$ is a closed immersion.

Claim. *For any point on D there exists a $V \in \mathcal{C}$ such that $P \in D(V)$ and an $\mathcal{X}$ and U as above such that the image of $\delta(D(V))$ equals $\delta(D) \cap (\mathcal{X} \times U)$.*

Suppose for the moment that we have this. Let E be the intersection of C with $\mathcal{X} \times U$. It is a subaffinoid and its points are the same as those of $D(V)$. It follows that $D(V)$ maps via a closed immersion onto the nilreduction of E. Since it is a bijection on closed points and $D(V)$ is reduced we conclude that it is an isomorphism. Thus the claim implies that ω is a local isomorphism. To prove the claim, suppose $\nu(P) = t$. Then using the analysis of Lemma A5.9 of [BGR], we see there exists an affinoid open neighborhood W of $w(P)$

in $\mathcal{W}$ such that the affinoid $Z(W,t) \in \mathcal{C}$ (this is the affinoid called $Z_1(W,t)$ in the last section). The choices $V = Z(W,t)$, $\mathcal{X} = W$ and $U = B[0,p^t]$ will fulfill the requirements of the claim. ∎

We now prove the theorem. Suppose $\mathcal{X}$ is an affinoid open in X_p whose image in $\mathcal{W}$ is an affinoid open W. The map from $D(W,t)$ to C^{red} factors through $C(\mathcal{X},t) := C^{red} \cap (\mathcal{X} \times B[0,p^t])$. Moreover, both $D(W,t)$ and $C(\mathcal{X},t)$ are affinoids and the map $f : D(W,t) \to C(\mathcal{X},t)$ is a bijection on closed points. It follows from the above that f is an isomorphism after the removal of finitely many points S in $D(W,t)$ and is a local isomorphism. Let $A = D(W,t) - S$, $B = C(\mathcal{X},t) - \omega(S)$ and let $\psi : B \to A$ be the inverse of $\omega|_A$. Now choose affinoid neighborhoods U of S in $D(W,t)$ and V of $\omega(S)$ in $C(\mathcal{X},t)$ such that ω restricts to an isomorphism $U \to V$. We can do this because S is finite and ω is a injection on closed points. Let $\phi : V \to U$ be the inverse of $\omega|_U$. It is clear that ψ and ϕ glue together into a morphism $\rho(\mathcal{X},t) : C(\mathcal{X},t) \to D(W,t)$ inverse to $\omega|_{D(W,t)}$. As $\mathcal{X}$ and t vary the affinoids $C(\mathcal{X},t)$ for an admissible open covering of C^{red} and the $\rho(\mathcal{X},t)$ glue together into an inverse of ω. ∎

7.6. Consequences of the relationship between D and C.

The first consequence of section 7.5 is that we now know that the eigencurve is a curve since D is by construction. Moreover, Theorems A and B of Chapter 1, section 1.5, follow from the results of sections 7.4 and 7.5. We also established

Corollary 7.6.1. *Every irreducible component of the reduced eigencurve C contains a classical point.*

Corollary 7.6.2. *Each irreducible component of C^{red} is generically isomorphic to a Fredholm hypersurface over $\mathcal{W}$.*

For the record, let us extract a slightly more precise statement that is useful in understanding the geometry of the rigid-analytic curve C. In the special case of arithmetic weight $\kappa = (\chi, k)$ where the slope σ of c is less than $k - 1$ and different from $(k-2)/2$ it follows from [C-BMF] that the $\mathbf{C}_p$-vector space $M_k(\mathbf{C}_p; \chi)_\sigma$ consists entirely of classical modular forms, and therefore it follows from standard facts that $\mathcal{H}$ acts semi-simply (with multiplicity one) on $M = M_{\kappa,\sigma}$. In particular, $\mu(c) = 1$. We therefore get:

Corollary 7.6.3. *Let c be a $\mathbf{C}_p$-valued point of the eigencurve $C = C_p$ of weight-character $w = \chi \otimes \eta_k \in \mathcal{W}$ and of slope $\sigma < k-1$, and $\sigma \neq (k-2)/2$. Then c is a smooth point of the curve C^{red} and the projection $C^{red} \to \mathcal{W}$ is locally an isomorphism in a neighborhood of c.*

One should note that if $\sigma \geq k - 1$ the point c may indeed be ramified in the projection mapping of C to weight space. We do not have to go far to see this phenomenon. Say that a point $c \in C$ of weight $w = \chi \otimes \eta_k$ is of **critical slope** if the slope of c is equal to $k - 1$. There are examples (cf. [C-OC]) of points c of critical slope ramified under the projection mapping of C to weight space.

From the previous discussion, we deduce:

Corollary 7.6.4. *If $\beta \in \mathbf{T}^*_{p,OF}$, and $d_\beta \in Z_\beta$, then the map $D \to Z_\beta$ is finite flat of degree at most $\mu(d_\beta)$ in a neighborhood of d_β. Moreover, if any of the points $d \in D$ above d_β satisfy the conditions of the previous corollary, they all do and this degree equals $\mu(d_\beta)$.* ∎

We may now prove Theorem G stated in section 1.5:

Proposition 7.6.5. *If $\{f_n\}$ is a sequence of normalized eigenforms of tame level 1 with Fourier coefficients in $\mathbf{C}_p$ such that f_n has weight $\kappa_n \in \mathcal{W}(\mathbf{C}_p)$, $v(a_p(f_n))$ is bounded independently of n and the sequence $\{f_n(q) \in \mathbf{C}_p[[q]](1/q)\}$ converges coefficientwise to a series $f(q)$. Then the sequence κ_n converges in $(\mathbf{Z}/(p-1)\mathbf{Z}) \times \mathbf{Z}_p$ to some weight $\kappa \in \mathcal{W}(\mathbf{C}_p)$ and $f(q)$ is the q-expansion of an overconvergent modular eigenform f of tame level 1, weight κ, and finite slope. Let f be a Katz modular eigenfunction (see section 2.2) of tame level 1, of finite slope, with (normalized) Fourier expansion*

$$\tilde{f} = q + a_2 q^2 + a_3 q^3 + \cdots$$

with coefficients $a_j \in \mathbf{C}_p$. Suppose that f is of accessible weight-character (cf. section 1.4) (χ, s) with $s \in \mathbf{Z}_p$. Then f is overconvergent if and only if f is the limit in the q-expansion topology of a sequence of classical, normalized, cuspidal-overconvergent (cf. section 3.6) modular eigenforms of tame level 1.

Proof. That the κ_n converge to a $\kappa \in \mathcal{W}(\mathbf{C}_p)$ follows from Theorem 2 of [S 3]. We will give two proofs of the rest of the first part of this proposition.

For the first proof note that because the slopes are bounded and the weights converge, eventually all the f_n correspond to points P_n on an affinoid of the form $D(V)$ for some $V \in \mathcal{C}_1$. Moreover in the notation of Theorem 7.1.1, $\eta_{P_n}(\mathcal{T})_n$ is the n-coefficient of $f_n(q)$. Since the images of Hecke operators generate the ring of functions on $D(V)$ over the weight space, it follows that the points P_n converge to a point P on $D(V)$ such that $\eta_P(\mathcal{T}_n)$ is the n-coefficient of $f(q)$. By Theorem 7.1.1, P correspond to an eigenform of weight κ with q-expansion $f(q)$ and finite slope.

For the second proof and the proof of the rest of this theorem, let us first consider the various natural topologies on sets of eigenforms. For every prime number $\ell \in \mathbf{P}$ we have a natural continuous mapping

$$T_\ell : \left(X_p \times (\mathbf{A}^1 \setminus \{0\})\right)(\mathbf{C}_p) \to \mathbf{C}_p$$

by putting $T_\ell(x, 1/u) := T_\ell(x)$ if $\ell \neq p$, where we view the Hecke operator T_ℓ as rigid analytic function on X_p, and putting $T_p(x, 1/u) := u$. On defines the **q-expansion topology** on the set $X_p \times (\mathbf{A}^1 \setminus \{0\})(\mathbf{C}_p)$ by means of these functions. Explicitly, let $\mathbf{P}$ be the set of all prime numbers, and $\mathbf{C}_p^{\mathbf{P}}$ the product of $\mathbf{C}_p$ with itself countably many times, indexed by $\mathbf{P}$ and given the natural product topology. Consider the mapping

$$T : X_p \times (\mathbf{A}^1 \setminus \{0\})(\mathbf{C}_p) \to \mathbf{C}_p^{\mathbf{P}},$$

where for any $\ell \in \mathbf{P}$, the ℓ-th entry of $T(c)$ is $T_\ell(c)$. By definition, the **q-expansion topology** on the set $X_p \times (\mathbf{A}^1 \setminus \{0\})(\mathbf{C}_p)$ is the topology this set inherits via the mapping T to the topological space $\mathbf{C}_p^{\mathbf{P}}$.

Lemma 7.6.6. *The natural topology that the set $X_p \times (\mathbf{A}^1 \setminus \{0\})(\mathbf{C}_p)$ inherits as the $\mathbf{C}_p$-valued points of the rigid analytic space $X_p \times (\mathbf{A}^1 \setminus \{0\})$ (which we will call the **rigid analytic topology**) is the same as its q-expansion topology.*

Proof. This follows from the fact that if $\mathcal{H}'$ is the Λ-algebra generated by the T_ℓ's (for $\ell \neq p$) (cf. Chapter 2), the natural ring homomorphism of $\mathcal{H}'$ to R_p is topologically dense. This fact can be regarded as given by a "standard argument" at least for those components of R_p which actually correspond to representations rather than pseudo-representations (for if you take such a component , e.g., a completion of R_p with respect to a maximal ideal m (call it R_m) and if you consider $R'_m \subset R_m$ which is the closure of the image of $\mathcal{H}'$ in R_m, one can see that (by continuity and Cebotarev) the image of the entire Galois group under the trace mapping is contained in R'_m. We may assume this is a local ring with the same residue field as R_m and then by standard Schur-type arguments you see that the universal representation over R_m descends to a representation r' over R'_m. Universality of R_m allows us to conclude that $R'_m = R_m$ (for if not, there would be two distinct ring-homomorphisms $R_m \to R_m$ inducing the same representation: the identity and the projection to R').

Now this argument works word-for-word (only is easier) for pseudo-representations, because we don't need "standard Schur-type arguments." The fact that the trace descends immediately implies (from the definition of pseudo-representation) that the entire pseudo-representation descends. ∎

The first part of our proposition is a consequence of this lemma. For Proposition 6.2.1 allows us to identify our sequence $\{f_n\}$ with a sequence of $\mathbf{C}_p$-valued points of (the nilreduction of) the eigencurve. But since the reduced eigencurve is a Zariski-closed rigid analytic subspace, its set of $\mathbf{C}_p$-valued points is a closed subset of $X_p \times (\mathbf{A}^1 \setminus \{0\})(\mathbf{C}_p)$ in the rigid analytic topology. By the lemma, this set is also closed in the q-expansion topology.

As for the second part, let $\mathcal{K}$ be the set of all Katz modular eigenfunctions of tame level 1, of finite slope, of arbitrary weight, but with Fourier expansion as hypothesized in our proposition. We view $\mathcal{K}$ as a topological space, given the q-expansion topology. Let $c : \mathcal{K} \to X_p \times \mathbf{A}^1(\mathbf{C}_p)$ be the mapping which assigns to a Katz modular eigenform $f \in \mathcal{K}$ the point $c(f) \in X_p \times \mathbf{A}^1(\mathbf{C}_p)$ whose coordinate in $X_p(\mathbf{C}_p)$ corresponds to the associated pseudo-representation of f, as given by the Gouvêa-Hida Theorem (see section 5.2) and whose coordinate in $\mathbf{A}^1(\mathbf{C}_p)$ is the reciprocal of its U_p-eigenvalue. The mapping c is a homeomorphism of the topological space $\mathcal{K}$ onto its image in $X_p \times (\mathbf{A}^1 \setminus \{0\})(\mathbf{C}_p)$. This may be seen from the above lemma by noting that the T_ℓ-eigenvalues (for all $\ell \neq p$) together with the U_p-eigenvalue of f determine the q-expansion of f (see the discussion about this in section 3.1).

Now take a Katz modular eigenform $f \in \mathcal{K}$ satisfying all the hypotheses of our proposition, and in particular which is of accessible weight-character. Suppose that f is overconvergent. The point $c(f)$ is then, by Theorem 6.2.1, a $\mathbf{C}_p$-valued point of the reduced eigencurve, and since f is of finite slope and accessible weight, using the argument in the proof of Corollary 7.6.3 one sees that f is the limit of classical eigenforms of tame level 1 (with normalized q-expansions). ∎

References.

[AM]   Atiyah, M. F., I. G. Macdonald: *Introduction to Commutative Algebra*, Addison-Wesley (1969).

[Bo]   Bosch, S.: Eine bemerkenswerte Eigenschaft der formellen Fasern affinoider Räume. Math. Ann. **229** (1977) 25 - 45

[BGR]   _____, Güntzer, U., and Remmert, R.: *Non-Archimedean Analysis*, Springer-Verlag, (1984).

[BL-FRI]   _____, and Lütkebohmert, W.: Formal and Rigid geometry, I. Rigid spaces. Math. Ann. **295** (1993) 291-317

[BL-FRII]   _____, and Lütkebohmert, W.: Formal and Rigid geometry, II. Flattening techniques. Math. Ann. **296** (1993) 403-429

[C-RLC]   Coleman, R.: Reciprocity laws on curves. Compositio **72** (1989) 205-235

[C-PPMF]   _____: A p-adic inner product on elliptic modular forms. Pp. 125-151 in the *Barsotti symposium in Algebraic Geometry* (1994) Academic Press.

[C-CO]   _____: Classical and overconvergent modular forms. Inventiones math. **124** (1996) 215-241.

[C-COHL]   _____: Classical and overconvergent modular forms of higher level. To appear in Le journal de théorie des nombres de Bordeaux.

[C-CCS]   _____: On the Coefficients of the Characteristic series of the U operator, Proc. of the NAS, Vol. 94, No. 21 (1997).

[C-BMF]   _____: P-adic Banach spaces and families of modular forms. Inventiones math. 127 (1997) 417-479.

[CGJ]   _____, Gouvea, F. and Jochnowitz, N.: E_2, Θ and U, 1 IMRN (1995) 23-41.

[CST]   _____, Stevens G. and Teitelbaum, J.: Numerical experiments on families of p-adic modular forms, R. Coleman, G. Stevens, J. Teitelbaum, in "Computational Perspectives on Number Theory, Proceedings of a Conference in Honor of A.O.L Atkin," AMS/IP studies in advanced mathematics, volume 7, (1998) 143-158.

[DDT]   Darmon, H., Diamond, F., Taylor, R.: Fermat's Last Theorem. pp. 1-107 in *Current Developments in Mathematics*, 1995. International Press.

[de J]   de Jong, J.: Crystalline Dieudonné module theory via formal and rigid geometry. Inst. Hautes Etudes Sci. Publ. Math. No. 82 (1995), 5-96 (1996).

[Ed]   Edixhoven, B.: The weight in Serre's conjectures on modular forms. Inventiones math. **109** (1992) 563-594

[Em]   Emerton, M.: 2-adic eigenforms of minimal slope. Preprint (1997)

[F]   Fontaine, J.-M.: Représentations p-adiques semi-stables. Exposé III: pp. 113-184 in *Périodes p-Adiques*. Astérisque **223** (1994)

[G-ApM]   Gouvêa, F.: Arithmetic of p-adic modular forms, SLN 1304, (1988).

[GM-FM]   Gouvêa, F., Mazur, B.: Families of modular eigenforms. Math. of Comp. **58** (1992) 793-805

[GM-CPS]   _____: On the characteristic power series of the U-operator. Annales de l'Institut Fourier **43** (1993) 301-321

[GM-ZD]   _____: On the Zariski-density of modular representations in "Computational Perspectives on Number Theory, Proceedings of a Conference in Honor of A.O.L Atkin," AMS/IP studies in advanced mathematics, volume 7 (1998).

[H-HA]   Hida, H.: On p-adic Hecke algebras for GL_2 over totally real fields, Ann. of Math. **128** (1988) 295-384.

[H-NO]   _____: Nearly Ordinary Hecke Algebras and Galois Representations of Several Variables, JAMI Inaugural Conference Proceedings, (supplement to) Amer. J. Math. (1990), 115-134.

[H-ET]   _____: *Elementary theory of L-functions and Eisenstein series*, London Mathematical Society (1993).

[I]   Iwasawa, K., Lectures on p-adic L functions, Ann. Math. Studies, Princeton University Press, 1972.

[JvP]   de Jong, J., van der Put, M.: Étale cohomology of rigid analytic spaces, Preprint of the University of Groningen, W-9506

[K-PMF]   Katz, N.: *P-adic properties of modular schemes and modular forms*, Modular Functions of one Variable III, SLN 350, (1972) 69-190.

[K-HC]   _____: Higher congruences between modular forms. Annals of Math. **101** (1975) 332-367

[K-LME]   _____: P-adic *L*-functions via Moduli of elliptic curves, in *Algebraic Geometry: Arcata 1974*, Robin Hartshorne, editor, AMS, Providence, RI (1975).

[KM]   Katz, N., Mazur, B.: *Arithmetic Moduli of Elliptic Curves*. Annals of Math. Stud. **108**, PUP (1985).

[L]   Lütkebohmert, W.: Formal-algebraic and rigid-analytic geometry. Math. Ann. **286** (1990) 341-371

[LvP]   Liu, Q., van der Put, M.: On one-dimensional separated rigid spaces. Preprint. Bordeaux (1994)

[M-DGR]   Mazur, B.: Deforming Galois representations. pp 385-487 in Galois groups over **Q** MSRI Publications **16** (1989) Springer-Verlag

[M-IF]   _____: An "infinite fern" in the deformation space of Galois representations, Collectanea Mathematica **48** 1-2 (1997) 155-193.

[M-DTR]   Mazur, B.: Deformation theory of Galois representations, pp. 243-311 in *Modular forms and Fermat's Last Theorem*, (eds.: G. Cornell, J. Silverman, G. Stevens), Springer (1997)).

[MW]   Mazur, B., Wiles, A.: Class fields of abelian extensions of **Q**. Inventiones math. **76** (1984) 179-330

[Mi]   Miyake, T.: *Modular Forms*, Springer (1989).

[Ny]   Nyssen, L.: Pseudo-représentations, Math. Ann. 306 (1996) 257-283.

[Ra]   Raynaud, M.: Revêtements de la droite affine en caractéristique $p > 0$, Inventiones math. (1994) **116** 425-462

[Ri]   Ribet, K.: Report on mod ℓ representations of $\mathrm{Gal}(\bar{\mathbf{Q}}/\mathbf{Q})$. Proceedings of Symposia in Pure Mathematics **55** (1994) 639-676

[Ro]   Rouquier, R.: Caractérisation des caractères et Pseudo-caractères. Journal of Algebra **180** (1996) 571-586

[Sch]   Schlessinger, M.: Functors on Artin Rings, Trans. AMS, **130** (1968) 208-222.

[S-ECC]   Serre, J-P.: Endomorphismes complètements continues des espaces de Banach p-adiques, Publ. Math. I.H.E.S., **12** (1962) 69-85.

[S-RD2]   _____: Sur les représentations modulaires de degré 2 de $\mathrm{Gal}(\bar{\mathbf{Q}}/\mathbf{Q})$. Duke Math. J. **54** (1987) 179-230.

[S-FMZ]   Serre, J. P., Formes modulaires et fonctiones zêta p-adiques, Modular Functions of one Variable III, SLN 350, (1972) 69-190.

[St]   Stevens, G., Overconvergent Modular Symbols and a Conjecture of Mazur, Tate, and Teitelbaum, to appear.

[T]   Taylor, R.: Galois representations associated to Siegel modular forms of low weight. Duke Math. J. **63** (1991) 281-332

[Wa]   Wan, D.: Dimension Variation of classical and p-adic modular forms, to appear in Inventiones Math.

[Wi]   Wiles, A.: On ordinary λ-adic representations associated to modular forms, Inventiones math. **94** (1988) 529-573.

DEPARTMENT OF MATHEMATICS, EVANS HALL, UNIVERSITY OF CALIFORNIA AT BERKELEY, BERKELEY, CA 94720, USA
coleman@math.berkeley.edu

DEPARTMENT OF MATHEMATICS, HARVARD UNIVERSITY, ONE OXFORD STREET, CAMBRIDGE, MA 02138, USA
mazur@math.harvard.edu

Geometric trends in Galois module theory

B. EREZ

to A. Fröhlich on his 80th birthday

Abstract

We review recent contributions by various authors to the study of group actions in arithmetic geometry related to L-functions. The results we present considerably extend previous work on the Galois module structure of rings of integers and units in algebraic number fields. In particular we present the solution of a generalized Fröhlich Conjecture relating the module structure of de Rham cohomology to epsilon constants on arithmetic schemes of arbitrary dimension, and we discuss new invariants attached to equivariant motives, which generalize the omega invariants introduced by Chinburg in connection with the Stark Conjectures.

Contents

Reprinted from 'Galois Representations in Arithmetic Algebraic Geometry',
edited by A. J. Scholl & R. L. Taylor. ©Cambridge University Press 1998

Introduction

Four talks–by D. Burns, T. Chinburg, G. Pappas and M.J. Taylor–were given at the Durham conference, which concerned results in what one may call geometric Galois module theory. In this survey article we attempt to explain the main results presented at the conference. We try to show how various lines of investigation, which originated in the work on the Galois module structure in number fields and which were inspired by recent advances in arithmetic geometry have converged to give a coherent body of results.

One result we will focus on is the proof of a generalized Fröhlich Conjecture. Recall that for a tamely ramified Galois extension of number fields N/K, with Galois group G, the ring of integers O_N in N is a projective $\mathbf{Z}G$-module. The original Fröhlich Conjecture, proved by Taylor in [T1], asserts that the obstruction to O_N being stably free is given by the signs of the constants in the functional equation of the Artin L-functions of symplectic representations of G. The generalized Fröhlich Conjecture concerns the relation between the epsilon constants and the Galois modules attached to a group action on an arithmetic scheme. Chinburg made a first fundamental advance towards finding the right formulation of the conjecture, by proving a result analogous to Taylor's theorem for varieties over a finite field in [C3]. Let (X, G) be a *tame* action of a finite group G on a projective scheme X over a commutative ring R. Chinburg showed how to define a homomorphism from the Grothendieck group $G_0(G, X)$ of coherent G-sheaves on X to the Grothendieck group $CT(RG)$ of finitely generated RG-modules which are cohomologically trivial for G: the refined Euler characteristic. If R is a field and X is a projective variety over R, such Euler characteristics were considered by Nakajima [Na2]; in this case $CT(RG)$ may be identified with the Grothendieck group $K_0(RG)$ of all finitely generated projective RG-modules. Suppose now that $R = \mathbf{F}_p$ is a field of prime order p and that X is a smooth variety over $\mathbf{F}_p$. Chinburg proved that a suitable combination $\Psi(X, G)$ of refined Euler characteristics of sheaves of differentials is characterised by the epsilon factors attached to the action (X, G). Then he translated this result into an equation inside the projective classgroup $Cl(\mathbf{Z}G)$, which is a strict analogue of Taylor's theorem. Indeed, mimicking the constructions performed in the case of number fields, he defined a root-number class $W_{X/Y}$ in $Cl(\mathbf{Z}G)$

which is determined by the signs at infinity of the epsilon constants $\epsilon(Y, V)$, for V ranging over the irreducible symplectic $\mathbf{C}G$-modules. He also introduced a ramification class $R_{X/Y}$, which depends on the epsilon constants of the branch locus of the covering X/Y, and showed the equality

$$\operatorname{Res}_\mathbf{Z}(\Psi(X, G)) = W_{X/Y} + R_{X/Y} ,$$

where $\operatorname{Res}_\mathbf{Z}$ sends a projective $\mathbf{F}_p G$-module to the (stable equivalence) class in $Cl(\mathbf{Z}G)$ of the $\mathbf{Z}G$-module obtained by restricting scalars from $\mathbf{F}_p$ to $\mathbf{Z}$. The corresponding conjecture for flat X over $\mathbf{Z}$ formulated in [CEPT2] involves a different class than $\Psi(X, G)$. This conjecture was proved in [CEPT2] for tame actions on arithmetic surfaces and for free actions in arbitrary dimensions. Again a suitable combination of refined Euler characteristics of sheaves of differentials was related to the generalization of the root-number class appearing in the original Fröhlich Conjecture. The case of tame actions in arbitrary dimensions is now also dealt with, see [CPT].

Another important set of results concerns some new invariants for equivariant motives which have been defined using recent work on L-values. The prototype of such invariants is the omega invariant introduced by Chinburg in [C1], in relation with Tate's work on the Stark Conjectures. For N/K a (not necessarily tamely ramified) Galois extension of number fields, with Galois group G, this is an invariant $\Omega(N/K, 3)$ in $Cl(\mathbf{Z}G)$, which encodes information on the Galois structure of the group of S-units in N, for S a large enough (finite, G-stable) set of places of N. The new invariants defined in [BF1] and [CKPS1] for Tate motives, encode information about the Galois structure of higher K-groups of S-integers. Moreover the very general approach of Burns and Flach sheds a new light on the work of Chinburg and Fröhlich, which had emphasized the analogies that exist between additive and multiplicative Galois module structure.

The contents of this paper can be divided into four parts. The first part consists of Sect. 1, in which we review various ways to obtain refined Euler characteristics attached to group actions. Such Euler characteristics arise from bounded complexes of finitely generated G-modules, which compute various kinds of sheaf cohomology and whose terms are of a restricted type–say projective or cohomologically trivial. If, for instance, the action (X, G) is tame, then the Čech complexes computing the cohomology of coherent G-sheaves are quasi-isomorphic to bounded complexes of finitely generated cohomologically trivial G-modules. The refined Euler characteristics arising from such complexes are one of our main objects of study in the second part of the paper, which consists of Sect. 2 to 4. There we present results whose common denominator is that they relate various refined Euler characteristics to classes defined in terms of L-values and epsilon constants. Sect. 2 contains results for tame actions on varieties over finite fields. Here, following

[CEPT5], we begin by indicating a new approach to Chinburg's results of [C3], in which Lefschetz-Riemann-Roch Theorems are used to compute the refined Euler characteristics. Such results are also used in the proof of the generalized Fröhlich Conjecture. In the second half of Sect. 2 we present a recent result from [CKPS2], which builds on work of Lichtenbaum to compute an equivariant Euler characteristic for the multiplicative group G_m in terms of L-values. Sect. 3 reports on the generalized Fröhlich conjecture, and Sect. 4 shows how one can characterize epsilon factors in terms of Galois modules, following [CEPT3] and [CEPT4]. Thus, as was explained above, in the second part we mainly deal with combinations of Euler characteristics for various sheaves of differentials. In contrast, in the third part, we describe results about the Euler characteristic of an arbitrary sheaf on certain surfaces equipped with a free action. Sect. 5 deals with a topic which has been initiated in [T2], where Taylor formulated the conjecture that certain orders defined using torsion points on elliptic curves admit a normal basis (see *loc. cit.* p. 433). We present Agboola's and Pappas' geometric approach to this problem, which generalizes previous work by Srivastav-Taylor and which comes from [P1]. In Sect 6, again following Pappas, we show how Deligne's Riemann-Roch theorem can be used to deduce that the refined Euler characteristic of the structure sheaf of a surface has order 1 or 2. In the last part we discuss the work of Burns-Flach and Burns on the omega invariants attached to equivariant motives mentioned at the beginning of the introduction.

Acknowledgements. This is a good time for reviewing our subject and I thank the organizers for having given me the oppurtunity to do this. (For other aspects of Galois module theory see the reviews [C-N,C,F,T] and [C-N,T].) This presentation is largely independent of the four talks given at the conference and has actually been influenced by results obtained after the conference. However I have greatly benefited from discussions with the speakers and I thank them for having shared their insights with me. Many thanks also go to A. Agboola and Ph. Cassou-Noguès for fruitful discussions, and to the referee, whose careful reading of the manuscript lead to the correction of many imprecisions.

1   Refined Euler characteristics and analytic classes

We are interested in classes arising from an action (X, G) of a finite group G on an arithmetic scheme X. For the main part of the paper we will be dealing with classes inside the projective classgroup $Cl(\mathbf{Z}G)$. Recall that this is the quotient of the Grothendieck group $K_0(\mathbf{Z}G)$ of projective $\mathbf{Z}G$-modules by the subgroup generated by the class of $\mathbf{Z}G$. The classgroup can also be identified with the subgroup of rank 0 elements in $K_0(\mathbf{Z}G)$. Moreover, Schanuel's lemma implies that one can identify $K_0(\mathbf{Z}G)$ with the Grothendieck group $CT(\mathbf{Z}G)$ of cohomologically trivial G-modules.

1.a Analytic classes

There are two kinds of classes in which we are interested. The first arises from considering homomorphisms from the group R_G of (virtual) complex representations to the multiplicative group of complex numbers, such as those defined by epsilon constants or L-value/regulator quotients. These define classes in the classgroup *via* Fröhlich's so-called Hom-description of the classgroup. Let E be a large enough (finite) Galois extension of the rationals. In particular assume that all representations of G are realized over E. Let $J(E)$ denote the group of ideles of E. Then R_G and $J(E)$ are modules for the Galois group of $E/\mathbf{Q}$. It can be shown that the group of Galois equivariant homomorphisms between these two modules surjects onto the classgroup:

$$cl : \mathrm{Hom}_{Gal(E/\mathbf{Q})}(R_G, J(E)) \to Cl(\mathbf{Z}G) \ .$$

(For this one uses Swan's result that every projective $\mathbf{Z}G$-module is locally free, see *e.g.* [F1]; it is sometimes more convenient to work with an algebraic closure $\overline{\mathbf{Q}}$ of the rationals in place of E.) Two examples of analytic classes: (a) Cassou-Noguès and Fröhlich have shown how to define a Galois equivariant homomorphism by using the root numbers attached to a Galois extension of number fields (see [F2]); (b) Tate's formulation of the Stark conjecture says that a certain L-value/regulator quotient defines a Galois equivariant homomorphism (see [Ta] Chap. I, 5.1). These two constructions are fundamental in the Galois module theory of number fields and have been greatly generalized. For instance, root number classes have been defined for tame actions on arithmetic schemes in [C3] 5.8, [CEPT2] 3.2 and also for motives endowed with a group action arising from base change by a Galois extension in [B1] Sect. 4. One can usually control the order of these analytic classes and locate them inside the classgroup. Thus they give a good measure for the triviality or otherwise of classes defined in other ways and that are related to them.

1.b Tame actions

The second kind of classes arise from taking Euler characteristics of perfect complexes, that is (complexes quasi-isomorphic to) bounded complexes with finitely generated and projective terms.

Let X be a projective scheme over a regular Noetherian commutative ring R, equipped with an action by a finite group G. We will say that the action of G on X is tame, if the order of the inertia group of each point is prime to the residue characteristic of the point. From [C3], [CE] and [CEPT1], Sect. 8 one has the Euler characteristic homomorphism

$$\chi = \chi^{CT} : G_0(G, X) \to CT(RG)$$

from the Grothendieck group of coherent G-sheaves on X to that of cohomologically trivial RG-modules. Indeed one shows that for any coherent G-sheaf F on X there is a perfect complex $(M^i)_i$ of RG-modules, well-defined up to quasi-isomorphism, which computes the cohomology of F and one puts $\chi(F) = \sum_i (-1)^i [M^i]$ (see [CEPT1] Sect. 8, where the case of actions by not necessarily constant finite affine group schemes is also considered; [Na1] contains a special case also noted independently by Kani). Recall that $G_0(G, X)$ (resp. $K_0(G, X)$) denotes the Grothendieck group of all coherent (resp. locally free) G-sheaves on X. For regular X the forgetful map defines an isomorphism from $K_0(G, X)$ to $G_0(G, X)$ and we will identify these two groups when X is regular.

1.c Perfect complexes in étale cohomology and omega invariants

Other kinds of perfect complexes of arithmetic interest arise as follows. The motivating example is Tate's construction of a four term exact sequence involving S-units and which represents a fundamental class coming from class field theory. Let N/K be a Galois extension of number fields with Galois group G and let S be a finite G-stable set of places of N. Denote by U the group of S-units in N, and by X the kernel of the augmentation/degree map on the free abelian group over S. Recall from [Ta] II.5 that if S is large enough (see *loc. cit.*), then there is an exact sequence of $\mathbf{Z}G$-modules

$$0 \to U \to A \to B \to X \to 0 \ ,$$

which represents Tate's fundamental class α in $\mathrm{Ext}^2_G(X, U)$ and where A and B are finitely generated and cohomologically trivial. By a homological lemma proved in–say–[Wa] 1.3, the class $[A] - [B]$ in $K_0(\mathbf{Z}G)$ only depends on U, X and α. In [C1] Chinburg showed that this class is independent from S as long as S is large enough, so the class only depends on the extension N/K. It is denoted $\Omega(N/K, 3)$. Note that it actually lies in the rank 0 subgroup of $K_0(\mathbf{Z}G)$ by the Dirichlet S-unit Theorem. Recall that this subgroup is naturally isomorphic to the classgroup $Cl(\mathbf{Z}G)$.

Independently, Pappas and Snaith saw how to use the results of Kahn in [K] to construct a four term sequence relating the third and second K-group of the S-integers in N, in the case that N/K does not ramify at infinity (see [Sn1] Ch. 7). A direct approach, using only K-theory, to get classes related to the Galois structure of higher K-groups does not seem possible. However one gets four term sequences for all extensions N/K which relate K_{2i-1} to K_{2i-2}, for any $i > 1$, by using étale cohomology and the Chern character homomorphisms (see [CKPS1] and the account in [BF2] 4.1). One of the basic ingredients in the construction is a lemma which we state in an imprecise form (see [Ka1] 4.17 or [BF1] 1.20).

Lemma 1.1 *For any prime p such that no place over p lies in S, the étale cohomology with compact support of a smooth constructible $\mathbf{Z}_p$-sheaf F on the ring of S-integers can be computed by a perfect complex of $\mathbf{Z}_p G$-modules.*

When G is abelian, this lemma was used by Burns and Flach in [BF1] to assign Galois structure invariants to the base change to N of a motive over K provided certain standard conjectures hold. These conjectures hold for the Tate motives $\mathbf{Z}(i)$, for example. The restriction to abelian G resulted from the use of determinants over the group ring of G. The case of arbitrary G and Tate motives was considered unconditionally by Chinburg, Pappas, Kolster and Snaith in [CKPS1]; they used a different method which involves the construction of certain four-term sequences of G-modules. The four-term sequence method was generalized to motives by Burns [B1]. An approach developed in [CKPS2] which does not rely on four-term sequences will be described in Sect. 1.d.

The four-term sequence approach of [CKPS1] for the Tate motive $\mathbf{Z}(i)$ proceeds in the following way. Let $C_p(i)$ be the cokernel of the injection $\mathbf{Z}_p(i) \to \sum_{v|\infty} i_{v*} i_v^* \mathbf{Z}_p(i)$, where $i_v : \mathrm{Spec} N_v \to \mathrm{Spec} O_{N,S}$ is the natural map. By Artin-Verdier duality, this sheaf has cohomology concentrated in degrees 0 and 1, thus one obtains an exact sequence of $\mathbf{Z}_p G$-modules

$$0 \to H^0(C_p(i)) \to M \to M' \to H^1(C_p(i)) \to 0 ,$$

with M and M' cohomologically trivial $\mathbf{Z}_p G$-modules. This four term sequence defines an element in $\mathrm{Ext}^2_{\mathbf{Z}_p G}(H^1(C_p(i)), H^0(C_p(i)))$. By taking products over p one gets a class E in $\mathrm{Ext}^2_{\hat{\mathbf{Z}} G}(\prod_p H^1(C_p(i)), \prod_p H^0(C_p(i)))$. By using results in [DF] on the Chern character homomorphisms, one is then lead to consider the following approximations to the K-groups of S-integers. One puts $K'_{2i-2} = \prod_p H^1(C_p(i))$ and shows that there is a finitely generated submodule K'_{2i-1} of $\prod_p H^0(C_p(i))$, which becomes equal to it after extending scalars from $\mathbf{Z}$ to $\hat{\mathbf{Z}}$. Hence we can view E as an element in

$$\mathrm{Ext}^2_{\hat{\mathbf{Z}} G}(K'_{2i-2} \otimes_{\mathbf{Z}} \hat{\mathbf{Z}}, K'_{2i-1} \otimes_{\mathbf{Z}} \hat{\mathbf{Z}}) = \mathrm{Ext}^2_{\mathbf{Z} G}(K'_{2i-2}, K'_{2i-1}) .$$

Because the groups K'_{2i-j} are finitely generated for $j = 1$ and 2 and cup product with E induces cohomology isomorphisms, E is represented by an exact sequence

$$0 \to K'_{2i-1} \to A_i \to B_i \to K'_{2i-2} \to 0$$

in which A_i and B_i are finitely generated cohomologically trivial $\mathbf{Z}G$-modules. The classes $[A_i] - [B_i]$ of $K_0(\mathbf{Z}G)$ are not of rank 0, hence one subtracts from them the free $\mathbf{Z}G$-module $\mathbf{Z}\Sigma_\infty$, which is the free abelian group on the set Σ_∞ of complex embeddings of N. The resulting classes are the generalized omega invariants of [CKPS1]:

$$\Omega_n(N/K) = [A_{n+1}] - [B_{n+1}] - [\mathbf{Z}\Sigma_\infty] \ .$$

The omega invariants of Burns and Flach are discussed briefly in the last section.

1.d   Nearly perfect complexes

In his work on the value at 1 of the zeta function of a surface X over a finite field, Lichtenbaum was lead to define a (numerical) Euler characteristic for the multiplicative group $\mathbf{G}_m$, viewed as a sheaf for the étale topology. Assume that the Brauer group $H^2(X, \mathbf{G}_m)$ is finite. Even under this assumption, one cannot expect $H^3(X, \mathbf{G}_m)$ to be finitely generated. However, this group is dual to a finitely generated group. Using this fact, Lichtenbaum was able define a numerical Euler characteristic $\chi(X, \mathbf{G}_m)$. The Euler characteristic $\chi(X, \mathbf{G}_a)$ of the additive sheaf $\mathbf{G}_a$ in the étale topology equals the usual numerical Euler characteristic of the coherent sheaf O_X. Lichtenbaum proved that (under the assumption that $H^2(X, \mathbf{G}_m)$ is finite), the quotient $\chi(X, \mathbf{G}_a)/\chi(X, \mathbf{G}_m)$ equals ± 1 times the leading term of the zeta function of X at $s = 1$; [Li2] 4.1 and [Li1] Sect. 3.

Now let (X, G) be a free action of the finite group G on the integral scheme X. Suppose R is a ring and that $F^\bullet$ is a complex of sheaves of R-modules for the étale topology on X/G, which is bounded below. Suppose that for each subgroup H of G, only finitely many of the cohomology groups $H^i(X/H, F^\bullet)$ are non-trivial. By arguments similar to those mentioned in Sect. 1.b, one can show that the étale hypercohomology $H^\bullet(X, F^\bullet)$ is isomorphic in the derived category to a bounded complex of RG-modules which are cohomologically trivial for G. For X a regular surface over a finite field, let $C^\bullet$ be the complex computing the cohomology of $\mathbf{G}_m$ on X. Assume still that the Brauer group $H^2(X, \mathbf{G}_m)$ is finite. For $i = 3$ or 4, let $L_i = H^{4-i}(X, \mathbf{G}_m)/H^{4-i}(X, \mathbf{G}_m)_{tor}$. Lichtenbaum's work on the cohomology of $\mathbf{G}_m$ in [Li2] Sect. 3.4 and 4, shows that for $i = 3$ or 4, there are isomorphisms

$$\tau_i : \mathrm{Hom}_{\mathbf{Z}}(L_i, \mathbf{Q}/\mathbf{Z}) \to H^i(X, \mathbf{G}_m)_{div} \ ,$$

where A_{div} is the subgroup of divisible elements of A. The same result is obtained by S. Saito in [SS] when X is a regular surface which is flat over $\mathbf{Z}$ with finite Brauer group.

Put $L_i = 0$ and $\tau_i = 0$ for $i \neq 3$ or 4. The triple $(C^\bullet, (L_i)_i, (\tau_i)_i)$ defines what is called a nearly perfect complex in [CKPS2]. In *loc. cit.* the authors show how to define an Euler characteristic class in $K_0(\mathbf{Z}G)$ for any nearly perfect complex. The class only depends on the quasi-isomorphism class of the complex. If $C^\bullet$ happens to have finitely generated cohomology, then L_i and τ_i are trivial and this new Euler characteristic coincides with the Euler characteristic of the perfect complex obtained from $C^\bullet$ by the usual approximating procedure.

2 Varieties over finite fields

In this section we review two ways of giving a module theoretic interpretation of results on zeta and L-functions for varieties over finite fields. The material presented in the first half of this section comes from [C3] (the presentation follows [CEPT5]). It relates the Galois structure of de Rham cohomology to ϵ-factors. In the second half of the section, following [CKPS1], we relate the Euler characteristics $\chi(X, \mathbf{G}_m)$, introduced at the end of the previous section, to L-values.

2.a Galois structure of de Rham cohomology

The work of Chinburg presented here represents the coming together of two lines of research, which take their origin in the Galois theory of the thirties. Namely Noether's work on the Normal Integral Basis problem and the work of Chevalley-Weil on the Galois structure of differentials (see [CE] for the references). The Galois structure of differentials for varieties over fields has been studied by various authors and, as we mentioned above, in his study of the question Nakajima already considered refined Euler characteristics. In his set-up, one knew how to calculate the character of the Euler characteristics. Projectivity then showed this was sufficient for their determination as modules. Following the approach suggested by the Fröhlich Conjecture, Chinburg shifted the emphasis from the explicit determination of the module structure of differentials to comparing Euler characteristics to analytic invariants associated to L-functions.

Let p be a prime number. In this subsection we consider a smooth projective variety X which is equidimensional of dimension d over the finite field k with $q = p^f$ elements. We assume X is equipped with an action by a finite group G, which is tame in the sense of Grothendieck and Murre. (This is stronger than the numerical tameness discussed in Sect. 1.b, see [C3], [CE] and [GM] for details; if one has resolutions of singularities, some of the results can be generalized to apply to the weaker notion of tameness in which one requires only that the order the inertia group of each point is relatively prime to the residue characteristic of the point.)

The zeta function $Z(X,t)$ satisfies the functional equation

$$Z(X,t) = \pm (q^d t^2)^{-e_X/2} Z(X, q^{-d} t^{-1}) \ .$$

Here e_X is the topological (or ℓ-adic) Euler characteristic, also equal to the self-intersection number of the diagonal Δ_X in $X \times X$. The valuation at p of the constant $\epsilon(X) := \pm q^{-d \cdot e_X/2}$ can be interpreted in terms of the de Rham cohomology of X. Namely, let $\Omega^i_{X/k}$ denote the sheaf of i-differentials on X and let $\chi(\Omega^i_{X/k})$ denote its (numerical) Euler characteristic, then

$$\frac{d}{2} \cdot e_X = \sum_{i=0}^{d} (-1)^i (d-i) \chi(\Omega^i_{X/k}) \ .$$

(see *e.g.* [CEPT5]; this is a consequence of the Hirzebruch-Riemann-Roch theorem and Serre duality).

Write $Y = X/G$ for the quotient of the action, then for any complex representation V of G there is defined an algebraic number $\epsilon(Y,V)$, which is the constant in the functional equation of the Artin L-function attached to the action (X,G) and V. Actually $\epsilon(Y,V)$ only depends on the character χ_V of V and we shall therefore write $\epsilon(Y, \chi_V) = \epsilon(Y,V)$. Let $v_p(-)$ denote the valuation at p. For any g in G, let

$$S(g) = - \sum_{\chi} v_p(\epsilon(Y,\chi)) \chi(g) \ ,$$

where the sum runs over the set of irreducible complex characters of G. Then, by the above, $-S(1) = v_p(\epsilon(X))$ equals $[k : \mathbf{F}_p] d \cdot e_X/2$ and thus has an expression in terms of de Rham cohomology. The first main result is an expression for $S(g)$ in terms of the Brauer traces of g acting on a virtual kG-module, which is a combination of refined Euler characteristics of the sheaves $\Omega^i_{X/k}$, as defined in Sect.1.b. Consider the class in $K_0(\mathbf{F}_p G)$

$$\Psi(X,G) = \sum_{i=0}^{d} (-1)^i (d-i) \mathrm{Res}^k_{\mathbf{F}_p} (\chi^{CT}(\Omega^i_{X/k})) \ ,$$

where Res denotes restricting scalars from k to $\mathbf{F}_p$. For N an $\mathbf{F}_p G$-module and g in G of order prime to p, let $BTr_g(N)$ denote the Brauer trace of g on N. Then we have (see [C3] Thm. 5.2)

Theorem 2.1 *For any g in G of order prime to p,*

$$S(g) = BTr_g(\Psi(X,G)) \ .$$

In [C3] Chinburg deduces the result from work of Milne and Illusie on the slopes of Frobenius in crystalline cohomology. A more direct approach is used in [CEPT5] to deal with curves and surfaces. It goes as follows. Once the ϵ-factors are connected to congruence Gauss sums by using the results of T. Saito in [ST], one can use Stickelberger's Theorem to determine the valuations $v_p(\epsilon(Y,\chi))$. The computation of the Brauer trace of g on $\Psi(X,G)$ is carried out by use of a Lefschetz-Riemann-Roch (LRR) theorem going back to Donovan (see [Do] and [BFQ]). (To use LRR one has to analyze in detail the geometry of a tame action. This analysis becomes combinatorially involved for varieties of dimension higher than two, which explains the restriction on dimension imposed in [CEPT5], but see [CPT] for a way to organize the computation in higher dimension.)

Let us indicate how the proof works for a curve X. The group $K_0(\mathbf{F}_pG)$ injects into the group $G_0(\mathbf{F}_pG)$ of all finitely generated $\mathbf{F}_pG$-modules and the Brauer trace is defined on the latter group (see [Se]). For simplicity of notation assume X is defined over $k = \mathbf{F}_p$ and assume that k is large enough. Let

$$f : X \to \text{Spec}\, k$$

denote the structure morphism. There is a direct image homomorphism

$$f_* : G_0(G, X) \to G_0(kG) \ ,$$

which factors over the refined Euler characteristic map. Also, for $g \neq 1$ in G let $< g >$ denote the subgroup of G generated by g, then there is a commutative diagram

$$
\begin{array}{ccc}
K_0(G, X) & \to & K_0(< g >, X) \\
f_* \downarrow & & \downarrow f_* \\
G_0(kG) & \to & G_0(k < g >)
\end{array}
$$

Assume that g has order prime to p, then the fixed point scheme X^g is regular and the inclusion $i : X^g \to X$ is a regular embedding. One has a restriction map $i^* : K_0(< g >, X) \to K_0(< g >, X^g)$, and an isomorphism $K_0(< g >, X^g) \cong K_0(X^g) \otimes G_0(kG)$. It follows that the conormal sheaf $\mathcal{N} = \mathcal{N}_{X^g|X}$ is locally free ([FuL]). Consider the class $\lambda_{-1}(\mathcal{N}) := [O_X] - [\mathcal{N}]$ in $K_0(< g >, X^g)$ (recall that X is a curve; in general one needs to consider the exterior powers of $\mathcal{N}$ up to $\dim(X)$). Let $W = W(k)$ denote the ring of Witt vectors of k (here $W = \mathbf{Z}_p$). One has a "localization" map

$$K_0(< g >, X^g) = K_0(X^g) \otimes G_0(kG) \to K_0(X^g) \otimes W$$

induced by the Brauer trace $BTr_g : G_0(kG) \to W$. The image of $\lambda_{-1}(\mathcal{N})$ under this map is invertible. The Lefschetz-Riemann-Roch Theorem asserts that the following diagram is commutative

$$
\begin{array}{ccc}
K_0(<g>,X) & \xrightarrow{\lambda_{-1}(\mathcal{N})^{-1} \cdot i^*} & K_0(X^g) \otimes W \\
f_* \downarrow & & \downarrow f_*^g \otimes W \\
G_0(kG) & \xrightarrow{BTr_g} & W = G_0(k) \otimes W
\end{array}
$$

This gives a way to compute the Brauer trace appearing in the theorem.

Let us calculate the Brauer trace $BTr_g(f_*(O_X))$. By LRR every closed point x which is ramified in the covering X/Y contributes to the computation. The conormal sheaf $\mathcal{N}$ at x is a one dimensional representation of G over k (large enough!)–say $k \cdot \chi_x$. So the contribution of x to $\lambda_{-1}(\mathcal{N})$ is $1 - \chi_x$. Hence:

$$
BTr_g(f_*(O_X)) = \sum_{x \in X^g} \frac{1}{1 - \chi_x(g)} .
$$

The expression $S(g)$ can be evaluated by using Deligne's expression for the ϵ-factors of curves as a product of local terms, which are essentially congruence Gauss sums by tameness (see [De2]). The valuation of such Gauss sums can be computed by Stickelberger's Theorem (see [F1] Thm. 27 or [La] IV.4). The theorem then results from the observation that if for an integer e the complex number $\omega \neq 1$ is such that $\omega^e = 1$, then there holds

$$
\frac{1}{1 - \omega} = -\frac{1}{e} \sum_{n=1}^{e-1} n\omega^n .
$$

Now, following Chinburg, we want to reformulate the above theorem as a result on classes in the classgroup $Cl(\mathbf{Z}G)$. Consider the map

$$
\mathrm{Res}_{\mathbf{Z}} : K_0(\mathbf{F}_p G) \to Cl(\mathbf{Z}G)
$$

obtained by composing restriction of scalars and considering the stable equivalence class. The result we want to state is that $\mathrm{Res}_{\mathbf{Z}}(\Psi(X, G))$ only depends on root number data at infinity for symplectic characters and on root number data over the branch locus b of the covering X/Y. We first factor $\mathrm{Res}_{\mathbf{Z}}$ over the map cl giving the Hom-description (see Sect. 1.b). Let Ω (resp. Ω_p) denote the absolute Galois group of $\mathbf{Q}$ (resp. $\mathbf{Q}_p$) and let $R_{G,p}$ be the group of representations of G over the algebraic closure $\overline{\mathbf{Q}}_p$ of $\mathbf{Q}_p$. There is an injection

$$
\Delta : K_0(\mathbf{F}_p G) \to \mathrm{Hom}_{\Omega_p}(R_{G,p}, \overline{\mathbf{Q}}_p^{\times}) ,
$$

obtained by writing a projective $\mathbf{F}_p G$-module $\overline{P}$ as the quotient of a projective module P by pP, and letting

$$
\overline{P} \mapsto (\chi \mapsto p^{-m_\chi}) ,
$$

where m_χ denotes the multiplicity of χ in $P \otimes \overline{\mathbf{Q}}_p$.

Next, let $(\overline{\mathbf{Q}})_p^\times = (\overline{\mathbf{Q}} \otimes_{\mathbf{Q}} \mathbf{Q}_p)^\times$ and choose an embedding h of $\overline{\mathbf{Q}}$ in $\overline{\mathbf{Q}}_p$. This gives an isomorphism

$$h^{*-1} : \operatorname{Hom}_{\Omega_p}(R_{G,p}, \overline{\mathbf{Q}}_p^\times) \cong \operatorname{Hom}_\Omega(R_G, (\overline{\mathbf{Q}})_p^\times) \ .$$

The last group maps into $\operatorname{Hom}_\Omega(R_G, J(\overline{\mathbf{Q}}))$, the group on which cl is defined. Then

$$\operatorname{Res}_{\mathbf{Z}} = cl \circ h^{*-1} \circ \Delta \ .$$

For any place v of $\overline{\mathbf{Q}}$ denote by $\epsilon_v(Y)(\overline{V})$ the image of $\epsilon(Y, V)$ under the injection of $\overline{\mathbf{Q}}^\times$ into the group of ideles $J(\overline{\mathbf{Q}})$, through $(\overline{\mathbf{Q}})_v^\times$. Note that for V symplectic $\epsilon(Y, V)$ is totally real and so has a well defined sign at each archimedean place of $\overline{\mathbf{Q}}$. These signs give an element $\operatorname{sign}(\epsilon_\infty(Y)(V))$ in $(\overline{\mathbf{Q}})_\infty^\times = (\overline{\mathbf{Q}} \otimes_{\mathbf{Q}} \mathbf{R})^\times$. The *root number class* attached to the covering X/Y is then the analytic class

$$W_{X/Y} = cl(\chi_V \mapsto h_\infty(\chi_V))$$

where the p-component of the idele $h_\infty(\chi_V)$, for V irreducible, is given by

$$h_\infty(\chi_V)_p = \begin{cases} 1 & \chi_V \text{ not symplectic} \\ 1 & p \text{ finite} \\ \operatorname{sign}(\epsilon_\infty(Y)(V)) & p = \infty \, , \ \chi_V \text{ symplectic} \end{cases}$$

Let as before b be the ramification divisor of X/Y. There exists a p-primary idele $|\epsilon_p(b)|_p(\chi_V)$ such that $|\epsilon_p(b)|_p(\chi_V) \cdot \epsilon_p(b)(\chi_V)$ is a p-unit idele. Then the *ramification class* is the analytic class

$$R_{X/Y} = cl(\chi_V \mapsto \epsilon_\infty(b) \cdot |\epsilon_p(b)|_p^{-1}(\chi_V)) \ .$$

The main result in [C3] for X of arbitrary dimension can now be stated (see *loc. cit.* Thm. 5.10).

Theorem 2.2 *With the above notation and assumptions, the following equality holds in the classgroup $Cl(\mathbf{Z}G)$,*

$$Res_{\mathbf{Z}}(\Psi(X, G)) = W_{X/Y} + R_{X/Y} \ .$$

We make some remarks on the proof, in particular we indicate why only ϵ-factors for symplectic representations and those supported on the branch locus appear in this result. Using the above factorization of $\operatorname{Res}_{\mathbf{Z}}$, Theorem 2.1 can be restated as the equality

$$\operatorname{Res}_{\mathbf{Z}}(\Psi(X, G)) = -cl(|\epsilon_p(Y)|_p) \ .$$

Here all characters and all of Y play a rôle. Let $U = Y \setminus b$ and let $\epsilon(U)$ be the
product of the $\epsilon_v(U)$ over all places v of $\mathbf{Q}$. Then $cl(\epsilon(U)) = 0$ since $\epsilon(U)$ is in
$\mathrm{Hom}_\Omega(R_G, \overline{\mathbf{Q}}^\times)$. Moreover $\epsilon(U)$ equals the product of the $\epsilon_v(U)$ for all places
v of $\mathbf{Q}$. In *loc. cit.* Thm. 5.12 Chinburg uses the connection between epsilon
factors and the determinant of the (opposite of) Frobenius on v-adic étale
cohomology to show that the right hand side of the last displayed equation
equals (up to sign) the analytic class defined by

$$|\epsilon_p(Y)|_p \cdot (\epsilon_p(U)\epsilon_\infty(U)) \ .$$

Using the additivity of epsilon factors under disjoint unions, this can be written as

$$|\epsilon_p(b)|_p|\epsilon_p(U)|_p \cdot \epsilon_p(U) \cdot \epsilon_\infty(b)^{-1}\epsilon_\infty(Y) \ .$$

In *loc. cit.* Thm. 5.13 Chinburg shows how to deduce that $cl(|\epsilon_p(U)|_p\epsilon_p(U)) = 0$ from the work of T. Saito in [ST] and Taylor's Fixed-Point Theorem for
group determinants (see [F1] Chap. II.6, Thm.10A). Thus U does not appear
anymore. Moreover for any V, all finite components of the idele $(\epsilon_\infty(Y) \cdot
h_\infty)(\chi_V)$ are trivial, and the infinite components are positive if V is symplec-
tic. From this one deduces that $cl(\epsilon_\infty(Y) \cdot h_\infty) = 0$ and thus only symplectic
characters come into the definition of $W_{X/Y}$. This then concludes our sketch
of the proof of the theorem.

2.b   L-values and the cohomology of $\mathbf{G}_m$ for surfaces

Let X be a smooth projective geometrically connected surface over a finite
field and assume that its Brauer group $H^2(X, \mathbf{G}_m)$ is finite. Let the finite
group G act freely on X so that the quotient morphism $X \to Y = X/G$ is
étale. We describe a result from [CKPS2] on the image of the Euler characteristic $\chi(X, \mathbf{G}_m)$ defined in the previous section, in the Grothendieck group
of finitely generated $\mathbf{Z}G$-modules, under the forgetful map

$$f : K_0(\mathbf{Z}G) \to G_0(\mathbf{Z}G) \ .$$

Let $G_0T(\mathbf{Z}G)$ be the Grothendieck group of all finite $\mathbf{Z}G$-modules, then we
also have the forgetful map

$$z : G_0T(\mathbf{Z}G) \to G_0(\mathbf{Z}G) \ .$$

By work of Queyrut, the group $G_0T(\mathbf{Z}G)$ admits a Hom-description, so that
one can define analytic classes in it as well. We shall define such a class by
using the leading terms at 1 of the L-functions attached to (X, G) and show
that it is equal to a class whose image under z is $f(\chi(X, \mathbf{G}_m))$.

Let c_V denote the leading term at $s = 1$ of the L-function $L(q^{-s}, V)$. Then the function

$$c_{X,G} : \chi_V \mapsto c_V$$

is Galois equivariant and thus defines a class in $G_0 T(\mathbf{Z}G)$ (see [CKPS2] Thm. 2.8). Let

$$\lambda : H^1(X, \mathbf{G}_m) \to H^1(X, \mathbf{G}_m)^D = \mathrm{Hom}_{\mathbf{Z}}(H^1(X, \mathbf{G}_m), \mathbf{Z})$$

denote the map induced by the intersection pairing on divisors. It has finite kernel and cokernel. Define classes in $G_0 T(\mathbf{Z}G)$ by

$$\begin{aligned}
\chi_T(X, \mathbf{G}_m) &= [H^0(X, \mathbf{G}_m)] - [H^1(X, \mathbf{G}_m)_{tor}] + [H^2(X, \mathbf{G}_m)] \\
&\quad + [\mathrm{coker}(\lambda)] - [H^3(X, \mathbf{G}_m)_{codiv}] + [H^4(X, \mathbf{G}_m)]
\end{aligned}$$

and

$$\chi_T(X, \mathbf{G}_a) = [H^0(X, O_X)] - [H^1(X, O_X)] + [H^2(X, O_X)] \; .$$

Then, $z(\chi_T(X, \mathbf{G}_m)) = f(\chi(X, \mathbf{G}_m))$ and a slight generalization of the work in [Li2] allows one to show that $c_{X,G} = \chi_T(X, \mathbf{G}_m) - \chi_T(X, \mathbf{G}_a)$. But because we assume the action is free, by work of Nakajima [Na2], $z(\chi_T(X, \mathbf{G}_a)) = 0$, so $f(\chi(X, \mathbf{G}_m)) = z(c_{X,G})$. The final part concerns identifying the right hand side in this last equality with a class only depending on the signs of the leading terms for symplectic characters. Let $L_{X,1}$ be the class defined in $G_0(\mathbf{Z}G)$ *via* the Hom-description by the element h of $\mathrm{Hom}_\Omega(R_G, J(\overline{\mathbf{Q}}))$ such that for irreducible χ

$$h(\chi)_v = \begin{cases} 1 & v \text{ finite} \\ 1 & v = \infty, \, \chi \text{ not symplectic} \\ \mathrm{sign}(c_{X,G}) & v = \infty, \, \chi \text{ symplectic} \end{cases}$$

Then $L_{X,1}$ has order at most two and $z(c_{X,G}) = L_{X,1}$. In summary

Theorem 2.3

$$f(\chi(X, \mathbf{G}_m)) = L_{X,1} \; .$$

3   The generalized Fröhlich Conjecture

Once results analogous to Taylor's theorem, for varieties over finite fields were known (in any dimension!), one started wondering about the possibility of formulating a generalized Fröhlich Conjecture for arithmetic schemes. Such a conjecture was formulated in [CEPT2] and proven there in many cases. A proof of that conjecture in general is contained in [CPT].

To arrive at a reasonable conjecture, the first thing to try was, of course, to see if one could define classes analogous to those defined by Chinburg in the case of–say–arithmetic surfaces. One had to find a combination of refined Euler characteristics, which was related to epsilon factors and possibly obtain a relation like Chinburg's $\Psi = W + R$. Classes analogous to the root number class and the ramification class could be defined by using Deligne's theory of ϵ-constants of representations of Weil-Deligne groups ([De2]). But, as we will see, the strict analogue of Ψ does not seem to be related nicely to ϵ-constants.

Let $\mathcal{X}$ be a regular projective scheme flat over Spec $\mathbf{Z}$, equidimensional of dimension $d+1$, which admits an action by the finite group G. Let $\mathcal{Y} = \mathcal{X}/G$. Assume that the action is tame in the sense that, for every closed point x of $\mathcal{X}$, the inertia subgroup at x has order relatively prime to the characteristic of the residue field of x. Then the branch locus b in $\mathcal{Y}$ is fibral and every coherent sheaf on $\mathcal{X}$ admits a refined Euler characteristic. Which combination of sheaves is related to a root number class? A root number class would have order at most two and it would lie inside the so-called kernel group $D(\mathbf{Z}G)$, that is the group of classes in the classgroup, which become trivial after extending scalars to one (hence any) maximal order in $\mathbf{Q}G$ containing $\mathbf{Z}G$. One can show that, even for surfaces, the class Ψ does not lie in the kernel group (see [C3] Rem. 3.9). The correct class in the present framework is the following. Note that in general the sheaf $\Omega^1_{\mathcal{X}/\mathbf{Z}}$ of absolute differentials on $\mathcal{X}$ is not locally free, however by the regularity of $\mathcal{X}$ we can identify the two groups $G_0(G, \mathcal{X})$ and $K_0(G, \mathcal{X})$. Thus it make sense to take exterior powers of classes in $G_0(G, \mathcal{X})$. We define the *Euler-de Rham class* in $Cl(\mathbf{Z}G)$, attached to the tame action $(\mathcal{X}, G)$, to be

$$\chi(\mathcal{X}, G) = \sum_{i=0}^{d}(-1)^i \chi^{CT}(\lambda^i([\Omega^1_{\mathcal{X}}])) \ .$$

It is shown in [CEPT2] Sect. 1.4.5 that if $(\mathcal{Z}, G)$ is a tame action on a variety of dimension $d+1$ over a finite field, which is fibered over a curve C, then the Euler-de Rham class defined using the relative differentials for $\mathcal{Z}/C$ equals the class $\Psi(\mathcal{Z}, G)$. To conclude the comparison of χ and Ψ over a finite field, we note that (up to sign) χ is the Euler characteristic of the top Chern class of Ω^1 whereas Ψ is the Euler characteristic of the penultimate Chern class (see [CPT]).

The root number class and the ramification class are defined by analogy with the case of varieties over finite fields and the case of extensions of number fields. Let $X = \mathcal{X} \otimes \mathbf{Q}$ be the generic fibre of $\mathcal{X}$. The local epsilon factors that come into the definition of the root number class are those attached, via a recipe due to Deligne, to the motives $X \otimes_G V$ whose realization are given by the cohomology groups $(H^*(X) \otimes V)^G$ (see [De2]). In fact the epsilon factor at v is modified by the epsilon factor of the fiber $\mathcal{X}_v/G$. One must also choose a fixed auxilary prime ℓ such that $\mathcal{X}_\ell$ is smooth together with an embedding $\overline{\mathbf{Q}}_\ell \hookrightarrow \mathbf{C}$ (see [CEPT2]). One can then define a root number class

$W_{\mathcal{X},\ell}$ and a ramification class $R_{\mathcal{X}}$. The ramification class is defined in terms of the epsilon factors of the fibers of the reduced branch locus of the quotient $\mathcal{X} \to \mathcal{Y} = \mathcal{X}/G$.

Theorem 3.1 *In addition to the above assumptions, also assume that the quotient $\mathcal{Y}$ is regular, with special fibers $\mathcal{Y}_p$ that are divisors with strict normal crossings and for each p assume that the multiplicities of the irreducible components of $\mathcal{Y}_p$ are prime to p. Then*

$$\chi(\mathcal{X}, G) = W_{\mathcal{X},\ell} + R_{\mathcal{X}} \ .$$

This was proved in [CEPT2] for surfaces and for free actions in arbitrary dimensions. The general case is the main theorem of [CPT]. Below we indicate a proof for free actions and outline the strategy for the general case. One of the reasons for the restrictions imposed on $\mathcal{Y}$ is the use of the results in [ST] on epsilon factors. Another reason is that they imply that a sheaf of relative log-differentials, which is used in the proof, is locally free. It is conjectured that the result holds without these restrictions. For the fact that Taylor's theorem is a special case of this result see [CEPT2] Thm. 4.13 (and also [P2]).

Assume now the action $(\mathcal{X}, G)$ is free. Then the quotient morphism $\pi : \mathcal{X} \to \mathcal{Y}$ is étale and the ramification class is trivial. Moreover, every coherent G-sheaf on $\mathcal{X}$ is of the form $\pi^* F$ for some coherent sheaf F on $\mathcal{Y}$. The idea is to reduce to the one dimensional case, where one can apply Taylor's result. Let $c_Y := (-1)^d c^d(\Omega^1_{Y/\mathbf{Q}})$ denote the fundamental class of the generic fiber Y of $\mathcal{Y}$. This is an element in the Chow group $CH^d(Y)$ represented by a 0-cycle on Y. Write c_Y as a linear combination of closed points of Y

$$c_Y = \sum_j e_j [P_j] \ .$$

Let $\overline{P}_j$ denote the Zariski closure of P_j in $\mathcal{Y}$. Then $O_{\overline{P}_j \times_{\mathcal{Y}} \mathcal{X}}$ is a G-stable order inside the G-Galois algebra whose spectrum is $P_j \times_Y X$. Let S_j be the normalization of $\overline{P}_j \times_{\mathcal{Y}} \mathcal{X}$. By comparing $\sum_j e_j [O_{S_j}]$ to $\sum_{i=0}^d (-1)^i \lambda^i(\Omega^1_{\mathcal{X}})$ in $K_0(G, \mathcal{X}) = G_0(G, \mathcal{X})$ and by using results of Thomason [Th] and Nakajima [Na1], [Na2], one can show that the Euler-de Rham class can be represented as

$$\chi(\mathcal{X}, G) = (-1)^d \sum_j e_j \cdot \chi^{CT}(S_j, G) \ .$$

Thus by applying Taylor's Theorem, one is able to relate the Euler-de Rham class to a combination of root number classes coming from number field extensions. The relation to the root number class $W_{\mathcal{X},l}$ comes from a product formula for the symplectic root numbers at infinity. This formula shows that the root numbers of the motives $X \otimes_G V$ over $\mathbf{R}$ for symplectic V can be obtained as the product of the symplectic root numbers of the points in the canonical cycle of Y. In the proof of the product formula, the root numbers of the real motive are calculated using a formula of Deligne, which expresses them in terms of Hodge numbers (see [Del]). The product of root numbers of the points in the canonical cycle is expressed in terms of the Euler numbers of the connected components of the real locus of Y. This uses the topological Lefschetz fixed point theorem and a result on cycles and characteristic classes due to Borel-Haefliger. The formula then results from a (long) calculation. For the details see [CEPT2] Sect. 5 and 6.

Let us now go back to the general case. Then not every G-sheaf on $\mathcal{X}$ is the pull-back of a sheaf on $\mathcal{Y}$. Still, as in the case of a free action, the strategy will be to study the pull-back of a locally free sheaf whose refined Euler characteristic is closely related to the Euler-de Rham class. Note that this class can be viewed as the Euler characteristic of the top Chern class of $\Omega_{\mathcal{X}}^1$:

$$\chi(\mathcal{X}, G) = (-1)^d \chi^{CT}(c^d(\Omega_{\mathcal{X}}^1)) \ .$$

Because of our assumptions on $\mathcal{Y}$, the sheaf

$$\Omega_{\mathcal{Y}}^1(\log) = \Omega_{\mathcal{Y}/\mathbf{Z}}^1(\log \mathcal{Y}_S^{red}/\log S)$$

of relative logarithmic differentials is locally free. Here S denotes the set consisting of those primes p at which the fibre $\mathcal{Y}_p$ is not smooth and those supporting the branch locus of $\mathcal{X}/\mathcal{Y}$ (recall that $\mathcal{X}/\mathcal{Y}$ is generically étale because the action is tame), and $\Omega_{\mathcal{Y}}^1(\log)$ is defined with respect to the morphisms of log-schemes $(\mathcal{Y}, \mathcal{Y}_S^{red}) \to (\operatorname{Spec} \mathbf{Z}, S)$. We put

$$\chi_1(\mathcal{X}, G) = (-1)^d \chi^{CT}(c^d(\pi^* \Omega_{\mathcal{Y}}^1(\log)))$$

and we let $\chi_2(\mathcal{X}, G) = \chi(\mathcal{X}, G) - \chi_1(\mathcal{X}, G)$. By using a Moving Lemma for Chern classes, $\chi_1(\mathcal{X}, G)$ can be written in terms of the refined Euler characteristics of one dimensional schemes flat over $\mathbf{Z}$ and hence can be related to symplectic root number classes by using Taylor's theorem. The relation to the root number class is achieved by using the product formula for the root numbers of real motives mentioned above and the product formula for epsilon factors of tame sheaves proved by Saito in [ST]. To deal with $\chi_2(\mathcal{X}, G)$ one notes that because it is the Euler characteristic of the difference of two classes in $K_0(G, \mathcal{X})$ which give the same class in $K_0(G, X)$, using [Th] Thm. 2.7, it can be written as

$$\chi_2(\mathcal{X}, G) = \sum_{p \in S} \operatorname{Res}_{\mathbf{Z}}(\chi^{CT}(F_p)) \ ,$$

where the F_p are elements of $G_0(G, \mathcal{X}_p^{red})$. Then one defines a class Ψ_p by using the Ψ-classes of Chinburg attached to the strata of the reduced fiber $\mathcal{X}_p^{red}$, such that

$$\operatorname{Res}_{\mathbf{Z}}(\chi^{CT}(F_p)) = (-1)^d \operatorname{Res}_{\mathbf{Z}}(\Psi_p) \ .$$

This equality is proved by showing, using LRR, that the Brauer traces of $\chi^{CT}(F_p)$ and Ψ_p (for elements different from the identity) are the same. The relation to root numbers is achieved by using the results of Chinburg on varieties over finite fields. For the details see [CEPT2] and [CPT].

4 Characterizing epsilon factors

So far we have seen how to determine (combinations of) refined Euler characteristics in terms of epsilon factors. Proving a second conjecture of Fröhlich, Cassou-Noguès and Taylor have shown how to characterize the signs of symplectic root numbers attached to tame Galois extensions of number fields, in terms of the hermitian Galois module structure of rings of integers supplied by the bilinear trace form. Here we describe analogous results for the cases of varieties over finite fields and of schemes over $\mathbf{Z}_p$. We also state a theorem generalizing a result of Serre, which asserts that for orthogonal representations the epsilon factors of curves are positive.

4.a Varieties over finite fields

Recall the notation of Sect. 2.a. In particular let $\epsilon(Y, V)$ denote the epsilon constant attached to an action (X, G) with quotient Y, and a complex representation V of G. In this subsection we consider real characters of even degree with trivial determinant, not just symplectic characters. Let R^r_G denote the subgroup of R_G which is generated by these characters. If the character of the representation V is real valued, then $\epsilon(Y, V)$ is real.

The following is an easy generalization, using Poincaré duality, of a result of Serre (proved in [FQ], see also [CEPT3] Thm. A).

Theorem 4.1 *If the dimension of X is even (resp. odd) and if V is a symplectic (resp. orthogonal) representation, then $\epsilon(Y, V)$ is positive.*

By the results of Chinburg described in Sect. 2.a, the p-adic valuations of the epsilon constants $\epsilon(Y, V)$ are integers, which are determined by the class $\Psi(X, G)$. To determine the sign of $\epsilon(Y, V)$ for χ_V in R^r_G, we introduce the (generalized) adelic hermitian classgroup $A(\mathbf{Z}G)$. Let E be a large enough Galois extension of the rationals and let $J_f(E)$ denote the group of finite ideles of E. The properties of $A(\mathbf{Z}G)$ that we need are:

(a) there is a surjection

$$\mu : \mathrm{Hom}_{Gal(E/\mathbf{Q})}(R^r_G, J_f(E))) \to A(\mathbf{Z}G) ,$$

analogous to the map giving the Hom-description of Sect. 1.a;

(b) $A(\mathbf{Z}G)$ contains a subgroup isomorphic to $\operatorname{Hom}_{Gal(E/\mathbf{Q})}(R_G^r, \mathbf{Q}^\times)$, called
     the group of *rational classes*.

(This last fact is a generalization of work of Cassou-Noguès and Taylor, see
[CEPT3] Thm. 3.2 and [F3] Cor. 3 to Thm. 17.) By composing the injection Δ of Sect. 2.a with the map induced on the Hom-groups by restricting
homomorphism from R_G to R_G^r and with the surjection μ, we obtain a map

$$\alpha : K_0(\mathbf{F}_p G) \to A(\mathbf{Z}G) \ .$$

We shall use a stratification $(U_i, G_i)_i$ of (X, G) and classes $\Psi(U_i, G_i)$ whose
image under the maps α_i, defined in the same way as α with G replaced by G_i,
are rational classes. These classes combine to determine the epsilon constants.
For simplicity assume that all irreducible components of the branch locus B
in X are in the same G-orbit. Then one can define locally closed U_i in X
with a number of desirable properties. Namely:

(1) all points of U_i have the same inertia group I_i, which is normal in the
     stabilizer D_i of U_i in G;

(2) U_i carries an action of $G_i = D/I_i$;

(3) $X = \coprod_i \coprod_{g \in G/D_i} gU_i$.

The epsilon factor of X can be expressed in terms of the epsilon factors of
(U_i, G_i). Thus, if we could characterize these epsilon factors in terms of Ψ-
invariants, we would be done. However, Chinburg's Ψ-invariants are only
defined for smooth projective varieties with a tame action. The idea then
is to define $\Psi(U_i, G_i)$ by inclusion-exclusion. For instance, if the difference
between the closure $\overline{U_i}$ of U_i and U_i is the union of two closed varieties Z_1
and Z_2, which intersect in P, then we would put

$$\Psi(U_i) = \Psi(\overline{U_i}) - \Psi(Z_1) - \Psi(Z_2) + \Psi(P) \ .$$

The surjection $D_i \to G_i$ and the inclusion $D_i \to G$ induce maps on the level
of the Hom-groups denoted Inf_i and Ind_i respectively. Denote by $\epsilon^r(X, G)$
the restriction to R_G^r of the map sending χ_V in R_G to $\epsilon(Y, V)$. The following
is essentially Thm. C of [CEPT3].

Theorem 4.2 *With the above notation*

(i) *The classes* $\alpha_i(\Psi(U_i, G_i))$ *are rational.*

(ii)
$$\epsilon^r(X, G) = \prod_i \mathrm{Ind}_i \circ \mathrm{Inf}_i(\alpha_i(\Psi(U_i, G_i))) \ .$$

Let us explain this result for curves. The U_i above are defined in terms of a stratification of the branch locus B of X/Y. If X is of dimension one, the branch locus is a collection of points, and epsilon factors for points are roots of unity. From this and the theorem one deduces the following.

Corollary 4.3 *Let X be of dimension one, then*

$$\epsilon^r(X, G) = \alpha([H^0(X, O_X)] - [H^1(X, O_X)]) .$$

4.b Arithmetic schemes

Here, following [CEPT4], we indicate how one can characterize symplectic epsilon constants, in terms of the intersection numbers of certain Pfaffian divisors with a sheaf constructed from logarithmic differentials. Although this does not transpire from our summary, the results we present rely heavily on the work of T. Saito in [ST].

Let $\mathcal{X}$ be a regular projective scheme flat over $\operatorname{Spec} \mathbf{Z}_p$, which is equidimensional of dimension $d + 1$. We assume that $\mathcal{X}$ is equipped with an action by a finite group G, which is such that the quotient $\mathcal{Y}$ is regular connected and such that the pair $(\mathcal{Y}, \mathcal{Y} \times \mathbf{Q}_p)$ is tame in the sense of [Ka2] (2.2) (*e.g.* $\mathcal{Y}$ semistable). Then the sheaf

$$\Omega^1_{\mathcal{Y}}(\log) = \Omega^1_{\mathcal{Y}/\mathbf{Z}_p}(\log \mathcal{Y}_p^{red} / \log \mathbf{F}_p) ,$$

of relative logarithmic differentials is locally free of $O_{\mathcal{Y}}$-rank d. Let

$$\lambda_{-1}(\Omega^1_{\mathcal{Y}}(\log)) = \sum_{i=0}^d (-1)^i \Lambda^i(\Omega^1_{\mathcal{Y}}(\log)) .$$

Assume that the action $(\mathcal{X}, G)$ is tame over a vertical divisor with strictly normal crossings. Then, for any symplectic representation V of G, imitating Fröhlich's construction in [F3] II.3, we can define a divisor $\mathbf{P}f(\mathcal{X}, G)(V)$ on $\mathcal{Y}$ by using the trace form at the local level. This divisor is called the *Pfaffian divisor*. Under the tameness assumption, for any symplectic representation $\tilde{V}$ of degree 0, the constant $\epsilon_0(\mathcal{X} \otimes_G \tilde{V})$ is (up to sign) an integral power of p. This power can be determined in terms of the Pfaffian divisor as

$$v_p(\epsilon_0(\mathcal{X} \otimes_G \tilde{V})) = d(\tilde{V}) := \deg(\lambda_{-1}(\Omega^1_{\mathcal{Y}}(\log)) \cdot \mathbf{P}f(\mathcal{X}, G)(\tilde{V})) .$$

(The right-hand side can be thought of as an intersection number; see *loc. cit.*)

This equality can be refined to get a statement about classes inside a classgroup. Let O denote the ring of integers in the cyclotomic field $\mathbf{Q}(e^{2i\pi/p})$.

As in the previous subsection, we consider the adelic hermitian classgroup $A(OG)$, which comes with a surjection

$$\alpha : \mathrm{Hom}_{Gal(E/\mathbf{Q})}(R_G^s, J_f(E)) \to A(OG) \ ,$$

and which contains a subgroup of rational classes. One shows that the image by α of the homomorphism sending χ_V to the idele with p-component $(-p)^{d(V)}$ and all other components 1, is a rational class c and thus corresponds to an element of $\mathrm{Hom}_{Gal(E/\mathbf{Q})}(R_G^s, \mathbf{Q}^\times)$ again denoted by c.

Theorem 4.4 *For all symplectic representations V of G*

$$c(V) = \epsilon_0(X \otimes_G (V - (dim(V) \cdot 1))) \ .$$

Corollary 4.5 *If $\mathcal{Y}$ has semistable reduction, then for any symplectic representation*

$$sign(c(V)) = sign(\epsilon_0(X \otimes_G V)) \ .$$

5   Normal bases for elliptic division orders

Let us begin by recalling what a normal basis for a Galois extension of rings is. Let Γ be a finite group and let R be a ring. A Γ-Galois extension of R is an R-algebra S on which Γ acts by ring automorphisms in such a way that $R = S^\Gamma$ and the S-algebra $S \otimes_R S$ is isomorphic to the algebra $\mathrm{Map}(\Gamma, S)$ of all maps from Γ to S, under the map $x \otimes y \mapsto (\gamma \mapsto x\gamma(y))$. In other words, $\mathrm{Spec}\, S$ is a principal homogeneous space/torsor for the constant group scheme Γ_R associated to the algebra $\mathrm{Map}(\Gamma, R)$. Then S has a normal basis, if S is isomorphic to $R\Gamma$ as an $R\Gamma$-module. Note that $R\Gamma$ is the dual R-algebra of $\mathrm{Map}(\Gamma, R)$.

We will be interested in the situation where S is replaced by an order whose spectrum is a torsor for a non-necessarily constant group scheme obtained as the kernel of an isogeny between abelian varieties. Let R be the ring of integers in a number field and let A be an abelian scheme on $T = \mathrm{Spec}\, R$. Consider an isogeny f from A to A, and let G denote its kernel. For any R-valued point of A, say $Q : T \to A$, we can form the pull-back $X_Q = A \times_A T$ of f along Q. This is an affine T-scheme which is a torsor for G. Let $\mathcal{A}$ (resp. $\mathcal{B}_Q$) denote the ring of G (resp. X_Q). These are both modules over the linear dual $\mathcal{B} = \mathrm{Hom}_R(\mathcal{A}, R)$. It can be shown that $\mathcal{B}_Q$ is a locally free module of finite rank over $\mathcal{B}$. Assume for simplicity that $\mathcal{A}$ is free over $\mathcal{B}$. Then the map from the group of R-valued points $A(R)$ to the classgroup of locally free $\mathcal{B}$-modules, which sends Q to the class of $\mathcal{B}_Q$, can be shown to be a homomorphism

$$\psi : A(R) \to Cl(\mathcal{B}) \ .$$

This is called the *class invariant homomorphism*. It has been first introduced by Waterhouse in [Wat] and it has been studied in relation with Galois module theory by Taylor and then Agboola, Cassou-Noguès, Srivastav *et al.* (see [T2] and the surveys [C-N,T], [BT] and [C-N,C,F,T]). It is clear that a point Q defines an element in the kernel if and only the order $\mathcal{B}_Q$ is free over $\mathcal{B}$, *i.e.* if and only if it admits a normal basis.

One of the main results about the class invariant homomorphism is that for $A = E$ of dimension 1 and f multiplication by the power of a prime $p > 3$, then any torsion point is in the kernel of ψ:

$$E(R)_{tor} \subset \ker(\psi) \ .$$

Srivastav-Taylor proved this for elliptic curves with complex multiplication, by finding explicit normal basis generators using modular functions (see [SrT]). Then, in [Ag2] Agboola was able to prove the result without assuming the curve has complex multiplication. In [Ag1] Agboola showed the result to be equivalent to the following geometric result on the restriction of torsion line bundles, which was proved by a new method and in complete generality by Pappas in [P1].

Theorem 5.1 *Let E be an abelian scheme of relative dimension 1 over a commutative ring R and let m be an integer coprime to 6. Let $\mathcal{L}$ be a torsion line bundle on E with trivial restriction to the 0-section. Then the restriction of $\mathcal{L}$ to the subscheme of m-torsion points E_m is a trivial line bundle.*

In [P1], Pappas shows how to construct examples of abelian schemes fibered over an affine curve defined over a finite field, for which the analogous result does not hold. The restriction to integers prime to 6 is necessary as shown by Bley-Klebel in [BlKl] and by Cassou-Noguès–Jehanne in [C-N,J].

6   The equivariant arithmetic genus

In the preceding sections we have insisted on the fact that the main results about refined Euler characteritics concern combinations of such. Not much is known about the Euler characteristics of individual sheaves. In this section we will see how, following [P2], one can obtain results on the refined Euler characteristic of one sheaf alone. Let $\mathcal{X}$ be an arithmetic surface, that is a projective scheme of dimension 2 flat over Spec $\mathbf{Z}$. Assume $\mathcal{X}$ supports a free action by a finite *abelian* group G and, as before, denote the quotient by $\pi : \mathcal{X} \to \mathcal{Y} = \mathcal{X}/G$. For simplicity we assume that $\mathcal{Y}$ is a local complete intersection. We want to show how to use Deligne's Riemann-Roch theorem for relative curves of [De3] to deduce that under the stated hypotheses

$$2 \cdot \chi^{CT}(O_\mathcal{X}) = 0 \ .$$

Let G^D denote the Cartier dual of G (viewed as a constant group scheme). Then because the action is free, and hence tame, $O_{\mathcal{X}}$ is locally free over $O_{\mathcal{Y}}G$ (see *e.g.* [CEPT1]). So we get a line bundle $\mathcal{L} = \mathcal{L}[\pi]$ on $\tilde{\mathcal{Y}} = \mathcal{Y} \times G^D$. Let $\tilde{h}$ denote the structure map of $\tilde{\mathcal{Y}}$ over G^D. Then,

$$\chi^{CT}(O_{\mathcal{X}}) = \det R\tilde{h}_*(\mathcal{L}[\pi]) \ ,$$

where $\det : K_0(G^D) \to \mathrm{Pic}(G^D)$ is the determinant homomorphism, and we identify $\mathrm{Pic}(G^D)$ with the quotient $Cl(\mathbf{Z}G)$ of $CT(\mathbf{Z}G)$. Recall from [De3], that to any two line bundles $\mathcal{L}$ and $\mathcal{M}$ on $\tilde{\mathcal{Y}}$ one can associate a line bundle $< \mathcal{L}, \mathcal{M} >$ on G^D. Also let $\omega = \omega_{\tilde{\mathcal{Y}}/G^D}$ denote the relative dualizing sheaf for $\tilde{h}$. Then Deligne's Riemann-Roch theorem (*loc. cit.* (7.5.1)) gives

$$\det R\tilde{h}_*(\mathcal{L}[\pi])^{\otimes 2} = < \mathcal{L}[\pi], \mathcal{L}[\pi] \otimes \omega^{-1} > \otimes \det R\tilde{h}_*(O_{\tilde{\mathcal{Y}}})^{\otimes 2} \ .$$

So we deduce that

$$2 \cdot \chi^{CT}(O_{\mathcal{X}}) = < \mathcal{L}, \mathcal{L} > \cdot < \mathcal{L}, \omega^{-1} > \cdot 2 \cdot \chi^{CT}(O_{\tilde{\mathcal{Y}}}) \ .$$

The last term in the right hand side is an induced class. Up to sign, the middle term equals the refined Euler characteristic of the structure sheaf of the inverse image by π of a divisor representing the relative dualizing sheaf. One can show that it is trivial by using the results of Taylor [T1] and Nakajima [Na2] mentioned in Sect. 3. So there remains

$$2 \cdot \chi^{CT}(O_{\mathcal{X}}) = < \mathcal{L}, \mathcal{L} > \ .$$

In order to prove that the right hand side is zero one writes it in terms of a $\mathbf{G}_m$-biextension of (G^D, G^D) and one uses the fact that there are no non-trivial biextensions on finite multiplicative group schemes over $\mathrm{Spec}\,\mathbf{Z}$. The latter fact can be obtained using Herbrand's Theorem (see [P2] or [Maz]).

It is interesting to note that in the approach of [P2], the "2-dimensional, absolute" Galois structure problem of determining $\chi^{CT}(O_{\mathcal{X}})$ is controlled by the "1-dimensional, relative" problem of determining the structure of a torsor over G^D. Actually Pappas' results are more general, here we only considered the simplest interesting case. [P2] also contains results connecting $\chi^{CT}(O_{\mathcal{X}})$ to values of the class invariant homomorphism considered in the previous section.

7   Equivariant motives

In this section I review the results on generalized omega invariants which I am aware of. These results fall under three headings: (i) Tate motives, (ii) equivariant motives obtained by base change by abelian extensions and (iii) motives obtained by base change by an arbitrary finite Galois extension.

Let N/K be a finite Galois extension of number fields, and let $G = Gal(N/K)$. Chinburg's invariants $\Omega(N/K, i)$ for $i = 1, 2, 3$ (see [C1], [C2]) arose from Tate's work on Stark's conjecture and on the cohomology of one-dimensional class field theory. The goal of extending Galois structure theory to the context of the generalizations of Stark's conjecture proposed by Lichtenbaum, Beilinson, Bloch-Kato, and others is mentioned at the end of the introduction of [C3].

The first major step towards this goal was achieved by Burns and Flach in [BF1]. Let M be a motive defined over the number field K and let M_N denote the base change of M by the finite Galois extension N/K. Burns and Flach assumed that the extension N/K is abelian, and used work of Bloch, Kato, Fontaine and Perrin-Riou. (See [Fo] for a presentation of the non-equivariant theory.) They defined a "fundamental line" $\Xi(M_N)$, which is a $\mathbf{Q}G$-module, which over the completions of $\mathbf{Q}$, conjecturally is isomorphic to some standard modules: $\mathbf{R}G$ over $\mathbf{R}$, and the determinant of a complex $R\Gamma_c(M_p)$ computing the cohomology with supports of the p-adic realization of M_N, over $\mathbf{Q}_p$. The definition of $\Xi(M_N)$ uses determinants of $\mathbf{Q}G$-modules. This is why one needs to assume G is abelian. The invariant $\Omega(N/K, M)$ defined by Burns and Flach is the class in $\mathrm{Pic}(\mathbf{Z}G)$ of an invertible module $\Xi(M_N)_{\mathbf{Z}}$ in the fundamental line. This module is defined as the intersection over all primes p of invertible $\mathbf{Z}_pG$-modules which are obtained from perfect complexes, *via* Lemma 1.1. For a general motive, the construction only works under some assumptions. However these are all satisfied for Tate motives and for the motive $h^1(X)(1)$ of an abelian variety X defined over K, such that the Tate-Shafarevich group of X_N is finite.

The problem of removing the restriction that N/K be abelian was studied by Chinburg, Kolster, Pappas and Snaith for the case of Tate motives. They developed a different approach from Burns and Flach based on the idea of using Chern character maps from K-theory to étale cohomology to construct four-term sequences of G-modules analogous to the Tate S-unit sequence. As described in Sect. 1.c, they obtained invariants $\Omega_n(N/K)$ for all finite Galois N/K and all Tate motives $\mathbf{Z}(n)$ for $n \geq 1$ which generalize Chinburg's $\Omega(N/K, 3) = \Omega_0(N/K)$ (multiplicative) invariant. In [CKPS1] the $\Omega_n(N/K)$ invariants are related to the Galois structure of K-groups under the condition that the Quillen-Lichtenbaum conjecture on the Chern character maps is true. A generalization of Stark's conjecture, namely the Lichtenbaum-Gross conjecture, implies that $\Omega_n(N/K)$ equals the root number class modulo the kernel group $D(\mathbf{Z}G)$. (This implication can in fact be proved in the function field case for $n \geq 2$). The relation between $\Omega_1(N/K)$ and the root number class for quaternionic extensions of the rationals is considered in [CKPS3].

In [BF2], Burns and Flach have compared their invariant for Tate motives, with the invariants of [CKPS1]: they show under very general assumptions that for $n \geq 0$ the image of their $\Omega(N/K, \mathbf{Q}(-n))$ by the involution on $\mathrm{Pic}(\mathbf{Z}G)$ induced by $g \mapsto g^{-1}$, equals $\Omega_n(N/K)$. Also $\Omega(N/K, \mathbf{Q}(1))$ becomes equal to the inverse of the image of Chinburg's $\Omega(N/K, 1)$ inside $\mathrm{Pic}(\mathbf{Z}[1/(2|G|)]G)$. (The invariant studied there *is* $\Omega_1(N/K)$ by [CKPS4].)

Burns has applied the four-term sequence approach of [CKPS1] to more general motives than Tate motives (see [B2]), and has related these generalizations to non-commutative generalizations of Iwasawa's Main Conjecture (see Kato's papers [Ka1] and [Ka3]). Burns has also applied the four-term sequence method to define analogues of Chinburg's $\Omega(N/K, 2)$ invariant for equivariant motives of the form M_N (see [B2]). Geometric class field theory leads naturally to complexes that are problematic from the point of view of the four-term sequence technique; this motivated the work of Chinburg, Kolster Pappas and Snaith on nearly perfect complexes which has been described in Sect. 1.d.

The depth and breadth of current research on values of L-functions associated to motives seems certain to be reflected in future work on the Galois module structure theory of motives.

References

[Ag1]   Agboola, A.: A geometric description of the class invariant homomorphism, J. Théor. Nombres Bordeaux **6**(1994), 273–280.

[Ag2]   Agboola, A.: Torsion points on elliptic curves and Galois module structure, Inventiones math. **123**(1996), 105–122.

[BFQ]   Baum, P., Fulton, W., Quart, G.: Lefschetz-Riemann-Roch for singular varieties, Acta Math. **143**(1979), 193–211.

[BlKl]   Bley, W., Klebel, M.: An infinite family of elliptic curves and Galois module structure, preprint Uni. Augsburg, 1996.

[B1]   Burns, D.: ϵ-constants and de Rham structure invariants associated to motives over number fields I: the general case, preprint King's College London, 1996.

[B2]   Burns, D.: Iwasawa theory and p-adic Hodge theory over noncommutative algebras I & II, preprint King's College London, 1997.

[BF1]   Burns, D., Flach, M.: Motivic L-functions and Galois module structures, Math. Ann. **305**(1996), 65–102.

[BF2]   Burns, D., Flach, M.: On Galois structure invariants associated to Tate motives, preprint King's College London, 1996.

[BT]   Byott, N., Taylor, M.J.: Hopf orders and Galois module structure, in "Group rings and classgroups", ed. by K.W. Roggenkamp and M.J. Taylor, DMV **18**, Birkhäuser, 1992, 153–210.

[C-N,C,F,T]   Cassou-Noguès, Ph., Chinburg, T., Fröhlich, A., Taylor, M.J.: L-functions and Galois modules (notes by D. Burns and N.P. Byott), in "L-functions and arithmetic", ed. by J. Coates and M.J. Taylor, London Math. Soc. Lecture Note Series **153**, C.U.P., Cambridge, 1991.

[C-N,T]   Cassou-Noguès, Ph., Taylor, M.J.: Structure galoisienne et courbes elliptiques, Journées Arithmétiques 1993, J. Théor. Nombres Bordeaux **7**(1995), 307–331.

[C-N,J]   Cassou-Noguès, Ph., Jehanne, A.: Espaces homogènes principaux et points de 2-division de courbes elliptiques, preprint Bordeaux, 1998.

[C1]   Chinburg, T.: On the Galois structure of algebraic integers and
            S-units, Inventiones math. **74**(1983), 321–349.

[C2]   Chinburg, T.: Exact sequences and Galois module structure,
            Ann. of Math. **121**(1985), 351–376.

[C3]   Chinburg, T. : Galois structure of de Rham cohomology of tame
            covers of schemes, Ann. of Math. **139**(1994), 443-490.

[CE]   Chinburg, T., Erez, B. : Equivariant Euler-Poincaré charac-
            teristics and tameness, 179–194, in Proceedings of Journées
            Arithmétiques 1991, Astérisque **209**, Société Math. de France,
            Paris, 1992.

[CEPT1]   Chinburg, T., Erez, B., Pappas, G., Taylor, M.J.: Tame actions
            of group schemes: integrals and slices, Duke Math. J. **82:2**(1996),
            269–302.

[CEPT2]   Chinburg, T., Erez, B., Pappas, G., Taylor, M.J.: ϵ-constants
            and the Galois structure of de Rham cohomology, Ann. of Math.
            146(1997), 411–473.

[CEPT3]   Chinburg, T., Erez, B., Pappas, G., Taylor, M.J.: On the
            ϵ-constants of a variety over a finite field, Am. J. of Math.
            119(1997), 503–522.

[CEPT4]   Chinburg, T., Erez, B., Pappas, G., Taylor, M.J.: On the ϵ-
            constants of arithmetic schemes, to appear in Math. Ann.

[CEPT5]   Chinburg, T., Erez, B., Pappas, G., Taylor, M.J.: Gauss sums
            and fixed point sets for curves and surfaces over finite fields, in
            preparation.

[CKPS1]   Chinburg, T., Kolster, M., Pappas, G., Snaith, V.: Galois struc-
            ture of K-groups of rings of integers, C.R. Acad. Sci. Paris,
            320(1995), Série I, 1435–1440.

[CKPS2]   Chinburg, T., Kolster, M., Pappas, G., Snaith, V.: Nearly per-
            fect complexes and Galois module structure, preprint Univ. of
            Pennsylvania, 1997.

[CKPS3]   Chinburg, T., Kolster, M., Pappas, G., Snaith, V.: Quaternionic
            exercises in K-theory Galois module structure, preprint McMas-
            ter Uni., 1994/95 No. 5.

[CKPS4]   Chinburg, T., Kolster, M., Pappas, G., Snaith, V.: Comparison of K-theory Galois module structure invariants, preprint McMaster Uni., 1995/96 No. 9.

[CPT]   Chinburg, T., Pappas, G., Taylor, M.J.: ε-constants and the Galois structure of de Rham cohomology II, preprint of Princeton Univ., 1997.

[De1]   Deligne, P.: Valeurs de fonctions L et périodes d'intégrales, Proc. Symp. Pure Math. **33**(1979), Part 2, 313–346.

[De2]   Deligne, P.: Les constantes des équations fonctionnelles des fonctions L, 501–597, LNM **349**, Springer-Verlag, Berlin, 1974.

[De3]   Deligne, P.: Le déterminant de la cohomologie, 93–177, Contemporary Math. **67**, AMS, Providence, 1987.

[Do]   Donovan, P.: The Lefschetz-Riemann-Roch formula, Bull. Soc. Math. France **97**(1969), 257–273.

[DF]   Dwyer, W.G., Friedlander, E.M.: Algebraic and étale K-theory, Trans. Am. Math. Soc. **292**(1)(1985), 247–280.

[Fo]   Fontaine, J.-M.: Valeurs spéciales des fonctions L des motifs, Sém. Bourbaki (1991-1992), Exp. 751, 205–249 Astérisque **206**, Soc. Math. de France, Paris, 1992

[F1]   Fröhlich, A.: *Galois Module Structure of Algebraic Integers*, Erg. der Math., 3. Folge, Bd. 1, Springer-Verlag, Heidelberg-New York-Tokyo, 1983.

[F2]   Fröhlich, A.: Some problems of Galois module structure for wild extensions, Proc. London Math. Soc. **37**(1978), 193–212.

[F3]   Fröhlich, A.: *Classgroups and Hermitian Modules*, Progress in Math. **48**, Birkhäuser, Boston-Basel-Stuttgart, 1984.

[FQ]   Fröhlich, A., Queyrut, J.: On the functional equation of the Artin L-function for characters of real representations, Inventiones math. **20**(1973), 125-138.

[FuL]   Fulton, W., Lang, S.: *Riemann-Roch Algebra*, Grund. math. Wiss. **277**, Springer-Verlag, New York-Berlin-Heidelberg, 1985.

[GM]   Grothendieck, A., Murre, J. P.: The tame fundamental group of
            a formal neighbourhood of a divisor with normal crossings on a
            scheme, Lect. Notes in Math. **208**, Springer-Verlag, Heidelberg,
            1971.

[K]   Kahn, B.: Descente galoisienne et K_2 des corps de nombres, K-
            theory **7**(1993), 55–100.

[Ka1]   Kato, K.: Iwasawa theory and p-adic Hodge theory, Kodai Math.
            J. **16**(1993), 1–31.

[Ka2]   Kato, K.: Class field theory, $\mathcal{D}$-modules, and ramification
            on higher dimensional schemes, part 1, Amer. Jour, Math.
            116(1994), 757–784.

[Ka3]   Kato, K.: Lectures on the approach to Iwasawa theory for Hasse-
            Weil L-functions via B_{dR}, Part I, 50–163, in "Arithmetic Alge-
            braic Geometry", LNM **1553**, Springer-Verlag, 1992.

[La]   Lang S.: *Algebraic number theory*, Addison-Wesley, Reading,
            1970.

[Li1]   Lichtenbaum, S.: Zeta-functions of varieties over finite fields
            at $s = 1$, in "Arithmetic and geometry–papers dedicated to
            I.R. Shafarevich...", Vol. I, Progress in Math. **35**, Birkhäuser,
            Boston-Basel-Stuttgart, 1983.

[Li2]   Lichtenbaum, S.: Behavior of the zeta-function of open surfaces
            at $s = 1$, in "Algebraic Number Theory–in honor of K. Iwasawa",
            Adv. Stud. in Pure Math. **17**(1989), 271–287.

[Maz]   Mazur, B.: Modular curves and the Eisenstein ideal, Publ. Math.
            IHES **47**(1977), 33-186.

[Na1]   Nakajima, S.: Galois module structure of cohomology groups
            for tamely ramified coverings of algebraic varieties, J. Number
            Theory **22**(1986), 115–123.

[Na2]   Nakajima, S.: On the Galois module structure of the cohomology
            groups of an algebraic variety, Inventiones math. **75**(1984), 1–8.

[P1]   Pappas, G.: On torsion line bundles and torsion points on
            abelian varieties, to appear in Duke Math. J.

[P2]   Pappas, G.: Galois modules and the theorem of the cube,
            preprint Princeton University, 1996.

[SS]   Saito, S: Arithmetic theory of arithmetic surfaces, Ann. of Math.
         129(1989), 547–589.

[ST]   Saito, T.: ϵ-factor of a tamely ramified sheaf on a variety, Inven-
         tiones math. **113**(1993), 389–417.

[Se]   Serre, J.-P. : Représentations linéaires des groupes finis, 3ème
         éd., Hermann Paris, 1978.

[Sn1]   Snaith, V.P.: *Galois Module Structure*, Fields Institute Mono-
         graphs **2**, Am. Math. Soc., Providence, 1994.

[Sn2]   Snaith, V.P.: Local fundamental classes derived from higher di-
         mensional K-groups, to appear in Proc. London Math. Soc.

[SrT]   Srivastav, A., Taylor, M.J.: Elliptic curves with complex mul-
         tiplication and Galois module structure, Inventiones math.
         99(1990), 165–184.

[Ta]   Tate, J.: *Les Conjectures de Stark sur les Fonctions L d'Artin
         en s = 0*, Progress in Math. **47**, Birkhäuser, Boston-Basel-
         Stuttgart, 1984.

[T1]   Taylor, M.J.: On Fröhlich's conjecture for rings of integers of
         tame extensions, Inventiones math. **63**(1981), 41–79.

[T2]   Taylor, M.J.: Mordell-Weil groups and the Galois module struc-
         ture of rings of integers, Illinois J. of Math. **32**(1988), 428–451.

[Th]   Thomason, R.W. : Algebraic K-theory of group scheme actions,
         Proc. of Moore Conference, Ann. of Math. Studies **113**, Prince-
         ton Univ. Press, Princeton 1987.

[Wa]   Wall, C.T.C.: Periodic projective resolutions, Proc. London
         Math. Soc. (3) **39**(1979), 509–553.

[Wat]   Waterhouse, W. : Principal homogeneous spaces of group scheme
         extensions, Trans. Amer. Math. Soc. **153**(1971), 181-189.

LABORATOIRE DE MATHÉMATIQUES PURES, UNIVERSITÉ BORDEAUX 1, 351,
COURS DE LA LIBÉRATION, F-33405 TALENCE CEDEX, FRANCE
erez@math.u-bordeaux.fr

Mixed elliptic motives

ALEXANDER GONCHAROV

Contents

1 Introduction

Summary. Let E be an elliptic curve over an arbitrary field k and $\mathcal{H}$ the motive $H^1(E)(1)$. We define complexes $B(E; n + 2)^\bullet$ and conjecture that they are quasiisomorphic to $RHom_{\mathcal{MM}_k}(\mathbb{Q}, Sym^n \mathcal{H}(1))$. If k is a number field this together with Beilinson's conjecture on regulators lead to a precise conjecture expressing the special values $L(Sym^n E, n + 1)$ via the classical

Reprinted from 'Galois Representations in Arithmetic Algebraic Geometry', edited by A. J. Scholl & R. L. Taylor. ©Cambridge University Press 1998

Eisenstein-Kronecker series. It can be considered as an elliptic analog of Zagier's conjecture.

We give a simple motivic interpretation of the elliptic polylogarithms and show how it together with the motivic formalism implies that the complexes $B(E; n + 2)^\bullet$ should map naturally to $RHom_{\mathcal{MM}_k}(\mathbb{Q}, Sym^n\mathcal{H}(1))$. When E degenerates to the nodal curve our complexes lead to the motivic complexes from [G1-2] reflecting the properties of the classical polylogarithms.

The complex $B(E; 3)^\bullet$ was constructed in [GL]. The groups similar to $H^1B(E; n + 2)^\bullet$ were also discussed in [W], [W2], where it was conjectured that they inject into $Ext^1_{\mathcal{MM}_k}(\mathbb{Q}, Sym^n\mathcal{H}(1))$.

We formulate several conjectures about the category of mixed elliptic motives and its motivic Galois group. Perhaps the most unexpected are conjecture 1.1 about a *small* category of mixed elliptic motives and its generalization to all mixed motives, conjecture 4.12. The others generalize some conjectures about the mixed Tate motives ([G1-2]). In the end we define the generalized Eisenstein-Kronecker series which should be related to $L(Sym^nE, n+m)$. For $n = 1, m = 2$ this was conjectured in [D3] and proved in [G3].

1. A triptych: special values of L-functions, motivic complexes and motivic Galois groups. Let E be an elliptic curve over a number field k and $L(Sym^nE, s)$ the L-function of the n-th symmetric power of $h^1(E)$. The seminal Beilinson conjecture relates the special values of the L-function of a motive X over a number field, considered up to a $\mathbb{Q}^*$-factor, with the volume of the image of certain pieces of the algebraic K-theory of X under the regulator maps. However for symmetric powers of elliptic curves one should be able to say more about the special values.

It is natural to adress the problem in the language of motives. Let $\mathcal{MM}_k$ (resp. $\mathcal{MM}_X$) be the (hypothetical) abelian $\mathbb{Q}$-category of all mixed motives over a field k (resp. all mixed motivic sheaves over a regular scheme X, [Be]). Let $\mathbb{Q}(-1) := h^2(P^1)$ be the Tate motive, $\mathbb{Q}(n) := \mathbb{Q}(1)^{\otimes n}$ for any integer n and $M(n) := M \otimes \mathbb{Q}(n)$.

Let E be an elliptic curve over a field k. Then $h^1(E)$ is a pure motive of weight 1. The cup product defines an isomorphism $\Lambda^2h^1(E) \longrightarrow \mathbb{Q}(-1)$. Set $\mathcal{H} := h^1(E)(1)$. It is a simple object of weight -1. The *elliptic motives* are the direct summands of the motives $\mathcal{H}^{\otimes n}(m)$. They form a rigid abelian tensor category $\mathcal{P}_E$.

Example. If E has no complex multiplication then $\mathcal{P}_E$ is equivalent to the category of finite dimensional rational GL_2-modules over $\mathbb{Q}$. The objects $S^n\mathcal{H}(m)$ are simple and mutually non isomorphic. Any simple object in $\mathcal{P}_E$ is isomorphic to one of them.

The category $\mathcal{M}_E$ of *mixed elliptic motives* is the smallest abelian tensor subcategory of $\mathcal{MM}_k$ which contains the elliptic motives and is closed under extensions.

For an integer a let $L^*(Sym^n E, a)$ be the leading coefficient of the Taylor expansion of the L-function at $s = a$. For a given pair of integers n, m there are the following intimately related problems:

Problem A. *Find explicit formulas for special values of the L-functions of pure elliptic motives, i.e.*

$$L^*(Sym^n E, n + m) \qquad E/k, \quad k \quad \text{is a number field}$$

Problem B. *Construct explicitly elliptic motivic complexes*

$$RHom_{\mathcal{MM}_k}(\mathbb{Q}(0), S^n \mathcal{H}(m)), \quad k \quad \text{is an arbitrary field} \qquad (1)$$

They may be non zero only if $-n - 2m$, the weight of the motive $Sym^n \mathcal{H}(m)$, is negative.

Problem C. *Find a precise description of the Galois group of the category of mixed elliptic motives.*

A weaker version of the Problem A concerns the special values of L-functions up to a $\mathbb{Q}^*$-factor. We will refer to it as Problem A*.

We are making sense of the motivic Ext's in a usual way. Namely, let $E^{(n)}$ be the kernel of the sum map

$$E^{n+1} \longrightarrow E, \qquad (x_1, ..., x_{n+1}) \longmapsto x_1 + ... + x_{n+1} \qquad (2)$$

For a group A living on E^{n+1} the notation A_{sgn} means the part which is alternating under the action of the group S_{n+1}. Beilinson's description of $Ext^i_{\mathcal{MM}_X}(\mathbb{Q}(0), \mathbb{Q}(n))$ implies (see lemma 3.4) that one should have

$$Ext^i_{\mathcal{MM}_k}(\mathbb{Q}(0), S^n \mathcal{H}(m)) = gr^\gamma_{n+m} K_{n+2m-i}(E^{(n)})_{sgn} \otimes \mathbb{Q} \qquad (3)$$

(Here γ is the γ-filtration on the K-groups). If k is a number field these groups are expected to be zero for $i > 1$. Beilinson's regulator map

$$Ext^1_{\mathcal{MM}_k}(\mathbb{Q}(0), S^n \mathcal{H}(m)) \longrightarrow Ext^1_{\mathbb{R}-\mathcal{MHS}}(\mathbb{R}(0), S^n \mathcal{H}(m))$$

is provided by the realization functor from $\mathcal{MM}_k$ to the category $\mathbb{R} - \mathcal{MHS}$ of mixed Hodge structures over $\mathbb{R}$.

Problem A is the deepest one. It is of arithmetic nature, while Problems B and C are geometric. We expect such a solution of Problem C that gives the desired answer for Problem B. This answer in the case of number fields should resolve Problem A*. The regulator map should be constructed first over $\mathbb{C}$, brining analysis into the picture and providing a key for Problem B.

In these problems one can consider the category of mixed motives generated by powers of any simple pure motive X. However the case of mixed elliptic motives seems to be especially interesting.

To smell the flavor of the problems let us look at similar questions for the category of mixed Tate motives over a field k (i.e. the rigid tensor subcategory of $\mathcal{MM}_k$ generated by the Tate motive $\mathbb{Q}(1)$).

Then Problem A is about special values $\zeta_k(n)$ of the Dedekind zeta function of a number field k. The ideal answer to Problem A* is given by Zagier's conjecture [Z2].

In Problem B we are looking for complexes $RHom_{\mathcal{MM}_k}(\mathbb{Q}(0), \mathbb{Q}(n))$ for an arbitrary field k. In [G1],[G2] we have constructed complexes $\mathcal{B}(\mathbb{Q}(n); k)^\bullet$:

$$\mathcal{B}_n(k) \longrightarrow \mathcal{B}_{n-1}(k) \otimes k^* \longrightarrow \dots \longrightarrow \mathcal{B}_2(k) \otimes \Lambda^{n-2}k^* \longrightarrow \Lambda^n k^* \qquad (4)$$

which reflect the properties of *classical n-logarithm function* $Li_n(z)$. Here $\mathcal{B}_n(k)$ is the quotient of $\mathbb{Z}[k^*]$ along a certain subgroup $\mathcal{R}_n$, which in the case $k = \mathbb{C}$ is the subgroup of all functional equations for the classical n-logarithm function. It is placed in degree 1. These complexes $\otimes_{\mathbb{Q}}$ were conjectured to be quasiisomorphic to $RHom_{\mathcal{MM}_k}(\mathbb{Q}, \mathbb{Q}(n))$.

A detailed exposition of this philosophy in the two simplest cases, for the motives $\mathbb{Q}(2)$ and $\mathcal{H}(1)$, is given in chapter 2.

According to the Tannakian formalism the category of mixed Tate motives a field k is equivalent to the category of finite dimensional representations of its motivic Galois group. The motivic Galois group is a semidirect product of $\mathbb{G}_m$ and a prounipotent algebraic group scheme $U(k)$ over $\mathbb{Q}$. Let $L(k)$ be the Lie algebra of $U(k)$. The action of $\mathbb{G}_m$ provides a grading $L(k)_\bullet = \oplus_{n \geq 1} L(k)_{-n}$ (it is negatively graded thanks to a weight argument).

The answer to the Problem C for mixed Tate motives which we have in mind is this. Let $I(k)_\bullet := \oplus_{n \geq 2} L(k)_{-n}$. It is an ideal, and $L(k)_\bullet / I(k)_\bullet = (k_{\mathbb{Q}}^*)^\vee$. (Here $V \to V^\vee$ is a duality between the inductive and projective limits of finite dimensional $\mathbb{Q}$-vector space.)

The Freeness Conjecture. ([G1],[G2]) $I(k)_\bullet$ *is a free graded pro-Lie algebra generated by the groups* $\mathcal{B}_n(k)^\vee$, $n \geq 2$, *sitting in degree n.*

For a more precise version see Conjecture 1.20 in [G2], where we proved that it is equivalent to the description (4) of the motivic complexes.

There are several other candidates for the motivic complexes, for example Bloch's beautiful complexes of higher Chow groups [Bl] and their versions. However the complexes (4) are the smallest possible and the only ones directly related just to the classical polylogarithms. For a "cycle" construction of the motivic Lie algebra see [BK].

In this paper I will formulate an elliptic analog of the freeness conjecture, see conjecture 1.1 below.

2. A small category of mixed elliptic motives and the freeness conjecture for its motivic Galois group. The Tannakian formalism tells us that there exists a pro-Lie algebra $L(E)$ in the tensor category $\mathcal{P}_E$ such that the category of mixed elliptic motives is equivalent to the category of modules over $L(E)$ in the category $\mathcal{P}_E$. There is a different tensor structure $\otimes'$ on the category of pure elliptic motives. Assume that E has no complex multiplication. Then $S^n \mathcal{H}(m) \otimes' S^{n'} \mathcal{H}(m') = S^{n+n'} \mathcal{H}(m + m')$. Denote by $\mathcal{P}_E^*$ the category of pure elliptic motives with *this* tensor structure.

Let J be the set of k-points of the jacobian of E.

Let V be a $\mathbb{Q}$-vector space and M an object of a category $\mathcal{C}$. Then if V is finite dimensional there is an object $V \boxtimes M$ of $\mathcal{C}$ such that $Hom_{\mathcal{C}}(N, V \boxtimes M) = V \otimes Hom_{\mathcal{C}}(N, M)$. If V is an inductive limit of finite dimensional $\mathbb{Q}$-vector spaces then we get an object of the category of Ind-objects in $\mathcal{C}$.

Let L, L^* be objects in $\mathcal{P}_E$ and $p : L \longrightarrow L^*$ is a projection. Suppose L has a structure of a Lie algebra in $\mathcal{P}_E$. Say that the quotient L^* is a Lie algebra with respect of the **both** tensor structures on the category of pure elliptic motives if there is a structure of a Lie algebra in $\mathcal{P}_E^*$ on L^* given by the commutator map $[,]^* : \Lambda^2_{\mathcal{P}_E^*} L \longrightarrow L^*$ and the following natural diagram is commutative:

$$
\begin{array}{ccc}
\Lambda^2_{\mathcal{P}_E} L & \xrightarrow{[,]} & L \\
\downarrow \Lambda^2 p & & \downarrow p \\
\Lambda^2_{\mathcal{P}_E^*} L & \xrightarrow{[,]^*} & L^*
\end{array}
$$

Then there are canonical maps of cohomology groups $H^\bullet_{\mathcal{P}_E^*}(L^*) \longrightarrow H^\bullet_{\mathcal{P}_E}(L)$.

Conjecture 1.1 *Assume that $k = \bar{k}$. Then*

*a) $L(E)$ has a unique quotient $L^*(E)$ which is a pro-Lie algebra with respect to the **both** tensor structures on the category of pure elliptic motives and such that one has*

$$H^\bullet_{\mathcal{P}_E^*}(L^*(E)) = H^\bullet_{\mathcal{P}_E}(L(E))$$

b) There is an ideal $I^*(E) \subset L^*(E)$ with an abelian quotient

$$L^*(E)/I^*(E) = (k^*)^\vee \boxtimes \mathcal{H} \oplus J^\vee \boxtimes \mathcal{H}$$

The pro-Lie algebra $I^*(E)$ should be free in the tensor category $\mathcal{P}_E^*$.

The Lie algebra $L^*(E)$, considered as a Lie algebra in $\mathcal{P}_E$, is the central theme of our story. The tensor category $\mathcal{P}_E^*$ at first seems to be only a tool which allows us to express the properties of the Lie algebra $L^*(E)$ in a neat form. See nevertheless the proof of the theorem 1.2 in s.4.4. The category $\mathcal{M}_E^*$ of finite dimensional modules over $L^*(E)$ in the category $\mathcal{P}_E$ is a subcategory of $\mathcal{M}_E$ which we call the *small category of mixed elliptic motives*. More about the Lie algebra $L^*(E)$ in s. 4.8. The conjectures about $L^*(E)$ imply a very specific structure for the elliptic motivic complexes.

So far the existence of such a Lie algebra $L^*(E)$ is a conjecture. However we proved

Theorem 1.2 *Let us assume for simplicity that E is not a CM curve. Then there exists a differential graded pro-Lie algebra $DL^*(E)$ in the category $\mathcal{P}_E^*$ such that for $(m,n) \neq (0,0)$ one has*

$$H^i_{\mathcal{P}_E^*}(DL^*(E))_{S^n\mathcal{H}(m)^\vee} = gr^\gamma_{n+m}K_{n+2m-i}(E^{(n)})_{sgn} \otimes \mathbb{Q}$$

Here V_M is the M-isotypical component of a simple object V in $\mathcal{P}_E^*$. The DG pro-Lie algebra $DL^*(E)$ is constructed in s.4.4. This result does not assume any conjecture on mixed motives.

Assuming the following vanishing conjecture (a version of the Beilinson-Soule conjecture; the left hand side is defined via (3))

$$Ext^i_{\mathcal{MM}_k}(\mathbb{Q}(0), S^n\mathcal{H}(m)) = 0 \quad \text{for } i < 0$$

one can show that $H^j_\partial(DL^*(E)) = 0$ for $j < 0$. Here ∂ is the differential in $\tilde{L}^*(E)$.

Conjecture 1.3 *Assume $k = \bar{k}$.*
   a) $H^j_\partial(DL^*(E)) = 0$ for $j \neq 0$.
   b) $H^0_\partial(DL^*(E))$ is isomorphic to the pro-Lie algebra $L^*(E)$ anticipated in conjecture 1.1.

Remark. Let G be a reductive group. Then there is a new tensor structure the category of finite dimensional G-modules given by $V_\lambda \otimes V_\mu \xrightarrow{=} V_{\lambda+\mu}$, where V_λ is the G-module with the highest weight λ. Notice that the category of all pure motives is equivalent to the category of finite dimensional modules over a pro-reductive group.

The Lie coalgebra $\mathcal{L}^*(E)$ has the following canonical filtration. Set

$$\mathcal{L}^*_{\leq n}(E) := \oplus_{0 \leq i \leq n} \tilde{\mathcal{L}}^*(E)_{S^i\mathcal{H}(m)} \boxtimes S^i\mathcal{H}(m)^\vee$$

It is clearly a Lie subcoalgebra of $\mathcal{L}^*(E)$ in the tensor category $\mathcal{P}^*(E)$. Therefore there is a natural filtration by the Lie subcoalgebras:

$$\mathcal{L}^*_{[0]}(E) := \mathcal{L}(k) \subset \mathcal{L}^*_{\leq 1}(E) \subset \ldots \subset \mathcal{L}^*_{\leq n}(E) \subset \ldots$$

$\mathcal{L}^*_{\leq n}(E)$ is also a Lie subcoalgebra of $\mathcal{L}(E)$. Let $\mathcal{M}^*_{\leq n}(E)$ be the category of comodules over $\mathcal{L}^*_{\leq n}(E)$ in the category $\mathcal{P}_E$. Then $\mathcal{M}^*_E$ is a tensor subcategory of $\mathcal{M}_E$. We will construct in chapter 5, assuming some conjectures on mixed Tate motives, a Lie coalgebra which should be isomorphic to $\mathcal{L}^*_{\leq 1}(E)$.

   3. The complexes $B(E,n)^\bullet$. In chapters 7-9 we study Problems A^* and B in the case $m = 1$. We construct a complex $B(E, n + 2)^\bullet$ which is conjecturaly quasiisomorphic to $RHom_{\mathcal{MM}_k}(\mathbb{Q}, Sym^n\mathcal{H}(1))$. When E is an elliptic curve over a number field k this leads to a precise conjecture on special values $L(Sym^nE, n + 1)$.

This complex is an *elliptic deformation* of the complex (4): if E is a nodal curve and $k = \bar{k}$, it is quasiisomorphic to (4).

Remark. A better notation for the group $B_{n+2}(E)$ and the complex $B(E, n+2)^\bullet$ would be $B_{S^n \mathcal{H}(1)}$ and $B(S^n \mathcal{H}(1))^\bullet$.

Let us assume first that $k = \bar{k}$. Let $\mathbb{Z}[X]$ be the free abelian group generated by a set X, $\{x\}$ are the generators. We will define subgroups

$$R_n(E/k) \subset \mathbb{Z}[E(k)], \quad \text{and set} \quad B_n(E) := \frac{\mathbb{Z}[E(k)]}{R_n(E/k)}$$

Put formally $B_0(E) = \mathbb{Z}$. By definition $R_1(E)$ is generated by the elements $\{x\} + \{y\} - \{x + y\}$ where $x, y \in E(k)$. So $B_1(E) = J(k)$, where J is the Jacobian of E.

Define a homomorphism $\delta : \mathbb{Z}[E(k)] \longrightarrow \mathbb{Z}[E(k)] \otimes J(k)$ by the formula $\{a\} \longmapsto \{a\} \otimes a$. We prove in chapter 7 that $\delta(R_n(E)) \subset R_{n-1}(E) \otimes J$, so we get a homomorphism $\delta : B_n(E) \longrightarrow B_{n-1}(E) \otimes J$. Consider the following cohomological complex

$$B_n(E) \longrightarrow B_{n-1}(E) \otimes J \longrightarrow \dots \longrightarrow B_1(E) \otimes \Lambda^{n-1} J \longrightarrow B_0(E) \otimes \Lambda^n J \quad (5)$$

where the differential is given by the formula

$$\delta : \{a\} \otimes b_1 \wedge b_2 \wedge \dots \wedge b_m \longmapsto \{a\} \otimes a \wedge b_1 \wedge b_2 \wedge \dots \wedge b_m$$

It is the complex $B(E, n+1)^\bullet$. We put it in degrees $[1, n+1]$. The complex is acyclic in the last two terms. To emphasize dependency on the ground field k we use a notation $B(E/k, n+1)^\bullet$. If k is not an algebraically closed field we postulate the Galois descent property: $B(E/k, n)^\bullet := \left(B(E/\bar{k}, n)^\bullet \right)^{Gal(\bar{k}/k)}$ The complexes $B(E/k, n)^\bullet$ for $n = 2, 3$ were constructed in [GL].

Conjecture 1.4 *a) There exists a canonical homomorphism*

$$H^i \left(B(E/k, n+2)^\bullet_\mathbb{Q} \right) \longrightarrow gr^\gamma_{n+1} K_{n+2-i}(E^{(n)})_{sgn} \otimes \mathbb{Q} \quad (6)$$

b) This homomorphism is an isomorphism.

Here "canonical" means compatibility with the regulator map, see below. The most nontrivial is the "surjectivity conjecture" from part b).

In [W] J. Wildeshaus, assuming standard conjectures on mixed motives, gave a conjectural inductive definition of groups similar to $H^1 B(E, n+2)^\bullet$ and formulated a conjecture similar to the part a) of conjecture (1.4) for $i = 1$. See also some new developments in this direction in the recent preprint [W2].

In chapter 9 we construct, assuming the standard conjectures, groups $\mathcal{B}_n(E)$. One has a surjective map $B_n(E) \longrightarrow \mathcal{B}_n(E)$ which is hope to be an isomorphism. These groups form a complex $\mathcal{B}(E;n)^{\bullet}$ similar to the complex (5). We show that the motivic formalism implies that (for $k = \bar{k}$) this complex maps to the standard cochain complex of the motivic Lie algebra $L(E)$, providing a morphism of complexes (for an arbitrary field k)

$$B(E/k, n+2)^{\bullet}_{\mathbb{Q}} \longrightarrow RHom_{\mathcal{M}\mathcal{M}_k}(\mathbb{Q}(0), S^n\mathcal{H}(1)) \qquad (7)$$

Conjecture 1.5 *The morphism (7) is a quasiisomorpism.*

This conjecture implies conjecture (1.4). See sections 4.8–4.9.

4. The Eisenstein-Kronecker series, groups $B_{n+1}(E)$ and regulators. The group $R_{n+1}(E/\mathbb{C})$ is a subgroup of the functional equations for the elliptic n-logarithm studied by Bloch for $n = 2$ [Bl] and by Beilinson and Levin [BL] in general. Its real version is the Eisenstein-Kronecker series known from the XIX-th century [We]. These are the special functions we need in our conjecture on $L(Sym^nE, n+1)$. Let us say a few words about them.

Let E be an elliptic curve over $\mathbb{C}$. Choose a holomorphic differential ω on $E(\mathbb{C})$. Let $\Gamma \in \mathbb{C}$ be the lattice of periods of ω. We will always normalize ω in such a way that $\Gamma := \mathbb{Z} \oplus \mathbb{Z}\tau$ where $Im\tau > 0$. The differential ω defines via the Abel-Jacobi map an isomorphism $E(\mathbb{C}) \longrightarrow \mathbb{C}/\Gamma$. The intersection form on $\Gamma = H_1(E(\mathbb{C}), \mathbb{Z})$ provides a pairing

$$(\cdot, \cdot) : E(\mathbb{C}) \times \Gamma \longrightarrow S^1; \qquad (z, \gamma) := exp(\frac{2\pi i(z\bar{\gamma} - \bar{z}\gamma)}{\tau - \bar{\tau}}) \qquad (8)$$

Consider the Eisenstein-Kronecker series

$$K_{i,j}(z; \tau) = \sum_{\gamma \in \Gamma \backslash 0} \frac{(z, \gamma)}{\gamma^i \bar{\gamma}^j}, \quad i, j \geq 1$$

There is a homomorphism $K_n : \mathbb{Z}[E(\mathbb{C})] \longrightarrow Sym^{n-2}H^1_B(E(\mathbb{C}), \mathbb{C})$ given by the formulas

$$\{z\} \longmapsto \sum_{a+b=n} K_{a,b}(z) \cdot (dz)^{a-1}(d\bar{z})^{b-1}$$

Theorem (8.2) below claims that it sends the subgroup $R_n(E(\mathbb{C}))$ to zero.

Suppose E is defined over $\mathbb{R}$. Choose a differential $\omega \in \Omega^1(E_{/\mathbb{R}})$ over $\mathbb{R}$. Then the lattice of periods Γ is invariant under the action of complex conjugation $z \longmapsto \bar{z}$ and the isomorphism $E(\mathbb{C}) \longrightarrow \mathbb{C}/\Gamma$ is compatible with complex conjugation.

For any lattice Γ one has $\bar{K}_{i,j}(z; \tau) = (-1)^{i+j}K_{j,i}(z; \tau)$. If a lattice Γ and a divisor P on $\mathbb{C}/\Gamma$ are invariant under complex conjugation, then $K_{i,j}(P; \tau) \in \mathbb{R}$. In this situation (+ means invariants under the complex conjugation acting on $E(\mathbb{C})$ and $\mathbb{R}(1)$) we find a map

$$K_n : B_n[E(\mathbb{C})]^+ \longrightarrow Sym^{n-2}H^1(E(\mathbb{C}), \mathbb{R}(1))^+$$

Restricting K_n to $Ker(B_n(E/\mathbb{C}) \to B_{n-1}(E/\mathbb{C}) \otimes J(\mathbb{C}))$ and assuming conjecture (1.4) we should get the regulator map

$$gr_{n-1}^\gamma K_{n-1}(E^{(n-2)})_{sgn} \otimes \mathbb{Q} \longrightarrow Sym^{n-2} H^1(E(\mathbb{C}), \mathbb{R}(1))^+$$

This together with Beilinson's conjecture on regulators implies a precise conjecture on $L(Sym^n E, n+1)$ for elliptic curves over number fields, see chapter 8.

5. The structure of the paper. In chapter 2 we recall explicit construction of the Bloch-Suslin complex ([S]) which essentially computes the group $RHom_{\mathcal{MM}_k}(\mathbb{Q}(0), \mathbb{Q}(2))$ and its elliptic analog ([GL]) which does the job for $RHom_{\mathcal{MM}_k}(\mathbb{Q}(0), \mathcal{H}(1))$. These constructions and results in the special case when k is a number field lead to explicit formulas for the special values of the Dedekind ζ-function and the L-function of elliptic curves at $s = 2$. In chapter 3 we spell out the basic formalism of motivic Galois groups and motivic Lie algebras. These chapters are expository.

In chapter 4 we prove theorem 1.1 and formaluate some conjectures on the structure of the motivic Lie algebra of the category of mixed elliptic motives. Let me mention three of them: conjecture 4.2 on the Ext groups in the category of mixed elliptic motives, conjecture 4.3 about the small motivic Lie algebra $L^*(E)$ and the freeness conjecture 4.10 for $L^*(E)$.

In chapter 7 we construct the complexes $B(E, n+2)^\bullet$ and $B^*(E, n+2)^\bullet$ (they are canonically quasiisomorphic). In chapter 8 a conjecture on $L(Sym^n E, n+1)$ is formulated. In chapter 9 we show that restriction of the Eisenstein-Kronecker series K_{n+1} to the $2n$-th power of the augmentation ideal in $\mathbb{Z}[E]$ is a real period of an explicitly constructed mixed elliptic motive. This motivic interpretaion seems to be quite different (and simpler) then the one suggested in [BL]. We show how this plus the motivic formalism of chapter 3 allow us to deduce conjecture 1.4a) from standard conjectures on mixed motives. In chapter 10 we define the single valued and multivalued elliptic Chow polylogarithms and introduce the generalized Eisenstein-Kronecker series. We hope to pursue further the material of this chapter in future.

Acknowledgement. Chapter 7–8 of this paper were written during my stay in the IHES in the Summer of 1995 and circulated as [G5]. The paper was finished when I enjoyed the hospitality of MPI(Bonn) and University Paris XI at Orsay. I am grateful to IHES and MPI and University Paris XI for hospitality and support. This work was partially supported by NSF grant DMS–9500010.

I am grateful to S. Bloch, A. Levin and V. Schechtman for useful discussions and to N. Schappacher for useful remarks about the text. I also grateful to the referee for useful remarks.

Leitfaden

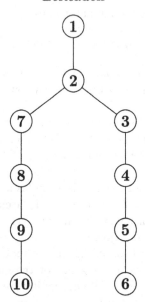

2   Two basic examples: a survey

1. The $\mathbb{Q}(1)$-story: the Bloch-Suslin complex and $\zeta_F(2)$. i) Let $R_{\mathbb{Q}(2)}(k)$ be the subgroup of $\mathbb{Z}[k^*]$ generated by elements $\sum_i(-1)^i\{r(x_1,...,\hat{x}_i,...,x_5)\}$, where x_i runs through all 5-tuples of distinct points over k on the projective line, and r is the cross ratio. The Bloch-Suslin complex $B(\mathbb{Q}(2);k)^\bullet$ for an arbitrary field k is defined as follows:

$$B_{\mathbb{Q}(2)}(k) \xrightarrow{\ \delta\ } \Lambda^2 k^*; \qquad B_{\mathbb{Q}(2)}(k) := \frac{\mathbb{Z}[k^*]}{R_{\mathbb{Q}(2)}(k)}; \qquad \delta : \{x\} \longmapsto (1-x) \wedge x$$

We put $B_{\mathbb{Q}(2)}(k)$ in degree 1. Then one should have a canonical isomorphism in the derived category

$$B(\mathbb{Q}(2);k)^\bullet \overset{qis}{=} RHom_{\mathcal{M}\mathcal{M}_k}(\mathbb{Q}(0),\mathbb{Q}(2))$$

The main reason for this are the following results describing the relation of the cohomology of the complex $B(\mathbb{Q}(2);k)^\bullet \otimes \mathbb{Q}$ with weight 2 motivic cohomology.

According to Matsumoto's theorem $H^2 B(\mathbb{Q}(2);k)^\bullet = K_2(k)$.

Suslin proved ([S], see also [DS]) that

$$H^1 B(\mathbb{Q}(2);k)^\bullet \otimes \mathbb{Q} = K_3^{ind}(k) \otimes \mathbb{Q} = gr_2^\gamma K_3(k) \otimes \mathbb{Q}$$

In fact, Suslin proved in [S] that the following sequence is exact:

$$0 \longrightarrow \tilde{T}or(k^*, k^*) \longrightarrow K_3^{ind}(k) \longrightarrow B_{\mathbb{Q}(2)}(k) \xrightarrow{\delta} \Lambda^2 k^* \longrightarrow K_2(k) \longrightarrow 0$$

Here $\tilde{T}or(k^*, k^*)$ is a nontrivial extension of $Tor(k^*, k^*)$ by $\mathbb{Z}/2\mathbb{Z}$.

Finally, there exists a canonical morphism in the derived category

$$B(\mathbb{Q}(2); k)^\bullet \otimes \mathbb{Q} \longrightarrow \tau_{\geq 1} RHom_{\mathcal{MM}_k}(\mathbb{Q}(0), \mathbb{Q}(2))$$

inducing the isomorphism on H^1 and H^2. The right hand side can be understood, for example, as the complex of Bloch's higher Chow groups. The existence of such a morphism follows also from the motivic philosophy and the motivic construction of the dilogarithm in [BGSV]. According to the Beilinson-Soulé vanishing conjecture one should have

$$gr_2^\gamma K_{4+i}(k) \otimes \mathbb{Q} = 0 \quad \text{for } i \geq 0$$

One can reformulate this by saying that the canonical morphism

$$RHom_{\mathcal{MM}_k}(\mathbb{Q}(0), \mathbb{Q}(2)) \longrightarrow \tau_{\geq 1} RHom_{\mathcal{MM}_k}(\mathbb{Q}(0), \mathbb{Q}(2))$$

is a quasiisomorphism

The Bloch-Suslin complex is definitely "the smallest possible" representative of the motivic complex $RHom_{\mathcal{MM}_k}(\mathbb{Q}(0), \mathbb{Q}(2))$. I think it is also "the best possible", but this is of cource matter of taste. It gives a solution of the Problem B for the motive $\mathbb{Q}(2)$ which we have in mind.

ii) Now let $k = \mathbb{C}$. Recall that the dilogarithm is the following multivalued analytic function of on $\mathbb{C}P^1 \setminus \{0, 1, \infty\}$):

$$Li_2(z) = -\int_0^z \log(1 - t)\frac{dt}{t}$$

It has a single-valued version, the Bloch-Wigner function:

$$\mathcal{L}_2(z) := ImLi_2(z) + \arg(1 - z) \cdot \log|z|$$

Let

$$\tilde{\mathcal{L}}_2 : \mathbb{Z}[\mathbb{C}^*] \longrightarrow \mathbb{R}, \quad \{z\} \longmapsto \mathcal{L}_2(z)$$

Then one can show that $\tilde{\mathcal{L}}_2(R_{\mathbb{Q}(2)}(\mathbb{C})) = 0$, so we get a well defined homomorphism

$$\tilde{\mathcal{L}}_2 : B_{\mathbb{Q}(2)}(\mathbb{C}) \longrightarrow \mathbb{R}$$

Theorem 2.1 *The restriction of the homomorphism $\tilde{\mathcal{L}}_2$ to the subspace $H^1 B(\mathbb{Q}(2); k)^\bullet \otimes \mathbb{Q} = K_3^{ind}(\mathbb{C}) \otimes \mathbb{Q}$ coincides with the Borel regulator on $K_3^{ind}(\mathbb{C}) \otimes \mathbb{Q}$.*

Here "coincides" means coincides up to a *known* rational factor depending on a normalization of the Borel class we choose. A simple direct proof of this result see in [G1].

Finally, if $k = F$ is a number field combining the two theorems above with the Borel theorem we get the solution of the Problem A* for $\zeta_F(2)$:

Theorem 2.2 *a) There exists elements* $y_1, ..., y_r \in H^1 B(\mathbb{Q}(2); F)^\bullet \otimes \mathbb{Q}$ *such that*

$$q \cdot \zeta_F(2) = \pi^{2(r_1+r_2)} d_F^{-1/2} det\left(\tilde{\mathcal{L}}_2(\sigma_i(y_j))\right)$$

for certain $q \in \mathbb{Q}^*$.

b) For any $y_1, ..., y_r \in H^1(B(\mathbb{Q}(2); F)^\bullet) \otimes \mathbb{Q}$ *one has the formula above with* $q \in \mathbb{Q}$.

To solve the Problem A one should be able to determine *effectively* the constant $q(y_1 \wedge ... \wedge y_r)$. I do not know how to do this. So we see that the analysis (the dilogarithm) indeed suggested a key for the solution of the Problem B, which leads to a solution of the Problem A*. The relation with the Problem C was explained in details in [G2] and will be scketched briefly in section 4.

Remark. The standard notation for the groups $B_{\mathbb{Q}(n)}(k)$ is $B_n(k)$.

2. The $\mathcal{H}(1)$-story: the elliptic deformation of the Bloch-Suslin complex and $L(E, 2)$ ([GL]) . Let us suppose that k is an algebraically closed field. Let I_E be the augmentaion ideal of the group algebra $\mathbb{Z}[E(k)]$, and I_E^4 its fourth power. Let B_3^* be the quotient of I_E^4 by the subgroup generated by the elements $(f) * (1 - f)^-$, where $*$ is the convolution in the group algebra $\mathbb{Z}[E]$, $f \in k(E)^*$, and $g^-(t) := g(-t)$. Then there is a map

$$\delta^*_{\mathcal{H}(1)} : B_3^*(E) \longrightarrow k^* \otimes J \qquad (9)$$

whose construction will be outlined below. The complex we get putting the left group in the degree 1 is called the complex $B^*(E, 3)^\bullet$. This is our candidate for $RHom_{\mathcal{MM}_k}(\mathbb{Q}(0), \mathcal{H}(1))$.

If $k = \mathbb{C}$ we can "define" $\delta^*_{\mathcal{H}(1)}$ as follows:

$$\sum n_i\{a_i\} \longmapsto \sum n_i\theta(a_i) \otimes a_i$$

where

$$\theta(z) = \prod_{n \in \mathbb{Z}}(1 - q^n z) := q^{1/12}z^{-\frac{1}{2}}\prod_{j \geq 0}(1 - q^j z)\prod_{j > 0}(1 - q^j z^{-1})$$

To make sense out of this formula for arbitrary field k we proceed as follows (see also s. 4.6 of [GL]).

i. The group $B_2(E)$. Let E be an elliptic curve over an arbitrary field k. In s.2.1-2.2 of [GL] we defined a group $B_2(E/k) = B_2(E)$ which fits in the following diagram

$$0$$

$$\uparrow$$

$$0 \longrightarrow \mathbb{G}_m(k) \longrightarrow B_2(E/k) \overset{p}{\longrightarrow} S^2 J(k) \longrightarrow 0$$

$$\uparrow \theta$$

$$\mathbb{Z}[E(k)\backslash 0]$$

such that $p \circ \theta : \{a\} \longmapsto a \cdot a$. here the horisontal sequence is exact, and the vertical is exact at least if $k = \bar{k}$. The map θ is defined modulo 6-torsion.

Moreover, if K is a local field there exists a canonical homomorphism

$$h : B_2(E/K) \longrightarrow \mathbb{R}$$

whose restriction to the subgroup $K^* \subset B_2(E/K))$ is given by $x \longmapsto \log|x|$, (see s. 2.3). Let h_K be the canonical local height. Then $-h_K$ is given by the composition

$$\mathbb{Z}[E(K)\backslash 0] \overset{\theta}{\longrightarrow} B_2(E/K) \overset{h}{\longrightarrow} \mathbb{R}$$

The group $B_2(E)$ appears naturally from the theory of biextensions. Namely, let $\mathcal{P}$ be the rigidified Poincare line bundle over $J \times J$: its fiber over $(0,0)$ is identified with k^*. The restriction of $\mathcal{P}$ to $x \times J$ (resp. $J \times y$) minus the zero section has canonical structure of a commutative algebraic group over k. It is the extension of J by $\mathbb{G}_m$ corresponding to x (resp y). This means that for every point $(x, y) \in J \times J$ we have a k^* torsor $T_{(x,y)}$ together with morphisms of k^*-torsors

$$T_{(x_1,y)} \otimes_{k^*} T_{(x_2,y)} \longrightarrow T_{(x_1+x_2,y)} \qquad T_{(x,y_1)} \otimes_{k^*} T_{(x,y_2)} \longrightarrow T_{(x,y_1+y_2)}$$

providing the group structure "in horizontal and vertical directions" such that the diagram

$$T_{(x_1,y_1)} \times T_{(x_2,y_1)} \times T_{(x_1,y_2)} \times T_{(x_2,y_2)} \longrightarrow T_{(x_1,y_1+y_2)} \times T_{(x_2,y_1+y_2)}$$

$$\downarrow \qquad\qquad\qquad\qquad\qquad\qquad \downarrow$$

$$T_{(x_1+x_2,y_1)} \times T_{(x_1+x_2,y_2)} \longrightarrow T_{(x_1+x_2,y_1+y_2)}$$

is commutative. Then by definition the k^*-torsor $p^{-1}(x \cdot y)$ is given by $T_{(x,y)}$. The horizontal sequence in the diagram above is just a different way to spell these properties of the torsors $T_{(x,y)}$. For the defenition of the map θ see [GL]. Roughly speacking it comes from a section of the restriction of the Poincare line bundle to the diagonal deleted at zero. Set $\{a\}_2 := \theta(\{a\}) \in B_2(E)$.

ii) The group $B_3^*(E)$. Consider the homomorphism $(J := J(k))$

$$\delta_{\mathcal{H}(1)} : \mathbb{Z}[E(k)] \longrightarrow B_2(E) \otimes J, \quad \{a\} \longrightarrow -\frac{1}{2}\{a\}_2 \otimes a$$

Let R_3^* be the subgroup of I^4 given by the elements $(f) * (1 - f)$ when $f \in k(E)^*$. Let $R_3 \subset \mathbb{Z}[E(k)]$ be the subgroup generated by R_3^* and the distribution relations

$$m \cdot (\{a\} - m \cdot \sum_{mb=a} \{b\}), \quad a \in E(k), \quad m = -1, 2$$

Then $\delta_{\mathcal{H}(1)}(R_3(E)) = 0$ by theorem (3.3) in [GL]. Setting

$$B_3(E) := \frac{\mathbb{Z}[E(k)]}{R_3(E)}$$

we get a homomorphism $\delta_{\mathcal{H}(1)} : B_3(E) \longrightarrow B_2(E) \otimes J$. We define $B_3^*(E)$ as the quotient of I_E^4 by the subgroup $R_3^*(E)$. There is a map $i : B_3^*(E) \longrightarrow B_3(E)$ provided by the inclusion $i : I_E^4 \hookrightarrow \mathbb{Z}[E]$. We define $\delta_{\mathcal{H}(1)}$ as a morphism making commutative the diagram

$$
\begin{array}{ccc}
B_3^*(E) & \xrightarrow{\delta^*_{\mathcal{H}(1)}} & k^* \otimes J \\
\downarrow & & \downarrow \\
B_3(E) & \xrightarrow{\delta_{\mathcal{H}(1)}} & B_2(E) \otimes J
\end{array}
$$

Let us consider the following complex

$$B(E;3)^{\bullet} : \qquad B_3(E) \xrightarrow{\delta_{\mathcal{H}(1)}} B_2(E) \otimes J \longrightarrow J \otimes \Lambda^2 J \longrightarrow \Lambda^3 J \qquad (10)$$

Here the middle arrow is $\{a\}_2 \otimes b \longmapsto a \otimes a \wedge b$ and the last one is the canonical projection. The complex is placed in degrees [1, 4]. It is acyclic in the last two terms. It was proved in [GL] that we get the commutative diagram:

$$
\begin{array}{ccccccc}
0 & & 0 & & & & \\
\downarrow & & \downarrow & & & & \\
B_3^*(E) & \xrightarrow{\delta^*_{\mathcal{H}(1)}} & k^* \otimes J & & & & \\
\downarrow & & \downarrow & & & & \\
B_3(E) & \xrightarrow{\delta_{\mathcal{H}(1)}} & B_2(E) \otimes J & \longrightarrow & J \otimes \Lambda^2 J & \longrightarrow & \Lambda^3 J \\
\downarrow & & \downarrow & & \downarrow = & & \downarrow = \\
S^3 J & \longrightarrow & S^2 J \otimes J & \longrightarrow & J \otimes \Lambda^2 J & \longrightarrow & \Lambda^3 J \\
\downarrow & & \downarrow & & & & \\
0 & & 0 & & & &
\end{array}
$$

where the vertical sequences are exact, and the bottom one is the Koszul complex, and thus also exact. So $B^*(E;3)^\bullet$ is canonically quasiisomorphic to $B(E;3)^\bullet$. If $k \neq \bar{k}$ we postulate the Galois descent:

$$B^*(E/k;3)^\bullet_{\mathbb{Q}} := B^*(E/\bar{k};3)^{\bullet \, Gal(\bar{k}/k)}_{\mathbb{Q}}$$

iii) The elliptic dilogarithm, algebraic K-theory and regulator on $K_2(E/\mathbb{C})$. Let $E(\mathbb{C}) = \mathbb{C}^*/q^{\mathbb{Z}}$ be the complex points of an elliptic curve E. Here $q := exp(2\pi i \tau), Im\tau > 0$.

$$\mathcal{L}_{2,q}(z) := \sum_{n \in \mathbb{Z}} \mathcal{L}_2(q^n z), \qquad \mathcal{L}_{2,q}(z^{-1}) = -\mathcal{L}_{2,q}(z)$$

(The series converges since $\mathcal{L}_2(z)$ has a singularity of type $|z| \log |z|$ near $z = 0$ and $\mathcal{L}_2(z) = -\mathcal{L}_2(z^{-1})$). When $k = \mathbb{C}$ the group $R_3(E)$ is the subgroup of functional equations for the elliptic dilogarithm. (The rigidity conjecutre tells us that $R_3(E)$ should give *all* of them). In particular the homomorphism

$$\tilde{\mathcal{L}}_{2,q} : \mathbb{Z}[E(\mathbb{C})] \longrightarrow \mathbb{R}, \quad \{a\} \longmapsto \mathcal{L}_{2,q}(a)$$

annihilates the subgroup $R_3(E/\mathbb{C})$.

Let $K(E)^-$ be the minus part of the action of the involution $x \to -x$, $x \in E$.

Theorem 2.3 [GL] *a) Let $k = \bar{k}$. Then there is a sequence*

$$0 \longrightarrow Tor(k^*, J) \longrightarrow gr_2^\gamma K_2(E)^- \longrightarrow B_3^*(E) \longrightarrow k^* \otimes J \longrightarrow gr_2^\gamma K_1(E)^- \to 0$$

It is exact modulo 2-torsion (and exact in $k^* \otimes J$).
b) Now let $k = \mathbb{C}$. Then the composition

$$gr_2^\gamma K_2(E)^- \longrightarrow B_3^*(E) \xrightarrow{\tilde{\mathcal{L}}_{2,q}} \mathbb{R}$$

coincides (up to a known multiple) with the Bloch-Beilinson regulator map.

Since we have the Galois descent on $gr_2^\gamma K_i(E)^-_{\mathbb{Q}}$, we immediately get

Corollary 2.4 *Let E be a curve over an arbitrary field k. Then*

$$H^1(B^*(E;3)^\bullet) = gr_2^\gamma K_2(E)^-_{\mathbb{Q}}, \qquad H^2(B^*(E;3)^\bullet) = gr_2^\gamma K_1(E)^-_{\mathbb{Q}}$$

Using the complex $B(E,3)$ instead of $B^*(E,3)$ the following lemma makes the conditions on the divisor P effectively computable (see a numerical example in s.1.3 of [GL]) .

Lemma 2.5 *Let E be an elliptic curve over a number field F. Then an F-rational divisor $P = \sum n_j(P_j)$ over $\bar{\mathbb{Q}}$ belongs to $Ker\left(B_3(E) \longrightarrow B_2(E) \otimes J\right)$ if and only if it satisfy the following conditions:*

a) $$\sum n_j P_j \otimes P_j \otimes P_j = 0 \quad in \quad S^3 J(\bar{\mathbb{Q}}) \tag{11}$$

b) For any valuation v of the field $F(P)$ generated by the coordinates of the points P_j

$$\sum n_j h_v(P_j) \cdot P_j = 0 \quad in \quad J(\bar{\mathbb{Q}}) \otimes \mathbb{R} \tag{12}$$

In particular for any field k the group of k-rational divisors on $E(\bar{\mathbb{Q}})$ satisfying the conditions a)-b) above maps surjectively to $gr_2^7 K_2(E_{/k})_{\bar{\mathbb{Q}}}^-$, and when k is a number field this map is compatible with the regulator map in the obvious way.

Combining lemma 2.5 with Beilinson's conjecture on regulators we come to an explicit formula for $L(E, 2)$, which we formulate only for $k = \mathbb{Q}$ leaving the general case to the reader.

Conjecture 2.6 *Let E be an elliptic curve over $\mathbb{Q}$. Then for any $\mathbb{Q}$-rational divisor P on $E(\bar{\mathbb{Q}})$ satisfying the conditions (11) and (12), and an integrality condition at each prime p where E has a split multiplicative reduction (see 57) one has*

$$\pi \tilde{\mathcal{L}}_{2,q}(P) = q \cdot L(E, 2) \quad for \; a \; certain \; q \in \mathbb{Q}$$

Adding to the game Beilinson's results on regulators for modular curves we proved this formula for modular elliptic curves over $\mathbb{Q}$ ([GL]):

Theorem 2.7 *Let E be a modular elliptic curve over $\mathbb{Q}$. Then there exists a $\mathbb{Q}$-rational divisor P on $E(\bar{\mathbb{Q}})$ satisfying the conditions of conjecture above such that*

$$\pi \tilde{\mathcal{L}}_{2,q}(P) = q \cdot L(E, 2) \quad for \; a \; certain \; q \in \mathbb{Q}^*$$

So for the motive $\mathcal{H}(1)$ we see the same kind of relationship between the Problems A* and B. It remains to see the role of the motivic Galois group of the category of mixed elliptic motives, i.e. to understand

Why the motivic complex $RHom_{\mathcal{M}\mathcal{M}_k}(\mathbb{Q}(0), \mathcal{H}(1))$ has the shape (9), how it reflects the structure of the motivic Galois group, and what it tells us about the motivic Galois group?

In particular, how to define the group $B_3^*(E)$ it terms of the motivic Lie algebra of the category $\mathcal{M}_E$? The answers are given in chapter 4 below.

3 Mixed motives and motivic Lie algebras

1. Categories of mixed motives. Let $\mathcal{MM}_X$ be the (hypothetical) abelian category of all mixed motivic sheaves over a regular scheme X. When $X = Spec(k)$, k is an arbitrary field, we get the category $\mathcal{MM}_k$. We will assume that it satisfies all the expected properties conjectured by Beilinson [Be]. In particular any object of $\mathcal{MM}_X$ has a canonical increasing weight filtration $W_{\bullet}$; morphisms are strictly compatible with $W_{\bullet}$. We will ignore the fact that existence of such an abelian category is not known yet.

Let $\pi : X \longrightarrow Spec(k)$ be the structure morphism. There are the Tate sheaves $\mathbb{Q}(n)_X := \pi^*\mathbb{Q}(n)_{Spec(k)}$ which we usually denote simply by $\mathbb{Q}(n)$. The basic conjecture is Beilinson's description of Ext's between them:

Conjecture 3.1

$$Ext^i_{\mathcal{MM}(X)}(\mathbb{Q}(0)_X, \mathbb{Q}(n)_X) = gr^\gamma_n K_{2n-i}(X) \otimes \mathbb{Q}$$

Consider the category of pure motives over a field k. One can have in mind Grothendieck's category of motives with morphisms given in terms of the Chow correspondences modulo *numerical* equivalence. We will assume that it is a semisimple abelian category.

Now let $\mathcal{P}$ be a rigid tensor subcategory of the category of pure motives, and $\mathcal{M}_{\mathcal{P}}$ the tensor category of mixed motives generated by $\mathcal{P}$. This means that $\mathcal{M}_{\mathcal{P}}$ is closed under extensions, contains $\mathcal{P}$ as a full subcategory and the weight graded quotients of any object of $\mathcal{M}_{\mathcal{P}}$ belong to $\mathcal{P}$.

Examples. 1. $\mathcal{P}$ is the category of pure Tate motives. $\mathcal{M}_{\mathcal{P}}$ is the category of mixed Tate motives.

2. $\mathcal{P}$ is the category of pure elliptic motives. Then $\mathcal{M}_{\mathcal{P}}$ is the category of mixed elliptic motives.

3. $\mathcal{P}$ is the category of all pure motives over k, so $\mathcal{M}_{\mathcal{P}}$ is the category of all mixed motives.

There is a canonical fiber functor

$$\Psi : \mathcal{M}_{\mathcal{P}} \longrightarrow \mathcal{P}, \quad M \longmapsto \oplus_{n\in\mathbb{Z}} gr^W_n M$$

Let us axiomatise this situation. Namely, let $\mathcal{P}$ be an abelian semisimple rigid tensor $\mathbb{Q}$-category . We will say that $\mathcal{M}_{\mathcal{P}}$ is a mixed category over $\mathcal{P}$ if it is an abelian rigid tensor $\mathbb{Q}$-category containing $\mathcal{P}$ as a full subcategory with the following properties:

1. Any object M of $\mathcal{M}_{\mathcal{P}}$ carries a canonical finite filtration $W_{\bullet}M$ (the weight filtration).

2. Morphisms are strictly compatible with $W_{\bullet}$.

3. The graded objects $gr^W_i M$ belong to $\mathcal{P}$.

4. $Hom_{\mathcal{M}_{\mathcal{P}}}(M, N)$ are finite dimensional.

Remark. We can assume also that $\mathcal{M_P}$ is an F-category where F is an arbitrary field of characteristic zero. This is essential when E is a curve with complex multiplication by $\mathcal{O}_F$, so $\mathcal{M}_E$ is an F-category. However to simplify a bit notations we will assume $F = \mathbb{Q}$ below.

Below we will assume that $\mathcal{M_P}$ is such a category, not necessarily of motivic origin.

2.   The fundamental Lie algebra of a mixed category over $\mathcal{P}$.
Set $H(\mathcal{M_P}) := End(\Psi)$. It is a cocommutative Hopf algebra in the tensor category $\mathcal{P}$.

Let $L(\mathcal{M_P})$ be the Lie algebra of all derivations of the functor Ψ:

$$L(\mathcal{M_P}) = Der(\Psi) = \{F \in End(\Psi)|F_{X \otimes Y} = F_X \otimes id_Y + id_X \otimes F_Y\}$$

It is a Lie algebra in the tensor category $Pro\mathcal{P}$. Its universal enveloping algebra is isomorphic to the Hopf algebra $H(\mathcal{M_P})$. The Tannakian formalism shows that the functor Ψ provides an equivalence between the category $\mathcal{M_P}$ and the category of $L(\mathcal{M_P})$-modules in the category $\mathcal{P}$.

A Lie coalgebra in a tensor category $\mathcal{C}$ is an object $\mathcal{L}$ together with a cobracket $\delta : \mathcal{L} \to \Lambda_{\mathcal{C}}^2 \mathcal{L}$ such that the composition

$$\mathcal{L} \xrightarrow{\delta} \Lambda_{\mathcal{C}}^2 \mathcal{L} \xrightarrow{\delta \otimes id - id \otimes \delta} \Lambda_{\mathcal{C}}^3 \mathcal{L}$$

is zero. The standard complex of $\mathcal{L}$ is defined as follows:

$$C_{\mathcal{C}}^{\bullet}(\mathcal{L}) := \mathcal{L} \xrightarrow{\delta} \Lambda_{\mathcal{C}}^2 \mathcal{L} \xrightarrow{\delta \otimes id - id \otimes \delta} \Lambda_{\mathcal{C}}^3 \mathcal{L} \longrightarrow \Lambda_{\mathcal{C}}^4 \mathcal{L} \longrightarrow \ldots$$

Here $\mathcal{L}$ placed in degree 1, and the differential has degree +1. We define the cohomology groups of a Lie coalgebra $\mathcal{L}$ setting $H_{\mathcal{C}}^{\bullet}(\mathcal{L}) := H^{\bullet}(C^{\bullet}(\mathcal{L}))$.

There is a duality $V \to V^{\vee}$, $(V^{\vee})^{\vee} = V$ between the inductive and projective limits of objects in the cateogory $\mathcal{P}$. Set $\mathcal{L}(\mathcal{M_P}) := L(\mathcal{M_P})^{\vee}$. It is a Lie coalgebra in the tensor category $Ind\mathcal{P}$.

Recall that for an $\mathbb{Q}$-vector space V which is an inductive limit of finite dimensional $\mathbb{Q}$-vector spaces and an object X of a category $\mathcal{C}$ there is an object $V \boxtimes X$ of $Ind\mathcal{C}$. Any pure motive W can be canonically decomposed into the isotypical components:

$$W = \oplus_M Hom_{\mathcal{M}}(M, W) \boxtimes M$$

where the sum is over the isomorphism classes of simple objects in $\mathcal{M_P}$. The standard complex of $\mathcal{L}(\mathcal{M_P})$ splits into a direct sum of isotypical components

$$\left(\mathcal{L}(\mathcal{M_P})\right)_M \xrightarrow{\delta} \left(\Lambda^2 \mathcal{L}(\mathcal{M_P})\right)_M \xrightarrow{\delta} \left(\Lambda^3 \mathcal{L}(\mathcal{M_P})\right)_M \xrightarrow{\delta} \ldots \qquad (13)$$

Notice that $\mathcal{L}(\mathcal{M_P})_M$ is zero unless the weight M is > 0. Therefore each of the complexes (13) has finite length.

3. The Galois group of a mixed category. Let $\Phi : \mathcal{M}_\mathcal{P} \longrightarrow Vect_\mathbb{Q}$ be composition of the fiber functor Ψ with a fiber functor φ from $\mathcal{P}$ to the category of finite dimensional $\mathbb{Q}$-vector spaces.

Then $G(\mathcal{M}_\mathcal{P}) := Aut^\otimes \Phi$ is a proalgebraic group scheme over $\mathbb{Q}$. It is the Galois group of the category $\mathcal{M}_\mathcal{P}$.

There are two canonical functors between the tensor categories $\mathcal{M}_\mathcal{P}$ and $\mathcal{P}$: the inclusion functor $\mathcal{P} \hookrightarrow \mathcal{M}_\mathcal{P}$, and the functor

$$gr^\bullet : \mathcal{M}_\mathcal{P} \longrightarrow \mathcal{P} \qquad gr^W : X \longmapsto \oplus_{n \in \mathbb{Z}} gr_n^W X$$

Their composition is the identity functor on $\mathcal{P}$. These functors obviously respect the fiber functor, and so lead to homomorphisms of groups $G(\mathcal{M}_\mathcal{P}) \longrightarrow G(\mathcal{P})$ and $G(\mathcal{P}) \longrightarrow G(\mathcal{M}_\mathcal{P})$. Thus the group $G(\mathcal{M}_\mathcal{P})$ is a semidirect product:

$$0 \longrightarrow U(\mathcal{M}_\mathcal{P}) \longrightarrow G(\mathcal{M}_\mathcal{P}) \longrightarrow G(\mathcal{P}) \longrightarrow 0$$

Passing to Lie algebras we get:

$$0 \longrightarrow LieU(\mathcal{M}_\mathcal{P}) \longrightarrow LieG(\mathcal{M}_\mathcal{P}) \longrightarrow LieG(\mathcal{P}) \longrightarrow 0$$

So $LieU(\mathcal{M}_\mathcal{P})$ is a pronilpotent Lie algebra in the category of $G(\mathcal{P})$-modules.

The category of finite dimensional $G(\mathcal{M}_\mathcal{P})$-modules is equivalent to the category of $U(\mathcal{M}_\mathcal{P})$-modules in the category of finite dimensional $G(\mathcal{P})$-modules. Since the group scheme $U(\mathcal{M}_\mathcal{P})$ is prounipotent, it is equivalent to the category of $LieU(\mathcal{M}_\mathcal{P})$-modules in the category of $G(\mathcal{P})$-modules. One can think about it as of the category $G(\mathcal{P})$-modules equipped with an action of the Lie algebra $LieU(\mathcal{M}_\mathcal{P})$ such that the action $LieU(\mathcal{M}_\mathcal{P}) \otimes V \longrightarrow V$ is a morphism of $G(\mathcal{P})$-modules.

Lemma 3.2 *Let M be a pure object. Then*

$$RHom_{LieU(\mathcal{M}_\mathcal{P})-mod}(\varphi(\mathbb{Q}(0)), \varphi(M)) \boxtimes M^\vee = \left(\left(\Lambda^\bullet \mathcal{L}(\mathcal{M}_\mathcal{P}) \right)_M, \partial \right)$$

4. A description of the fundamental Lie coalgebra. (Compare with [BMS] and [BGSV]). Let A, B be simple objects of the category $\mathcal{P}$. Let us say that an object M of $\mathcal{M}_\mathcal{P}$ is n-framed by A, B if we are given nonzero homomorphisms

$$v_A : A \to gr_0^W M, \quad f_B : gr_{-n}^W M \to B$$

Consider the finest equivalence relation on the set of all objects n-framed by A, B such that $(M; v_A; f_B)$ and $(M'; v'_A; f'_B)$ are equivalent if there is a morphism $M \to M'$ respecting the frames. Denote by $\mathcal{A}(A, B)$ the set of equivalence classes of mixed motives n-framed by A, B.

Let us define an addition law setting

$$(M; v_A; f_B) + (M'; v_A'; f_B') := (M \oplus M'; (v_A, v_A'); f_B + f_B')$$

The neutral element is $A \oplus B$ with the obviuos framing. Indeed, the equivalence between $(M \oplus A \oplus B; (v_A, id_A); (f_B + id_B))$ and $(M; v_A; f_B)$ is provided by the natural morphisms $M \longleftarrow M \oplus A \hookrightarrow M \oplus A \oplus B$.

The inversion is given by $-(M; v_A; f_B) := (M; v_A; -f_B)$, and one also has $(M; v_A; -f_B) = (M; -v_A; f_B)$. Let us prove the first claim. The (A, B)-framed object object $W_{\leq 0} M / Ker f_B$ is equivalent to $(M; v_A; f_B)$. So we may assume without loss of generality that $W_{\leq 0} M = M, W_{<-n} M = 0$, and moreover $gr_0^W M = A, gr_{-n}^W M = B$. There is a morphism

$$i_- : W_{<0} M \hookrightarrow M \oplus M, m \longmapsto (m, -m)$$

Then

$$M \oplus M \longrightarrow \frac{M \oplus M}{i_-(M)} =: N_1$$

is a morphism of (A, B)-framed objects. We get an extension

$$0 \longrightarrow W_{<0} N_1 \longrightarrow N_1 \longrightarrow A \oplus A \longrightarrow 0$$

Let us consider the extension $0 \longrightarrow W_{<0} N_1 \longrightarrow N \longrightarrow A \longrightarrow 0$ induced by the morphism $j_- : A \to A \oplus A, a \longmapsto (a, -a)$. Then there is a well defined morphism $\alpha : A \to N$ given by $a \longmapsto (n_a, -n_a)$ where $n_a \in N$ is any element projecting to a. So we get a morphism of (A, B)-framed objects $A \oplus B \xrightarrow{(\alpha, \beta)} N$, where $\beta : B \hookrightarrow N$ is the natural inclusion.

It is easy to check that $\mathcal{A}(A, B)$ is an abelian group. Since $\mathcal{M}_\mathcal{P}$ is a $\mathbb{Q}$-category, it is a $\mathbb{Q}$-vector space, and in fact an inductive limit of finite dimensional F-vector spaces.

If B is a simple object then $\mathcal{A}(\mathbb{Q}(0), B) \boxtimes B^\vee$ is an object of $\mathcal{P}$ defined by the isomorphism class of B up to a *unique* isomorphism. Set

$$H'(\mathcal{M}_\mathcal{P}) := \oplus \mathcal{A}(\mathbb{Q}(0), B) \boxtimes B^\vee$$

The sum is over all isomorphism classes of simple objects B in $\mathcal{P}$, but only finitely many terms are non zero.

The tensor product of mixed motives provides $H'(\mathcal{M}_\mathcal{P})$ with a structure of a commutative algebra in the tensor category $\mathcal{P}$:

$$\mathcal{A}(\mathbb{Q}(0), B_1) \boxtimes B_1^\vee \otimes \mathcal{A}(\mathbb{Q}(0), B_2) \boxtimes B_2^\vee \to \mathcal{A}(\mathbb{Q}(0), B_1 \otimes B_2) \boxtimes B_1^\vee \otimes B_2^\vee$$

Let us define the coproduct. Let C be a simple pure object of weight $-k$. Let $c \in Hom_\mathcal{P}(gr_{-k}^W M, C)$ and $c^* \in Hom_\mathcal{P}(C, gr_{-k}^W M)$. Let

$$v(c^*) : \quad \mathbb{Q}(0) \longrightarrow C \otimes C^\vee \xrightarrow{c^* \otimes id} gr_{-k}^W M \otimes C^\vee = gr_0^W(M \otimes C^\vee)$$

(The first arrow is the canonical morphism). Set

$$\alpha(c) := (M; v_{\mathbb{Q}(0)}; c) \in \mathcal{A}(\mathbb{Q}(0), C) \boxtimes C^\vee$$

$$\alpha^*(c^*) := (M \otimes C^\vee; v(c_i^*); f_B \otimes id_{C^\vee}) \in \mathcal{A}(\mathbb{Q}(0), C^\vee \otimes B) \boxtimes C \otimes B^\vee$$

Define a natural map

$$\nu_C : Hom_{\mathcal{P}}(gr_{-k}^W M, C) \otimes Hom_{\mathcal{P}}(C, gr_{-k}^W M) \longrightarrow \qquad (14)$$

$$\left(\mathcal{A}(\mathbb{Q}(0), C) \boxtimes C^\vee \right) \otimes \left(\mathcal{A}(\mathbb{Q}(0), C^\vee \otimes B) \boxtimes C \otimes B^\vee \right)$$

by setting

$$\nu_C : c \otimes c^* \longmapsto \alpha(c) \otimes \alpha^*(c^*)$$

There is a canonical element in (14) defined as follows. Let $c_1, ..., c_m$ be a basis of the $\mathbb{Q}$-vector space $Hom_{\mathcal{P}}(gr_{-k}^W M, C)$ and $c_1^*, ..., c_m^*$ the dual basis of $Hom_{\mathcal{P}}(C, gr_{-k}^W M)$, i.e. $c_i \circ c_j^* = \delta_{i,j} Id_C$. Then $\sum_i c_i \otimes c_i^*$ does not depend on the choice of basis c_i. By definition

$$\Delta : \quad (M; v_{\mathbb{Q}(0)}; f_B) \boxtimes B^\vee \longmapsto \oplus_C \nu_C(\sum_i c_i \otimes c_i^*) \quad \in$$

$$\oplus_C \left(\mathcal{A}(\mathbb{Q}(0), C) \boxtimes C^\vee \right) \otimes \left(\mathcal{A}(\mathbb{Q}(0), C^\vee \otimes B) \boxtimes C \otimes B^\vee \right)$$

The sum is over all isomorphism classes of simple objects in $\mathcal{P}$ whose weights are between 0 and $wt(B)$. It turnes out to be a finite sum.

Notice that $V := \oplus V_B \boxtimes B^\vee$ is a pro (resp. ind) object in $\mathcal{P}$ if and only if V_B is pro (resp. ind) $\mathbb{Q}$-vector space for each isomorphism class of simple objects in $\mathcal{P}$. There is a duality between the pro- and ind-objects: if V is a pro (resp. ind) object in $\mathcal{P}$, then $V^\vee := \oplus V_B^\vee \boxtimes B$ is an ind (resp. pro) object in $\mathcal{P}$, and the canonical map $V \to (V^\vee)^\vee$ is an isomorphism.

For any simple object B of $\mathcal{P}$ the B-isotypical component of $H'(\mathcal{M}_\mathcal{P})$ is an ind-object in the category $\mathcal{P}$. This means that $H'(\mathcal{M}_\mathcal{P})$ itself is an ind-object in the category $\mathcal{P}$. On the other hand $H(\mathcal{M}_\mathcal{P})$ is a pro-object in $\mathcal{P}$.

Let $E \in End(\Psi)$ and E_M the corresponding endomorphism of $\Psi(M)$. There is a natural map $\varphi : H'(\mathcal{M}_\mathcal{P}) \to H(\mathcal{M}_\mathcal{P})^\vee$ given by

$$\varphi\left((M; v_{\mathbb{Q}(0)}; f_B) \boxtimes B^\vee \right)(E_M) := \quad f_B E_M v_{\mathbb{Q}(0)}(\mathbb{Q}(0));$$

Here we extended f_B to a morphism $\Psi(M) \to B$ postulating that its restriction to $gr_{-l}^W M$ for $l \neq n$ is zero. We will always extend the framing morphisms in a similar way.

Theorem 3.3 *a)* $H'(\mathcal{M}_\mathcal{P})$ *is dual to* $H(\mathcal{M}_\mathcal{P})$.

b) Δ *provides* $H'(\mathcal{M}_\mathcal{P})$ *with a structure of a commutative Hopf algebra in the category* $\mathcal{P}$ *which is canonically isomorphic to* $H(\mathcal{M}_\mathcal{P})^\vee$.

Proof. a) According to the Tannaka theory (which remains valid for fiber functors with values in the semisimple tensor categories) the fiber functor Ψ provides an equivalence between the category $\mathcal{M}_\mathcal{P}$ and the category $H(\mathcal{M}_\mathcal{P})$-mod of finite dimensional $H(\mathcal{M}_\mathcal{P})$-modules in the category $\mathcal{P}$. So we will work with the category $H(\mathcal{M}_\mathcal{P})$-mod.

Let us show that φ is injective. Suppose that $\varphi\big((M; v_{\mathbb{Q}(0)}; f_B) \boxtimes B^\vee\big) = 0$. We may assume that $gr_p^W M = 0$ for $p > 0$ and $p < -n$ and $gr_0^W M = \mathbb{Q}(0), gr_{-n}^W M = B$. Consider the cyclic submodule $M' := H(\mathcal{M}_\mathcal{P}) \cdot \Psi(\mathbb{Q}(0))$. Then its weights are bigger then $-n$, since otherwise $\varphi\big((M; v_{\mathbb{Q}(0)}; f_B) \boxtimes B^\vee\big) \neq 0$. So we get a morphism $M' \oplus B \to M$ which obviously respects the frames. On the other hand there is canonical projection $M' \oplus B \to \mathbb{Q}(0) \oplus B$ providing by the projection $M' \to gr_0^M = \mathbb{Q}(0)$. Thus $(M; v_{\mathbb{Q}(0)}; f_B)$ is equivalent to $\mathbb{Q}(0) \oplus B$ (with the natural frame).

Now let us show that φ is surjective. Let $f \in H(\mathcal{M}_\mathcal{P})_B^\vee$ be a subobject isomorphic to $B^\vee$. Denote by $Ker f$ the corresponding subobject of $H(\mathcal{M}_\mathcal{P})_B$ such that $H(\mathcal{M}_\mathcal{P})_B/ker f = B$. Consider the $H(\mathcal{M}_\mathcal{P})$-module: $\mathbb{Q}(0) \oplus H(\mathcal{M}_\mathcal{P})_B/Ker f$ whith the action of $H(\mathcal{M}_\mathcal{P})$ provided by f, the nontrivial part of the action is a morphism $H(\mathcal{M}_\mathcal{P})_B \otimes \mathbb{Q}(0) \to H(\mathcal{M}_\mathcal{P})_B/Ker f$, and the obvious framing. The map φ sends it to f.

The part b) follows easily from the definitions and the part a). The theorem is proved.

Remark. A coalgebra similar to $H'(\mathcal{M}_\mathcal{P})$ was considered in [BMS]. However that coalgebra is *not* isomorphic to ours, it is bigger then $H'(\mathcal{M}_\mathcal{P})$.

4 Conjectures on the motivic Galois group

1. A conjecture on Ext's in the category of mixed elliptic motives.

Lemma 4.1

$$RHom_{\mathcal{M}\mathcal{M}_k}(\mathbb{Q}(0), Sym^n \mathcal{H}(m)) = RHom_{\mathcal{M}\mathcal{M}_{E^{(n)}}}(\mathbb{Q}(0), \mathbb{Q}(n+m))[n]_{sgn} \tag{15}$$

Proof. Let $p : E^{(n)} \longrightarrow Spec(k)$ be the canonical projection. Then we should have the motivic Leray spectral sequence

$$E_2^{p,q} = Ext_{\mathcal{M}\mathcal{M}_k}^p\big(\mathbb{Q}(0), R^q p_* \mathbb{Q}(n+m)\big)$$

degenerating at E_2 and abbuting to $Ext_{\mathcal{M}\mathcal{M}_{E^{(n)}}}^{p+q}\big(\mathbb{Q}(0), \mathbb{Q}(n+m)\big)$. Since

$$H^i(E^{(n)})_{sgn} = 0 \quad \text{for} \quad i \neq n; \qquad H^n(E^{(n)})_{sgn} = Sym^n h^1(E)$$

we get (15). Conjecture (3.1) tells us that the cohomology of the elliptic motivic complexes are given by the formula

$$R^i Hom_{\mathcal{MM}_k}(\mathbb{Q}(0), Sym^n\mathcal{H}(m)) \otimes \mathbb{Q} = gr^\gamma_{n+m}K_{n+2m-i}(E^{(n)})_{sgn} \otimes \mathbb{Q} \quad (16)$$

Let us choose a fiber functor $\varphi : \mathcal{P}_E \to Vect_\mathbb{Q}$. Set $H := \varphi(\mathcal{H})$ Then $\varphi(\mathbb{Q}(1))$ corresponds to the one-dimensional representation $det : GL_2(H) \to \mathbb{G}_m$ and $S^m\mathcal{H}(n)$ to the GL_2-module $S^m H(n)$, both with the trivial $LieU(\mathcal{M}_E)$-action.

The following conjecture plays a crucial role.

Conjecture 4.2 *Let E be an elliptic curve over an algebraically closed field k The canonical morphism*

$$RHom_{\mathcal{M}_E}(\mathbb{Q}(0), S^n\mathcal{H}(m)) \longrightarrow RHom_{\mathcal{MM}_k}(\mathbb{Q}(0), S^n\mathcal{H}(m))$$

is a quasiisomorphism. In particular the Ext groups in the category $\mathcal{M}_E$ should be isomorphic to the Ext groups (between the same objects) in the category $\mathcal{MM}_k$.

Remark. This conjecture should not be true if k is not an algebraically closed field. A similar conjecture for the category of mixed Tate motives is expected to be valid for all fields k.

Question. Consider the mixed category $\mathcal{M}_{\{X_i\}}$ generated by given simple pure motives X_i. When could we expect that the Ext's in the category $\mathcal{M}_X$ coinside with the Ext's (between the same objects) in the category $\mathcal{MM}_k$?

2. The main hero: a Lie coalgebra $\mathcal{L}^*(E)$. Let us assume for simplicity that E is an ellliptic curve over a field k without complex multiplication. Then any simple object of the category $\mathcal{P}_E$ of pure elliptic motives is isomorphic to just one of the objects $S^n\mathcal{H}(m)$.

$\mathcal{P}_E$ is a rigid tensor category. Let us define a new tensor structure $\otimes'$ on this category setting

$$S^n\mathcal{H}(m) \otimes' S^{n'}\mathcal{H}(m') \xrightarrow{=} S^{n+n'}\mathcal{H}(m+m')$$

where the morphism is given by the usual tensor product followed by the projection to the $S^{n+n'}\mathcal{H}(m+m')$ component. We get an abelian tensor category $\mathcal{P}_E^*$ which will be called *the reduced tensor category of elliptic motives.*

Remark. $\mathcal{P}_E^*$ is not a rigid tensor category since there is no dual object for $S^n\mathcal{H}(m)$ if $n > 0$.

The mixed Tate motives are contained in the category of mixed elliptic motives, and we have a canonical functor $\mathcal{M}_T(k) \longrightarrow \mathcal{M}_E$. Let $\pi^* : \mathcal{L}(k) \to \mathcal{L}(E)$ be the corresponding morphism of Lie coalgebras.

Let $\mathcal{L}$ be a Lie coalgebra in the tensor category $\mathcal{P}_E$. We say that $\mathcal{L}^*$ is a Lie subcoalgebra of $\mathcal{L}$ with respect to the *both* tensor structures, $\mathcal{P}_E^*$ and

$\mathcal{P}_E$, if $\mathcal{L}^*$ has a structure of a Lie coalgebra in $\mathcal{P}_E^*$ given by the coproduct $\delta^* : \mathcal{L}^* \longrightarrow \Lambda^2_{\mathcal{P}_E^*} \mathcal{L}^*$ and the diagram

$$
\begin{array}{ccc}
\mathcal{L} & \xrightarrow{\delta} & \Lambda^2_{\mathcal{P}_E} \mathcal{L} \\
\uparrow i & & \uparrow \Lambda^2 i \\
\mathcal{L}^* & \xrightarrow{\delta^*} & \Lambda^2_{\mathcal{P}_E^*} \mathcal{L}^*
\end{array}
$$

is commutative. Then we get a morphism of complexes

$$
\begin{array}{ccccccc}
\mathcal{L} & \xrightarrow{\delta} & \Lambda^2_{\mathcal{P}_E} \mathcal{L} & \xrightarrow{\delta} & \Lambda^3_{\mathcal{P}_E} \mathcal{L} & \xrightarrow{\delta} & \dots \\
\uparrow i & & \uparrow \Lambda^2 i & & \uparrow \Lambda^3 i & & \\
\mathcal{L}^* & \xrightarrow{\delta^*} & \Lambda^2_{\mathcal{P}_E^*} \mathcal{L}^* & \xrightarrow{\delta^*} & \Lambda^3_{\mathcal{P}_E^*} \mathcal{L}^* & \xrightarrow{\delta^*} & \dots
\end{array}
$$

and thus a morphism of cohomology groups.

$$
i^* : H^*_{\mathcal{P}_E^*}(\mathcal{L}^*) \longrightarrow H^*_{\mathcal{P}_E}(\mathcal{L})
$$

Conjecture 4.3 *There exists a Lie subcoalgebra $\mathcal{L}^*(E) \subset \mathcal{L}(E)$ which enjoys the following properties:*
  1. $\mathcal{L}^*(E)$ is also a Lie coalgebra in the reduced tensor category $\mathcal{P}_E^*$.
  2. The natural morphism of cohomology groups

$$
H^{\bullet}_{\mathcal{P}_E^*}(\mathcal{L}^*(E)) \longrightarrow H^{\bullet}_{\mathcal{P}_E}(\mathcal{L}(E))
$$

is an isomorphism.
  3. π^* maps $\mathcal{L}(k)_{\mathbb{Q}(n)^\vee}$ isomorphically onto $\mathcal{L}^*(E)_{\mathbb{Q}(n)^\vee}$.

A construction of a Lie coalgebra in the tensor category $\mathcal{P}_E^*$ which hypothetically satisfies all these properties is given in s. 4.5.

Remarks. 1. A Lie subcoalgebra of $\mathcal{L}(E)$ is usually not a Lie coalgebra in the tensor category $\mathcal{P}_E^*$.

2. $H^{\bullet}_{\mathcal{P}_E^*}(\mathcal{L}^*(E))$ are quite different from $H^{\bullet}_{\mathcal{P}_E}(\mathcal{L}^*(E))$. The reason is clear from the following example:

$$
(V \boxtimes \mathcal{H}) \otimes (V \boxtimes \mathcal{H}) = (V \otimes V) \boxtimes S^2\mathcal{H} \oplus (V \otimes V) \boxtimes S^2\mathcal{H}
$$

$$
(V \boxtimes \mathcal{H}) \otimes' (V \boxtimes \mathcal{H}) = (V \otimes V) \boxtimes S^2\mathcal{H}
$$

Consider $\mathcal{L}^*(E)$ as a Lie coalgebra in $\mathcal{P}_E$. Let $\mathcal{M}_E^*$ be the category of comodules over it in the category $\mathcal{P}_E$. If $\mathcal{L}^*(E)$ is a Lie subcoalgebra of $\mathcal{L}(E)$, then $\mathcal{M}_E^*$ is a subcategory of $\mathcal{M}_E$. We will call it the *small category of mixed elliptic motives*.

Warning. The natural functor $\mathcal{P}_E^* \to \mathcal{M}_E^*$ is not a tensor functor.

The weight consideration shows that the picture of nonzero isotypical components of $\mathcal{L}^*(E)$ looks as follows:

A boldface point with coordinates (m, n) corresponds to nonzero isotypical component $\mathcal{L}^*(E)_{S^n \mathcal{H}(m)^\vee}$. For example $\mathcal{L}^*(E)_{\mathbb{Q}(0)} = 0$, $\mathcal{L}^*(E)_{S^2 \mathcal{H}(-1)^\vee} = 0$.

3. Corollaries. According to conjectures (4.3) and (4.2) one has the following quasiisomorphisms:

$$\left(\Lambda^\bullet_{\mathcal{P}^*_E} \mathcal{L}^*(E), \partial\right)_{S^n \mathcal{H}(m)^\vee} \overset{4.3}{=} RHom_{\mathcal{M}_E}(\mathbb{Q}(0), S^n \mathcal{H}(m)) \overset{4.2}{=} \tag{17}$$

$$RHom_{\mathcal{M}\mathcal{M}_k}(\mathbb{Q}(0), S^n \mathcal{H}(m)) \tag{18}$$

So we get the basic formula

$$H^i(\Lambda^\bullet_{\mathcal{P}^*_E} \mathcal{L}^*(E))_{S^n \mathcal{H}(m)^\vee} = gr^\gamma_{n+m} K_{n+2m-i}(E^{(n)})_{sgn} \otimes \mathbb{Q} \tag{19}$$

Using this we immediately conclude that

$$\mathcal{L}^*(E)_{\mathbb{Q}(1)^\vee} = k^* \boxtimes \mathbb{Q}(1)^\vee, \quad \mathcal{L}^*(E)_{\mathcal{H}^\vee} = J \boxtimes \mathcal{H}^\vee$$

More generally, let $J_{S^{2m+1}\mathcal{H}(-m)}$ be the image of the subgroup of homologically trivial cycles in $CH^m(E^{(2m-1)})_{sgn}$ under the Abel-Jacobi map. Then

$$\mathcal{L}^*(E)_{S^{2n+1}\mathcal{H}(-n)^\vee} = J_{S^{2n+1}\mathcal{H}(-n)} \boxtimes S^{2n+1}\mathcal{H}(-n)^\vee, \quad n > 0$$

Indeed, if H is a pure motive of weight 1 then

$$C^\bullet(\mathcal{L}^*(E))_H = \mathcal{L}^*(E)_H = H^1_{\mathcal{P}^*(E)}(\mathcal{L}^*(E))_H \overset{2}{=} H^1(\mathcal{L}^*(E))_H$$

Recall that $B_2(k)$ (denoted by $B_{\mathbb{Q}(1)}(k)$ in chapter 2) is the Bloch group of a field k. Let $B_3(k)$ be its analog for the classical trilogarithm introduced in [G1-2]).

Lemma 4.4 *(17) together with the results of [G2], [GL] implies that the lower left corner of the diagram looks as folows (all groups are tensored by* $\mathbb{Q}$*):*

$$J_{S^3\mathcal{H}(-1)} \quad \bullet \quad \bullet \qquad \bullet \qquad \bullet \qquad \bullet$$
$$\bullet \quad \bullet \qquad \bullet \qquad \bullet \qquad \bullet$$
$$J \quad B_3^*(E) \qquad \bullet \qquad \bullet \qquad \bullet$$
$$0 \quad k^* \qquad B_2(k) \quad B_3(k) \quad \bullet$$

Proof. The structure of the bottom row follows from 3) and the main results in [G2]. The $\mathcal{H}(1)^\vee$-isotypical component of the standard complex of $\mathcal{L}^*(E)$ is

$$\mathcal{L}^*(E)_{\mathcal{H}(1)^\vee} \longrightarrow \mathcal{L}^*(E)_{\mathbb{Q}(1)^\vee} \otimes \mathcal{L}^*(E)_{\mathcal{H}^\vee} = k^* \otimes J \boxtimes \mathcal{H}(1)^\vee$$

The motivic interpretation of the group $B_3^*(E)$ as a group of $\mathcal{H}(1)$-framed mixed elliptic motives (see theorem 9.15 below) leads to a morphism of complexes

$$\begin{array}{ccc} B_3^*(E) \boxtimes \mathcal{H}(1)^\vee & \longrightarrow & k^* \otimes J \boxtimes \mathcal{H}(1)^\vee \\ \downarrow & & \downarrow= \\ \mathcal{L}^*(E)_{\mathcal{H}(1)^\vee} & \longrightarrow & k^* \otimes J \boxtimes \mathcal{H}(1)^\vee \end{array}$$

Thanks to the main result of [GL] and property 2) this map is a quasiisomorphism. So $\mathcal{L}^*(E)_{\mathcal{H}(1)^\vee} = B_3^*(E) \boxtimes \mathcal{H}(1)^\vee$.

The Lie coalgebra $\mathcal{L}(E)$ has a much more complicated structure, see s.4.8 below for the simplest example: description of $\mathcal{L}(E)_{\mathbb{Q}(1)^\vee}$.

Zero cycles on $E^{(n)}$. The only way how the object $S^n\mathcal{H}$ could appear as a direct summand of a tensor product of n simple objects of negative weights *in the small tensor category* $\mathcal{P}_E^*$ is this:

$$S^n\mathcal{H} = \mathcal{H} \otimes' \ldots \otimes' \mathcal{H} \quad (n \text{ times})$$

Set

$$\mathcal{L}^*(E)_{S^2\mathcal{H}^\vee} := B_{S^2\mathcal{H}}^* \boxtimes S^2\mathcal{H}^\vee \tag{20}$$

The $S^2\mathcal{H}^\vee$- isotypic component of the cochain complex $\Lambda_{\mathcal{P}_E^*}^\bullet \mathcal{L}^*(E)$ looks as follows:

$$\left(\mathcal{L}^*(E) \longrightarrow \Lambda_{\mathcal{P}_E^*}^2 \mathcal{L}^*(E)\right)_{S^2\mathcal{H}^\vee} = \left(B_{S^2\mathcal{H}}^*(E) \xrightarrow{\delta_{S^2\mathcal{H}}} \Lambda^2 J\right) \boxtimes S^2\mathcal{H}^\vee$$

Similarly the $S^n\mathcal{H}^\vee$-isotypic component of $\Lambda_{\mathcal{P}_E^*}^\bullet \mathcal{L}^*(E)$ is $(\ldots \longrightarrow \Lambda^n J) \boxtimes S^n\mathcal{H}^\vee$. Therefore we get

$$Ext^n(\mathbb{Q}(0), S^n\mathcal{H}) = CH^n(E^{(n)})_{sgn} = \text{a quotient of } \Lambda^n J \tag{21}$$

The last equality indeed follows from the Bloch's theorem on zero cycles on abelian varieties [Bl5].

Similarly we deduce that for $m > 0$ one should have

$$Ext_{\mathcal{M}_E}^{n+m}(\mathbb{Q}(0), S^n \mathcal{H}(m)) \quad = \quad \text{a quotient of} \quad \Lambda^n J \otimes_{\mathbb{Q}} K_m^M(k)$$

4. The duality between DG Lie coalgebras and commutative DG algebras. Let $\mathcal{L}^\bullet$ be a DG Lie coalgebra in a tensor category $\mathcal{C}$, which is not supposed to be a rigid tensor category. This means that we have a complex $(\mathcal{L}^\bullet, \partial)$ and a cobracket

$$\delta : \mathcal{L}^\bullet \to \Lambda_{\mathcal{C}}^2(\mathcal{L}^\bullet)$$

which is a morphism of complexes satisfying the co Jacobi identity. Let

$$C(\mathcal{L}^\bullet) := \mathcal{L}^\bullet[-1] \oplus S_{\mathcal{C}}^2(\mathcal{L}^\bullet[-1]) \oplus S_{\mathcal{C}}^3(\mathcal{L}^\bullet[-1]) \oplus \dots$$

be the free super commutative algebra (without unit) generated by $\mathcal{L}^\bullet[-1]$. There are two cohomological differentials on $C(\mathcal{L}^\bullet)$: the first is provided by the differential in $\mathcal{L}^\bullet$, and the second comes via the Leibniz rule from the cobracket δ. Their sum is a differential providing a structure of a commutative DG algebra on $C(\mathcal{L}^\bullet)$. For example, if $\mathcal{L}^\bullet = \mathcal{L}$ is concentrated in degree 0, then $S_{\mathcal{C}}^n(\mathcal{L}[-1]) = \Lambda_{\mathcal{C}}^n \mathcal{L}[-n]$ and we get the standard cochain complex of the Lie coalgebra:

$$C(\mathcal{L}) := \mathcal{L}[-1] \oplus \Lambda_{\mathcal{C}}^2(\mathcal{L})[-2] \oplus \Lambda_{\mathcal{C}}^3(\mathcal{L})[-3] \oplus \dots$$

Let $\mathcal{L}^\bullet$ be a DG Lie coalgebra and $A^\bullet$ is a DG commutative algebra. Define $MC(Hom_k(\mathcal{L}^\bullet[-1], A^\bullet))$ as the set of all degree zero elements in $\omega \in Hom_k(\mathcal{L}^\bullet[-1], A^\bullet)$ satisfying the Maurer-Cartan equation $d\omega + \frac{1}{2}\omega \wedge \omega = 0$. Here $\omega \wedge \omega$ is the composition

$$\mathcal{L}^\bullet[-1] \to S_{\mathcal{C}}^2(\mathcal{L}^\bullet[-1]) \to S_{\mathcal{C}}^2 A^\bullet \to A^\bullet$$

If we think of ω as of element in $(\mathcal{L}^\bullet[-1])^\vee \otimes A^\bullet$, we get the usual formula for $\omega \wedge \omega$. Then one has

$$MC(Hom_k(\mathcal{L}^\bullet[-1], A^\bullet)) \quad = \quad Hom_{DGCom}(C(\mathcal{L}), A)$$

We get a functor

$$C : DGCoLie \to DGCom$$

There is a functor in the opposite direction:

$$F : DGCom \to DGCoLie$$

satisfying

$$Hom_{DGcoLie}(\mathcal{L}^\bullet, F(A^\bullet)) \quad = \quad MC(Hom_k(\mathcal{L}^\bullet, A^\bullet[1]))$$

Namely, $F(A^\bullet)$ is the cofree Lie supercoalgebra $\mathcal{F}(A^\bullet[1])$ generated by the complex $A^\bullet[1]$. There is a canonical projection $p : \mathcal{F}(A^\bullet[1]) \longrightarrow A^\bullet[1]$. According to the universality property of the cofree Lie coalgebras to define the differential on $\mathcal{F}(A^\bullet[1])$ it is sufficient to define its projection $\mathcal{F}(A^\bullet[1]) \to A^\bullet[1]$. By definition it is the sum of the differential in $A^\bullet[1]$ and the product $\Lambda^2(A^\bullet[1]) = S^2(A^\bullet)[2] \to A^\bullet[1]$.

The functors C and F are obviously adjoint. So there are canonical morphisms

$$A^\bullet \longrightarrow C \circ F(A^\bullet) \qquad \mathcal{L}^\bullet \longrightarrow F \circ C(\mathcal{L}^\bullet)$$

Theorem 4.5 *These morphisms are quasiisomorphisms.*

This theorem was proved by Quillen [Q] when $\mathcal{C}$ is the category of $\mathbb{Q}$-vector spaces, but it is true for an arbitrary tensor category $\mathcal{C}$.

The functor F has another description via the bar construction. Namely, let $B(A^\bullet)$ be the reduced bar complex of a DG commutative algebra $A^\bullet$. It has a structure of a Hopf algebra. Let $B_0(A^\bullet)$ be the augmentation ideal of $B(A^\bullet)$. Then the space $B_0(A^\bullet)/(B_0(A^\bullet) \cdot B_0(A^\bullet))$ of the indecomposable elements has a natural structure of a DG Lie coalgebra (a good reference for the bar construction is ch. 2 of [BK]).

5. A cycle construction of $\mathcal{L}^*(E)$ (Compare with [Bl4]). Let G_n be the semidirect product of the symmetric group S_n and $(\mathbb{Z}/2\mathbb{Z})^n$. Let *sgn* be the one dimensional alternating representation of G_n where a generator of each factor $\mathbb{Z}/2\mathbb{Z}$ acts by -1 and the restriction to S_n is the alternating representation. The group G_n acts naturally on E^n.

The idea of the construction. Let $\Gamma_X(n)$ be a motivic complex on a regular scheme X, i.e., it is a *complex* of $\mathbb{Q}$-vector spaces quasiisomorphic to $RHom_{\mathcal{MM}_X}(\mathbb{Q}(0), \mathbb{Q}(n))$. We will need also a canonical morphism of complexes $\Gamma_X(m) \times \Gamma_Y(n) \to \Gamma_{X \times Y}(m+n)$. We will take below the complex of Bloch's Higher Chow groups on X as a concrete version of $\Gamma_X(n)$ to work with. Then

$$\mathcal{N}^*(E) := \oplus_{2m+n>0, n \geq 0} \Gamma_{E^n}(m+n)_{sgn} \boxtimes S^n \mathcal{H}(m)^\vee$$

has a natural structure of a commutative DG algebra (without unit) in the reduced tensor category of pure elliptic motives. Namely, the product is provided by the natural morphism of complexes

$$\Gamma_{E^n}(m+n)_{sgn} \times \Gamma_{E^{n'}}(m'+n')_{sgn} \longrightarrow \Gamma_{E^{n+n'}}(m+m'+n+n')_{sgn}$$

(take the external product of the complexes on $E^{n+n'}$ and project it onto the signum part with respect to the action of the group $G_{n+n'}$). It remains to apply the functor F!

Now let us spell out the details for the complexes of Bloch's Higher Chow groups. Let $\mathcal{Z}^n(E^m, k)$ be the $\mathbb{Q}$-vector space of codimension n cycles with $\mathbb{Q}$ coefficients on $E^m \times (\mathbb{P}^1 \backslash \{1\})^k$ which are skewsymmetric with respect to the action of G_k on $(\mathbb{P}^1 \backslash \{1\})^k$ and meet the faces of the coordinate cube defined by setting some of the coordinates equal to 0 or ∞ properly. Set $\mathcal{Z}^{m+n}(E^n, c)_{sgn} := (\mathcal{Z}^{m+n}(E^n, c) \otimes sgn)^{G_n}$. We define a complex $\mathcal{Z}^{m+n}(E^n, *)_{sgn}$ placing

$$\mathcal{Z}^{m+n}(E^n, c)_{sgn} \quad \text{in degree} \quad 2n + m - c$$

The differential is the alternating sum of the face maps:

$$\mathcal{Z}^{m+n}(E^n, c)_{sgn} \longrightarrow \mathcal{Z}^{m+n}(E^n, c - 1)_{sgn}$$

A $\otimes'$-product of $RHom$'s

$$RHom_{\mathcal{MM}_k}(\mathbb{Q}(0), S^n \mathcal{H}(m)) \otimes' RHom_{\mathcal{MM}_k}(\mathbb{Q}(0), S^{n'} \mathcal{H}(m')) \longrightarrow \quad (22)$$

$$RHom_{\mathcal{MM}_k}(\mathbb{Q}(0), S^{n+n'} \mathcal{H}(m + m'))$$

is provided by the usual tensor product

$$RHom(\mathbb{Q}, A) \otimes RHom(\mathbb{Q}, B) \longrightarrow RHom(\mathbb{Q}, A \otimes B)$$

and the canonical projection $S^n \mathcal{H}(m) \otimes S^{n'} \mathcal{H}(m') \to S^{n+n'} \mathcal{H}(m + m')$.

Lemma 4.6 *a) One has a quasiisomorphism*

$$\mathcal{Z}^{m+n}(E^n, *)_{sgn} \quad = \quad RHom_{\mathcal{MM}_k}(\mathbb{Q}(0), S^n \mathcal{H}(m))$$

b) The product of cycles followed by the projection to the $G_{n+n'}$-alternating part provides a morphism of complexes

$$\mathcal{Z}^{m+n}(E^n, *)_{sgn} \otimes \mathcal{Z}^{m'+n'}(E^{n'}, *')_{sgn} \longrightarrow \mathcal{Z}^{m+n+m'+n'}(E^{n+n'}, * + *')_{sgn}$$

which coincides in the derived category with the $\otimes'$-product of $RHom$'s (22).

Proof. Follows from lemma (4.1) and the results of Bloch [Bl2], [BK]. Set

$$\mathcal{N}^*(E) := \oplus_{2m+n>0} \mathcal{Z}^{m+n}(E^n, *)_{sgn} \boxtimes S^n \mathcal{H}(m)^\vee$$

There is a commutative product on $\mathcal{N}^*(E)$ given by

$$\left(\mathcal{Z}^{m+n}(E^n, *)_{sgn} \boxtimes S^n \mathcal{H}(m)^\vee \right) \otimes \left(\mathcal{Z}^{m'+n'}(E^{n'}, *')_{sgn} \boxtimes S^{n'} \mathcal{H}(m')^\vee \right)$$

$$\left(\mathcal{Z}^{m+n+m'+n'}(E^{n+n'}, * + *')_{sgn} \boxtimes S^{n+n'} \mathcal{H}(m + m')^\vee \right)$$

Proposition 4.7 $\mathcal{N}^*(E)$ *is a commutative DG algebra (without unit) in the reduced tensor category of pure elliptic motives.*

Thus setting $\tilde{\mathcal{L}}^*(E) := F\mathcal{N}^*(E)$ and using theorem 4.5 we get a proof of theorem 1.2.

Conjecture 4.8 $H^i(F\mathcal{N}^*(E)) = 0$ *if* $i \neq 0$.

$H^0(F\mathcal{N}^*(E))$ is a Lie coalgebra in the tensor category $\mathcal{P}_E^*$ which is our candidate for $\mathcal{L}^*(E)$.

Proposition 4.9 *Assuming the conjecture (4.8) one has for* $(n, m) \neq (0, 0)$:

$$\left(H_{\mathcal{P}_E^*}^*(H^0(\tilde{\mathcal{L}}^*(E))) \right)_{S^n\mathcal{H}(m)^\vee} = RHom_{\mathcal{MM}_k}(\mathbb{Q}(0), S^n\mathcal{H}(m)) \boxtimes S^n\mathcal{H}(m)^\vee$$

Proof. This folows immediately from theorem 4.5 and lemma 4.6.

6. The Freeness conjecture for mixed elliptic motives. The Lie coalgebras in this paper are *Ind*-objects in the category of pure motives. Denote by $L(E)$, $L^*(E)$, $I^*(E)$ the Lie algebras dual to the Lie coalgebras $\mathcal{L}(E)$, $\mathcal{L}^*(E)$, $\mathcal{I}^*(E)$. Set

$$I^*(E) := \oplus_{n+m>1} L^*(E)_{S^n\mathcal{H}(m)}$$

It is clear from the picture above that $I^*(E)$ is an ideal in $L^*(E)$ and

$$L^*(E)/I^*(E) = (k^*)^\vee \boxtimes \mathbb{Q}(1) \oplus J^\vee \boxtimes \mathcal{H} \tag{23}$$

is an abelian Lie algebra.

Conjecture 4.10 $I^*(E)$ *is a free Lie algebra in the tensor category* $\mathcal{P}_E^*$.

A Lie algebra L in $\mathcal{P}_E^*$ is free if and only if $H_{\mathcal{P}_E^*}^i(L) = 0$ for $i > 1$.

Remark. According to the property 3) this conjecture implies the freeness conjecture for mixed Tate motives.

7. The vector space $\mathcal{L}(E)_{\mathbb{Q}(1)^\vee}$. It follows from (19) that one should have

$$\mathcal{L}(E)_{S^{2n+1}\mathcal{H}(-n)^\vee} = J_{S^{2n+1}\mathcal{H}(-n)} = CH^{n+1}(E^{2n+1})_{sgn}$$

Let M be a simple object of $\mathcal{P}_E$. According to H.Weyl's theorem $Hom_{\mathcal{P}_E}(M, \otimes^m\mathcal{H})$ is an irreducible G_m-module. Denote it by $\rho_M^{(m)}$ (we will omit (m) sometimes). Set

$$CH^{2n+2}(E^{4n+2})_{\rho_{\mathbb{Q}(1)^\vee}} := \left(CH^{2n+2}(E^{4n+2}) \otimes \rho_{\mathbb{Q}(1)^\vee} \right)_{G_{4n+2}} \subset S^2(J_{S^{2n+1}\mathcal{H}(-n)})$$

For any integer $n \geq 0$ one can define an abelian group $B_{\mathbb{Q}(1)}^{(n)}$ together with the following exact sequence:

$$0 \longrightarrow k_{\mathbb{Q}}^* \longrightarrow B_{\mathbb{Q}(1)}^{(n)} \longrightarrow CH^{2n+2}(E^{4n+2})_{\rho_{\mathbb{Q}(1)^\vee}} \longrightarrow 0 \tag{24}$$

For $n = 0$ this is the group $B_{\mathbb{Q}(1)}(E)$ we discussed in chapter 2. In general the extension 24 comes in a similar way from the biextension related to codimension $n + 1$ cycles in E^{2n+1} (see [Bl3]).

Let us take the sum of all these extensions:

$$0 \longrightarrow k^* \otimes \mathbb{Q}[t] \longrightarrow \oplus_{n \geq 0} B_{\mathbb{Q}(1)}^{(n)} \longrightarrow \oplus_{n \geq 0} CH^{2n+2}(E^{4n+2})_{\rho_{\mathbb{Q}(1)}\vee} \boxtimes \mathbb{Q}(1)^{\vee} \longrightarrow 0$$

There is a homomorphism $\alpha : k^* \otimes \mathbb{Q}[t] \longrightarrow k^*$, $a \otimes t^n \longmapsto a$.

Theorem 4.11

$$\mathcal{L}_{\mathbb{Q}(1)^{\vee}} = \frac{\oplus_{n \geq 0} B_{\mathbb{Q}(1)}^{(n)}}{ker\,\alpha}$$

Using the identification

$$\oplus_{n \geq 0} \left(\Lambda^2 \mathcal{L}(E)_{S^{2n+1}\mathcal{H}(-n)^{\vee}} \right)_{\mathbb{Q}(1)^{\vee}} = \oplus_{n \geq 0} CH^{2n+2}(E^{4n+2})_{\rho_{\mathbb{Q}(1)}\vee} \boxtimes \mathbb{Q}(1)^{\vee}$$

we get a canonical homomorphism, provided by (24)

$$\mathcal{L}(E)_{\mathbb{Q}(1)^{\vee}} \longrightarrow \oplus_{n \geq 0} \left(\Lambda^2 \mathcal{L}(E)_{S^{2n+1}\mathcal{H}(-n)^{\vee}} \right)_{\mathbb{Q}(1)^{\vee}}$$

This homomorphism gives the cobracket δ. Its kernel is isomorphic to k^*.

Notice how simple is $\mathcal{L}^*(E)_{\mathbb{Q}(1)^{\vee}} = k_{\mathbb{Q}}^*$ and how complicated is $\mathcal{L}(E)_{\mathbb{Q}(1)^{\vee}}$!

8. The small motivic Galois group ? Let $\mathcal{M}_k$ be the abelian category of all pure motives over a field k. It should have a natural small tensor structure. Namely, if $\text{char}\,k = 0$ then $\mathcal{M}_k$ should be a neutral tannakian tensor category, so one should have a canonical equivalence

$$\psi : \mathcal{M}_k \longrightarrow G(\mathcal{M}_k) - mod$$

between the category of all pure motives and the category of finite dimensional $\mathbb{Q}$-rational representations of a pro-reductive group scheme $G(\mathcal{M}_k)$ over $\mathbb{Q}$. The l-adic cohomology fiber functor on $\mathcal{M}_k$ is given by $\psi \otimes \mathbb{Q}_l$. (The case $\text{char}\,k \neq 0$ we left to the reader).

Let $G(\mathcal{M}_k)^s$ be the maximal split over $\mathbb{Q}$ quotient of the proreductive group $G(\mathcal{M}_k)$. According to E. Cartan's theory the irreducible $G(\mathcal{M}_k)^s$-modules are the highest weight modules V_λ, where λ is the highest weight. We define the new tensor structure $\otimes'$ on the category of $G(\mathcal{M}_k)^s - mod$ by $V_\lambda \otimes' V_\beta := V_{\lambda+\beta}$. The category $G(\mathcal{M}_k)^s - mod$ is a subcategory of $G(\mathcal{M}_k) - mod$.

We conjecture that there exists a small tensor structure on the category $G(\mathcal{M}_k) - mod$ which coincides with the small tensor structure $\otimes'$ defined above on the subcategory $G(\mathcal{M}_k)^s - mod$. Combining it with the equivalence ψ we would get a small tensor structure on the category of pure motives. Denote this new tensor category by $\mathcal{M}_k^*$.

Conjecture 4.12 *Assume* $k = \bar{k}$. *Then there exists a Lie subcoalgebra* $\mathcal{L}^*(\mathcal{M}_k) \subset \mathcal{L}(\mathcal{M}_k)$ *which is a Lie coalgebra for* **both** *tensor structures on the category of pure motives over* k *such that the canonical morphism*

$$H^{\bullet}_{\mathcal{M}_k^{\bullet}} \mathcal{L}^*(\mathcal{M}_k) \longrightarrow H^{\bullet}_{\mathcal{M}_k}(\mathcal{L}(\mathcal{M}_k)$$

is an isomorphism.

9. Some evidence for conjecture 4.12.

Let X be a regular projective curve over an algebraically closed field k. Set $h_1(X) := h^1(X)(1)$ and $F := k(X)$. The Gersten complex $K_2(F) \longrightarrow k^* \otimes \mathbb{Z}[X(k)]$ provides an sequence

$$K_2(k) \hookrightarrow K_2(F) \longrightarrow k^* \otimes \mathbb{Z}[X(k)] \overset{id \otimes s}{\longrightarrow} k^* \longrightarrow 0$$

Here $s : \mathbb{Z}[X(k)] \to \mathbb{Z}$ is the augmentation map, the left arrow can be proved to be an inclusion and the right one is surjective by the Weil reciprocity law.

Let $\mathbb{Z}[X(k)]_0$ be the subgroup of the degree zero divisors.

Lemma 4.13 *Let us assume the rigidity conjecture for* $K_3^{ind}(F) \otimes \mathbb{Q}$. *Then the complex*

$$\frac{K_2(F)}{K_2(k)} \longrightarrow k^* \otimes \mathbb{Z}[X(k)]_0 \qquad (25)$$

placed in degrees $[1,2]$ *and tensored by* $\mathbb{Q}$ *is quasiisomorphic to* $RHom_{\mathcal{M}\mathcal{M}_k}(\mathbb{Q}(0), h_1(X))$.

Proof. The standard motivic Leray spectral sequence argument shows that one expects to have an exact sequence

$$RHom_{\mathcal{M}\mathcal{M}_k}(\mathbb{Q}(0), \mathbb{Q}(2)) \hookrightarrow RHom_{\mathcal{M}\mathcal{M}_X}(\mathbb{Q}(0), \mathbb{Q}(2))[-1] \longrightarrow$$

$$RHom_{\mathcal{M}\mathcal{M}_k}(\mathbb{Q}(0), \mathbb{Q}(1))[-2] = k_{\mathbb{Q}}^*[-3]$$

The complex

$$\left(B_2(F) \longrightarrow \Lambda^2 F^* \longrightarrow k^* \otimes \mathbb{Z}[X(k)] \right) \otimes \mathbb{Q}$$

placed in degrees $[1,3]$ is quasiisomorphic to $RHom_{\mathcal{M}\mathcal{M}_X}(\mathbb{Q}(0), \mathbb{Q}(2))$. It contains the subcomplex $B_2(k) \longrightarrow \Lambda^2 k^*$ computing $RHom_{\mathcal{M}\mathcal{M}_k}(\mathbb{Q}(0), \mathbb{Q}(2))$ modulo torsion. The map $B_2(F)/B_2(k) \longrightarrow \Lambda^2 F^*$ is supposed to be injective modulo torsion by the rigidity conjecture for $K_3^{ind}(F) \otimes \mathbb{Q}$. Taking the quotient we get a complex

$$\frac{K_2(F)}{K_2(k)} \longrightarrow k^* \otimes \mathbb{Z}[X(k)]$$

Our complex is its subcomplex, and the quotient is isomorphic to k^*. So we get the lemma.

Let $P(X)$ be the subgroup of principal divisors in $\mathbb{Z}[X(k)]_0$. The complex (25) containes a subcomplex $k^* \otimes P_X \overset{\text{id}}{\to} k^* \otimes P_X$. Taking the quotient we get a complex

$$\frac{K_2(F)}{K_2(k) + \{k^*, F^*\}} \longrightarrow k^* \otimes J_X(k) \qquad (26)$$

Now let us put

$$\mathcal{L}^*(\mathcal{M}_k)_{h_1(X)(1)} := \frac{K_2(F)}{K_2(k) + \{k^*, F^*\}}$$

Then the complex (26) provides the cobracket

$$\mathcal{L}^*(\mathcal{M}_k)_{h_1(X)(1)} \boxtimes h_1(X)(1)^\vee \longrightarrow \left(k_{\mathbb{Q}}^* \boxtimes \mathbb{Q}(1)^\vee\right) \otimes \left(J_X(k)_{\mathbb{Q}} \boxtimes h_1(X)^\vee\right)$$

This complex by definition is isomorphic to the complex (26).

Finally, one can define an embedding $\mathcal{L}^*(\mathcal{M}_k)_{h_1(X)(1)} \hookrightarrow \mathcal{L}(\mathcal{M}_k)_{h_1(X)(1)}$ where we think about $\mathcal{L}(\mathcal{M}_k)$ as of the Lie coalgebra given by the framed mixed motives. Namely, we can associate to a triple (X, f_1, f_2) where f_1, f_2 are rational functions on X a $h_1(X)$-framed mixed motive in such a way that we get a morphism of complexes

$$\begin{array}{ccc} \mathcal{L}(\mathcal{M}_k)_{h_1(X)(1)} \boxtimes h_1(X)(1)^\vee & \longrightarrow & \left(k_{\mathbb{Q}}^* \boxtimes \mathbb{Q}(1)^\vee\right) \otimes \left(J_X(k)_{\mathbb{Q}} \boxtimes h_1(X)^\vee\right) \\ i \uparrow & & \uparrow = \\ \mathcal{L}^*(\mathcal{M}_k)_{h_1(X)(1)} \boxtimes h_1(X)(1)^\vee & \longrightarrow & \left(k_{\mathbb{Q}}^* \boxtimes \mathbb{Q}(1)^\vee\right) \otimes \left(J_X(k)_{\mathbb{Q}} \boxtimes h_1(X)^\vee\right) \end{array}$$

(See section 9.2 below where the case of an elliptic curve is considered. The general case is completely similar).

We will see in the next section that the existence of a similar construction of $\mathcal{L}^*(\mathcal{M}_k)_{h_1(X)(2)}$ boils down to the following

Conjecture 4.14 *Assume* $k = \bar{k}$. *Then*

$$\frac{K_3^M(F)}{k^* \cdot K_2(F)} \otimes \mathbb{Q} = 0$$

This conjecture is a special case of conjecture 5.4. It is proved for curves of genus one and two using the Basss-Tate lemma in proposition 5.5.

5 Towards the Lie coalgebra $\mathcal{L}_{\leq 1}^*(E)$

We construct, assuming some conjectures on mixed **Tate** motives, a Lie coalgebra $\tilde{\mathcal{L}}_{\leq 1}^*(E)$ which should be isomorphic to $\mathcal{L}_{\leq 1}^*(E)$. For this we need to recall some basic things about the specialization.

1. The specialization homomorphism and motivic cohomology of a curve X. Let X be a smooth irreducible variety over a field k, $Y \subset X$ a smooth divisor, N_Y the normal bundle to Y and $\dot{N}_Y := N_Y \backslash$ zero section. Then one should have the specialization functor $s_Y : \mathcal{M}_T(X \backslash Y) \to \mathcal{M}_T(\dot{N}_Y)$ and hence a homomorphism of graded Lie algebras $s_Y : L(\dot{N}_Y) \to L(X \backslash Y)$. The Lie algebra $L(\dot{N}_Y)$ is a central extension :

$$0 \longrightarrow \mathbb{Q}_Y(1) \longrightarrow L(\dot{N}_Y) \longrightarrow L(Y) \longrightarrow 0$$

where $\mathbb{Q}_Y(1)$ is a one-dimensional space sitting in degree -1 (the "monodromy" around Y). Dualizing we get an extension of the Lie coalgebras

$$0 \longrightarrow \mathcal{L}(Y) \longrightarrow \mathcal{L}(\dot{N}_Y) \longrightarrow \mathbb{Q}_Y(-1) \longrightarrow 0$$

Let us restrict our attension to the generic points of X and Y. Then we are in the special case of the following general situation. Let K be a field with a discrete valuation v, the residue field k. Then one should have a Lie coalgebra $\mathcal{L}(\dot{N}_v)$ given as an extension

$$0 \longrightarrow \mathcal{L}(k) \longrightarrow \mathcal{L}(\dot{N}_v) \stackrel{v}{\longrightarrow} \mathbb{Q}_Y(-1) \longrightarrow 0 \qquad (27)$$

and a specialization homomorphism of Lie coalgebras

$$s_v : \mathcal{L}(K) :\longrightarrow \mathcal{L}(\dot{N}_v) \qquad (28)$$

such that the composition $\mathcal{L}(K)_{-1} = K^* \longrightarrow \mathcal{L}(\dot{N}_v)_{-1} \longrightarrow \mathbb{Q}_Y(-1)$ coincides with the valuation homomorphism $v : K^* \to \mathbb{Z}$.

Then there is an exact sequence of complexes

$$0 \longrightarrow \Lambda^\bullet \mathcal{L}(k) \longrightarrow \Lambda^\bullet \mathcal{L}(\dot{N}_v) \stackrel{p_v}{\longrightarrow} \Lambda^{\bullet-1} \mathcal{L}(k)(-1)[-1] \longrightarrow 0 \qquad (29)$$

Composing the morphism coming from (28) with the projection map p_v in (29) we get the residue map (which is a morphism of complexes)

$$r_v : \Lambda^\bullet \mathcal{L}(K) :\longrightarrow \Lambda^{\bullet-1} \mathcal{L}(k)(-1)[-1] \qquad (30)$$

Lemma 5.1 *The residue map r_v vanishes on $\Lambda^\bullet \mathcal{L}(K)_{\geq 2}$.*

Proof. Clear from (27) since $\mathbb{Q}(-1)$ is sitting in the degree +1.

Let us assume that we are always given such a specialization data. In particular, let F be the field of rational functions on a regular curve X over an arbitrary field k, x is a closed point of X, $v(x)$ the corresponding valuation on F and $k(x)$ the residue field. Then one should have a specialization morphism $\mathcal{L}(F) \longrightarrow \mathcal{L}(\dot{N}_{v(x)})$ and thus a morphism of complexes

$$\Lambda^\bullet \mathcal{L}(F) \longrightarrow \Lambda^{\bullet-1} \mathcal{L}(k(x))(-1)[-1]$$

Taking the sum of these morphisms over the set X_1 of all closed points of X we get a bicomplex

$$\Lambda^\bullet \mathcal{L}(F) \longrightarrow \bigoplus_{x \in X_1} \Lambda^{\bullet-1} \mathcal{L}(k(x))[-1] \tag{31}$$

Denote by $\Lambda(X)^\bullet$ the total complex associated with this bicomplex.

If X is a regular curve over a closed field k we get

$$\Lambda^\bullet \mathcal{L}(F) \longrightarrow \Lambda^{\bullet-1} \mathcal{L}(k) \otimes \mathbb{Q}[X(k)][-1] \tag{32}$$

Then $\Lambda(X)^\bullet = \oplus_{n \geq 1} \Lambda(X, n)^\bullet$ where $\Lambda(X, n)^\bullet$ in the case $k = \bar{k}$ looks as follows:

$$\mathcal{L}(F)_n \longrightarrow \bigoplus \mathcal{L}(F)_{i_1} \wedge \mathcal{L}(F)_{i_2} \longrightarrow \bigoplus \mathcal{L}(F)_{i_1} \wedge \mathcal{L}(F)_{i_2} \wedge \mathcal{L}(F)_{i_3} \longrightarrow \cdots$$

$$\downarrow r_1 \qquad\qquad\qquad \downarrow r_2$$

$$\mathcal{L}(k)_{n-1} \otimes \mathbb{Q}[X(k)] \longrightarrow \bigoplus \mathcal{L}(k)_{j_1} \wedge \mathcal{L}(k)_{j_2} \otimes \mathbb{Q}[X(k)] \longrightarrow \cdots$$

Here the summation is over $i_1 \leq \ldots \leq i_k$, $i_1 + \ldots + i_k = n$ and $j_1 \leq \ldots \leq j_k$, $j_1 + \ldots + j_k = n - 1$.

Conjecture 5.2 *The complex $\Lambda(X, n)^\bullet$ is quasiisomorphic to $RHom_{\mathcal{MM}_X}(\mathbb{Q}(0), \mathbb{Q}(n))$.*

Let $s : \mathbb{Q}[X(k)] \longrightarrow \mathbb{Q}$ be the augmentation map $\{x\} \longmapsto 1$. Consider a morphism of complexes

$$id \otimes s : \quad \Lambda^{\bullet-1} \mathcal{L}(k) \otimes \mathbb{Q}[X(k)][-1] \quad \longrightarrow \quad \Lambda^{\bullet-1} \mathcal{L}(k)[-1]$$

Conjecture 5.3 (A strong reciprocity law). *There exists a homomorphism of complexes*

$$R : \Lambda(X)^\bullet \longrightarrow \Lambda^{\bullet-1} \mathcal{L}(k)[-2]$$

which on the subcomplex $\Lambda^{\bullet-1} \mathcal{L}(k) \otimes \mathbb{Q}[X(k)][-2] \subset \Lambda(X)^\bullet$ coincides with $id \otimes s$.

Taking the cohomology we get the reciprocity law for K-groups which tells us that the composition

$$gr_n^\gamma K_m(F)_{\mathbb{Q}} \xrightarrow{res} gr_{n-1}^\gamma K_{m-1}(k)_{\mathbb{Q}} \otimes \mathbb{Q}[X(k)] \xrightarrow{id \otimes s} gr_{n-1}^\gamma K_{m-1}(k)_{\mathbb{Q}}$$

is zero.

The motivic Leray spectral sequence for the structural morphism $\pi : X \to \operatorname{Spec}(k)$ should degenerate in the E_2-term, which looks as follows:

$$Ext_{\mathcal{MM}_k}^i(\mathbb{Q}(0), h^j(E)(n))) \longrightarrow Ext_{\mathcal{MM}_X}^{i+j}(\mathbb{Q}(0), \mathbb{Q}(n)))$$

Since $h^0(X)(n) = \mathbb{Q}(n)$ and $h^2(X)(n) = \mathbb{Q}(n-1)$, the group $Ext^{n-1}_{\mathcal{M}\mathcal{M}_k}(\mathbb{Q}(0), h^1(X)(n)))$ is the middle cohomology of the complex

$$0 \longrightarrow Ext^i_{\mathcal{M}\mathcal{M}_k}(\mathbb{Q}(0), \mathbb{Q}(n))) \xrightarrow{\pi^*} Ext^i_{\mathcal{M}\mathcal{M}_E}(\mathbb{Q}(0), \mathbb{Q}(n))) \xrightarrow{R}$$

$$Ext^{i-2}_{\mathcal{M}\mathcal{M}_k}(\mathbb{Q}(0), \mathbb{Q}(n-1))) \longrightarrow 0$$

which is exact in the other terms. We should have morphisms on the level of complexes:

$$RHom_{\mathcal{M}\mathcal{M}_k}(\mathbb{Q}(0), \mathbb{Q}(n)) \xrightarrow{\pi^*} RHom_{\mathcal{M}\mathcal{M}_E}(\mathbb{Q}(0), \mathbb{Q}(n))$$

$$RHom_{\mathcal{M}\mathcal{M}_E}(\mathbb{Q}(0), \mathbb{Q}(n)) \xrightarrow{R} RHom_{\mathcal{M}\mathcal{M}_k}(\mathbb{Q}(0), \mathbb{Q}(n-1))[-2]$$

Therefore we expect a quasiisomorphism

$$\mathrm{Ker}\Big(\frac{\Lambda(X,n)^\bullet}{\Lambda^\bullet_{(n)}\mathcal{L}(k)} \xrightarrow{R} \Lambda^{\bullet-1}_{(n-1)}\mathcal{L}(k)[-2]\Big) = RHom_{\mathcal{M}\mathcal{M}_k}(\mathbb{Q}(0), h^1(X)(n)) \tag{33}$$

(The subscript (n) means the degree n part of the complex).

$\mathcal{L}(k)$ is a Lie subcoalgebra of $\mathcal{L}(F)$. Let

$$\bar{\mathcal{L}}(F) := \mathcal{L}(F)/\mathcal{L}(k)$$

be the quotient Lie coalgebra.

Conjecture 5.4 *a)* $H^1_{(n)}(\bar{\mathcal{L}}(F)) = 0$ *if* $n \neq 1$.
b) $H^i(\bar{\mathcal{L}}(F)) = 0$ *if* $i > 2$.

Here $H^i_{(n)}$ is the degeee n part of H^i.

Remark. By the rigidity conjecture $gr^\gamma_n K_{2n-1}(k) \otimes \mathbb{Q} \overset{?}{=} gr^\gamma_n K_{2n-1}(F) \otimes \mathbb{Q}$. So if for a field K one ha s $H^1_{(n)}\mathcal{L}(K)) = gr^\gamma_n K_{2n-1}(K) \otimes \mathbb{Q}$, as expected, the n $H^1_{(n)}(\mathcal{L}(k)) = H^1_{(n)}(\mathcal{L}(F))$ for $n \neq 1$, so we get the first statement of the conjecture.

2. The Lie coalgebra $\tilde{\mathcal{L}}^*_{\leq 1}(E)$.

Proposition 5.5 *Let F be the field of rational functions on a curve X of genus ≤ 2 over an algebraically closed field k. $H^n_{(n)}(\bar{\mathcal{L}}(F)) = 0$ if $n > 2$.*

Proof. If $X = P^1$ the statement is trivial. If X is a curve of genus 1 or 2 then it is a hyperelliptic curve, so F is a degree two extension of $k(P^1)$. We have to show that $K^M_n(F) = \{k^*, F^*, F^*, ..., F^*\}$. Let $K \subset F$ be a degree two extension of fields. Then according to the Bass-Tate lemma [BT] one has $K^M_n(F) = K^M_{n-1}(K) \otimes F^*$. Let $p : E \to P^1$ be a degree two projection. Set $K := p^*k(P^1)$. It is known that $K_2(k(P^1)) = \{k^*, k(P^1)\}$. Thus $K^M_n(F) = \{k^*, ..., k^*, K^*, F^*\}$, which is even stronger then the result we need.

Now let F be the field of rational functions on an elliptic curve E over an algebraically closed field k.

Definition 5.6 $\tilde{\mathcal{L}}_{\leq 1}^{*}(E)_{\mathcal{H}} := J_{\mathbb{Q}}$ and $\tilde{\mathcal{L}}_{\leq 1}^{*}(E)_{\mathcal{H}(n-1)} := H_{(n)}^{2}(\bar{\mathcal{L}}(F))$ *for* $n \geq 2$.
Set

$$\tilde{\mathcal{L}}_{\leq 1}^{*}(E) := \quad \oplus_{i \geq 1} \mathcal{L}(k)_i \boxtimes \mathbb{Q}(i)^{\vee} \quad \bigoplus \quad \oplus_{n \geq 1} \tilde{\mathcal{L}}_{\leq 1}^{*}(E)_{\mathcal{H}(n-1)} \boxtimes \mathcal{H}(n-1)^{\vee}$$

We will also need a slightly bigger object $\hat{\mathcal{L}}_{\leq 1}^{*}(E)$ which sits in the exact sequence

$$0 \longrightarrow \tilde{\mathcal{L}}_{\leq 1}^{*}(E) \longrightarrow \hat{\mathcal{L}}_{\leq 1}^{*}(E) \longrightarrow \mathbb{Q} \boxtimes \mathcal{H}^{\vee} \longrightarrow 0$$

such that the $\mathcal{H}^{\vee}$-component of $\hat{\mathcal{L}}_{\leq 1}^{*}(E)$ is $\mathrm{Pic}(\mathrm{J})_{\mathbb{Q}} \boxtimes \mathcal{H}^{\vee}$.

Theorem 5.7 *a)* $\hat{\mathcal{L}}_{\leq 1}^{*}(E)$ *has a structure of a Lie coalgebra in* $\mathcal{P}_E^*$ *and* $\mathcal{L}_{[0]}^*(E)$ *is its Lie subcoalgebra.*

b) Let us assume the conjecture 5.3. Then $\tilde{\mathcal{L}}_{\leq 1}^{*}(E)$ *is a Lie coalgebra in* $\mathcal{P}^*(E)$.

c) If we assume in addition conjectures 5.2 and 5.4, then the $\mathcal{H}(n-1)^{\vee}$*- isotypical component of* $\Lambda_{\mathcal{P}^{\bullet}(E)}^{\bullet} \tilde{\mathcal{L}}_{\leq 1}^{*}(E)$ *is quasiisomorphic to* $RHom_{\mathcal{MM}_k}(\mathbb{Q}(0), \mathcal{H}(n-1))$.

Proof. Consider $\mathcal{L}(k) \subset \mathcal{L}(F)$ as a decreasing filtration $G^{\bullet}$ on $\mathcal{L}(F)$:

$$G^0 \mathcal{L}(F) = \mathcal{L}(k), \quad G^1 \mathcal{L}(F) = \mathcal{L}(F)$$

It induces a decreasing filtration $G^{\bullet}$ on $\Lambda^{\bullet} \mathcal{L}(F)$ such that

$$gr_G^m(\Lambda^{\bullet} \mathcal{L}(F)) = \Lambda^m \mathcal{L}(k) \otimes \Lambda^{\bullet} \bar{\mathcal{L}}(F) \tag{34}$$

Let us extend this filtration to the complex $\Lambda(E, n)$ so that

$$gr_G^m(\Lambda^{\bullet-1} \mathcal{L}(k)[-1] \otimes \mathbb{Q}[E(k)]) = \Lambda^m \mathcal{L}(k)[-1] \otimes \mathbb{Q}[E(k)]$$

Embedding $\Lambda^{\bullet} \mathcal{L}(k)$ as a subcomplex into $\Lambda^{\bullet} \mathcal{L}(F)$ we notice that it is annihilated by the residue homomorphism. Therefore we get a subcomplex in the complex $\Lambda(E, n)$. Let us take the quotient of $\Lambda(E, n)$ by this subcomplex and consider the induced filtration $G^{\bullet}$ on it.

We are going to compute the spectral sequence corresponding to this filtration and show that, assuming conjecture 5.4, its E_1-term provides us the standard cochain complex of a Lie coalgebra $\hat{\mathcal{L}}_{\leq 1}^{*}(E)$ in the $\otimes'$-category $\mathcal{P}_E^*$. All higher differentials will vanish.

The complex $gr_G^m(\Lambda(E, n)^{\bullet})$ is the tensor product of $\Lambda^m \mathcal{L}(k)$ with the complex

$$
\begin{array}{ccccccc}
\bar{\mathcal{L}}^{*}(F) & \longrightarrow & \Lambda^2 \bar{\mathcal{L}}^{*}(F) & \longrightarrow & \Lambda^3 \bar{\mathcal{L}}^{*}(F) & \longrightarrow & \dots \\
\downarrow & & \downarrow & & \downarrow & & \\
\mathbb{Q}[E(k)] & \longrightarrow & 0 & \longrightarrow & 0 & \longrightarrow & \dots
\end{array}
$$

placed in degrees ≥ 2. According to conjecture 5.4 and definition 5.6 the last complex is quasiisomorphic to

$$\left(\text{Pic}(J)_{\mathbb{Q}} \quad \bigoplus \quad \oplus_{j\geq 1}\tilde{\mathcal{L}}^*(E)_{\mathcal{H}(j)}\right)[-3]$$

Indeed, the top line is quasiisomorphic to $\bar{F}_{\mathbb{Q}}^*[-2] \quad \bigoplus \quad \oplus_{j\geq 1}\tilde{\mathcal{L}}^*(E)_{\mathcal{H}(j)}[-3]$. The vertical arrow is zero on the second term and sends $\bar{F}_{\mathbb{Q}}^*$ isomorphically to I_E^2. It remains to notice that $\text{Pic}(J)_{\mathbb{Q}} = \mathbb{Q}[E(k)]/I_E^2$. So the $\mathcal{H}(n)^{\vee}$-part of the E_1-term is a direct sum

$$\oplus_{1\leq i\leq n-1}(\Lambda^{\bullet}\mathcal{L}(k))_{\mathbb{Q}(i)} \otimes \tilde{\mathcal{L}}_{\leq 1}^*(E)_{\mathcal{H}(n-i)} \quad \bigoplus \quad (\Lambda^{\bullet}\mathcal{L}(k))_{\mathbb{Q}(n)} \otimes \text{Pic}(E)_{\mathbb{Q}}$$

which is just the $\mathcal{H}(n)^{\vee}$-part of $\Lambda_{\mathcal{P}_E}^{\bullet}\hat{\mathcal{L}}_{\leq 1}^*(E)$. The differential in the E_1-term provides us with a homomorphism

$$\mathcal{L}_{\leq 1}^*(E)_{\mathcal{H}(n)} \longrightarrow \oplus_{1\leq k\leq n}\mathcal{L}_{\leq 1}^*(E)_{\mathcal{H}(n-k)} \otimes \mathcal{L}(k)_{\mathbb{Q}(k)} \quad \bigoplus \quad \mathcal{L}(k)_{\mathbb{Q}(n)} \otimes \text{Pic}(E)_{\mathbb{Q}}$$

Consider it as a $\mathcal{H}(n)^{\vee}$-part of the cobracket $\mathcal{L}_{\leq 1}^*(E) \longrightarrow \Lambda_{\mathcal{P}_E}^2\mathcal{L}_{\leq 1}^*(E)$. Then it is easy to see that we get a Lie coalgebra and the differentials in E_1 are just the same as in $\Lambda_{\mathcal{P}_E}^{\bullet}\hat{\mathcal{L}}_{\leq 1}^*(E)$.

Example. The $\mathcal{H}(2)^{\vee}$-component of the E_1 term looks as follows:

$$\begin{array}{ccc} \tilde{\mathcal{L}}_{\leq 1}^*(E)_{\mathcal{H}(2)} & \longrightarrow & \tilde{\mathcal{L}}_{\leq 1}^*(E)_{\mathcal{H}(1)} \otimes k_{\mathbb{Q}}^* \\ \downarrow & & \downarrow \\ \mathcal{L}(k)_{\mathbb{Q}(2)} \otimes \text{Pic}(E)_{\mathbb{Q}} & \longrightarrow & \Lambda^2 k^* \otimes \text{Pic}(E)_{\mathbb{Q}} \end{array}$$

The complex $\Lambda_{\mathcal{P}_E}^{\bullet}\hat{\mathcal{L}}_{\leq 1}^*(E)$ should be quasiisomorphic to

$$RHom_{\mathcal{MM}_E}(\mathbb{Q}(0), \mathbb{Q}(n+1))/\pi^* RHom_{\mathcal{MM}_k}(\mathbb{Q}(0), \mathbb{Q}(n+1))$$

Here $\pi : E \to \text{Spec}(k)$ is the structure morphism.

According to a very special case of conjecture 5.3 the composition

$$\hat{\mathcal{L}}_{\leq 1}^*(E)_{\mathcal{H}(n)} \longrightarrow \mathcal{L}(k)_{\mathbb{Q}(n)} \otimes \text{Pic}(E)_{\mathbb{Q}} \xrightarrow{id\otimes\delta} \mathcal{L}(k)_{\mathbb{Q}(n)}$$

is zero. For $n = 1$ the statement follows from the Weil reciprocity law. For $n = 2$ it can be deduced from the rigidity conjecture, and I am sure this is true in general. This provides the statement b) of the theorem. The statements a) and c) has already been proved above.

The theorem is proved.

Remark. Let X be a curve of genus 2. Then it is a hyperelliptic curve. Therefore $k(X)$ is a degree 2 extension of $k(P^1)$, and thus we can use the Bass-Tate lemma to prove an analog of proposition 5.5. This means the following. Consider the mixed category generated by $h^1(X)$. Then we can construct the

$\mathbb{Q}$-vector spaces $\mathcal{L}^*_{\leq 1}(X)_{h^1(X)(2)}$ and $\mathcal{L}^*_{\leq 1}(X)_{h^1(X)(3)}$ just as before, and get the coproduct map

$$\mathcal{L}^*_{\leq 1}(X)_{h^1(X)(3)} \longrightarrow \mathcal{L}^*_{\leq 1}(X)_{h^1(X)(2)} \otimes k^* \oplus B_2(k)_{\mathbb{Q}} \otimes J(X)_{\mathbb{Q}}$$

Here $J(X)$ is the k-points of the Jacobian of X and, of cource, $J(X)_{\mathbb{Q}} = \mathcal{L}^*_{\leq 1}(X)_{h^1(X)(1)}$ and $B_2(k)_{\mathbb{Q}} = \mathcal{L}^*_{\leq 1}(X)_{\mathbb{Q}(2)}$. So we are getting a subcoalgebra Lie of $\mathcal{L}^*_{\leq 1}(X)$ with the expected cohomology. This provide some evidence for the existence of a small motivic Lie coalgebra with a similar list of properties for the category generated by a genus two curve!

It would be very interesting to know whether proposition 5.5 is true for an arbitrary curve over k. If it is true, one might hope for the existence of a small motivic Galois group for the whole category of mixed motives.

6 Reflections on elliptic motivic complexes

1. Some hypothesis on elliptic motivic complexes. Suppose that E is defined over an algebraically closed field k. I conjecture that:

a) For any m, n such that $n \geq 0, n + 2m > 0$ there exists an abelian group $B^*_{S^n \mathcal{H}(m)}$ together with homomorphisms

$$\delta_1 : B^*_{S^n \mathcal{H}(m)} \longrightarrow B^*_{S^{n-1} \mathcal{H}(m)} \otimes J, \quad n + 2m > 0$$

$$\delta_2 : B^*_{S^n \mathcal{H}(m)} \longrightarrow B^*_{S^n \mathcal{H}(m-1)} \otimes k^* \quad m \geq 1$$

For $m \geq 1$ one should have

$$B^*_{S^{2m+1} \mathcal{H}(-m)} = J_{S^{2m+1} \mathcal{H}(-m)} \quad \text{and} \quad B^*_{\mathbb{Q}(m)} = B_m(k)$$

where $B_m(k)$ are the groups introduced in [G1-2]. For instance $B^*_{\mathcal{H}} = J$ and $B^*_{\mathbb{Q}(1)} = k^*$.

b) One should get a bicomplex $\Gamma^*_{S^n \mathcal{H}(m)}$ of the following shape.
For $m > 0$:

$$
\begin{array}{ccccccc}
B^*_{S^n \mathcal{H}(m)} & \longrightarrow & B^*_{S^n \mathcal{H}(m-1)} \wedge k^* & \longrightarrow \dots \longrightarrow & B^*_{S^n \mathcal{H}(1)} \wedge \Lambda^{m-1} k^* \\
\downarrow & & \downarrow & \dots & \downarrow \\
B^*_{S^{n-1} \mathcal{H}(m)} \wedge J & \longrightarrow & B^*_{S^{n-1} \mathcal{H}(m-1)} \wedge J \wedge k^* & \longrightarrow \dots \longrightarrow & B^*_{S^{n-1} \mathcal{H}(1)} \wedge J \wedge \Lambda^{m-1} k^* \\
\downarrow & & \downarrow & \dots & \downarrow \\
\dots & & \dots & \dots & \dots \\
\downarrow & & \downarrow & \dots & \downarrow \\
B^*_{\mathbb{Q}(m)} \wedge \Lambda^n J & \longrightarrow & B^*_{\mathbb{Q}(m-1)} \wedge \Lambda^n J \wedge k^* & \longrightarrow \dots \longrightarrow & \Lambda^n J \wedge \Lambda^m k^*
\end{array}
$$

Here the horizontal differentials are $b \wedge x \longmapsto \delta_1(b) \wedge x$ and the vertical ones are $b \wedge x \longmapsto \delta_2(b) \wedge x$.

For $m = 0$:

$$B^*_{S^n \mathcal{H}} \longrightarrow B^*_{S^{n-1} \mathcal{H}} \wedge J \longrightarrow \dots \longrightarrow B^*_{S^2 \mathcal{H}} \wedge \Lambda^{n-2} J \longrightarrow \Lambda^n J$$

For $m < 0$, set $m' := -m$. Then

$$B^*_{S^n \mathcal{H}(-m')} \longrightarrow B^*_{S^{n-1} \mathcal{H}(-m')} \otimes J \longrightarrow \dots \longrightarrow B^*_{S^{2m'+1} \mathcal{H}(-m')} \otimes \Lambda^{n-2m'-1} J$$

c) The total complex associated with the bicomplex $\Gamma^*_{S^n \mathcal{H}(m)}$ (abusing notations I will denote it also by $\Gamma^*_{S^n \mathcal{H}(m)}$) after tensoring with $\mathbb{Q}$ should be quasiisomorphic to $RHom_{\mathcal{MM}_k}(\mathbb{Q}(0), S^n \mathcal{H}(m))$.

Example. $\Gamma^*_{\mathcal{H}(m)}$ is the total complex associated with the bicomplex

$$
\begin{array}{ccccccccc}
B^*_{\mathcal{H}(m)} & \to & B^*_{\mathcal{H}(m-1)} \wedge k^* & \to \dots \to & B^*_{\mathcal{H}(2)} \wedge \Lambda^{m-2} k^* & \to & B^*_3(E) \wedge \Lambda^{m-1} k^* \\
\downarrow & & \downarrow & \dots & \downarrow & & \downarrow \\
B^*_{\mathbb{Q}(m)} \wedge J & \to & B^*_{\mathbb{Q}(m-1)} \wedge J \wedge k^* & \to \dots \to & B^*_{\mathbb{Q}(2)} \wedge J \wedge \Lambda^{m-2} k^* & \to & J \wedge \Lambda^m k^*
\end{array}
$$

d) There should be a variation $P_{n,m}$ of *mixed elliptic motives* framed by $\mathbb{Q}(0)$ and $S^n \mathcal{H}(m)$ over a certain *finite dimensional* variety $X(n,m)$ over k. The groups $B^*_{S^n \mathcal{H}(m)}$ should come from it as follows. The variation $P_{m,n}$ provides a homomorphism

$$\tilde{l}_{n,m} : \mathbb{Q}[X(n,m)(k)] \to \mathcal{A}(\mathbb{Q}(0), S^n \mathcal{H}(m))$$

where $l_{n,m}(\{x\})$ for $x \in X_{n,m}$ is the class of the framed mixed motive $P_{n,m}(x)$. ($P_{n,m}(x)$ is the fiber at x of the variation $P_{n,m}$). Passing to the coalgebra Lie we get a homomorphism

$$l_{n,m} : \mathbb{Q}[X(n,m)(k)] \to \mathcal{L}(E)_{S^n \mathcal{H}(m)^\vee}$$

Set $B_{S^n \mathcal{H}(m)} := Im(l_{n,m})$. The group $B^*_{S^n \mathcal{H}(m)}$ is the largest $\mathbb{Q}$-subspace of $B_{S^n \mathcal{H}(m)}$ such that the restriction of the coproduct δ to the group $B^*_{S^n \mathcal{H}(m)}$ has non zero components only in

$$\mathcal{L}_{S^{n-1} \mathcal{H}(m)^\vee} \wedge J \boxtimes \mathcal{H}^\vee \quad \oplus \quad \mathcal{L}_{S^n \mathcal{H}(m-1)^\vee} \wedge k^* \boxtimes \mathbb{Q}(1)^\vee, \quad \text{if } m > 1$$

and in

$$\mathcal{L}_{S^{n-1} \mathcal{H}(m)^\vee} \wedge J \boxtimes \mathcal{H}^\vee, \quad \text{if } m \leq 1$$

By definition δ_1 and δ_2 are the components of the restriction of δ to $B^*_{S^n \mathcal{H}(m)} \boxtimes S^n \mathcal{H}(m)^\vee$. The restriction of δ to $B_{S^n \mathcal{H}(m)} \boxtimes S^n \mathcal{H}(m)^\vee$ may be more complicated.

The periods of the $\mathbb{R}$-Hodge realization of the variation $P_{n,m}$ are the new transcendental functions denoted $\mathcal{P}_{n,m}$ which is needed to get the special values $L(S^n\mathcal{H}, m+n)$.

Example. $P_{n,1}$ is the (motivic) elliptic polylogarithm sheaf on $E\backslash 0$, and $\mathcal{P}_{n,1}$ are the Eisenstein-Kronecker series from 1.5. The group $\mathcal{B}_{S^n\mathcal{H}(1)}$ (resp. $\mathcal{B}^*_{S^n\mathcal{H}(1)}$) should coincide with the group

$$\mathcal{B}_{n+2}(E) = \frac{\mathbb{Q}[E(k]]}{\mathcal{R}_{n+2}(E)}, \quad \mathcal{B}^*_{n+2}(E) = \frac{I_E^{2n+2}}{\mathcal{R}_{n+2}(E)}$$

which will be defined later on. Here I_E is the augmentation ideal of the group algebra $\mathbb{Q}[E(k)]$.

2. The structure of $L^*(E)$ and elliptic motivic complexes. Let us define the $\mathbb{Q}$-vector spaces $C^*_{S^n\mathcal{H}(m)}$ by setting

$$H^1(I^*(E)) = \oplus C^*_{S^n\mathcal{H}(m)} \boxtimes S^n\mathcal{H}(m)^\vee$$

Let δ^* be the the cobracket in the Lie coalgebra $\mathcal{L}^*(E)$. Conjecture 4.10 means that

$$\delta^* : C^*_{S^n\mathcal{H}(m)} \longrightarrow C^*_{S^n\mathcal{H}(m-1)} \wedge k^* \quad \oplus \quad C^*_{S^{n-1}\mathcal{H}(m)} \wedge J \tag{35}$$

Lemma 6.1 *The $S^n\mathcal{H}(m)^\vee$-isotypical component of the Serre-Hochschild spectral sequence for the ideal $I^*(E) \subset L^*(E)$ computing cohomology of $L^*(E)$ collapses to the total complex associated with a bicomplex (which should be tensored $\boxtimes S^n\mathcal{H}(m)^\vee$):*

$$
\begin{array}{ccccccc}
C^*_{S^n\mathcal{H}(m)} & \to & C^*_{S^n\mathcal{H}(m-1)} \otimes k^* & \to \cdots \to & C^*_{S^n\mathcal{H}(-[\frac{n-1}{2}])} \otimes S^{m+[\frac{n-1}{2}]}k^* \\
\downarrow & & \downarrow & & \downarrow \\
C^*_{S^{n-1}\mathcal{H}(m)} \otimes J & \to & C^*_{S^{n-1}\mathcal{H}(m-1)} \otimes J \otimes k^* & \to \cdots \to & C^*_{S^{n-1}\mathcal{H}(-[\frac{n-2}{2}])} \otimes J \otimes S^{m+[\frac{n-2}{2}]}k^* \\
\downarrow & & \downarrow & & \downarrow \\
\cdots & & \cdots & & \cdots \\
\downarrow & & \downarrow & & \downarrow \\
C^*_{\mathcal{H}(m)} \otimes \Lambda^{n-1}J & \to & C^*_{\mathcal{H}(m-1)} \otimes \Lambda^{n-1}J \otimes k^* & \to \cdots \to & C^*_{\mathcal{H}(1)} \otimes \Lambda^{n-1}J \otimes \Lambda^{m-1}k^* \\
\downarrow & & \downarrow & & \downarrow \\
C^*_{\mathbb{Q}(m)} \otimes \Lambda^n J & \to & C^*_{\mathbb{Q}(m-1)} \otimes \Lambda^n J \otimes k^* & \to \cdots \to & \Lambda^n J \otimes \Lambda^m k^*
\end{array}
$$

The length of the rows decreases when we are going down. Here is how it looks for $n = 4, m = 1$:

$$
\begin{array}{ccccc}
C^*_{S^4\mathcal{H}(1)} & \to & C^*_{S^4\mathcal{H}} \otimes k^* & \to & C^*_{S^4\mathcal{H}(-1)} \otimes \Lambda^2 k^* \\
\downarrow & & \downarrow & & \downarrow \\
C^*_{S^3\mathcal{H}(1)} \otimes J & \to & C^*_{S^3\mathcal{H}} \otimes J \otimes k^* & \to & C^*_{S^3\mathcal{H}(-1)} \otimes J \otimes \Lambda^2 k^* \\
\downarrow & & \downarrow & & \\
C^*_{S^2\mathcal{H}(1)} \otimes \Lambda^2 J & \to & C^*_{S^2\mathcal{H}} \otimes \Lambda^2 J \otimes k^* & & \\
\downarrow & & & & \\
C^*_{\mathcal{H}(1)} \otimes \Lambda^3 J & & & & \\
\downarrow & & & & \\
k^* \otimes \Lambda^2 J & & & &
\end{array}
$$

Proof. The E_1 term of this spectral sequence is

$$
E_1^{p,q} = C^p\big(k^* \oplus J, \; H^q_{\mathcal{P}^{\bullet}_E}(\mathcal{I}^*(E))\big)_{S^n\mathcal{H}(m)^\vee}
$$

Since $H^q_{\mathcal{P}^{\bullet}_E}(\mathcal{I}^*(E))$ is zero for $q > 1$ the only non zero rows are $E_1^{\bullet,0}$ and $E_1^{\bullet,1}$. Moreover $E_1^{p,0}$ is zero unless $p = n + m$, and

$$
E_1^{m+n,0} = \Lambda^n J \otimes \Lambda^m k^*
$$

Further, $E_1^{p,1}$ is non zero only if $0 \le p < m + n$. If so, then

$$
E_1^{p,1} = \oplus_{a+b=p} \; C^*_{S^{n-a}\mathcal{H}(m-b)} \otimes \Lambda^a J \otimes \Lambda^b k^*
$$

So the E_1 term of the spectral sequence gives us precisely the groups in the bicomplex. The differential $d_1 : E_1 \to E_1$ provides all the differentials exept the last two in the right bottom corner targeting to $\Lambda^n J \otimes \Lambda^m k^*$. These two we get from the differential d_2. The lemma is proved.

The structure of the quotient $L^*(E)/[I^*(E), I^*(E)]$. There is an exact sequence of Lie algebras

$$
0 \to \frac{I^*(E)}{[I^*(E), I^*(E)]} \to \frac{L^*(E)}{[I^*(E), I^*(E)]} \to \frac{L^*(E)}{I^*(E)} \to 0 \tag{36}
$$

The action of $L^*(E)$ on $I^*(E)$ leads to the action of $L^*(E)/I^*(E)$ on $I^*(E)/[I^*(E), I^*(E)]$. The Lie algebra structure on $L^*(E)/[I^*(E), I^*(E)]$ is determined by this action.

The inclusion $L^*(E)_{\mathbb{Q}(1)^\vee} \oplus L^*(E)_{\mathcal{H}(1)^\vee} \hookrightarrow L^*(E)$ provides a canonical splitting $s : L^*(E)/I^*(E) \to L^*(E)/[I^*(E), I^*(E)]$ as an extension of $\mathbb{Q}$-vector spaces. Recall that $L^*(E)/I^*(E) = (k^*_{\mathbb{Q}})^\vee \boxtimes \mathbb{Q}(1) \oplus J^\vee_{\mathbb{Q}} \boxtimes \mathcal{H}$ is an abelian Lie algebra. So to define the Lie algebra structure on $L^*(E)/[I^*(E), I^*(E)]$ we need to know the $\mathbb{Q}$-vector spaces $C^*_{S^n\mathcal{H}(m)}$ and the homomorphisms (35). That is precisely what the following two conjectures are doing.

Conjecture 6.2 *i) There is a canonical map $B^*_{S^n\mathcal{H}(m)} \hookrightarrow C^*_{S^n\mathcal{H}(m)}$ such that* $\delta^*|_{B^*_{S^n\mathcal{H}(m)}} = \delta_1 + \delta_2$.

ii) $B^*_{S^n\mathcal{H}(m)} = C^*_{S^n\mathcal{H}(m)}$ if $n = 0, 1$. Otherwise one has an exact sequence

$$0 \to B^*_{S^n\mathcal{H}(m)} \to C^*_{S^n\mathcal{H}(m)} \to B^*_{S^n\mathcal{H}(-[\frac{n-1}{2}])} \otimes S^{m+[\frac{n-1}{2}]}k^* \to 0 \qquad (37)$$

Set $\bar{C}^*_{S^n\mathcal{H}(m)} := C^*_{S^n\mathcal{H}(m)}/B^*_{S^n\mathcal{H}(m)}$. Then δ^* induces a map

$$\bar{C}^*_{S^n\mathcal{H}(m)} \longrightarrow \bar{C}^*_{S^n\mathcal{H}(m-1)} \otimes k^* \quad \oplus \quad \bar{C}^*_{S^{n-1}\mathcal{H}(m)} \otimes J \qquad (38)$$

Now we can formulate the second part of conjecture (4.10).

Conjecture 6.3 *The second component of the map (38) is zero, and the first component coinsides with the homomorphism*

$$B^*_{S^n\mathcal{H}(-[\frac{n-1}{2}])} \otimes S^{m+[\frac{n-1}{2}]}k^* \longrightarrow B^*_{S^n\mathcal{H}(-[\frac{n-1}{2}])} \otimes S^{m+[\frac{n-1}{2}]-1}k^* \otimes k^* \qquad (39)$$

given by the identity × the Koszul differential.

Theorem 6.4 *Assume all the conjectures of this section. Then the complex $\Gamma^*_{S^n\mathcal{H}(m)}$ is quasiisomorphic to $RHom_{\mathcal{M}_E}(\mathbb{Q}(0), S^n\mathcal{H}(m))$*

Proof. The complex $RHom_{\mathcal{M}_E}(\mathbb{Q}(0), S^n\mathcal{H}(m))$ is quasiisomorphic to the standard complex $C^\bullet(\mathcal{L}(E))$ of the Lie coalgebra $\mathcal{L}(E)$. According to the property 1) there is a morphism of complexes

$$C^\bullet_{\mathcal{P}^*_E}(\mathcal{L}^*(E)) \longrightarrow C^\bullet_{\mathcal{P}_E}(\mathcal{L}(E))$$

which is a quasiisomorphism thanks to the property 2).

The part ii) of conjecture (6.2) just means that there is a canonical embedding of complexes

$$j : \Gamma^*_{S^n\mathcal{H}(m)} \hookrightarrow C^\bullet_{\mathcal{P}^*_E}(\mathcal{L}^*(E))$$

We are going to show that it is a quasiisomorphism.

Using the part i) of conjecture (6.2) we see that the quotient of the bicomplex we got from the spectral sequence along the bicomplex $\Gamma^*_{S^n\mathcal{H}(m)}$ looks as follows. The two bottom rows become zero, and each of the remaining rows is $B^*_?$ times a Koszul complex. For instance in the case $n = 4, m = 1$ the quotient is

$$
\begin{array}{ccccc}
B^*_{S^4\mathcal{H}(-1)} \otimes S^2 k^* & \to & B^*_{S^4\mathcal{H}(-1)} \otimes k^* \otimes k^* & \to & B^*_{S^3\mathcal{H}(-1)} \otimes \Lambda^2 k^* \\
0 \downarrow & & 0 \downarrow & & 0 \downarrow \\
B^*_{S^3\mathcal{H}(-1)} \otimes J \otimes S^2 k^* & \to & B^*_{S^3\mathcal{H}(-1)} \otimes J \otimes k^* \otimes k^* & \to & B^*_{S^3\mathcal{H}(-1)} \otimes J \otimes \Lambda^2 k^* \\
0 \downarrow & & 0 \downarrow & & \\
B^*_{S^2\mathcal{H}} \otimes \Lambda^2 J \otimes k^* & \to & B^*_{S^2\mathcal{H}} \otimes \Lambda^2 J \otimes k^* & &
\end{array}
$$

7   The complexes $B(E, n)^\bullet$ and $B^*(E, n)^\bullet$

1.   An auxillary complex. Let A be an abelian group. Consider the following complex

$$\mathbb{Q}[A] \xrightarrow{\delta} \mathbb{Q}[A] \otimes A_\mathbb{Q} \xrightarrow{\delta} \mathbb{Q}[A] \otimes \Lambda^2 A_\mathbb{Q} \xrightarrow{\delta} \mathbb{Q}[A] \otimes \Lambda^3 A_\mathbb{Q} \xrightarrow{\delta} \dots \qquad (40)$$

The differential is defined by the formula

$$\delta : \{a\} \otimes b_1 \wedge b_2 \wedge \dots \wedge b_m \longmapsto \{a\} \otimes a \wedge b_1 \wedge b_2 \wedge \dots \wedge b_m \qquad (41)$$

It is infinite if $A_\mathbb{Q} := A \otimes \mathbb{Q}$ is infinite dimensional. Let I_A^k be the k-th degree of the augmentation ideal $I_A \subset \mathbb{Z}[A]$.

Lemma 7.1 $\delta(I_A^k) \subset I_A^{(k-1)} \otimes A$.

Proof. I_A^k is generated by the elements

$$(\{a_1\} - \{0\}) * (\{a_1\} - \{0\}) * \dots * (\{a_k\} - \{0\})$$

Clearly

$$\delta(\{a_1\} - \{0\}) * \dots * (\{a_k\} - \{0\}) = \sum_i \prod_{j \neq i} (\{a_j\} - \{0\}) * \{a_i\} \otimes a_i$$

So the complex (40) has the "diagonal" filtration by subcomplexes

$$I_A^n \xrightarrow{\delta} I_A^{n-1} \otimes A_\mathbb{Q} \xrightarrow{\delta} I_A^{n-2} \otimes \Lambda^2 A_\mathbb{Q} \xrightarrow{\delta} I_A^{n-3} \otimes \Lambda^3 A_\mathbb{Q} \xrightarrow{\delta} \dots$$

Each graded quotient is isomorphic to the Koszul complex

$$S^n A_\mathbb{Q} \longrightarrow S^{n-1} A_\mathbb{Q} \otimes A_\mathbb{Q} \longrightarrow \dots \longrightarrow A_\mathbb{Q} \otimes \Lambda^{n-1} A_\mathbb{Q} \longrightarrow \Lambda^m A_\mathbb{Q}$$

2. The groups $B_n(E)$. Recall that $B_0(E) := \mathbb{Z}$ and $B_1(E) = J(k)$. We will usually write J for $J(k)$. The group $B_2(E)$ is the group discussed in s. 2.1. In particular if $k = \bar{k}$ it is a quotient of $\mathbb{Z}[E(k)]$.

Recall that

$$E^{(n-1)} = \{(x_1, \dots, x_n) \subset E^n | \sum x_i = 0\}$$

The group S_n acts naturally on $E^{(n-1)}$. Let $p_i : E^{(n-1)} \to E$ be the projection to the i-th factor. We will use the "coordinate notations" denoting $p_i^* f$ by $f(x_i)$ etc. Let us define the following diagram:

$$K_n^M \left(k(E^{(n-1)}) \right)$$

$$\mu_n \nearrow$$

$$S^n \left(k(E)^* \right)$$

$$\beta_n \searrow$$

$$I_E^{2n}$$

by setting

$$\bar{\mu}_n : f_1 \circ \dots \circ f_n \longmapsto \sum_{\sigma \in S_n} (-1)^{|\sigma|} \{p^*_{\sigma(1)} f_1, \dots, p^*_{\sigma(n)} f_n\} = \sum_{\sigma \in S_n} \{p^*_1 f_{\sigma(1)}, \dots, p^*_n f_{\sigma(n)}\}$$

For $n \geq 2$ set

$$\beta_n : S^n k(E)^* \longrightarrow I^{2n}_{E(k)} \qquad f_1 \circ \dots \circ f_n \longmapsto (f_1) * (f_2) * \dots * (f_n)$$

Definition 7.2 *Let* $k = \bar{k}$ *and* $n \geq 3$. *Then* $R_n(E/k)$ *is the subgroup of* $\mathbb{Z}[E]$ *generated by* $\beta_{n-1}(Ker\bar{\mu}_{n-1})$ *and "distribution relations"* :

$$\{a\}_n - m^{n-2} \cdot \sum_{mb=a} \{b\}_n, \qquad a \in E(k), \qquad m = -1, 2 \qquad (42)$$

Example $R_3(E)$ is generated by the elements $(1-f)*(f)^-$, where $\{x\}^- :=$ $\{-x\}$, and (42).

Remark. It would be more natural to add to the subgroup $R_n(E/k)$ the distribution relations for all $m \in \mathbb{Z}$, $m \neq 0$. However we will get the same group, and for our purposes the definition above is technically more convenient.

Definition 7.3 *Let* $k = \bar{k}$.

$$B_n(E) := \frac{\mathbb{Z}[E(k)]}{R_n(E/k)}$$

Theorem 7.4 $\delta(R_n(E)) \subset R_{n-1}(E) \otimes J$.

The proof consists of two independent parts. For a more dificult one see proof of the theorem (7.9) below. The easy one follows from

Lemma 7.5 *For any* $m \in \mathbb{Z}$, $m \neq 0$, *one has*

$$\delta_n\left(\{a\}_n - m^{n-2} \cdot \sum_{mb=a} \{b\}_n\right) = 0 \quad \text{in the group} \quad B_{n-1}(E) \otimes J \qquad (43)$$

Proof. For $n = 2$ this is done in [GL]. The general case follows by induction:

$$\delta\left(\{a\}_n - m^{n-2} \cdot \sum_{mb=a} \{b\}_n\right) = \{a\}_{n-1} \otimes a - m^{n-3} \cdot \sum_{mb=a} \{b\}_{n-3} \otimes mb =$$

$$\left(\{a\}_{n-1} - m^{n-3} \cdot \sum_{mb=a} \{b\}_{n-1}\right) \otimes a = 0$$

So we get a homomorphism $\delta : B_n(E) \longrightarrow B_{n-1}(E) \otimes J$ and thus the following complex $B(S^{n-2}\mathcal{H}(1))^\bullet$:

$$B_n(E) \xrightarrow{\delta} B_{n-1}(E) \otimes J \xrightarrow{\delta} \dots \xrightarrow{\delta} B_2(E) \otimes \Lambda^{n-2}J \xrightarrow{\delta} J \otimes \Lambda^{n-1}J \xrightarrow{\delta} \Lambda^n J$$
$$(44)$$

Here the very left group sits in degree one. The differential is defined by the formula (41) and has degree +1.

If k is not an algebraically closed we postulate the Galois descent:

$$B(E/k, n+2)_{\mathbb{Q}}^{\bullet} := (B(E/\bar{k}, n+2)_{\mathbb{Q}}^{\bullet})^{Gal(\bar{k}/k)}$$

Let us also define the groups $B_n^*(E)$ for $\bar{k} = k$:

$$B_n^*(E) := Im\left(I^{2n-2} \longrightarrow B_n(E)\right)$$

Here the map is induced by the natural inclusion $I^{2n-2} \hookrightarrow \mathbb{Z}[E]$.

It follows from the lemma (7.1) that $\delta_n(B_n^*(E)) \subset B_{n-1}^*(E) \otimes J$. So we can consider the following subcomplex $B^*(E, n)^{\bullet}$ of the complex (44).

$$B_n^*(E) \xrightarrow{\delta} B_{n-1}^*(E) \otimes J \xrightarrow{\delta} \dots \xrightarrow{\delta} k^* \otimes \Lambda^{n-2}J$$

Proposition 7.6 *The canonical morphism of complexes*

$$B^*(E, n+2)_{\mathbb{Q}}^{\bullet} \longrightarrow B(E, n+2)_{\mathbb{Q}}^{\bullet}$$

is a quasiisomorphism.

Proof. This morphism is injective by the definition of the groups $B_n^*(E)$. It follows immediately from the lemma below that the quotient is isomorphic to the Koszul complex

$$S^n J_{\mathbb{Q}} \longrightarrow S^{n-1}J_{\mathbb{Q}} \otimes J_{\mathbb{Q}} \longrightarrow \dots \longrightarrow S^2 J_{\mathbb{Q}} \otimes \Lambda^{n-2}J_{\mathbb{Q}} \longrightarrow J_{\mathbb{Q}} \otimes \Lambda^{n-1}J_{\mathbb{Q}} \longrightarrow \Lambda^n J_{\mathbb{Q}}$$

Lemma 7.7
$$B_n(E)/B_n^*(E) \otimes \mathbb{Q} = S^n J_{\mathbb{Q}}$$

Proof. We may assume that k is algebraically closed. We need to study the quotient of the group $\mathbb{Q}[E(k)]/I^{2n-2}$ by the subgroup generated by the distribution relations. Notice that

$$\mathbb{Q}[E(k)]/I^{2n-2} = \mathbb{Q} \oplus J_{\mathbb{Q}} \oplus S^2 J_{\mathbb{Q}} \oplus \dots \oplus S^{2n-3} J_{\mathbb{Q}}$$

where the isomorphism is given by $i = (i_0, ..., i_{2n-3})$ where

$$i_m : \mathbb{Q}[E(k)] \longrightarrow S^m J_{\mathbb{Q}}; \qquad \{a\} \longmapsto a^m$$

Let us denote by DR_n the subgroup generated by the distribution relations (42). Then the homomorphism

$$i_{\hat{n}} = (i_0, ..., \hat{i_n}, ..., i_{2n-3}) : DR_n \longrightarrow \mathbb{Q} \oplus J_{\mathbb{Q}} \oplus \dots \oplus \hat{S}^n J_{\mathbb{Q}} \oplus \dots \oplus S^{2n-3} J_{\mathbb{Q}}$$

is surjective. Indeed,

$$i_k\Big(\{a\}_n - m^{n-2} \cdot \sum_{mb=a} \{b\}_n\Big) = (1 - m^{n-k}) \cdot a^k$$

In particular $i_n(DR_n) = 0$. The lemma and hence the proposition are proved.

3. **The complex $B^*(E/k, n+1)^\bullet_{\mathbb{Q}}$ is an** *elliptic deformation* **of the complex** $B(Spec(k), n)^\bullet_{\mathbb{Q}}$. Let me recall that the complex $B(Spec(k), n)^\bullet$ looks as follows:

$$\mathcal{B}_n(k) \longrightarrow \mathcal{B}_{n-1}(k) \otimes k^* \longrightarrow \ldots \longrightarrow \mathcal{B}_2(k) \otimes \Lambda^{n-2}k^* \longrightarrow \Lambda^n k^*$$

where $\mathcal{B}_n(k)$ is the quotient of $\mathbb{Z}[k^*]$ along a certain subgroup $\mathcal{R}_n$.

There is an important difference between these two complexes. The complex $B(Spec(k), n)^\bullet_{\mathbb{Q}}$ is defined directly in terms of a field k, while to define the complex $B(E/k, n+1)^\bullet_{\mathbb{Q}}$ we have to go to the algebraic closure of k and then postulate the Galois descent property. So in general they can only be quasiisomorphic.

Suppose $k = \bar{k}$. When E degenerates to $\big(P^1, \{0\} \cup \{\infty\}\big)$ the complex $B(E/k, n+1)^\bullet_{\mathbb{Q}}$ degenerates to a complex quasiisomorphic to $B(Spec(k), n)^\bullet_{\mathbb{Q}}$ (for $n = 2$ see s.3.4 in [GL]).

In a sense the elliptic situation is simpler. For example our definition of the group $R_n(E)$ is not inductive and more explicit then the definition of the group $\mathcal{R}_n(k)$ in [G2]. In fact when E degenerates to $\big(P^1, \{0\} \cup \{\infty\}\big)$ our definition suggests a new way of defining of the groups $\mathcal{B}_n(k)$ from [G1]-[G2].

4. **Proof of the theorem (7.4).** Let $\sigma \in S^n$ be a permutation and $a_i \in E$. Consider the following codimension p cycle in $E^{(n-1)}$:

$$X(\sigma; a_1, ..., a_p) := \{(x_1, ..., x_n) \subset E^{(n-1)} | x_{\sigma(1)} = a_1, ..., x_{\sigma(p)} = a_p\}$$

It is a product of elliptic curves E. Let us define a homomorphism

$$\mu_{n;p} : \Lambda^p \mathbb{Z}[E] \otimes S^{n-p}k(E)^* \longrightarrow \coprod_{X \in (E^{(n-1)})^p} \Lambda^{n-p}(k(X))^*$$

by setting

$$\{a_1\} \wedge \ldots \wedge \{a_p\} \otimes f_{p+1} \circ \ldots \circ f_n \longmapsto$$
$$\sum_{\sigma \in S_n} (-1)^{|\sigma|} f_{p+1}(x_{\sigma(p+1)}) \wedge \ldots \wedge f_n(x_{\sigma(n)})|_{X(\sigma;a_1,...,a_p)}$$

Denote by $\tilde{\mathcal{D}}^p_{(n)}$ the image of the homomorphism $\mu_{n;p-1}$.

The elements of type $f_1(x_1) \wedge \ldots \wedge f_m(x_m)$ in $\Lambda^m k(E^{(m-1)})^*$ and their linear combinations may be called the *decomposable* elements; this suggests the notation.

Let F be an arbitrary field with a discrete valuation v and the residue class F_v. The group of units U has a natural homomorphism $U \longrightarrow F_v^*$, $u \mapsto \bar{u}$. Choose a uniformizer $\pi \in F^*$, $\mathrm{ord}_v \pi = 1$. There is canonical homomorphism

$$\Lambda^n(F^*) \xrightarrow{\partial_v} \Lambda^{n-1}(F_v^*)$$

uniquely defined by the properties

$$\partial_v(u_1 \wedge ... \wedge u_n) = 0; \qquad \partial_v(\pi \wedge u_2 \wedge ... \wedge u_n) = (\bar{u}_2, ..., \bar{u}_n)$$

Consider the following complex on $E^{(n-1)}$:

$$\Lambda^n(k(E^{(n-1)})) \xrightarrow{\partial} \coprod_{X \in (E^{(n-1)})^1} \Lambda^{n-1}(k(X)) \xrightarrow{\partial} ... \xrightarrow{\partial} \coprod_{X \in (E^{(n-1)})^{n-1}} k(X)^*$$

$$(45)$$

Here $\partial := \sum_X \partial_{v(X)}$ where $v(X)$ is the valuation corresponding to the irreducible divisor X.

Lemma 7.8 *The groups $\tilde{\mathcal{D}}_{(n)}^p$ form a subcomplex in the complex (45)*

Proof. Clear from the definitions.

Let us define for $p < n$ a homomorphism

$$\tilde{\beta}_n^{(p+1)} : \tilde{\mathcal{D}}_{(n)}^{p+1} \longrightarrow I_E^{2(n-p+1)} \otimes \Lambda^p J$$

We define $\tilde{\beta}_n^{(p+1)}$ first on space of decomposable elements on the subvariety $x_1 = a_1, ..., x_p = a_p$ by the formula

$$\tilde{\beta}_n^{(p+1)} : f_{p+1}(x_{p+1}) \wedge ... \wedge f_n(x_n)|_{x_1=a_1,...,x_p=a_p} \longmapsto$$

$$(f_{p+1}) * ... * (f_n) * (a_1 + ... + a_p) \otimes a_1 \wedge ... \wedge a_p$$

It extends uniquely to the space of all decomposable elements assuming the skewsymmetry with respect to the action S_n. In particulary it is defined on $\tilde{\mathcal{D}}_{(n)}^p$.

If $p = n - 1$ then $\tilde{\mathcal{D}}_{(n)}^n = k^* \otimes \mathbb{Z}[E^{(n-1)}]^{sgn}$ and we have a homomorphism

$$\tilde{\beta}_n^{(n)} : k^* \otimes \mathbb{Z}[E^{(n-1)}]^{sgn} \longrightarrow k^* \otimes \Lambda^{n-1} J; \qquad x \otimes (a_1, ..., a_n) \longmapsto x \otimes a_1 \wedge ... \wedge a_{n-1}$$

Finally, one can define a homomorphism

$$I_E^4 \otimes \Lambda^{n-2} J \xrightarrow{\delta} k^* \otimes \Lambda^{n-1} J$$

Namely, there is a homomorphism

$$B_3(E) \otimes \Lambda^{n-2} J \xrightarrow{\delta} B_3(E) \otimes \Lambda^{n-1} J \qquad \{a\}_3 \otimes b_1 \wedge ... \longmapsto \{a\}_2 \otimes a \wedge b_1 \wedge ...$$

Let $B_n^{(k)}(E)$ be the subgroup of $B_n(E)$ generated by k-th degree of the augmentation ideal I_E^k. The restriction of this homomorphism to $B_3^{(4)}(E) \otimes \Lambda^{n-2}J \subset B_3(E) \otimes \Lambda^{n-2}J$ lands in $k^* \otimes \Lambda^{n-1}J$. Indeed, the composition

$$I_E^4 \otimes \Lambda^{n-2}J \xrightarrow{\delta} B_{\mathbb{Q}(1)}(E) \otimes \Lambda^{n-1}J \longrightarrow S^2 J \otimes \Lambda^{n-1}J$$

is equal to zero.

Theorem 7.9 *The maps $\tilde{\beta}_n^{(i)}$ provide a homomorphism of complexes*

$$
\begin{array}{ccccccccc}
\tilde{\mathcal{D}}_{(n)}^1 & \xrightarrow{\ \partial\ } & \tilde{\mathcal{D}}_{(n)}^2 & \xrightarrow{\ \partial\ } & \cdots & \xrightarrow{\ \partial\ } & \tilde{\mathcal{D}}_{(n-1)}^n & \xrightarrow{\ \partial\ } & \tilde{\mathcal{D}}_{(n)}^n \\
\downarrow \tilde{\beta}_n^{(1)} & & \downarrow \tilde{\beta}_n^{(2)} & & & & \downarrow \tilde{\beta}_n^{(n-1)} & & \downarrow \tilde{\beta}_n^{(n)} \\
I_E^{2n} & \xrightarrow{\ \delta\ } & I_E^{2n-2} \otimes J & \xrightarrow{\ \delta\ } & \cdots & \xrightarrow{\ \delta\ } & I_E^4 \otimes \Lambda^{n-2}J & \xrightarrow{\ \delta\ } & k^* \otimes \Lambda^{n-1}J
\end{array}
$$

Proof. We will do in details the commutativity of the left square, at the same time proving the theorem (7.4). The commutativity of the other squares except the very right one is completely similar.

The commutativity of the right square is a more subtle statement. For $n = 2$ it was already proved in [GL], s.3. The general case $n > 2$ is similar.

Consider an element

$$f_1(x_1) \wedge \ldots \wedge f_n(x_n) \in \Lambda^n k(E^{(n-1)}) \tag{46}$$

Let $v_a(f)$ be the order of zero of the function $f(t)$ at $t = a$.

The part of the coboundary (in the complex $\tilde{\mathcal{D}}_{(n)}^\bullet$) of the element (46) sitting on the divisor $x_1 = a_1$ is equal to

$$\sum_{a_1 \in E(k)} v_{a_1}(f_1) \cdot f_2(x_2) \wedge \ldots \wedge f_n(x_n), \quad x_2 + \ldots + x_n = -a_1 \tag{47}$$

Let t_a be the shift by a on the group $E(k)$, so $t_a\{b\} = \{a + b\}$ and $t_a f(x) = f(x - a)$ (sic). Then $(t_a f) = (f) * (a)$. Then it can be written as

$$\sum_{a_1 \in E(k)} v_{a_1}(f_1) \cdot f_2(y_2) \wedge \ldots \wedge t_{a_1} f_n(y_n), \quad y_2 + \ldots + y_n = 0$$

Applying the homomorphism $\tilde{\beta}_n^{(2)}$ to the element (47) we get

$$\sum_{a_1 \in E(k)} v_{a_1}(f_1) \cdot (f_2) * \ldots * (f_n) * (a_1) \otimes a_1 \tag{48}$$

On the other hand writing

$$(f_1) * \ldots * (f_n) = \sum_{a_i \in E(k)} v_{a_1}(f_1) \cdot \ldots \cdot v_{a_n}(f_n) \cdot \{a_1 + \ldots + a_n\}$$

we get

$$\delta((f_1)*...*(f_n)) = \sum_{a_i \in E(k)} v_{a_1}(f_1) \cdot ... \cdot v_{a_n}(f_n) \cdot \{a_1+...+a_n\} \otimes (a_1+...+a_n) \quad (49)$$

Collecting the terms with a_1 in the right factor we get just what needed: the formula (48). Taking into consideration the coboundaries of the element (46) sitting on the divisors $x_p = a_p$ we will get the other terms (with a_p) in (49).

To prove the commutativity of the right square we replace it by a bigger diagram

$$\tilde{\mathcal{D}}^n_{(n-1)} \xrightarrow{\partial} \tilde{\mathcal{D}}^n_{(n)}$$

$$\downarrow \tilde{\beta}^{(n-1)}_n \qquad\qquad \downarrow \tilde{\beta}^{(n)}_n$$

$$B_3(E) \otimes \Lambda^{n-2}J \xrightarrow{\delta} B_{\mathbb{Q}(1)}(E) \otimes \Lambda^{n-1}J$$

and then prove its commutativity in a way similar to the proof of theorem 4.5 in [GL]. The theorem is proved.

Consider the Gersten complex for the Milnor K-theory on $E^{(n-1)}$:

$$K_n^M(k(E^{(n-1)})) \xrightarrow{\partial} \coprod_{X \in (E^{(n-1)})^1} K_{n-1}^M(k(X)) \xrightarrow{\partial} ... \xrightarrow{\partial} \coprod_{X \in (E^{(n-1)})^{n-1}} k(X)^*$$

The Gersten complex is obtained from (45) by factorisation along the subgroup generated by the Steinberg relations. Let

$$\mathcal{D}^\bullet_{(n)}: \qquad \mathcal{D}^1_{(n)} \longrightarrow \mathcal{D}^2_{(n)} \longrightarrow ... \longrightarrow \mathcal{D}^n_{(n)}$$

be the image of the complex $\tilde{\mathcal{D}}_{(n)}$ in the Gersten complex. In other words

$$\mathcal{D}^p_{(n)} = \tilde{\mathcal{D}}^p_{(n)}/St^p_{(n)}$$

where $St^p_{(n)}$ is the intersection of $\tilde{\mathcal{D}}^p_{(n)}$ with the subgroup generated by the Steinberg relations in $\coprod_{X \in (E^{(n-1)})^{p-1}} \Lambda^{n-p+1}k(X)^*$.

Lemma 7.10 $\beta_n^{(p)}(St^p_{(n)}) = 0$ in $B^*_{n+2-p}(E) \otimes \Lambda^{p-1}J$.

Proof. Consider the subvariety $X(id; a_1, ..., a_{p-1})$ in $E^{(n-1)}$ and its projection

$$p: X(id; a_1, ..., a_{p-1}) \longrightarrow E^{n-p} \qquad p: (x_1, ..., x_n) \longmapsto (x_p, ..., x_{n-1})$$

The subgroup of Steinberg relations is generated by

$$\{f(x_p, ..., x_{n-1}), (1-f)(x_p, ..., x_{n-1}), g_1(x_p, ..., x_{n-1}), ...\}$$

Notice that $x_n = -a_1 - ... - a_{p-1} - x_p - ... - x_{n-1}$. This means that an element $\sum_j\{f_p^{(j)}(x_p), ..., f_n^{(j)}(x_n)\}$ of the subgroup of Steinberg relations can be written as

$$p^* \sum_j \{f_p^{(j)}(y_p), ..., t_{-a_1-...-a_p}f_n^{(j)}(y_n)\}$$

on

$$E^{(n-p)} = \{(y_p, ..., y_n)|y_p + ... + y_n = 0\}$$

Now the lemma follows immediately.

Theorem 7.11 *There is a canonical homomorphism of complexes*

$$
\begin{array}{ccccccc}
\mathcal{D}^1_{(n)} & \longrightarrow & \mathcal{D}^2_{(n)} & \longrightarrow & ... & \longrightarrow & \mathcal{D}^n_{(n)} \\[2mm]
\downarrow \bar{\beta}_n^{(1)} & & \downarrow \bar{\beta}_n^{(2)} & & & & \downarrow \bar{\beta}_n^{(n)} \\[2mm]
B^*_{n+1}(E) & \overset{\delta}{\longrightarrow} & B^*_n(E) \otimes J & \overset{\delta}{\longrightarrow} & ... & \overset{\delta}{\longrightarrow} & k^* \otimes \Lambda^{n-1}J
\end{array}
$$

Proof. Follows immediately from lemma (7.10) and theorem (7.9).

8 The regulator integrals, Eisenstein-Kronecker series and a conjecture on $L(Sym^nE, n+1)$

1. Beilinson's conjecture in the case of $L(Sym^n h^1(E), n+1)$. Let E be an elliptic curve over a number field k. According to the general Beilinson conjecture on regulators the special value $L(Sym^n h^1(E), n+1)$ should be equal, up to standard factors, to the (co)volume of a certain lattice obtained as follows.

The n-th Deligne complex $\mathbb{R}(n)_\mathcal{D}$ on a regular variety X can be defined as the total complex associated with the following bicomplex:

$$
\begin{array}{ccccccccc}
\mathcal{A}^0_X(n-1) & \overset{d}{\longrightarrow} & ... & \overset{d}{\longrightarrow} & \mathcal{A}^n_X(n-1) & \overset{d}{\longrightarrow} & \mathcal{A}^{n+1}_X(n-1) & \overset{d}{\longrightarrow} & ... \\[4mm]
& & & & \uparrow \pi_n & & \uparrow \pi_n & & \\[4mm]
& & & & \Omega^n_X & \overset{\partial}{\longrightarrow} & \Omega^{n+1}_X & \overset{\partial}{\longrightarrow} &
\end{array}
$$

$$\tag{50}$$

Here $(\mathcal{A}^*_X, d)$ is the C^∞-De Rham complex, $\pi_n : \mathcal{A}^m_X \otimes \mathbb{C} \longrightarrow \mathcal{A}^m_X(n-1)$ is the projection induced by $\pi_n : \mathbb{C} \longrightarrow \mathbb{R}(n-1)$, $z \longmapsto z + (-1)^{(n-1)}\bar{z}$, the group $\mathcal{A}^0_X(n-1)$ is placed in degree 1 and (Ω^*_X, ∂) is the De Rham complex of holomorphic forms with logarithmic singularities at infinity.

One has the regulator map

$$r_{Be} : H^n_{\mathcal{M}}\left(E^{(n-1)}, \mathbb{Q}(n)\right)_{sgn} \longrightarrow H^n_{\mathcal{D}}\left(E^{(n-1)} \otimes_{\mathbb{Q}} \mathbb{R}, \mathbb{R}(n)\right)_{sgn} \qquad (51)$$

The right hand side is a group of purely topological origin:

$$H^n_{\mathcal{D}}\left(E^{(n-1)} \otimes_{\mathbb{Q}} \mathbb{R}, \mathbb{R}(n)\right)_{sgn} = H^{n-1}_B\left(E^{(n-1)} \otimes_{\mathbb{Q}} \mathbb{C}, \mathbb{R}(n-1)\right)^+_{sgn} =$$

$$\left(\oplus_{\sigma: k \hookrightarrow \mathbb{C}} H^{n-1}_B(E^{(n-1)}_\sigma(\mathbb{C}), \mathbb{R}(n-1))_{sgn}\right)^+ =$$

$$\left(\oplus_{\sigma: k \hookrightarrow \mathbb{C}} Sym^{n-1} H^1_B(E_\sigma(\mathbb{C}), \mathbb{R}(1))\right)^+$$

Here $+$ means invariants of the complex conjugation acting on σ's and on the coefficients $\mathbb{R}(n-1)$.

The image of the regulator map (51) is conjectured to be a lattice. $H^n_{\mathcal{D}}\left(E^{(n-1)} \otimes_{\mathbb{Q}} \mathbb{R}, \mathbb{Q}(n)\right)_{sgn}$ gives another lattice in $H^n_{\mathcal{D}}\left(E^{(n-1)} \otimes_{\mathbb{Q}} \mathbb{R}, \mathbb{R}(n)\right)_{sgn}$. The covolume of the lattice Imr_{Be} measured with respect to the second lattice should coincide (up to standard factors) with the special value of our L-function at $s = n + 1$.

2. The Eisenstein-Kronecker series. Let me recall their definition:

$$K_{i,j}(z; \tau) = \sum_{\gamma \in \Gamma \backslash 0} \frac{(z, \gamma)}{\gamma^i \bar{\gamma}^j}, \quad i, j \geq 1$$

For the relation with the elliptic polylogarithms see [BL], [Z].

Lemma 8.1 a) *For any lattice Γ one has $\bar{K}_{i,j}(z; \tau) = (-1)^{i+j} K_{j,i}(z; \tau)$*

b) *Suppose that the lattice Γ and a divisor P on $\mathbb{C}/\Gamma$ are invariant under the complex conjugation. Then $K_{i,j}(P; \tau) \in \mathbb{R}$.*

Proof. Clear.

Consider for each $n > 2$ a homomorphism

$$K_n : \mathbb{Z}[E(\mathbb{C})] \longrightarrow Sym^{n-2} H^1(E(\mathbb{C}), \mathbb{C})$$

$$\{z\} \longmapsto \sum_{a+b=n} K_{a,b}(z; \tau)(dz)^{a-1}(d\bar{z})^{b-1}$$

Theorem 8.2 $K_n(R_n(E(\mathbb{C}))) = 0.$

We will prove it in s 6.3 below. So we get a homomorphism

$$K_n : B_n[E(\mathbb{C})] \longrightarrow Sym^{n-2} H^1(E(\mathbb{C}), \mathbb{C})$$

This means that $R_n(E(\mathbb{C}))$ is a subgroup of functional equations for the Eisenstein-Kronecker series.

Lemma 8.3 *Suppose that E is defined over $\mathbb{R}$ and the lattice Γ was defined using a real differential ω. Then*

$$K_n : B_n[E(\mathbb{C})]^+ \longrightarrow Sym^{n-2}H^1(E(\mathbb{C}), \mathbb{R}(1))^+$$

Here $+$ means invariants of the complex involution acting on both $\mathbb{R}(1)$ and $E(\mathbb{C})$.

Proof. Follows from lemma (8.1).

3. Computation of the regulator integral. The main result of this section is due to Deninger (see [D1], s.6). Our presentation is technically simpler since working with distributions we avoid the convergence problems.

For any functions $f_1, ..., f_n$ on a manifold X consider the following $(n-1)$-form with values in $\mathbb{R}(n-1) := (2\pi i)^{n-1}\mathbb{R}$

$$r_n(f_1, ..., f_n) := \tag{52}$$

$$\text{Alt}_n \sum_{j \geq 0} C_j \cdot \log|f_1| d\log|f_2| \wedge ... \wedge d\log|f_{2j+1}| \wedge di \arg f_{2j+2} \wedge ... \wedge di \arg f_n$$

Here $C_j := \frac{1}{(2j+1)!(n-2j-1)!}$ and Alt_n is the operation of alternation of f_i's. Then

$$dr_n(f_1, ..., f_n) = \pi_n(d\log f_1 \wedge ... \wedge d\log f_n)$$

This just means that the pair

$$(r_n(f_1, ..., f_n), \quad d\log f_1 \wedge ... \wedge d\log f_n)$$

is an n-cycle in the Deligne complex (50). It is the product in the real Deligne cohomology of the 1-cocycles $(\log|f_i|, d\log f_i)$. Set

$$\omega_{p,q} := \left(\sum \overset{(-)}{dz_1} \wedge...\wedge \overset{(-)}{dz_{p+q}}\right)^{(p,q)} \tag{53}$$

where $\overset{(-)}{dz_i}$ means either dz_i or $d\bar{z}_i$ and the sum is taken over all possible terms. For example

$$\omega_{2,0} = dz_1 \wedge dz_2, \quad \omega_{1,1} = dz_1 \wedge d\bar{z}_2 + d\bar{z}_1 \wedge dz_2, \quad \omega_{0,2} = d\bar{z}_1 \wedge d\bar{z}_2$$

The forms $\pi_n\omega_{p,q}$ for $p \geq q, p + q = n - 1$, form a basis over $\mathbb{R}$ in $H_B^{n-1}(E_{/\mathbb{R}}^{(n-1)}, \mathbb{R}(n-1))_{sgn}$.

We can represent elements in $H_B^{n-1}(E_{/\mathbb{R}}^{(n-1)}, \mathbb{R}(n-1))_{sgn}$ by their cup product with forms $\pi_n\omega_{p,q}$:

$$\frac{1}{(2\pi i)^{(n-1)}} \int_{E^{(n-1)}(\mathbb{C})} \sum_{\sigma \in S_n} (-1)^{|\sigma|} r_n(p^*_{\sigma(1)}f_1, ..., p^*_{\sigma(n)}f_n) \wedge \pi_n\omega_{p,q}$$

Theorem 8.4

$$\int_{E^{(n-1)}(\mathbb{C})} \sum_{\sigma \in S_n} (-1)^{|\sigma|} r_n(p^*_{\sigma(1)} f_1, ..., p^*_{\sigma(n)} f_n) \wedge \omega_{p,q} = c_n \frac{Im\tau}{\pi} \cdot K_{p+1,q+1}(f_1 * ... * f_n)$$

where $c_n \in \mathbb{Q}^*$.

The constant c_n can be obtained from the proof below.

Proof. It consists of several reductions of the integral.

Step 1. The form $\omega_{p,q}$ is skew invariant under the action of the group S_n. So the integral is equal to

$$n! \cdot \int_{E^{(n-1)}(\mathbb{C})} r_n(p^*_1 f_1, ..., p^*_n f_n) \wedge \omega_{p,q}$$

Step 2. Let

$$\alpha_n(f_1, ..., f_n) := \sum_{i=1}^{n} (-1)^i \log|f_i| d\log|f_1| \wedge ... d\widehat{\log}|f_i| ... \wedge d\log|f_n| \qquad (54)$$

Lemma 8.5 *For any functions $f_1, ..., f_n$*

$$\int_{E^{(n-1)}(\mathbb{C})} r_n(f_1, ..., f_n) \wedge \omega_{p,q} = b_n \cdot \int_{E^{(n-1)}(\mathbb{C})} \alpha_n(f_1, ..., f_n) \wedge \omega_{p,q}$$

where $b_n \in \mathbb{Q}^*$ is a (computable) constant.

Proof. One always have either $d\log f_i \wedge \omega_{p,q} = 0$ or $d\log \bar{f}_i \wedge \omega_{p,q} = 0$. So we can replace everywhere $di \arg f$ by $\pm d\log|f|$.

Step 3.

$$\int_{E^{(n-1)}(\mathbb{C})} \alpha_n(f_1, ..., f_n) \wedge \omega_{p,q} = n \cdot \int_{E^{(n-1)}(\mathbb{C})} \log|f_n| d\log|f_1| \wedge ... \wedge d\log|f_{n-1}| \wedge \omega_{p,q}$$

$$(55)$$

Indeed, $\log|f_1| d\log|f_2| + \log|f_2| d\log|f_1| = d(\log|f_1| \cdot \log|f_2|)$ and so by the Stokes theorem

$$\int_{E^{(n-1)}(\mathbb{C})} d(\log|f_1| \cdot \log|f_2|) d\log|f_3| \wedge ... \wedge d\log|f_n| \wedge \omega_{p,q} = 0$$

Step 4. Let us compute the right integral in (55).

Lemma 8.6

$$\log|f(z)| = -\frac{Im\tau}{2\pi} \sum_{\gamma \in \Gamma \setminus 0} v_a(f) \frac{(z-a, \gamma)}{|\gamma|^2} + C_f \qquad (56)$$

where C_f is a certain constant.

Proof. One can get a proof applying $\partial\bar{\partial}$ to the both parts of (55). The constant C_f can be computed from the decomposition of f on the product of theta functions using the formula in s. 18 ch. VIII of [We]. According to step 5 it does not play any role in our considerations.

Step 5. By the Stokes formula

$$\int_{E^{(n-1)}(\mathbb{C})} C_f \cdot d\log|f_1| \wedge ... \wedge d\log|f_{n-1}| \wedge \omega_{p,q} = 0$$

we see that one can neglect the constants C_f.

Step 6. We may suppose that in (55) the form $\omega_{p,q}$ is written in variables $z_1, ..., z_{n-1}$. Then for each summand in $\omega_{p,q}$ one can replace $d\log|f_i|$ by $1/2 \cdot \partial\log f_i$ or $1/2 \cdot \bar{\partial}\log f_i$ depending whether in this summand appeared $d\bar{z}_i$ or dz_i.

For example for the summand $dz_1 \wedge ... \wedge dz_p \wedge d\bar{z}_{p+1} \wedge ... \wedge d\bar{z}_{n-1}$ we will have the integral

$$\frac{n}{2^{n-1}} \cdot \int_{E^{(n-1)}(\mathbb{C})} \log|f_n(z_n)|\bar{\partial}\log f_1(z_1) \wedge ... \wedge \bar{\partial}\log f_p(z_p)\wedge$$

$$\partial\log f_{p+1}(z_{p+1}) \wedge ... \wedge \partial\log f_{n-1}(z_{n-1}) \wedge dz_1 \wedge ... \wedge dz_p \wedge d\bar{z}_{p+1} \wedge ... \wedge d\bar{z}_{n-1}$$

Differentiating the distributions we get

$$\partial\log f(\bar{z}) = \sum_{\gamma\in\Gamma\backslash 0} v_a(f)\frac{(z-a,\gamma)\cdot\bar{\gamma}}{|\gamma|^2}$$

$$\bar{\partial}\log f(z) = -\sum_{\gamma\in\Gamma\backslash 0} v_a(f)\frac{(z-a,\gamma)\cdot\gamma}{|\gamma|^2}$$

The condition $z_1 + ... + z_n = 0$ just mean that we compute the value of the convolution of one variable distributions:

$$\log|f_n(z)| * \frac{\bar{\partial}\log f_2(\bar{z})}{\partial\bar{z}} * ... * \frac{\bar{\partial}\log f_p(\bar{z})}{\partial\bar{z}} * \frac{\partial\log f_p(z)}{\partial z} * ... * \frac{\partial\log f_{n-1}(z)}{\partial z}$$

at 0. Using the fact that the Fourier transform sends the convolution to the product and the formulas above we get the theorem.

Recall we have a homomorphism $K_{n+1} : \mathbb{Z}[E(\mathbb{C})] \longrightarrow Sym^{n-1}H_B^1(E(\mathbb{C}), \mathbb{C})$ which is constructed as follows

$$\{z\} \longmapsto \sum_{i=0}^{n-1} K_{a,b}(z) \cdot (dz)^{a-1}(d\bar{z})^{b-1} \in Sym^{n-1}H_B^1(E(\mathbb{C}), \mathbb{C})$$

Theorem (8.2) claims that it sends the subgroup $R_{n+1}(E(\mathbb{C}))$ to zero.

4. Proof of the theorem (8.2). For any function f on a manifold one has $d\log|f| \wedge d\log|1-f|$. So a non zero term in the integral (8.5) could be

$$\int_{E^{(n-1)}(\mathbb{C})} (\log|f|d\log|1-f| - \log|1-f|d\log|f|) \wedge d\log|f_3| \wedge ... \wedge d\log|f_n| \wedge \omega_{p,q}$$

One always has

$$(\log|f|d\log|1-f| - \log|1-f|d\log|f|) \wedge \omega_{p,q} =$$

$$\pm i \cdot (\log|f|darg(1-f) - \log|1-f|dargf) \wedge \omega_{p,q}$$

Further one has, even in the sence of the distribution,

$$d\mathcal{L}_2(f) = \log|f|darg(1-f) - \log|1-f|dargf$$

So we can rewrite the integral as

$$\pm i \cdot \int_{E^{(n-1)}(\mathbb{C})} d(\mathcal{L}_2(f) \wedge d\log|f_3| \wedge ... \wedge d\log|f_n| \wedge \omega_{p,q}) = 0$$

It is zero by the Stokes formula.

Now using theorem (8.4) relating Eisenstein-Kronecker series to the regulator integral we come to the proof of the theorem.

Theorem (8.2) and lemma (8.1) imply

Theorem 8.7 *The Eisenstein-Kronecker map K_{n+1} provides a homomorphism*

$$K_{n+1}: B_{n+1}[E(\bar{\mathbb{Q}})]^{Gal(\bar{\mathbb{Q}}/F)} \longrightarrow \left(\oplus_{\sigma:F\hookrightarrow\mathbb{C}} Sym^{n-1}H_B^1(E_\sigma(\mathbb{C}), \mathbb{R}(1))\right)^+$$

Combining these results with conjecture (1.4), our construction of the elliptic motivic complexes presented in s. 4.1 and Beilinson's conjecture on regulators we come to conjecture (8.8) about L-function of $Sym^{n-1}h^1(E)$ at $s = n$ for an arbitrary elliptic curve E over a number field.

5. A conjecture on $L(Sym^nE, n+1)$ In this section we will assume for simplicity that E be an elliptic curve over $\mathbb{Q}$. We will left to reader as an easy exercise to generalise all the discussion to the case of an elliptic curve over an arbitrary field F.

Conjecture (1.4) together with Beilinson's conjecture on regulators imply a precise conjecture on $L(Sym^nE, n+1)$

For any divisor $P = \sum n_s(P_s)$ on $E(\mathbb{C})$ put $K_{i,j}(P) := \sum n_s K_{i,j}(P_s)$.

The integrality condition . Suppose E has a split multiplicative reduction at p with N-gon as a special fibre. Let L be a finite extention of $\mathbb{Q}_p$ of degree $n = ef$ and $\mathcal{O}_L$ the ring of integers in L. Let E^0 be the connected component of the Néron model of E over $\mathcal{O}_L$. Let us fix an isomorphism

$E^0_{F_{p^f}} = \mathbb{G}_m / F_{p^f}$. It provides a bijection between $\mathbb{Z}/eN\mathbb{Z}$ and the components of $E_{F_{p^f}}$. For a divisor P such that all its points are defined over L denote by $d(P; \nu)$ the degree of the restriction of the flat extension of a divisor P to the ν'th component of the (eN)-gon.

Let $B_{n+1}(x)$ be the $(n+1)$-th Bernoulli polynomial. The integrality condition at p is the following condition on a divisor P, provided by the work of Schappaher and Scholl ([SS]). For a certain (and hence for any, see s. 3.3 in [GL]) extention L of $\mathbb{Q}_p$ such that all points of the divisor P are defined over L one has ($[L : \mathbb{Q}_p] = ef$):

$$\sum_{\nu \in \mathbb{Z}/(eN)\mathbb{Z}} d(P; \nu) B_{n+1}(\frac{\nu}{eN}) = 0 \tag{57}$$

Let C_n be the conductor of the system of the l-adic representations related to $Sym^n h^1(E)$. Set

$$\beta_{2l+1} = C_{2l+1}^{-(l+1)} \cdot (Im\tau)^{(l+1)(l+2)}, \qquad \beta_{2l} = C_{2l}^{-\frac{2l+1}{2}} \cdot \pi^{-2l} (Im\tau)^{(l+1)^2}$$

Conjecture 8.8 *a) For any elliptic curve over $\mathbb{Q}$ there exist $[\frac{n}{2}]+1$ $\mathbb{Q}$-rational divisors P_a on $E(\bar{\mathbb{Q}})$ such that*

$$L(Sym^n h^1(E), n+1) \sim_{\mathbb{Q}^*} \beta_n det|K_{b,n+2-b}(P_a; \tau)| \tag{58}$$

($1 \leq a, b \leq [\frac{n}{2}] + 1$), and the divisors P_a satisfy the following two conditions:

i) $$\delta(P_a) = 0 \quad in \quad B_{n+1}(E) \otimes J(\bar{\mathbb{Q}})_{\mathbb{Q}} \tag{59}$$

ii) the integrality condition: at each prime p where E has a split multiplicative reduction with special fibre a Neron N-gon

$$\sum_{\nu \in \mathbb{Z}/(eN)\mathbb{Z}} d(P; \nu) B_{n+1}(\frac{\nu}{eN}) = 0 \tag{60}$$

b) For any $[\frac{n}{2}]+1$ $\mathbb{Q}$-rational divisors P_a on $E(\bar{\mathbb{Q}})$, satisfying the conditions above the right hand side of (58) is equal to $q \cdot L(S^n h^1(E), n+1)$ where q is a rational number, perhaps equal to 0.

In [W] J.Wildeshaus, assuming standard conjectures about mixed motives, formulated a conjecture similar to the part b) of the conjecture (8.8) (an elliptic analog of the weak version of Zagier's conjecture).

For $n = 2$ the formula (58) was proved for modular elliptic curves over $\mathbb{Q}$ in [GL]. Formula (58), even without precise conditions on the divisors P_a, is the most nontrivial part of the conjecture for $n > 2$ (see also s.8 in [G4]). An efficient way to formulate the conditions on the divisors P_a without referring to the definition of the subgroups $R_n(E)$ is given in the chapter 7.

When E degenerates to the nodal curve, the conjecture on $L(Sym^2E, 3)$ leads to Zagier's conjecture [Za2] at $s = 3$, which was proved in [G1]-[G2]. This gives a credit for the conjecture (8.8).

The key condition (59) is obviously satisfied if P_a are (multiples of) torsion divisors. The determinants from (58) for torsion divisors where considered by Deninger ([De2], s.5) (and inspired by the Eisenstein symbol of Beilinson [Be]). They work well for CM elliptic curves. However Mestre and Schappacher [SM] deduced from a result of Serre that for a given non CM elliptic curve over $\mathbb{Q}$ for all $n > n_0$ the determinant is always zero for any $\mathbb{Q}$-rational torsion divisors P_a. So to get the L-values one has to consider the non torsion divisors.

6. A more explicit form of the conditions on the divisors P_a. Let $P = \sum n_i P_i$ and $k(P)$ be the field generated by the points P_i. Let us denote by h_v the canonical local height related to a valuation v of a number field K. Let K_v be the completion of a number field K corresponding to v. Recall the height homomorphism $h_v : B_2(E(K_v)) \longrightarrow \mathbb{R}$. If v is a non archimedean valuation then the target of this homomorphism is $(\log p) \cdot \mathbb{Q}$.

Let us consider a homomorphism

$$d_m : B_{n+1}(E) \longrightarrow B_m(E) \otimes S^{n+1-m} J_{\mathbb{Q}}, \qquad \{a\}_{n+1} \longmapsto \{a\}_{n+1-m} \otimes a \cdot \ldots \cdot a$$

We will need the following pairs of homomorphisms. If $m = 2$:
i) A homomorphism

$$p_2 \otimes id : B_2(E) \otimes S^{n-1} J_{\mathbb{Q}} \longrightarrow S^2 J_{\mathbb{Q}} \otimes S^{n-1} J_{\mathbb{Q}}$$

where $p_2 : \{a\}_2 \longmapsto a \cdot a$.

For any valuation v of we have
ii) The height homomorphism

$$h_v \otimes id : B_2(E) \otimes S^{n-1} J_{\mathbb{Q}} \longrightarrow \mathbb{R} \otimes_{\mathbb{Q}} S^{n-1} J_{\mathbb{Q}}$$

where v is any valuation of the field $K(P)$.

For $m > 2$:
iii) The Bernoulli homomorphism, defined for any bad prime where E has a split multiplicative reduction with the Néron N-gon:

$$Ber_m : B_m(E) \otimes S^{n+1-m} J_{\mathbb{Q}} \longrightarrow S^{n+1-m} J_{\mathbb{Q}},$$

$$\{a\}_m \otimes b_1 \cdot \ldots \cdot b_{n+1-m} \longmapsto \sum_\nu d(a, \nu) B_m(\frac{\nu}{N}) \cdot b_1 \cdot \ldots \cdot b_{n+1-m}$$

iv) The Eisenstein-Kronecker homomorphism

$$K_m \otimes id : B_m(E) \otimes S^{n+1-m} J_{\mathbb{Q}} \longrightarrow Sym^{n-2} H^1(E(\mathbb{C}), \mathbb{R}(1))^+ \otimes S^{n+1-m} J_{\mathbb{Q}}$$

Remarks. 1. In formulas above $J_{\mathbb{Q}}$ means $J(\bar{\mathbb{Q}})_{\mathbb{Q}}$. However for a given divisor P we land in $Gal(\bar{\mathbb{Q}}/\mathbb{Q})$-invariant part of powers of $J(k(P))_{\mathbb{Q}}$.

2. Let us consider the Eisenstein-Kronecker homomorphism only on the kernel of the Bernoulli homomorphism. Then Beilinson's conjecture on regulators means that it should land in

$$\left(\text{the regulator lattice in } Sym^{n-2} H^1(E(\mathbb{C}), \mathbb{R}(1))^+ \right) \otimes S^{n+1-m} J_{\mathbb{Q}}$$

So by the Mordell-Weil theorem if $k(P)$ is a given number field then the target group is a finite dimensional $\mathbb{Q}$-vector space.

3. If v is a p-adic valuation of the field $k(P)$, then $(\log p)^{-1} \cdot h_v(P_i) \in \mathbb{Q}$, and so the target of the height homomorphism is a finite dimensional $\mathbb{Q}$-vector space.

4. If the a divisor P is in the kernel of the height homomorphisms for all archimedean valuations but one then thanks to the product formula it is sent to zero by all of them. In particulary if $k(P) = \mathbb{Q}$ we can forget the archimedian valuation.

Composing each of these homomorphism with the appropriate map d_m we get homomorphisms

$$(p_2 \otimes id) \circ d_2, \quad (h_v \otimes id) \circ d_2, \quad Ber_m \circ d_m, \quad (K_m \otimes id) \circ d_m \qquad (61)$$

here $m > 2$.

Proposition 8.9 *The condition $\delta(P) = 0$ in the group $B_n(E) \otimes J_{\mathbb{Q}}$ implies that all of the homomorphisms (61) are equal to zero.*

Proof. Clear.

The height condition is the crucial one. If it is satisfied, then for a given field $k(P_a)$ the other conditions should give only a finite number of conditions on the divisors P_a.

If a $\mathbb{Q}$-rational divisor P is sent to zero by all of the homomorphisms (61) then this essentially means that $\delta(P) = 0$ in the group $B_n(E) \otimes J(\bar{\mathbb{Q}})_{\mathbb{Q}}$. To see this we write the homomorphism d_m as a composition of homomorphisms $\delta \otimes id$

$$B_{n+1}(E) \to B_n(E) \otimes J_{\mathbb{Q}} \to (B_{n-1}(E) \otimes J_{\mathbb{Q}}) \otimes J_{\mathbb{Q}} \to \ldots \to B_m(E) \otimes J_{\mathbb{Q}}^{\otimes n+1-m} \qquad (62)$$

followed by the projection

$$B_m(E) \otimes J_{\mathbb{Q}}^{\otimes n+1-m} \longrightarrow B_m(E) \otimes S^{n+1-m} J_{\mathbb{Q}}$$

Let us spell the details in the first interesting case: $n = 3$.

Proposition 8.10 *Let us assume that the Bloch-Beilinson regulator r_{Be} : $K_2(E)_\mathbb{Z} \longrightarrow \mathbb{R}$ is injective. Then if for $n = 3$ a $\mathbb{Q}$-rational divisor P belongs to the kernel of homomorphisms (61) then $\delta(P) = 0$ in the group $B_3(E) \otimes J_\mathbb{Q}$*

Proof. Consider the homomorphism

$$d_2 : B_4(E) \longrightarrow B_2(E) \otimes S^2 J, \qquad \{a\}_4 \longmapsto \{a\}_2 \otimes a \cdot a \qquad (63)$$

Suppose an element $P \in B_4(E)$ is in the kernel of the homomorphism $(p_2 \otimes id) \circ d_2$. Then $P \in B_4^*(E)$ and $f(P) \in (\bar{\mathbb{Q}}^* \otimes S^2 J(\bar{\mathbb{Q}}))^{Gal(\bar{\mathbb{Q}}/\mathbb{Q})}$.

The image of the divisor P under this homomorphism belongs to the subgroup $k(P)^* \otimes S^2 J(k(P))$. The hight condition is a way to say that it is equal to zero.

Let us write the map d_2 as a composition

$$B_4(E) \longrightarrow B_3(E) \otimes J \longrightarrow (B_2(E) \otimes J) \otimes J \longrightarrow B_2(E) \otimes S^2 J$$

Notice that

$$Ker\Big(B_3(E) \otimes J \to (B_2(E) \otimes J) \otimes J\Big) \otimes \mathbb{Q} = Ker\Big(B_3(E) \to B_2(E) \otimes J\Big) \otimes J \otimes \mathbb{Q}$$

Consider the homomorphism

$$B_3(E(\mathbb{C})) \otimes J(\mathbb{C}) \longrightarrow J(\mathbb{C}) \otimes \mathbb{R}, \qquad b \otimes \{a\}_3 \longmapsto K_{2,1}(a) \otimes b \qquad (64)$$

There is a homomorphism $K_2(E(\mathbb{C})) \longrightarrow B_3(E(\mathbb{C}))$ such that the following diagram is commutative (see [GL]):

$$
\begin{array}{ccc}
K_2(E(\mathbb{C})) & \xrightarrow{r_{Be}} & \mathbb{R} \\
\downarrow & & \downarrow Id \\
B_3(E(\mathbb{C})) & \xrightarrow{K_3} & \mathbb{R}
\end{array}
$$

So assuming the injectivity of the regulator $K_2(E)_\mathbb{Z} \longrightarrow \mathbb{R}$ we see that (64) should be injective on $Ker\Big(B_3(E) \longrightarrow B_2(E) \otimes J\Big) \otimes J \otimes \mathbb{Q}$.

9 The complexes $\mathcal{B}(E; n)^\bullet$ and motivic elliptic polylogarithms

1. In this chapter $k = \bar{k}$, and all abelian groups are tensored by $\mathbb{Q}$, so we work with the corresponding $\mathbb{Q}$-vector spaces. For instance $J := J \otimes \mathbb{Q}$ etc.

Let

$$\mathbb{Q}[E(k)] \xrightarrow{\delta} \mathbb{Q}[E(k)] \otimes J, \qquad \{a\} \longmapsto \{a\} \otimes a$$

Theorem 9.1 *Let us assume standard conjectures on mixed motives. Then there exist canonical homomorphisms* $l_n : \mathbb{Q}[E(k)] \to \mathcal{L}(E)_{S^{n-2}\mathcal{H}(1)^\vee}$ *such that the following digram is commutative:*

$$
\begin{array}{ccc}
\mathbb{Q}[E(k)] & \xrightarrow{\delta} & \mathbb{Q}[E(k)] \otimes J \\[2mm]
l_{n-1} \downarrow & & \downarrow l_{n-2} \otimes id \\[2mm]
\mathcal{L}(E)_{S^{n-2}\mathcal{H}(1)^\vee} & \xrightarrow{\partial} & \mathcal{L}(E)_{S^{n-3}\mathcal{H}(1)^\vee} \otimes J
\end{array}
$$

Proof. The proof is based on the existence and basic properties of the motivic elliptic polylogarithms of Beilinson and Levin [BL]. For any nonzero point $a \in E(k)$ let $G_a^{(1)}$ be an element of $Ext^1_{\mathcal{M}_E}(\mathbb{Q}(0), \mathcal{H})$ wich corresponds to $a \in J$ under the isomorphism $Ext^1_{\mathcal{M}_E}(\mathbb{Q}(0), \mathcal{H}) = J$. Set $G_a^{(m)} := S^m(G_a^{(1)})$. The motivic elliptic $(n-1)$-logarithm at a is a mixed elliptic motive $El_{n-1}(a)$ which provides a certain extension class in $Ext^1_{\mathcal{M}_E}(\mathcal{H}, G_a^{(n-1)}(1))$. In particular its weight graded quotients are

$$
\mathcal{H}, \mathbb{Q}(1), \mathcal{H}(1), ..., S^{(n-1)}\mathcal{H}(1)
$$

Therefore it has canonical framing and so defines an element of $\mathcal{A}(\mathcal{H}, S^{(n-1)}\mathcal{H}(1))$. After tensoring it by $\mathcal{H}$ and twisting by $\mathbb{Q}(-1)$ we can introduce a natural framing by $\mathbb{Q}$ and $S^{(n-2)}\mathcal{H}(1)$ (since $S^{(n-1)}\mathcal{H} \otimes \mathcal{H} = S^{(n)}\mathcal{H} \oplus S^{(n-2)}\mathcal{H}$). Therefore we picked up an element $l_{n-1}(a) \in \mathcal{A}(\mathbb{Q}, S^{(n-2)}\mathcal{H}(1))$.

The commutativity of the diagram follows from the properties of the elliptic polylogarithms ([BL]). The crucial point is this. Since $W_{\leq -3}El_{n-1}(a)$ is a symmetric power of the motive $G_a^{(1)}$, The projection to $\mathcal{L}(E)$ of the elements in $\mathcal{A}(S^{(k)}\mathcal{H}(1), S^{(n-1)}\mathcal{H}(1))$ coming from the canonical framing by $S^{(k)}\mathcal{H}(1)$ and $S^{(n-1)}\mathcal{H}(1))$ of the motive $El_{n-1}(a)$ are zero provided $0 \leq k < n - 2$. So projecting the coproduct of $El_{n-1}(a)$ to $\Lambda^2 \mathcal{L}(E)$ the only nonzero contribution we get is given by the component of the coproduct coming from $\mathcal{A}(\mathcal{H}, S^{(n-2)}\mathcal{H}(1)) \otimes \mathcal{A}(S^{(n-2)}\mathcal{H}(1), S^{(n-1)}\mathcal{H}(1))$. The fact that it is equal to $l_{n-2}(a) \wedge a$ follows immediately from the basic property of the elliptic polylogarithm motive (see [BL]).

Definition 9.2

$$
\mathcal{R}_n(E) = Ker l_{n-1}, \qquad \mathcal{B}_n(E) = \frac{\mathbb{Q}[E(k)]}{\mathcal{R}_n(E)}
$$

Theorem (9.1) implies that δ provides a well defined homomorphism $\delta : \mathcal{B}_n(E) \longrightarrow \mathcal{B}_{n-1}(E) \otimes J$. So we get a complex $\mathcal{B}(E; n)^\bullet$:

$$
\mathcal{B}_n(E) \longrightarrow \mathcal{B}_{n-1}(E) \otimes J \longrightarrow ... \longrightarrow \mathcal{B}_2(E) \otimes \Lambda^{n-2}J \longrightarrow J \otimes \Lambda^{n-1}J \longrightarrow \Lambda^n J
$$

Set

$$r_n(J) := Ker\left(J \otimes \Lambda^{n-1}J \longrightarrow \Lambda^n J\right)$$

The theorem (9.1) and this definition immediately imply that there exists canonical homomorphism of complexes

$$
\begin{array}{ccccccc}
\mathcal{B}_n(E) & \longrightarrow & \mathcal{B}_{n-1}(E) \otimes J & \longrightarrow \dots \longrightarrow & \mathcal{B}_2(E) \otimes \Lambda^{n-2}J & \longrightarrow & r_n(J) \\
l_{n-1} \downarrow & & l_{n-2} \otimes id \downarrow & & \downarrow l_1 \otimes id & & \| \\
\mathcal{L}(E)_{S^{n-2}\mathcal{H}(1)^\vee} & \longrightarrow & \mathcal{L}(E)_{S^{n-3}\mathcal{H}(1)^\vee} \otimes J & \longrightarrow \dots \longrightarrow & \mathcal{L}(E)_{\mathbb{Q}(1)^\vee} \otimes \Lambda^{n-2}J & \longrightarrow & r_n(J)
\end{array}
$$

Lemma 9.3 *The bottom complex is a subcomplex of the $S^{n-2}\mathcal{H}(1)^\vee$-isotypical component of the standard cochain complex of the Lie coalgebra $\mathcal{L}(E)$.*

So if $k = \bar{k}$, we get a canonical injective morphism of the complexes

$$\mathcal{B}(E; n+1)^\bullet \longrightarrow \left(\Lambda^\bullet \mathcal{L}(E), \partial\right)_{S^{n-1}\mathcal{H}(1)^\vee} \tag{65}$$

Let $\mathcal{K}_{n-1}^M$ be the sheaf of Milnor K-groups. Combining this morphism with the canonical morphism from the right hand side to $RHom_{\mathcal{MM}_k}(\mathbb{Q}(0), S^{n-1}\mathcal{H}(1))$ provided by the inclusion functor $\mathcal{M}_E \to \mathcal{MM}_k$ we get

Corollary 9.4 *Let us assume the formalism of mixed motives. Then*
a) there exists canonical homomorphisms

$$H^i(\mathcal{B}(E; n+1)_{\mathbb{Q}}^\bullet) \longrightarrow gr_n^\gamma K_{n+1-i}(E^{(n-1)})_{sgn} \otimes \mathbb{Q} \tag{66}$$

b) The homomorphism for $i = 1$ is injective.

Indeed, thanks to (65) this is true if $k = \bar{k}$. The general case follows since we have the descent property both for rational K-theory and, (by definition), for the complexes $\mathcal{B}(E; n+1)_{\mathbb{Q}}^\bullet$.

Remark. We do not expect a morphism of complexes (65) exist if k is not algebraically closed. The reason is this. If k is not algebraically closed we have postulated the Galois descent for the complexes $\mathcal{B}(E; n+1)^\bullet$. On the other hand the standard complex of the Lie algebra $\mathcal{L}(E/k)$ should not satisfy the Galois decent.

I hope a stronger result should be valid:

Conjecture 9.5 *a)*

$$gr_n^\gamma K_{n+1-i}(E^{(n-1)})_{sgn} \otimes \mathbb{Q} = H^{i-1}(E^{(n-1)}, \mathcal{K}_n^M)_{sgn} \otimes \mathbb{Q} \tag{67}$$

b) There exists a canonical isomorpism in the derived category

$$\mathcal{B}(E; n+1)_{\mathbb{Q}}^{\bullet} = RHom_{\mathcal{MM}_k}(\mathbb{Q}(0), S^n\mathcal{H}(1))$$

in particular

$$H^i\left(\mathcal{B}(E, n+1)_{\mathbb{Q}}^{\bullet}\right) = gr_n^{\gamma}K_{n+1-i}(E^{(n-1)})_{sgn} \otimes \mathbb{Q} \qquad (68)$$

c) Let $k = \bar{k}$. Then the homomorphism of complexes (65) is a quasiisomorphism.

If the conjecture (4.2) is correct, then b) is equivalent to c).
According to the lemma (9.3) this conjecture implies conjecture (1.4).
Part a) of the conjecture is trivial for $n = 2$.

Lemma 9.6 *For $n = 3$ and $k = \bar{k}$ the part a) of the conjecture (9.5) follows from the rigidity conjecture for $K_3^{ind}(k)$.*

Proof. The statement boils down to $Ker\beta = Im\alpha$ in the sgn-part of the diagram

$$
\begin{array}{ccc}
\longrightarrow \quad B_2(k(E^{(2)})) \otimes k(E^{(2)})^* & \longrightarrow & \Lambda^3 k(E^{(2)})^* \\
\quad\quad\quad\quad\quad\quad\quad \downarrow \alpha & & \downarrow \\
\amalg_{Y \in (E^{(2)})^1} B_2(k(Y)) & \xrightarrow{\;\beta\;} & \amalg_{Y \in (E^{(2)})^1} \Lambda^2 k(Y)^* \\
\quad\quad\quad\quad\quad \downarrow & & \downarrow \\
\quad\quad\quad\quad\quad 0 & & \cdots
\end{array}
$$

Let $\beta = \amalg_Y \beta_Y$. Then $Ker\beta_Y = K_3^{ind}(k(Y))_{\mathbb{Q}}$. By the rigidity conjecture any point $y \in Y$ provides an isomorphism $K_3^{ind}(k)_{\mathbb{Q}} = K_3^{ind}(k(Y))_{\mathbb{Q}}$. So $Ker\beta/Im\alpha$ is a subgroup of

$$Coker\left(K_3^{ind}(k)_{\mathbb{Q}} \otimes k(E^{(2)})^* \longrightarrow \coprod_{Y \in (E^{(2)})^1} K_3^{ind}(k)_{\mathbb{Q}}\right) = CH^1(E^{(2)})_{sgn} \otimes \mathbb{Q} = 0$$

The lemma is proved.

2. Motivic realization of elliptic polylogarithms. Let $f = (f_1, ..., f_n)$ be an n-tuple of rational functions on E.

Motivation. Consider the following multivalued analytic function on

$$\{n - \text{tuples of rational functions on} \quad E(\mathbb{C})\} \times H_{n-1}(E^{(n-1)}(\mathbb{C}))_{sgn_n}$$

$$P(E^{(n-1)}; f; \gamma) := \int_{\gamma} \text{Alt}_{(x_1,...,x_n)} \log f_1(x_1)) d\log f_2(x_2) \wedge ... \wedge d\log f_n(x_n) \qquad (69)$$

where γ is a cycle representing a nontrivial class in $H_{n-1}(E^{(n-1)}(\mathbb{C}))_{sgn_n}$. The subscript sgn_n means the skewsymmetric part with respect to the permutations. (A better way to define this function is given by formula (74) below.) We will show that this function is a period of a mixed elliptic motive.

Choose a coordinate z on $\mathbb{P}^1$. Let Δ_n be the coordinate cube in $(\mathbb{P}^1; 0 \cup \infty)^n$, i.e. union of $2n$ hyperplanes $z_i = 0$, $z_i = \infty$. Notice that $(\mathbb{P}^1)^n \backslash \Delta_n = (\mathbb{G}_m)^n$.

Let us define a codimension n cycle

$$Z(E^{(n-1)}; f) \subset E^{(n-1)} \times (\mathbb{P}^1)^n$$

as follows:

$$Z(E^{(n-1)}; f) := \text{Alt}_{(x_1, \dots, x_n)}\{x_1, \dots, x_n; f_1(x_1), \dots, f_n(x_n)\} \cup E^{(n-1)} \times \{1, \dots, 1\}$$

Here we use the coordinate system $z_1, \dots, z_n$, i.e. $z_i = f_i(x_i)$. The group S_n acts on $(\mathbb{G}_m)^n$ by permutations. So we get an action of the group $S_n \times S_n$ on $E^{(n-1)} \times (\mathbb{G}_m)^n$. In this chapter we mark by the subscript sgn the skewsymmetric part under this action. Consider the following mixed motive

$$h(E^{(n-1)}; f) := H^n(E^{(n-1)} \times (\mathbb{G}_m)^n, Z^0(E^{(n-1)}; f))(n)_{sgn}$$

where $Z^0(E^{(n-1)}; f) := Z(E^{(n-1)}; f) \backslash \left(Z(E^{(n-1)}; f) \cap \Delta_n \right)$.

More precisely, $h(E^{(n-1)}; f)(-n) := R^n p_! \mathcal{F}(E; f)_{sgn}$ where $\mathcal{F}(E; f)$ is the following mixed motivic sheaf on $E^{(n-1)} \times (\mathbb{P}^1)^n$. Take the constant sheaf on the complement to $Z(E^{(n-1)}; f) \cup E^{(n-1)} \times \Delta_n$ in $E^{(n-1)} \times (\mathbb{P}^1)^n$; extend it by j_* to the divisor $E^{(n-1)} \times \Delta_n$ and then by $j_!$ to $Z(E^{(n-1)}; f)$.

Lemma 9.7 $h(E^{(n-1)}; f)$ *is a mixed elliptic motive framed by* $\mathbb{Q}(0)$ *and* $\text{Sym}^{n-1}\mathcal{H}(1)$.

Proof. i) $\mathbb{Q}(0)$-*component of the frame.* Let us prove that

$$gr_{2n}^W H^n(E^{(n-1)} \times (\mathbb{G}_m)^n, Z^0(E^{(n-1)}; f))_{sgn} = \mathbb{Q}(-n)$$

Notice that $H^n(\mathbb{G}_m)_{sgn}^n = H^n(\mathbb{G}_m)^n$ and $H^i(\mathbb{G}_m)_{sgn}^n = 0$ for $i < n$. So the projection $E^{(n-1)} \times (\mathbb{G}_m)^n \longrightarrow (\mathbb{G}_m)^n$ induces an isomorphism

$$\mathbb{Q}(-n) = gr_{2n}^W H^n((\mathbb{G}_m)^n)_{sgn} \xrightarrow{\sim} gr_{2n}^W H^n(E^{(n-1)} \times (\mathbb{G}_m)^n)_{sgn}$$

The canonical morphism

$$H^n(E^{(n-1)} \times (\mathbb{G}_m)^n, Z^0(E^{(n-1)}; f))_{sgn} \longrightarrow H^n(E^{(n-1)} \times (\mathbb{G}_m)^n)_{sgn}$$

induces an isomorphism after taking gr_{2n}^W. Indeed, there is an exact sequence

$$H^{n-1} Z^0(E^{(n-1)}; f) \longrightarrow H^n(E^{(n-1)} \times (\mathbb{G}_m)^n, Z^0(E^{(n-1)}; f))_{sgn} \longrightarrow$$

$$H^n(E^{(n-1)} \times (\mathbb{G}_m)^n)_{sgn} \longrightarrow H^n Z^0(E^{(n-1)}; f)$$

and

$$gr_{2n}^W H^{n-1} Z^0(E^{(n-1)}; f) = gr_{2n-1}^W H^{n-1} Z^0(E^{(n-1)}; f) = 0$$

since $Z^0(E^{(n-1)}; f)$ is an open regular variety of dimension $n-1$.

ii) $Sym^{n-1}\mathcal{H}(1)$-*component of the frame.* One has

$$H^n(E^{(n-1)} \times (\mathbb{P}^1)^n, Z(E^{(n-1)}; f))_{sgn} = Sym^{n-1}h^1(E) \tag{70}$$

Indeed, $H^n(E^{(n-1)} \times (\mathbb{P}^1)^n)_{sgn} = 0$. So there is an exact sequence

$$H^{n-1}(E^{(n-1)} \times (\mathbb{P}^1)^n)_{sgn} \overset{\alpha}{\longrightarrow} H^{n-1}(Z(E^{(n-1)}; f))_{sgn} \overset{\beta}{\longrightarrow}$$

$$H^n(E^{(n-1)} \times (\mathbb{P}^1)^n, Z(E^{(n-1)}; f))_{sgn} \longrightarrow 0$$

Further,

$$H^{n-1}(E^{(n-1)} \times (\mathbb{P}^1)^n)_{sgn} = Sym^{n-1}h^1(E)$$

$$H^{n-1}(Z(E^{(n-1)}; f))_{sgn} = Sym^{n-1}h^1(E) \oplus Sym^{n-1}h^1(E)$$

and α is injective.

The restriction map induces an isomorphism

$$gr_{n-1}^W H^n(E^{(n-1)} \times (\mathbb{P}^1)^n, Z(E^{(n-1)}; f))_{sgn} \longrightarrow \tag{71}$$

$$gr_{n-1}^W H^n(E^{(n-1)} \times (\mathbb{G}_m)^n, Z^0(E^{(n-1)}; f))_{sgn}$$

because $(Z(E^{(n-1)}; f)$ is regular of dimension $n-1$).

$$W_{n-1} H^{n-1} Z^0(E^{(n-1)}; f) = W_{n-1} H^{n-1} Z(E^{(n-1)}; f)$$

Combining (70) and (71) we get the $Sym^{n-1}\mathcal{H}(1)$-component of the frame. The lemma is proved.

Finally, we show that $h(E^{(n-1)}; f)$ is a mixed *elliptic* motive by induction using the following basic observation: the intersection of the cycle $Z(E^{(n-1)}; f_1, ..., f_n)$ with any codimension 1 face of the cube Δ_n is a sum of cycles of form $Z(E^{(n-2)}; g_1, ..., g_{n-1})$. For example

$$Z(E^{(n-1)}; f_1, ..., f_n) \cap \{z_1 = 0\} = \sum_{x \in E} m_x(f_1) \cdot Z(E^{(n-2)}; f_2, ..., f_n)$$

where $m_x(f)$ is the multiplicity of zero at x. The lemma is proved.

Remark 9.8 *One can apply the same construction to n arbitrary functions* $f_1(z_1, ..., z_n), ..., f_n(z_1, ..., z_n)$ *on* $E^{(n-1)}$. *However it is not clear whether the motive* $h(E^{(n-1)}; f_1, ..., f_n)$ *is a mixed elliptic motive in general.*

The functions $f_1, ..., f_n$ on the E define a map

$$f : E^{(n-1)} \longrightarrow (\mathbb{P}^1)^n, \quad (x_1, ..., x_n) \longmapsto (f_1(x_1), ..., f_n(x_1))$$

Let $E_f^{(n-1)}$ be the image of this map and $\tilde{E}_f^{(n-1)} := E_f^{(n-1)} \backslash \left(E_f^{(n-1)} \cap \Delta \right)$. Consider the following motive:

$$\tilde{h}(E^{(n-1)}; f) := H^n((\mathbb{G}_m)^n, \tilde{E}_f^{(n-1)})(n)_{sgn} \tag{72}$$

Lemma 9.9 *Suppose that* $f_* : H_{n-1}(E^{(n-1)})_{sgn} \to H_{n-1}(E_f^{(n-1)})_{sgn}$ *is a nonzero map. Then*
  a) $\tilde{h}(E^{(n-1)}; f)$ *is a* $(\mathbb{Q}(0), Sym^{n-1}\mathcal{H}(1))$*-framed mixed elliptic motive.*
  b) $\tilde{h}(E^{(n-1)}; f) = h(E^{(n-1)}; f)$ *as framed motives.*
  c) If $f_* = 0$ *then* $h(E^{(n-1)}; f) = 0$.

Proof. a) is very similar to the proof of the lemma above. For instance the $Sym^{n-1}h^1(E)$ part of the framing comes from isomorphism

$$W_{n-1}H^n((\mathbb{P}^1)^n\backslash\Delta_n, \tilde{E}_f^{(n-1)}) \longrightarrow Sym^{n-1}h^1(E)$$

Namely, $H^i(P^n)_{sgn} = 0$, so there is an isomorphism

$$H^{n-1}(E_f^{(n-1)})_{sgn} \longrightarrow H^n((\mathbb{P}^1)^n, E_f^{(n-1)})_{sgn}$$

Combining it with $f^* : H^{n-1}(E_f^{(n-1)})_{sgn} \longrightarrow H^{n-1}(E^{(n-1)})_{sgn}$ we get a morphism

$$H^n((\mathbb{P}^1)^n, E_f^{(n-1)})_{sgn} \longrightarrow H^{n-1}(E^{(n-1)})_{sgn} = Sym^{n-1}h^1(E) \tag{73}$$

  b) The projection

$$E^{(n-1)} \times (\mathbb{P}^1)^n \longrightarrow (\mathbb{P}^1)^n$$

induces a morphism respecting the frames.
  c) is clear from the construction. The lemma is proved.
  The period corresponding to this framing is exactly the function (69). Indeed, consider the differential form

$$\omega_{\Delta_n} := d\log(z_1) \wedge ... \wedge d\log(z_n)$$

in $(\mathbb{CP}^1)^n\backslash\Delta_n$. Let Γ be a relative n-chain in $\mathbb{CP}^n$ which bounds an $(n-1)$-cycle γ, $[\gamma] \in H_{n-1}((\mathbb{CP}^1)^n, E_f^{(n-1)})_{sgn}$. Then

$$L_n(E^{(n-1)}; f; \gamma) = \int_\Gamma \omega_{\Delta_n} \tag{74}$$

The Stokes formula shows that the integrals (74) and (69) coinside.

Proposition 9.10 *The $\mathbb{R}$-valued period of the Hodge realization of $h(E^{(n-1)}; f)$ is given by*

$$\mathcal{L}(E^{(n-1)}; f) := \int_{E^{(n-1)}(\mathbb{C})} \mathrm{Alt}_{x_1,\ldots,x_n} r_n(f_1, \ldots, f_n) \wedge \omega_{p,q}$$

This integral coincides with the one computed by means of the Eisenstein-Kronecker series, as was explained before.

Recall that $\mathcal{A}(\mathbb{Q}, Sym^{n-1}\mathcal{H}(1))$ is the group of mixed elliptic motives framed by $\mathbb{Q}(0)$ and $Sym^{n-1}\mathcal{H}(1)$.

Lemma 9.11 *For any $\lambda_i \in k^*$ one has the equality of framed motives*

$$h(E^{(n-1)}; f_1, \ldots, f_n) = h(E^{(n-1)}; \lambda_1 \cdot f_1, \ldots, \lambda_n \cdot f_n)$$

Proof. The action of an element $g = (\lambda_1, \ldots, \lambda_n) \in \mathbb{G}_m^n$ on $P^n \backslash \Delta_n = \mathbb{G}_m^n$ gives a morphism of motives $h(E^{(n-1)}; \lambda_1 \cdot f_1, \ldots, \lambda_n \cdot f_n) \to h(E^{(n-1)}; f_1, \ldots, f_n)$ which obviously preserves the framing.

Theorem 9.12 *There is a well defined homomorphism of abelian groups*

$$m_n^* : S^n k(E)^* \longrightarrow \mathcal{A}(\mathbb{Q}(0), Sym^{n-1}\mathcal{H}(1)) \qquad f_1 \circ \ldots \circ f_n \longmapsto h(E^{(n-1)}; f)$$

It is zero if one of the functions f_i is a constant; one has $m_n^*(Ker \mu_n) = 0$.

Proof.

Below a generalization of the construction above is given. Let D^0 be the group of degree zero divisors on E. For any $d := (d_1, \ldots, d_n)$ let us construct a mixed elliptic motive $h(E^{(n-1)}; d) := h(E^{(n-1)}; d_1, \ldots, d_n)$ framed by $\mathbb{Q}(0)$ and $Sym^{n-1}\mathcal{H}(1)$.

Let $\mathcal{P}$ be the rigidified Poincaré line bundle over $J \times J$. For any two degree zero divisors d_1, d_2 with disjoint support there is an element

$$< d_1, d_2 > \in \mathcal{P}_{[d_1],[d_2]}$$

where $[d_i] \in J$ is the class of a degree zero divisor d_i.

Consider the following $(n-1)$-cycle

$$Z(E^{(n-1)}; d) \subset E^{(n-1)} \times \mathcal{P}^n \qquad (75)$$

$$c(E^{(n-1)}; d) := \mathrm{Alt}_{(x_1,\ldots,x_n)}(x_1, \ldots, x_n; < d_1, (x_1) - (0) >; \ldots; < d_n, (x_n) - (0) >)$$

Here $x_1 + \ldots + x_n = 0$ and $(x_1, \ldots, x_n)$ belongs to Zariski open part of $E^{(n-1)}$ where the divisors d_i and $(x_i) - (0)$ are disjoint. and set

$$h(E^{(n-1)}; d) := H^n\Big(E^{(n-1)} \times \mathcal{P}^n, Z(E^{(n-1)}; d)\Big)(n)$$

Theorem 9.13 *a)* $h(E^{(n-1)}; d)$ *is a mixed elliptic motive framed by* $\mathbb{Q}(0)$ *and* $Sym^{n-1}\mathcal{H}(1)$.

b) There is a well defined homomorphism of abelian groups

$$S^n D^0 \longrightarrow \mathcal{A}(\mathbb{Q}(0), Sym^{n-1}\mathcal{H}(1)) \qquad d_1 \circ \ldots \circ d_n \longmapsto h(E^{(n-1)}; d)$$

Proof. Restriction to a fiber of the Poincare line bundle provides an isomorphism

$$gr_2^W H^1(\mathcal{P}) \longrightarrow H^1(\mathbb{G}_m) = \mathbb{Q}(-1)$$

and thus we get a first part of the framing:

$$\mathbb{Q}(-n) \longrightarrow gr_{2n}^W H^n(\mathcal{P}^n) \xrightarrow{\pi^*} gr_{2n}^W H^n(E^{(n-1)} \times \mathcal{P}^n)$$

where $\pi : E^{(n-1)} \times \mathcal{P}^n \to \mathcal{P}^n$ is the natural projection.

The second part of the framing comes from the fundamental cycle of $E^{(n-1)}$ just as before. The rest is straitforward.

3. Motivic construction of the complex $B^*(E; 3)$. Recall the convolution map $\beta_2 : S^2 k(E)^* \longrightarrow B_3^*(E), f_1 \circ f_2 \longmapsto (f_1) * (f_2)$. We are going to show that the diagram

$$
\begin{array}{ccc}
S^2 k(E)^* & & \\
\downarrow \beta_2 & \searrow^{m_2^*} & \\
B_3^*(E) & \xrightarrow{l_2^*} & \mathcal{L}^*(E)_{\mathcal{H}(1)}
\end{array}
$$

provides a well defined homomorphism

$$l_2^* : B_3^*(E) \longrightarrow \mathcal{L}^*(E)_{\mathcal{H}(1)}$$

(Here $\mathcal{L}^*(E)_{\mathcal{H}(1)^\vee} = \mathcal{L}^*(E)_{\mathcal{H}(1)} \boxtimes \mathcal{H}(1)^\vee$, so $\mathcal{L}^*(E)_{\mathcal{H}(1)}$ is a $\mathbb{Q}$-vector space.)

Consider the map

$$\mu_2 : S^2 k(E)^* \to \frac{K_2(K(E)_-}{\{k^*, k(E)^*\}_-}, \qquad f_1 \circ f_2 \mapsto \{f_1(x), f_2(-x)\} - \{f_1(-x), f_2(x)\}$$

According to theorem 3.9 in [GL] one has

Theorem 9.14 $\mu_2(Ker\beta_2) = 0$

It remains to use that $l_2^*(f * (1 - f)) = 0$ by theorem 9.12.

Theorem 9.15 *We get a commutative diagram*

$$
\begin{array}{ccc}
B_3^*(E)_\mathbb{Q} & \xrightarrow{\delta_{\mathcal{H}(1)}^*} & (k^* \otimes J)_\mathbb{Q} \\
& & \\
l_2^* \downarrow & & \downarrow = \\
& & \\
\mathcal{L}^*(E)_{\mathcal{H}(1)} & \xrightarrow{\delta} & \mathcal{L}^*(E)_{\mathbb{Q}(1)} \otimes \mathcal{L}^*(E)_{\mathcal{H}}
\end{array}
$$

To prove this theorem we need only to compute $\delta h(E; f_1, f_2)$, which is left to the reader.

10 Elliptic Chow polylogarithms and generalized Eisenstein-Kronecker series

1. Elliptic Chow polylogarithms. *The single valued version.* Let C be a codimension n cycle in $E^k \times (\mathbb{P}^1)^l$, $k + l = 2n - 1$, skewinvariant under the action of $G_k \times G_l$.

Recall the forms $\pi_n \omega_{p,q}$ (see (53)), which for $p \geq q, p + q = n - 1$ form a basis over $\mathbb{R}$ in $H_B^{n-1}(E_{/\mathbb{R}}^{n-1}, \mathbb{R}(n-1))_{sgn_n}$. We represent elements in $H_B^{n-1}(E_{/\mathbb{R}}^{n-1}, \mathbb{R}(n-1))_{sgn_n}$ by their cup product with the forms $\pi_n \omega_{p,q}$:

The single valued elliptic Chow polylogarithm is a function

$$\mathcal{P}_{k,l} : \mathcal{Z}^n(E^k, l) \longrightarrow H_B^{n-1}(E_{/\mathbb{R}}^{n-1}, \mathbb{R}(n-1))_{sgn_n} \qquad k + l = 2n - 1$$

defined as follows. Let $\pi : C \to (\mathbb{P}^1)^l$ and $p : C \to E^k$. If $k > 0$:

$$< \mathcal{P}_{k,l}(C), \omega_{p,q} > := \int_C \pi^* r_{k-1}(z_1, ..., z_k) \wedge p^* \omega_{p,q} \quad p \geq q, p + q = n - 1,$$

Here we integrate over the nonsingular part of the complex points of the cycle C. The integral is always convergent (see [G6]). For example

$$\mathcal{P}_{2,1}(C) := \text{Alt}_{(x_2, x_3)} \int_C \pi_1^* \log |z_1| \pi_2^* \omega \wedge \pi_3^* \bar{\omega}$$

The multivalued elliptic Chow polylogarithm, denoted $P_{k,l}(C)$, is defined as follows. Let $(x_1, ..., x_k, z_{k+1}, ..., z_{k+l})$ be the coordinates on $E^k \times (\mathbb{P}^1)^l$. Assume $l \neq 0$. Let π_i (resp. p_j) be the projection of C to the i-th coordinate $\mathbb{P}^1$ (resp. j-th factor E) in $E^k \times (\mathbb{P}^1)^l$.

i). *Assume $k \leq n$. Then*

$$P_{k,l}(C) :=$$

$$\text{Alt}_{(G_k \times G_l)} \int_{p_1^* \gamma \times ... \times p_k^* \gamma \times \pi_{k+1} \delta \times ... \times \pi_{k+l} \delta} \log z_{n+1} d \log z_{n+2} \wedge ... \wedge d \log z_{2n-1}$$

ii). *Assume $n < k < 2n - 1$. Then*

$$P_{k,l}(C) :=$$

$$\text{Alt}_{(G_k \times G_l)} \int_{p_1^* \gamma \times ... \times p_n^* \gamma} p_{n+1}^* \omega \wedge ... \wedge p_k^* \omega \wedge (\log z_{k+1} d \log z_{k+2} \wedge ... \wedge d \log z_{2n-1})$$

Example 1. The multivalued elliptic Chow dilogarithms:

$$P_{0,3}(C) := \text{Alt}_{(G_3)} \int_{\pi_1^* \delta} \log z_2 d \log z_3$$

$$P_{1,2}(C) := \text{Alt}_{(G_1 \times G_2)} \int_{p_1^* \gamma} \log z_2 d \log z_3$$

$$P_{2,1}(C) := \mathrm{Alt}_{(G_2 \times G_1)} \int_{p_1^* \gamma} p_2^* \omega \cdot \log z_3$$

Example 2. The multivalued elliptic Chow trilogarithms:

$$P_{0,5}(C) := \mathrm{Alt}_{(G_5)} \int_{\pi_1^* \delta \times \pi_2^* \delta} \log z_3 d \log z_4 \wedge d \log z_5$$

$$P_{1,4}(C) := \mathrm{Alt}_{(G_1 \times G_4)} \int_{p_1^* \gamma \times \pi_2^* \delta} \log z_3 d \log z_4 \wedge d \log z_5$$

$$P_{2,3}(C) := \mathrm{Alt}_{(G_2 \times G_3)} \int_{p_1^* \gamma \times p_2^* \gamma} \log z_3 d \log z_4 \wedge d \log z_5$$

$$P_{3,2}(C) := \mathrm{Alt}_{(G_3 \times G_2)} \int_{p_1^* \gamma \times p_2^* \gamma} p_3^* \omega \wedge \log z_4 \wedge d \log z_5$$

$$P_{4,1}(C) := \mathrm{Alt}_{(G_4 \times G_1)} \int_{p_1^* \gamma \times p_2^* \gamma} p_3^* \omega \wedge p_4^* \omega \cdot \log z_5$$

The multivalued elliptic Chow polylogarithms are periods of mixed motives, which are easy to write down.

2. Some interesting cycles. Let $\mathbb{L}_n(a)$ be the codimension n cycle in $(\mathbb{P}^1)^{2n-1}$ responsible for the classical n-logarithm (see [Bl6] and [BK]):

$$\mathbb{L}_n(a) := \{x_1, ..., x_{k-1}, 1-x_1, 1-x_2/x_1, ..., 1-x_{k-1}/x_{k-2}, 1-a/x_{k-1}\} \in \mathcal{Z}^n(2n-1)$$

Consider the following cycle in $\mathcal{Z}^n(E^{(n-k-1)}, k+n)$:

$$\mathrm{Alt}_{(x_1,...,x_{n-k})}\Big(x_1, ..., x_{n-k}, \mathbb{L}_k(f_1(x_1)), f_2(x_2), ..., f_{n-k}(x_{n-k})\Big) \qquad (76)$$

Notice that we are using this $E^{(n-k-1)}$, not E^{n-k-1}.

Examples of cycles.

$$\text{in} \quad \mathcal{Z}^2(E, 2): \quad \mathrm{Alt}_{(x_1,x_2)}\{x_1, f_1(x_1), f_2(x_2)\}$$

$$\text{in} \quad \mathcal{Z}^2(3): \quad \{z_1, 1-z_1, 1-a/z_1\}$$

$$\text{in} \quad \mathcal{Z}^3(E^{(2)}, 3): \quad \mathrm{Alt}_{(x_1,x_2,x_3)}\{x_1, x_2, f_1(x_1), f_2(x_2), f(x_3)\}$$

$$\text{in} \quad \mathcal{Z}^3(E, 4): \quad \mathrm{Alt}_{(x_1,x_2)}\{x_1, 1-x_1, 1-f_1(x_1)/x_1, f_2(x_2)\}$$

$$\text{in} \quad \mathcal{Z}^3(5): \quad \{z_1, z_2, 1-z_1, 1-z_2/z_1, 1-a/z_2\}$$

I think the single valued elliptic Chow polylogarithm on these cycles should provide the new transcendental functions needed for $L(Sym^{n-k-1}E, n)$.

3. The generalized Eisenstein-Kronecker series and $L(Sym^n E, n+m)$ for $m \geq 1$. Conjecture 2.1 in [G2] for the field $k(E^{(n)})$ tells us that

$H^{n+1}_{\mathcal{M}}(Speck(E^{(n)}), \mathbb{Q}(n+m))$ is generated by the sums of the elements of form

$$\sum_i \{f_0^{(i)}(x)\}_m \otimes f_1^{(i)}(x) \wedge ... \wedge f_n^{(i)}(x) \quad \text{in} \quad \mathcal{B}_m(k(E^{(n)})) \otimes \Lambda^n k(E^{(n)})^*$$

satisfying the condition

$$\sum_i \{f_0^{(i)}(x)\}_{m-1} \wedge f_0^{(i)}(x) \wedge f_1^{(i)}(x) \wedge ... \wedge f_n^{(i)}(x) \quad \text{in} \quad \mathcal{B}_{m-1}(k(E^{(n)})) \otimes \Lambda^{n+1} k(E^{(n)})^*$$

$$(77)$$

Definition 10.1 $\mathcal{D}^{0,1}_{(n,m)}$ *is the subgroup of* $\mathcal{B}_m(k(E^{(n)})) \otimes \Lambda^n k(E^{(n)})^*$ *generated by the elements*

$$\text{Alt}_{(x_0,...,x_n)}\{f(x_0)\}_m \otimes g_1(x_1) \wedge ... \wedge g_n(x_n) \tag{78}$$

where $f, g_k \in k(E)^*$, $x_0 + ... + x_n = 0$.

Denote by X the element (78). Set

$$\partial(X) := \text{Alt}_{(x_0,...,x_n)}\{f_0(x_0)\}_{m-1} \wedge f_0(x_0) \wedge g_1(x_1) \wedge ... \wedge g_n(x_n)$$

It belongs to the group $\mathcal{B}_{m-1}(\mathbb{C}(E^{(n)})) \otimes \Lambda^{n+1}\mathbb{C}(E^{(n)})^*$.

$$r(X) := \sum_{i=1}^n (-1)^i \sum_{a \in E(k)} v_a g_i \cdot \text{Alt}_{(x_0,...,x_n)}\{f_0(x_0)\}_m \otimes g_1(x_1) \wedge ... \hat{g}_i ... \wedge g_n(x_n)|_{x_i=a}$$

Here $v_a g$ is the valuation of g at the point a. Let $d_{n,m} := \partial + r$. $r(X)$ lies in the sum of the groups $\mathcal{B}_m(\mathbb{C}(X_{i,a}))^* \otimes \Lambda^{n-1}\mathbb{C}(X_{i,a})^*$ where $X_{i,a}$ is the divisor $x_i = a$.

Conjecture 10.2 *a) There exists a map*

$$\text{Ker} d_{n,m} \longrightarrow H^{n+1}_{\mathcal{M}}(E^{(n)}, \mathbb{Q}(n+m))_{sgn}$$

which in the case $k = \mathbb{C}$ *commutes with the regulator map.*
   b) One might hope that this map is surjective.

Part a) of this conjecture can be deduced from standard conjectures. If $m = 1$ this is exactly conjecture discussed in chapter 5. If $n = 1$ the conjecture follows from the conjecture 2.1 in [G2], see also conjecture 8 in [G4] and $n = 1, m = 2$ it is proved in [G3]. In general the main argument for the hope expressed in the part b) is simplicity of the ansatz used to define the elements (78).
   Element (78) provides a cycle of type (76), as was explained above. The value of the regulator on the element which lies in $\text{Ker} d_{n,m}$ should coinside

with the value of the elliptic Chow polylogarithm $\mathcal{P}_{n,m}$ on the corresponding cycle. Thus to get the generalized Eisenstein-Kronecker series responsible for the special values $L(Sym^n E, n+m)$ we evaluate the Chow polylogarithm $\mathcal{P}_{n,m}$ on the cycle (76) using the Fourier decomposition of $\mathcal{L}_k(f(x))$ and then the same method as in chapter 6. This boils down to computation of the following regulator integral $(p+q=n)$

$$r(\{f\}_m \otimes g_1 \wedge ... \wedge g_n; \bar{\omega}_{p,q}) := \int_{E^n(\mathbb{C})} \log|f|^{m-2} \cdot \alpha_2(1-f, f) \wedge \alpha_n(g_1, ..., g_n) \wedge \omega_{p,q}$$

Set

$$r(\{f\}_m \otimes g_1 \wedge ... \wedge g_n) :=$$

$$\sum_{p+q=n} r(\{f\}_m \otimes g_1 \wedge ... \wedge g_n; \omega_{q,p})(dz)^p (d\bar{z})^q \in Sym^n H^1(E(\mathbb{C}), \mathbb{C})$$

Let

$$\left(\mathbb{Z}[E] \otimes \mathbb{Z}[E] \otimes \mathbb{Z}[E]\right)_E = \left(\mathbb{Z}[E \times E \times E]\right)_E$$

be the abelian group generated by the elements $\{x, y, z\}$ where $x, y, z \in E(k)$, subject to the relations $\{x, y, z\} = \{x+c, y+c, z+c\}$ for any $c \in E(k)$. Define

$$\beta_{n,m} : \{f\}_m \otimes g_1 \circ ... \circ g_n \longmapsto \left(\mathbb{Z}[E] \otimes \mathbb{Z}[E] \otimes \mathbb{Z}[E]\right)_E$$

$$\{f\}_m \otimes g_1 \circ ... \circ g_n \longmapsto (1-f) \otimes (f) \otimes (g_1) * ... * (g_n)$$

Consider the following functions where $p+q = n, m \geq 2$:

$$K_{n,m}^{p,q}(x, y, z) := (\frac{Im\tau}{\pi})^m \times$$

$${\sum_{\gamma_0 + ... + \gamma_m = 0}}' \frac{(x, \gamma_0)(y, \gamma_1 + ... + \gamma_{m-1})(z, \gamma_m) \cdot (\bar{\gamma}_m(\gamma_0 - \gamma_1) + \gamma_m(\bar{\gamma}_0 - \bar{\gamma}_1)) \cdot \gamma_m^{p-1} \cdot \bar{\gamma}_m^{q-1}}{|\gamma_0|^2 \cdot ... \cdot |\gamma_{m-1}|^2 |\gamma_m|^{2n}}$$

I will call them the generalized Eisenstein-Kronecker series. For $n=1$ this is the functions $K_{m+1}(x, y, z)$ defined in [G4], see also [G3]. For $n=1, m=2$ this function was considered by Deninger [D3].

Conjecture 10.3 *There exists a variation of mixed elliptic motives over* $(E \times E \times E)/E$ *such that its real periods are given by the generalized Eisenstein-Kronecker series.*

Define a homomorphism

$$K_{n,m} : \left(\mathbb{Z}[(E \times E \times E)(\mathbb{C})]\right)_E \longrightarrow Sym^n H^1(E(\mathbb{C}), \mathbb{C})$$

$$\{x, y, z\} \longmapsto \sum_{p+q=n} K_{n,m}^{p,q}(x, y, z)(dz)^p (d\bar{z})^q$$

Theorem 10.4 *Assume*

$$\text{Alt}_{(x_0,\dots,x_n)}\Big(\{f(x_0)\}_{m-1} \otimes f(x_0) \wedge g_1(x_1) \wedge \dots \wedge g_n(x_n)\Big) = 0$$

in the group $\mathcal{B}_{m-1}(\mathbb{C}(E^{(n)})) \otimes \Lambda^{n+1}\mathbb{C}(E^{(n)})^*$. *Then*

$$r(\{f\}_m \otimes g_1 \wedge \dots \wedge g_n) = K_{n,m} \circ \beta_{n,m}\Big(\{f\}_m \otimes g_1 \circ \dots \circ g_n\Big)$$

The proof is completely similar to the proof of theorems 8.4 and theorems 3.4 and 4.7 in [G3], and thus is omitted.

REFERENCES

[B1] Beilinson A.A.: *Higher regulators and values of L-functions*, VINITI, 24 (1984), 181–238 (in Russian); English translation: J. Soviet Math. 30 (1985), 2036–2070.

[B2] Beilinson A.A.: *Higher regulators of modular curves*, Contemp. Math. 55, 1–34 (1986)

[B3] Beilinson A.A.: *Height pairings between algebraic cycles*, Lect. Notes in Math. 1289, (1987), 1–26.

[BD1] Beilinson A.A., Deligne P.: *Polylogarithms and regulators*, To appear.

[BD2] Beilinson A.A., Deligne P.: *Interpretation motivique de la conjecture de Zagier*, in Symp. in Pure Math., v. 55, part 2, (1994), 23–41

[BL] Beilinson A.A., Levin A.M.: *Elliptic polylogarithms*, Symposium in pure mathematics, 1994, vol 55, part 2, 101–156.

[BMS] Beilinson A.A., MacPherson R., Schechtman V.V.: *Notes on motivic cohomology*, Duke math. Journal, 54 (1987), 679–710.

[BGSV] Beilinson A.A., Goncharov A.A., Schechtman V.V., Varchenko A.N.: *Aomoto dilogarithms, mixed Hodge structures and motivic cohomology*, the Grothiendieck Feschtrift, Birkhäuser, vol 86, 1990, p. 135–171.

[Bl1] Bloch S.: *Higher regulators, algebraic K-theory and zeta functions of elliptic curves*, Lect. Notes U.C. Irvine, 1977.

[Bl2] Bloch S.: *Algebraic cycles and higher K-theory*, Advances in Math. vol. 61 (1986) 267–304.

[Bl3] Bloch S.: *Cycles and biextensions*, Contemporary Math. 1989, vol 83, 19–31

[Bl4] Bloch S.: *Remarks on mixed elliptic motives*, To appear in the Proceedings of the conference on regulators in Jerusalem, 1998

[Bl5] Bloch S.: *Some elementary theorems about algebraic cycles on abelian varieties*, Inventiones Math, 37 (1976), 215–228.

[Bl6] Bloch S.: *Algebraic cycles and the Lie algebra of mixed Tate motives* Amer. J. of Math., 4 (1991), 771–791.

[BK] Bloch S., Kriz I.: *Mixed Tate motives*, Annals of mathematics, 1994, vol. 140, N3, 557–605.

[De] Deligne P.: *Valeurs de fonction L et périodes des intégrales*, In Proc. Symp. Pure Math., 33 (1979), 313–346.

[De2] Deligne P.: *A quoi servent les motifs?*, in Symp. in Pure Math., v. 55, part 1, (1994), 143–163.

[D1] Deninger, C.: *Higher regulators and Hecke L-series of imaginary quadratic fields I*, Invent. Math. 96, 1–96 (1989)

[D2] Deninger, C.: *Higher regulators and Hecke L-series of imaginary quadratic fields II*, Ann. of Math, 132, N1 (1990), 131–158.

[D3] Deninger, C.: *Higher order operations in Deligne cohomology*. Inventiones Math. 122, N1, (1995), 289–316.

[DS] Dupont J., Sah S.H.: *Scissors congruences II*, J. Pure Appl. Algebra, v. 25, (1982), 159–195.

[G1] Goncharov A.B.: *Geometry of configurations, polylogarithms and motivic cohomology*, Advances in Math. vol 144, N2, (1995), 279–312.

[G2] Goncharov A.B.: *Polylogarithms and motivic Galois groups*, Symposium in pure mathematics, 1994, vol 55, part 2, 43–96.

[G3] Goncharov A.B.: *Deninger's conjecture on special values of L-functions of an elliptic curve at s = 3*, Special volume dedicated to Manin's 60th birthday, Plenum, 1997 (alg-geom e-preprint, Preprint MPI January 1996.)

[G4] Goncharov A.B.: *Polylogarithms in arithmetic and geometry*, Proc. ICM-94 in Zurich, vol 1, 374–387.

[G5] Goncharov A.B.: *The Eisenstein-Kronecker series and* $L(Sym^2E, 3)$, Preprint September 1995.

[G6] Goncharov A.B.: *Chow polylogarithms and regulators.* Math. Res. Letters, N1 (1995) 95–112

[GL] Goncharov A.B., Levin A.M.: *Zagier's conjecture on* $L(E, 2)$, Inventiones Math. (1998).

[J] de Jeu R.: *Zagier's conjecture and wedge complexes in algebraic K-theory*, Comp. Math. 96, N2 (1995) 197–247.

[MS] Mestre J.-F., Schappacher, N.: *Series de Kronecker et fonctions L des puissances symmetriques de courbes elliptiques sur* $\mathbb{Q}$, In: Arithmetic algebraic geometry. Van de Geer. G., Oort, F., Steenbrink, J. (eds) (Prog. Math., vol. 89, 209–245). Birkhauser 1991.

[Q] Quillen D. : *Rational homotopy theory*, Ann. of Math, 90 (1969) 204–295.

[SS] Schappacher N., Scholl, A.: *The boundary of the Eisenstein symbol*, Math. Ann. 1991, 290, 303–321.

[S] Suslin A.A.: K_3 *of a field and Bloch's group*, Proceedings of the Steklov Institute of Mathematics 1991, Issue 4.

[We] Weil A.: *Elliptic functions according to Eisenstein and Kronecker*, Ergebnisse der Mathematik, 88, Springer 77.

[W] Wildeshaus J.: *On an analog of Zagier's conjecture for elliptic curves*, Duke Math. Journal, (1997) vol. 87 N2, 355–407.

[W2] Wildeshaus J.: *On the generalized Eisenstein symbol*, Preprint, 1997.

[Z] Zagier D.: *The Bloch-Wigner-Ramakrishnan polylogarithm function*, Math. Ann. 286, 613–624 (1990)

[Z2] Zagier D.: *Polylogarithms, Dedekind zeta functions and the algebraic K-theory of fields*, Arithmetic Algebraic Geometry (G.v.d.Geer, F.Oort, J.Steenbrink, eds.), Prog. Math., Vol 89, Birkhauser, Boston, 1991, pp. 391–430.

DEPARTMENT OF MATHEMATICS, BROWN UNIVERSITY, PROVIDENCE, RI 02912, USA
sasha@math.brown.edu

On the Satake isomorphism

BENEDICT H. GROSS

In this paper, we present an expository treatment of the Satake transform. This gives an isomorphism between the spherical Hecke algebra of a split reductive group G over a local field and the representation ring of the dual group $\hat{G}$.

If one wants to use the Satake isomorphism to convert information on eigenvalues for the Hecke algebra to local L-functions, it has to be made quite explicit. This was done for $G = GL_n$ by Tamagawa, but the results in this case are deceptively simple, as all of the fundamental representations of the dual group are minuscule. Lusztig discovered that, in the general case, certain Kazhdan-Lusztig polynomials for the affine Weyl group appear naturally as matrix coefficients of the transform. His results were extended by S. Kato.

We will explain some of these results below, with several examples.

CONTENTS

1. The algebraic group $\underline{G}$

Throughout the paper, we let F be a local field with ring of integers A. We fix a uniformizing parameter π for A, and let q be the cardinality of the residue field $A/\pi A$.

Reprinted from 'Galois Representations in Arithmetic Algebraic Geometry',
edited by A. J. Scholl & R. L. Taylor. ©Cambridge University Press 1998

Let $\underline{G}$ be a connected, reductive algebraic group over F. We will assume throughout that $\underline{G}$ is split over F. Then $\underline{G}$ is the general fibre of a group scheme (also denoted $\underline{G}$) over A with reductive special fibre; in the semi-simple case $\underline{G}$ is a Chevalley group scheme over A. We fix a maximal torus contained in a Borel subgroup $\underline{T} \subset \underline{B} \subset \underline{G}$, all defined over A, and define the Weyl group of $\underline{T}$ by $W = N_G(\underline{T})/\underline{T}$.

Define the characters and co-characters of $\underline{T}$ by

$$X^{\bullet} = X^{\bullet}(\underline{T}) = \mathrm{Hom}(\underline{T}, \mathbb{G}_m)$$
$$X_{\bullet} = X_{\bullet}(\underline{T}) = \mathrm{Hom}(\mathbb{G}_m, \underline{T})$$

These are free abelian group of rank $l = \dim(\underline{T})$, which are paired into $\mathrm{Hom}(\mathbb{G}_m, \mathbb{G}_m) = \mathbb{Z}$. The first contains the roots Φ of $\underline{G}$: the characters of $\underline{T}$ occuring in the adjoint representation on $\mathrm{Lie}\,(\underline{G})$, and the second contains the coroots [1, Ch.6,§1].

The subset Φ^+ of positive roots which occur in the representation on $\mathrm{Lie}\,(\underline{B})$ satisfies: $\Phi = \Phi^+ \cup -\Phi^+$. It determines a root basis $\Delta \subset \Phi^+$ of positive, indecomposable roots. When $\underline{G}$ is of adjoint type, the elements of Δ give a $\mathbb{Z}$-basis of $X^{\bullet}$.

The root basis determines a positive Weyl chamber P^+ in $X_{\bullet}(\underline{T})$, defined by

$$P^+ = \{\lambda \in X_{\bullet} : \langle \lambda, \alpha \rangle \geq 0 \text{ all } \alpha \in \Phi^+\} \qquad (1.1)$$
$$= \{\lambda \in X_{\bullet} : \langle \lambda, \alpha \rangle \geq 0 \text{ all } \alpha \in \Delta\}$$

Let

$$(1.2) \qquad\qquad 2\rho = \sum_{\Phi^+} \alpha \quad \text{in } X^{\bullet}(T)$$

Then, for all λ in P^+, the half-integer $\langle \lambda, \rho \rangle$ is non-negative.

There is a partial ordering on P^+, written $\lambda \geq \mu$, if the difference $\lambda - \mu$ can be written as the sum of positive coroots. If $\check{\alpha}$ is a basic coroot in $\check{\Delta}$, then

$$(1.3) \qquad\qquad \langle \check{\alpha}, \rho \rangle = 1$$

Hence $\lambda \geq \mu$ implies that $\langle \lambda - \mu, \rho \rangle$ is a non-negative integer.

Let $\hat{G}$ be the complex dual group of $\underline{G}$. This is a connected, reductive group over $\mathbb{C}$ whose root datum is dual to $\underline{G}$. If we fix a maximal torus in a Borel subgroup $\hat{T} \subset \hat{B} \subset \hat{G}$, there is an isomorphism

$$(1.4) \qquad\qquad X^{\bullet}(\hat{T}) \simeq X_{\bullet}(\underline{T})$$

which takes the positive roots corresponding to $\hat{B}$ to the positive coroots corresponding to B. The elements λ in $P^+ \subset X^{\bullet}(\hat{T})$ index the irreducible representations V_λ of the group $\hat{G}$: λ is the highest weight for $\hat{B}$ on V_λ. Let

$\chi_\lambda = \text{Trace}(V_\lambda)$ be the character of V_λ, viewed as an element of $\mathbb{Z}[X^\bullet(\hat{T})]$. Then χ_λ lies in the subring

$$R(\hat{G}) = \mathbb{Z}[X^\bullet(\hat{T})]^W$$

fixed by the Weyl group. If we write

$$\chi_\lambda = \sum m_\lambda(\mu) \cdot [\mu],$$

then $m_\lambda(\mu) = m_\lambda(w\mu)$. Hence it suffices to determine the integers $m_\lambda(\mu)$ for μ in P^+, as these weights represent the orbits of the Weyl group on $X^\bullet(\hat{T})$. A simple result is that the integer $m_\lambda(\mu) = \dim V_\lambda(\mu)$ is non-zero if and only if $\lambda \geq \mu$ in P^+ [5, pg.202-203].

2. The Gelfand pair (G, K)

We define compact and locally compact topological groups by taking the A- and F-rational points of the group scheme $\underline{G}$:

(2.1) $$K = \underline{G}(A) \subset G = \underline{G}(F)$$

Then K is a hyperspecial maximal compact subgroup of G [17, 3.8.1]. Similarly, we have the locally compact, closed subgroups

(2.2) $$T = \underline{T}(F) \subset B = \underline{B}(F) \subset G = \underline{G}(F).$$

We let $N = \underline{N}(F)$, where $\underline{N}$ is the unipotent radical of $\underline{B}$. Then

(2.3) $$B = T \ltimes N, \quad \text{and}$$

(2.4) $$\det(\text{ad}(t)|\, \text{Lie}(N)) = 2\rho(t).$$

The Hecke ring $\mathcal{H} = \mathcal{H}(G, K)$ is by definition the ring of all locally constant, compactly supported functions $f : G \to \mathbb{Z}$ which are K-bi-invariant: $f(kx) = f(xk') = f(x)$ for all k, k' in K. The multiplication in $\mathcal{H}$ is by convolution

(2.5) $$f \cdot g\,(z) = \int_G f(x) \cdot g(x^{-1}z)dx$$

where dx is the unique Haar measure on G giving K volume 1. We will see below that the product function $f \cdot g$ also takes values in $\mathbb{Z}$. The characteristic function of K is the unit element of $\mathcal{H}$.

Each function f in $\mathcal{H}$ is constant on double cosets KxK; since it is also compactly supported it is a finite linear combination of the characteristic functions $\text{char}(KxK)$ of double cosets. Hence these characteristic functions give a $\mathbb{Z}$-basis for $\mathcal{H}$.

For any $\lambda \in X_\bullet(\underline{T}) = \text{Hom}(\mathbb{G}_m, \underline{T})$ we have the element $\lambda(\pi)$ in $T(F)$. Since $\lambda(A^*) \subset T(A) \subset K$, the double coset $K\lambda(\pi)K$ depends only on λ, not on the choice of uniformizing element. Here we view λ multiplicatively, so $(\lambda + \mu)(\pi) = \lambda(\pi) \cdot \mu(\pi)$.

Proposition 2.6. (cf. [17, p.51]) *The group G is the disjoint union of the double cosets $K\lambda(\pi)K$, where λ runs through the co-characters in P^+.*

This is a refinement of the Cartan decomposition: $G = KTK$; for $\underline{G} = GL_n$ it is proved by the theory of elementary divisors [13, pg.56-57]. It follows that the elements

$$(2.7) \qquad c_\lambda = \mathrm{char}(K\lambda(\pi)K) \qquad \lambda \in P^+$$

give a $\mathbb{Z}$-basis for $\mathcal{H}$, and multiplication is determined by the products

$$(2.8) \qquad c_\lambda \cdot c_\mu = \sum n_{\lambda,\mu}(\nu) \cdot c_\nu, \text{ with } n_{\lambda,\mu}(\nu) \in \mathbb{Z}.$$

To obtain an explicit formula for the integers $n_{\lambda,\mu}(\nu)$, we write

$$\nu(\pi) = t$$
$$K\lambda(\pi)K = \coprod x_i K$$
$$K\mu(\pi)K = \coprod y_j K$$

Then

$$
\begin{aligned}
n_{\lambda,\mu}(\nu) &= (c_\lambda \cdot c_\mu)(t) \\
&= \int_G c_\lambda(x) c_\mu(x^{-1}t) dx \\
&= \sum_i \int_{x_i K} c_\mu(x^{-1}t) dx \\
&= \sum_i \int_K c_\mu(kx_i^{-1}t) dk \\
&= \sum_i c_\mu(x_i^{-1}t) \\
&= \#\{(i,j) : \nu(\pi) \in x_i y_j K\}
\end{aligned}
$$

Since we can take $x_i = \lambda(\pi)$ and $y_j = \mu(\pi)$, this shows that $n_{\lambda,\mu}(\lambda + \mu) \geq 1$. In fact, we will see later that $n_{\lambda,\mu}(\lambda + \mu) = 1$ and that $n_{\lambda,\mu}(\nu) \neq 0$ implies that $\nu \leq (\lambda + \mu)$. Therefore

$$(2.9) \qquad c_\lambda \cdot c_\mu = c_{\lambda+\mu} + \sum_{\nu < (\lambda+\mu)} n_{\lambda,\mu}(\nu) \cdot c_\nu$$

The most important property of $\mathcal{H}$ is not evident from this calculation. It is the fact that $n_{\lambda,\mu}(\nu) = n_{\mu,\lambda}(\nu)$, or in other words:

Proposition 2.10. *The Hecke ring $\mathcal{H}$ is commutative.*

This is equivalent to the statment that (G, K) is a Gelfand pair. It is usually proved via Gelfand's Lemma (cf. [6, pg.279], which requires the existence of an anti-involution of G which fixes each double coset. For $\underline{G} = GL_n$ one takes $g \mapsto {}^t g$, which fixes the torus $\underline{T}$ of diagonal matrices. In general, there is an involution σ of $\underline{G}$ over A which acts as -1 on $\underline{T}$ (cf. [1, Ch.8,§2]), and one takes $g \mapsto \sigma(g^{-1})$.

A more involved proof of commutativity is via the Satake transform, which injects $\mathcal{H}$ into the commutative ring $R(\hat{G}) \otimes \mathbb{Z}[q^{-1/2}, q^{1/2}]$. We will discuss this transform in the next section.

One case when $\mathcal{H}$ is obviously commutative is when $\underline{G} = \underline{T}$ is a torus! We then have an exact sequence of locally compact groups

$$(2.11) \qquad o \to \underline{T}(A) \to \underline{T}(F) \xrightarrow[\gamma]{} X_{\bullet}(\underline{T}) \to 0$$

with $\gamma(t)$ the cocharacter satisfying

$$(2.12) \qquad \langle \gamma(t), \chi \rangle = \operatorname{ord}(\chi(t))$$

for all characters χ in $X^{\bullet}(T) = \operatorname{Hom}(\underline{T}, \mathbb{G}_m)$. The choice of uniformizing parameter π gives a splitting of this sequence: map λ in $X_{\bullet}(\underline{T})$ to $\lambda(\pi)$ in T.

Since each $\underline{T}(A)$ double coset is a single right coset, we have

$$c_\lambda \cdot c_\mu = c_{\lambda+\mu}$$

in $\mathcal{H}$. This agrees with (2.9), as there are no elements ν in $P^+ = X_{\bullet}(\underline{T})$ with $\nu < \lambda + \mu$. In other words, we have an isomorphism of rings

$$\mathcal{H}_T \simeq \mathbb{Z}[X_{\bullet}(\underline{T})] \qquad (2.13)$$
$$c_\lambda \longleftrightarrow [\lambda]$$

3. The Satake transform

The Satake transform gives a ring homomorphism

$$\mathcal{S} : \mathcal{H}_G \longrightarrow \mathcal{H}_T \otimes \mathbb{Z}[q^{1/2}, q^{-1/2}],$$

with image in the invariants for the Weyl group. It is defined by an integral of a type much studied by Harish-Chandra.

Fix $\underline{T} \subset \underline{B} = \underline{T} \cdot \underline{N} \subset \underline{G}$ over A, and let dn be the unique Haar measure on the unipotent group $N = \underline{N}(F)$ which gives $\underline{N}(A) = N \cap K$ volume 1. Let

$$(3.1) \qquad \delta : B \longrightarrow \mathbb{R}_+^*$$

be the modular function of B, defined by the formula

$$(3.2) \qquad d(bnb^{-1}) = \delta(b) \cdot dn$$

Then δ is trivial on N, so defines a character $\delta : T \longrightarrow \mathbb{R}_+^*$. Let $\delta^{1/2}$ be the positive square-root of this character; if $t = \mu(\pi)$ with μ in $X_{\bullet}(\underline{T})$ we have

$$\delta^{1/2}(t) = |\det(\operatorname{ad} t | \operatorname{Lie}(N))|^{1/2}$$
$$= |\pi^{\langle \mu, 2\rho \rangle}|^{1/2} \qquad (3.3)$$
$$= q^{-\langle \mu, \rho \rangle}.$$

In particular, $\delta^{1/2}$ takes values in the subgroup $q^{(1/2)\mathbb{Z}}$. If $\rho \in X^{\bullet}(\underline{T})$ then $\delta^{1/2}$ takes values in the subgroup $q^{\mathbb{Z}}$.

For f in $\mathcal{H}_G$, we define $\mathcal{S}f$, the Satake transform, as a function on T by the integral

$$(3.4) \qquad \mathcal{S}f(t) = \delta(t)^{1/2} \cdot \int_N f(tn)dn$$

Then $\mathcal{S}f$ is a function on $T/T \cap K = X_*(T)$ with values in $\mathbb{Z}[q^{1/2}, q^{-1/2}]$. The main result is that the image lies in the subring

$$(3.5) \qquad \left(\mathcal{H}_T \otimes \mathbb{Z}[q^{1/2}, q^{-1/2}] \right)^W = R(\hat{G}) \otimes \mathbb{Z}[q^{1/2}, q^{-1/2}]$$

of W-invariants, and that furthermore (cf. [11],[3, p.147]),

Proposition 3.6. *The Satake transform gives a ring isomorphism* $\mathcal{S} : \mathcal{H}_G \otimes \mathbb{Z}[q^{1/2}, q^{-1/2}] \simeq R(\hat{G}) \otimes \mathbb{Z}[q^{1/2}, q^{-1/2}]$. *If ρ is an element of $X^*(\underline{T})$, then the Satake transform gives a ring isomorphism*

$$\mathcal{H}_G \otimes \mathbb{Z}[q^{-1}] \simeq R(\hat{G}) \otimes \mathbb{Z}[q^{-1}].$$

Without giving the full proof, we give a sample calculation that illustrates the main idea. Assume that λ and μ lie in P^+; we wish to evaluate $\mathcal{S}(c_\lambda)$ on the element $t = \mu(\pi)$ of T.

Recall that c_λ is the characteristic function of the double coset $K\lambda(\pi)K = \coprod x_i K$. Since $G = BK$, we may assume each $x_i = t(x_i)n(x_i)$ lies in $B = TN$. Then

$$\begin{aligned} \mathcal{S}(c_\lambda)(t) &= \delta(t)^{1/2} \int_N c_\lambda(tn)dn \\ &= q^{-\langle \mu, \rho \rangle} \sum_i \int_{N \cap t^{-1}x_i K} dn \qquad\qquad (3.7) \\ &= q^{-\langle \mu, \rho \rangle} \#\{i : t(x_i) \equiv \mu(\pi) \quad (\mathrm{mod}\ \underline{T}(A))\} \end{aligned}$$

Hence the transform counts the number of single cosets where the "diagonal entries" $t(x_i)$ have the same valuation as $\mu(\pi)$. In particular, as we can take $x_i = \lambda(\pi)$, we have

$$(3.8) \qquad\qquad \mathcal{S}(c_\lambda)(\lambda(\pi)) \neq 0.$$

In fact, one has

$$\mathcal{S}(c_\lambda)(\lambda(\pi)) = q^{\langle \lambda, \rho \rangle}.$$

Moreover, if μ is in P^+ and $\mathcal{S}(c_\lambda)(\mu(\pi)) \neq 0$, then $\mu \leq \lambda$. Therefore we have

$$(3.9) \qquad\qquad \mathcal{S}(c_\lambda) = q^{\langle \lambda, \rho \rangle}\chi_\lambda + \sum_{\mu < \lambda} a_\lambda(\mu)\chi_\mu$$

in $R(\hat{G}) \otimes \mathbb{Z}[q^{-1/2}, q^{1/2}]$. From this the isomorphism of rings follows easily.

If one takes the scaled basis

$$(3.10) \qquad\qquad \varphi_\lambda = q^{\langle \lambda, \rho \rangle} \cdot \chi_\lambda$$

of $R(\hat{G}) \otimes \mathbb{Z}[q^{-1/2}, q^{1/2}]$, one finds that

$$(3.11) \qquad \mathcal{S}(c_\lambda) = \varphi_\lambda + \sum_{\mu < \lambda} b_\lambda(\mu)\varphi_\mu$$

with coefficients $b_\lambda(\mu)$ in $\mathbb{Z}$. Inversely, for all λ in P^+ we have the identity:

$$(3.12) \qquad q^{\langle \lambda, \rho \rangle} \operatorname{Trace}(V_\lambda) = \varphi_\lambda = \mathcal{S}(c_\lambda) + \sum_{\mu < \lambda} d_\lambda(\mu)\mathcal{S}(c_\mu)$$

with integers $d_\lambda(\mu)$.

If λ is a minuscule weight for $\hat{G}$, there are no elements μ in P^+ with $\mu < \lambda$. In this case we obtain

$$(3.13) \qquad q^{\langle \lambda, \rho \rangle} \operatorname{Trace}(V_\lambda) = \mathcal{S}(c_\lambda)$$

This pleasant situation occurs for <u>all</u> fundamental representations $V_\lambda = \bigwedge^i \mathbb{C}^n$ of $\hat{G}(\mathbb{C}) = GL_n(\mathbb{C})$, in the case when $G = GL_n$. This gives the formula of Tamagawa [15], [13, pg.56-62]:

$$(3.14) \qquad q^{i(n-i)/2} \operatorname{Trace}(\overset{i}{\bigwedge} \mathbb{C}^n) = \mathcal{S}\Big(\operatorname{char}(K \begin{pmatrix} \overset{\pi}{\ddots} & & \\ & \pi_1 & \\ & & \overset{\ddots}{1} \end{pmatrix} K)\Big)$$

$$\overset{i \text{ times}}{\nwarrow} \qquad \underset{n-i \text{ times}}{\nearrow}$$

Similarly, one obtains the transform of minuscule λ corresponding to real forms $G_{\mathbb{R}}$ of G with Hermitian symmetric spaces. The term $\langle \lambda, \rho \rangle$ is half of the complex dimension of $G_{\mathbb{R}}/K_{\mathbb{R}}$. For example, when $\underline{G} = GSp_{2n}$ and $\hat{G}(\mathbb{C}) = CSpin_{2n+1}(\mathbb{C})$ one has a minuscule weight λ corresponding to the spin representation V_λ of dimension 2^n. We find [8, 2.1.3]

$$(3.15) \qquad q^{\frac{n(n+1)}{4}} (\operatorname{Trace} V_\lambda) = \mathcal{S}\Big(\operatorname{char}(K \begin{pmatrix} \overset{\pi}{\ddots} & & \\ & \pi_1 & \\ & & \overset{\ddots}{1} \end{pmatrix} K)\Big)$$

$$\overset{n \text{ times}}{\nwarrow} \qquad \underset{n \text{ times}}{\nearrow}$$

4. Kazhdan-Lusztig polynomials

If one wishes to determine $\mathcal{S}(c_\lambda)$ for non-minuscule λ, one has to calculate the integers $b_\lambda(\mu)$ in (3.11), or equivalently the integers $d_\lambda(\mu)$ of the inverse matrix in (3.12). These depend on λ, μ, and the cardinality q of the residue field of A; in fact, we will see that $d_\lambda(\mu) = P_{\mu,\lambda}(q)$ is a polynomial in q with non-negative integer coefficients. Lusztig realized that $P_{\mu,\lambda}(q)$ is a Kazhdan-Lusztig polynomial for the affine Weyl group of $\hat{G}$ [9]. His work was completed by S. Kato [7], and we review it below. We assume in this section that $\underline{G}$ is a group of adjoint type.

For any μ in $X_\bullet(T)$, we let

$$(4.1) \qquad \hat{P}(\mu) = \sum_{\mu = \sum n(\alpha^\vee)\alpha^\vee} q^{-\sum n(\alpha^\vee)}$$

be the polynomial in q^{-1} which counts the number of expressions of μ as a non-negative sum of positive coroots. If μ cannot be expressed as such a sum, $\hat{P}(\mu) = 0$. Since we include the empty sum, when $\mu = 0$ we have $\hat{P}(0) = 1$. In all cases,

$$(4.2) \qquad q^{\langle \mu, \rho \rangle} \cdot \hat{P}(\mu)$$

is a polynomial in q with integral coefficients. If μ is in P^+ and $\mu \geq 0$, the constant coefficient of $q^{\langle \mu, \rho \rangle} \hat{P}(\mu)$ is equal to 1. Let

$$(4.3) \qquad 2\rho^\vee = \sum_{(\Phi)^+} \alpha^\vee \qquad \text{be the sum of all positive coroots.}$$

Proposition 4.4. *The coefficient $d_\lambda(\mu)$ appearing in the Satake isomorphism is given by the formula*

$$d_\lambda(\mu) = P_{\mu,\lambda}(q) = q^{\langle \lambda - \mu, \rho \rangle} \sum_W \varepsilon(\sigma)\hat{P}\big(\sigma(\lambda + \rho^\vee) - (\mu + \rho^\vee)\big),$$

where $\varepsilon(\sigma) = \det(\sigma | X_\bullet(T))$ is the sign character on the Weyl group W.

Kato shows that the polynomial $P_{\mu,\lambda}(q)$ defined by the alternating sum in Proposition 4.4 is a Kazhdan-Lusztig polynomial for the affine Weyl group W_a of $\hat{G}$. It is associated to the pair of elements $w_\mu \leq w_\lambda$ in the extended affine Weyl group $\tilde{W}_a = X^\bullet(\hat{T}) \rtimes W$ of maximal length in the double cosets $W\mu W$ and $W\lambda W$ respectively. These elements have lengths: $\ell(w_\mu) = \langle \mu, 2\rho \rangle + \dim(G/B)$ and $\ell(w_\lambda) = \langle \lambda, 2\rho \rangle + \dim(G/B)$. The general theory of Kazhdan-Lusztig polynomials then implies that $P_{\mu,\lambda}(q)$ has non-negative integer coefficients, and has degree strictly less than $\langle \lambda - \mu, \rho \rangle$ in q [9, pg.215].

If we set $q = 1$, $\hat{P}(\mu)$ becomes the partition function, and $P_{\mu,\lambda}(1) = \dim V_\lambda(\mu)$ by a formula of Kostant (cf. [5, pg.421]). More generally, R. Brylinski [2] has shown how to obtain $P_{\mu,\lambda}(q)$ from the action of a principal SL_2 in $\hat{G}$ on the space $V_\lambda(\mu)$.

Assume $\mu \leq \lambda$ in P^+. Then $P_{\mu,\lambda}(q)$ has constant coefficient $= 1$. In particular,

$$(4.5) \qquad \dim V_\lambda(\mu) = 1 \Rightarrow P_{\mu,\lambda}(q) = 1.$$

A non-trivial case is due to Lusztig [9, p.226]. Assume G is simple and λ is the highest coroot (= the highest weight of the adjoint representation of $\hat{G}$).

Then $0 \leq \lambda$ in P^+ and

$$(4.6) \qquad P_{0,\lambda}(q) = \sum_{i=1}^{l} q^{m_i - 1}$$

where $m_1, m_2, \ldots, m_l$ are the exponents of G [1, ch.5,§6].

5. Examples

We first treat the case $\underline{G} = PGL_2$. Then $\hat{G} = SL_2(\mathbb{C})$ and $X_\bullet(T) = X^\bullet(\hat{T}) = \mathbb{Z} \cdot \chi$, where χ is the character of the standard representation on $\mathbb{C}^2$. If $\lambda = n\chi$ and $\mu = m\chi$ are elements of $P^+(m, n \geq 0)$, then $\lambda \geq \mu$ if and only if

$$(5.1) \qquad \begin{cases} n \geq m \\ n \equiv m \pmod{2}. \end{cases}$$

Since $\dim V_\lambda(\mu) = 1$, we have $d_\lambda(\mu) = 1$ in this case. If we use the traditional notation

$$(5.2) \qquad T_{\pi^m} = \mathrm{char}(K(\begin{smallmatrix} \pi^m \\ & 1 \end{smallmatrix})K)$$

for c_μ in $\mathcal{H}$ we obtain the well-known formula [12, p.73]

$$(5.3) \qquad q^{\frac{n}{2}}\left(\mathrm{Trace\, Sym}^n(\mathbb{C}^2) \right) = \mathcal{S}\left(\sum_{\substack{m \leq n \\ m \equiv n \pmod{2}}} T_{\pi^m} \right).$$

The simplest adjoint group with a fundamental co-weight λ which is not minuscule is $G = PSp_4 = SO_5$. Then $\hat{G} = Sp_4(\mathbb{C})$ and V_λ is the 5-dimensional orthogonal representation. We have $0 \leq \lambda$ in P^+, and $P_{0,\lambda}(q) = 1$. Hence

$$\begin{aligned} q^2 \cdot \mathrm{Trace}(V_\lambda) &= \mathcal{S}(c_\lambda) + \mathcal{S}(c_0) \\ &= \mathcal{S}(c_\lambda) + 1. \end{aligned}$$

When $G = G_2$, so $\hat{G} = G_2(\mathbb{C})$, neither of the fundamental co-weights λ_1, λ_2 are minuscule. If $V_1 = V_{\lambda_1}$ is the seven-dimensional representation and $V_2 = V_{\lambda_1}$ is the fourteen-dimensional adjoint representation, we have

$$(5.5) \qquad 0 \leq \lambda_1 \leq \lambda_2 \qquad \text{in } P^+.$$

The polynomials $P_{\mu,\lambda}(q)$ are given by

$$(5.6) \qquad \begin{cases} P_{0,\lambda_1}(q) = 1 \\ P_{\lambda_1,\lambda_2}(q) = 1 \\ P_{0,\lambda_2}(q) = 1 + q^4 \end{cases}$$

as the exponents for G_2 are $m_1 = 1, m_2 = 5$. Hence we find

$$(5.7) \qquad \begin{cases} q^3 \cdot \mathrm{Trace}(V_1) = \mathcal{S}(c_{\lambda_1}) + 1 \\ q^5 \cdot \mathrm{Trace}(V_2) = \mathcal{S}(c_{\lambda_2}) + \mathcal{S}(c_{\lambda_1}) + (1 + q^4). \end{cases}$$

6. *L*-functions

One application of the Satake isomorphism is in the calculation of *L*-functions. This is based on the fact that the characters

$$(6.1) \qquad\qquad \omega : R(\hat{G}) \otimes \mathbb{C} \longrightarrow \mathbb{C}$$

of the representation ring are indexed by the semi-simple conjugacy classes s in the dual group $\hat{G}(\mathbb{C})$. The value of the character ω_s on χ_λ in $R(\hat{G})$ is given by

$$(6.2) \qquad\qquad \omega_s(\chi_\lambda) = \chi_\lambda(s) = \mathrm{Trace}(s|V_\lambda)$$

Since we have a fixed isomorphism

$$\mathcal{S} : \mathcal{H}_G \otimes \mathbb{C} \simeq R(\hat{G}) \otimes \mathbb{C}$$

we see that to any complex character of the Hecke algebra $\mathcal{H}_G$, we can associate a semi-simple class s in $\hat{G}(\mathbb{C})$, its Satake parameter.

Let π be an irreducible, smooth complex representation of G. The space π^K of K-fixed vectors has dimension ≤ 1. When $\dim(\pi^K) = 1$, we say π is unramified; in this case π gives a character of $\mathcal{H}_K$:

$$(6.3) \qquad\qquad \mathcal{H}_K \longrightarrow \mathrm{End}(\pi^K) = \mathbb{C}.$$

We let $s = s(\pi)$ be the Satake parameter of this character.

Proposition 6.4. (cf. [3, ch. III]) *The map $\pi \longrightarrow s(\pi)$ gives a bijection between the set of isomorphism classes of unramified irreducible representations of G and the set of semi-simple conjugacy classes in $\hat{G}(\mathbb{C})$.*

We write $\pi(s)$ for the unramified representation with Satake parameter s in $\hat{G}(\mathbb{C})$. It is known that $\pi(s)$ is tempered, so lies in the support of the Plancherel measure, if and only if s lies in a compact subgroup of $\hat{G}(\mathbb{C})$. Macdonald has determined the Plancherel measure on the unramified unitary dual [10, Ch.V].

If $\pi = \pi(s)$ is an unramified representation of G, and V is a complex, finite-dimensional representation of $\hat{G}(\mathbb{C})$, we define the local *L*-function $L(\pi, V, X)$ in $\mathbb{C}[[X]]$ by the formula.

$$(6.5) \qquad\qquad L(\pi, V, X) = \det(1 - sX|V)^{-1}.$$

When π is tempered, the eigenvalues of s on V have absolute value 1, so $L(\pi, V, X)$ has no poles in the disc $|X| < 1$. In general, we have

$$(6.6) \qquad\qquad \det(1 - sX|V) = \sum_{k=0}^{\dim V} (-1)^k \, \mathrm{Tr}(s|\textstyle\bigwedge^k V) X^k.$$

Hence, if we can write the elements $\mathrm{Tr}(\wedge^k V)$ in $R(\hat{G})$ as polynomials in the elements $S(c_\lambda)$, we can calculate the local *L*-function from the eigenvalues of c_λ in $\mathcal{H}$ acting on π^K.

For example, let $\underline{G} = GL_n$, so $\hat{G} = GL_n(\mathbb{C})$. We take $V = \mathbb{C}^n$, the standard representation. Let α_i in $\mathbb{C}$ be the eigenvalue of the elements c_{λ_i} on π^K. Then by Tamagawa's formula (3.14):

$$(6.7) \qquad L(\pi, V, X) = \left(\sum_{k=0}^{n} (-1)^k q^{-k(n-k)/2} \cdot \alpha_k \cdot X^k \right)^{-1}$$

Equivalently, one has [13, pg.61]

$$L(\pi, V, q^{\frac{n-1}{2}} \cdot X) = \left(\sum_{k=0}^{n} (-1)^k q^{k(k-1)/2} \cdot \alpha_k \cdot X^k \right)^{-1}.$$

For $n = 2$, this gives

$$L(\pi, V, q^{\frac{1}{2}} \cdot X) = (1 - \alpha_1 X + q\alpha_2 X^2)^{-1}$$

Next, consider the case of $G = SO_5$, so $\hat{G} = Sp_4(\mathbb{C})$. First we want to consider the L-function of the standard representation $V_{\lambda_1} = \mathbb{C}^4$, with minuscule weight. We have $\bigwedge^2 V_{\lambda_1} = V_{\lambda_2} \oplus \mathbb{C}$, where V_{λ_2} is the other fundamental representation (of dimension 5). Let α_i be the eigenvalues of c_{λ_i} on π^K.

Then by our previous calculation of the Satake transform:

$$q^{3/2} \operatorname{Tr}(s|V_{\lambda_1}) = \alpha_1$$
$$q^2 \operatorname{Tr}(s|V_{\lambda_2}) = \alpha_2 + 1.$$

Hence

$$q^2 \operatorname{Tr}(s| \bigwedge^2 V_{\lambda_1}) = \alpha_2 + 1 + q^2,$$

and we find (cf.[14], [16]):

$$(6.8) \quad L(\pi, V_{\lambda_1}, q^{\frac{3}{2}} X) = (1 - \alpha_1 X + (q\alpha_2 + q + q^3) X^2 - q^3 \alpha_1 X^3 + q^6 X^4)^{-1}$$

Now consider the L-function of the representation V_{λ_2}. We have $\bigwedge^2 V_{\lambda_2} \simeq \bigwedge^3 V_{\lambda_2} = V_{2\lambda_1}$, the adjoint representation of dimension 10. Using the fact that $V_{\lambda_1}^{\otimes 2} = V_{2\lambda_1} + \bigwedge^2 V_{\lambda_1}$, we find the relation

$$\chi_{2\lambda_1} = (\chi_{\lambda_1})^2 - \chi_{\lambda_2} - 1 \qquad \text{in } R(\hat{G}).$$

Hence

$$\operatorname{Tr}(s|V_{2\lambda_1}) = \operatorname{Tr}(s|V_{\lambda_1})^2 - \operatorname{Tr}(s|V_{\lambda_2}) - 1$$
$$= \frac{\alpha_1^2}{q^3} - \frac{(\alpha_2 + 1)}{q^2} - 1$$

and the L-function of V_{λ_2} is given by

$$
\begin{aligned}
L(\pi, V_{\lambda_2}, q^2 X) = &(1 - (\alpha_2 + 1)X + (q\alpha_1^2 - q^2\alpha_2 - q^2 - q^4)X^2 \\
&- (q^3\alpha_1^2 - q^4\alpha_2 - q^4 - q^6)X^3 \\
&+ q^6(\alpha_2 + 1)X^4 - q^{10}X^5)^{-1}
\end{aligned}
\tag{6.9}
$$

Finally, consider the case when $G = G_2$, so $\hat{G} = G_2(\mathbb{C})$. Let V be the 7-dimensional representation of $\hat{G}$, with weight λ_1, and let λ_2 be the weight of the 14-dimensional representation. We have

$$q^3\chi_1(s) = \alpha_1 + 1$$
$$q^5\chi_2(s) = \alpha_2 + \alpha_1 + 1 + q^4$$

where α_i is the eigenvalue of c_{λ_i}. On the other hand,

$$\mathrm{Tr}(V) = \chi_1$$

$$\mathrm{Tr}(\overset{2}{\bigwedge} V) = \chi_2 + \chi_1$$

$$\mathrm{Tr}(\overset{3}{\bigwedge} V) = \chi_1^2 - \chi_2$$

in $R(\hat{G})$, and $\bigwedge^k V \simeq \bigwedge^{7-k} V$ for all k. Hence we find

$$
\begin{aligned}
L(\pi, V, q^3 X) =& \Big(1 - (\alpha_1 + 1)X + (q\alpha_2 + (q + q^3)\alpha_1 + (q + q^3 + q^5))X^2 \\
& - \big(q^3\alpha_1^2 + (2q^3 - q^4)\alpha_1 - q^4\alpha_2 + (q^3 - q^4 - q^8) \big)X^3 \\
& \qquad\qquad\qquad\qquad\qquad\qquad\qquad\qquad\qquad\qquad (6.10) \\
& + \ldots - q^{21}X^7 \Big)^{-1}
\end{aligned}
$$

7. The trivial representation

One interesting unramified representation π of G is the trivial representation. Then $\pi = \pi^K$ affords a representation of $\mathcal{H}$, and c_λ acts by multiplication by

$$(7.1) \qquad \deg(c_\lambda) = \# \text{ of single } K\text{-cosets in } K\lambda(\pi)K.$$

This integer is given by a polynomial in q, with leading term $q^{\langle \lambda, 2\rho \rangle}$. Let $P_\lambda \subset \underline{G}$ be the standard parabolic subgroup defined by the co-character λ. We have

$$(7.2) \qquad \mathrm{Lie}(P_\lambda) = \mathrm{Lie}(T) + \bigoplus_{\langle \lambda, \alpha \rangle \geq 0} \mathrm{Lie}(G)_\alpha$$

and

$$(7.3) \qquad \dim(G/P_\lambda) = \#\{\alpha \in \Phi : \langle \lambda, \alpha \rangle < 0\}$$

If $\lambda = 0$ we find $P_\lambda = G$; if λ is regular we find $P_\lambda = B$. Let

$$\ell : \tilde{W}_a = X^\bullet(\hat{T}) \rtimes W \longrightarrow \mathbb{Z}$$

be the length function on the extended affine Weyl group, defined in [9, pg.209]. The following is a simple consequence of the Bruhat-Tits decomposition of G [17, 3.3.1].

Proposition 7.4. *For all λ in P^+, we have:*

$$\deg(c_\lambda) = \sum_{W\lambda W} q^{\ell(y)} \Big/ \sum_W q^{\ell(w)} = \frac{\#(G/P_\lambda)(q)}{q^{\dim(G/P_\lambda)}} \cdot q^{\langle\lambda,2\rho\rangle}.$$

Moreover, λ is a minuscule co-weight if and only if

$$\deg(c_\lambda) = \#(G/P_\lambda)(q)$$

It is also known that the Satake parameter of the trivial representation is the conjugacy class $s = \rho(q) = 2\rho(q^{1/2})$ in $\hat{G}(\mathbb{C})$. Equivalently, if

$$(7.5) \qquad s_0 = \begin{pmatrix} q^{1/2} & \\ & q^{-1/2} \end{pmatrix} \text{ in } SL_2(\mathbb{C})$$

is the Satake parameter of the trivial representation of PGL_2, then s is the image in $\hat{G}(\mathbb{C})$ of s_0 in a principal SL_2.

This gives a check on our various formulas. For example, when $G = G_2$ we found

$$q^3\chi_1(s) = \alpha_1 + 1$$
$$q^5\chi_2(s) = \alpha_2 + \alpha_1 + 1 + q^4$$

On the trivial representation, we find

$$\alpha_1 = \deg(c_{\lambda_1}) = q^6 + q^5 + q^4 + q^3 + q^2 + q$$
$$\alpha_2 = \deg(c_{\lambda_2}) = q^{10} + q^9 + q^8 + q^7 + q^6 + q^5$$

Since

$$V_1 = S^6(\mathbb{C}^2)$$
$$V_2 = S^{10}(\mathbb{C}^2) + S^2(\mathbb{C}^2)$$

as representations of the principal SL_2 in G_2, we find

$$q^3\chi_1(s) = q^3(q^3 + q^2 + q + 1 + q^{-1} + q^{-2} + q^{-3})$$
$$= q^6 + q^5 + q^4 + q^3 + q^2 + q + 1$$
$$q^5\chi_2(s) = q^{10} + q^9 + q^8 + q^7 + 2q^6 + 2q^5 + 2q^4 + q^3 + q^2 + q + 1$$

which checks!

One consequence of Proposition 7.4 is that the degrees of Hecke operators are quite large. For example, if $G = E_8$ and $\lambda \neq 0$, $\deg(c_\lambda) > q^{58}$.

8. Normalizing the Satake isomorphism

One unpleasant, but necessary, feature in the Satake isomorphism is the presence of the irrationalities $q^{m/2}$. As already noted, these are not needed in the case when $\rho \in X^*(T)$. When the derived group G' of G is simply-connected, we will see how they can be removed by a choice of normalization.

Let Y be the quotient torus G/G'. The exact sequence:

$$(8.1) \qquad 1 \longrightarrow G' \longrightarrow G \longrightarrow Y \longrightarrow 1$$

induces an exact sequence

(8.2) $$0 \longrightarrow X^\bullet(Y) \longrightarrow X^\bullet(T) \longrightarrow \mathrm{Hom}(\mathbb{Z}\check{\Phi}, \mathbb{Z}) \longrightarrow 0$$

Since $\langle \check{\alpha}, \rho \rangle$ is integral for all coroots, there is a class ρ_Y in $\frac{1}{2}X^\bullet(Y)$ with

(8.3) $$\rho \equiv \rho_Y \qquad \mathrm{mod}\ X^\bullet(T)$$

The class ρ_Y is well-determined in the quotient group $\frac{1}{2}X^\bullet(Y)/X^\bullet(Y)$. A choice of representative in $\frac{1}{2}X^\bullet(Y)$ will be called a normalization.

Since $X^\bullet(Y) = X_\bullet(\hat{Z})$, where $\hat{Z}$ is the connected center of $\hat{G}$, a normalization ρ_Y gives us a central element

(8.4) $$z = \rho_Y(q) = 2\rho_Y(q^{1/2}) \quad \text{in } \hat{Z}(\mathbb{C}) \subset \hat{G}(\mathbb{C}).$$

We adopt the convention that the normalized Satake parameter of an unramified representation π is given by:

(8.5) $$s' = z \cdot s(\pi) \quad \text{in } \hat{G}(\mathbb{C}).$$

Of course, this depends on the choice of ρ_Y. It has the advantage that the relation between eigenvalues of the Hecke operators c_λ on π^K and the traces $\chi_\lambda(s')$ is now algebraic, involving only integral powers of q. Indeed, $\chi_\lambda(s') = q^{\langle \lambda, \rho_Y \rangle}\chi_\lambda(s)$, so

(8.6) $$q^{\langle \lambda, \rho - \rho_Y \rangle}\chi_\lambda(s') = (c_\lambda|\pi^K) + \sum_{\mu < \lambda} d_\lambda(\mu)(c_\mu|\pi^K).$$

Note that $\rho - \rho_Y$ is an element of $X^\bullet(T)$.

One example, discussed by Deligne [4, pg.99-101], is for $G = GL_2$. Then $\rho = \frac{1}{2}\alpha = \frac{1}{2}(e_1 - e_2)$ is not in $X^\bullet(T) = \mathbb{Z}e_1 + \mathbb{Z}e_2$. The normalization

(8.7) $$\rho_Y = \tfrac{1}{2}(e_1 + e_2) = (\det)^{1/2}$$

gives the Hecke parameter $s' = q^{1/2} \cdot s$, and the normalization

(8.8) $$\rho_Y = -\tfrac{1}{2}(e_1 + e_2) = (\det)^{-1/2}$$

gives the Tate parameter $s' = q^{-1/2} \cdot s$.

For the trivial representation $\pi = \mathbb{C}$ of $GL_2(F)$, the Satake parameter is the element

$$s = \begin{pmatrix} q^{1/2} & 0 \\ 0 & q^{-1/2} \end{pmatrix} \quad \text{in } SL_2(\mathbb{C}).$$

The normalized Hecke parameter is the element

$$s' = \begin{pmatrix} q & 0 \\ 0 & 1 \end{pmatrix} \quad \text{in } GL_2(\mathbb{C}),$$

and the normalized Tate parameter is the element

$$s' = \begin{pmatrix} 1 & 0 \\ 0 & q^{-1} \end{pmatrix} \quad \text{in } GL_2(\mathbb{C}).$$

One may choose to take a normalization even when ρ lies in $X^\bullet(T)$. For example, when $G = GL_n$ the normalization $\rho_Y = (\det)^{\frac{n-1}{2}}$ is popular, as in the formula following (6.7). In the global situation, the choice of normalization often depends on the infinity type, and is chosen to render the parameters s'_p integral. The Hecke normalization does this for holomorphic forms of weight 2 on the upper half-plane.

Acknowledgements

I would like to thank R. Taylor, for explaining his calculations in [16, 3.5], and M. Reeder and J.-P. Serre for their comments on a draft of this paper.

References

1. N. Bourbaki, *Groupes et algèbres de Lie*, Ch. 5-8, Hermann, Paris, 1968.
2. R. K. Brylinski, *Limits of weight spaces, Lusztig's q-analogs, and fiberings of adjoint orbits*, JAMS **2** (1989) 517-533.
3. P. Cartier, *Representations of p-adic groups*, In: Automorphic forms, representations, and L-functions, Proc. Symp. AMS **33** (1979) 111-155.
4. P. Deligne, *Formes modulaires et représentations de GL(2)*, Springer Lecture Notes **349** (1973) 55-106.
5. W. Fulton, and J. Harris, *Representation theory*, Springer GTM 129 (1991).
6. B. Gross, *Some applications of Gelfand pairs to number theory*, Bull. AMS **24** (1991) 277-301.
7. S. I. Kato, *Spherical functions and a q-analog of Kostant's weight multiplicity formula*, Invent. Math. **66** (1982) 461-468.
8. R. Kottwitz, *Shimura varieties and twisted orbital integrals*, Math. Ann. **269** (1984) 287-300.
9. G. Lusztig, *Singularities, character formulas, and a q-analog of weight multiplicities*, Astérisque **101** (1983) 208-227.
10. I. Macdonald, *Spherical functions on a group of p-adic type*, Ramanujan Inst. Publ., Madras, 1971.
11. I. Satake, *Theory of spherical functions on reductive algebraic groups over p-adic fields*, IHES **18** (1963) 1-69.
12. J.-P. Serre, *Trees*, Springer, 1980.
13. G. Shimura, *Arithmetic theory of automorphic functions*, Princeton Univ. Press, 1971.
14. G. Shimura, *On modular corespondences for Sp(n, ℤ) and their congruence relations*, Proc. NAS **49** (1963) 824-828.
15. T. Tamagawa, *On the ζ-function of a division algebra*, Ann. of Math. **77** (1963) 387-405.
16. R. Taylor, *Galois representations associated to Siegel modular forms of low weight*, Duke Math. J. **63** (1991) 281-332.
17. J. Tits, *Reductive groups over local fields*, In: Automorphic forms, representations, and L-functions, Proc. Symp. AMS **33** (1979) 29-69.

DEPARTMENT OF MATHEMATICS, HARVARD UNIVERSITY, ONE OXFORD STREET, CAMBRIDGE, MA 02138, USA

E-mail address: gross@math.harvard.edu

Open problems regarding rational points on curves and varieties.

B. MAZUR

Contents

Reprinted from 'Galois Representations in Arithmetic Algebraic Geometry',
edited by A. J. Scholl & R. L. Taylor. ©Cambridge University Press 1998

PART I. Questions of Uniformity

The organizers of the Durham conference on Arithmetic Algebraic Geometry, in July 1996 asked me to give a general lecture in the conference on my work with Lucia Caporaso and Joe Harris [C-H-M 1,2]. I was more than happy to oblige: there had been some recent progress in the subject, thanks to Hassett, Pacelli, and Abramovich, and I was thankful for an opportunity to report on this work. I also used the opportunity to discuss some specific "applications" of these ideas to the study of rational isogenies and to list some related open problems. Part I of this article is the written version of that lecture. I am grateful to J.-P. Serre to include in this text some portions of a letter he wrote to me (cf. the Appendix to part I), and to Brian Conrad and Richard Taylor for conversations leading to the material in section 8 below.

One of the goals of [C-H-M 1,2] was to set up something which might be called a "machine" which shows that a conjecture due to Lang [L], which on first view seems purely "qualitative" implies conjectures which assert that there are uniform upper bounds for numbers of rational points.

Our first chore, then, is to describe Lang's conjecture.

1. Rational points on varieties of general type

Our base field K will always be a number field; i.e., a field of finite degree over $\mathbf{Q}$. Let V be a reduced, irreducible, positive-dimensional variety over K.

If V is proper and smooth of dimension d, let ω denote its canonical line bundle; that is, $\omega = \Omega^d$ is the sheaf of holomorphic differential d-forms. We will say that such a V is of **general type** if there is a positive integer N such that the rational mapping of V to projective space effected by considering the global sections of ω^N yields a birational isomorphism between V and its image in projective space. In older language, this last statement is sometimes expressed by saying that V has "enough pluricanonical forms". More generally, a proper, reduced, irreducible variety V over K is of *general type* if some desingularization of V is so (and equivalently: if all are so). In full generality, if V is noncomplete, V is of *general* type if some completion of V is so (and equivalently: if all are so). The property of being of general type or not is a birational invariant.

Examples: Smooth proper curves are of general type if and only if their genus is > 1. Symmetric d-th powers of curves are of general type provided $d <$ genus. Smooth hypersurfaces in projective space are so if and only if the difference between their degree and their dimension is > 2.

Lang's Conjecture: If V is a variety of general type over a number field K, there is a subvariety $Z \subset V$ of dimension smaller than the dimension of V, defined over K, such that $V - Z$ has at most a finite number of L-rational points over any number field extension L of K.

For any variety V over K, consider the set $\mathcal{Z}$ of all subvarieties $Z \subset V$ that have the property enunciated by the conjecture, i.e., that there are only a finite number of L-rational points in the complement of Z for every finite extension field L of K. One sees directly that the set $\mathcal{Z}$ is closed under finite intersection. By the noetherian property of V, the intersection, Z_0, of all the subvarieties $Z \in \mathcal{Z}$, is equal to the intersection of some finite subcollection of subvarieties $Z \in \mathcal{Z}$ and is therefore again in $\mathcal{Z}$. The subvariety Z_0, by construction, is then the *smallest* subvariety of V whose complement contains at most a finite number of L-rational points for any L of finite degree over K. We shall refer to this Z_0 as the **Lang locus** of V. Lang's Conjecture then amounts to the assertion that if V is of general type, its Lang locus is a "proper" subvariety, i.e., a subvariety of dimension smaller than that of V.

Example: By Faltings' theorem ("Mordell conjecture for curves") Lang's Conjecture is true for smooth curves of genus > 1 (the Lang locus of such a curve is empty). Also for symmetric d-th powers of curves with $d <$ genus (where the Lang locus is related to the "gonality" of the curve; cf. [A-H,D-F, H-S]). For a surface S of general type, if the Lang locus $Z \subset S$ is of dimension smaller than the dimension of S (as would follow from Lang's Conjecture), Z would, of necessity, be the union of all curves of geometric genus 0 and 1 lying on S. In particular, Lang's Conjecture would then imply that there are only a finite number of curves of geometric genus 0 and 1 lying on S. To the best of my knowledge this latter statement is still not known. Lang's Conjecture is still unknown for smooth quintic surfaces in $\mathbf{P}^3$.

Dependence upon the ground field? Let V be a variety of general type, and let Z be its Lang locus. How does the cardinality of the set of L-rational points on $V - Z$ vary with the field extension L? Is there, for example, an upper bound for this cardinality dependent only upon the *degree* of L over K?

2. An illustrative case

To give a sense of our original method, let us consider the simple case of the following ("isotrivial") family of hyperelliptic curves of genus > 1:

$$V_t : t \cdot y^2 = g(x)$$

where $g(x) \in K[x]$ is a polynomial of degree ≥ 5 with no multiple roots.

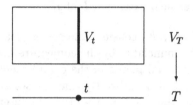

For each $t \in K - \{0\}$, we get a smooth curve V_t over K of fixed genus > 1, and therefore by Falting's Theorem, $|V_t(K)|$ is finite. Let us show, as illustration of our method, that (assuming Lang's Conjecture) there is a uniform upper bound B to the cardinalities of $V_t(K)$ for all $t \in K - \{0\}$.

On the one hand, thinking of Faltings' Theorem, one might imagine that the rational points are fairly scarce in V_T: there are only finitely many K-rational points on each fiber. But, of course, V_T is a rational variety over K (for any choice of coordinates x, y we may solve for t) and therefore the K-rational points are, in fact, all over the place; they are dense in the total space V_T. In a word, the rational points, fiber-by-fiber, are not "correlated" (I'm not using the term "correlated" in any rigorous sense). But, as we will see in a moment, Lang's Conjecture implies that these K-rational points satisfy a certain kind of "two-point correlation". To visualize this, consider what we might call the "two-point correlation space" for $V_T = V \times_T V$, i.e., the family parametrized by T which for each given $t \in T$ is the product of V_t with itself:

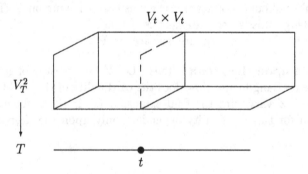

The variety $V_T{}^2$ is a threefold, whose equations are:

$$t \cdot y^2 = g(x), \qquad t \cdot v^2 = g(u),$$

and we immediately see that we have a dominating mapping of $V_T{}^2$ onto the surface W given by the equation

$$W : z^2 = g(x) \cdot g(u)$$

by sending $x \mapsto x$, $u \mapsto u$, and $z \mapsto t \cdot v \cdot y$.

Now the surface W is of general type, and therefore by Lang's Conjecture, its K-rational points lie in a "proper" subvariety (in particular, a curve), whose inverse image in the threefold $V_T{}^2$ is therefore a subvariety of dimension ≤ 2 which contains all the K-rational points of $V_T{}^2$. In other words, Lang's Conjecture tells us that there are some nontrivial algebraic relations satisfied by the locus of couples of K-rational points on the same fiber. Let, then, $S \subset V_T{}^2$ be a surface containing the Zariski closure of the set of K-rational points of $V_T{}^2$ (we may assume that every component of S is two-dimensional, and for simplicity of discussion we will do so).

Step 1: Consider the points $t \in T$ for which the entire fiber V_t^2 of $V_T{}^2$ is contained in S. There are at most a finite number of such points $\{t_1, t_2, \ldots, t_\nu\}$. It suffices to find an upper bound B to the cardinalities of the sets $V_t(K)$ for all $t \in K - \{0\}$ other than this finite set of points. Relabel the complement of the finite set $\{t_1, t_2, \ldots, t_\nu\}$ in $\mathbf{A}^1 - \{0\}$ as T and restrict our family to this parameter space T so that we may now assume that S contains no component which is equal to a fiber of $V_T \to T$.

Step 2: Consider the projection $V_T{}^2 \to V_T$ to the first coordinate, and let $\pi : S \to V_T$ be the restriction of this projection to S. Now suppose that π is finite. Then our bound B can be taken to be the "generic" degree of π. For, take a value $t \in T(K)$. Then either V_t has no K-rational points at all (and then B would be a perfectly good upper bound for the cardinality of $V_t(K)$) or else there is a point $u \in V_t(K)$. Fix such a point u, and consider the fiber above u for the projection $V_T{}^2 \to V_T$. This fiber is $\{u\} \times V_t$ and therefore is isomorphic over K to V_t. But the set of its K-rational points is contained in S, and therefore B is an upper bound for its cardinality. We are now ready to consider the general situation.

Step 3: Consider the locus $C \subset V_T$ consisting of those points for which the mapping $\pi \colon S \to V_T$ is not finite, and let B_1 denote the "generic" degree of π when restricted to the inverse image of the complement of C. Let $\phi \colon C \to T$ denote the restriction to C of the projection $f : V_T \to T$. Now the mapping ϕ is finite, for if it weren't finite at some point then it would have the property that S contains the entire fiber above that point. Let B_2 denote the generic degree of ϕ. I claim that the uniform bound B that we see'k may be taken to be the maximum of B_1 and B_2. For let $t \in T(K)$. If the cardinality of $V_t(K)$ is $\leq B_2$, then we are done. If it is $> B_2$ then there is a K-rational point $u \in V_t$ which does not lie in C, and hence the argument of Step 2 bounds the cardinality of $V_t(K)$ by B_1.

3. Finer uniformity?

If we knew the Lang locus of the surface W of general type given by the equation $z^2 = g(x) \cdot g(u)$, finer information could be obtained for the example

$t \cdot y^2 = g(x)$ of the previous section. There are, of course, the "evident" curves of genus 0 lying on W, i.e.,

- $x = u$, $z = g(x)$;

- $x = u$, $z = -g(x)$;

- $x = x_0$, for $x_0 \in \bar{K}$ such that $g(x_0) = 0$;

- $u = u_0$, for $u_0 \in \bar{K}$ such that $g(u_0) = 0$.

If $V_T \to T$ is a family defined over K for which these curves comprise the full Lang locus of W, i.e., for which these are *all* the curves of genus ≤ 1 lying on W, then the above argument gives the following more stringent uniformity result (conditional, of course, on Lang's Conjecture): For every number field extension L/K there is a finite set of curves defined over L,

$$\mathcal{C}(L) = \{C_i; i = 1, \ldots, m\}$$

such that for all $t \in L - \{0\}$ if V_t is not isomorphic over L to a curve in $\mathcal{C}(L)$ then there is at most *one* antipodal pair of non-Weierstrass L-rational points in $V_t(L)$. Here "antipodal" means that the pair of points is stable under the hyperelliptic involution of V_t.

But the above remark is of limited use at present, because to my knowledge, for no polynomial $g(x)$ has the Lang locus of the corresponding surface W (of the construction above) been computed. Can one compute some example?

4. The general "correlation theorem"

The illustrative case we have just discussed might suggest to us that we could put the following theorem to good use.

THE FIBERED POWER THEOREM. *Let T be an irreducible variety over* **C**, *and let $f : V_T \to T$ be a smooth proper family of varieties of general type, and let*

$$f_N \colon V_T^N \to T$$

denote the N-th fiber power of V_T over T. Then there exists a positive integer N such that V_T^N dominates a variety W of general type (by a rational mapping, $\phi \colon V_T^N - \to W$).

ADDENDA. *If the family V_T/T is of "maximal variation" then V_T^N is itself of general type. If [the fibers are curves, and if] $V_T \to T$ is defined over* *a number field* K, *then* W *and the rational mapping* $\phi : V_T{}^N \to W$ *may be taken to be defined over* K.

Remarks. This theorem was proved in the case of fiber dimension one in [C-H-M 1] and was conjectured to be true in general, in that paper. Brendan Hassett proved this in case the fiber dimension is two [H] and Abramovich [A 2] has recently shown the theorem in full generality (see also his earlier note: [A 1]). We might mention in passing that the bracketed requirement in the second Addendum above (that the fibers are curves) is not necessary if one appeals to the proof of the theorem in [A 2].

Here are a few remarks which may evoke something of the flavor of the proof. First, although we have assumed that our family is smooth in the statement of the theorem, in the actual proof of the theorem we must compactify V_T and therefore must deal with possible singularities. Second, let us loosely use the word "positive" to mean "having many pluricanonical sections" and the word "negative" to mean "not necessarily having many pluricanonical sections". Our goal, of course, is to show that a desingularization of a compactification of some fiber power $V_T{}^N$ is "positive". Now the base, T, is "negative" (which may hinder us in our goal). The fibers of $V_T \to T$ are, of course, "positive" (which will help us) and therefore, the idea of increasing N sounds like a good idea to counteract the possibly "negativity" of the base. A problem which one encounters, however, is that as N gets larger the singularities of the compactification will also get more complicated, and this is a hindrance. In a word, it is a race (as N goes to infinity) between the good effect of the high power of the fiber and the bad effect of the complicating singularities. Happily, the good effect wins out. In the proof for fiber dimension one in [C-H-M 1] we make strong use of the stable compactification of families of curves, which have very mild controllable singularities. In the proof for families of surfaces due to Hassett, again it is the known structure of the moduli space for stable surfaces of general type that is relied on. In the proof of the full theorem due to Abramovich, first a general positivity result of Kollàr is used, and the idea of de Jong's alterations [dJ] is adapted to this situation to control the problem due to singularities (Abramovich finds a Galois cover $Y \to V_T$ where Y is a composition of curves with at worst nodal singularities).

5. Uniformity statements for the number of rational point on curves

I hope the "illustrative example" of section **2** is sufficient to suggest that the theorem given in the previous section can be used to show that Lang's Conjecture implies the existence of uniform upper bounds for the numbers of rational points of certain varieties, as these varieties vary in algebraic families. Restricting attention to curves, at the moment, Lang's Conjecture implies

the three statements which I have labelled A, B, and C below. In these
statements, g will denote a number > 1, and "curve" will mean smooth
proper curve.

A. There is a number $N(g)$ with the following property. For each number
field K, there is a finite set $\mathcal{C}(g, K)$ of curves of genus g defined over K such
that if C is any curve of genus g defined over K with more than $N(g)$ K-
rational points, then C is isomorphic over K to one of the curves in $\mathcal{C}(g, K)$.
Call the curves in $\mathcal{C}(g, K)$ "K- exceptional curves of genus g".

Questions. Assume A to be true, i.e., that there is such a number $N(g)$.
Let us take $N(g)$ to be the smallest positive number for which **A** is true, and
let us take $\mathcal{C}(g, K)$ to be the corresponding sets of "K- exceptional" curves.

1) Find lower bounds for $N(g)$. To show $N(g) >$ some number B you
need only find some fixed number field K and an infinity of K- isomorphism
classes of curves over K of genus g each of which possesses more than B points
rational over K. In this direction, both Mestre and Brumer have shown (by
independent constructions) that $N(g) \geq 16(g+1)$ for all $g > 1$. For particular
small g, i.e., $g = 2, 3$, one has slightly better lower estimates for $N(g)$.

2) Does the number of K-exceptional curves of genus g depend only on
(g and) the degree of K over $\mathbf{Q}$?

For the proof that Lang's Conjecture implies **A** see [C-H-M 1]; it uses
the fiber power theorem only for fiber dimension equal to one. Statement
A implies (in conjunction with Faltings' Theorem applied to each of the K-
exceptional curves of genus g) that there is a uniform upper bound for the
number of K-rational points of any curve of genus g defined over K. By
an intricate (and elegant) argument, Pacelli has shown (in [P]) that Lang's
Conjecture implies such a uniformity, where the bound depends only upon
the degree d of K (see also [A-V]). Explicitly, Lang's Conjecture implies:

B. (Pacelli): For integers $d \geq 1$ and $g > 1$ there is a number $N(d, g)$
with the following property. Given any number field K of degree $\leq d$, and
curve C of genus g defined over K, the number of K-rational points on C is
$\leq N(d, g)$.

To prove this result, Pacelli studies the loci (F_n, in her terminology) given
by the Zariski closure of the set of what she calls (indefinitely) "prolongable"
points in the symmetric d-the power of the n-th fiber power of families $V_T \to T$
of curves of genus $g \geq 1$. In an appropriate context she shows that if n is
appropriately large, the irreducible components of maximal dimension in F_n
are of general type. Since, by construction, these F_n have Zariski dense sets
of points rational over a fixed number field, Lang's conjecture implies that F_n
is empty, giving the uniformity that is sought.

Remark and further questions. With Pacelli's statement **B** in mind, let us return to the question we had raised when we first formulated Lang's Conjecture in section **1**. Namely, given a variety of general type V with Lang locus Z, how does the cardinality of the set of K-rational points of $V - Z$ vary with K? More specifically, is there an upper bound for this cardinality that depends only on the degree of K. Statement **B** is an affirmation of this, in the case where V is a variety of general type of dimension 1. Returning to our original question we might wonder whether the correlation- theoretic techniques we have been discussing, suitably strengthened, may be used to show that Lang's Conjecture "improves itself" for varieties of general type of all dimensions, in the sense that it implies that the set of K-rational points of $V - Z$ depends only on the degree of K (where Z is the Lang locus of V) for V varying through a "bounded" family of varieties of general type over K. In this connection, see [A-V] Theorems 1.5-1.7.

If C is a curve defined over the field of complex numbers, by the *gonality* of C let us mean the minimal degree δ of any nonconstant function on C, i.e., it is the minimal degree of any nonconstant mapping from the Riemann surface C to $\mathbf{P}^1$. We refer the reader to [A 3] and to [Y], [L-Y] for discussion and results regarding gonality of modular curves.

Recall that if V is a variety over K, a point of degree $\leq d$ on $V_{/K}$ is an L-rational point of V where L/K is some field extension of degree $\leq d$.

Using the correlation theorem that Abramovich proved (i.e., for all fiber dimensions), Abramovich showed that Lang's Conjecture implies:

C. (Abramovich): For integers d, $e \geq 1$ and $g > 1$ there is a number $N(d, e, g)$ with the following property. Given any number field K of degree $\leq d$, and curve C defined over K, of genus g and of gonality $> 2e$, the set of points of C of degree $\leq e$ over K is less than or equal to $N(d, e, g)$.

6. Counting K-rational isogenies

If E is an elliptic curve over K, by a K-isogeny one means a surjective homomorphism $E \to E'$ where E' is an elliptic curve over K and the homomorphism is defined over K. Let $N(K, E)$ denote the cardinality of the set of j-invariants of all the elliptic curves E' which are K-isogenous to E. It is known, for example, by a result of Kenku that for any elliptic curve E over $\mathbf{Q}$, $N(\mathbf{Q}, E) \leq 8$. It is also easy to see that the (maximum) value $N(\mathbf{Q}, E) = 8$ is reached infinitely often: we have $N(\mathbf{Q}, E) = 8$, for example, for any elliptic curve E over $\mathbf{Q}$ which has no complex multiplications (over $\mathbf{C}$) and which admits a $\mathbf{Q}$-rational isogeny whose kernel is cyclic of order 12; there are infinitely many such E's since $X_0(12)$ is of genus 0.

Proposition 1. *Statement* **A** *of section* **5** *(and therefore, also, Lang's Conjecture) implies that there is a "universal" natural number N such that for any number field K, the set*

$$\mathcal{J}(K) := \begin{pmatrix} \text{the set of } j\text{-invariants of elliptic curves} \\ E \text{ over } K \text{ for which } N(K, E) > N \end{pmatrix}$$

is finite.

Remarks and further questions.

1. Can one take $N = 20$ in Proposition 1? Without appeal to Lang's Conjecture, finiteness of $\mathcal{J}(K)$ is not yet known. Is there an upper bound for the cardinality of $\mathcal{J}(K)$ which depends only upon the degree of K?

2. A related question, however, replacing *K-rational isogenies* by *K-rational torsion points* is now known, by the recent work of Merel [Me]; namely, there *is* an integer M which has the property that for all number fields K the set of K-isomorphism classes of elliptic curves defined over K with Mordell-Weil torsion (over K) of order $> M$ is finite. One can, in fact, take $M = 18$. Lang's Conjecture implies the further fact that the cardinality of this finite set of K-isomorphism classes of elliptic curves with large Mordell-Weil torsion admits an upper bound depending only on the degree of K. Explicitly,

Proposition 2. *Let $d \geq 1$ be an integer. Lang's Conjecture implies that there is a finite bound $\mathcal{M}(d)$ such that for any number field K of degree d the number of K-isomorphism classes of elliptic curves having Mordell-Weil torsion (over K) of order > 18 is $\leq \mathcal{M}(d)$.*

Proof: The number $M = 18$ is chosen to guarantee that the modular curves that parametrize elliptic curves with rational torsion of order $> M$ are all of genus > 1. Merel (cf [Me]) has shown that the order of K-rational torsion in the Mordell-Weil group of any elliptic curve over K is bounded by a quantity $U(d)$ which depends only on the degree d. It follows that the "exceptions", i.e., the set of K-isomorphism classes of elliptic curves with Mordell-Weil torsion (over K) of order $> M$ is of cardinality no greater than the cardinality of the set of K-rational points on the disjoint union of the following set $\mathcal{S}(d)$ of modular curves: For integers m_1, m_2, denote by $X(m_1, m_2)$ the modular curve which classifies pairs (F, α) where F is an elliptic curve, and α is an injective homomorphism of the product of two cyclic groups $\mathbf{Z}/m_1 \cdot \mathbf{Z} \times \mathbf{Z}/m_2 \cdot \mathbf{Z}$ into F. Now let $\mathcal{S}(d)$ denote the disjoint union of these curves $X(m_1, m_2)$ where (m_1, m_2) runs through all pairs of integers with $1 \leq m_1 \cdot m_2$ and such that $M \leq m_1 \cdot m_2 \leq U(d)$. Clearly, the cardinality of $\mathcal{S}(d)$ admits a finite upper bound depending only on d, and the genus of any curve which is a member of $\mathcal{S}(d)$ is > 1 and admits a finite upper bound depending only on d. It follows (from Pacelli's Theorem quoted above) , using Lang's Conjecture, that the total number of K-rational points on the disjoint union of the curves in $\mathcal{S}(d)$ admits a finite upper bound depending only on d.

Proposition 1 will be proved in section 10 below.

7. The modular curves $X(E; p)$

Let p be a prime number. Our proof of Proposition 1 rests on the properties of certain models (over number fields K) of the modular curves $X(p)_{/\mathbf{C}}$ for $p \geq 7$. Recall that $X(p)$ is the Riemann surface which is the compactification of

$$Y(p) := (\text{upper half-plane})/\Gamma(p)$$

where $\Gamma(p) \subset \mathrm{PSL}_2(\mathbf{Z})$ is the kernel of the projection $\mathrm{PSL}_2(\mathbf{Z})$ to $\mathrm{PSL}_2(\mathbf{F}_p)$ and the action of $\Gamma(p)$ is via the natural action of $\mathrm{PSL}_2(\mathbf{Z})$ on the upper half-plane. The group $\mathrm{PSL}_2(\mathbf{F}_p)$ then acts naturally on $X(p)$. The quotient space of this action is the projective line $\mathbf{P}^1$ which we view as parametrized by the elliptic modular function

$$j: (\text{upper half-plane}) \to \mathbf{P}^1$$

and we denote by the same letter j the projection

$$j: X(p) \to \mathbf{P}^1.$$

The genus g_p of $X(p)$, for $p \geq 7$, satisfies the formula

$$84 \cdot (g_p - 1) = |\mathrm{PSL}_2(\mathbf{F}_p)| \cdot (7 - \frac{42}{p}) = \frac{1}{2}(p^2 - 1)(7p - 42)$$

as can be computed from Prop. 1.40 of [Sh] and is therefore > 1.

The full automorphism group of $X(p)$ for $p \geq 7$ is $\mathrm{PSL}_2(\mathbf{F}_p)$ and the following paraphrase of this will be particularly useful to us:

Proposition. *If $p \geq 7$, any automorphism of $X(p)$ preserves j.*

The appendix below contains an elegant argument due to Serre which proves this Proposition (as Serre remarked, this result is very likely in the literature, but it seems difficult to find an adequate reference for it).

For $p \geq 7$, and any elliptic curve E over a number field K there is an affine curve $Y(E; p)$ defined over K and its smooth complete model $X(E; p)$ whose underlying Riemann surfaces are isomorphic to the disjoint union of $p - 1$ copies of $Y(p)$ and of $X(p)$ respectively. For any scheme U over K the U- valued points of $Y(E; p)$ are naturally identified with isomorphism classes of pairs $(F_{/U}, \alpha)$ where $F_{/U}$ is an elliptic curve over U (i.e., an abelian scheme over U of dimension 1) and where

$$\alpha: E[p]_{/U} \to F[p]$$

is an isomorphism of finite flat group schemes over U. Here $E[p]_{/U}$ is the pullback to U of the kernel of multiplication by p, viewed as endomorphism of E and $F[p]$ is the kernel of multiplication by p, viewed as endomorphism of $F_{/U}$. The above property characterizes the curves $Y(E;p) \subset X(E;p)$ over K.

Let us take a moment to discuss the fact that the curve $X(E;p)$ is not connected. To any point $x \in X(E;p)$ over a K-scheme U, we may associate an element of $\mathbf{F}_p^*$ as follows: the Weil pairing identifies the "wedge-squares" over $\mathbf{F}_p$ of $E[p]_{/U}$ and of $F[p]_{/U}$ with the group scheme μ_p over U. Thus $\wedge^2(\alpha)$ induces an automorphism

$$\wedge^2(\alpha) \colon \mu_p \to \mu_p,$$

and since the automorphism group of μ_p is canonically $\mathbf{F}_p^*$, $\wedge^2(\alpha)$ identifies with a well-defined element (call it $\delta(x)$) in $\mathbf{F}_p^*$. For each element $u \in \mathbf{F}_p^*$, there is a (geometrically irreducible) component $X(E;p)_u$ of $X(E;p)$ characterized by the property that if $x \in X(E;p)_u$, then $\delta(x) = u$. The components $X(E;p)_u$ are each defined over K and are isomorphic over $\mathbf{C}$ to $X(p)$.

The multiplicative group $\mathbf{F}_p^*$ acts on $X(E;p)$ by the following rule. If $\lambda \in \mathbf{F}_p^*$ then λ sends the pair $(F_{/U}, \alpha)$ representing a U-valued point in $X(E;p)$ to $(F_{/U}, \lambda \cdot \alpha)$. Clearly, then, λ sends the component $X(E;p)_u$ to $X(E;p)_{\lambda^2 \cdot u}$. It follows that among the $p - 1$ components $X(E;p)_u$ (as u ranges through $\mathbf{F}_p^*$) there are at most two distinct K-isomorphism classes of geometrically irreducible curves (of genus g_p) represented, e.g., $X(E;p)_1$ and $X(E;p)_w$ for w a choice of quadratic nonresidue modulo p. The curves $X(E;p)_1$ and $X(E;p)_w$ are twists of one another over K and become isomorphic over the splitting field $K(E;p)$ of the K-Galois module $E[p]$. Moreover, the curve $X(E;p)_1$ has at least one K-rational point (namely the point represented by the pair $(E[p]_K, \alpha)$ where α is the identity mapping.

8. Digression: When are the different components of $X(E;p)$ isomorphic over K?

The results of this sections is an account of some conversations with Brian Conrad and Richard Taylor and I am grateful for their permission to include it in this article.

Since we have isomorphisms over K between $X(E;p)_v$ and $X(E;p)_{vu^2}$ for any $u, v \in \mathbf{F}_p^*$, to answer the question in the title of this section it suffices, for $E_{/K}$ an elliptic curve, and p an odd prime, to fix a quadratic non-residue mod p, $w \in \mathbf{F}_p^*$ and to give "necessary and sufficient" criteria for $X(E;p)_1$ and $X(E;p)_w$ to be isomorphic curves over K. First note that for every prime $\ell \neq p$ we have a natural "Hecke operator",

$$T_\ell \colon X(E;p)_v \dashrightarrow X(E;p)_{v\ell},$$

i.e., it is the correspondence of degree $\ell+1$ going from the curve $X(E;p)_v$ over K to the curve $X(E;p)_{v\ell}$ and characterized by the following property. Let x be a $\bar{K}$-valued point of $X(E;p)_v$. Suppose that x corresponds to the isomorphism class of pairs $(F_{\bar{K}}, \alpha)$ where $F_{\bar{K}}$ is an elliptic curve over $\bar{K}$ and where $\alpha: E[p]_{\bar{K}} \to F[p]$ is an isomorphism of determinant $v \in \mathbf{F}_p^*$. For every cyclic subgroup $C \subset F$ of order ℓ, consider the $\bar{K}$-valued point of $X(E;p)_{v\ell}$ corresponding to the isomorphism class $(F/C, \alpha_C)$ where $\alpha_C: E[p]_{\bar{K}} \to (F/C)[p]$ is the composition of α with the isogeny $F \to F/C$. Since the determinant of α_C is equal to $v\ell$, the point x_C lies on $X(E;p)_{v\ell}$. We have the formula

$$T_\ell(x) = \sum_C x_C.$$

Proposition 1. *(the genus zero cases) Suppose that $p = 3$, or 5. The curves $X(E;p)_v$ are isomorphic to $\mathbf{P}^1$ over K for all $v \in \mathbf{F}_p^*$.*

Proof: Since $p = 3$, or 5 the curves $X(E;p)_v$ are of genus 0, and since $X(E;p)_1$ has a K-rational point it follows that $X(E;p)_1$ is isomorphic to $\mathbf{P}^1$ over K. Since 2 is a quadratic non-residue mod 3 and mod 5, to prove the proposition it suffices to show that $X(E;p)_w$ is isomorphic to $\mathbf{P}^1$ for w equal to 2 mod p ($p = 3, 5$). But the Hecke correspondence T_2 is a K-rational correspondence of degree 3 from $X(E;p)_1$ to $X(E;p)_2$. The image of any K-rational point on $X(E;p)_1$ is a(n effective) K-rational divisor of degree 3 on the genus zero curve $X(E;p)_2$. Since $X(E;p)_1$ has a K-rational point, it follows that $X(E;p)_2$ has a K-rational divisor of odd degree, and is therefore isomorphic to $\mathbf{P}^1$ as was to be proved.

Here is the answer to the title question in this section.

Proposition 2. *Let E be an elliptic curve over K and $p \geq 7$ a prime number. Denote by*

$$\rho: \mathrm{Gal}(\bar{K}/K) \to \mathrm{Aut}(E[p]) \cong \mathrm{GL}_2(\mathbf{F}_p)$$

the natural continuous Galois representation on p-torsion points of E. Then the curves $X(E;p)_1$ and $X(E;p)_w$ are isomorphic over K if and only if either:

a. *the image of ρ is contained in a Cartan subgroup of $\mathrm{GL}_2(\mathbf{F}_p)$;*

b. *$p \equiv -1$ mod 4 and the image of ρ is contained in the normalizer of a (split) Cartan subgroup;*

c. *$p \equiv 1$ mod 4 and the image of ρ is contained in the normalizer of a nonsplit Cartan subgroup.*

Proof: First consider the set $\mathrm{Isom}_{\mathbf{C}}(X(E;p)_1, X(E;p)_w)$ of isomorphisms $X(E;p)_1 \to X(E;p)_w$ over $\mathbf{C}$, which may be viewed as a torsor over the group $\mathrm{Aut}_{\mathbf{C}}(X(E;p)_1)$. Identifying $X(E;p)_1$ (taken over $\mathbf{C}$) with the Riemann surface $X(p)$ and applying the Proposition of section 7 above, we see that $\mathrm{Isom}_{\mathbf{C}}(X(E;p)_1, X(E;p)_w)$ is naturally a torsor over $\mathrm{PSL}_2(\mathbf{F}_p)$. It follows that any isomorphism $\iota\colon X(E;p)_1 \to X(E;p)_w$ may be given as follows. There is an automorphism $r\colon E[p] \to E[p]$ of determinant w such that

$$\iota(F,\alpha) = (F, \alpha \cdot r).$$

Moreover such an automorphism r is unique up to multiplication by ± 1. The correspondence $\iota \leftrightarrow \{\pm r\}$ gives us a natural bijection of the sets:

$$\mathrm{Isom}_{\mathbf{C}}(X(E;p)_1, X(E;p)_w) \cong \mathrm{Aut}(E[p])_w/\{\pm 1\},$$

where $\mathrm{Aut}(E[p])_w \cong \mathrm{GL}_2(\mathbf{F}_p)_w$ denotes the $\mathrm{SL}_2(\mathbf{F}_p)$-coset of automorphisms of determinant w. One easily computes the necessary and sufficient condition for such an isomorphism in $\mathrm{Isom}_{\mathbf{C}}(X(E;p)_1, X(E;p)_w)$ to be defined over K. Namely, it is necessary and sufficient that conjugation by elements of ρ stabilize the set $\{\pm r\}$. Consequently, there exist an isomorphism $X(E;p)_1 \cong X(E;p)_w$ over K if and only if there exists an element $r \in \mathrm{GL}_2(\mathbf{F}_p)$ whose determinant is a nonquadratic residue mod p, and such that the image of ρ stabilizes the two-element set $\{\pm r\}$. Such an element r, whose determinant w is a nonquadratic residue mod p, is semi-simple.

At this point it suffices to list cases. First suppose that r is diagonalizable. Then after multiplication by a scalar we may take $\{\pm r\}$ to be the set of diagonal (2×2) matrices with diagonal elements $\{\pm[1,w]\}$; the group of elements stabilizing this set (under conjugation) is either the group of diagonal matrices (in which case we are in case **a**) or the normalizer of the group of diagonal matrices if $w = -1$. But $w = -1$ is allowed as a possibility only if -1 is a nonresidue, i.e. if we are in case **b**. of the Proposition. Next, suppose that r is not diagonalizable, in which case the $\mathbf{F}_p$-algebra E generated by r in the matrix algebra $M_2(\mathbf{F}_p)$ is a maximal commutative subfield, and the hypothesis that the image of ρ stabilizes the two-element set $\{\pm r\} \subset E$ (under the action of conjugation) implies that the image of ρ is contained in the normalizer of the corresponding (nonsplit) Cartan subgroup. Again, we have the possibility that the image of ρ is contained in the (nonsplit) Cartan subgroup itself (in which case we are in case **a**) or there is an element g in the image of ρ such that $grg^{-1} = -r$. Since conjugation by g is a field automorphism of E it follows that $u := r^2 \in \mathbf{F}_p$, and since E is a field, we know that $u \in \mathbf{F}_p$ is a quadratic non-residue. Since $w = \det r = -u^2$ is also a quadratic nonresidue, we have that -1 is a quadratic residue, i.e., we are in case **c**. This establishes the Proposition.

Idle Question. Are there occasions when $X(E;p)_1$ is not isomorphic to $X(E;p)_w$ over K, and yet the jacobians of these two curves are isomorphic over K? Note that we have a good number of homomorphisms defined over K between these jacobians: choose any prime number $\ell \equiv w$ mod p and consider the homomorphism induced from the Hecke correspondence T_ℓ.

9. The isomorphism class of the curve $X(E;p)$ over K

Fix the prime $p \geq 7$, and the elliptic curve E over K. Note that twisting E by a quadratic character χ over K does not change the isomorphism class of $X(E;p)$ over K; i.e., there is a canonical isomorphism

$$X(E;p) \cong X(E \otimes \chi; p)$$

given by the rule

$$[F, \alpha] \mapsto [F \otimes \chi, \alpha \otimes \chi].$$

Also, if $[F, \alpha]$ represents a K-rational point of $X(E;p)$, and if the determinant of α is $u \in \mathbf{F}_p^*$ then for each $w \in \mathbf{F}_p^*$ there is a canonical isomorphism over K

$$X(F;p)_w \cong X(E;p)_{u \cdot w}.$$

Lemma. *Let $u \in \mathbf{F}_p^*$. Let E be an elliptic curve over K such that there is some K-rational point of $X(E;p)_u$ represented by a pair $[F, \alpha]$ where F is an elliptic curve without complex multiplication (i.e., such that $j(F) \neq 0, 1728$). Let E' be an elliptic curve and u' an element in $\mathbf{F}_p^*$ such that*

$$X(E;p)_u \cong X(E';p)_{u'}.$$

Then there is a quadratic character χ over K and a $\mathrm{Gal}(\bar{K}/K)$-equivariant isomorphism $\beta \colon E[p] \to E'[p] \otimes \chi$ such that the pair $[E' \otimes \chi, \alpha']$ represents a K-rational point of $X(E;p)_{u \cdot u'^{-1}}$.

Proof: Let $h \colon X(E;p)_u \to X(E';p)_{u'}$ be the K-isomorphism, and let $[F', \alpha']$ be the image of $[F, \alpha]$ under h. By the previous discussion, we have that $j(F) = j(F')$ and since these j-invariants are distinct from 0 and 1728 we have that F' is a twist of F by a (quadratic) character over K. Let χ be this character. Since $\alpha \colon E[p] \to F[p]$ and $\alpha' \colon E'[p] \to F'[p] = F \otimes \chi[p]$ are $\mathrm{Gal}(\bar{K}/K)$-equivariant isomorphisms of determinants u and u' respectively, if we denote by α' again the induced isomorphism $E' \otimes \chi[p] \to F[p]$, we see that

$$\beta = \alpha'^{-1} \cdot \alpha$$

does what we want.

Corollary. *Let $p \geq 7, u \in \mathbf{F}_p^*$, and E an elliptic curve over K such that there is some K-rational point of $X(E;p)_u$ represented by a pair $[F, \alpha]$ where F is an elliptic curve without complex multiplication. Then the set of j-invariants* *of elliptic curves E' over K for which there is an element in $u' \in \mathbf{F}_p^*$ such that $X(E;p)_u \cong X(E';p)_{u'}$ is finite.*

Proof: By the previous lemma, any E' having the above property must be a quadratic twist of some elliptic curve representing some K-rational point on $X(E;p)$. Since $p \geq 7$, the curve $X(E;p)$ is a finite union of curves each of genus > 1 and therefore has only a finite number of K-rational points by Faltings' Theorem.

10. Proof of Proposition 1

Let p be a prime number > 7. Any K-isogeny $\phi\colon E \to E'$ of degree prime to p gives rise to a K-rational point

$$[\phi] = (E', \phi\colon E[p] \to E'[p]) \in Y(E;p).$$

More precisely, if the degree of the K-isogeny $\phi\colon E \to E'$ is δ (assumed prime to p) then the point $[\phi]$ lies in the component $Y(E;p)_u \subset X(E;p)_u$ where u is the reduction of δ modulo p.

Let $N(K, E, p)$ denote the number of j-invariants of elliptic curves which are K-isogenous to E via K-isogenies of degree prime to p. Since the set of these j-invariants is contained in the image (in the j-line) of the set $Y(E;p)(K)$ (under the natural projection $(E', \phi) \mapsto j(E')$) we have:

$$N(K, E, p) \leq \operatorname{card}\{Y(E;p)(K)\} \leq \operatorname{card}\{X(E;p)(K)\}.$$

To prove Proposition 1 let us assume Statement **A** of section 5. Let $N(g)$ be the number given to us in Statement **A** for the genus $g > 1$. Put

$$M(g) := (p-1) \cdot \max(3 \cdot |\mathrm{PSL}_2(\mathbf{F}_p)|, N(g_p))$$

Choose any two distinct prime numbers p_1 and p_2 both ≥ 7 ($p_1 = 7$ and $p_2 = 11$ will do). We will prove Proposition 1 with

$$N = M(g_{p_1}) \cdot M(g_{p_2}).$$

If we are given any K-isogeny $\phi\colon E \to E'$ of degree d, writing $d = d_o \cdot (p_1)^a$ with d_o relatively prime to p_1 we may factor the isogeny ϕ as in the commutative diagram of K-isogenies below, where the arrows are labelled by their degrees, and in particular, the horizontal isogenies are of degree prime to p_1 and the vertical ones are of degree prime to p_2:

$$
\begin{array}{ccc}
E & \xrightarrow{\ d_0\ } & E_1' \\
{\scriptstyle p_1^a}\big\downarrow & \searrow^{d} & \big\downarrow{\scriptstyle p_1^a} \\
E_1 & \xrightarrow{\ d_0\ } & E'
\end{array}
$$

Since the isogeny $E \to E'$ is determined by the pair of isogenies $E \to E_1$ and $E \to E_1'$, we get that $N(K, E) \leq N(K, E, p_1) \cdot N(K, E, p_2)$. It follows then that if $N(K, E) > M(g_{p_1}) \cdot M(g_{p_2})$ we must have either $N(K, E, p_1) > M(g_{p_1})$ or $N(K, E, p_2) > M(g_{p_2})$, i.e., we have that one of the curves $X(E, p)_u$ (for $p = p_1$ or p_2, and for some u) has the property that

$$\operatorname{card}\{X(E, p)_u(K)\} > \max(3 \cdot |\mathrm{PSL}_2(\mathbf{F}_p)|, N(g_p))$$

It follows from this inequality that

a. The curve $X(E, p)_u$ is a "K-exceptional curve of genus g_p", and

b. The set $Y(E, p)_u(K)$ contains a point whose associated elliptic curve has j-invariant different from 0 and 1728.

Statement **A** together with the corollary above guarantee that there are at most a finite number of j-invariants of elliptic curves satisfying a) and b), which concludes the proof of Proposition 1.

Appendix: The automorphism group of $X(p)$. (Copied from a letter of J.-P. Serre, June 26, 1996)

Let p be a prime number. Let $X(p)$ denote the quotient of the extended upper half plane $\mathbf{H}^*$ under the action of the congruence subgroup $\Gamma(p) = \ker(\mathrm{PSL}_2(\mathbf{Z}) \to \mathrm{PSL}_2(\mathbf{F}_p))$. Let $G = \mathrm{PSL}_2(\mathbf{F}_p)$ which acts faithfully on the Riemann surface $X(p)$. Denote by A the group of all automorphisms of $X(p)$ and put $m = [A : G]$. The genus g_p of $X(p)$, for $p \geq 7$, satisfies the formula

$$84 \cdot (g_p - 1) = |\mathrm{PSL}_2(\mathbf{F}_p)| \cdot (7 - 42/p),$$

and therefore, by the well known Hurwitz bound for the order of the automorphism group of a Riemann surface of given genus, we have $m < 7$.

Lemma. *The subgroup G is normal in A.*

Proof: The natural action of G on A/G via left multiplication is trivial. This can be seen by first noting that any element $s \in G$ of order p acts trivially on A/G (since $p > m$). It then follows that the action of (all of) G on A/G is trivial, since G is generated by its elements of order p. Therefore G is normal in A.

If $e = 2, 3$, or p, call X_e the finite subset of points of $X = X(p)$ which have ramification index e under the mapping $X \to X/G = \mathbf{P}^1$. Any element $t \in A$ normalizes G and therefore stabilizes the subsets X_e for $e = 2, 3$, or p and also induces an automorphism t' of $\mathbf{P}^1$. This automorphism fixes the three image points of the sets X_e ($e = 2, 3$, and p) and is therefore the identity on $\mathbf{P}^1$. It follows that A acts trivially on X/G and consequently $A = G$.

PART II. Speculations about the topology of rational points: a further up-date

1. The topological closure of the set of rational points in the real locus

A few years ago I formulated some conjectures about the topology of rational points [Ma 1,2,3] the original motivation for them being to examine the question of whether or not "**Z** is Diophantinely definable in **Q**". That question had been raised by mathematical logicians because an affirmative answer to it would imply (using Matjasevic's work) the non-existence of any algorithmic solution to to the problem of determining if a polynomial in many variables with rational coefficients has a rational solution.

Here is the list of those conjectures:

XConjecture 1. Let V be a smooth variety over **Q** such that $V(\mathbf{Q})$ is Zariski-dense. Then the topological closure of $V(\mathbf{Q})$ in $V(\mathbf{R})$ is equal to a union of connected components in $V(\mathbf{R})$.

Conjecture 2. Let V be any variety over **Q**. Then the topological closure of $V(\mathbf{Q})$ in $V(\mathbf{R})$ is homeomorphic to a finite simplical complex.

Conjecture 3. Let V be any variety over **Q**. Then the topological closure of $V(\mathbf{Q})$ in $V(\mathbf{R})$ possesses at most a finite number of connected components.

Conjecture 4. **Z** is not Diophantinely definable in **Q**.

Each of the conjectures in the above list implies the subsequent ones.

The reason for the "**X**" in front of Conjecture 1 above is that recently, a counter-example to it was constructed by Colliot-Thélène, Skorobogatov, and Swinnerton-Dyer [C-T, S, S-D]. In the light of their counter-example, they "repair" **XConjecture 1** by making the following modification of it:

Conjecture A (of C-T, S and S-D): Let V be a smooth integral variety over **Q** and U a connected component of the real locus $V(\mathbf{R})$ such that $V(\mathbf{Q}) \cap U$ is Zariski-dense in V. Then $V(\mathbf{Q}) \cap U$ is topologically dense in U.

Remark. This is Conjecture 5 in the article [C-T, S, and S-D]; the authors also formulate a strengthened version of it (Conjecture 4 of [C-T, S, and S-D]) as follows:

Conjecture B (of C-T, S, and S-D): Let V be a smooth integral variety over **Q** and U a connected component of the real locus $V(\mathbf{R})$. If W is the topological closure of $V(\mathbf{Q}) \cap U$ in U, then there is a Zariski-closed set $Y \subset V$ defined over **Q** such that W is a (finite) union of connected components of $Y(\mathbf{R})$.

Either of these Conjectures of C-T, S and S-D imply Conjectures 2,3 and 4 above; and they both avoid the (counter)-examples that C-T, S and S-D construct. The formulation of the "repaired conjectures" by C-T, S and S-D focusses on a phenomenon illustrated by their counter-examples; namely, that there can be a good deal of "possible variation" in the Zariski-closures of $V(\mathbf{Q}) \cap U$ as U ranges through the connected components of the real locus of V.

How much variation is there in the Zariski-closures of $V(\mathbf{Q}) \cap U$? To focus on this, I found it useful to formulate **Question I** below, an affirmative or negative solution of which would be interesting. First, an affirmative answer to Question I still implies **Conjecture 4**. Second, Question I concerns the topology of K-rational points of a variety V defined over K, where K is *any* number field (not just $\mathbf{Q}$) and concerns the placement of these K-rational points in the topological space of S-adic points of V where S is *any* finite set of places (not just a real archimedean prime) of K.

2. S-adic topological closure

Let V be any variety defined over a number field K. Let S be a finite set of places of K, and consider

$$K_S = \prod_{v \in S} K_v$$

viewed as locally compact topological ring. Let $V(K_S)$ denote the topological space of K_S-rational points of K. For every point $p \in V(K_S)$ define $W(p) \subset V$ to be the subvariety defined over K which is the intersection of the Zariski-closures of the subsets $V(K) \cap U$ where U ranges through all open neighborhoods of p in $V(K_S)$. By the noetherian property of V, $W(p)$ is, in fact, the Zariski-closure of $V(K) \cap U$ for some open neighborhood of p which is "small enough".

Question I. As p ranges through the points of $V(K_S)$ are there only a finite number of distinct subvarieties $W(p)$?

Proposition. *An affirmative answer to Question I implies Conjecture 4.*

Proof: If Conjecture 4 is false, then there is a variety V over $\mathbf{Q}$ and a $(\mathbf{Q}\text{-})$ rational function f on V such that the image of $V(\mathbf{Q})$ under f is the subset $\mathbf{Z} \subset \mathbf{Q}$. For each $N \in \mathbf{Z}$ choose a rational point $p_N \in V(\mathbf{Q})$ such that $f(p_N) = N$ and note that $W(p_N)$ is a nonempty subvariety contained in the fiber of f over $N \in \mathbf{Q}$; the existence of such an f thus implies that there are an infinite number of distinct $W(p)$'s and therefore provides an instance where Question I has a negative answer. More generally, an affirmative answer to Question I implies that there is no rational function f such that $f(V(\mathbf{Q}))$ is an infinite subset of $\mathbf{Z} \subset \mathbf{Q}$.

I will recall the construction of [C-T, S, S-D] below. Their construction exhibits certain varieties V over $\mathbf{Q}$ possessing a number of distinct Zariski-closures of $V(\mathbf{Q}) \cap U$ for distinct connected components $U \subset V(\mathbf{R})$.

3. Isotrivial families that are trivialized by finite étale extensions of the base

The basic idea of [C-T, S, S-D] is to examine isotrivial families V_T of varieties over a parameter space T which is a smooth projective variety, the family becoming trivial over a finite étale cover of T. The variety V of the previous paragraph is the "total space" V_T. More specifically, T will be taken to be an elliptic curve E and the finite étale cover of T which trivializes the family will be given by an isogeny $E' \to E$. We now prepare for the construction.

Let E' and E be elliptic curves over K, and let $\pi \colon E' \to E$ be an isogeny of degree two defined over a number field K. The kernel of the isogeny π is a cyclic group of order two which we denote by $C = \{1, c\}$ where c is a point of order two in $E'(K)$. We shall ignore the group structures of E' and E for a moment and think of E' as C-torsor over the curve E. As such it is classified by an element, denote it γ, in the étale cohomology group $H^1(E, C)$. For any K-algebra R and any R-valued point $e \in E(R)$ the isogeny π determines a C-torsor $T_e = \pi^{-1}(e)$ whose associated cohomology class γ_e in the étale cohomology group $H^1(Spec(R), C)$ is simply the pullback of γ via the section e. Alternatively, we might form the exact sequence of sheaves of abelian groups for the étale topology,

$$0 \to C \to E' \to E \to 0,$$

and form the corresponding long exact sequence of cohomology over $Spec(R)$,

$$0 \to H^0(Spec(R), C) \to E'(R) \to E(R) \to H^1(Spec(R), C),$$

and think of γ_e as the image of e under the coboundary mapping

$$\delta \colon E(R) \to H^1(Spec(R), C).$$

From this we see that the C-torsor T_e classified by γ_e depends only on the coset of $E(R)$ modulo the image of $E'(R)$ and, in fact, the set of isomorphism classes $\mathcal{E}(R)$ of C-torsors "T_e obtained from this process as e ranges through the elements of $E(R)$ is in natural one-one correspondence with $E(R)/\pi E'(R)$.

4. Comparing global to local

First we consider the global context, and take $R = K$, a number field. We have

$$\mathcal{E}(K) = E(K)/\pi E'(K)$$

and since K is a number field, $\mathcal{E}(K)$, which is a quotient of $E(K)/2 \cdot E(K)$, is a finite abelian group of exponent 2.

Now let S be a finite set of places (possibly including archimedean places) of K. We will be interested in the locally compact topological space (group, in fact)

$$E(K_S) = \prod_{v \in S} E(K_v).$$

Put $R = K_S$ so that $\mathcal{E}(R)$ is again a finite group of exponent 2. As e ranges through the compact group $E(R) = E(K_S)$, the isomorphism class of the C-torsor T_e is "locally constant" and is constant "precisely" on the cosets of $E(K_S)$ modulo the open subgroup of finite index $\pi\{E'(K_S)\} \subset E(K_S)$.

It is easy to see that one can find finite sets of places S for which the restriction of the natural mapping

$$\mathcal{E}(K) \to \mathcal{E}(K_S)$$

is injective. When S is such a finite set of places, let us simply say that S is **large enough**.

5. Twisting varieties by π:

Let $Y_{/K}$ be a variety, equipped with a K-rational involution ι. We view the group C as acting on Y (in a K-rational way) by having the generator of C act on Y as the involution ι. The particular case of this that is used in [C-T, S, S-D] is when Y is an elliptic curve over K and the involution ι is given by $y \mapsto -y$.

We will twist the "constant" family over E,

$$Y \times E \to E, \tag{1}$$

by the isogeny π, to get an isotrivial family which we will denote

$$Y \tilde{\times} E \to E \tag{2}$$

Explicitly, $Y \tilde{\times} E$ is obtained by passing to the quotient of the product $Y \times E'$ by the diagonal action of C; let

$$u: Y \times E' \to Y \tilde{\times} E \tag{3}$$

denote this quotient morphism. Then u is an étale morphism of degree two; we may view $Y \times E'$ as a C-torsor over $Y \tilde{\times} E$. Given an R-valued point e of $E(R)$ the fiber Y_e of the twisted family (2) over the point e is the variety over R obtained by twisting the pullback of Y over R via that cohomology class in $H^1(\mathrm{Spec}(R), \mathrm{Aut}(Y/R))$ which is the image of γ_e under the natural homomorphism

$$H^1(\mathrm{Spec}(R), C) \to H^1(\mathrm{Spec}(R), \mathrm{Aut}(Y/R)).$$

In the case where $R = K$ is our number field, the $\bar{K}$-rational points of Y_e admit the following somewhat more explicit description. Let $\tilde{e}' \in E'(\bar{K})$ be a choice of one of the two inverse images of $e \in E(K) \subset E(\bar{K})$, the other choice being $\tilde{e}' + c$. Then $Y_e(\bar{K})$, the set of $\bar{K}$-rational points of Y_e, may be naturally identified with the set of couples

$$\{(y, \tilde{e}'), (\iota(y), \tilde{e}' + c)\} \tag{4}$$

of points in $Y(\bar{K}) \times E'(\bar{K})$, this identification being equivariant with respect to Galois action.

The above discussion gives rise to the following straightforward

Proposition 1. *(Topological local constancy of fibers) If S is large enough (in the sense of section 2) then for points $e \in E(K)$ the K-isomorphism class of the variety Y_e (i.e., the fiber of (2) over the point e) is determined by the coset modulo the open subgroup $\pi E'(K_S) \subset E(K_S)$ of the image of e under the inclusion $E(K) \subset E(K_S)$.*

To get a more precise picture of the placement of K-rational points in $Y \tilde{\times} E$ note that for every $e \in E(K)$ we have a degree-two covering $u_e: Y_e \times E' \to Y \tilde{\times} E$ defined over K, which on $\bar{K}$-rational points sends $(y_e, e') \in Y_e \times E'(\bar{K})$ to the image of either of the points $(y, \tilde{e}' + e')$ or $(\iota(y), \tilde{e}' + e' + c)$ under the mapping $u: Y \times E' \to Y \tilde{\times} E$, where y_e is represented by the couple $\{(y, \tilde{e}'), (\iota(y), \tilde{e}' + c)\}$ as in (4) above. We have a cartesian diagram of K-rational morphisms,

$$
\begin{array}{ccc}
Y_e \times E' & \xrightarrow{u_e} & Y \tilde{\times} E \\
\pi' \downarrow & & \downarrow \\
E & \xrightarrow{\tau_e} & E
\end{array}
$$

where π' is the composition of projection to the second factor and the isogeny π; the unmarked vertical morphism is the natural projection; and τ_e is translation by e, i.e., $\tau_e(y) = y + e$. We see that the fiber of $Y \tilde{\times} E \to E$ over any point $e + \pi(e')$ in the coset $e + \pi E'(K)$ is isomorphic over K to Y_e (and, more precisely, is the image $Y_e \times \{e'\}$ under the mapping u_e).

Now choose a system $\{e_j\}_{j=1,\dots,\nu}$ of coset representatives of $E(K)$ modulo $\pi E'(K)$. The above discussion shows

Proposition 2. *(Topological local constancy of the set of K-rational points) If S is large enough (in the sense of section 2) the set $(Y\tilde{\times}E)(K)$ of K-rational points of the family $Y\tilde{\times}E$ breaks up into the disjoint union of ν subsets, the i-th subset ($i = 1,\ldots,\nu$) being equal to the image under the morphism u_{e_i} of the set $Y_{e_i}(K) \times E'(K)$, of K-rational points of $Y_{e_i} \times E'$. Under the natural projection $Y\tilde{\times}E \to E$, this "i-th" subset maps into the coset of $\pi E'(K)$ containing e_i in $E(K)$.*

We now examine the topological closure of $(Y\tilde{\times}E)(K)$ in the topological space $(Y\tilde{\times}E)(K_S)$ when S is a finite set of places of K, which is "large enough" (which means that each $i = 1,\ldots,\nu$ also determines a distinct coset of $E(K_S)$ modulo $\pi E'(K_S)$; we refer to this coset as the "i-th coset" of $E(K_S)$ modulo $\pi E'(K_S)$. Of course, it is not necessarily the case that all cosets of $E(K_S)$ modulo $\pi E'(K_S)$ are included in this enumeration. Let $U_i \subset (Y\tilde{\times}E)(K_S)$ be the inverse image of the i-th coset of $E(K_S)$ modulo $\pi E'(K_S)$ under the natural mapping

$$(Y\tilde{\times}E)(K_S) \to E(K_S),$$

so that each U_i is an open compact subset of the compact topological space $(Y\tilde{\times}E)(K_S)$. Let $W_i \subset Y_{e_i}$ denote the Zariski closure of the subset of K-rational points $Y_{e_i}(K)$ in Y_{e_i}. By Proposition 2, the subset of K-rational points of our family $Y\tilde{\times}E$ whose image in $(Y\tilde{\times}E)(K_S)$ is contained in U_i is equal to

$$u_{e_i}(W_i(K) \times E'(K)) \subset U_i.$$

In particular, for each of the U_i, the Zariski-closure of the set of K-rational points of $Y\tilde{\times}E$ which are contained in U_i is equal to $Z_i := u_{e_i}(W_i \times E')$. The fun, then, is to find examples where the set of K-rational points of Y_{e_i}, and hence also the Zariski closures W_i, differ substantially — differ even in dimension — for distinct i's. In diagram 1 overleaf we are schematically envisioning a case where $\nu = 2$, W_1 consists of four points, and $W_2 = Y_{e_2}$. The cross-hatching in U_1 and U_2 stand for Z_1 and Z_2 respectively. Let us now prepare more specifically to describe the kind of (counter-)examples given in [C-T, S, S-D]. Their construction hinges on taking $K = \mathbf{Q}$, $S =$ the real prime, and finding 2-isogenies $E' \to E$ of elliptic curves over $\mathbf{Q}$ with the property that

(*) 1. The real locus $E(\mathbf{R})$ has two components; the image of $E'(\mathbf{R})$ in $E(\mathbf{R})$ is $E^o(\mathbf{R})$ the connected component of $E(\mathbf{R})$ containing the identity element.

(*) 2. The Mordell-Weil group $E(\mathbf{Q})$ is infinite and is dense in $E(\mathbf{R})$.

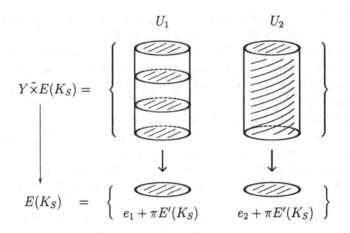

$$Y \tilde{\times} E(K_S) =$$

U_1 U_2

$$E(K_S) = \left\{ \begin{array}{cc} & \\ e_1 + \pi E'(K_S) & e_2 + \pi E'(K_S) \end{array} \right\}$$

Diagram 1

(*) **3.** The natural mapping

$$E(\mathbf{Q})/\text{image}\, E'(\mathbf{Q}) \to E(\mathbf{R})/\text{image}\, E'(\mathbf{R})$$

is an isomorphism of groups of order two.

In such a situation, one can apply Proposition 1 to get that the family $Y \tilde{\times} E \to E$ has only two isomorphism classes of $\mathbf{Q}$-rational fibers, the ones lying over points in $E^o(\mathbf{R})$ being $\mathbf{Q}$-isomorphic to Y, and the ones lying over the other connected component of $E(\mathbf{R})$, call it $E^1(\mathbf{R})$, being $\mathbf{Q}$-isomorphic to a single specific twist Y_e of Y. Now take Y to be an elliptic curve over $\mathbf{Q}$ and take the involution ι to be $y \mapsto -y$, but consider only such elliptic curves Y with the added property that $Y(\mathbf{Q})$ is infinite and $Y_e(\mathbf{Q})$ is finite. Applying Proposition 2 one sees that the intersection of $(Y \tilde{\times} E)(\mathbf{Q})$ with the union of connected components of $(Y \tilde{\times} E)(\mathbf{R})$ lying over $E^o(\mathbf{R})$ is Zariski dense in $Y \tilde{\times} E$ while the Zariski closure of the intersection of $(Y \tilde{\times} E)(\mathbf{Q})$ with the union of connected components of $(Y \tilde{\times} E)(\mathbf{R})$ lying over $E^1(\mathbf{R})$ is (non-empty, and is) a finite union of (elliptic) curves.

6. " C.-T., S. and Sw.-D." examples

Here is a specific choice of example, following the recipe of C.-T., S. and Sw.-D. which produces a surface "out of" elliptic curves of low conductor which have the desired property. For the base curve of the surface take the elliptic curve E displayed below, which admits a 2-isogeny $E' \to E$ over $\mathbf{Q}$.

$$\begin{array}{rrl} E': & y^2 + xy & = x^3 + 4x + 1 \\ E: & y^2 + xy & = x^3 - x. \end{array}$$

The curves E, E' are the two elliptic curves of conductor 65 and the pair of them satisfy (*)**1,2,3**. This is the example of smallest conductor (of a pair of elliptic curves) satisfying (*) occurring in Cremona's tables [Cr].

One computes that $E(\mathbf{R})$ is disconnected and $E'(\mathbf{R})$ is connected. Condition **1** of (*) then follows. The point of order two in $E(\mathbf{Q})$ is given by $x = 0, y = 0$, and lies on the connected component of $E(\mathbf{R})$ not containing the identity. The Mordell-Weil group $E(\mathbf{Q})$ is generated by this point of order two, together with the point $(x, y) = (1, -1)$ which is of infinite order. This gives Condition **2** of (*).

Letting $E'/\mathbf{Z}$ and $E/\mathbf{Z}$ denote the associated Néron models over $\operatorname{Spec} \mathbf{Z}$, we have an exact sequence of group schemes over $\mathbf{Z}$

$$0 \to \mu_{2/\mathbf{Z}} \to E'_{/\mathbf{Z}} \to E_{/\mathbf{Z}} \to 0$$

where $\mu_{2/\mathbf{Z}}$ is the finite flat group scheme over $\operatorname{Spec} \mathbf{Z}$ which is the kernel of multiplication by 2 in $\mathbf{G}_{\mathrm{m}/\mathbf{Z}}$. The exactness of this sequence may be checked fairly directly using the information given in Cremona's listing for conductor 65. Passing to the diagram of long exact sequence of flat cohomology groups, one gets the commutative diagram

$$
\begin{array}{ccccc}
E'(\mathbf{Q}) & \longrightarrow & E(\mathbf{Q}) & \longrightarrow & H^1(\operatorname{Spec} \mathbf{Z}, \mu_2) \\
\downarrow & & \downarrow & & {\scriptstyle j}\downarrow \\
E'(\mathbf{R}) & \longrightarrow & E(\mathbf{R}) & \longrightarrow & H^1(\operatorname{Spec} \mathbf{R}, \mu_2) & \longrightarrow & 0
\end{array}
$$

where the horizontal rows are exact. One computes that the vertical mapping labelled j is an isomorphism of groups of order two. This gives Condition **3** of (*).

The unique nontrivial element in the group $H^1(\operatorname{Spec} \mathbf{Z}, \mu_2)$ "classifies" the μ_2-torsor given by the extension $\operatorname{Spec} \mathbf{Z}[\sqrt{-1}]$ over $\operatorname{Spec} \mathbf{Z}$. Denote by χ the quadratic Dirichlet character (of discriminant 4) attached to this extension.

7. An explicit example

To complete the construction of an example, we need only take our "fiber" Y to be an elliptic curve over $\mathbf{Q}$ which has the property that the Mordell-Weil rank of Y is 0 while the Mordell-Weil rank of $Y \otimes \chi$, i.e., of Y twisted by χ, is positive. Of course, we take our ι to be the involution $y \mapsto -y$. There is no difficulty finding such an elliptic curve: $Y = X_0(17) \otimes \chi$, for example, has the property that its Mordell-Weil group over $\mathbf{Q}$ is of positive rank, while $Y \otimes \chi = X_0(17)$ has finite Mordell-Weil group (it is cyclic of order 4).

We therefore focus on the surface V over $\mathbf{Q}$ given by $V = Y \tilde{\times} E \to E$ (in the notation of section **5**.) where E is the elliptic curve of conductor 65 whose equations is $y^2 + xy = x^3 - x$, and where $Y = X_0(17) \otimes \chi$. The real locus of V consists of two components, U_1 (which projects onto the component of $E(\mathbf{R})$ containing the identity element) and U_2 (which projects onto the other topological component of $E(\mathbf{R})$). Both U_1 and U_2 are compact two-manifolds of genus one.

The $\mathbf{Q}$-rational points of V meeting U_1 are topologically dense in U_1 and therefore the Zariski-closure of $V(\mathbf{Q}) \cap U_1$ in V is equal to V.

In contrast, the Zariski-closure of $V(\mathbf{Q}) \cap U_2$ in V is the image of $X_0(17)(\mathbf{Q}) \times E'(\mathbf{R})$ in V. In other words the Zariski-closure of $V(\mathbf{Q}) \cap U_2$ is the the union of three elliptic curves! (one of which is isomorphic to E' (projecting by a map of degree 2 to the base) and the other two are isomorphic to E). Our example, then, is of the sort envisioned in Diagram 3 (with $K = \mathbf{Q}$, and $K_S = \mathbf{R}$, and each of the cosets $e_1 + \pi E'(K)$ and $e_2 + \pi E'(K)$ are topologically homeomorphic to the circle.

References

[A 1] Abramovich, D.: Uniformité des points rationnels des courbes algébriques sur les extensions quadratiques et cubiques, C.R. Acad. Sci. Paris **321** (1995) 755-758

[A 2] Abramovich, D.: A high fibered power of a family of varieties of general type dominates a variety of general type. Invent. math. **128** (1997) 481-494

[A 3] Abramovich, D.: A linear lower bound on the gonality of modular curves. International Mathematics Research Notices, **20** (1996) 1005-1013

[A-H] Abramovich, D., Harris, J.: Abelian varieties and curves in $W_d(C)$. Comp. Math. **78** (1991) 227-238

[A-V] Abramovich, D., Voloch, J.F.: Lang's conjecture, fibered powers and uniformity, New York J. Math **II** (1996) 20-34

[C-H-M 1] Caporaso, L., Harris, J., Mazur, B.: Uniformity of rational points. Journal of the A.M.S. **10** (1997) 1-35; a more expository version available by anonymous ftp:
ftp://ftp.math.harvard.edu/pub/uniformityofrationalpoints.tex

[C-H-M 2] Caporaso, L., Harris, J., Mazur, B.: How many rational points can a curve have? Proceedings of the Texel Conference, Progress in Math. **129** Birkhauser, Boston (1995) 13-31

[C-T, S, S-D] Colliot-Thélène, J.-L., Skorobogatov, A.N., Swinnerton-Dyer, P.: Double fibres and double covers: paucity of rational points. To appear.

[Cr] Cremona, J.: Algorithms for modular elliptic curves. Cambridge Univ. Press (1992)

[D-F] Debarre, O., Fahlaoui, R.: Abelian varieties in $W_d^r(C)$ and points of bounded degree on algebraic curves. Compositio Math. **88** (1993) 235-249

[dJ] de Jong, J.: Smoothness, semistability, and alterations. Publ.Math. I.H.E.S. **83** (1996) 51-93

[H-S] Harris, J., Silverman, J.: Bi-elliptic curves and symmetric products. Proc. Amer. Math. Soc. **112** (1991)347-356

[H] Hassett, B.: Correlation for surfaces of general type. Duke Math. J. **85** (1996) 95-107

[L] Lang, S.: Hyperbolic diophantine analysis. Bull. A.M.S. **14** (1986) 159-205

[L-Y] Li, P., Yau, S.-T.: A new conformal invariant and its applications to the Willmore conjecture and the first eigenvalue of compact surfaces. Inventiones math. **69** (1982) 269-291

[Ma 1] Mazur, B.: The topology of rational points. Journ. Exp. Math. **1** (1992) 35-45

[Ma 2] Mazur, B.: Questions of decidability and undecidability in number theory. J. Symbolic Logic **59** (1994) no. 2, 353-371

[Ma 3] Mazur, B.: Speculation about the topology of rational points: an update. Astérisque **228** (1995) 165-181

[Me] Merel, L.: Bornes pour la torsion des courbes elliptiques sur les corps des nombres. Inventiones Math. **124** 437-449

[P] Pacelli, P.: Uniform boundedness for rational points. Duke Math. J. **88** (1997) 77-102

[Sh] Shimura, G.: Introduction to the Arithmetic Theory of Automorphic Functions. Princeton Univ. Press, 1971

[Y] Yau, S.-T.: An application of eigenvalue estimate to algebraic curves defined by congruence subgroups. Math. Research letters **3** (1996) 167-172

DEPARTMENT OF MATHEMATICS, HARVARD UNIVERSITY, ONE OXFORD STREET, CAMBRIDGE, MA 02138, USA
mazur@math.harvard.edu

Models of Shimura varieties in mixed characteristics

BEN MOONEN

Contents

Introduction

At the 1996 Durham symposium, a series of four lectures was given on Shimura varieties in mixed characteristics. The main goal of these lectures was to discuss some recent developments, and to familiarize the audience with some of the techniques involved. The present notes were written with the same goal in mind.

It should be mentioned right away that we intend to discuss only a small number of topics. The bulk of the paper is devoted to models of Shimura varieties over discrete valuation rings of mixed characteristics. Part of the discussion only deals with primes of residue characteristic p such that the group G in question is unramified at p, so that good reduction is expected. Even at such primes, however, many technical problems present themselves, to begin with the "right" definitions.

Reprinted from 'Galois Representations in Arithmetic Algebraic Geometry',
edited by A. J. Scholl & R. L. Taylor. ©Cambridge University Press 1998

There is a rather large class of Shimura data—those called of pre-abelian type—for which the corresponding Shimura variety can be related, if maybe somewhat indirectly, to a moduli space of abelian varieties. At present, this seems the only available tool for constructing "good" integral models. Thus, if we restrict our attention to Shimura varieties of pre-abelian type, the construction of integral canonical models (defined in §3) divides itself into two parts:

Formal aspects. If, for instance, we have two Shimura data which are "closely related", then this should have consequences for the existence of integral canonical models. Loosely speaking, we would like to show that if one of the two associated Shimura varieties has an integral canonical model, then so does the other. Most of such "formal" results are discussed in §3.

Constructing models for Shimura varieties of Hodge type. By definition, these are the Shimura varieties that can be embedded into a Siegel modular variety. As we will see, the existence of an integral canonical model is essentially a problem about smoothness, which therefore can be studied using deformation theory. We are thus led to certain deformation problems for p-divisible groups. These can be dealt with using techniques of Faltings, which are the subject of §4. This is not to say that we can now easily prove the existence of integral canonical models. Faltings's results only apply under some assumptions, and in the situation where we want to use them, it is not at all clear that these are satisfied. To solve this, Adrian Vasiu has presented an ingenious, but technically complicated strategy. We will discuss this in §5. Unfortunately, it seems that Vasiu's program has not yet been brought to a successful end. We hope that our presentation of the material can help to clarify what technical points remain to be settled.

I have chosen to include quite a bit of "basic material" on Shimura varieties, which takes up sections 1 and 2. Most of this is a review of Deligne's papers [De1] and [De3]. I also included some examples and some references to fundamental work that was done later, such as the generalization of the theory to mixed Shimura varieties. The main strategy of [De3] is explained in some detail, since it will reappear in our study of integral models.

The only new result in the first two sections concerns the existence of canonical models for Shimura varieties which are not of abelian type. It was pointed out to me by J. Wildeshaus that the argument as it is found in the literature is not complete. We will discuss this in section 2, and we present an argument to complete the proof.

In §3 we take up the study of integral models of Shimura varieties. The first major problem here is to set up good definitions. We follow the pattern laid out by Milne in [Mi3], defining an integral canonical model as a smooth model which satisfies a certain Néron extension property. The main difficulty is to decide what class of "test schemes" to use. We explain why the class used by Milne in loc. cit. leads to unwanted results, and we propose to use a smaller class of test schemes which (at least for $p > 2$ and ramification $e < p - 1$) avoids this. Our definition differs from the one used by Vasiu in [Va2].

In the rest of §3 we prove a number of "formal" results about integral canonical models, and, inspired by Deligne's approach in [De3], we develop the notion of a connected Shimura variety in the p-adic setting. The main result of this section is Cor. 3.23. It says, roughly, that in order to prove the existence of integral canonical models for all Shimura varieties of pre-abelian type at primes of characteristic $p > 2$ where the group in question is unramified, it suffices to show that certain models obtained starting from an embedding into a Siegel modular variety, are formally smooth. As we will explain, there are finitely many primes that may cause additional problems if the group has simple factors of type A_ℓ. We give full proofs of most statements. Although the reader may find some details too cumbersome, we think that they are quite essential, and that only by going through all arguments we are able to detect some unexpected problems. Some of our results were also claimed in [Va2], but most proofs given here were obtained independently (see also remark 3.24).

In §4 we study deformation theory of p-divisible groups with given Tate classes. The main results are based on a series of remarks in Faltings's paper [Fa3], of which we provide detailed proofs.

In §5 we attempt to follow Vasiu's paper [Va2]. Our main goal here is to explain Vasiu's strategy, and to explain which technical problems remain to be solved. This section consists of two parts. Up until Thm. 5.8.3, we prove most statements in detail. This leads to a result about the existence of integral canonical models under a certain additional hypothesis (5.6.1). After that we indicate a number of statements that should allow to remove this hypothesis. It is in this part of Vasiu's work that, to our understanding, further work needs to be done before the main result (see 5.9.6) can be accepted as a solid theorem.[1]

---

[1] After completing our manuscript we received new versions of Vasiu's work (A. Vasiu, *Integral canonical models of Shimura varieties of Preabelian type*, third version, July 15,

The last section contains a hodgepodge of questions and results, due to various people. We will try to give references in the main text. The main topic here is no longer the existence of integral canonical models per se. Instead, we discuss some results about the local structure of (examples of) such models, in relation to conjectures of Coleman and Oort.

There are some interesting related topics for which we did not find place in this article. Among the casualties are the recent work [RZ] of Rapoport and Zink, examples of bad reduction (see, e.g., [R2]), the Newton polygon stratification of A_g in characteristic p (for an overview, see [Oo1], [Oo2]) and the study of isocrystals with additional structure as in [Ko1], [Ko3], [RR].

Acknowledgements. In preparing this paper I benefited a lot from discussions with Y. André, D. Blasius, C. Deninger, B. Edixhoven, O. Gabber, J. de Jong, G. Kings, E. Landvogt, F. Oort, A. Vasiu, A. Werner and J. Wildeshaus. I thank them all cordially. Also I wish to thank the referee for several useful comments.

Notations. *Superscripts and subscripts:* ^{0} denotes connected components for the Zariski topology, $^+$ connected components for other (usually analytic) topologies. A superscript $^-$ (as in $G(\mathbb{Q})_+^-$ for example) denotes the closure of a subset of a topological space. If G is an algebraic group then ^{ad} (adjoint group), ^{ab} (maximal abelian quotient), ^{der} (derived group) have the usual meaning, $G(\mathbb{R})_+$ denotes the pre image of $G^{\text{ad}}(\mathbb{R})^+$ under the adjoint map, and in case G is defined over $\mathbb{Q}$ we write $G(\mathbb{Q})_+$ for the intersection of $G(\mathbb{Q})$ and $G(\mathbb{R})_+$ inside $G(\mathbb{R})$. For fields, ^{ab} denotes the maximal abelian extension. A superscript ^{p} usually denotes a structure "away from p"; a subscript $_p$ something "at p".

If (X, λ) is a g-dimensional principally polarized abelian scheme over a basis S then λ gives rise to a Weil pairing $e^\lambda \colon X[n] \times X[n] \to \mu_{n,S}$. Write $\psi_n \colon (\mathbb{Z}/n\mathbb{Z})^{2g} \times (\mathbb{Z}/n\mathbb{Z})^{2g} \to (\mathbb{Z}/n\mathbb{Z})$ for the standard symplectic form. By a Jacobi level n structure on (X, λ) we mean an isomorphism $\eta \colon X[n] \xrightarrow{\sim} (\mathbb{Z}/n\mathbb{Z})_S$ such that there exists an isomorphism $\alpha \colon (\mathbb{Z}/n\mathbb{Z})_S \xrightarrow{\sim} \mu_{n,S}$ with $\alpha \circ \psi_n \circ (\eta \times \eta) = e^\lambda$. We write $A_{g,1,n}$ for the (coarse) moduli scheme over

1997, UC at Berkeley, and *Ibid.*, December 1997, UC at Berkeley.) We have not yet had the opportunity to study this work in detail, and we therefore cannot say whether it can take away all doubts we have about the arguments in [Va2]. We strongly recommend the interested reader to consult Vasiu's original papers.

$\mathrm{Spec}(\mathbb{Z}[1/n])$ of principally polarized, g-dimensional abelian varieties with a Jacobi level n structure. If $n \geq 3$ then it is a fine moduli scheme.

Let $\mathbb{S} := \mathrm{Res}_{\mathbb{C}/\mathbb{R}} \mathbb{G}_{m,\mathbb{C}}$. We write $\mu \colon \mathbb{G}_{m,\mathbb{C}} \to \mathbb{S}_{\mathbb{C}}$ for the cocharacter which on complex points is given by $\mathbb{C}^* \ni z \mapsto (z,1) \in \mathbb{C}^* \times \mathbb{C}^* \cong (\mathbb{C} \otimes_{\mathbb{R}} \mathbb{C})^*$. The natural inclusion $w \colon \mathbb{G}_{m,\mathbb{R}} \to \mathbb{S}$ is called the weight cocharacter.

We write $\mathbb{A}_f$ for the ring of finite adèles of $\mathbb{Q}$ and $\mathbb{A}_L$ for the ring of (full) adèles of a number field L. We refer to [Pi], 0.6, for an explanation of when a subgroup $K \subseteq G(\mathbb{A}_f)$ (where G is an algebraic group over $\mathbb{Q}$) is called neat, and for some basic properties concerning this notion.

Abbreviations: H.S. for Hodge structure, V.H.S. for variation of Hodge structure, d.v.r. for discrete valuation ring, p.p.a.v. for principally polarized abelian variety, i.c.m. for integral canonical model (see 3.3), a.t.s. for admissible test scheme (see 3.5), e.e.p. for extended extension property (see 3.20).

§1 Shimura varieties

1.1 Recall ([De2]) that a pure Hodge structure of weight n with underlying $\mathbb{Q}$-vector space V is given by a homomorphism of algebraic groups $h \colon \mathbb{S} \to \mathrm{GL}(V)_{\mathbb{R}}$ such that the weight cocharacter $h \circ w \colon \mathbb{G}_m \to \mathrm{GL}(V)_{\mathbb{R}}$ maps z to $z^{-n} \cdot \mathrm{id}_V$. The Tate twist $\mathbb{Q}(1)$ corresponds to the norm character $\mathrm{Nm} \colon \mathbb{S} \to \mathbb{G}_{m,\mathbb{R}}$. An element $v \in V$ is called a Hodge class (in the strict sense) if v is purely of type $(0,0)$ in the Hodge decomposition $V_{\mathbb{C}} = \oplus V^{p,q}$. In other words: the Hodge classes are the *rational* classes $v \in V$ which, as elements of $V_{\mathbb{R}}$, are invariant under the action of $\mathbb{S}$ given by h.

The Mumford-Tate group $\mathrm{MT}(V)$ of V is defined as the smallest algebraic subgroup of $\mathrm{GL}(V) \times \mathbb{G}_m$ which is defined over $\mathbb{Q}$ and such that $h \times \mathrm{Nm} \colon \mathbb{S} \to \mathrm{GL}(V)_{\mathbb{R}} \times \mathbb{G}_{m,\mathbb{R}}$ factors through $\mathrm{MT}(V)_{\mathbb{R}}$. In Tannakian language $\mathrm{MT}(V)$ is the automorphism group of the forgetful fibre functor $\langle V, \mathbb{Q}(1) \rangle^{\otimes} \to \mathsf{Vec}_{\mathbb{Q}}$, where $\langle V, \mathbb{Q}(1) \rangle^{\otimes} \subset \mathsf{Hdg}_{\mathbb{Q}}$ is the Tannakian subcategory generated by V and $\mathbb{Q}(1)$. Concretely, this means that for every tensor space

$$V(r_1, r_2; s) := V^{\otimes r_1} \otimes (V^*)^{\otimes r_2} \otimes \mathbb{Q}(s),$$

the Hodge classes in $V(r_1, r_2; s)$ are precisely the invariants under the natural action of $\mathrm{MT}(V)$.

In more classical language one would define a Hodge class to be a rational class $v \in V$ which is purely of type $(n/2, n/2)$ in the Hodge decomposition. Clearly there are in general more Hodge classes in this sense than in the "strict" sense, but the difference is only a matter of weights. If we define the

Hodge group Hg(V) (sometimes called the special Mumford-Tate group) to be the kernel of the second projection map MT(V) $\twoheadrightarrow \mathbb{G}_m$, then the Hodge classes (in the more general sense) of a tensor space $V^{\otimes r_1} \otimes (V^*)^{\otimes r_2}$ are precisely the invariants of Hg(V). All in all, the Hodge group contains essentially the same information as the Mumford-Tate group, except that it does not keep track of the weight.

The main principle that we want to stress here is the following: if $h\colon \mathbb{S} \to$ GL(V)$_\mathbb{R}$ defines a Hodge structure on the $\mathbb{Q}$-vector space V, and if we are given tensors $t_1, \dots, t_k$ in spaces of the form $V(r_1, r_2; s)$, then there is an algebraic group $G \subseteq$ GL(V) (depending on the classes t_i) such that

$$t_1, \dots, t_k \quad \text{are Hodge classes} \quad \Longleftrightarrow \quad h \quad \text{factors through } G_\mathbb{R}.$$

1.2 To illustrate the usage of Mumford-Tate groups, let us discuss some examples pertaining to Hodge classes on abelian varieties. There are at least two reasons why abelian varieties are special:

(i) Riemann's theorem tells us that there is an equivalence of categories

$$\{\text{complex abelian varieties}\} \xrightarrow{\ \text{eq.}\ } \{\text{polarizable } \mathbb{Z}\text{-H.S. of type } (0,1) + (1,0)\},$$

sending X to H$^1(X, \mathbb{Z})$. (This should really be done covariantly, using H$_1$; as we shall later always work with cohomology we phrase everything in terms of H^1.) This result has some important variants, in that polarized abelian varieties are in equivalence with polarized $\mathbb{Z}$-H.S. of type $(0,1) + (1,0)$, abelian varieties up to isogeny correspond to polarizable $\mathbb{Q}$-H.S. of type $(0,1) + (1,0)$, and if S is a smooth variety over $\mathbb{C}$ then abelian schemes over S correspond to polarizable $\mathbb{Z}$-V.H.S. of type $(0,1) + (1,0)$ over S. (See [De2], section 4.4.) Furthermore, all cohomology of X and of its powers X^m, can be expressed directly in terms of H$^1(X, \mathbb{Z})$: we have natural isomorphisms of Hodge structures

$$\mathrm{H}^k(X^m, \mathbb{Z}) \cong \bigwedge^k \left(\oplus^m \mathrm{H}^1(X, \mathbb{Z}) \right).$$

(ii) Let $V := \mathrm{H}^1(X, \mathbb{Q})$, and write Hg($X$) := Hg($V$). Choose a polarization of X. The corresponding Riemann form φ is a Hodge class in Hom$\left(V^{\otimes 2}, \mathbb{Q}(-1)\right) = V(0,2;-1)$, hence it is invariant under Hg(X). This means that Hg(X) is contained in the symplectic group Sp(V, φ). Next we remark that, because of the above equivalence of categories,

$$\mathrm{End}(X) \otimes_\mathbb{Z} \mathbb{Q} =: \mathrm{End}^0(X) \cong \{\text{Hodge classes in End}(V)\} = \mathrm{End}(V)^{\mathrm{Hg}(X)}.$$

We conclude that $\mathrm{Hg}(X)$ is contained in the centralizer of $\mathrm{End}^0(X)$ inside $\mathrm{Sp}(V, \varphi)$, and that the commutant of $\mathrm{Hg}(X)$ in $\mathrm{End}(V)$ equals $\mathrm{End}^0(X)$. These observations become even more useful if we remark that for abelian varieties of a given dimension, the Albert classification (see [Mu2], section 21) gives a finite list of possible types for the endomorphism algebra $\mathrm{End}^0(X)$. When combined with other properties of the Hodge group, knowing $\mathrm{End}^0(X)$ is in some cases sufficient to determine $\mathrm{Hg}(X)$ and its action on V. This then enables us—at least in principle—to determine the Hodge ring of all powers of X. In general, however, the endomorphism algebra does not determine the Hodge group.

Example 1. Main references: [Ri], [Haz], [Ku], [Se2], [Ch]. Let X be a simple abelian variety of dimension 1 or of prime dimension. Then the Hodge group is equal to the centralizer of $\mathrm{End}^0(X)$ in $\mathrm{Sp}(V, \varphi)$. (This does not depend on the choice of the polarization.) The Hodge ring of every power of X is generated by divisor classes; in particular, the Hodge conjecture is true for all powers of X.

Example 2. Main references: [We], [MZ2]. Suppose k is an imaginary quadratic field, acting on X by endomorphisms. If σ and τ are the two complex embeddings of k, then $\mathrm{H}^0(X, \Omega_X^1) = V^{1,0}$ is a module over $k \otimes_{\mathbb{Q}} \mathbb{C} = \mathbb{C}^{(\sigma)} \times \mathbb{C}^{(\tau)}$, hence it decomposes as $V^{1,0} = V^{1,0}(\sigma) \oplus V^{1,0}(\tau)$. Suppose that the dimensions $n_\sigma = \dim V^{1,0}(\sigma)$ and $n_\tau = \dim V^{1,0}(\tau)$ are equal. This implies that $\dim(X)$ is even, say $\dim(X) = 2n$. The 1-dimensional k-vector space

$$W_k := \bigwedge_k^{2n} V$$

can be identified with a subspace of $\wedge_{\mathbb{Q}}^{2n} V = \mathrm{H}^{2n}(X, \mathbb{Q})$. Moreover, the condition that $n_\sigma = n_\tau$ implies that $W_k \subset \mathrm{H}^{2n}(X, \mathbb{Q})$ consists of Hodge classes. This construction was first studied by Weil in [We]; we call W_k the space of Weil classes with respect to k. Weil showed that for a generic abelian variety X with an action of k (subject to the condition $n_\sigma = n_\tau$), the nonzero classes in W_k are exceptional, i.e., they do not lie in the $\mathbb{Q}$-subalgebra $\mathcal{D}^{\cdot}(X) \subset \oplus \mathrm{H}^{2i}(X, \mathbb{Q})$ generated by the divisor classes.

The construction of Weil classes works in much greater generality. They play a role in Deligne's proof of "Hodge = absolute Hodge" for abelian varieties. In [MZ2] the space W_F of Weil classes w.r.t. the action of an arbitrary field $F \hookrightarrow \mathrm{End}^0(X)$ is studied. We find here criteria, purely in terms of F, $\mathrm{End}^0(X)$ and the action of F on the tangent space $V^{1,0}$, of when W_F contains Hodge classes, and of when these Hodge classes are exceptional.

Example 3. Main references: [Mu1], [MZ1], [Ta]. Let X be an abelian fourfold with $\text{End}^0(X) = \mathbb{Q}$. Then either $\text{Hg}(X)$ is the full symplectic group $\text{Sp}_{8,\mathbb{Q}}$, in which case the Hodge ring of every power of X is generated by divisor classes, or $\text{Hg}(X)$ is isogenous to a $\mathbb{Q}$-form of $\text{SL}_2 \times \text{SL}_2 \times \text{SL}_2$, where the representation V is the tensor product of the standard 2-dimensional representations of the three factors. Both possibilities occur. In the latter case, the Hodge ring of X is generated by divisor classes, but for X^2 this is no longer true: $\text{H}^4(X^2, \mathbb{Q})$ contains exceptional classes. These are not of the same kind as in example 2, i.e., they are not Weil classes with respect to the action of a field on X^2.

In case X is defined over a number field, we have "the same" two possibilities for the image of the Galois group acting on the Tate module. In particular, knowing $\text{End}^0(X)$ here is not sufficient to prove the Mumford-Tate conjecture. Known by many as "the Mumford example", this is actually the lowest dimensional case where the Mumford-Tate conjecture for abelian varieties remains, at present, completely open. Mumford's example can be generalized to abelian varieties of dimension 4^k, see [Ta].

1.3 Guided by the considerations in 1.1, we can make sense of the problem to study Hodge structures with "a given collection of Hodge classes". How one translates this in purely group-theoretical terms is explained with great clarity in [De3], especially section 1.1. Here we summarize the most important points.

Fix an algebraic group $G_{\mathbb{R}}$ over $\mathbb{R}$, and consider the space $\text{Hom}(\mathbb{S}, G_{\mathbb{R}})$ of homomorphisms of algebraic groups $h: \mathbb{S} \to G_{\mathbb{R}}$. Its connected components are the $G(\mathbb{R})^+$-conjugacy classes. Given one such component X^+, and fixing a representation $\rho_{\mathbb{R}}: G_{\mathbb{R}} \to \text{GL}(V_{\mathbb{R}})$, we obtain a family of $\mathbb{R}$-Hodge structures on $V_{\mathbb{R}}$, parametrized by X^+. From an algebro-geometric point of view, the natural conditions to impose on this family are:

(a) the weight decomposition $V_{\mathbb{R}} = \oplus_{n \in \mathbb{Z}} V_{\mathbb{R}}^n$ does not depend on $h \in X^+$,

(b) there is a complex structure on X^+ such that the family of Hodge structures on each $V_{\mathbb{R}}^n$ is a polarizable $\mathbb{R}$-V.H.S. over X^+.

Now an important fact is that (a) and (b) can be expressed directly in terms of $G_{\mathbb{R}}$ and X^+, and that, at least for faithful representations, they do not depend on $\rho_{\mathbb{R}}$. If (a) and (b) are satisfied for some (equivalently: every) faithful representation $\rho_{\mathbb{R}}$, then the complex structure in (b) is unique and X^+ is a hermitian symmetric domain. (For all this, see [De3], 1.13–17.) By adding a $\mathbb{Q}$-structure on $G_{\mathbb{R}}$, one is led to the following definition.

1.4 Definition. A Shimura datum is a pair (G, X) consisting of a connected reductive group G over $\mathbb{Q}$, and a $G(\mathbb{R})$-conjugacy class $X \subset \operatorname{Hom}(\mathbb{S}, G_{\mathbb{R}})$, such that for all (equivalently: for some) $h \in X$,

(i) the Hodge structure on $\operatorname{Lie}(G)$ defined by $\operatorname{Ad} \circ h$ is of type $(-1, 1) + (0, 0) + (1, -1)$,

(ii) the involution $\operatorname{Inn}\big(h(i)\big)$ is a Cartan involution of $G_{\mathbb{R}}^{\mathrm{ad}}$,

(iii) the adjoint group G^{ad} does not have factors defined over $\mathbb{Q}$ onto which h has a trivial projection.

In this definition we have followed [De3], section 2.1. Pink has suggested (cf. [Pi]) to allow not only $G(\mathbb{R})$-conjugacy classes $X \subset \operatorname{Hom}(\mathbb{S}, G_{\mathbb{R}})$ but also finite coverings of such. We will not use this generalization in this paper.

There are some other conditions that sometimes play a role. For instance, condition (i) implies that the weight cocharacter $h \circ w \colon \mathbb{G}_{m,\mathbb{C}} \to G_{\mathbb{C}}$ (for which we sometimes simply write w) does not depend on $h \in X$, and one could require that it is defined over $\mathbb{Q}$. It turns out, however, that the theory works well without this assumption, and that there are rather natural examples where it is not satisfied.

1.5 Let (G, X) be a Shimura datum, and let K be a compact open subgroup of $G(\mathbb{A}_f)$. We set

$$Sh_K(G, X)_{\mathbb{C}} = G(\mathbb{Q}) \backslash X \times G(\mathbb{A}_f)/K,$$

where $G(\mathbb{Q})$ acts diagonally on $X \times \big(G(\mathbb{A}_f)/K\big)$. If $X^+ \subseteq X$ is a connected component, and if $g_1, \ldots, g_m$ are representatives in $G(\mathbb{A}_f)$ for the finite set $G(\mathbb{Q})_+ \backslash G(\mathbb{A}_f)/K$, then we can rewrite $Sh_K(G, X)_{\mathbb{C}}$ as a disjoint sum

$$Sh_K(G, X)_{\mathbb{C}} = \coprod_{i=1,\ldots,m} \Gamma_i \backslash X^+,$$

where Γ_i is the image of $G(\mathbb{Q})_+ \cap g_i K g_i^{-1}$ inside $G^{\mathrm{ad}}(\mathbb{Q})^+$, which is an arithmetic subgroup. By the results of Baily and Borel in [BB], the quotients $\Gamma_i \backslash X^+$ have a natural structure of a quasi-projective algebraic variety. For compact open subgroups $K_1 \subseteq K_2$, the natural map

$$Sh(K_1, K_2) \colon Sh_{K_1}(G, X)_{\mathbb{C}} \longrightarrow Sh_{K_2}(G, X)_{\mathbb{C}}$$

is algebraic. We thus obtain a projective system of (generally non-connected) algebraic varieties $Sh_K(G, X)_{\mathbb{C}}$, indexed by the compact open subgroups $K \subset G(\mathbb{A}_f)$. This system, or its limit

$$Sh(G, X)_{\mathbb{C}} = \varprojlim_{K} Sh_K(G, X)_{\mathbb{C}},$$

(which exists as a scheme, since the transition maps are finite) is called the Shimura variety defined by the datum (G, X).

1.6   We will briefly recall some basic definitions and results. For further discussion of these topics, see [De1], [De3], [Mi2].

1.6.1   The group $G(\mathbb{A}_f)$ acts continuously on $Sh(G, X)_{\mathbb{C}}$ from the right. The continuity here means that the action of an element $g \in G(\mathbb{A}_f)$ is obtained as the limit of a system of isomorphism $\cdot g \colon Sh_K(G, X)_{\mathbb{C}} \xrightarrow{\sim} Sh_{g^{-1}Kg}(G, X)_{\mathbb{C}}$, see [De3], 2.1.4 and 2.7, or [Mi2], II.2 and II.10. On "finite levels", the $G(\mathbb{A}_f)$-action gives rise to Hecke correspondences: for compact open subgroups K_1, $K_2 \subset G(\mathbb{A}_f)$ and $g \in G(\mathbb{A}_f)$, set $K' = K_1 \cap gK_2g^{-1}$; then the Hecke correspondence T_g from $Sh_{K_1}(G, X)_{\mathbb{C}}$ to $Sh_{K_2}(G, X)_{\mathbb{C}}$ is given by

$$T_g \colon \quad Sh_{K_1}(G, X)_{\mathbb{C}} \xleftarrow{\;Sh(K', K_1)\;} Sh_{K'}(G, X)_{\mathbb{C}} \xrightarrow{\;\cdot g\;} Sh_{K_2}(G, X)_{\mathbb{C}}.$$

1.6.2   A morphism of Shimura data $f \colon (G_1, X_1) \to (G_2, X_2)$ is given by a homomorphism of algebraic groups $f \colon G_1 \to G_2$ defined over $\mathbb{Q}$ which induces a map from X_1 to X_2. Such a morphism induces a morphism of schemes

$$Sh(f) \colon Sh(G_1, X_1)_{\mathbb{C}} \longrightarrow Sh(G_2, X_2)_{\mathbb{C}}.$$

If $f \colon G_1 \to G_2$ is a closed immersion then so is $Sh(f)$. (See [De1], 1.14–15.)

1.6.3   Let (G, X) be a Shimura datum. Associated to $h \in X$, we have a cocharacter $h \circ \mu \colon \mathbb{G}_{m, \mathbb{C}} \to G_{\mathbb{C}}$, whose $G(\mathbb{C})$-conjugacy class is independent of $h \in X$. The reflex field $E(G, X) \subset \mathbb{C}$ is defined as the field of definition of this conjugacy class. It is a finite extension of $\mathbb{Q}$. If $f \colon (G_1, X_1) \to (G_2, X_2)$ is a morphism of Shimura data, then $E(G_2, X_2) \subseteq E(G_1, X_1) \subset \mathbb{C}$.

1.6.4   A point $h \in X$ is called a special point if there is a torus $T \subseteq G$, defined over $\mathbb{Q}$, such that $h \colon \mathbb{S} \to G_{\mathbb{R}}$ factors through $T_{\mathbb{R}}$. In this case $(T, \{h\})$ is a Shimura datum, and the inclusion $T \hookrightarrow G$ gives a morphism $(T, \{h\}) \to (G, X)$. A point $x \in Sh_K(G, X)_{\mathbb{C}}$ is called a special point if it is of the form $x = [h, gK]$ with h special. (This does not depend on the choice of the representative (h, gK) for x.) Here we follow [De3]; the definition in [De1], 3.15, is more restrictive.

1.6.5   Consider a triplet $(G^{\mathrm{ad}}, G', X^+)$ consisting of an adjoint group G^{ad} over $\mathbb{Q}$, a covering G' of G^{ad}, and a $G^{\mathrm{ad}}(\mathbb{R})^+$-conjugacy class $X^+ \subset \mathrm{Hom}(\mathbb{S}, G^{\mathrm{ad}}_{\mathbb{R}})$

such that the conditions (i), (ii) and (iii) in 1.4 are satisfied. Let $\tau(G')$ be the linear topology on $G^{\mathrm{ad}}(\mathbb{Q})$ for which the images in $G^{\mathrm{ad}}(\mathbb{Q})$ of the congruence subgroups in $G'(\mathbb{Q})$ form a fundamental system of neighbourhoods of the identity. The connected Shimura variety $Sh^0(G^{\mathrm{ad}}, G', X^+)_{\mathbb{C}}$ is defined as the projective system

$$Sh^0(G^{\mathrm{ad}}, G', X^+)_{\mathbb{C}} = \varprojlim_{\Gamma} \Gamma\backslash X^+,$$

where Γ runs through the arithmetic subgroups of $G^{\mathrm{ad}}(\mathbb{Q})$ which are open in $\tau(G')$. It comes equipped with an action of the completion $G^{\mathrm{ad}}(\mathbb{Q})^{+\wedge}$ of $G^{\mathrm{ad}}(\mathbb{Q})^+$ for the topology $\tau(G')$.

Given a Shimura datum (G, X) and a connected component $X^+ \subseteq X$, we obtain a triplet $(G^{\mathrm{ad}}, G^{\mathrm{der}}, X^+)$ as above. The associated connected Shimura variety $Sh^0(G^{\mathrm{ad}}, G^{\mathrm{der}}, X^+)_{\mathbb{C}}$ is the connected component of $Sh(G, X)_{\mathbb{C}}$ containing the image of $X^+ \times \{e\} \subset X \times G(\mathbb{A}_f)$. In particular, we see that this component only depends on G^{ad}, G^{der} and $X^+ \subset X$. In the sequel, when working with connected Shimura varieties, we will usually omit G^{ad} from the notation. For lack of better terminology, we will refer to a pair (G', X^+) as above as "a pair defining a connected Shimura variety".

1.6.6 Let G be a reductive group over a number field L. Write $\rho\colon \widetilde{G} \to G^{\mathrm{der}}$ for the universal covering (in the sense of algebraic groups) of its derived group. By [Del], Prop. 2.2 and [De3], Cor. 2.0.8, $G(L) \cdot \rho\widetilde{G}(\mathbb{A}_L)$ is a closed subgroup of $G(\mathbb{A}_L)$ with abelian quotient $\pi(G) := G(\mathbb{A}_L)/G(L) \cdot \rho\widetilde{G}(\mathbb{A}_L)$. (Note: $\mathbb{A}_L$ is the ring of full adèles of L.) Consequently, the set of connected components $\pi_0\pi(G)$ is also an abelian group.

Now let (G, X) be a Shimura datum. If $K \subset G(\mathbb{A}_f)$ is a compact open subgroup then $Sh_K(G, X)_{\mathbb{C}}$ is a scheme of finite type over $\mathbb{C}$. For K getting smaller, its number of connected components generally increases. Deligne proves in [De3], 2.1.3 that

$$\pi_0\big(Sh_K(G, X)_{\mathbb{C}}\big) \cong G(\mathbb{A}_f)/G(\mathbb{Q})_+ \cdot K \cong \overline{\pi}_0\pi(G)/K,$$

where $\overline{\pi}_0\pi(G) := \pi_0\pi(G)/\pi_0G(\mathbb{R})_+$. Passing to the limit one finds that the $G(\mathbb{A}_f)$-action on $Sh(G, X)_{\mathbb{C}}$ makes $\pi_0\big(Sh(G, X)_{\mathbb{C}}\big)$ a principal homogeneous space under $\overline{\pi}_0\pi(G) \cong G(\mathbb{A}_f)/G(\mathbb{Q})_+^-$.

1.6.7 Given a Shimura datum (G, X), we can define some other data as follows. Write $X^{\mathrm{ad}} \subset \mathrm{Hom}(\mathbb{S}, G^{\mathrm{ad}}_{\mathbb{R}})$ for the $G^{\mathrm{ad}}(\mathbb{R})$-conjugacy class containing the image of X under the map $\mathrm{Hom}(\mathbb{S}, G_{\mathbb{R}}) \to \mathrm{Hom}(\mathbb{S}, G^{\mathrm{ad}}_{\mathbb{R}})$. The map

$X \to X^{\mathrm{ad}}$ is not necessarily an isomorphism, but every connected component of X maps isomorphically to its image. The pair $(G^{\mathrm{ad}}, X^{\mathrm{ad}})$ is a Shimura datum, called the adjoint Shimura datum. Similarly, the image X^{ab} of X in $\mathrm{Hom}(\mathbb{S}, G^{\mathrm{ab}}_{\mathbb{R}})$ is a $G^{\mathrm{ab}}(\mathbb{R})$-conjugacy class (necessarily a single point), and we have a Shimura datum $(G^{\mathrm{ab}}, X^{\mathrm{ab}})$.

Another construction that is sometimes useful is the following. Suppose the group G is of the form $G = \mathrm{Res}_{F/\mathbb{Q}}(H)$, where F is a totally real number field and H is an absolutely simple algebraic group over F. (Such is the case, for example, if G is a simple adjoint group.) Now take an extension $F \subset F'$ of totally real number fields, and set $G_2 = \mathrm{Res}_{F'/\mathbb{Q}}(H_{F'})$. There is a unique $G_2(\mathbb{R})$-conjugacy class $X_2 \subset \mathrm{Hom}(\mathbb{S}, G_{2,\mathbb{R}})$ such that the natural homomorphism $G \to G_2$ gives a closed immersion of Shimura data $(G, X) \hookrightarrow (G_2, X_2)$.

1.7 One might ask "how many" Shimura varieties there are. A possible approach is to begin by classifying the Shimura varieties of adjoint type. These are products of Shimura varieties $Sh(G, X)_{\mathbb{C}}$, where G is a $\mathbb{Q}$-simple adjoint group. The group $G_{\mathbb{R}}$ is an inner form of a compact group, of one of the types A, B, C, $\mathrm{D}^{\mathbb{R}}$, $\mathrm{D}^{\mathbb{H}}$, E_6 or E_7, and given $G_{\mathbb{R}}$, the possibilities for X are classified in terms of special nodes in the Dynkin diagram. We refer to [De3], sections 1.2 and 2.3 for more details.

Given $(G^{\mathrm{ad}}, X^{\mathrm{ad}})$, we can list all possibilities for G^{der}. As we have seen, the pair (G^{der}, X^+) consisting of G^{der} and a connected component $X^+ \subseteq X$, determines the connected components of the Shimura variety. In particular, the "toric part" $(G^{\mathrm{ab}}, X^{\mathrm{ab}})$ does not contribute to the geometry of $Sh(G, X)_{\mathbb{C}}$, in the sense that it has no effect on $Sh^0_{\mathbb{C}}$, but only on $\pi_0(Sh(G, X)_{\mathbb{C}})$. Finally, let us remark that "toric" Shimura data are in bijective correspondence to pairs (Y, μ) consisting of a free $\mathbb{Z}$-module Y of finite rank with a continuous action of $\mathrm{Gal}(\overline{\mathbb{Q}}/\mathbb{Q})$ (the cocharacter group of the torus), together with an element $\mu \in Y$.

1.8 The definition of a Shimura variety is set up in such a way that that if $\xi \colon G_{\mathbb{R}} \to \mathrm{GL}(V_{\mathbb{R}})$ is a representation, then we obtain a (direct sum of) polarizable $\mathbb{R}$-VHS $\mathcal{V}(\xi)_{\mathbb{R}}$ over X with underlying local system $X \times V_{\mathbb{R}}$. If $V_{\mathbb{R}} = V \otimes_{\mathbb{Q}} \mathbb{R}$ for a $\mathbb{Q}$-vector space V, and if the weight $\xi \circ w \colon \mathbb{G}_{m,\mathbb{R}} \to \mathrm{GL}(V)_{\mathbb{R}}$ is defined over $\mathbb{Q}$, then $\mathcal{V}(\xi)_{\mathbb{R}}$ comes from a polarizable $\mathbb{Q}$-VHS $\mathcal{V}(\xi)$. Under some conditions on $G/\mathrm{Ker}(\xi)$, this VHS descends, for K sufficiently small, to a $\mathbb{Q}$-VHS on $Sh_K(G, X)$. (It suffices if the center of $G/\mathrm{Ker}(\xi)$ is the almost direct product of a $\mathbb{Q}$-split torus and a torus T for which $T(\mathbb{R})$ is compact.)

One expects (see [De5], [Mi4]) that these variations of Hodge structure are the Betti realizations of families of motives, and that Shimura varieties, at least those for which the weight is defined over $\mathbb{Q}$, have an interpretation (depending on the choice of a representation ξ) as moduli spaces for motives with certain additional structures. What is missing, at present, is a sufficiently good theory of motives. In certain cases, however, the dictionary between abelian varieties and certain Hodge structures (see 1.2 above) leads to a modular interpretation of $Sh(G, X)$. Let us briefly review some facts and terminology.

1.8.1 *Siegel modular varieties.* Let φ denote the standard symplectic form on $\mathbb{Q}^{2g}$, and set $G = \mathrm{CSp}_{2g}$. The homomorphisms $h\colon \mathbb{S} \to G_\mathbb{R}$ which determine a $\mathbb{Q}$-H.S. of type $(-1,0)+(0,-1)$ on $\mathbb{Q}^{2g}$ such that $\pm\varphi$ is a polarization, form a single $G(\mathbb{R})$-conjugacy class $\mathfrak{H}_g^\pm$. It can be identified with the Siegel double-$\frac{1}{2}$-space. The pair $(\mathrm{CSp}_{2g}, \mathfrak{H}_g^\pm)$ is a Shimura datum with reflex field $\mathbb{Q}$. The associated Shimura variety is often referred to as the Siegel modular variety.

For $K \subset G(\mathbb{A}_f)$ a compact open subgroup, $Sh_K(\mathrm{CSp}_{2g}, \mathfrak{H}_g^\pm)_\mathbb{C}$ is a moduli space for g-dimensional complex p.p.a.v. with a level K-structure (as defined, for instance, in [Ko2], §5). Here a couple of remarks should be added. The interpretation of $Sh_K(\mathrm{CSp}_{2g}, \mathfrak{H}_g^\pm)_\mathbb{C}$ in terms of abelian varieties up to *isomorphism* depends on the choice of a lattice $\Lambda \subset \mathbb{Q}^{2g}$. This choice also determines the "type" of the polarization; if we want to work with principally polarized abelian varieties then we must choose Λ such that $\varphi_{|\Lambda}$ has discriminant 1 (e.g., $\Lambda = \mathbb{Z}^{2g}$). For further details see [De1], §4. In the sequel, we identify $Sh(\mathrm{CSp}_{2g}, \mathfrak{H}_g^\pm)_\mathbb{C}$ and $\varprojlim_n A_{g,1,n} \otimes \mathbb{C}$.

1.8.2 *Shimura varieties of PEL and of Hodge type.* By definition, a Shimura datum (G, X) (as well as the associated Shimura variety) is said to be of Hodge type, if there exists a closed immersion of Shimura data $j\colon (G, X) \hookrightarrow (\mathrm{CSp}_{2g}, \mathfrak{H}_g^\pm)$ for some g. If this holds, the Shimura variety $Sh(G, X)_\mathbb{C} \hookrightarrow Sh(\mathrm{CSp}_{2g}, \mathfrak{H}_g^\pm)_\mathbb{C}$ has an interpretation in terms of abelian varieties with certain "given Hodge classes". The precise formulation of such a modular interpretation is usually rather complicated.

This is already the case for Shimura varieties of PEL type (see [De1], 4.9–14, [Ko2]). Loosely speaking, these are the Shimura varieties parametrizing

abelian varieties with a given algebra acting by endomorphisms.[2] Recall (1.2)
that endomorphisms of an abelian variety are particular examples of Hodge
classes. On finite levels one thus looks at abelian varieties with a *Polarization*,
certain given *Endomorphisms*, and a *Level* structure.

Shimura varieties of PEL type are more special in that they represent a
moduli problem that can be formulated over an arbitrary basis. For more
general Shimura varieties of Hodge type we can only do this if we assume the
Hodge conjecture.

In 2.10, we will introduce two more classes of Shimura varieties: those of
abelian and of pre-abelian type. Among these classes we have the following
inclusions

$$
\left(\begin{array}{c} \text{Sh. var. of} \\ \text{PEL type} \end{array} \right) \subset \left(\begin{array}{c} \text{Sh. var. of} \\ \text{Hodge type} \end{array} \right) \subset \left(\begin{array}{c} \text{Sh. var. of} \\ \text{abelian type} \end{array} \right) \subset
$$

$$
\subset \left(\begin{array}{c} \text{Sh. var. of} \\ \text{pre-ab. type} \end{array} \right) \subset \left(\begin{array}{c} \text{general} \\ \text{Sh. var.} \end{array} \right)
$$

All inclusions are strict; for the first one see 1.2. (A priory, the Shimura
variety corresponding to a "Mumford example" could have a different realization for which it is of PEL type. By looking at the group involved over $\mathbb{R}$,
one easily shows that this does not happen.)

1.9 *Compactifications; mixed Shimura varieties.* This is a whole subject in
itself, and we cannot say much about it here. We will briefly indicate some
important statements, referring to the literature for details.

The first compactification to mention is the Baily-Borel (or minimal) compactification, for which we write $Sh_K(G, X)^*_{\mathbb{C}}$. (References: [BB], see also [Br],
§4 for a summary.) It was constructed by Baily and Borel in the setting of
locally symmetric varieties. If $\Gamma \backslash X^+$ is a component of $Sh_K(G, X)_{\mathbb{C}}$, say with
K neat so that $\Gamma \backslash X^+$ is non-singular, then its Baily-Borel compactification
is given as a quotient $\Gamma \backslash X^*$. Here X^* is the Satake compactification of X^+;
as a set it is the union of X^+ and its (proper) rational boundary components,
which themselves are again hermitian symmetric domains. It is shown in
[BB] that $\Gamma \backslash X^*$ has a natural structure of a normal projective variety. The

---

[2]For the reader who has not worked with Shimura varieties before, it may be instructive
to read Shimura's paper [Sh]. Here certain Shimura varieties of PEL type are written
down "by hand". Both for understanding Shimura's paper and for understanding the
abstract Deligne-formalism we are presenting here, it is a good exercise to translate the
considerations of [Sh] to the "(G, X) language".

stratification of X^* by its boundary components F induces a stratification of $\Gamma\backslash X^*$ by locally symmetric varieties $\Gamma_{\overline{F}}\backslash F$.

As the referee pointed out to us, it is worth noticing that the Baily-Borel compactification $Sh_K(G,X)^*_{\mathbb{C}}$ is not simply defined as the disjoint union of the $\Gamma\backslash X^*$. Instead, one starts with a Satake compactification of the whole of X at once, and one defines $Sh_K(G,X)^*_{\mathbb{C}}$ as a suitable quotient. Thus, for example, if $(G,X) = (\mathrm{GL}_2, \mathfrak{H}^{\pm})$, one does not adjoin the points of $\mathbb{P}^1(\mathbb{Q})$ to $\mathfrak{H}^+$ and $\mathfrak{H}^-$ separately but one works with $\mathbb{P}^1(\mathbb{Q})\cup\mathfrak{H}^+\cup\mathfrak{H}^-$. We refer to [Pi], Chap. 6 for further details.

The Baily-Borel compactification is canonical. In particular, it is easy to show (see [Br], p. 90) that it descends to a compactification of the canonical model $Sh_K(G,X)$ (to be discussed in the next section). In general, $Sh_K(G,X)^*$ is singular along the boundary.

Next we have the toroidal compactifications[3] studied in the monograph [Aea]. These are no longer canonical, as they depend on the choice of a certain cone decomposition. We will reflect this in our notation, writing $Sh_K(G,X;\mathcal{S})_{\mathbb{C}}$ for the toroidal compactification corresponding to a K-admissible partial cone decomposition $\mathcal{S}$ as in [Pi], Chap. 6. From the construction, we obtain a natural stratification of the boundary. For suitable choices of $\mathcal{S}$ (and K neat), one obtains a projective non-singular scheme $Sh_K(G,X;\mathcal{S})_{\mathbb{C}}$ such that the boundary is a normal crossings divisor—in this case one speaks of a smooth toroidal compactification.

Although both the Baily-Borel and the toroidal compactifications were initially studied in the setting of locally symmetric varieties, it was realized that they should be tied up with the theory of degenerating Hodge structures (e.g., see [Aea], p. iv). For certain Shimura varieties this was done by Brylinski in [Br], using 1-motives. Subsequently, Pink developed a general theory of mixed Shimura varieties and studied compactifications in this setting. Similar ideas, but in a less complete form, were presented by Milne in [Mi2]. It seems that several important ideas can actually be traced back to Deligne.

The main results of [Pi] include the following statements (some of which had been known before for pure Shimura varieties or some special mixed Shimura varieties). We refer to loc. cit. for definitions, more precise statements and of course for the proofs.

(i) Let $Sh_K(G,X)_{\mathbb{C}}$ be a pure Shimura variety. It has a canonical model

---

[3]Here we indulge in the customary abuse of terminology to call these compactifications, even though $Sh_K(G,X;\mathcal{S})$ is compact only if the cone decomposition $\mathcal{S}$ satisfies some conditions.

$Sh_K(G,X)$ over the reflex field $E(G,X)$ (see §2 below). The Baily-Borel compactification descends to a compactification $Sh_K(G,X)^*$ of this canonical model. The boundary has a stratification by finite quotients of (canonical models of) certain pure Shimura varieties; each such stratum is a finite union of the natural boundary components in the Baily-Borel compactification. Which pure Shimura varieties occur in this way can be described directly in terms of the Shimura datum (G,X).

(ii) Next consider K-admissible cone decompositions $\mathcal{S}$ for (P,X). If $\mathcal{S}$ satisfies certain conditions (such $\mathcal{S}$ always exists if K is neat) then the following assertions hold. The toroidal compactification $Sh_K(P,X;\mathcal{S})_\mathbb{C}$ descends to a compactification $Sh_K(P,X;\mathcal{S})$ of the canonical model. It is a smooth projective scheme, and the boundary is a normal crossings divisor. The boundary has a stratification by finite quotients of (canonical models of) certain other mixed Shimura varieties; each such stratum is a finite union of the strata of $Sh_K(P,X;\mathcal{S})$ as a toroidal compactification. The natural morphism $\pi\colon Sh_K(P,X;\mathcal{S}) \to Sh_K(P,X)^*$ is compatible with the stratifications. If $Sh_K(P',X')$ is a mixed Shimura variety of which a finite quotient occurs as a boundary stratum $\mathcal{C} \subset Sh_K(P,X;\mathcal{S})$, then the restriction of π to $\mathcal{C}$ is induced by the canonical morphism of $Sh_K(P',X')$ to the associated pure Shimura variety.

Furthermore, Pink proves several results about the functoriality of the structures in (i) and (ii).

To conclude this section, let us remark that in some cases (modular curves: Deligne and Rapoport, [DR]; Hilbert modular surfaces: Rapoport, [R1]; Siegel modular varieties: Chai and Faltings, [FC]) we even have smooth compactifications of Shimura varieties over (an open part of) Spec($\mathbb{Z}$) or the ring of integers of a number field. As Chai and Faltings remark in the introduction to [FC], many of their ideas also apply to Shimura varieties of PEL type; they conclude: "...and as our ideas usually either carry over directly, or we are lead to hard new problems which require new methods, we leave these generalizations to the reader."

§2 Canonical models of Shimura varieties.

2.1 Before turning to more recent developments, we will discuss some aspects of the theory of canonical models of Shimura varieties (over number fields). Our motivation for doing so is twofold.

(i) For "most" Shimura varieties, the existence of a canonical model was shown by Deligne in his paper [De3]. As we will see, the same strategy of proof is useful in the context of integral canonical models.

(ii) The existence of canonical models in general, i.e., including the cases where the group G has factors of exceptional type, was claimed in [Mi1] (see also [La], [Bo2], [MS], [Mi2]) as a consequence of the Langlands conjecture on conjugation of Shimura varieties. It was pointed out to us by J. Wildeshaus that the argument given there is not complete. Below we will explain this in more detail, and we correct the proof.

2.2 Recall (1.6.6) that for a reductive group G over a global field K of characteristic 0 we have set $\pi(G) = G(\mathbb{A}_K)/G(K) \cdot \rho\widetilde{G}(\mathbb{A}_K)$. We have the following constructions, for which we refer to [De3], section 2.4.

(a) Given a finite field extension $K \subset L$, there is a norm homomorphism $\mathrm{Nm}_{L/K} \colon \pi(G_L) \longrightarrow \pi(G)$.

(b) If T is a torus over K and M is a $G(\overline{K})$-conjugacy class of homomorphisms $T_{\overline{K}} \to G_{\overline{K}}$ which is defined over K, then there is associated to M a homomorphism $q_M \colon \pi(T) \longrightarrow \pi(G)$.

If (G, X) is a Shimura datum with reflex field $E = E(G, X) \subset \mathbb{C}$, we use this to define a reciprocity homomorphism

$$r_{(G,X)} \colon \mathrm{Gal}(\overline{\mathbb{Q}}/E) \longrightarrow \overline{\pi}_0\pi(G) = G(\mathbb{A}_f)/G(\mathbb{Q})_+^-$$

as follows. Global class field theory provides us with an isomorphism

$$(2.2.1) \qquad\qquad \mathrm{Gal}(\overline{\mathbb{Q}}/E)^{\mathrm{ab}} \xrightarrow{\ \sim\ } \pi_0\pi(\mathbb{G}_{\mathrm{m},E}) \,.$$

Applying (b) to the conjugacy class $M = \{h \circ \mu \colon \mathbb{G}_{\mathrm{m},\mathbb{C}} \to G_{\mathbb{C}} \mid h \in X\}$, which (by definition) is defined over E, we obtain a map $q_M \colon \pi(\mathbb{G}_{\mathrm{m},E}) \longrightarrow \pi(G_E)$. From (a) we get $\mathrm{Nm}_{E/\mathbb{Q}} \colon \pi(G_E) \longrightarrow \pi(G)$. Combining these maps we can now define the reciprocity map as

$(2.2.2)$

$$r_{(G,X)} \colon \mathrm{Gal}(\overline{\mathbb{Q}}/E) \twoheadrightarrow \mathrm{Gal}(\overline{\mathbb{Q}}/E)^{\mathrm{ab}} \xrightarrow{(2.2.1)} \pi_0\pi(\mathbb{G}_{\mathrm{m},E})$$

$$\xrightarrow{\pi_0(\mathrm{Nm}_{E/\mathbb{Q}} \circ q_M)} \pi_0\pi(G) \twoheadrightarrow \overline{\pi}_0\pi(G) \,.$$

For a Shimura datum $(T, \{h\})$ where T is a torus, the reciprocity map can be described more explicitly: if v is a place of E dividing p then

$$r_{(T,\{h\})} \colon \mathrm{Gal}(\overline{\mathbb{Q}}/E) \longrightarrow \overline{\pi}_0\pi(T) = T(\mathbb{A}_f)/T(\mathbb{Q})^-$$

sends a geometric Frobenius element $\Phi_v \in \mathrm{Gal}(\overline{\mathbb{Q}}/E)$ at v to the class of the element $\mathrm{Nm}_{E/\mathbb{Q}}\big(h(\pi_v)\big) \in T(\mathbb{Q}_p) \hookrightarrow T(\mathbb{A}_f)$, where π_v is a uniformizer at v.

2.3 If $(T, \{h\})$ is a Shimura datum with T a torus, then for every compact open subgroup $K \subset T(\mathbb{A}_f)$ the Shimura variety $Sh_K(T, \{h\})_{\mathbb{C}}$ consists of finitely many points. To define a model of it over the reflex field $E = E(T, \{h\})$ it therefore suffices to specify an action of $\mathrm{Gal}(\overline{\mathbb{Q}}/E)$. Write $Sh_K(T, \{h\})$ for the model over E determined by the rule that $\sigma \in \mathrm{Gal}(\overline{\mathbb{Q}}/E)$ acts on $Sh_K(T, \{h\})_{\mathbb{C}}$ by sending $[h, tK]$ to $[h, r_{(T, \{h\})}(\sigma) \cdot tK]$. It is clear that the transition maps $Sh_{(K', K)}$ descend to E, and we define the canonical model of $Sh(T, \{h\})_{\mathbb{C}}$ to be

$$Sh(T, \{h\}) = \varprojlim_K Sh_K(T, \{h\}) \,.$$

2.4 Definition. Let (G, X) be a Shimura datum.

(i) A model of $Sh(G, X)_{\mathbb{C}}$ over a field $F \subset \mathbb{C}$ is a scheme S over F together with a continuous action of $G(\mathbb{A}_f)$ from the right and a $G(\mathbb{A}_f)$-equivariant isomorphism $S \otimes_F \mathbb{C} \xrightarrow{\sim} Sh(G, X)_{\mathbb{C}}$.

(ii) Let $F \subset \mathbb{C}$ be a field containing $E(G, X)$. A weakly canonical model of $Sh(G, X)$ over F is a model S over F such that for every closed immersion of Shimura data $i \colon (T, \{h\}) \hookrightarrow (G, X)$ with T a torus, the induced morphism $Sh(T, \{h\})_{\mathbb{C}} \hookrightarrow Sh(G, X)_{\mathbb{C}} \cong S_{\mathbb{C}}$ descends to a morphism $Sh(T, \{h\}) \otimes_E EF \hookrightarrow S \otimes_F EF$, where $E = E(T, \{h\})$, and where $Sh(T, \{h\})$ is the model defined in 2.3.

(iii) A canonical model of $Sh(G, X)$ is a weakly canonical model over the reflex field $E(G, X)$.

It should be noticed that if S is a model of $Sh(G, X)$ over the field $F \subset \mathbb{C}$, then we have an action of $G(\mathbb{A}_f) \times \mathrm{Gal}(\overline{F}/F)$ on $S_{\overline{F}}$ (i.e., two commuting actions of $G(\mathbb{A}_f)$ and $\mathrm{Gal}(\overline{F}/F)$.)

2.5 Let $f \colon (G_1, X_1) \to (G_2, X_2)$ be a morphism of Shimura data, and suppose there exist canonical models $Sh(G_1, X_1)$ and $Sh(G_2, X_2)$. Then, as shown in [Del], section 5, the morphism $Sh(f)$ descends uniquely to a morphism $Sh(G_1, X_1) \to Sh(G_2, X_2) \otimes_{E(G_2, X_2)} E(G_1, X_1)$, which we will also denote $Sh(f)$. In particular, it follows that a canonical model, if it exists, is unique up to isomorphism. (The isomorphism is also unique, since the isomorphism $Sh(G, X) \otimes_E \mathbb{C} \xrightarrow{\sim} Sh(G, X)_{\mathbb{C}}$ is part of the data.)

2.6 If $Sh(G, X)$ is a canonical model of a Shimura variety, then the Galois group $\mathrm{Gal}(\overline{\mathbb{Q}}/E)$ acts on the set of connected components of $Sh(G, X)_{\mathbb{C}}$, which, as recalled in 1.6.6, is a principal homogeneous space under $\pi_0 \pi(G)$.

Deligne proves in [De3], section 2.6, that the homomorphism $\text{Gal}(\overline{\mathbb{Q}}/E) \to \pi_0\pi(G)$ describing the Galois action on $\pi_0\big(Sh(G,X)_{\mathbb{C}}\big)$ is equal to the homomorphism $r_{(G,X)}$ defined above. (Strictly speaking, this is only true up to a sign: in [De3] the Galois action on $\pi_0(Sh_{\mathbb{C}})$ is described to be $r_{(G,X)}$; Milne pointed out in [Mi3], Remark 1.10, that the reciprocity law is given by $r_{(G,X)}$, not its inverse.)

2.7 An important technique for proving the existence of canonical models is the reduction to a problem about connected Shimura varieties. To explain this, let us assume that $Sh(G,X)$ is a canonical model of the Shimura variety associated to the datum (G,X), and let us inventory the available structures. As in all of this section, we are mainly repeating things from Deligne's paper [De3].

The group $G(\mathbb{A}_f)$ acts continuously on $Sh_{\mathbb{C}} = Sh(G,X)_{\mathbb{C}}$ from the right. If Z denotes the center of G then $Z(\mathbb{Q})^- \subset G(\mathbb{A}_f)$ acts trivially. Write $G^{\text{ad}}(\mathbb{Q})_1 := G^{\text{ad}}(\mathbb{Q}) \cap \text{Im}\big(G(\mathbb{R}) \to G^{\text{ad}}(\mathbb{R})\big)$. The action of G^{ad} on G by inner automorphisms induces (by functoriality) a left action of $G^{\text{ad}}(\mathbb{Q})_1$ on $Sh(G,X)_{\mathbb{C}}$. For $g \in G(\mathbb{Q})$, the action of g through $G^{\text{ad}}(\mathbb{Q})_1$ coincides with the one of g^{-1} considered as an element of $G(\mathbb{A}_f)$. In total we therefore obtain a continuous left action of the group

$$\Gamma := \big(G(\mathbb{A}_f)/Z(\mathbb{Q})^-\big) \underset{G(\mathbb{Q})/Z(\mathbb{Q})}{*} G^{\text{ad}}(\mathbb{Q})_1 = \big(G(\mathbb{A}_f)/Z(\mathbb{Q})^-\big) \underset{G(\mathbb{Q})_+/Z(\mathbb{Q})}{*} G^{\text{ad}}(\mathbb{Q})^+$$

(converting the operation of $G(\mathbb{A}_f)$ to a left action). The group Γ operates transitively on $\pi_0(Sh_{\mathbb{C}})$. For any connected component of $Sh_{\mathbb{C}}$, the stabilizer of this component is the subgroup

$$\big(G(\mathbb{Q})_+^-/Z(\mathbb{Q})^-\big) \underset{G(\mathbb{Q})_+/Z(\mathbb{Q})}{*} G^{\text{ad}}(\mathbb{Q})^+ \cong G^{\text{ad}}(\mathbb{Q})^{+\wedge} \quad (\text{rel. } \tau(G^{\text{der}})),$$

where the completion $G^{\text{ad}}(\mathbb{Q})^{+\wedge}$ is taken relative to the topology $\tau(G^{\text{der}})$. The profinite set $\pi_0(Sh_{\mathbb{C}})$ is a principal homogeneous space under the abelian group

$$G(\mathbb{A}_f)/G(\mathbb{Q})_+^- \cong \pi_0\pi(G).$$

(Cf. 1.6.5 and 1.6.6.)

From now on we fix a connected component $X^+ \subset X$, and we write $Sh_{\mathbb{C}}^0 = Sh^0(G^{\text{der}}, X^+)_{\mathbb{C}}$ for the corresponding connected Shimura variety, to be identified with a connected component of $Sh(G,X)_{\mathbb{C}}$. We have an action of the Galois group $\text{Gal}(\overline{\mathbb{Q}}/E)$ on $Sh_{\mathbb{C}}$. As mentioned in 2.6, it acts on $\pi_0(Sh_{\mathbb{C}})$

through the reciprocity homomorphism $r_{(G,X)}$. The subgroup $\mathcal{E}_E(G^{\text{der}}, X^+) \subset$ $\Gamma \times \text{Gal}(\overline{\mathbb{Q}}/E)$ which fixes the connected component $Sh_{\mathbb{C}}^0$ is an extension

$$0 \longrightarrow G^{\text{ad}}(\mathbb{Q})^{+\wedge} \longrightarrow \mathcal{E}_E(G^{\text{der}}, X^+) \longrightarrow \text{Gal}(\overline{\mathbb{Q}}/E) \longrightarrow 0.$$

With these notations, we have the following important remarks.

(i) The extension $\mathcal{E}_E(G^{\text{der}}, X^+)$ depends only on the pair (G^{der}, X^+); in particular this justifies the notation. (See [De3], section 2.5.)

(ii) Galois descent (see also 2.15 below) tells us that it is equivalent to give a model of $Sh(G, X)_{\mathbb{C}}$ over E or to give a scheme S over $\overline{\mathbb{Q}}$ with a continuous action of $\Gamma \times \text{Gal}(\overline{\mathbb{Q}}/E)$ and a Γ-equivariant isomorphism $S \otimes_{\overline{\mathbb{Q}}} \mathbb{C} \stackrel{\sim}{\rightarrow} Sh(G, X)_{\mathbb{C}}$.

(iii) Write $e \in \pi_0(Sh_{\mathbb{C}})$ for the class of the connected component $Sh_{\mathbb{C}}^0$. To give a $\overline{\mathbb{Q}}$-scheme S as in (ii), which, in particular, comes equipped with a Γ-equivariant isomorphism $\pi_0(S) \cong \pi_0(Sh_{\mathbb{C}})$, is equivalent to giving its connected component S^e corresponding to e together with a continuous action of $\mathcal{E}_E(G^{\text{der}}, X^+)$. The idea here is that we can recover S from S^e by "induction" from $\mathcal{E}_E(G^{\text{der}}, X^+)$ to $\Gamma \times \text{Gal}(\overline{\mathbb{Q}}/E)$. (See [De3], section 2.7.)

2.8 Definition. (i) Let (G', X^+) be a pair defining a connected Shimura variety with reflex field E, let $F \subset \overline{\mathbb{Q}}$ be a finite extension of E, and write $\mathcal{E}_F(G', X^+)$ for the extension of $\text{Gal}(\overline{\mathbb{Q}}/F)$ by $G^{\text{ad}}(\mathbb{Q})^{+\wedge}$ (completion for the topology $\tau(G')$) described in [De3], Def. 2.5.7. Then a weakly canonical model of the connected Shimura variety $Sh^0(G', X^+)_{\mathbb{C}}$ over F consists of a scheme S over $\overline{\mathbb{Q}}$ together with a continuous left action of the group $\mathcal{E}_F(G', X^+)$ and an isomorphism $i\colon S \otimes_{\overline{\mathbb{Q}}} \mathbb{C} \stackrel{\sim}{\rightarrow} Sh^0(G', X^+)_{\mathbb{C}}$ such that the following conditions are satisfied.

(a) The action of $\mathcal{E}_F(G', X^+)$ on S is semi-linear, i.e., compatible with the canonical action on $\text{Spec}(\overline{\mathbb{Q}})$ through the quotient $\text{Gal}(\overline{\mathbb{Q}}/F)$.

(b) The isomorphism i is equivariant w.r.t. the action of $G^{\text{ad}}(\mathbb{Q})^{+\wedge} \subset$ $\mathcal{E}_F(G', X^+)$ (which by (i) acts linearly on S).

(c) Given a special point $h \in X^+$, factoring through a subtorus $h\colon \mathbb{S} \rightarrow$ $H_{\mathbb{C}} \subset G_{\mathbb{C}}^{\text{ad}}$ defined over $\mathbb{Q}$, let $E^{(h)}$ denote the field of definition of the cocharacter $h \circ \mu$. Delinge defines in loc. cit., 2.5.10, an extension $0 \rightarrow H(\mathbb{Q}) \rightarrow$ $\mathcal{E}_F(h) \rightarrow \text{Gal}(\overline{\mathbb{Q}}/EE^{(h)}) \rightarrow 0$, for which there is a natural homomorphism $\mathcal{E}_F(h) \rightarrow \mathcal{E}_F(G', X^+)$. Then we require that the point in $Sh^0(G', X^+)_{\mathbb{C}}$ defined by h is defined over $\overline{\mathbb{Q}}$ and is fixed by $\mathcal{E}_F(h)$.

(ii) A canonical model of the connected Shimura variety $Sh^0(G', X^+)_{\mathbb{C}}$ is a weakly canonical model over the reflex field E.

Although our formulation of condition (c) in (i) is a little awkward, it should be clear that this definition is just an attempt to formalize the above remarks. In fact, these remarks lead to the following result (= [De3], Prop. 2.7.13).

2.9 Proposition. *Let* (G, X) *be a Shimura datum, and choose a connected component* X^+ *of* X. *If* $Sh(G, X)$ *is a weakly canonical model of* $Sh(G, X)_{\mathbb{C}}$ *over* $F \supseteq E(G, X)$, *then the connected component* $Sh^0(G, X)_{\overline{\mathbb{Q}}}$ *determined by the choice of* X^+ *is a weakly canonical model of* $Sh^0(G^{der}, X^+)_{\mathbb{C}}$ *over* F. *Conversely, if there exists a weakly canonical model of* $Sh^0(G^{der}, X^+)_{\mathbb{C}}$ *over* F, *then it is obtained in this way from a weakly canonical model of* $Sh(G, X)_{\mathbb{C}}$.

2.10 The main result of [De3] is the existence of canonical models for a large class of Shimura varieties (see below). Since the strategy of proof also works for other statements about Shimura varieties, let us present it in an abstract form (following [Mi2], II.9). So, suppose we want to prove a statement $\mathcal{P}(G, X)$ about Shimura varieties.

(a) Prove $\mathcal{P}(\mathrm{CSp}_{2g}, \mathfrak{H}_g^{\pm})$ using the interpretation of $Sh(\mathrm{CSp}_{2g}, \mathfrak{H}_g^{\pm})_{\mathbb{C}}$ as a moduli space.

(b) For a closed immersion $i \colon (G_1, X_1) \hookrightarrow (G_2, X_2)$, prove the implication $\mathcal{P}(G_2, X_2) \Longrightarrow \mathcal{P}(G_1, X_1)$.

(c) Find a statement $\mathcal{P}^0(G', X^+)$ for pairs (G', X^+) defining a connected Shimura variety, such that, for any connected component $X^+ \subseteq X$, we have $\mathcal{P}(G, X) \Longleftrightarrow \mathcal{P}^0(G^{der}, X^+)$.

(d) Given pairs (G_i', X_i^+), $i = 1, \ldots, m$, prove that $\forall i \; \mathcal{P}^0(G_i', X_i^+) \Longrightarrow \mathcal{P}^0(\prod_i G_i', \prod_i X_i^+)$.

(e) For an isogeny $G' \to G''$, prove that $\mathcal{P}^0(G', X^+) \Longrightarrow \mathcal{P}^0(G'', X^+)$.

Roughly speaking, the class of Shimura varieties of abelian type is the largest class for which (a)–(e) suffice to prove statement $\mathcal{P}$. (As we will see below, this is not completely true: we may have to modify the strategy a bit, and even then it is not clear whether we obtain property $\mathcal{P}$ for all Shimura varieties of abelian type.) More precisely, a Shimura datum (G, X) is said to be of abelian type if there exists a Shimura datum (G_2, X_2) of Hodge type and an isogeny $G_2^{der} \twoheadrightarrow G^{der}$ which induces an isomorphism $(G_2^{ad}, X_2^{ad}) \overset{\sim}{\to} (G^{ad}, X^{ad})$. Deligne has analysed which simple Shimura data belong to this class. He showed that if (G, X) is of abelian type with G simple over $\mathbb{Q}$, then the following two conditions hold:

(i) The adjoint datum (G^{ad}, X^{ad}) is of type A, B or C, or of type $D^{\mathbb{R}}$, or of type $D^{\mathbb{H}}$ (cf. [De3], section 1.2), and

(ii) For a datum $(G^{\mathrm{ad}}, X^{\mathrm{ad}})$ of type A, B, C or $D^{\mathbb{R}}$, let $G^{\sharp}$ denote the universal covering of G^{ad}; for $(G^{\mathrm{ad}}, X^{\mathrm{ad}})$ of type $D_{\ell}^{\mathbb{H}}$, let $G^{\sharp}$ be the double covering of G^{ad} which is an inner form of (a product of copies of) $SO(2\ell)$, cf. ibid., 2.3.8, and notice that the case $D_{4}^{\mathbb{H}}$ is defined to exclude the case $D_{4}^{\mathbb{R}}$. Then G^{der} is a quotient of $G^{\sharp}$.

Conversely, if (G', X^{+}) is a pair defining a connected Shimura variety such that (i) and (ii) hold, then there exists a Shimura datum (G, X) of abelian type with $G^{\mathrm{der}} = G'$, $X^{+} \subseteq X$.

Finally, we define (G, X) to be of pre-abelian type if condition (i) holds. We see that, as far as connected Shimura varieties is concerned, this class is only slightly larger than that of data of abelian type.

2.11 Let us check steps (a)–(e) above for the statement

$$\mathcal{P}(G, X): \quad \text{there exists a canonical model for } Sh(G, X)_{\mathbb{C}}.$$

(a) The scheme $\varprojlim_{n} A_{g,1,n} \otimes \mathbb{Q}$ is a canonical model for $Sh(CSp_{2g}, \mathfrak{H}_{g}^{\pm})_{\mathbb{C}}$. Given the definitions as set up above, this boils down to a theorem of Shimura and Taniyama—see [De1], section 4. (Needless to say, the theorem of Shimura and Taniyama historically came first. The definition of a canonical model was modelled after a number of examples, including the Siegel modular variety.)

(b) This is shown in ibid., section 5. We should note here that, using a modular interpretation, one can prove $\mathcal{P}(G, X)$ more directly for Shimura varieties of Hodge type. This was indicated in the introduction of [De3], and carried out in detail in [Br].

For steps (c)–(e), let us work with the statement

$$\mathcal{P}^{0}(G', X^{+}): \quad \text{there exists a canonical model for } Sh^{0}(G', X^{+})_{\mathbb{C}}.$$

The (d) and (e) follow easily from the definitions (cf. [De3], 2.7.11) As for (c), we see that our strategy is not completely right: to prove $\mathcal{P}^{0}(G', X^{+})$, we want to take a Shimura datum (G_{2}, X_{2}) of Hodge type (for which we know $\mathcal{P}(G_{2}, X_{2})$ by (a) and (b)) with $(G_{2}^{\mathrm{der}}, X_{2}^{+}) \cong (G', X^{+})$, and then we can apply Prop. 2.9. The problem here is that this only gives the existence of a weakly canonical model of $Sh^{0}(G', X^{+})_{\mathbb{C}}$ over $E(G_{2}, X_{2})$, which in general is a proper field extension of $E(G^{\mathrm{ad}}, X^{+})$. (Notice that (G_{2}, X_{2}) is required to be of Hodge type—without this condition there would be no problem.) Thus we see that our "naive" strategy has to be corrected. This is done in two steps.

First one assumes that G is $\mathbb{Q}$-simple, and one considers the maximal covering $G^{\natural} \to G^{\mathrm{ad}}$ (as in 2.10) which occurs as the semi-simple part in a Shimura datum of Hodge type. As explained, the Shimura data (G_2, X_2) of Hodge type with $G_2^{\mathrm{der}} \cong G^{\natural}$ in general have $E(G^{\mathrm{ad}}, X^+) \subsetneq E(G_2, X_2)$. Deligne shows, however, that by "gluing in" a suitable toric part, the field extension $E(G_2, X_2)$ can be made in almost every "direction"; for a precise statement see [De3], Prop. 2.3.10. Finally one shows that this is enough to guarantee the existence of a canonical model of $Sh^0(G', X^+)_{\mathbb{C}}$; one proves (ibid., Cor. 2.7.19): if for every finite extension $F \subset \overline{\mathbb{Q}}$ of $E = E(G', X^+)$, there exists another finite extension $E \subseteq F' \subset \overline{\mathbb{Q}}$ which is linearly disjoint from F, and such that $Sh^0(G', X^+)_{\mathbb{C}}$ has a weakly canonical model over F', then it has a canonical model.

Putting everything together, one obtains the following result.

2.12 Theorem. (Deligne, [De3]) *Let (G, X) be a Shimura datum, and let $(G^{\mathrm{ad}}, X^{\mathrm{ad}}) \cong (G_1, X_1) \times \cdots \times (G_m, X_m)$ be the decomposition of its adjoint datum into simple factors. Suppose that, using the notations of 2.10, G^{der} is a quotient of $G_1^{\natural} \times \cdots \times G_m^{\natural}$. Then there exists a canonical model of $Sh(G, X)$.*

Notice that it is not clear whether this statement covers all data (G, X) of abelian type.

To extend this result to arbitrary Shimura data, additional arguments are needed. Since eventually we want to apply a Galois descent argument, it would be useful if we could first descend $Sh(G, X)_{\mathbb{C}}$ to a scheme over $\overline{\mathbb{Q}}$. Faltings has shown that this can be done using a rigidity argument.

2.13 Theorem. (Faltings, [Fa1]) *Let G be a semi-simple algebraic group over $\mathbb{Q}$, $K_{\infty} \subseteq G(\mathbb{R})$ a maximal compact subgroup, and $\Gamma \subset G(\mathbb{Q})$ a neat arithmetic subgroup. If $X = G(\mathbb{R})/K_{\infty}$ is a hermitian symmetric domain, then the locally symmetric variety $\Gamma \backslash X$ (with its unique structure of an algebraic variety) is canonically defined over $\overline{\mathbb{Q}}$. The special points on $\Gamma \backslash X$ are defined over $\overline{\mathbb{Q}}$. If $\Gamma_1, \Gamma_2 \subset G(\mathbb{Q})$ are neat arithmetic subgroups, $\gamma \in G(\mathbb{Q})$ an element with $\gamma \Gamma_1 \gamma^{-1} \subseteq \Gamma_2$, then the natural morphism $\gamma: \Gamma_1 \backslash X \to \Gamma_2 \backslash X$ is also defined over $\overline{\mathbb{Q}}$.*

Next we have to recall Langlands's conjecture on the conjugation of Shimura varieties (now a theorem, due to work of Borovoi, Deligne, Milne, and Milne-Shih, among others). We will not go into details here; the interested reader can consult [Bo1], [Bo2], [Mi1], [MS].

2.14 Theorem. (Borovoi, Deligne, Milne, Shih, ...) *Given a Shimura datum* (G, X), *a special point* $x \in X$, *and a* $\tau \in \mathrm{Aut}(\mathbb{C})$, *one can define a Shimura datum* $({}^{\tau,x}G, {}^{\tau,x}X)$, *a special point* ${}^\tau x \in {}^{\tau,x}X$, *and an isomorphism* $G(\mathbb{A}_f) \overset{\sim}{\to} {}^{\tau,x}G(\mathbb{A}_f)$, *denoted* $g \mapsto {}^{\tau,x}g$, *satisfying the following conditions (writing* $\mathcal{T}(g)$ *for the action of an element* $g \in G(\mathbb{A}_f)$ *on* $Sh(G, X)_{\mathbb{C}})$

(i) *There is a unique isomorphism* $\varphi_{\tau,x}$: ${}^{\tau}Sh(G, X)_{\mathbb{C}} \overset{\sim}{\to} Sh({}^{\tau,x}G, {}^{\tau,x}X)_{\mathbb{C}}$ *with* $\varphi_{\tau,x}({}^{\tau}[x, 1]) = [{}^{\tau}x, 1]$ *and with* $\varphi_{\tau,x} \circ {}^{\tau}\mathcal{T}(g) = \mathcal{T}({}^{\tau,x}g) \circ \varphi_{\tau,x}$ *for all* $g \in G(\mathbb{A}_f)$.

(ii) *If* $x' \in X$ *is another special point then there is an isomorphism* $\varphi(\tau; x, x')$: $Sh({}^{\tau,x}G, {}^{\tau,x}X)_{\mathbb{C}} \overset{\sim}{\to} Sh({}^{\tau,x'}G, {}^{\tau,x'}X)_{\mathbb{C}}$ *such that* $\varphi(\tau; x, x') \circ \varphi_{\tau,x} = \varphi_{\tau,x'}$ *and such that* $\varphi(\tau; x, x') \circ \mathcal{T}({}^{\tau,x}g) = \mathcal{T}({}^{\tau,x'}g) \circ \varphi(\tau; x, x')$ *for all* $g \in G(\mathbb{A}_f)$.

As explained in [La], section 6 (see also [Mi2], section II.5), using the theorem one obtains a "pseudo" descent datum from $\mathbb{C}$ to $E = E(G, X)$ on $Sh(G, X)_{\mathbb{C}}$. By this we mean a collection of isomorphisms

$$\{f_\tau \colon {}^{\tau}Sh(G, X)_{\mathbb{C}} \overset{\sim}{\longrightarrow} Sh(G, X)_{\mathbb{C}}\}_{\tau \in \mathrm{Aut}(\mathbb{C}/E)}$$

satisfying the cocycle condition $f_{\sigma\tau} = f_\sigma \circ {}^{\sigma}f_\tau$. At several places in the literature (e.g., [La], section 6, [Mi2], p. 340, [MS], §7) it is asserted that "by descent theory" this gives a model of $Sh(G, X)_{\mathbb{C}}$ over E. (Due to the properties of the f_τ, this model would then be a canonical model.) We think that this argument is not complete—let us explain why.

2.15 To descend a scheme $X_{\mathbb{C}}$ from $\mathbb{C}$ to a number field $E \subset \mathbb{C}$, it does in general not suffice to give a collection of isomorphisms $\{f_\tau \colon {}^{\tau}X_{\mathbb{C}} \overset{\sim}{\to} X_{\mathbb{C}}\}_{\tau \in \mathrm{Aut}(\mathbb{C}/E)}$ with $f_{\sigma\tau} = f_\sigma \circ {}^{\sigma}f_\tau$ (or, what is the same, a homomorphism of groups $\alpha \colon \mathrm{Aut}(\mathbb{C}/E) \to \mathrm{Aut}(X_{\mathbb{C}})$ sending τ to a τ-linear automorphism of $X_{\mathbb{C}}$). For instance, using the fact that $\mathbb{Q}$ is an injective object in the category of abelian groups, we easily see that there exist non-trivial group homomorphisms $c \colon \mathrm{Aut}(\mathbb{C}/E) \to \mathbb{Q}$. Taking $X_{\mathbb{C}} = \mathbb{A}^1_{\mathbb{C}}$, on which we let $\tau \in \mathrm{Aut}(\mathbb{C}/E)$ act as the τ-linear translation over $c(\tau)$, we get an example of a non-effective "pseudo" descent datum. The same remarks apply if we replace $\mathbb{C}$ by $\overline{\mathbb{Q}}$. (Thus, for example, [Mi4], Lemma 3.23 is not correct as it stands.)

In this context it seems useful to remark the following. Given a $\overline{\mathbb{Q}}$-scheme $X_{\overline{\mathbb{Q}}}$, one might expect that a descent datum on $X_{\overline{\mathbb{Q}}}$ relative to $\overline{\mathbb{Q}}/E$ can be expressed as a collection of isomorphisms $\{\varphi_\tau \colon {}^{\tau}X_{\overline{\mathbb{Q}}} \overset{\sim}{\to} X_{\overline{\mathbb{Q}}}\}_{\tau \in \mathrm{Gal}(\overline{\mathbb{Q}}/E)}$ for which, apart from the cocycle condition $\varphi_{\sigma\tau} = \varphi_\sigma \circ {}^{\sigma}\varphi_\tau$, a certain "continuity condition" holds. To see why a continuity condition should enter, one must realize that a scheme such as $\mathrm{Spec}(\overline{\mathbb{Q}} \otimes_E \overline{\mathbb{Q}})$ is not a disjoint union of copies of

$\mathrm{Spec}(\overline{\mathbb{Q}})$ indexed by $\mathrm{Gal}(\overline{\mathbb{Q}}/E)$ (which would not be a quasi-compact scheme), but rather a projective limit $\mathrm{Spec}(\overline{\mathbb{Q}} \otimes_E \overline{\mathbb{Q}}) = \varprojlim_F \mathrm{Spec}(\overline{\mathbb{Q}})^{\mathrm{Gal}(F/E)}$, where F runs through the finite Galois extensions of E in $\overline{\mathbb{Q}}$. (In other words: this is $\mathrm{Gal}(\overline{\mathbb{Q}}/E)$ as a pro-finite group scheme.) It seems though that it is not so easy to formulate the desired continuity condition directly. Even if one succeeds in doing this, however, it should be remarked that descent data relative to $\overline{\mathbb{Q}}/E$ are not necessarily effective (cf. [SGA1], Exp. VIII).

Since we are really only interested in *effective* descent data relative to $\overline{\mathbb{Q}}/E$, we take a slightly different approach. Let us call a (semi-linear) action $\alpha\colon \mathrm{Gal}(\overline{\mathbb{Q}}/E) \to \mathrm{Aut}(X')$ of $\mathrm{Gal}(\overline{\mathbb{Q}}/E)$ on a $\overline{\mathbb{Q}}$-scheme X' continuous if it is continuous as an action of a locally compact, totally disconnected group (see [De3], section 2.7). Since the Galois group is actually compact, the following statement is then a tautology.

2.15.1 *The functor $X \mapsto X' = X \otimes_E \overline{\mathbb{Q}}$ gives an equivalence of categories*

$$\begin{pmatrix} \text{quasi-projective} \\ \text{schemes } X \text{ over } E \end{pmatrix} \xrightarrow{\text{eq.}} \begin{pmatrix} \text{quasi-projective schemes } X' \text{ over } \overline{\mathbb{Q}} \\ \text{with a continuous semi-linear action} \\ \text{of } \mathrm{Gal}(\overline{\mathbb{Q}}/E) \end{pmatrix}.$$

We thus see that, in order to prove the existence of canonical models in the general case, we need to show that Theorem 2.14 provides us with a *continuous* Galois action on $Sh(G,X)_{\overline{\mathbb{Q}}}$. For this we will use the following lemma.

2.16 Lemma. *Let (G,X) be a Shimura datum, $K \subset G(\mathbb{A}_f)$ a compact open subgroup, and let $S = \Gamma\backslash X^+$ be a connected component of $Sh_K(G,X)_{\mathbb{C}}$. Then we can choose finitely many special points $x_1,\dots,x_n \in S^0$ such that S has no non-trivial automorphisms fixing the x_i.*

Proof. Let $j\colon S \hookrightarrow S^*$ denote the Baily-Borel compactification. Every automorphism of S extends to an automorphism of S^*. There exists an ample line bundle $\mathcal{L}$ on S^* such that $\alpha^*\mathcal{L} \cong \mathcal{L}$ for every $\alpha \in \mathrm{Aut}(S)$. In fact, if G has no simple factors of dimension 3 then we can take $\mathcal{L} := j_*\Omega_S^d$, where $d = \dim(S)$. In the general case one has to impose growth conditions at infinity: using the terminology of [BB] we can take for $\mathcal{L}$ the subsheaf of $j_*\Omega_S^d$ (now taken in the analytic sense) of automorphic forms which are integral at infinity. (So $\mathcal{L}$ is the bundle $\mathcal{O}(1)$ corresponding to the projective embedding of S^* as in loc. cit., §10. Mumford showed in [Mu3] that if $\overline{S}$ is a smooth toroidal compactification and $\pi\colon \overline{S} \to S^*$ is the canonical birational morphism, then $\pi^*\mathcal{L}$ is the sheaf $\Omega_{\overline{S}}^d(\log \partial\overline{S})$.)

Let P be the "doubled" Hilbert polynomial of $\mathcal{L}$, given by $P(x) = P_{\mathcal{L}}(2x)$. Recall from [FGA], Exposé 221, p. 20, that the scheme $\text{Hom}(S^*, S^*)^P$ given by

$$\text{Hom}(S^*, S^*)^P(T) = \{g \colon S^* \times_{\mathbb{C}} T \longrightarrow S^* \times_{\mathbb{C}} T \mid \chi\big((\mathcal{L} \otimes_{\mathcal{O}_T} g^*\mathcal{L})^{\otimes n}\big) = P(n)\}$$

is of finite type. With the obvious notations, it follows that $\text{Aut}(S^*)^P$ is a scheme of finite type, being a locally closed subscheme of $\text{Hom}(S^*, S^*)^P$. The lemma now follows from the following two trivial remarks:

(i) if $\alpha \in \text{Aut}(S^*)$ fixes all special points of S then $\alpha = \text{id}$,

(ii) if $x_1, \dots, x_n$ are special points of S^0 then

$$\text{Aut}(S^*; x_1, \dots, x_n)^P := \{\alpha \in \text{Aut}(S^*)^P \mid \alpha(x_i) = x_i \quad \text{for all } i = 1, \dots, n\}$$

is a closed subgroup scheme of $\text{Aut}(S^*)^P$. $\qquad\qquad\qquad\qquad\qquad\qquad\square$

2.17 We now complete the argument showing that $Sh(G, X)_{\mathbb{C}}$ has a canonical model. Obviously, the first step is to use Theorem 2.13, so that we obtain a model $Sh(G, X)_{\overline{\mathbb{Q}}}$ over $\overline{\mathbb{Q}}$. We claim that the "pseudo" descent datum $\{f_\tau \colon {}^\tau Sh(G, X)_{\mathbb{C}} \xrightarrow{\sim} Sh(G, X)_{\mathbb{C}}\}_{\tau \in \text{Aut}(\mathbb{C}/E)}$ considered in 2.14 induces a semi-linear action of $\text{Gal}(\overline{\mathbb{Q}}/E)$ on $Sh(G, X)_{\overline{\mathbb{Q}}}$, which is functorial. We can show this using the special points: if $Sh(T, \{h\})_{\mathbb{C}} \hookrightarrow Sh(G, X)_{\mathbb{C}}$ is a 0-dimensional sub-Shimura variety, then the canonical model $Sh(T, \{h\})$ over E' $= E(T, \{h\})$ gives rise to a collection of isomorphisms $\{\tilde{f}_\sigma \colon {}^\sigma Sh(T, \{h\})_{\mathbb{C}} \xrightarrow{\sim} Sh(T, \{h\})_{\mathbb{C}}\}_{\sigma \in \text{Aut}(\mathbb{C}/E')}$, and for $\sigma \in \text{Aut}(\mathbb{C}/E')$, the two maps f_σ and $\tilde{f}_\sigma$ are equal on ${}^\sigma Sh(T, \{h\})_{\mathbb{C}}$. Using the fact that the special points on $Sh(G, X)_{\mathbb{C}}$ are defined over $\overline{\mathbb{Q}}$ for the $\overline{\mathbb{Q}}$-structure $Sh(G, X)_{\overline{\mathbb{Q}}}$, one now checks that the f_σ induce a system

$$\{\varphi_\tau \colon {}^\tau Sh(G, X)_{\overline{\mathbb{Q}}} \xrightarrow{\sim} Sh(G, X)_{\overline{\mathbb{Q}}}\}_{\tau \in \text{Gal}(\overline{\mathbb{Q}}/E)}$$

with $\varphi_{\sigma\tau} = \varphi_\sigma \circ {}^\sigma\varphi_\tau$. What we shall use is that the action on the special points agrees with the one obtained from the canonical models $Sh(T, \{h\})$.

Now for the continuity of the Galois action on $Sh(G, X)_{\overline{\mathbb{Q}}}$. First let us remark that it suffices to prove that the semi-linear Galois action on each of the $Sh_K(G, X)$ is continuous, since the transition morphisms then automatically descend. Here we may even restrict to "levels" Sh_K where K is neat. Furthermore, it suffices to show that there is an open subgroup of $\text{Gal}(\overline{\mathbb{Q}}/E)$ which acts continuously. In fact, if we assume this then $Sh_K(G, X)$ descends to a finite Galois extension F of E. On the model $Sh_K(G, X)_F$ thus obtained we still have a Galois descent datum relative to F/E, and since this is now a finite Galois extension, the descent datum is effective.

Since $Sh_K(G,X)_{\overline{\mathbb{Q}}}$ is a $\overline{\mathbb{Q}}$-scheme of finite type, there exists a finite extension E' of E and a model $S_{E'}$ of $Sh_K(G,X)$ over E'. This model gives rise to semi-linear action of $\mathrm{Gal}(\overline{\mathbb{Q}}/E')$ on $Sh_K(G,X)_{\overline{\mathbb{Q}}}$, which we can describe as a collection of automorphisms

$$\{\psi_\tau \colon {}^\tau Sh_K(G,X)_{\overline{\mathbb{Q}}} \xrightarrow{\sim} Sh_K(G,X)_{\overline{\mathbb{Q}}}\}_{\tau \in \mathrm{Gal}(\overline{\mathbb{Q}}/E)}.$$

Observe that $\varphi_\tau \circ (\psi_\tau)^{-1}$ is a $\overline{\mathbb{Q}}$-*linear* automorphism of $Sh_K(G,X)_{\overline{\mathbb{Q}}}$, and that $\{\tau \in \mathrm{Gal}(\overline{\mathbb{Q}}/E) \mid \varphi_\tau = \psi_\tau\}$ is a subgroup of $\mathrm{Gal}(\overline{\mathbb{Q}}/E)$.

At this point we apply Lemma 2.16. It gives us special points $x_1, \ldots, x_n \in Sh_K(G,X)_{\overline{\mathbb{Q}}}$ such that there are no automorphisms of $Sh_K(G,X)_{\overline{\mathbb{Q}}}$ fixing all x_i. For each x_i, choose a closed immersion $j_i \colon (T_i, \{h_i\}) \hookrightarrow (G,X)$ and an element $g_i \in G(\mathbb{A}_f)$ such that x_i lies in $g_i \cdot Sh(T_i, \{h_i\})_{\overline{\mathbb{Q}}} \subseteq Sh(G,X)_{\overline{\mathbb{Q}}}$. Let $K_i := j_i^{-1}(K) \subset T_i(\mathbb{A}_f)$. There exists a finite extension E'' of E', containing the reflex fields $E(x_i)$, such that the x_i are all E''-rational on the chosen model $S_{E'}$ and such that furthermore all points of $Sh_{K_i}(T_i, \{h_i\})$ are rational over E'' (for every $i = 1, \ldots, n$). It now follows from what was said above that the two Galois actions on $Sh_K(G,X)_{\overline{\mathbb{Q}}}$, given by the φ_τ and the ψ_τ, respectively, are the same when restricted to $\mathrm{Gal}(\overline{\mathbb{Q}}/E'')$. This finishes the proof of the following theorem.

2.18 Theorem. *Let (G,X) be a Shimura datum. Then there exists a canonical model $Sh(G,X)$ of the associated Shimura variety.*

2.19 Remark. In [Pi], the notion of a canonical model is generalized to the mixed case, and the existence of such canonical models is proven for arbitrary mixed Shimura varieties. Pink's proof essentially reduces the problem to statement 2.18; once we have 2.18, the mixed case does not require any further corrections.

2.20 Remark. There is also a theory of a canonical models for automorphic vector bundles on Shimura varieties. The interested reader is referred to [Ha] and [Mi2].

§3 Integral canonical models

3.1 Let (G,X) be a Shimura datum with reflex field $E = E(G,X)$, and let v be a prime of E dividing $p > 0$. We want to study models of the Shimura variety $Sh(G,X)$ over the local ring $\mathcal{O}_{E,(v)}$ of E at v. In our personal view, the theory of such models is still in its infancy. How to set up the definitions, what properties to expect, etc., are dictated by the examples where the Shimura variety represents a moduli problem that can be formulated in mixed characteristics (notably Shimura varieties of PEL type). Even in the case where G is unramified over $\mathbb{Q}_p$, this leaves open some subtle questions.

Some of the rules of the game become clear already from looking at Siegel modular varieties. We have seen that the canonical model in this case can be identified with the projective limit $\varprojlim_n A_{g,1,n} \otimes \mathbb{Q}$. Fixing a prime number p, we see that, for constructing a model over $\mathbb{Z}_{(p)}$, we run into problems at the levels $A_{g,1,n}$ with $p \mid n$. By contrast, if we only consider levels with $p \nmid n$, then we have a natural candidate model, viz. $\varprojlim_{p \nmid n} A_{g,1,n} \otimes \mathbb{Z}_{(p)}$, which has all good properties we can expect.

Returning to the general case, this suggests the following set-up. Let (G,X), E and v be as above. We fix a compact open subgroup $K_p \subset G(\mathbb{Q}_p)$, and we consider

$$Sh_{K_p}(G,X) = \varprojlim_{K^p} Sh_{K_p \times K^p}(G,X)\,,$$

where K^p runs through the compact open subgroups of $G(\mathbb{A}_f^p)$. It is this scheme $Sh_{K_p}(G,X)$, the quotient of $Sh(G,X)$ for the action of K_p, of which we shall study models. Notice that we can expect to find a smooth model (to be made precise in a moment) only for special choices of K_p.

3.2 Definition. Let (G,X) be a Shimura datum, $E = E(G,X)$, v a finite prime of E dividing p, and let K_p be a compact open subgroup of $G(\mathbb{Q}_p)$. Let $\mathcal{O}$ be a discrete valuation ring which is faithfully flat over $\mathcal{O}_{(v)}$. Write F for the quotient field of $\mathcal{O}$.

(i) An integral model of $Sh_{K_p}(G,X)$ over $\mathcal{O}$ is a faithfully flat $\mathcal{O}$-scheme $\mathcal{M}$ with a continuous action of $G(\mathbb{A}_f^p)$ and a $G(\mathbb{A}_f^p)$-equivariant isomorphism $\mathcal{M} \otimes F \cong Sh_{K_p}(G,X) \otimes_E F$.

(ii) An integral model $\mathcal{M}$ of $Sh_{K_p}(G,X)$ over $\mathcal{O}$ is said to be smooth (respectively normal) if there exists a compact open subgroup $C \subset G(\mathbb{A}_f^p)$, such that for every pair of compact open subgroups $K_1^p \subseteq K_2^p \subset G(\mathbb{A}_f^p)$ contained in C, the canonical map $\mathcal{M}/K_1^p \to \mathcal{M}/K_2^p$ is an étale morphism between smooth (resp. normal) schemes of finite type over $\mathcal{O}$.

It should be clear that an integral model $\mathcal{M}$, if it exists, is by no means unique. For example, given one such model, we could delete a $G(\mathbb{A}_f^p)$-orbit properly contained in the special fibre, or we could blow up in such an orbit, to obtain a different integral model. To arrive at the notion of an integral canonical model, we will impose the condition that $\mathcal{M}$ satisfy an "extension property", similar to the Néron mapping property in the theory of Néron models (cf. [BLR], section 1.2). This idea was first presented by Milne in [Mi3]. As we shall see, one of the main difficulties in this approach is to find a good class of "test schemes" for which the extension property should hold. Given a base ring $\mathcal{O}$, we will work with a class of $\mathcal{O}$-schemes that we call "admissible test schemes over $\mathcal{O}$", abbreviated "a.t.s.". We postpone the precise definition of the class that we will work with until 3.5.

3.3 Definition. Let (G, X), E, v, K_p, $\mathcal{O}$ and F be as in 3.2.

(i) An integral model $\mathcal{M}$ of $Sh_{K_p}(G, X)$ over $\mathcal{O}$ is said to have the extension property if for every admissible test scheme S over $\mathcal{O}$, every morphism $S_F \to \mathcal{M}_F$ over F extends uniquely to an $\mathcal{O}$-morphism $S \to \mathcal{M}$.

(ii) An integral canonical model of $Sh_{K_p}(G, X)$ at the prime v is a separated smooth integral model over $\mathcal{O}_{(v)}$ which has the extension property. A local integral canonical model is a separated smooth integral model over $\mathcal{O}_v := \mathcal{O}_{(v)}^{\wedge}$ having the extension property.

3.4 Comments. A definition in this form was first given by Milne in [Mi3]. As admissible test schemes over $\mathcal{O}$ he used all regular $\mathcal{O}$-schemes S for which S_F is dense in S. Later it was seen that this is not the right class to work with (cf. [Mi4], footnote on p. 513); the reason for this is the following. One wants to set up the theory in such a way that $\varprojlim_{p \nmid n} A_{g,1,n} \otimes \mathbb{Z}_{(p)}$ is an integral canonical model for the Siegel modular variety. Using Milne's definition, this boils down to [FC], Cor. V.6.8, which, however, is false as it stands. Recall that this concerns the following question: suppose given a regular scheme S with maximal points of characteristic 0, a closed subscheme $Z \hookrightarrow S$ of codimension at least 2, and an abelian scheme over the complement $U = S \backslash Z$. Does this abelian scheme extend to an abelian scheme over S? In loc. cit. it is claimed that the answer is "yes"—this is not correct in general. A counterexample, due to Raynaud-Ogus-Gabber, is discussed in [dJO], section 6. Let us try to explain the gist of the example, referring to loc. cit. for details.

As base scheme we take $S = \mathrm{Spec}(R)$, where $R = W(\overline{\mathbb{F}}_p)[\![x, y]\!]/((xy)^{p-1} - p)$. There exists a primitive pth root of unity ζ_p in R. Let $s \in S$ be the closed point, and set $U_1 = D(x)$, $U_2 = D(y)$, $U = S \backslash \{s\} = U_1 \cup U_2$, $U_{12} = D(xy) = U_1 \cap U_2$. We obtain a finite locally free group scheme G_U of rank p^2 over U by gluing

$$G_1 = \left(\mu_p \times \mathbb{Z}/p\mathbb{Z}\right)_{U_1} \quad \text{and} \quad G_2 = \left(\mu_p \times \mathbb{Z}/p\mathbb{Z}\right)_{U_2}$$

via the isomorphism

$$\varphi\colon G_{1|U_{12}} \xrightarrow{\sim} G_{2|U_{12}} \quad \text{given by the matrix} \quad \begin{pmatrix} 1 & 0 \\ \beta & 1 \end{pmatrix},$$

where $\beta\colon \mu_p \xrightarrow{\sim} \mathbb{Z}/p\mathbb{Z}$ (over U_{12}) maps $\zeta_p \in \Gamma(U_{12}, \mu_p)$ to $\bar{1} \in \Gamma(U_{12}, \mathbb{Z}/p\mathbb{Z})$. One easily sees from the construction that we have an exact sequence

$$0 \longrightarrow (\mathbb{Z}/p\mathbb{Z})_U \xrightarrow{\gamma_U} G_U \longrightarrow \mu_{p,U} \longrightarrow 0,$$

and that this extension is not trivial.

The group scheme G_U extends uniquely to a finite locally free group scheme G over S. Also, the homomorphism γ_U extends uniquely to a homomorphism $\gamma\colon (\mathbb{Z}/p\mathbb{Z})_S \to G$, which, however, is not a closed immersion. (The whole point!) To get the desired example, one only has to embed G into an abelian scheme X over S (using the theorem [BM3], Thm. 3.1.1 by Raynaud), and take $Y_U := X_U/(\mathbb{Z}/p\mathbb{Z})_U$, where $(\mathbb{Z}/p\mathbb{Z})_U$ is viewed as a subgroup scheme of X_U via γ_U and the chosen embedding $G \hookrightarrow X$.

To understand what is going on, the following remarks may be of help. One can show that the fibre G_s is isomorphic to $\alpha_p \times \alpha_p$. There is a blowing up $\pi\colon \widetilde{S} \to S$ with center in s such that $(\mathbb{Z}/p\mathbb{Z})_U \hookrightarrow G_U$ extends to a closed flat subgroup scheme $N \hookrightarrow G_{\widetilde{S}}$. Over $\widetilde{S}$, the abelian scheme Y_U extends to the abelian scheme $Y_{\widetilde{S}} := X_{\widetilde{S}}/N$. When restricted to the exceptional fibre E, we have $Y_{\widetilde{S}|E} \cong (X_s \times E)/N_E$, where $N_E \hookrightarrow (\alpha_p \times \alpha_p)_E$ is a non-constant subgroup scheme isomorphic to α_p. Therefore, we cannot blow down $Y_{\widetilde{S}}$ to an abelian scheme over S.

In order to guarantee that $\varprojlim_{p\nmid n} A_{g,1,n} \otimes \mathbb{Z}_{(p)}$ is an i.c.m., we want our a.t.s. to satisfy the following condition. (Here $\mathcal{O}$ is a d.v.r. with field of fractions F and S is an $\mathcal{O}$-scheme.)

(3.4.1)   for every closed subscheme $Z \hookrightarrow S$, disjoint from S_F and of codimension at least 2 in S, every abelian scheme over the complement $U = S \setminus Z$ extends to an abelian scheme over S.

On the other hand, we want that an integral canonical model, if it exists, is unique up to isomorphism. Thus we want it to be an a.t.s. over $\mathcal{O}_{(v)}$ itself. The notion that we will work with in this paper is the following.

3.5 Definition. Let $\mathcal{O}$ be a discrete valuation ring. We call an $\mathcal{O}$-scheme S an admissible test scheme (a.t.s.) over $\mathcal{O}$ if every point of S has an open neigbourhood of the form $\text{Spec}(A)$, such that there exist $\mathcal{O} \subseteq \mathcal{O}' \subseteq A_0 \subseteq A$, where

—$\mathcal{O} \subseteq \mathcal{O}'$ is a faithfully flat and unramified extension of d.v.r. with $\mathcal{O}'/(\pi)$ separable over $\mathcal{O}/(\pi)$,

—A_0 is a smooth $\mathcal{O}'$-algebra, and where

—$\text{Spec}(A) \to \text{Spec}(A_0)$ is a pro-étale covering.

We write $\text{ATS}_\mathcal{O}$ for the class of a.t.s. over $\mathcal{O}$.

We want to stress that this should be seen as a working definition, see also the remarks in 3.9 below. Clearly, a smooth model of a Shimura variety over $\mathcal{O}$ belongs to $\text{ATS}_\mathcal{O}$. In particular, we have unicity of integral canonical models:

3.5.1 Proposition. *Let (G, X) be a Shimura datum, v a prime of its reflex field E dividing the rational prime p, and let K_p be a compact open subgroup of $G(\mathbb{Q}_p)$. If there exists an integral canonical model of $Sh_{K_p}(G, X)$ over $\mathcal{O}_{(v)}$, then it is unique up to isomorphism.*

Furthermore, we have the following properties.

(3.5.2) If $S \in \text{ATS}_\mathcal{O}$ then S is a regular scheme, formally smooth over $\mathcal{O}$. (To prove that the local rings of S are noetherian, we can follow the arguments of [Mi3], Prop. 2.4.)

(3.5.3) If $\mathcal{O} \subseteq \mathcal{O}'$ is an unramified faithfully flat extension of d.v.r., then $S \in \text{ATS}_{\mathcal{O}'} \Rightarrow S \in \text{ATS}_\mathcal{O}$, and $S \in \text{ATS}_\mathcal{O} \Rightarrow (S \otimes_\mathcal{O} \mathcal{O}') \in \text{ATS}_{\mathcal{O}'}$.

Next we investigate whether (3.4.1) holds. For this we use the following two lemmas.

3.6 Lemma. (Faltings) *Let $\mathcal{O}$ be a d.v.r. of mixed characteristics $(0, p)$ with $p > 2$. Suppose that the ramification index e satisfies $e < p - 1$. Then every regular formally smooth $\mathcal{O}$-scheme S satisfies condition (3.4.1).*

Proof (sketch). As mentioned above, some statements in [FC], section V.6, are not correct. The mistake can be found on p. 182: the map $p^{-\dim(G)} \cdot \text{trace}_{G[p^{n+1}]/G[p^n]}$ is not a splitting of the map $\mathcal{O}_{G[p^n]} \subset \mathcal{O}_{G[p^{n+1}]}$, as claimed. Most arguments in the rest of the section are correct however, and with some modifications we can use them to prove the lemma. Let us provisionally write $RFS_\mathcal{O}$ for the class of regular, formally smooth $\mathcal{O}$-schemes. For $S \in RFS_\mathcal{O}$, we have the following version of [FC], Thm. V.6.4'.

3.6.1 *Let S be a regular, formally smooth $\mathcal{O}$-scheme ($\mathcal{O}$ as above, with $e < p - 1$), and let $U \hookrightarrow S$ be the complement of a closed subscheme $Z \hookrightarrow S$ of codimension at least 2. Then every p-divisible group $\mathcal{G}_U$ over U extends uniquely to a p-divisible group $\mathcal{G}$ over S.*

The only step in the proof of [FC], Thm. V.6.4' that we have to correct is the one showing the existence of an extension $\mathcal{G}$ in case $\dim(S) = 2$ (loc. cit., top of p. 183). So we may assume $S = \mathrm{Spec}(R) \hookleftarrow U = S \setminus \{s\}$, where R is a 2-dimensional regular local ring, and where s is the closed point of S. The $\mathcal{G}_{U,n} := \mathcal{G}_u[p^n]$ extend uniquely to an inductive system of finite flat group schemes $\{\mathcal{G}_n; i_n \colon \mathcal{G}_n \to \mathcal{G}_{n+1}\}$. (See [FC], Lemma V.6.2.) We have to prove that the sequences

$$(3.6.2) \qquad\qquad 0 \longrightarrow \mathcal{G}_n \xrightarrow{i_n} \mathcal{G}_{n+1} \xrightarrow{p^n} \mathcal{G}_1 \longrightarrow 0$$

are exact. That i_n is a closed immersion needs to be checked only on the closed fibre. The formal smoothness of R over $\mathcal{O}$ guarantees that there exists an unramified faithfully flat extension of d.v.r. $\mathcal{O} \subset \mathcal{O}'$ such that S has a section over $\mathcal{O}'$ with s contained in the image. Pulling back to $\mathcal{O}'$, it then follows from [Ra1], Cor. 3.3.6, that i_n is a closed immersion. Finally, this implies that $\mathcal{G}_{n+1}/\mathcal{G}_n$ is a finite flat extension of $\mathcal{G}_{U,1}$, and because of the unicity of such an extension it follows that (3.6.2) is exact.

It remains to be checked that, using 3.6.1 to replace [FC] Thm. V.6.4', all steps in the proof of ibid., Thm. 6.7 go through for $S \in RFS_\mathcal{O}$. One has to note that in carrying out the various reduction steps, we stay within the class $RFS_\mathcal{O}$. At some points one furthermore needs arguments similar to the above ones, i.e., taking sections over an extension $\mathcal{O}'$ and using [Ra1], Cor. 3.3.6. We leave it to the reader to verify the details. $\square$

3.7 Lemma. *Let (G, X) be a Shimura datum, and let v be a prime of $E(G, X)$ dividing p. Assume that $G_{\mathbb{Q}_p}$ is unramified (see 3.11 below). Then v is an unramified prime (in the extension $E(G, X) \supset \mathbb{Q}$).*

Proof. See [Mi4], Cor. 4.7. $\square$

3.8 Corollary. *Notations as in 3.2. If $p > 2$ then every $S \in \mathrm{ATS}_{\mathcal{O}_{(v)}}$ satisfies
(3.4.1). In particular, if $p > 2$ then $\varprojlim_{p \nmid n} A_{g,1,n} \otimes \mathbb{Z}_{(p)}$ is an integral canonical
model of $Sh_{K_p}(\mathrm{CSp}_{2g,\mathbb{Q}}, \mathfrak{H}_g^{\pm})$, where $K_p = \mathrm{CSp}_{2g}(\mathbb{Z}_p)$.*

Proof. We can follow Milne's proof of [Mi3], Thm. 2.10, except that we have
to modify the last part of the proof in the obvious way. Notice that the group
CSp_{2g} is unramified everywhere, so that Lemmas 3.6 and 3.7 apply. □

3.9 Remarks. (i) We do not know whether the corollary is also true for
$p = 2$. (Note that in the example in 3.4, the base scheme S is not an a.t.s.
over W or $W[\zeta_p]$.) This is one of the reasons why we do not pretend that
Def. 3.5 is in its final form.

(ii) Our definitions differ from those used in [Va2]. Vasiu's definition of
an integral canonical model is of the above form, but the class $\mathrm{ATS}_{\mathcal{O}}$ he
works with is the class of all regular schemes S over $\mathrm{Spec}(\mathcal{O})$, for which the
generic fibre S_F is Zariski dense and such that condition (3.4.1) holds. As
we have seen above, this contains the class we are working with if $p > 2$ and
$e(\mathcal{O}) < p - 1$.

It seems to us that Vasiu's definition is more difficult to work with. For
example, it is not clear to us whether his notion of an a.t.s. is a local one,
and whether it satisfies $S \in \mathrm{ATS}_{\mathcal{O}} \Rightarrow (S \otimes_{\mathcal{O}} \mathcal{O}') \in \mathrm{ATS}_{\mathcal{O}'}$. (This *is* important
for some of the constructions.) On the other hand, if we want that the
extension property is preserved under extension of scalars from $\mathcal{O}_{(v)}$ to $\mathcal{O}_v$ or
to $W\left(\overline{\kappa(v)}\right)$, then this forces us to work with a class $\mathrm{ATS}_{\mathcal{O}}$ which is not "too
small". Here we should draw a comparison with the theory of Néron models:
we note that the proof of [BLR], Thm. 7.2.1 (ii) makes essential use of Weil's
theorem, ibid. Thm. 4.4.1, for which we see no analogue in the context of
Shimura varieties. This may help to explain why we set up the Def. 3.5 the
way we did.

3.10 Proposition. *Let (G, X), $E = E(G, X)$, v and K_p be as in 3.2.*

*(i) There exists an integral canonical model of $Sh_{K_p}(G, X)$ at v if and only
if there exists a local integral canonical model.*

*(ii) Suppose that $p > 2$ and that the prime v is (absolutely) unramified.
Write B for the fraction field of $W(\overline{\mathbb{F}}_p)$, and choose an embedding $\mathcal{O}_v \hookrightarrow
W(\overline{\mathbb{F}}_p)$, where $\mathcal{O}_v = \mathcal{O}_{(v)}^{\wedge}$ is the completed local ring of $\mathcal{O}_E$ at v. Suppose
there exists a smooth integral model $\overline{\mathcal{M}}$ for $Sh_{K_p}(G, X) \otimes B$ over $W(\overline{\mathbb{F}}_p)$
having the extension property. Then there exists an integral canonical model
of $Sh_{K_p}(G, X)$ over $\mathcal{O}_{(v)}$.*

Proof. (i) In the "only if" direction this readily follows from (3.5.3). For the converse, suppose that $\mathcal{M}^\natural$ is a local integral canonical model of $Sh_{K_p}(G, X)$ over $\mathcal{O}_v$. We have $\mathcal{M}^\natural = \varprojlim \mathcal{M}^\natural_{K^p}$, where K^p runs through the compact open subgroups of $G(\mathbb{A}_f^p)$. Write $S' = \mathrm{Spec}(\mathcal{O}_v) \to S = \mathrm{Spec}(\mathcal{O}_{(v)})$ and $\eta' = \mathrm{Spec}(E_v) \to \eta = \mathrm{Spec}(E)$. Also write $S'' = S' \times_S S'$, $\eta'' = \eta' \times_\eta \eta'$, and write p_i ($i = 1, 2$) for the ith projection $S'' \to S'$ (resp. $\eta'' \to \eta'$). On the generic fibre $\mathcal{M}^\natural \otimes E_v$ we have an effective descent datum relative to $\eta' \to \eta$. If we consider $p_1^*(\mathcal{M}^\natural \otimes E_v) \to \eta''$ as a η'-scheme via p_2: $eta'' \to \eta'$, then this descent datum is equivalent to giving a morphism $p_1^*(\mathcal{M}^\natural \otimes E_v) \to \mathcal{M}^\natural \otimes E_v$ over η'. (Here we ignore the cocycle condition for a moment.) Since $p_1^* \mathcal{M}^\natural$, considered as a S'-scheme via p_2: $S'' \to S'$, is an a.t.s. over S', and since $\mathcal{M}^\natural$ was assumed to have the extension property, the descent datum on $\mathcal{M}^\natural \otimes E_v$ extends to one on $\mathcal{M}^\natural$ relative to $S' \to S$. (It is clear that the extended descent datum again satisfies the cocycle condition, $\mathcal{M}^\natural$ being separated.)

By the arguments of [BLR], pp. 161–162, the extended descent datum is effective. (We can work with each of the $\mathcal{M}^\natural_{K^p}$ separately, and since $\mathcal{M}^\natural$ is a smooth model, we may furthermore restrict our attention to those $\mathcal{M}^\natural_{K^p}$ which are smooth over $\mathcal{O}_v$.) Thus we obtain a smooth model $\mathcal{M}$ over $\mathcal{O}_{(v)}$. It remains to be shown that this model again has the extension property. This follows easily from property (3.5.3) and the fact that descent data for morphisms are effective ([BLR], Prop. D.4(b) in section 6.2).

(ii) The descent from $\overline{\mathcal{M}}$ to a local i.c.m. $\mathcal{M}^\natural$ is done following the same argument. By (i) this suffices.   □

3.11 From now on, we will concentrate on the case where $K_p \subset G(\mathbb{Q}_p)$ is a hyperspecial subgroup. This means that there exists a reductive group scheme $\mathcal{G}_{\mathbb{Z}_p}$ over $\mathbb{Z}_p$ (uniquely determined by K_p) with generic fibre $G_{\mathbb{Q}_p}$ such that $K_p = \mathcal{G}(\mathbb{Z}_p)$. Hyperspecial subgroups of $G(\mathbb{Q}_p)$ exist if and only if $G_{\mathbb{Q}_p}$ is unramified, i.e., quasi-split over $\mathbb{Q}_p$ and split over an unramified extension. For more on hyperspecial subgroups we refer to [Ti], [Va2].

One can show ([Va2], Lemma 3.13) that the group $\mathcal{G}_{\mathbb{Z}_p}$ is obtained by pull-back from a group scheme $\mathcal{G}$ over $\mathbb{Z}_{(p)}$. This suggests that we define an integral Shimura datum to be a pair $(\mathcal{G}, X)$, where $\mathcal{G}$ is a reductive group scheme over $\mathbb{Z}_{(p)}$, and where, writing $G = \mathcal{G}_{\mathbb{Q}}$, the pair (G, X) is a Shimura datum in the sense of 1.4[4]. To $(\mathcal{G}, X)$ we associate the Shimura variety

---

[4]We hasten to add that one has to be careful about morphisms: if we have two pairs $(\mathcal{G}_1, X_1)$ and $(\mathcal{G}_2, X_2)$ plus a morphism f: $(G_1, X_1) \to (G_2, X_2)$ such that $f(K_{p,1}) \subseteq K_{p,2}$ then it is *not* true in general that f extends to a morphism $\tilde{f}$: $\mathcal{G}_1 \to \mathcal{G}_2$; cf. [BT], 1.7 and

$Sh(\mathcal{G}, X) := Sh_{K_p}(G, X)$, where of course $K_p := \mathcal{G}(\mathbb{Z}_p)$.

Suppose $G_{\mathbb{Q}_p}$ is unramified. Whether there exists an integral canonical model of $Sh_{K_p}(G, X)$ does not depend on the choice of the hyperspecial subgroup $K_p \subset G(\mathbb{Q}_p)$. This is a consequence of the fact that the hyperspecial subgroups of $G(\mathbb{Q}_p)$ are conjugate under $G^{\mathrm{ad}}(\mathbb{Q}_p)$, see [Va2], 3.2.7.

3.12 Examples. (i) Let $(T, \{h\})$ be a Shimura datum with T a torus. The group $T_{\mathbb{Q}_p}$ is unramified precisely if the character group $X^*(T)$ is unramified at p as a $\mathrm{Gal}(\overline{\mathbb{Q}}/\mathbb{Q})$-module. If this is the case then $T_{\mathbb{Q}_p}$ extends uniquely to a torus $\mathcal{T}$ over $\mathbb{Z}_p$, and $K_p := \mathcal{T}(\mathbb{Z}_p)$ is the unique hyperspecial subgroup of $T(\mathbb{Q}_p)$. Let $K^p \subset T(\mathbb{A}_f^p)$ be a compact open subgroup. It follows from the description given in 2.2 and 2.3 that $Sh_{K_p \times K^p}(T, \{h\}) = \mathrm{Spec}(L_1 \times \cdots \times L_r)$ for certain number fields $L_i \supset E$ which are unramified above p. Now set $\mathcal{M}_{K_p \times K^p} = \mathrm{Spec}(\mathcal{O}_1 \times \cdots \times \mathcal{O}_r)$, where $\mathcal{O}_i$ is the normalization of $\mathcal{O}_{(v)}$ in L_i. Then $\varprojlim_{K^p} \mathcal{M}_{K_p \times K^p}$ is an integral canonical model of $Sh_{K_p}(T, \{h\})$ over $\mathcal{O}_{(v)}$.

(ii) If (G, X) defines a Shimura variety of PEL type, then we can use the modular interpretation of $Sh(G, X)$ to study integral canonical models. As mentioned before, the precise formulation of a moduli problem requires a lot of data, and we refer to [Ko2] for details. We remark that the Shimura varieties that we are interested in, in general only form an open subscheme of the moduli space studied in loc. cit., section 5. The arguments given there (see also [LR], §6) show that, for primes p satisfying suitable conditions which imply the existence of a hyperspecial subgroup $K_p \subset G(\mathbb{Q}_p)$, the Shimura variety $Sh_{K_p}(G, X)$ has an i.c.m. over $\mathcal{O}_{(v)}$ for all primes v of $E(G, X)$ above p.

3.13 Remark. If there exists an i.c.m. $\mathcal{M}$ for $Sh(\mathcal{G}, X)$, then one expects that each "finite level" $\mathcal{M}_{K^p}$ is a quasi-projective $\mathcal{O}_{(v)}$-scheme. This is certainly the case for the examples in 3.8 and 3.12. Moreover, one easily checks that the quasi-projectivity is preserved under all constructions presented in this section.

3.14 Our next goal is to show that if $G_{\mathbb{Q}_p}$ is unramified, then we can adapt [De3], 2.1.5–8 (which we summarized in 1.6.5) to the present context. The connected component of $Sh_{K_p}(G, X)_{\overline{\mathbb{Q}}}$ containing the image of $X^+ \times \{e\}$ is the projective limit $\varprojlim \Gamma \backslash X^+$, where $\Gamma = \mathrm{Im}\big([G^{\mathrm{der}}(\mathbb{Q})^+ \cap (K_p \times K^p)] \to G^{\mathrm{ad}}(\mathbb{Q})^+\big)$ for some compact open subgroup $K^p \subset G^{\mathrm{der}}(\mathbb{A}_f^p)$. (Here we use [De3], 2.0.13.)

Formalizing this, we are led to consider pairs $(\mathcal{G}', X^+)$ consisting of a semi-simple group $\mathcal{G}'$ over $\mathbb{Z}_{(p)}$ and a $\mathcal{G}^{\mathrm{ad}}(\mathbb{R})^+$-conjugacy class of homomorphisms $h\colon \mathbb{S} \to G_{\mathbb{R}}^{\mathrm{ad}}$ (writing $G' = \mathcal{G}'_{\mathbb{Q}}$, $G^{\mathrm{ad}} = \mathcal{G}_{\mathbb{Q}}^{\mathrm{ad}} := (\mathcal{G}')_{\mathbb{Q}}^{\mathrm{ad}}$) such that conditions (i), (ii) and (iii) in 1.4 are satisfied. For such a pair we define the topology $\tau(\mathcal{G}')$ on $\mathcal{G}^{\mathrm{ad}}(\mathbb{Z}_{(p)})$ as the linear topology having as a fundamental system of neighbourhoods of 1 the images of the $\{p, \infty\}$-congruence subgroups $\mathcal{G}'(\mathbb{Z}_{(p)}) \cap K^p$, where K^p is a compact open subgroup of $\mathcal{G}'(\mathbb{A}_f^p)$. We then write $Sh^0(\mathcal{G}', X^+)_{\mathbb{C}} := \varprojlim \Gamma \backslash X^+$, where Γ runs through the $\{p, \infty\}$-arithmetic subgroups of $\mathcal{G}^{\mathrm{ad}}(\mathbb{Z}_{(p)})$ which are open in $\tau(\mathcal{G}')$.

On $Sh^0(\mathcal{G}', X^+)_{\mathbb{C}}$ we have a continuous action of $\mathcal{G}^{\mathrm{ad}}(\mathbb{Z}_{(p)})^{+\wedge}$ (completion rel. $\tau(\mathcal{G}')$), and by 2.13, these data are all canonically defined over $\overline{\mathbb{Q}}$. (Even over a much smaller field, as we shall see next.) For an integral Shimura datum $(\mathcal{G}, X)$ and a connected component $X^+ \subseteq X$, the corresponding connected component of $Sh(\mathcal{G}, X)_{\overline{\mathbb{Q}}}$ is a scheme with continuous $\mathcal{G}^{\mathrm{ad}}(\mathbb{Z}_{(p)})^{+\wedge}$-action, isomorphic to $Sh^0(\mathcal{G}^{\mathrm{der}}, X^+)_{\overline{\mathbb{Q}}}$. Note that $Sh^0(\mathcal{G}', X^+)$ is an integral scheme (use [EGA], IV, Cor. 8.7.3).

3.15 Lemma. *Let (G, X) be a Shimura datum, $E = E(G, X)$, v a prime of E dividing p. Assume that $G_{\mathbb{Q}_p}$ is unramified, and let $K_p \subset G(\mathbb{Q}_p)$ be a hyperspecial subgroup. Then the connected components of $Sh_{K_p}(G, X)$ are defined over an abelian extension $\widetilde{E}$ of E which is unramified above p.*

Proof. First we prove this under the additional assumption that G^{der} is simply connected. The $G(\mathbb{C})$-conjugacy class of homomorphisms $\mu_x\colon \mathbb{G}_{m,\mathbb{C}} \to G_{\mathbb{C}}$ (for $x \in X$) gives rise to a well-defined cocharacter $\mu^{\mathrm{ab}}\colon \mathbb{G}_{m,\mathbb{C}} \to G_{\mathbb{C}}^{\mathrm{ab}}$, which has field of definition $E(G^{\mathrm{ab}}, X^{\mathrm{ab}}) \subseteq E$. Writing $T_E = \mathrm{Res}_{E/\mathbb{Q}} \mathbb{G}_{m,E}$, we get a homomorphism

$$\rho = \underset{E/\mathbb{Q}}{\mathrm{Nm}} \circ \mu_E^{\mathrm{ab}}\colon T_E \to G^{\mathrm{ab}},$$

inducing a map $\rho(\mathbb{A}/\mathbb{Q})\colon \mathbb{A}_E^*/E^* \to G^{\mathrm{ab}}(\mathbb{A})/G^{\mathrm{ab}}(\mathbb{Q}) = \pi(G^{\mathrm{ab}})$. The assumption that G^{der} is simply connected implies (see [De1], 2.7) that $\pi_0 \pi(G)$ is a quotient of $\pi(G^{\mathrm{ab}})$. Moreover, the action of $\mathrm{Gal}(\overline{\mathbb{Q}}/E)^{\mathrm{ab}}$ on $\pi_0(Sh(G, X))$ factors through $\pi_0 \rho(\mathbb{A}/\mathbb{Q})\colon \pi_0(T_E) \to \pi_0 \pi(G^{\mathrm{ab}})$. By class field theory it therefore suffices to show that the image under $\rho(\mathbb{Q}_p)$ of $C_p := \prod_{v|p} \mathcal{O}_v^* \subset T_E(\mathbb{Q}_p)$ in $G^{\mathrm{ab}}(\mathbb{Q}_p)$ is contained in $K_p^{\mathrm{ab}} := \mathrm{Im}\big(K_p \subset G(\mathbb{Q}_p) \to G^{\mathrm{ab}}(\mathbb{Q}_p)\big)$.

The fact that $G_{\mathbb{Q}_p}$ is unramified implies ([Mi4], Cor. 4.7) that T_E is unramified over $\mathbb{Q}_p$, so it extends to a torus $\mathcal{T}_E$ over $\mathbb{Z}_p$. Clearly, $C_p = \mathcal{T}_E(\mathbb{Z}_p)$. Write $\mathcal{G}$ for the extension of $G_{\mathbb{Q}_p}$ to a reductive group scheme over $\mathbb{Z}_p$ with $K_p = \mathcal{G}(\mathbb{Z}_p)$. The map ρ extends to a homomorphism $\mathcal{T}_E \to \mathcal{G}^{\mathrm{ab}}$ over $\mathbb{Z}_p$, hence we are done if we show that $\mathcal{G}^{ab}(\mathbb{Z}_p) = K_p^{ab}$, i.e., $\mathcal{G}(\mathbb{Z}_p)$ maps surjectively to $\mathcal{G}^{ab}(\mathbb{Z}_p)$. Again using that G^{der} is simply connected we have $H^1(\mathbb{Q}_p, G^{der}) = \{1\}$, hence $G(\mathbb{Q}_p) \twoheadrightarrow G^{ab}(\mathbb{Q}_p)$. For $s \in \mathcal{G}^{ab}(\mathbb{Z}_p)$ we thus can lift the corresponding $s_\eta \in G^{ab}(\mathbb{Q}_p)$ to $\tilde{s}_\eta \in G(\mathbb{Q}_p)$. Taking the Zariski closure of the image of $\tilde{s}_\eta$ inside $\mathcal{G}$ then gives the desired $\mathbb{Z}_p$-valued point $\tilde{s}$ of $\mathcal{G}$ mapping to s.

The general case is reduced to the previous one. An easy generalization of [MS], Application 3.4 shows that there exists a morphism of Shimura data $f \colon (G_1, X_1) \to (G, X)$ such that $f^{der} \colon G_1^{der} \to G^{der}$ is the universal covering of G^{der}, such that $E(G_1, X_1) = E(G, X)$, and such that there is a hyperspecial subgroup $\widetilde{K}_p \subset G_1(\mathbb{Q}_p)$ with $f(\widetilde{K}_p) \subseteq K_p$. This suffices to prove the lemma, since the components of $Sh_{\widetilde{K}_p}(G_1, X_1)$ map surjectively to components of $Sh_{K_p}(G, X)$ and since all components have the same field of definition (being permuted transitively under the $G(\mathbb{A}_f)$-action). □

3.16 Consider a pair $(\mathcal{G}', X^+)$ as in 3.14. Write $G' = \mathcal{G}'_{\mathbb{Q}}$, and write $\widetilde{E}$ for the maximal subfield of $E(G^{ad}, X^{ad})^{ab}$ which is unramified above p. The lemma implies that the connected Shimura variety $Sh^0(\mathcal{G}', X^+)$ has a well-defined "canonical" model over $\widetilde{E}$. Indeed, we can choose an integral Shimura datum $(\mathcal{G}, X)$ with $\mathcal{G}' = \mathcal{G}^{der}$, $X^+ \subseteq X$ and $E(\mathcal{G}, X) = E(\mathcal{G}^{ad}, X^{ad})$, and take $Sh^0(\mathcal{G}^{der}, X^+)_{\widetilde{E}}$ (which makes sense, grace to the lemma) as the desired model. That this does not depend on the chosen pair $(\mathcal{G}, X)$ follows from the facts in 2.7.

3.17 Definition. Write $Sh^0(\mathcal{G}', X^+)_{\widetilde{E}}$ for the model over $\widetilde{E}$ just defined, and let w be a prime of $\widetilde{E}$ above p. We adapt Def. 3.2 to connected Shimura varieties, replacing E by $\widetilde{E}$ and $G(\mathbb{A}_f)$ by $\mathcal{G}(\mathbb{Z}_{(p)})^{+\wedge}$. Then an integral canonical model (resp. local i.c.m.) for $Sh^0(\mathcal{G}', X^+)_{\widetilde{E}}$ at w is a separated smooth integral model over $\mathcal{O}_{(w)}$ (resp. $\mathcal{O}_w$) which has the extension property.

Of course, the point of this definition is that a Shimura variety can be recovered from the (or rather: some) corresponding connected Shimura variety by an "induction" procedure. This will enable us to follow the same strategy as in 2.10. We consider the properties

$$\mathcal{P}(\mathcal{G}, X; v) : \quad \text{there exists an i.c.m. for } Sh(\mathcal{G}, X) \text{ over } \mathcal{O}_{(v)}$$

(for $(\mathcal{G}, X)$ an integral Shiumura datum, v a prime of $E = E(G, X)$ above p), and

$$\mathcal{P}^0(\mathcal{G}', X^+; w) : \quad \text{there exists a local i.c.m. for } Sh^0(\mathcal{G}', X^+)_{\widetilde{E}} \text{ over } \mathcal{O}_{(w)}$$

(for $(\mathcal{G}', X^+)$, $\widetilde{E}$ and w as above). Using the induction technique of [De3], Lemma 2.7.3 and Prop. 3.10, we can prove the following statement. We leave the details of the proof to the reader.

3.18 Proposition. *Notations as above, with* $\mathcal{G}' = \mathcal{G}^{\mathrm{der}}$, $X^+ \subseteq X$. *Suppose that v and w restrict to the same prime of $E \cap \widetilde{E}$. Then* $\mathcal{P}(\mathcal{G}, X; v) \Longleftrightarrow \mathcal{P}^0(\mathcal{G}^{\mathrm{der}}, X^+; w)$.

From now on we restrict our attention to the case $p > 2$. Recall that it is implicit in our notations that we are working at a prime where the group is unramified, since $\mathcal{G}$ and $\mathcal{G}'$ are supposed to be *reductive* group schemes over $\mathbb{Z}_{(p)}$. Write $\mathcal{P}(\mathcal{G}, X)$ for "$\mathcal{P}(\mathcal{G}, X; v)$ holds for all primes v of E above p", and similarly for $\mathcal{P}^0(\mathcal{G}', X^+)$. We have shown that statements (a) and (c) in 2.10 hold. Furthermore, statement (d) is almost trivially true. By contrast, it is not at all obvious how to prove (b). The only thing we get more or less for free is a good *normal* model.

3.19 Proposition. *Let* $i \colon (G_1, X_1) \hookrightarrow (G_2, X_2)$ *be a closed immersion of Shimura data such that there exist hyperspecial subgroups $K_{j,p} \subset G_j(\mathbb{Q}_p)$ with $i(K_{1,p}) \subseteq K_{2,p}$. Suppose there exists an i.c.m. $\mathcal{M}$ for $\mathrm{Sh}_{K_{2,p}}(G_2, X_2)$ over $\mathcal{O}_{E_2,(v)}$. If w is a prime of $E_1 = E(G_1, X_1)$ above v then there exists a normal integral model $\mathcal{N}$ of $\mathrm{Sh}_{K_{1,p}}(G_1, X_1)$ over $\mathcal{O}_{E_1,(w)}$ which has the extension property (see Def. 3.3).*

Proof. Let $\mathcal{G}_j$ ($j = 1, 2$) denote the extension of G_j to a reductive group scheme over $\mathbb{Z}_{(p)}$ with $\mathcal{G}_j(\mathbb{Z}_p) = K_{j,p}$. Write $\mathcal{K}$ for the set of pairs (K_1^p, K_2^p) of compact open subgroups $K_j^p \subset G_j(\mathbb{A}_f^p)$ such that $i(K_1^p) \subset K_2^p$, partially ordered by $(K_1^p, K_2^p) \preceq (L_1^p, L_2^p)$ iff $K_1^p \supseteq L_1^p$ and $K_2^p \supseteq L_2^p$. Given $(K_1^p, K_2^p) \in \mathcal{K}$, we have a morphism

$$i(K_1^p, K_2^p) \colon \mathrm{Sh}_{K_1^p}(\mathcal{G}_1, X_1) \longrightarrow \mathrm{Sh}_{K_2^p}(\mathcal{G}_2, X_2) \hookrightarrow \mathcal{M}_{K_2^p} \otimes \mathcal{O}_{E_1,(w)}\,.$$

Write $N(K_1^p, K_2^p)$ for the (scheme-theoretical) image of $i(K_1^p, K_2^p)$, and let $\mathcal{N}(K_1^p, K_2^p)$ be its normalization. For fixed K_1^p we set

$$N_{K_1^p} = \varprojlim_{K_2^p} N(K_1^p, K_2^p)\,, \quad \mathcal{N}_{K_1^p} = \varprojlim_{K_2^p} \mathcal{N}(K_1^p, K_2^p)\,, \quad \mathcal{M}_{K_1^p} = \varprojlim_{K_2^p} \mathcal{M}_{K_2^p}\,,$$

where the limits run over all K_2^p such that $(K_1^p, K_2^p) \in \mathcal{K}$. Also we set

$$N := \varprojlim_{K_1^p} N_{K_1^p}\,, \quad \mathcal{N} := \varprojlim_{K_1^p} \mathcal{N}_{K_1^p}\,.$$

First we show that, for $K_1^p \supseteq L_1^p$ sufficiently small, the canonical morphism $\mathcal{N}_{L_1^p} \to \mathcal{N}_{K_1^p}$ is étale. For this, we take compact open subgroups $C_j^p \subset G_j(\mathbb{A}_f^p)$ with $i(C_1^p) \subseteq C_2^p$, and such that for all $K_j^p \supseteq L_j^p$ contained in C_j^p ($j = 1, 2$), the transition morphisms $Sh_{L_1^p}(\mathcal{G}_1, X_1) \to Sh_{K_1^p}(\mathcal{G}_1, X_1)$ and $\mathcal{M}_{L_2^p} \to \mathcal{M}_{K_2^p}$ are étale morphisms of smooth schemes over E_1 and $\mathcal{O}_{E_2,(v)}$ respectively. One checks that for all such $K_1^p \supseteq L_1^p$, the morphism $t \colon \mathcal{M}_{L_1^p} \to \mathcal{M}_{K_1^p}$ is again étale, of degree $[K_1^p : L_1^p]$. It follows that $N_{L_1^p} \to N_{K_1^p}$ is a pull-back of t, hence étale. Now $N_{K_1^p}$ has finitely many irreducible components, being a scheme-theoretical image of $Sh_{K_1^p}(\mathcal{G}_1, X_1)$, and the normalization of $N_{K_1^p}$ is just $\mathcal{N}_{K_1^p}$. Using this remark, it follows that $\mathcal{N}_{L_1^p} \to \mathcal{N}_{K_1^p}$ is étale, so that $\mathcal{N}$ is a normal model of $Sh(\mathcal{G}_1, X_1)$ over $\mathcal{O}_{E_1,(w)}$.

That $\mathcal{N}$ has the extension property is seen as follows. We consider an $S \in$ ATS$_{\mathcal{O}}$ (with $\mathcal{O} = \mathcal{O}_{E_1,(w)}$) and a morphism $\alpha_{E_1} \colon S_{E_1} \to \mathcal{N}_{E_1}$ on the generic fibre. The fact that $\mathcal{O}_{E_1,(w)}$ is an unramified extension of $\mathcal{O}_{E_2,(v)}$ implies, using (3.5.3), that $\mathcal{M} \otimes \mathcal{O}_{E_1,(w)}$ has the extension property over $\mathcal{O}_{E_1,(w)}$, hence α_E extends to a morphism

$$\alpha \colon S \longrightarrow N \hookrightarrow \mathcal{M} \otimes \mathcal{O}_{E_1,(w)}.$$

Now fix $(K_1^p, K_2^p) \in \mathcal{K}$, and set

$$\widetilde{S} = \widetilde{S}(K_1^p, K_2^p) := S \underset{N(K_1^p, K_2^p)}{\times} \mathcal{N}(K_1^p, K_2^p) \xrightarrow{\ \rho\ } S.$$

Then $\widetilde{S}$ is integral over S, since ρ is a pull-back of the normalization map $\mathcal{N}(K_1^p, K_2^p) \to N(K_1^p, K_2^p)$. On the generic fibre, ρ is an isomorphism. Since S is a normal scheme (being an a.t.s.), it follows that ρ is an isomorphism, hence α lifts to $\tilde{\alpha} \colon S \to \mathcal{N}$. $\qquad\Box$

3.20 Remark. Suppose $\mathcal{M}$ is an integral model of a Shimura variety over a d.v.r. $\mathcal{O}$. We will say that $\mathcal{M}$ has the extended extension property (e.e.p.), if it satisfies the condition that for every $S = \mathrm{Spec}(\mathcal{O}_1)$ with $\mathcal{O} \subset \mathcal{O}_1$ a faithfully flat extension of d.v.r., setting $F = \mathrm{Frac}(\mathcal{O}_1)$, every morphism $\alpha_F \colon \mathrm{Spec}(F) \to \mathcal{M}_F$ over $\mathcal{O}$ extends to an $\mathcal{O}$-morphism $\alpha \colon S \to \mathcal{M}$.

It follows from the Néron-Ogg-Shafarevich criterion that the model of the Siegel modular variety as in 3.8 enjoys the e.e.p. Also it is clear that in the situation of 3.19, we have the implication "$\mathcal{M}$ has the e.e.p. $\Rightarrow \mathcal{N}$ has the e.e.p.", if $\mathcal{N}$ is the model constructed in the proof.

3.21 The last step in our strategy is statement (e). So, we consider a pair (G', X^+) defining a connected Shimura variety and an isogeny $\pi \colon G' \twoheadrightarrow G''$.

We assume that G' (hence also G'') is unramified over $\mathbb{Q}_p$, so that π extends to an isogeny $\pi\colon \mathcal{G}' \to \mathcal{G}''$ of semi-simple groups over $\mathbb{Z}_{(p)}$.

Let us also assume that there exists an i.c.m. $\mathcal{M}$ of $Sh^0(\mathcal{G}', X^+)_{\widetilde{E}}$ over $\mathcal{O}_{(w)}$, where w is a prime of the field $\widetilde{E}$ (as in 3.16) above p. We want to show that there exists an i.c.m. $\mathcal{N}$ of $Sh^0(\mathcal{G}'', X^+)_{\widetilde{E}}$ over $\mathcal{O}_{(w)}$.

3.21.1 Set

$$\Delta := \mathrm{Ker}\left[\mathcal{G}^{\mathrm{ad}}(\mathbb{Z}_{(p)})^{+\wedge}\quad \mathrm{rel.}\ \tau(\mathcal{G}') \longrightarrow \mathcal{G}^{\mathrm{ad}}(\mathbb{Z}_{(p)})^{+\wedge}\quad \mathrm{rel.}\ \tau(\mathcal{G}'')\right].$$

This is a finite group which acts freely on $Sh^0(\mathcal{G}', X^+)_{\widetilde{E}}$. The canonical morphism $Sh(\pi)\colon Sh^0(\mathcal{G}', X^+)_{\widetilde{E}} \to Sh^0(\mathcal{G}'', X^+)_{\widetilde{E}}$ is a quotient morphism for this action. (Cf. [De3], 2.7.11 (b).) Since $\mathcal{M}$ has the extension property, the action of Δ on $\mathcal{M}_{\widetilde{E}}$ extends uniquely to an action on $\mathcal{M}$. The natural candidate for an i.c.m. of $Sh^0(\mathcal{G}'', X^+)_{\widetilde{E}}$ is the quotient $\mathcal{N} := \mathcal{M}/\Delta$.

3.21.2 Problem. *Consider a faithfully flat extension of d.v.r. $\mathbb{Z}_{(p)} \subseteq \mathcal{O}$. Let Δ be a finite (abstract) group acting on a faithfully flat $\mathcal{O}$-scheme $\mathcal{M}$ which is locally noetherian and formally smooth over $\mathcal{O}$. Assume the action of Δ on the generic fibre of $\mathcal{M}$ is free. Under what further conditions does it follow that the action of Δ on all of $\mathcal{M}$ is free?*

3.21.3 It follows from a result of Edixhoven ([Ed1], Prop. 3.4) that, under the previous assumptions, the action of Δ on all of $\mathcal{M}$ is free if p does not divide the order of Δ. On the other hand, if p does divide $|\Delta|$, then extra assumptions are needed.

Example 1: take $\mathcal{O} = \mathbb{Z}_p[\zeta_p]$, $\mathcal{M} = \mathrm{Spec}(\mathcal{O}[\![x]\!])$ with the automorphism of order p given by $x \mapsto \zeta_p \cdot x - (\zeta_p - 1)$. In this case, the action of $\mathbb{Z}/p\mathbb{Z}$ on the generic fibre is free (note that $x - 1$ is a unit in $\mathcal{O}[\![x]\!]$), but the action on the special fibre is trivial.

In order to avoid examples of this kind, we can add the assumption that $p > 2$ and $e(\mathcal{O}/\mathbb{Z}_{(p)}) < p - 1$. (In the situation where we want to use it, this holds anyway.) That this is not a sufficient condition is shown by the following example that was communicated to us by Edixhoven.

Example 2: write Λ for the $\mathbb{Z}_p$-module $\mathbb{Z}_p \oplus \mathbb{Z}_p[\zeta_p]$, and consider the automorphism of order p given by $(x, y) \mapsto (x, \zeta_p \cdot y)$. This induces a $\mathbb{Z}_p$-linear automorphism of order p on $\mathbb{P}(\Lambda) = \mathbb{P}_{\mathbb{Z}_p}^{p-1}$. On the generic fibre there are (geometrically) p fixed points. On the special fibre there is an $\mathbb{F}_p$-rational line of fixed points. By removing the closure of the fixed points in the generic fibre we obtain a $\mathbb{Z}_p$-scheme $\mathcal{M}$ with a $\mathbb{Z}/p\mathbb{Z}$-action as in 3.21.2, such that the action is *not* free on the special fibre.

3.21.4 Proposition. *In the situation of 3.21, suppose that (i) the action of Δ on $\mathcal{M}$ is free, and (ii) $\mathcal{M}$ has the extended extension property (see 3.20). Then $\mathcal{N} := \mathcal{M}/\Delta$ is an i.c.m. of $Sh^0(\mathcal{G}'', X^+)_{\widetilde{E}}$ over $\mathcal{O}_{(w)}$.*

Proof.   Condition (i) implies that $\mathcal{N}$ is a smooth model, so it remains to be shown that it has the extension property. Consider an a.t.s. S over $\mathcal{O}_{(w)}$ plus a morphism $\alpha_{\widetilde{E}} \colon S_{\widetilde{E}} \to \mathcal{N}_{\widetilde{E}}$. Let

$$ T_{\widetilde{E}} := (S_{\widetilde{E}} \underset{\mathcal{N}_{\widetilde{E}}}{\times} \mathcal{M}_{\widetilde{E}}) \xrightarrow{\ \beta_{\widetilde{E}}\ } \mathcal{M}_{\widetilde{E}}, $$

and write T for the integral closure of S in the fraction ring of $T_{\widetilde{E}}$. We have a canonical morphism $\rho \colon T \to S$. If $U \subseteq S$ is an open subscheme such that $\alpha_{\widetilde{E}|U_{\widetilde{E}}}$ extends to $\alpha_U \colon U \to \mathcal{N}$, then $\rho^{-1}(U) \cong U \times_{\mathcal{N}} \mathcal{M}$, so that $\rho^{-1}(U) \to U$ is étale.

We now first consider the special case where $S = \mathrm{Spec}(A)$ for some d.v.r. A which is faithfully flat over $\mathcal{O}_{(w)}$. It then follows from the e.e.p. of $\mathcal{M}$ that $\beta_{\widetilde{E}}$ extends to a morphism $\beta \colon T \to \mathcal{M}$ which is equivariant for the action of Δ. On quotients this gives the desired extension α of $\alpha_{\widetilde{E}}$.

Back to the general case, it follows from the special case, the remarks preceding it and the Zariski-Nagata purity theorem of [SGA1] Exp. X, 3.1, that $\rho \colon T \to S$ is étale, so that $T \in \mathrm{ATS}_{\mathcal{O}_{(w)}}$. This again gives an extension β of $\beta_{\widetilde{E}}$ and, on quotients, an extension α as desired. $\square$

3.21.5   For a reductive group G over $\mathbb{Q}$, define δ_G as the degree of the covering $\widetilde{G} \to G^{\mathrm{ad}}$. (In other words: δ_G is the "connectedness index" of the root system of $G_{\overline{\mathbb{Q}}}$.) By definition, δ_G depends only on G^{ad}. We claim that, in the situation of 3.21 and 3.21.1, the order of Δ is invertible in $\mathbb{Z}[1/\delta]$, where $\delta = \delta_{G'} = \delta_{G''}$. To prove this, we need some facts and notations. We write $\rho_1 \colon \widetilde{G} \to G'$ and $\rho_2 \colon \widetilde{G} \to G''$ for the canonical maps from the universal covering. Writing $\Gamma_1 := \rho_1 \widetilde{G}(\mathbb{A}_f^p) \cap \mathcal{G}'(\mathbb{Z}_{(p)})$, $\Gamma_2 := \rho_2 \widetilde{G}(\mathbb{A}_f^p) \cap \mathcal{G}''(\mathbb{Z}_{(p)})$, we have

$$ \mathcal{G}^{\mathrm{ad}}(\mathbb{Z}_{(p)})^{+\wedge} \quad \mathrm{rel.}\ \tau(\mathcal{G}') = \rho_1 \widetilde{G}(\mathbb{A}_f^p) \underset{\Gamma_1}{*} \mathcal{G}^{\mathrm{ad}}(\mathbb{Z}_{(p)})^+ , $$

and similarly for $\mathcal{G}^{\mathrm{ad}}(\mathbb{Z}_{(p)})^{+\wedge}$ rel. $\tau(\mathcal{G}'')$. (Cf. [De3], (2.1.6.2).)

Write $\mathcal{K} := \mathrm{Ker}(\rho_2 \colon \widetilde{\mathcal{G}} \to \mathcal{G}'')$. We claim there is an exact sequence

$$ (3.21.6) \qquad\qquad \mathcal{K}(\mathbb{A}_f^p) \xrightarrow{\ t\ } \Delta \xrightarrow{\ u\ } \Gamma_2/\rho_2 \widetilde{\mathcal{G}}(\mathbb{Z}_{(p)}) . $$

Here the map t sends an element $g \in \widetilde{\mathcal{G}}$ with $\rho_2(g) = e_{\mathcal{G}''}$ to the element $\rho_1(g) *_{\Gamma_1} e_{\mathcal{G}^{\mathrm{ad}}}$, which obviously lies in Δ. The map u sends an element $x *_{\Gamma_1} y \in \Delta \subset \rho_1 \widetilde{G}(\mathbb{A}_f^p) *_{\Gamma_1} \mathcal{G}^{\mathrm{ad}}(\mathbb{Z}_{(p)})^+$ to $\pi(x)$ mod $\rho_2 \widetilde{\mathcal{G}}(\mathbb{Z}_{(p)})$; notice that $x *_{\Gamma_1} y \in \Delta$ means that $(\pi(x), y) = (\gamma^{-1}, \mathrm{ad}(\gamma))$ for some $\gamma \in \Gamma_2$. If $x *_{\Gamma_1} y \in \mathrm{Ker}(u)$ then we can take $\gamma = \rho_2(g)$ for some $g \in \widetilde{\mathcal{G}}(\mathbb{Z}_{(p)})$, in which case $x *_{\Gamma_1} y = (x \cdot \rho_1(g)) *_{\Gamma_1} e_{\mathcal{G}^{\mathrm{ad}}} \in \mathrm{Im}(t)$. This proves the exactness of (3.21.6).

It follows from the definitions that every element of $\mathcal{K}(\mathbb{A}_f^p)$ has a finite order dividing δ. On the other hand, $\Gamma_2/\rho_2 \widetilde{\mathcal{G}}(\mathbb{Z}_{(p)})$ is a subgroup of $\mathrm{H}^1_{\mathrm{fppf}}(\mathbb{Z}_{(p)}, \mathcal{K})$, in which again all elements are killed by δ. This proves our claim that $|\Delta| \in \mathbb{Z}[1/\delta]^*$.

For simple groups G, the number δ_G is given by $\delta(\mathrm{A}_\ell) = \ell + 1$, $\delta(\mathrm{B}_\ell) = 2$, $\delta(\mathrm{C}_\ell) = 2$, $\delta(\mathrm{D}_\ell) = 4$, $\delta(\mathrm{E}_6) = 3$, $\delta(\mathrm{E}_7) = 2$. (The other three simple types have $\delta = 1$ but do not occur as part of a Shimura datum.) In particular, we see that $|\Delta|$ is invertible in $\mathbb{Z}[1/6]$ if G does not contain factors of type A_ℓ.

After the technical problems encountered in our discussion of steps (b) and (e), the good news is that we can prove the converse of (e).

3.22 Proposition. *Consider the situation as in the first paragraph of 3.21, and assume that $Sh^0(\mathcal{G}'', X^+)_{\widetilde{E}}$ has an i.c.m. $\mathcal{N}$ over $\mathcal{O}_{(w)}$. Then the normalization $\mathcal{M}$ of $\mathcal{N}$ in the fraction field of $Sh^0(\mathcal{G}', X^+)_{\widetilde{E}}$ is an i.c.m. of $Sh^0(\mathcal{G}', X^+)_{\widetilde{E}}$ over $\mathcal{O}_{(w)}$.*

Proof. First we remark that the action of the group Δ on $\mathcal{M}_{\widetilde{E}}$ extends to an action on $\mathcal{M}$ and that $\mathcal{M}/\Delta \overset{\sim}{\to} \mathcal{N}$. (We have a map $\mathcal{M}/\Delta \to \mathcal{N}$ which is an isomorphism on generic fibres; now use that $\mathcal{N}$ is normal.) We claim that the action of Δ on $\mathcal{M}$ is free. On the generic fibre we know this. The important point now is that the purity theorem applies, so that possible fixed points must occur in codimension 1. So, suppose Δ has fixed points. Without loss of generality we may assume that Δ is cyclic of order p (cf. 3.21.3). Restricting to a suitable open part $\mathrm{Spec}(A) \subset \mathcal{M}$, we then obtain a nontrivial automorphism of order p of the $\mathcal{O}_{(w)}$-module A which (using purity and the fact that the action is free on the generic fibre) is the identity modulo p. But now we have the following fact from algebra, probably well-known and in any case not difficult to prove: if R is a principal ideal domain, $p > 2$ a prime number with $(p) \neq R$, M a flat R-module, and α an R-module automorphism of M with $\alpha^p = \mathrm{id}_M$ and $(\alpha \bmod p \colon M/pM \to M/pM) = \mathrm{id}_{M/pM}$, then $\alpha = \mathrm{id}_M$. Applying this fact we obtain a contradiction, and it follows that $\mathcal{M}$ is a smooth model.

For the extension property, consider an $S \in \mathrm{ATS}_{\mathcal{O}_{(w)}}$ and a morphism $\alpha_{\widetilde{E}} \colon S_{\widetilde{E}} \to \mathcal{M}_{\widetilde{E}}$. The projection $\beta_{\widetilde{E}} \colon S_{\widetilde{E}} \to \mathcal{N}_{\widetilde{E}}$ of $\alpha_{\widetilde{E}}$ to $\mathcal{N}$ extends to a morphism $\beta \colon S \to \mathcal{N}$. Set $T := S \times_{\mathcal{N}} \mathcal{M}$, then $T \to S$ is a finite étale Galois covering with group Δ. The section $T_{\widetilde{E}} \leftarrow S_{\widetilde{E}}$ on the generic fibres (corresponding to $\alpha_{\widetilde{E}}$) therefore extends to a section on all of S (recall that S and T are flat over $\mathcal{O}_{(w)}$ and normal), which means that $\alpha_{\widetilde{E}}$ extends to a morphism α. $\qquad\square$

Combining all the results in this section, we arrive at the following conclusion.

3.23 Corollary. *Fix a prime number $p > 2$. Let (H, Y) be a Shimura datum of pre-abelian type with $p \nmid \delta_H$, and let v be a prime of $E(H, Y)$ above p. Suppose that for each simple factor $(G^{\mathrm{ad}}, X^{\mathrm{ad}})$ of the adjoint datum $(H^{\mathrm{ad}}, Y^{\mathrm{ad}})$, there exist:*

(i) a Shimura datum (G, X) covering $(G^{\mathrm{ad}}, X^{\mathrm{ad}})$,

(ii) a closed immersion $i \colon (G, X) \hookrightarrow (\mathrm{CSp}_{2g}, \mathfrak{H}_g^{\pm})$,

(iii) a prime w of $E(G, X)$ such that v and w restrict to the same prime of $E(G^{\mathrm{ad}}, X^{\mathrm{ad}})$,

(iv) a hyperspecial subgroup $K_p \subset G(\mathbb{Q}_p)$ with $i(K_p) \subseteq \mathrm{CSp}_{2g}(\mathbb{Z}_p)$,

such that the normal model $\mathcal{N}$ of $Sh_{K_p}(G, X)$ constructed in 3.19 is a formally smooth $\mathcal{O}_{(w)}$-scheme. Then for every hyperspecial subgroup $L_p \subset H(\mathbb{Q}_p)$ there exists an integral canonical model of $Sh_{L_p}(H, Y)$ over $\mathcal{O}_{(v)}$.

3.24 Remark. In this section, we have tried to follow the strategy of [De3] very closely, adapting results to the p-adic context whenever possible. We wish to point out that our presentation of the above material is very different from the treatment in Vasiu's paper [Va2]. In particular, our definitions are different (see 3.9), and models of connected Shimura varieties (which play a central role in our discussion) do not appear in [Va2]. Vasiu claims 3.23 (using his definitions) *without* the condition that $p \nmid \delta_H$. We were not able to understand his proof of this (in which one step is postponed to a future publication). It seems to us that at several points the arguments are incomplete, and that Vasiu's proof furthermore contains some arguments which are not correct as they stand.

3.25 Remark. We should mention Morita's paper [Mor]. (See also Carayol's paper [Ca].) Of particular interest, in connection with the material discussed in this section, are the following two aspects.

(i) Morita proves that certain Shimura varieties (of dimension 1) have good reduction by relating them to other Shimura varieties which are of Hodge type. (The example is classical—see §6, "Modèles étranges" in Deligne's Bourbaki paper [De3].) In Morita's method of proof we recognize several results that have reappeared in this section in an abstract and somewhat more general form.

(ii) The Shimura varieties in question $(M_0' = M_{\mathbb{Z}_p^\times \times K_\mathfrak{p}}'(G', X'))$ in the notations of [Ca]) are shown to have good reduction at certain primes $\mathfrak{p}$ of the reflex field. This includes cases where the group $G_{\mathbb{Q}_p}'$ in question is ramified. Thus we see that good reduction is possible also if the group K_p (the "level at p") is not hyperspecial.

§4 Deformation theory of p-divisible groups with Tate classes

In the next section, we will try to approach the smoothness problem appearing in 3.23 using deformation theory. The necessary technical results are due to Faltings and are the suject of the present section. Here we work out some details of a series of remarks in Faltings's paper [Fa3].

4.1 To begin with, let us recall a result from crystalline Dieudonné theory. For an exposition of this theory, we refer to the work of Berthelot-Messing and Berthelot-Breen-Messing ([BM1], [BBM], [BM3]); some further results can be found in [dJ].

Let k be a perfect field of characteristic $p > 2$, let $W = W(k)$ be its ring of infinite Witt vectors, and write σ for the Frobenius automorphism of W. We will be working with rings of the form $A = W[\![t_1, \ldots, t_n]\!]$. For such a ring, set $A_0 = k[\![t_1, \ldots, t_n]\!]$, $\mathfrak{m} = \mathfrak{m}_A = (p, t_1, \ldots, t_n)$, $J = J_A = (t_1, \ldots, t_n)$, let $e_A \colon A \to W$ be the zero section, and define a Frobenius lifting ϕ_A by $\phi_A = \sigma$ on W, $\phi_A(t_i) = t_i^p$.

With these notations we have the following fact: the category of p-divisible groups over $\mathrm{Spf}(A)$ is equivalent to that of p-divisible groups over $\mathrm{Spec}(A)$ (see [dJ], Lemma 2.4.4), and these categories are equivalent to the category of 4-tuples $(M, \mathrm{Fil}^1, \nabla, F)$, where

—M is a free A-module of finite rank,

—$\mathrm{Fil}^1 \subset M$ is a direct summand,

—$\nabla \colon M \to M \otimes \hat{\Omega}_{A/W}^1$ is an integrable, topologically quasi-nilpotent connection,

—$F: M \to M$ is a ϕ_A-linear horizontal endomorphism,

such that, writing $\widetilde{M} = M + p^{-1}\mathrm{Fil}^1$,

(4.1.1) F induces an isomorphism $F: \phi_A^* \widetilde{M} \overset{\sim}{\longrightarrow} M$,   and

(4.1.2) $\mathrm{Fil}^1 \otimes_A A_0 = \mathrm{Ker}(F \otimes \mathrm{Frob}_{A_0}: M \otimes_A A_0 \longrightarrow M \otimes_A A_0)$.

(Here, as often in the sequel, we write ϕ_A^*- for $-\otimes_{A,\phi_A} A$.) Notice that (4.1.1) implies that there is a ϕ_A^{-1}-linear endomorphism $V: M \to M$ such that

(4.1.3) $F \circ V = p \cdot \mathrm{id}_M = V \circ F$.

This equivalence is an immediate corollary to [Fa2], Thm. 7.1. One also obtains it by combining the following results:

—the description of a Dieudonné crystal on $\mathrm{Spf}(A_0)$ in terms of a 4-tuple (M, ∇, F, V), see [BBM], [BM3], [dJ],

—the Grothendieck-Messing deformation theory of p-divisible groups, see [Me],

—the results of de Jong, saying that over formal $\mathbb{F}_p$-schemes satisfying certain smoothness conditions, the crystalline Dieudonné functor for p-divisible groups is an equivalence of categories, see [dJ].

If $(M, \mathrm{Fil}^1, \nabla, F)$ corresponds to a p-divisible group $\mathcal{H}$ over A then

$$\mathrm{rk}_A(M) = \mathrm{height}(\mathcal{H}) \quad , \quad \mathrm{rk}_A(\mathrm{Fil}^1) = \dim(\mathcal{H}).$$

4.2 The 4-tuples $(M, \mathrm{Fil}^1, \nabla, F)$ form a category $\mathrm{MF}_{[0,1]}^\nabla(A)$ similar to the category $\mathcal{MF}_{[0,1]}^\nabla(A)$ as in [Fa2], except that we are working here with p-adically complete, torsion-free modules, rather than with p-torsion modules. More generally, let us write $\mathrm{MF}_{[a,b]}^\nabla(A)$ for the category of 4-tuples $(M, \mathrm{Fil}^\cdot, \nabla, F)$, where M and ∇ are as in 4.1, where F is a ϕ_A-linear endomorphism $M \otimes A[1/p] \to M \otimes A[1/p]$, and where $\mathrm{Fil}^\cdot$ is a descending filtration of M such that

Fil^{i+1} is a direct summand of Fil^i, $\mathrm{Fil}^a M = M$, $\mathrm{Fil}^{b+1}M = 0$,

and such that, writing $\widetilde{M} = \sum_{i=a}^b p^{-i}\mathrm{Fil}^i M$,

F induces an isomorphism $F: \phi_A^* \widetilde{M} \overset{\sim}{\longrightarrow} M$.

The arguments of [Fa2], Thm. 2.3, show that for $p > 2$ and $0 \leq b - a \leq p-1$, the category $\mathrm{MF}_{[a,b]}^\nabla(A)$ is independent, up to canonical isomorphism, of the chosen Frobenius lifting ϕ_A. Every morphism in $\mathrm{MF}_{[a,b]}^\nabla(A)$ (the definition of which, we hope, is clear) is strictly compatible with the filtrations (cf. [Wi], Prop. 1.4.1(i), in which the subscript "lf" should be replaced by "tf").

For $a' \leq a$ and $b' \geq b$, we have a natural inclusion $\mathrm{MF}^{\nabla}_{[a,b]}(A) \subseteq \mathrm{MF}^{\nabla}_{[a',b']}(A)$. We will write $\mathrm{MF}^{\nabla}_{[,]}(A)$ for the union of these categories, i.e., $M \in \mathrm{MF}^{\nabla}_{[,]}(A)$ means that $M \in \mathrm{MF}^{\nabla}_{[a,b]}(A)$ for *some* a and b.

The Tate object $A(-n) \in \mathrm{MF}^{\nabla}_{[n,n]}(A)$ is given by the A-module A with $\nabla = \mathrm{d}$, $\mathrm{Fil}^n = A\{n\} \supset \mathrm{Fil}^{n+1} = (0)$ and $F(a) = p^n \cdot \phi_A(a)$.

4.3 Before we turn to the deformation theory of p-divisible groups, we need to discuss some properties of 4-tuples $(M, \mathrm{Fil}^1, \nabla, F)$ as in 4.1.

4.3.1 The connection ∇ induces a connection $\widetilde{\nabla}$ on $\phi^*_A \widetilde{M}$ (not on $\widetilde{M}$ itself): if $m \in M$ and $\nabla(m) = \sum m_\alpha \otimes \omega_\alpha$, then $\widetilde{\nabla}(m \otimes 1) = \sum (m_\alpha \otimes 1) \otimes \mathrm{d}\phi_A(\omega_\alpha)$. One checks that this gives a well-defined integrable connection $\widetilde{\nabla}$. The horizontality of F can be expressed by saying that $\widetilde{\nabla}$ is the pull-back of ∇ via $F \colon \phi^*_A \widetilde{M} \xrightarrow{\sim} M$.

4.3.2 Given (M, Fil^1, F) satisfying (4.1.1) and (4.1.2), there is at most one connection ∇ for which F is horizontal. Indeed, the difference of two such connections ∇ and ∇' is a linear form $\delta \in \mathrm{End}(M) \otimes \hat{\Omega}^1_{A/W}$ satisfying $\mathrm{Ad}(F)(\delta) = \delta$. Here $\mathrm{Ad}(F)(\delta) = (F \otimes \mathrm{id}) \circ \tilde{\delta} \circ F^{-1}$, where $\tilde{\delta} = \widetilde{\nabla} - \widetilde{\nabla}'$. One checks that if $\delta \in J^t \mathrm{End}(M) \otimes \hat{\Omega}^1_{A/W}$, then $\mathrm{Ad}(F)(\delta) \in p \cdot J^{t+1} \mathrm{End}(M) \otimes \hat{\Omega}^1_{A/W}$, so that $\mathrm{Ad}(F)(\delta) = \delta$ implies $\delta = 0$. Similar arguments show that any connection δ for which F is horizontal, is integrable and topologically quasi-nilpotent.

4.3.3 Suppose $A = W[\![t_1, \ldots t_n]\!]$ and $B = W[\![u_1, \ldots u_m]\!]$ are two rings of the kind considered above. Let $f \colon A \to B$ be a W-homomorphism. If $\mathcal{H}$ is a p-divisible group over A corresponding to the 4-tuple $\mathbb{D} = (M, \mathrm{Fil}^1_M, \nabla_M, F_M)$, then the pull-back $f^*\mathcal{H}$ corresponds to a 4-tuple $f^*\mathbb{D} = (N, \mathrm{Fil}^1_N, \nabla_N, F_N)$ described as follows:

(i) $N = f^*M := M \otimes_{A,f} B$, $\mathrm{Fil}^1_N = f^*\mathrm{Fil}^1_M$, $\nabla_N = f^*\nabla_M$.

(ii) To describe F_N, we have to take into account that f may not be compatible with the two chosen Frobenius liftings ϕ_A and ϕ_B. First we use the connection ∇_M to construct an isomorphism

$$c = c(\phi_B \circ f, f \circ \phi_A) \colon \phi^*_B f^* M \xrightarrow{\sim} f^* \phi^*_A M,$$

which, using multi-index notations

$$\nabla(\partial)^{\underline{i}} = \nabla(\partial t_1)^{i_1} \cdots \cdots \nabla(\partial t_n)^{i_n} \quad , \quad z^{\underline{i}} = z_1^{i_1} \cdots \cdots z_n^{i_n} \quad , \quad \text{etc.},$$

is given (for $m \in M$) by

$$c(m \otimes 1) = \sum_{\underline{i}} \nabla(\partial)^{\underline{i}}(m) \otimes p^{|\underline{i}|} \cdot \frac{z^{\underline{i}}}{\underline{i}!},$$

where $z_i = \big(\phi_B \circ f(t_i) - f \circ \phi_A(t_i)\big)/p$. Then one defines F_N as the composition

$$F_N: \quad \phi_B^* N = \phi_B^* f^* M \xrightarrow{\ \ \underset{\sim}{c}\ \ } f^* \phi_A^* M \xrightarrow{\ f^* F_M\ } f^* M = N.$$

4.4 Theorem. (Faltings) *Let $A = W[\![t_1, \ldots t_n]\!]$ and consider a p-divisible group $\mathcal{H}$ over A with filtered Dieudonné crystal $\mathbb{D}(\mathcal{H}) = (\mathcal{M}, \mathrm{Fil}^1_{\mathcal{M}}, \nabla_{\mathcal{M}}, F_{\mathcal{M}})$. Write $H = e_A^* \mathcal{H}$, which has Dieudonné module $\mathbb{D}(H) = (M, \mathrm{Fil}^1_M, F_M) = e_A^*(\mathcal{M}, \mathrm{Fil}^1_{\mathcal{M}}, F_{\mathcal{M}})$. Assume that $\mathcal{H}$ is a versal deformation of H in the sense that the Kodaira-Spencer map*

$$\kappa: \ W\partial t_1 + \cdots + W\partial t_n \longrightarrow \mathrm{Hom}_W\big(\mathrm{Fil}^1_M, M/\mathrm{Fil}^1_M\big)$$

is surjective.

Next consider a ring $B = W[\![u_1, \ldots u_m]\!]$ and a 3-tuple $\mathbb{E}' = (\mathcal{N}, \mathrm{Fil}^1_{\mathcal{N}}, F_{\mathcal{N}})$ satisfying (4.1.1) and (4.1.2), and such that $\mathbb{E}' \otimes_{B, e_B} W \cong \mathbb{D}(H)$. Then there exists a W-homomorphism $f: A \to B$ such that $\mathbb{E}'$ is isomorphic to the pull-back of $(\mathcal{M}, \mathrm{Fil}^1_{\mathcal{M}}, F_{\mathcal{M}})$. In particular, $\mathbb{E}'$ can be completed to a filtered Dieudonné crystal $\mathbb{E}$ by setting $\nabla_{\mathcal{N}} = f^* \nabla_{\mathcal{M}}$, and therefore corresponds to a deformation of H.

Proof. For every W-homomorphism $f_1: A \to B$ there is an isomorphism of filtered B-modules $g_1: f_1^*(\mathcal{M}, \mathrm{Fil}^1_{\mathcal{M}}) \xrightarrow{\sim} (\mathcal{N}, \mathrm{Fil}^1_{\mathcal{N}})$, which is unique up to an element of $\mathrm{Aut}(\mathcal{N}, \mathrm{Fil}^1_{\mathcal{N}})$. The map g_1 induces an isomorphism $\tilde{g}_1: f_1^* \widetilde{\mathcal{M}} \xrightarrow{\sim} \widetilde{\mathcal{N}}$. By induction on $n \geq 1$ we may assume that the two Frobenii

$$F_{\mathcal{N}}: \ \phi_B^* \widetilde{\mathcal{N}} \xrightarrow{\ \sim\ } \mathcal{N}$$

and (with $c_1 = c(\phi_B \circ f_1, f_1 \circ \phi_A)$ as in 4.3.3)

$$F_{\mathcal{N}}': \ \phi_B^* \widetilde{\mathcal{N}} \xrightarrow[\sim]{\phi_B^* \tilde{g}_1^{-1}} \phi_B^* f_1^* \widetilde{\mathcal{M}} \xrightarrow[\sim]{c_1} f_1^* \phi_A^* \widetilde{\mathcal{M}} \xrightarrow[\sim]{f_1^* F_{\mathcal{M}}} f_1^* M \xrightarrow[\sim]{g_1} \mathcal{N}$$

are congruent modulo J_B^n. (For $n = 1$ this is so by our assumptions.) Because B is J_B-adically complete, it suffices to show that we can modify f_1 and g_1 such that the new $F_{\mathcal{N}}'$ is congruent to $F_{\mathcal{N}}$ modulo J_B^{n+1}.

314   Models of Shimura varieties

Consider an $f_2\colon A \to B$ which is congruent to f_1 modulo J_B^n. Notice that $f_1 \circ \phi_A \equiv f_2 \circ \phi_A$ and $\phi_B \circ f_1 \equiv \phi_B \circ f_2$ modulo J_B^{n+1}, so we have canonical isomorphisms

$$\phi_B^* f_1^* \widetilde{\mathcal{M}} \otimes B/J_B^{n+1} \cong \phi_B^* f_2^* \widetilde{\mathcal{M}} \otimes B/J_B^{n+1}$$

and

$$f_1^* \phi_A^* \widetilde{\mathcal{M}} \otimes B/J_B^{n+1} \cong f_2^* \phi_A^* \widetilde{\mathcal{M}} \otimes B/J_B^{n+1}.$$

Next we choose an isomorphism $h\colon f_2^*(\mathcal{M}, \mathrm{Fil}_{\mathcal{M}}^1) \xrightarrow{\sim} f_1^*(\mathcal{M}, \mathrm{Fil}_{\mathcal{M}}^1)$ which reduces to the canonical isomorphism modulo J_B^n, and we set $g_2 = g_1 \circ h$.

The first important remark is that, given the above identifications, the two maps

$$c_i\colon \phi_B^* f_i^* \widetilde{\mathcal{M}} \xrightarrow{\sim} f_i^* \phi_A^* \widetilde{\mathcal{M}} \qquad (i = 1, 2)$$

are equal modulo J_B^{n+1}. One can check this using the description of the maps c_i given in 4.3.3 and using that $f_i(t_j) \in J_B$.

Write ν for the automorphism of $\mathcal{N}$ such that $F_{\mathcal{N}} = \nu \circ F_{\mathcal{N}}'$. The induction hypothesis gives us that $\nu \equiv \mathrm{id}_{\mathcal{N}} \bmod J^n \cdot \mathrm{End}(\mathcal{N})$. It follows from the previous remarks that we are done if we can choose f_2 and h such that the diagram

(4.4.1)

$$
\begin{array}{ccccc}
f_1^* \phi_A^* \widetilde{\mathcal{M}} \otimes B/J^{n+1} & \xrightarrow{f_1^* F_{\mathcal{M}}} & f_1^* \mathcal{M} \otimes B/J^{n+1} & \xrightarrow{\nu \circ g_1} & \mathcal{N} \otimes B/J^{n+1} \\
\wr \| & & & & \| \\
f_2^* \phi_A^* \widetilde{\mathcal{M}} \otimes B/J^{n+1} & \xrightarrow{f_2^* F_{\mathcal{M}}} & f_2^* \mathcal{M} \otimes B/J^{n+1} & \xrightarrow{g_2} & \mathcal{N} \otimes B/J^{n+1}
\end{array}
$$

commutes. Note that the diagram is commutative modulo J_B^n and that, given f_2, we can still change h (and consequently g_2) by an element of $\mathrm{Aut}(f_2^* \mathcal{M}, f_2^* \mathrm{Fil}_{\mathcal{M}}^1)$.

The composition $g_1^{-1} \circ \nu^{-1} \circ g_2$ induces a W-linear map

$$\xi\colon f_1^* \mathrm{Fil}_{\mathcal{M}}^1 \otimes B/J \overset{\mathrm{can}}{\cong} f_2^* \mathrm{Fil}_{\mathcal{M}}^1 \otimes B/J \longrightarrow f_1^*(\mathcal{M}/\mathrm{Fil}_{\mathcal{M}}^1) \otimes J^n/J^{n+1},$$

which is independent of the choice of h. Similarly, $f_1^* F_{\mathcal{M}} \circ (f_2^* F_{\mathcal{M}})^{-1}$ induces a W-linear map

$$\eta\colon f_1^* \mathrm{Fil}_{\mathcal{M}}^1 \otimes B/J \longrightarrow f_1^*(\mathcal{M}/\mathrm{Fil}_{\mathcal{M}}^1) \otimes J^n/J^{n+1}.$$

The assumption that $\mathcal{H}$ is a versal deformation of H now implies that we can choose f_2 such that $\eta = \xi$. This means precisely that we can modify g_2 by something in $\mathrm{Aut}(f_2^* \mathcal{M}, f_2^* \mathrm{Fil}_{\mathcal{M}}^1)$ such that the diagram (4.4.1) commutes. This proves the induction step. $\qquad \square$

4.5 Let H be a p-divisible group over W, with special fibre H_0. Write $n = \dim(H_0) \cdot \dim(H_0^D)$, and let $A = W[\![t_1, \ldots, t_n]\!]$. The formal deformation functor of H_0 is pro-represented by A (see [Il]), where we may choose the coordinates such that H corresponds to the zero section e_A. Write $\mathcal{H}$ for the universal p-divisible group over A, and let $\mathbb{D}(\mathcal{H}) = (\mathcal{M}, \mathrm{Fil}^1_{\mathcal{M}}, \nabla_{\mathcal{M}}, F_{\mathcal{M}})$ be its filtered Dieudonné crystal. We will use the previous result to give a more explicit description of $\mathbb{D}(\mathcal{H})$.

Let $(M, \mathrm{Fil}^1_M, F_M) = e_A^* \mathbb{D}(\mathcal{H})$ be the filtered Dieudonné module of H. Choose a complement M' for $\mathrm{Fil}^1_M \subseteq M$. Inside the reductive group $\mathrm{GL}(M)$ over W, consider the parabolic subgroup of elements g with $gM' = M'$, and let U be its unipotent radical. Notice that U is (non-canonically) isomorphic to $\mathbb{G}^n_{a,W}$. Let $\widehat{U} = \mathrm{Spf}(B)$ be the formal completion of U along the identity, and choose coordinates $B \cong W[\![u_1, \ldots, u_n]\!]$ such that e_B gives the identity section. Over B we define a filtered Dieudonné crystal $\mathbb{E} = (\mathcal{N}, \mathrm{Fil}^1_{\mathcal{N}}, \nabla_{\mathcal{N}}, F_{\mathcal{N}})$ as follows. We set

$$\mathcal{N} = M \otimes_W B, \quad \mathrm{Fil}^1_{\mathcal{N}} = \mathrm{Fil}^1_M \otimes_W B, \quad F_{\mathcal{N}} = g \cdot (F_M \otimes \phi_B),$$

where $g \colon \mathcal{N} \xrightarrow{\sim} \mathcal{N}$ is the "universal" automorphism, i.e., the automorphism given by the canonical B-valued point of $U \subset \mathrm{GL}(M)$. At this point we apply the theorem. This gives us a connection $\nabla_{\mathcal{N}}$ and a W-homomorphism $f \colon A \to B$ such that $\mathbb{E} \cong f^* \mathbb{D}(\mathcal{H})$.

We claim that the map f is an isomorphism. Since A and B are formally smooth W-algebras of the same dimension, it suffices for this to show that $\mathbb{E}$ is a versal deformation of $(M, \mathrm{Fil}^1_M, F_M)$. Now we have an isomorphism

(4.5.1)
$$\widetilde{\mathcal{N}} \otimes_{B, \phi_B} B/\phi(J_B) = \left((\widetilde{\mathcal{N}} \otimes_{B, e_B} W) \otimes_{W, \sigma} W \right) \otimes_W B/\phi(J_B)$$
$$\overset{\mathrm{can}}{\cong} (\widetilde{M} \otimes_{W, \sigma} W) \otimes_W B/\phi(J_B).$$

On the left hand term we have the connection $\widetilde{\nabla}$; on the right we take $1 \otimes \mathrm{d}$. It follows easily from the definition of $\widetilde{\nabla}$ that (4.5.1) is horizontal modulo J_B^{p-1}. Composing with the isomorphisms $F_{\mathcal{N}}$ and F_M then gives a horizontal isomorphism

$$\bar{g} \colon \mathcal{N} \otimes_B B/J^{p-1} \xleftarrow{\sim} M \otimes_W B/J^{p-1},$$

which, as is clear from the constructions, is just the reduction modulo J^{p-1} of the automorphism g. Since $p > 2$, it then follows from the choice of U that $\mathbb{E}$ is a versal deformation.

4.6   Our next goal is to redo some of the above constructions for p-divisible groups with given Tate classes. We keep the notations of 4.5. For $r_1, r_2 \in \mathbb{Z}_{\geq 0}$ and $s \in \mathbb{Z}$, set

$$M(r_1, r_2; s) := M^{\otimes r_1} \otimes (M^*)^{\otimes r_2} \otimes W(s),$$

with its induced structure of an object of $\mathsf{MF}_{[,]}(W)$. We will refer to any direct sum of such objects as a tensor space $T = T(M)$ obtained from M.

We assume given a polarization $\psi \colon M \otimes_W M \to W(-1)$, i.e., a morphism in $\mathsf{MF}_{[0,2]}(W)$ which on modules is given by a perfect symplectic form. We let $\mathrm{CSp}(M, \psi)$ act on the Tate twist $W(-1)$ through the multiplier character. Then we consider a closed reductive subgroup $\mathcal{G} \subseteq \mathrm{CSp}(M, \psi)$ such that

(4.6.1)
$$\begin{array}{l} \text{there exists a tensor space } T \text{ and an element } t \in T \text{ such that} \\ L = W \cdot t \text{ is a subobject of } T \text{ in } \mathsf{MF}_{[,]}(W) \text{ isomorphic to } W(0), \\ \text{and such that } \mathcal{G} \subseteq \mathrm{CSp}(M, \psi) \text{ is the stabilizer of the line } L. \end{array}$$

4.7 Remark.   In [Fa3], Faltings gives an argument which shows that, for (4.6.1) to hold, it suffices if the Lie algebra $\mathfrak{g} \subset \mathrm{End}(M)$ is a subobject in $\mathsf{MF}_{[-1,-1]}(W)$. Since, by assumption, $\mathcal{G}$ is a smooth group, an easy argument then shows that (4.6.1) is equivalent to the condition that $\mathfrak{g}$ is stable under the Frobenius on $\mathrm{End}(M)$.

4.8   We can now construct a "universal" deformation of H such that the Tate class t remains a Tate class. The procedure is essentially the same as in 4.5. First, however, we have to find the right unipotent subgroup $U_\mathcal{G} \subset \mathcal{G}$. For this we use the canonical decomposition $M = M^0 \oplus \mathrm{Fil}^1$ defined by Wintenberger in [Wi]. The corresponding cocharacter

$$\mu \colon \mathbb{G}_{m,W} \longrightarrow \mathrm{GL}(M); \quad \mu(z) = \begin{cases} \mathrm{id} & \text{on} \quad M^0 \\ z^{-1} \cdot \mathrm{id} & \text{on} \quad \mathrm{Fil}^1 M \end{cases}$$

factors through $\mathcal{G}$. In 4.5 we now take $M' = M^0$, and we set $U_\mathcal{G} = U \cap \mathcal{G}$. Then $U_\mathcal{G}$ is a smooth unipotent subgroup of $\mathcal{G}$, whose Lie algebra is a complement of $\mathrm{Fil}^0 \mathfrak{g} \subseteq \mathfrak{g}$. (Here we use that $\mathcal{G}$ is reductive.)

Taking formal completions of $U_\mathcal{G} \hookrightarrow U$ along the origin corresponds, on rings, to a surjection

$$B = W[\![u_1, \dots, u_n]\!] \twoheadrightarrow C = W[\![v_1, \dots, v_q]\!],$$

where $q = \dim_W(\mathfrak{g}/\mathrm{Fil}^0 \mathfrak{g})$. We set

$$\mathcal{P} = M \otimes_W C, \quad \mathrm{Fil}^1_\mathcal{P} = \mathrm{Fil}^1_M \otimes_W C, \quad F_\mathcal{P} = h \cdot (F_M \otimes \phi_C),$$

where $h\colon \mathcal{P} \overset{\sim}{\to} \mathcal{P}$ is the universal element of $U_{\mathcal{G}}$. As in 4.5, applying Theorem 4.4 gives a connection $\nabla_{\mathcal{P}}$ and a homomorphism $f_{\mathcal{G}}\colon A \to C$ such that

$$\mathbb{E}_{\mathcal{G}} := (\mathcal{P}, \mathrm{Fil}_{\mathcal{P}}^1, \nabla_{\mathcal{P}}, F_{\mathcal{P}})$$

is the Dieudonné crystal of a deformation $\mathcal{H}_{\mathcal{G}} = f_{\mathcal{G}}^* \mathcal{H}$ of H over $\mathrm{Spf}(C) = \widehat{U_{\mathcal{G}}}$.

From the fact that $F_{\mathcal{P}}$ is horizontal w.r.t. $\nabla_{\mathcal{P}}$, one can derive that $\nabla_{\mathcal{P}}$ is of the form $\nabla_{\mathcal{P}} = \mathrm{d} + \beta$ with $\beta \in \mathfrak{g}_C \otimes \hat{\Omega}^1_{C/W} \subseteq \mathrm{End}(M) \otimes \hat{\Omega}^1_{C/W}$. It follows that if we extend the space T to an object $\mathcal{T} \in \mathrm{MF}_{[a,b]}^{\nabla}(C)$ by applying to $\mathcal{P} = M \otimes C$ the same linear algebra construction as was used to obtain T from M, then the line $L \subset T$ extends to a subobject $\mathcal{L} \subset \mathcal{T}$ in $\mathrm{MF}_{[a,b]}^{\nabla}(C)$.

To finish, let us prove that, conversely, every deformation of H over a ring $D = W[\![x_1, \ldots, x_r]\!]$ such that the tensor t deforms as a Tate class (i.e., the line $L \subset T$ extends to an inclusion $\mathcal{L} \subset \mathcal{T}$ in $\mathrm{MF}_{[a,b]}^{\nabla}(D)$), can be obtained by pullback from $\mathcal{H}_{\mathcal{G}}$. The map $\mathrm{End}(M) \to T/L$ obtained by sending $\alpha \in \mathrm{End}(M)$ to the evaluation at t of the induced $T(\alpha) \in \mathrm{End}(T)$ is a morphism in $\mathrm{MF}_{[,]}(W)$, hence strictly compatible with the filtrations. It follows that

(4.8.1)   if $T(\alpha)$ maps L into $\mathrm{Fil}^0 T$, then $\alpha \in \mathrm{Fil}^0 \mathrm{End}(M) + \mathfrak{g}$.

To prove the claim we can now follow the same reasoning as in 4.4, making use of (4.8.1). Alternatively, it follows from what we did in 4.5 that our deformation of H over D is obtained by pulling back the universal deformation $\mathcal{H}_B$ over B via a homomorphism $\pi\colon B \to D$. It then suffices to show that $\pi_n\colon B/J_B^n \to D/J_D^n$ factors via C/J_C^n for every n. For this we can argue by induction, and because of the way we have chosen $U_{\mathcal{G}}$ and U, the induction step easily follows from (4.8.1). This proves:

4.9 Proposition. *Notations and assumptions as above. We have a formally smooth deformation space $\widehat{U_{\mathcal{G}}} = \mathrm{Spf}(C) \hookrightarrow \widehat{U}$ of relative dimension equal to $\dim_W(\mathfrak{g}/\mathrm{Fil}^0\mathfrak{g})$ which parametrizes the deformations of H such that the horizontal continuation of t remains a Tate class.*

§5 Vasiu's strategy for proving the existence of integral canonical models

After our excursion to deformation theory, we return to the problem of the existence of integral canonical models. Our aim in this section is to explain the principal ideas in Vasiu's paper [Va2] (which is a revised version of part of [Va1]). We add that, to our understanding, some technical points are not treated correctly in loc. cit, to the effect that the main conclusions remain conjectural.

5.1   Consider a closed immersion of Shimura data i: $(G, X) \hookrightarrow (\mathrm{CSp}_{2g}, \mathfrak{H}_g^{\pm})$. Let v be a prime of $E = E(G, X)$ above $p > 2$, and assume that there is a hyperspecial subgroup $K_p \subset G(\mathbb{Q}_p)$ with $i(K_p) \subset C_p := \mathrm{CSp}_{2g}(\mathbb{Z}_p)$. (In particular, $G_{\mathbb{Q}_p}$ is unramified.) Write $\mathcal{A} := \varprojlim_{p \nmid n} \mathsf{A}_{g,1,n} \otimes \mathbb{Z}_{(p)}$ which, as we have seen, is an i.c.m. of $Sh_{C_p}(\mathrm{CSp}_{2g}, \mathfrak{H}_g^{\pm})$ over $\mathbb{Z}_{(p)}$, and let $\mathcal{N} \twoheadrightarrow N \subset \mathcal{A} \otimes \mathcal{O}_{(v)}$ be the normal integral model constructed in the proof of 3.19 (so $\mathcal{N}$ is the normalization of N). Choose embeddings $\overline{\mathbb{Q}} \subset \overline{E_v} = \overline{\mathbb{Q}}_p \subset \mathbb{C}$, and write $\overline{\mathcal{N}} \to \overline{N} \subset \overline{\mathcal{A}}$ for the base-change of $\mathcal{N}$, N and $\mathcal{A}$ to $\overline{W} := W\big(\overline{\kappa(v)}\big)$. Let $\tilde{x}_0 \in \overline{\mathcal{N}}$ be a closed point mapping to $x_0 \in \overline{N} \subset \overline{\mathcal{A}}$.

We would like to show that $\overline{\mathcal{N}}$ is formally smooth at $\tilde{x}_0$. If this holds (for all $\tilde{x}_0$) then $\mathcal{N}$ is an i.c.m. of $Sh_{K_p}(G, X)$ over $\mathcal{O}_{(v)}$. To achieve this, we would like to use Prop. 4.9. This is a reasonable idea: over our Shimura variety we have certain Hodge classes, which, by a results of Blasius and Wintenberger, give crystalline Tate classes (in the sense used in §4). The corresponding formal deformation space of p-divisible groups with these Tate classes is formally smooth and has a dimension equal to that of $\overline{\mathcal{N}}$. Arguing along these lines one could hope to prove that $\overline{\mathcal{N}}$ is formally smooth at $\tilde{x}_0$.

We see at least two obstacles in this argument: (i) in §4 we started from a p-divisible group over a ring of Witt-vectors, and (ii) we need a reductive group $\mathcal{G} \subset \mathrm{GL}(M)$ (the generic fibre of which should essentially be our group G). To handle these problems, we will first try to prove the formal smoothness of $\mathcal{N}$ under an additional hypothesis (5.6.1). In rough outline, the argument runs as follows. We start with a lifting of x_0 to a V-valued point of $\overline{N}$, where V is a purely ramified extension of $\overline{W}$. If (X, λ) is the corresponding p.p.a.v. over V then the associated filtered Frobenius crystal can be described as a module M over some filtered ring R_e. We have a closely related ring $\tilde{R}_e$ which is a projective limit of nilpotent PD-thickenings of V/pV and on which we have sections i_0: $\mathrm{Spec}(\overline{W}) \hookrightarrow \mathrm{Spec}(\tilde{R}_e)$ and i_π: $\mathrm{Spec}(V) \hookrightarrow \mathrm{Spec}(\tilde{R}_e)$. We will construct a deformation $(\tilde{X}, \tilde{\lambda})$ over $\tilde{R}_e$ which corresponds to an $\tilde{R}_e$-valued point of $\overline{N}$ and such that $i_\pi^*(\tilde{X}, \tilde{\lambda}) = (X, \lambda)$. Then $i_0^*(\tilde{X}, \tilde{\lambda})$ will give a lifting of x_0 to a $\overline{W}$-valued point of $\overline{N}$, which takes care of problem (i).

To construct $(\tilde{X}, \tilde{\lambda})$, we will use the Grothendieck-Messing deformation theory; the essential problem is to find the right Hodge filtration on the module $\tilde{M} := M \otimes_{R_e} \tilde{R}_e$. (We remark that the filtration on M cannot be used directly for this purpose: M is filtered free over the filtered ring R_e, whereas the desired Hodge filtration should be a direct summand.) One of the key steps in the argument is to show that the Zariski closure of a certain reductive group $G_{1,R_e[1/p]} \hookrightarrow \mathrm{GL}(M[1/p])$ inside $\mathrm{GL}(M)$ is a reductive group scheme—this will also take care of problem (ii). To achieve this, we have to keep track of Hodge classes on X in various cohomological realizations. At a crucial point we use a result of Faltings which permits to compare étale and crystalline classes with integral coefficients.

Once we have shown that the closure of $G_{1,R_e[1/p]}$ is reductive, an argument about reductive group schemes leads to the definition of the desired Hodge filtration on $\widetilde{M}$. After checking that it has the right properties, this brings us in a situation where the deformation theory of § 4 can be applied. The formal smoothness of $\widetilde{\mathcal{N}}$ at $\tilde{x}_0$ is then a relatively simple consequence of Prop. 4.9.

Sections 5.2 and 5.5 contain the necessary definitions and a brief description of the crystalline theory with values in R_e-modules. In 5.6 the argument that we just sketched is carried out, resulting in Thm. 5.8.3. What then remains to be shown is that there exist "enough" Shimura data for which (5.6.1) is satisfied. Vasiu's strategy to solve this problem is discussed briefly from 5.9 on.

5.2 Let $\mathcal{O}$ be a d.v.r. with uniformizer π and field of fractions F. Let W be a finite dimensional F-vector space with a non-degenerate symplectic form ψ. Write $F(-n)$ for the vector space F on which $\mathrm{CSp}(W, \psi)$ acts through the nth power of the multiplier character, and consider tensor spaces $W(r_1, r_2; s) := W^{\otimes r_1} \otimes (W^*)^{\otimes r_2} \otimes F(s)$. The fact that $\psi \in W(0, 2; -1)$ is non-degenerate implies that there exists a class $\psi^* \in W(2, 0; 1)$ such that $\langle \psi, \psi^* \rangle = 1 \in F = W(0, 0; 0)$.

Consider a faithfully flat $\mathcal{O}$-algebra R and a free R-module M with a given identification $M \otimes_R R[1/\pi] = W \otimes_F R[1/\pi]$. An element $t \in W(r_1, r_2; s)$ is said to be M-integral if, with the obvious notations, $t \otimes 1$ lies in the subspace $M(r_1, r_2; s)$ of $W(r_1, r_2; s) \otimes_F R[1/\pi]$. For example, ψ and ψ^* are both M-integral precisely if ψ induces a perfect form $\psi_M: M \times M \to R$.

If $t \in W(r_1, r_2; s)$ then we shall say that t is of type $(r_1, r_2; s)$ and has degree $r_1 + r_2$. In the sequel we shall often use a notation $T(W)$ for direct sums of spaces $W(r_1, r_2; s)$, and we call such a space a "tensor space obtained from W".

5.3 Definition. Let $G \subset \mathrm{CSp}(W, \psi)$ be a reductive subgroup, and consider a collection $\{t_\alpha\}_{\alpha \in \mathcal{J}}$ of G-invariants in spaces $T_\alpha(W)$. We say that $\{t_\alpha\}_{\alpha \in \mathcal{J}}$

is a well-positioned family of tensors for the group G over the d.v.r. $\mathcal{O}$ if, for every R and M as above, we have

$$
\begin{array}{c}
\psi, \psi^* \text{ and } \{t_\alpha\} \\
\text{are } M\text{-integral}
\end{array}
\;\Longrightarrow\;
\begin{array}{c}
\text{the Zariski closure of } G_{R[1/\pi]} \text{ inside } \mathrm{CSp}(M, \psi_M) \\
\text{is a reductive group scheme over } R
\end{array}\;.
$$

If in addition there exists an $\mathcal{O}$-lattice $M \subset W$ such that ψ, ψ^* and all t_α are M-integral, then we say that $\{t_\alpha\}_{\alpha \in J}$ is a *very* well-positioned family of tensors.

5.4 Remarks. (i) One should *not* think of a well-positioned family of tensors as some special family of tensors which cut out the group G (i.e., such that G is the subgroup of $\mathrm{CSp}(W, \psi)$ leaving invariant all t_α), since G may be strictly contained in the group cut out by the t_α. We only use the well-positioned families of tensors to guarantee that certain models of G are again reductive groups.

(ii) For general reductive $G \subset \mathrm{CSp}(W, \psi)$, the main difficulty with this notion is to prove the existence of (very) well-positioned families of tensors. We will come back to this point in 5.9 below.

5.5 Consider a purely ramified extension of d.v.r. $\overline{W} = W(\overline{\mathbb{F}}_p) \subset V$. Write $\mathrm{Frac}(\overline{W}) = K_0 \subset K = \mathrm{Frac}(V)$, fix $K \subset \overline{K} \hookrightarrow \mathbb{C}$, and let $e = e(V/\overline{W}) = [K : K_0]$. Suppose we have a p.p.a.v. (X, λ) over V. We write

$$
\begin{aligned}
&\mathrm{H}^1_{\mathrm{B},\mathbb{Z}} := \mathrm{H}^1_{\mathrm{B}}(X(\mathbb{C}), \mathbb{Z}), \quad \mathrm{H}^1_{\mathrm{B}} := \mathrm{H}^1_{\mathrm{B},\mathbb{Z}} \otimes \mathbb{Q}, \\
&\mathrm{H}^1_{\mathrm{dR},K} := \mathrm{H}^1_{\mathrm{dR}}(X_K/K), \quad \mathrm{H}^1_{\mathrm{dR},\mathbb{C}} := \mathrm{H}^1_{\mathrm{dR}}(X_{\mathbb{C}}/\mathbb{C}) = \mathrm{H}^1_{\mathrm{dR},K} \otimes_K \mathbb{C} \quad \text{and} \\
&\mathrm{H}^1_{\mathrm{\acute{e}t},\mathbb{Z}_p} := \mathrm{H}^1_{\mathrm{\acute{e}t}}(X_{\overline{K}}, \mathbb{Z}_p), \quad \mathrm{H}^1_{\mathrm{\acute{e}t}} := \mathrm{H}^1_{\mathrm{\acute{e}t},\mathbb{Z}_p} \otimes \mathbb{Q}_p.
\end{aligned}
$$

Let $T_{\mathrm{B}} = T(\mathrm{H}^1_{\mathrm{B}})$ be a tensor space as in 5.2, obtained from $\mathrm{H}^1_{\mathrm{B}}$. We adopt the notational convention that T_{dR}, $T_{\mathrm{\acute{e}t}}$ etc. stand for "the same" tensor space built from the corresponding first cohomology group $\mathrm{H}^1_{\mathrm{dR}}$, $\mathrm{H}^1_{\mathrm{\acute{e}t}}$ etc. In each case, $T_?$ naturally comes equipped with additional structures (Hodge structure/filtration/Galois action/$\cdots$), where we interpret $F(n)$ as a Tate twist. (Cf. [De4], Sect. 1 and Sect. 5.5.8 below.) In each theory, the polarization λ gives rise to a symplectic form on $\mathrm{H}^1_?$, which, if there is no risk of confusion, we denote by ψ without further indices.

5.5.1 Choose a uniformizer π of V, and write $g = T^e + a_{e-1}T^{e-1} + \cdots + a_0$ for its minimum polynomial over K_0, which is an Eisenstein polynomial. The PD-hull (compatible with the standard PD-structure on (p)) of $\overline{W}[T] \twoheadrightarrow$

$\overline{W}[T]/(g) = V$ is the ring S_e obtained from $\overline{W}[T]$ by adjoining all $T^{en}/n!$. Let $I := (p, g) = (p, T^e) \subset S_e$. We define R_e as the p-adic completion of S_e and $\widetilde{R}_e$ as the completion of S_e w.r.t. the filtration by the ideals $I^{[n]}$. (Thus $\widetilde{R}_e$ is the nilpotent PD-hull of $\overline{W}[T] \twoheadrightarrow V/pV$.) Notice that these rings only depend on the ramification index e, which justifies the notation. We can identify $\widetilde{R}_e$ (resp. R_e) with the subring of $K_0[T]$ consisting of all formal power series $\sum a_n \cdot T^n$ such that all $\lfloor n/e \rfloor! \cdot a_n$ are integral (resp. the coefficients $\lfloor n/e \rfloor! \cdot a_n$ are integral and p-adically convergent to zero for $n \to \infty$). On R_e we have

—a filtration by the ideals $\mathrm{Fil}^n(R_e) := (g)^{[n]}$,

—a σ-linear Frobenius endomorphism $\phi = \phi_{R_e}$ given by $T \mapsto T^p$,

—a continuous action of $\mathrm{Gal}(\overline{\mathbb{Q}}_p/K_0)$, commuting with ϕ and respecting the filtration.

5.5.2 Next we briefly recall the definition of the ring A_{crys} as in [Fo]. (In [Fa3] and [Va2] the notation $B^+(V)$ is used.) Write $\mathcal{O}_C$ for the p-adic completion of the integral closure of $\mathbb{Z}_p$ in $\overline{K} = \overline{\mathbb{Q}}_p$, and let C $(= \mathbb{C}_p)$ be its fraction field. Let

$$R_{\mathcal{O}_C} := \varprojlim(\mathcal{O}_C/p\mathcal{O}_C \leftarrow \mathcal{O}_C/p\mathcal{O}_C \leftarrow \cdots \leftarrow \mathcal{O}_C/p\mathcal{O}_C \leftarrow \cdots),$$

where the transition maps are given by $x \mapsto x^p$. It is a perfect ring of characteristic p. Choose a sequence of elements $\pi^{(n)} \in \mathcal{O}_C$ with $\pi^{(1)} = \pi$ (the chosen uniformizer of V) and $(\pi^{(m+1)})^p = \pi^{(m)}$, and set $\underline{\pi} = (\pi^{(1)} \bmod p, \pi^{(2)} \bmod p, \ldots) \in R_{\mathcal{O}_C}$. There is a surjective homomorphism $\theta \colon W(R_{\mathcal{O}_C}) \twoheadrightarrow \mathcal{O}_C$ whose kernel is the principal ideal generated by $\xi := g([\underline{\pi}])$, where $[\underline{\pi}]$ is the Teichmüller representative of $\underline{\pi}$ and where g is the polynomial as in 5.5.1 (see [Fa3], sect. 4).

Define A_{crys} as the p-adic completion of the PD-hull of $W(R_{\mathcal{O}_C}) \twoheadrightarrow \mathcal{O}_C$, compatible with the canonical PD-structure on (p). Then A_{crys} is a $\overline{W}$-algebra which comes equipped with

—a filtration by the ideals $\mathrm{Fil}^n(A_{\mathrm{crys}}) := (\xi)^{[n]}$,

—a σ-linear Frobenius endomorphism $\phi = \phi_{A_{\mathrm{crys}}}$,

—a continuous action of $\mathrm{Gal}(\overline{\mathbb{Q}}_p/K_0)$ commuting with ϕ and respecting the filtration.

There is a $\mathbb{Z}_p$-linear homomorphism $\mathbb{Z}_p(1) \hookrightarrow \mathrm{Fil}^1(A_{\mathrm{crys}})$. We let β (called t in [Fo]) denote the image of a generator of $\mathbb{Z}_p(1)$; we have $\phi(\beta) = p \cdot \beta$.

5.5.3 We have ring homomorphisms

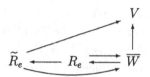

where the sections $R_e \to \overline{W}$ and $\tilde{R}_e \to \overline{W}$ are given by $T \mapsto 0$, and where $\tilde{R}_e \to V$ is given by $T \mapsto \pi$. Also we have a homomorphism

$$\iota \colon R_e \hookrightarrow A_{\mathrm{crys}}$$

given by $T \mapsto [\underline{\pi}]$ (hence $g \mapsto \xi$). This map is strictly compatible with the filtrations and induces ([Fa3], sect. 4) an isomorphism $\mathrm{gr}^{\cdot}(R_e) \otimes_{\overline{W}} \mathcal{O}_C \overset{\sim}{\to} \mathrm{gr}^{\cdot}(A_{\mathrm{crys}})$. Also, ι is compatible with the Frobenii, but in general *not* with the Galois-actions.

5.5.4 Let H be a p-divisible group over V. Its Dieudonné crystal can be described ([BBM], Thm. 1.2.7) as a free R_e-module $M = M(H)$ of finite rank with an integrable, topologically nilpotent connection ∇ on M as a $\overline{W}[T]$-module. On M we have

— a filtration by R_e-submodules $\mathrm{Fil}^{\cdot}_M \subset M$,

— a ϕ_{R_e}-linear horizontal endomorphism F,

such that

— there exists a basis $m_1^{(0)}, \ldots, m_{r_0}^{(0)}, m_1^{(1)}, \ldots, m_{r_1}^{(1)}$ such that $\mathrm{Fil}^j M = \sum_{a+b=j} \mathrm{Fil}^a R_e \cdot m_i^{(b)}$,

— the connection satisfies Griffiths transversality,

— F is divisible by p^j on $\mathrm{Fil}^j M$, and we can choose the basis $\{m_i^{(j)}\}$ as above in such a way that the elements $F(m_i^{(j)})/p^j$ form a new R_e-basis of M. (Modules M with these additional structures are the objects of a category $\mathrm{MF}^{\nabla}_{[0,1]}(V)$, analogous to the categories considered in §4; see [Fa3], §3.)

The Dieudonné module of the special fibre $H_0 := H \otimes_V \overline{\mathbb{F}}_p$ is a free $\overline{W}$-module M_0 with a σ-linear Frobenius endomorphism F_0. We have a canonical isomorphism of Frobenius crystals

$$(5.5.5) \qquad\qquad M_0 \cong M \otimes_{R_e} R_e/T \cdot R_e.$$

(Recall that $R_e/T \cdot R_e \overset{\sim}{\to} \overline{W}$.) On the other hand, the reduction $(H \bmod p)$ on $\mathrm{Spec}(V/pV)$ is isogenous, via some power of Frobenius, to $H_0 \otimes_{\overline{\mathbb{F}}_p} V/pV$, so that

$$(5.5.6) \qquad\qquad M \otimes_{R_e} R_e[1/p] \cong M_0 \otimes_{\overline{W}} R_e[1/p]$$

as Frobenius crystals.

Although the homomorphism ι from 5.5.3 is in general not compatible with the Galois actions, there is a canonical action of $\mathrm{Gal}(\overline{\mathbb{Q}}_p/K)$ on $M \otimes_{R_e,\iota} A_{\mathrm{crys}}$. (Here one uses that M is a crystal over R_e, see [Fa3], §4.) We also define

$$\mathrm{H}^1_{\text{ét},\mathbb{Z}_p}(H) := T_p(H)^*,$$

which is a free $\mathbb{Z}_p$-module with $\mathrm{Gal}(\overline{\mathbb{Q}}_p/K)$-action.

5.5.7 Theorem. (Faltings, [Fa3]) *There is a functorial injection*

$$\rho\colon M(H) \otimes_{R_e} A_{\mathrm{crys}} \hookrightarrow \mathrm{H}^1_{\text{ét},\mathbb{Z}_p}(H) \otimes_{\mathbb{Z}_p} A_{\mathrm{crys}},$$

which after extension of scalars to $B_{\mathrm{crys}} := A_{\mathrm{crys}}[1/\beta]$, and using the isomorphism $M \otimes_{R_e} R_e[1/p] \cong M_0 \otimes_{\overline{W}} R_e[1/p]$ gives back Faltings's comparison isomorphism $\tilde{\rho}\colon M_0 \otimes_{\overline{W}} B_{\mathrm{crys}} \xrightarrow{\sim} \mathrm{H}^1_{\text{ét}}(H) \otimes_{\mathbb{Q}_p} B_{\mathrm{crys}}$ of [Fa2]. The map ρ is compatible with the Frobenii, filtrations and Galois actions on both sides. Its cokernel is annihilated by $\beta \in A_{\mathrm{crys}}$.

5.5.8 We shall try to be precise about Tate twists and polarization forms. We have

—$\mathbb{Z}_B(1) := 2\pi i \cdot \mathbb{Z} \subset \mathbb{C}$, with H.S. purely of type $(-1,-1)$,

—$K_{\mathrm{dR}}(1) := K$ with filtration $\mathrm{Fil}^{-1} = K \supset \mathrm{Fil}^0 = (0)$ (similarly for other fields than K),

—$\mathbb{Z}_p(1) := \varprojlim \mu_{p^n}(\overline{K})$ as $\mathrm{Gal}(\overline{K}/K)$-module (similarly for other fields). Fixing $i \in \mathbb{C}$ with $i^2 = -1$ we have generators $2\pi i$ for $\mathbb{Z}_B(1)$ and 1 for $K_{\mathrm{dR}}(1)$. Also, the choice of i determines a generator $\left(\exp(2\pi i/p^n)\right)_{n\in\mathbb{N}}$ for $\mathbb{Z}_p(1)$ over $\mathbb{C}$. Via the chosen embedding $\overline{K} \hookrightarrow \mathbb{C}$ this gives a generator ζ for $\mathbb{Z}_p(1)$ over $\overline{K}$. We have comparison isomorphisms (over $\mathbb{C}$)

$$\mathbb{Z}_B(1) \otimes_{\mathbb{Z}} \mathbb{Z}_p \xrightarrow{\sim} \mathbb{Z}_p(1) \quad \text{and} \quad \mathbb{Z}_B(1) \otimes_{\mathbb{Z}} \mathbb{C} \xrightarrow{\sim} \mathbb{C}_{\mathrm{dR}}(1),$$

(see [De4], Sect. 1) mapping generators to generators.

If (X,λ) is a p.p.a.v. over $\mathbb{C}$ then λ gives rise to a perfect symplectic form $\psi_B\colon \mathrm{H}^1_B \times \mathrm{H}^1_B \longrightarrow \mathbb{Z}_B(-1)$. For de Rham and étale cohomology we have an analogous statement, and the various forms $\psi_?$ are compatible via the comparison isomorphisms. Using the chosen generators for Tate objects, we can view the forms $\psi_?$ as "ordinary" bilinear forms with values in the coefficient ring corresponding to $? \in \{B, \mathrm{dR}, \text{ét}\}$. This is consistent with our usage of the notation $F(n)$ in 5.2.

For crystalline cohomology (or Dieudonné modules) with values in R_e, the Tate twist is given by

$-R_e(-1) := R_e$ with filtration $\mathrm{Fil}^j(R_e(-1)) = \mathrm{Fil}^{j-1}R_e$ and Frobenius $F = p \cdot \phi_{R_e}$.

(This should be thought of as an object of a category $\mathsf{MF}(V)$, see [Fa3]. We have $R_e(-1) \cong M(\widehat{\mathbb{G}}_\mathrm{m})$.) The generator $\zeta \in \mathbb{Z}_p(1)$ over $\overline{K}$ determines a generator ζ^* of $\mathbb{Z}_p(-1)$ and an element $\beta \in A_\mathrm{crys}$ as in 5.5.2 (see [Fo], 1.5.4). On Tate twists, the comparison map of Thm. 5.5.7 is (after a suitable normalization) the map $\delta \colon R_e(1) \otimes_{R_e} A_\mathrm{crys} \xrightarrow{\sim} \mathbb{Z}_p(1) \otimes_{\mathbb{Z}_p} A_\mathrm{crys}$ with $1 \otimes 1 \mapsto \zeta \otimes \beta$. Here we see a factor β entering. In other words: if (X, λ) is a p.p.a.v. over V with associated p-divisible group $H = X[p^\infty]$, then λ gives polarization forms

$$\psi_\mathrm{\acute et} \colon \mathrm{H}^1_{\mathrm{\acute et}, \mathbb{Z}_p} \times \mathrm{H}^1_{\mathrm{\acute et}, \mathbb{Z}_p} \longrightarrow \mathbb{Z}_p(-1) \quad \text{and} \quad \psi_\mathrm{crys} \colon M(H) \times M(H) \longrightarrow R_e(-1),$$

so that under the map ρ from 5.5.7 we have $\delta \circ (\psi_\mathrm{crys} \otimes 1) = (\psi_\mathrm{\acute et} \otimes \beta) \circ (\rho \times \rho)$.

5.6 *Vasiu's strategy—first part.* We return to the situation considered in 5.1. We shall first try to prove the formal smoothness of $\overline{\mathcal{N}}$ under the following assumption. Here we recall that we write CSp_{2g} for the Chevalley group scheme $\mathrm{CSp}(\mathbb{Z}^{2g}, \psi)$, where ψ is the standard symplectic form on $\mathbb{Z}^{2g}$.

(5.6.1)   There is a collection of tensors $\{t_\alpha \in (\mathbb{Z}^{2g}_{(p)})(r_\alpha, r_\alpha; 0)\}_{\alpha \in \mathcal{J}_1}$ of degrees $2r_\alpha \le 2(p-2)$ such that this collection is very well-positioned over the d.v.r. $\mathbb{Z}_{(p)}$ for the group G (considered as a subgroup of $\mathrm{CSp}_{2g,\mathbb{Q}}$ via the given closed embedding i).

Also, we shall consider a larger collection $\{t_\alpha\}_{\alpha \in \mathcal{J}}$ (with $\mathcal{J}_1 \subset \mathcal{J}$) of tensors which, together with the tensor ψ, cut out the group G. (The t_α with $\alpha \in \mathcal{J} \setminus \mathcal{J}_1$ again of types $(r_\alpha, r_\alpha, 0)$ but not neccesarily $\mathbb{Z}^{2g}_{(p)}$-integral, nor of degree $\le 2(p-2)$.)

5.6.2 Consider triplets $(X_\mathbb{C}, \lambda_\mathbb{C}, \theta^p)$ consisting of a g-dimensional p.p.a.v. over $\mathbb{C}$ with a compatible system of Jacobi level n structures for all n with $p \nmid n$ (which we represent by the single symbol θ^p). The modular interpretation

$$Sh_{C_p}(\mathrm{CSp}_{2g,\mathbb{Q}}, \mathfrak{H}^\pm_g)(\mathbb{C}) \xrightarrow{\sim} \{(X_\mathbb{C}, \lambda_\mathbb{C}, \theta^p)\}/\cong$$

is given as follows. If $(h, \gamma) \in \mathfrak{H}^\pm_g \times \mathrm{CSp}_{2g}(\mathbb{A}_f)$, then we can view γ as an isomorphism $\gamma \colon \mathbb{Q}^{2g} \otimes_\mathbb{Q} \mathbb{A}_f \xrightarrow{\sim} \hat{\mathbb{Z}}^{2g} \otimes_{\hat{\mathbb{Z}}} \mathbb{A}_f$. For $X_\mathbb{C}$ we take the abelian variety determined by the lattice $\Lambda := \mathbb{Q}^{2g} \cap \gamma^{-1}(\hat{\mathbb{Z}}^{2g})$ and the Hodge structure h.

There is a unique $q \in \mathbb{Q}^*$ such that $q \cdot \psi$ is the Riemann form of a principal polarization $\lambda_{\mathbb{C}}$ on $X_{\mathbb{C}}$, and the system of level structures is given by the isomorphism $\gamma \colon \Lambda \otimes (\prod_{\ell \neq p} \mathbb{Z}_\ell) \xrightarrow{\sim} \prod_{\ell \neq p} \mathbb{Z}_\ell^{2g}$. One checks that this gives a well-defined bijection as claimed.

By construction of the model $\mathcal{N}$, there exists a purely ramified extension $\overline{W} \subset V$ as in 5.5 such that the closed point $\tilde{x}_0 \in \mathcal{N}$ lifts to a V-valued point $x \colon \mathrm{Spec}(V) \to \mathcal{N}$. Considered as a point of $\overline{\mathcal{A}}$ it corresponds to a p.p.a.v. with a system of level structures (X, λ, θ^p) over V. We shall use the notations and assumptions of 5.5; in particular we obtain a triplet $(X_{\mathbb{C}}, \lambda_{\mathbb{C}}, \theta^p)$ over $\mathbb{C}$ via base-change over the chosen embedding $V \subset K \hookrightarrow \mathbb{C}$. The corresponding point of $Sh_{C_p}(\mathrm{CSp}_{2g,\mathbb{Q}}, \mathfrak{H}_g^\pm)(\mathbb{C})$ can be represented by a pair $(h, e) \in \mathfrak{H}_g^\pm \times \mathrm{CSp}_{2g}(\mathbb{A}_f^p)$. In particular, we get an identification $\mathrm{H}_{\mathrm{B}}^1(X(\mathbb{C}), \mathbb{Z}) \otimes \mathbb{Z}_{(p)} \cong \mathbb{Z}_{(p)}^{2g}$. The fact that x factors through $\overline{N}$ now implies that the tensors t_α as in (5.6.1) correspond to Hodge classes $t_{\alpha,\mathrm{B}}$ on $X_{\mathbb{C}}$ which for $\alpha \in \mathcal{J}_1 \subset \mathcal{J}$ are integral w.r.t. the $\mathbb{Z}_{(p)}$-lattice $\mathrm{H}_{\mathrm{B},\mathbb{Z}}^1 \otimes \mathbb{Z}_{(p)} \subset \mathrm{H}_{\mathrm{B}}^1$. Notice that ψ gives an isomorphism $(\mathrm{H}_{\mathrm{B},\mathbb{Z}}^1)^* \cong \mathrm{H}_{\mathrm{B},\mathbb{Z}}^1(1)$, so that it is no restriction to assume that all t_α live in spaces $(\mathbb{Z}^{2g})(r_\alpha, r_\alpha; 0)$.

By [De4], Prop. 2.9, the de Rham realizations $t_{\alpha,\mathrm{dR}}$ $(\alpha \in \mathcal{J})$ are defined over a finite extension of K. Possibly after replacing V by a finite extension, we may therefore assume that the $t_{\alpha,\mathrm{dR}}$ are defined over K (i.e., they are elements of tensor spaces of the form $\mathrm{H}_{\mathrm{dR},K}^1(r_\alpha, r_\alpha; 0)$). In particular, we obtain a subgroup $G_{1,K} \subset \mathrm{CSp}(\mathrm{H}_{\mathrm{dR},K}^1, \psi)$ such that $G_{1,K} \otimes_K \mathbb{C} = G \otimes_{\mathbb{Q}} \mathbb{C}$.

Write $t_{\alpha,\text{ét}}$ for the (p-adic) étale realization of $t_{\alpha,\mathrm{B}}$, which is an element of some tensor space $T_{\alpha,\text{ét}}$. For $\alpha \in \mathcal{J}_1$, the class $t_{\alpha,\text{ét}}$ is $\mathrm{H}_{\text{ét},\mathbb{Z}_p}^1$-integral. Since the $t_{\alpha,\mathrm{B}}$ are Hodge classes on $X_{\mathbb{C}}$, the $t_{\alpha,\mathrm{dR}}$ and $t_{\alpha,\text{ét}}$ correspond to each other via the p-adic comparison isomorphism. More precisely, we have the following result, which was obtained independently by Blasius (see [Bl]) and Wintenberger. A simplified proof was given by Ogus in [Og].

5.6.3 Theorem. (Blasius, Wintenberger) *Let $\mathcal{O}$ be a complete d.v.r. of characteristic $(0, p)$ with perfect residue field. Set $F = \mathrm{Frac}(\mathcal{O})$, and let X_F be an abelian variety over F with good reduction over $\mathcal{O}$. Let $t_{\mathrm{dR}} \in \mathrm{H}_{\mathrm{dR}}^1(r_1, r_2; s)$ and $t_{\text{ét}} \in \mathrm{H}_{\text{ét}}^1(r_1, r_2; s)$ be the de Rham component and the p-adic étale component respectively of an absolute Hodge class on X. Under the comparison isomorphism*

$$\gamma \colon \mathrm{H}_{\mathrm{dR}}^1(X_F/F) \otimes_F B_{\mathrm{dR}} \xrightarrow{\sim} \mathrm{H}_{\text{ét}}^1(X_{\overline{F}}, \mathbb{Q}_p) \otimes_{\mathbb{Q}_p} B_{\mathrm{dR}}$$

we have $\gamma(t_{\mathrm{dR}} \otimes 1) = t_{\text{ét}} \otimes \beta^{-s}$.

We remark that in [Bl] and [Og], this results is only proven under the additional assumption that X_F is obtained via base-change from an abelian variety over a number field. It can be shown that this condition, which appears in the proof of a version of Deligne's "Principle B", is superfluous. In [Va2], Vasiu shows this by using a trick of Lieberman. One can also remark that the "Principle B" is needed only in the situation where we have a family of abelian varieties $\mathfrak{X} \to S$ over a variety S over $\overline{\mathbb{Q}}$, such that X_F occurs as the fibre over an F-valued point of S. (The variety S constructed in [De4], Sect. 6 is a component of a Shimura variety.) In this situation, the arguments given in [Bl], Sect. 3 and [Og], Prop. 4.3 suffice.

We also remark that the factor β^{-s} appears because we choose $K_{\mathrm{dR}}(1) \otimes B_{\mathrm{dR}} \xrightarrow{\sim} \mathbb{Q}_p(1) \otimes B_{\mathrm{dR}}$ to be the map $1 \otimes 1 \mapsto \zeta \otimes \beta^{-1}$. (In [Bl] a different normalization is used.)

5.6.4 Set $H := X[p^\infty]$, where X is as in 5.6.2. Notice that $\mathrm{H}^1_{\text{ét},\mathbb{Z}_p}(H) \cong \mathrm{H}^1_{\text{ét},\mathbb{Z}_p}(X_{\overline{K}})$. There are well-defined F_0-invariants $\tau_{\alpha,\mathrm{crys}} \in M_0[1/p](r_\alpha, r_\alpha; 0)$ such that $\tilde{\rho}(\tau_{\alpha,\mathrm{crys}} \otimes 1) = t_{\alpha,\text{ét}} \otimes 1$. Using (5.5.6), we then obtain horizontal F-invariant classes $t_{\alpha,\mathrm{crys}} \in M[1/p](r_\alpha, r_\alpha; 0)$. Also we have polarization forms ψ_{crys} and $\psi_{\text{ét}}$ as already mentioned in 5.5.8.

We claim that, writing $T_{\alpha,\mathrm{crys}}$ for the tensor spaces obtained from $M = M(H)$, the $t_{\alpha,\mathrm{crys}}$ lie in $\mathrm{Fil}^0(T_{\alpha,\mathrm{crys}}[1/p])$. To see this, we use that $t_{\alpha,\mathrm{crys}}$ is a lifting of $t_{\alpha,\mathrm{dR}}$ in the following sense. By (5.5.5) and the isomorphism $M_0 \otimes_{\overline{W}} K \cong \mathrm{H}^1_{\mathrm{dR},K}$ from [BO], we have $M \otimes_{R_e} K \xrightarrow{\sim} \mathrm{H}^1_{\mathrm{dR},K}$. Combining this with the isomorphism $R_e(-1) \otimes_{R_e} K \xrightarrow{\sim} K_{\mathrm{dR}}(-1)$ by $1 \otimes 1 \mapsto 1$, we obtain maps $M(H)(r_1, r_2; s) \otimes_{R_e} K \xrightarrow{\sim} \mathrm{H}^1_{\mathrm{dR},K}(r_1, r_2; s)$. The functoriality of the map ρ in Theorem 5.5.7 implies that $t_{\alpha,\mathrm{crys}} \otimes 1 \mapsto t_{\alpha,\mathrm{dR}}$ and $\psi_{\mathrm{crys}} \otimes 1 \mapsto \psi_{\mathrm{dR}}$. That $t_{\alpha,\mathrm{crys}} \in \mathrm{Fil}^0 T_{\alpha,\mathrm{crys}}$ now follows from the fact that $\mathrm{Fil}^1 M(H)$ is the inverse image of $\mathrm{Fil}^1 \mathrm{H}^1_{\mathrm{dR},K}$ under the map $M(H) \to M(H) \otimes_{R_e} K \cong \mathrm{H}^1_{\mathrm{dR},K}$. In this way we see that ψ_{crys} and the $t_{\alpha,\mathrm{crys}}$ are crystalline Tate classes, in the sense that they are horizontal, in Fil^0 and invariant under Frobenius.

We will be able to exploit assumption (5.6.1) by using the following supplement to Thm. 5.5.7.

5.6.5 Theorem. (Faltings, [Fa3]) *Suppose that $r \leq (p-2)$, and consider Tate classes*

$$t_{\mathrm{crys}} \in M(H)(r, r; 0) \otimes_{R_e} R_e[1/p] \quad \text{and} \quad t_{\text{ét}} \in \mathrm{H}^1_{\text{ét},\mathbb{Z}_p}(H)(r, r; 0) \otimes_{\mathbb{Z}_p} \mathbb{Q}_p,$$

with $\rho(t_{\mathrm{crys}} \otimes 1) = (t_{\text{ét}} \otimes 1)$. Then t_{crys} is $M(H)$-integral if and only if $t_{\text{ét}}$ is $\mathrm{H}^1_{\text{ét},\mathbb{Z}_p}(H)$-integral.

It follows from this result that the $t_{\alpha,\text{crys}}$ with $\alpha \in \mathcal{J}_1$ are $M(H)$-integral classes. (Notice that we assumed these classes to have degree $\leq 2(p-2)$, as required in Faltings's theorem.)

5.6.6 We are now ready to prove one of the key steps in the argument. Since at this point we were not able to follow [Va2], we present our own explanation of what is going on.

The tensors $\tau_{\alpha,\text{crys}}$ cut out a subgroup G_1 of $\text{CSp}(M_0[1/p], \psi_{M_0})$. By Thm. 5.6.3 we have $G_1 \otimes_{K_0} K = G_{1,K}$ (where the latter is the group cut out by the $t_{\alpha,\text{dR}}$ that was introduced in 5.6.2 above), so that G_1 is reductive. What we would like to show now is that the Zariski closure $\mathcal{G}_{1,R_e}$ of $G_{1,R_e[1/p]}$ inside $\text{CSp}(M(H), \psi_{\text{crys}})$ is a reductive group scheme over R_e. Here we use the identification (5.5.6) to identify $G_{1,R_e[1/p]}$ as a subgroup of $\text{CSp}(M(H), \psi_{\text{crys}})$.

Obviously, we will try to achieve our goal by using (5.6.1). If we look at Def. 5.3 then we see that we already know (grace to Thm. 5.6.5) that ψ_{crys}, ψ_{crys}^* and the $t_{\alpha,\text{crys}}$ with $\alpha \in \mathcal{J}_1$ are $M(H)$-integral, and therefore it only remains to show that there exists an isomorphism $\mathbb{Q}^{2g} \otimes R_e[1/p] \xrightarrow{\sim} M(H)[1/p]$ such that ψ and the t_α ($\alpha \in \mathcal{J}$) are sent to ψ_{crys} and the $t_{\alpha,\text{crys}}$ respectively. The first step is that we have an isomorphism $\mathbb{Q}^{2g} \otimes \mathbb{Q}_p \xrightarrow{\sim} \text{H}^1_{\text{ét}}$ sending ψ and the t_α to their étale realizations $\psi_{\text{ét}}$ and $t_{\alpha,\text{ét}}$.

Notice that we now only have to consider rings with p inverted. Since the $t_{\alpha,\text{crys}}$ were obtained from the $\tau_{\alpha,\text{crys}}$ using (5.5.6), it suffices (and will actually be easier) to show that there exists an isomorphism $\nu\colon \text{H}^1_{\text{ét}} \otimes_{\mathbb{Q}_p} K_0 \xrightarrow{\sim} M_0[1/p]$ such that $\psi_{\text{ét}} \mapsto \psi_{M_0}$ and $t_{\alpha,\text{ét}} \mapsto \tau_{\alpha,\text{crys}}$. (This isomorphism is of course not required to have any "meaning".)

By what was explained before, we can compare $\text{H}^1_{\text{ét}}$ and $M_0[1/p]$ after extension of scalars to the ring B_{crys}, in such a way that the tensors $t_{\alpha,\text{ét}}$ and the $\tau_{\alpha,\text{crys}}$ correspond. We have to be a little more careful about the polarization forms, since ψ_{crys} and $\psi_{\text{ét}}$ correspond to each other only up to a factor β (see 5.5.8). The ring to work over therefore is $B_{\text{crys}}[\sqrt{\beta}]$, since the factor $\sqrt{\beta}$ allows us to modify the isomorphism $M_0[1/p] \otimes B_{\text{crys}} \cong \text{H}^1_{\text{ét}} \otimes B_{\text{crys}}$ in such a way that the forms ψ_{crys} and $\psi_{\text{ét}}$ do correspond. This does not affect the tensors t_α, since these are of type $(r_\alpha, r_\alpha; 0)$. In any case, we see that there exists a field extension $K_0 \subset \Omega$ such that the desired comparison isomorphism exists after extension of scalars to Ω. Writing $\mathfrak{V} = (\text{H}^1_{\text{ét}}; \psi, \{t_{\alpha,\text{ét}}\}_{\alpha \in \mathcal{J}})$ and $\mathfrak{V}' = (M_0[1/p]; \psi_{M_0}, \{\tau_{\alpha,\text{crys}}\}_{\alpha \in \mathcal{J}})$ the torsor $\mathcal{I}som(\mathfrak{V}, \mathfrak{V}')$ is therefore non-empty. Since the automorphism group of the system $\mathfrak{V}$ is precisely $G_{\mathbb{Q}_p}$, the obstruction for finding ν then is a class in $\text{H}^1\big(\text{Gal}(\overline{\mathbb{Q}}_p/K_0), G(\overline{\mathbb{Q}}_p)\big)$. Now the fact that K_0 is a field of dimension ≤ 1, together with [Se2], Thm. 1 in Chap. III, §2.2, proves that the obstruction vanishes, whence the existence of an isomorphism ν as desired. By applying (5.6.1) this gives the following statement.

5.6.7 Proposition. *The Zariski closure $\mathcal{G}_{1,R_e}$ of $G_{1,R_e[1/p]}$ inside the scheme* $\mathrm{CSp}(M(H), \psi_{\mathrm{crys}})$ *is a reductive group scheme over R_e.*

It will now rapidly become clear why 5.6.7 is important. For this, we set

$$\widetilde{M} := M \otimes_{R_e} \widetilde{R}_e, \quad \text{and } M_V := M \otimes_{R_e} V = \mathrm{H}^1_{\mathrm{dR}}(X/V),$$

using the maps $R_e \subset \widetilde{R}_e \to V$ from 5.5.3. Also we set

$$\mathcal{G}_{1,\widetilde{R}_e} := \mathcal{G}_1 \times_{R_e} \widetilde{R}_e, \quad \mathcal{G}_{1,V} := \mathcal{G}_1 \times_{R_e} V,$$

which are reductive groups over $\widetilde{R}_e$ and V respectively. We write $\mathrm{Fil}^1(M_V) = \mathrm{Fil}^1(M) \otimes_{R_e} V$ for the Hodge filtration on M_V.

5.7 Lemma. *(i) There exists a complement M'_V for $\mathrm{Fil}^1(M_V) \subset M_V$ such that the cocharacter $\mu\colon \mathbb{G}_{\mathrm{m},V} \to \mathrm{GL}(M_V)$ given by*

$$\mu(z) = \begin{cases} \mathrm{id} & \text{on } M'_V \\ z^{-1} \cdot \mathrm{id} & \text{on } \mathrm{Fil}^1(M_V) \end{cases}$$

factors through $\mathcal{G}_{1,V}$.

(ii) The cocharacter $\mu\colon \mathbb{G}_{\mathrm{m},V} \to \mathcal{G}_{1,V}$ lifts to a cocharacter $\widetilde{\mu}\colon \mathbb{G}_{\mathrm{m},\widetilde{R}_e} \to \mathcal{G}_{1,\widetilde{R}_e}$.

We will admit this lemma, referring to [Va2], sect. 5.3 for a proof. We remark that the reductiveness of $\mathcal{G}_{1,V}$ and $\mathcal{G}_{1,\widetilde{R}_e}$ is used in an essential way.

The cocharacter $\widetilde{\mu}$ yields a direct sum decomposition $\widetilde{M} = \widetilde{M}' \oplus \widetilde{M}''$ with $\widetilde{M}'' \otimes_{\widetilde{R}_e} V = \mathrm{Fil}^1(M_V)$. Since $\widetilde{R}_e$ is a projective limit of nilpotent PD-thickenings of V/pV, we can apply the Grothendieck-Messing deformation theory of abelian varieties (see [Me], in particular Chap. V). This gives us a formal p.p.a.v. $(\mathfrak{X}, \boldsymbol{\lambda})$ over $\mathrm{Spf}(\widetilde{R}_e)$ (with the I-PD-adic topology on $\widetilde{R}_e$), the de Rham cohomology of which is given by $\mathrm{H}^1_{\mathrm{dR}}(\mathfrak{X}/\widetilde{R}_e) = \widetilde{M}$ with Hodge filtration $\widetilde{M}''$ and Gauß-Manin connection induced from the connection on $M(H)$ as a crystal. The fact that we have a polarization on $\mathfrak{X}$ implies that $(\mathfrak{X}, \boldsymbol{\lambda})$ algebraizes to a p.p.a.v. $(\widetilde{X}, \widetilde{\lambda})$ over $\mathrm{Spec}(\widetilde{R}_e)$. We have $(\widetilde{X}, \widetilde{\lambda}) \otimes_{\widetilde{R}_e} V = (X, \lambda)$. Since we only consider level n structures with $p \nmid n$, the system of level structures θ^p extends to a system $\widetilde{\theta}^p$ on $(\widetilde{X}, \widetilde{\lambda})$. We claim that the morphism $\mathrm{Spec}(\widetilde{R}_e) \to \overline{\mathcal{A}}$ corresponding to $(\widetilde{X}, \widetilde{\lambda}, \widetilde{\theta}^p)$ factors through $\overline{N} \subset \overline{\mathcal{A}}$. To prove this, we will use the following lemma.

5.8 Lemma. *Notations as above. Let $R := \mathbb{C}[\![z]\!]$ with its (z)-adic topology, and let $y\colon \operatorname{Spf}(R) \to \mathcal{A} \otimes \mathbb{C}$ be a morphism corresponding to a (formal) p.p.a.v. (Y, μ, η^p) over $\operatorname{Spf}(R)$. Let $i_0\colon \operatorname{Spec}(\mathbb{C}) \to \operatorname{Spf}(R)$ be the unique $\mathbb{C}$-valued point (given by $z \mapsto 0$), and assume that $y_0 := y \circ i_0$ is a point of $Sh(G, X)_{\mathbb{C}} \hookrightarrow \mathcal{A} \otimes \mathbb{C}$. As in 5.6.2, we obtain de Rham classes $t_{\alpha,\mathrm{dR},0} \in T_{\alpha,\mathrm{dR},0}$ for $\alpha \in \mathcal{J}$, where the subscript "$_0$" refers to the fact that these are classes on the special fibre Y_0. Assume that the formal horizontal continuations of the classes $t_{\alpha,\mathrm{dR},0}$ over $\operatorname{Spf}(R)$ remain inside $\operatorname{Fil}^0 T_{\alpha,\mathrm{dR}}$. Then y factors through $Sh(G, X)$.*

Proof (sketch). There exists a p.p.a.v. $(\widetilde{Y}, \widetilde{\mu})$ over an algebraic curve S such that the formal completion at some non-singular point $s_0 \in S$ gives back (Y, μ). (In this sketch of the argument we will forget about the level structures.) Over some open disc $U \hookrightarrow S^{\mathrm{an}}$ around s_0, we can choose a symplectic basis of $\mathrm{H}^1_{\mathrm{dR}}(\widetilde{Y}_U/U)$. By virtue of the Hodge filtration, this gives rise to a map $q\colon U \to \mathfrak{H}_g^{\vee}$, where $\mathfrak{H}_g^{\vee}$ (the compact dual of $\mathfrak{H}_g$) is the domain parametrizing g-dimensional subspaces $\operatorname{Fil}^1 \subset \mathbb{C}^{2g}$ which are totally isotropic for the standard symplectic form ψ. The point $q(s_0)$ lies on a subvariety $\check{X} \subset \mathfrak{H}_g^{\vee}$ (where $\check{X}$ is the compact dual of the hermitian symmetric domains $X^+ \subseteq X$ as in the given Shimura datum) parametrizing those flags for which the horizontal continuations $\tilde{t}_{\alpha,\mathrm{dR}}$ of the $t_{\alpha,\mathrm{dR},0}$ remain in the filtration step Fil^0. By consideration of the Taylor series development of the map q at s_0 one shows that q maps U into $\check{X}$, and this implies the assertion. $\square$

5.8.1 Proposition. *The morphism $\tilde{x}\colon \operatorname{Spec}(\widetilde{R}_e) \to \overline{\mathcal{A}}$ factors through $\overline{N} \subset \overline{\mathcal{A}}$.*

Proof (sketch). It suffices to show that the generic point of $\operatorname{Spec}(\widetilde{R}_e)$ maps to $Sh(G, X)$. Consider the homomorphism $j\colon \widetilde{R}_e \hookrightarrow \mathbb{C}[\![z]\!]$ with $T \mapsto z + \sigma(\pi)$ (using the chosen embedding $K_0 \subseteq K \xrightarrow{\sigma} \mathbb{C}$). Note that if we set $(Y, \mu, \eta^p) := j^*(\widetilde{X}, \widetilde{\lambda}, \widetilde{\theta}^p)$, then $(Y_0, \mu_0, \eta_0^p) = \sigma^*(X, \lambda, \theta)$. The de Rham classes $j^* t_{\alpha,\mathrm{dR}}$ on Y are formally horizontal, since $\partial/\partial z \cdot (j^* t_{\alpha,\mathrm{dR}}) = j^*(\partial/\partial T \cdot t_{\alpha,\mathrm{dR}}) = 0$. The claim now follows by applying the lemma. $\square$

5.8.2 The rest of the argument is easy. Pulling back $(\widetilde{X}, \widetilde{\lambda}, \widetilde{\theta}^p)$ via the morphism $\operatorname{Spec}(\overline{W}) \hookrightarrow \operatorname{Spec}(\widetilde{R}_e)$ we obtain a lifting of the closed point x_0 that we started off with, to a $\overline{W}$-valued point of $\overline{N}$. If H_1 is the corresponding p-divisible group over $\overline{W}$ then, by specialization, we have a collection $\{t_{\alpha,\mathrm{crys},1}\}_{\alpha \in \mathcal{J}}$ of crystalline Tate classes on H_1. Writing $M_1 := M(H_1) = M \otimes_{\widetilde{R}_e} \overline{W}$ and $\mathcal{G} := \mathcal{G}_{1,\widetilde{R}_e} \times_{\widetilde{R}_e} \overline{W} \hookrightarrow \operatorname{CSp}(M_1, \psi_1)$, the group $\mathcal{G}$ is reductive and is precisely the group fixing all tensors $t_{\alpha,\mathrm{crys},1}$. This brings us in a situation where we can apply the results of §4. Writing $\overline{\mathcal{N}}^\wedge$, $\overline{N}^\wedge$ and $\overline{\mathcal{A}}^\wedge$ for the formal completions at $\tilde{x}_0$ and x_0, and using the notations of 4.5–4.9, the same reasoning as in 5.8.1 above shows that the composition $\mathrm{Spf}(C) = \widehat{U_{\mathcal{G}}} \hookrightarrow \widehat{U} \cong \overline{\mathcal{A}}^\wedge$ factors through $\overline{N}^\wedge$. Now C and $\widehat{\mathcal{O}}_{\overline{N},x_0}$ are local $\overline{W}$-algebras of the same dimension, hence $\widehat{U_{\mathcal{G}}} \hookrightarrow \overline{N}^\wedge$ is dominant onto a component of $\overline{N}^\wedge$ and lifts to a morphism $\widehat{U_{\mathcal{G}}} \hookrightarrow \overline{\mathcal{N}}^\wedge$. Then $\widehat{\mathcal{O}}_{\overline{\mathcal{N}},\tilde{x}_0} \twoheadrightarrow C$ is a surjective homomorphism between local domains of the same dimension, hence an isomorphism. This concludes the proof of the following result.

5.8.3 Theorem. *In the situation of 5.1, assume that (5.6.1) holds. Then the model $\mathcal{N}$ is an integral canonical model of $Sh_{K_p}(G,X)$ over $\mathcal{O}_{(v)}$.*

5.9 *Vasiu's strategy—second part.* We continue our discussion of the paper [Va2]. What remains to be done to complete Vasiu's program is to show that, in the situation of Corollary 3.23, there exists a covering (G,X) with $i\colon (G,X) \hookrightarrow (\mathrm{CSp}_{2g},\mathfrak{H}_g^\pm)$ for which the assumption (5.6.1) holds. This is a highly non-trivial problem, and it is not clear to us if one can expect to solve this with the definition of a well-positioned family of tensors as in 5.3. The presentation of this material as it is presently available is too sketchy to convince us of the correctness of all arguments[5]; we will indicate by a marginal symbol $\boxed{?}$ statements of which we have not seen a complete proof.

In the rest of this section we shall only indicate the main line of Vasiu's arguments, without much further explanation.

5.9.1 Let W be a finite dimensional vector space over a field F of characteristic zero, and consider a semi-simple subgroup $G \subset \mathrm{GL}(W)$. On Lie algebras we have $\mathfrak{gl}(W) = \mathfrak{g} \oplus \mathfrak{g}^\perp$, where $\mathfrak{g}^\perp$ is the orthogonal of $\mathfrak{g} := \mathrm{Lie}(G)$ w.r.t. the form $(A_1,A_2) \mapsto \mathrm{Tr}(A_1 A_2)$ on $\mathfrak{gl}(W)$. Write $\pi_\mathfrak{g}$ for the projector onto $\mathfrak{g}$; we view $\pi_\mathfrak{g}$ as an element of $W(2,2;0)$. Next we consider the Killing form $\beta_\mathfrak{g}\colon \mathfrak{g} \times \mathfrak{g} \to F$. Since G is semi-simple, the form $\beta_\mathfrak{g}$ is non-degenerate, so that there exists a form $\beta_\mathfrak{g}^*\colon \mathfrak{g}^* \times \mathfrak{g}^* \to F$ with $\langle \beta_\mathfrak{g}, \beta_\mathfrak{g}^* \rangle = 1$. Using the direct sum decomposition $\mathfrak{gl}(W) = \mathfrak{g} \oplus \mathfrak{g}^\perp$ and the induced isomorphism $\mathfrak{gl}(W)^* = \mathfrak{g}^* \oplus (\mathfrak{g}^\perp)^*$, we can view $\beta_\mathfrak{g}$ and $\beta_\mathfrak{g}^*$ as elements of $W(2,2;0)$. Clearly the tensors $\pi_\mathfrak{g}$, $\beta_\mathfrak{g}$ and $\beta_\mathfrak{g}^*$ are G-invariant. Even better: if G is the derived subgroup of a reductive group $H \subset \mathrm{GL}(W)$ then $\pi_\mathfrak{g}$, $\beta_\mathfrak{g}$ and $\beta_\mathfrak{g}^*$ are

---

[5] As remarked before, we strongly encourage the reader to read Vasiu's original papers, some versions of which appeared after we completed this manuscript.

also H-invariant.

Finally we define an integer $s(\mathfrak{g}, W)$. For this we fix an algebraic closure $\bar{F}$ of F and we choose a Cartan subalgebra $\mathfrak{t} \subset \mathfrak{g}_{\bar{F}}$. For a root $\alpha \in R(\mathfrak{g}_{\bar{F}}, \mathfrak{t})$, let $\mathfrak{s}_\alpha \subset \mathfrak{g}_{\bar{F}}$ denote the Lie subalgebra (isomorphic to $\mathfrak{sl}_2$) generated by $\mathfrak{g}^\alpha$ and $\mathfrak{g}^{-\alpha}$. We write

$$d_\alpha := \max\{\dim(Y) \mid Y \subset W_{\bar{F}} \text{ is an irreducible } \mathfrak{s}_\alpha\text{-submodule}\},$$

and we define $s(\mathfrak{g}, W) := \max\{d_\alpha \mid \alpha \in R(\mathfrak{g}, \mathfrak{t})\}$. If Ξ is the set of weights occurring in the $\mathfrak{g}$-module W then $d_\alpha = 1 + \max\{\alpha^\vee(\xi) \mid \xi \in \Xi\}$. So, if $\alpha = \alpha_1, \alpha_2, \ldots, \alpha_r$ is a basis of $R(\mathfrak{g}, \mathfrak{t})$ and if W is irreducible with highest weight $\varpi = n_1 \cdot \varpi_1 + \cdots + n_r \cdot \varpi_r$, where ϖ_i is the fundamental dominant weight corresponding to α_i, then $d_\alpha = d_{\alpha_1} = 1 + n_1$.

5.9.2 Claim. *Let W be a finite dimensional $\mathbb{Q}$-vector space with a non-degenerate symplectic form ψ. If $G \subset \mathrm{CSp}(W, \psi)$ is a semi-simple subgroup and if $p \geq s(\mathfrak{g}, W)$ then $\{\pi_\mathfrak{g}, \beta_\mathfrak{g}, \beta_\mathfrak{g}^*\}$ is a well-positioned family of tensors for the group G over the d.v.r. $\mathbb{Z}_{(p)}$.*

If one tries to prove a statement like this then a priori one would have to consider an arbitrary faithfully flat $\mathbb{Z}_{(p)}$-algebra R and a free R-module M with $M \otimes_R R[1/p] \cong W \otimes_\mathbb{Q} R[1/p]$. Since we are dealing with a finite collection of tensors, however, one easily reduces to the case that R is of finite type over $\mathbb{Z}_{(p)}$. Also we may replace R by a faithfully flat covering, since taking a Zariski closure of something quasi-compact commutes with flat base-change. This allows one to reduce to the case that R is a complete local noetherian ring.

It should be noted that in general the Zariski closure of $G_{R[/p]}$ inside $\mathrm{GL}(M)$ is *not* a subgroup scheme, even if R is a regular local ring. We refer to the work [BT] of Bruhat and Tits, especially loc. cit., 3.2.15, for further theory and a very instructive example.

5.9.3 Corollary. *Assume 5.9.2 to hold. Consider a closed immersion of Shimura data $i \colon (G, X) \hookrightarrow (\mathrm{CSp}_{2g, \mathbb{Q}}, \mathfrak{H}_g^\pm)$. Let p be a prime number, with $p \geq 5$. Assume that the Zariski closure of G inside $\mathrm{CSp}_{2g, \mathbb{Z}_{(p)}}$ is reductive and that the tensors $\pi_{\mathfrak{g}^{\mathrm{der}}}$, $\beta_{\mathfrak{g}^{\mathrm{der}}}$ and $\beta_{\mathfrak{g}^{\mathrm{der}}}^*$ are $\mathbb{Z}_{(p)}^{2g}$-integral. Then condition (5.6.1) is satisfied. In particular: for every prime v of $E = E(G, X)$ above p and every hyperspecial subgroup $K_p \subset G(\mathbb{Q}_p)$, there exists an i.c.m. of $Sh_{K_p}(G, X)$ over $\mathcal{O}_{E,(v)}$.*

Up to one technical detail, we can derive this corollary from the previous claim by the following argument. By the results of [De3], Sect. 1.3, all highest weights in the representation $i\colon G \hookrightarrow \mathrm{GL}(\mathbb{Q}^{2g})$ are miniscule in the sense of [Bou], Chap. VIII, §7, n° 3. It follows from this that $s(\mathfrak{g}^{\mathrm{der}}, \mathbb{Q}^{2g}) = 2$. Since $p \geq 5$, the set of tensors $\{\pi_{\mathfrak{g}^{\mathrm{der}}}, \beta_{\mathfrak{g}^{\mathrm{der}}}, \beta^*_{\mathfrak{g}^{\mathrm{der}}}\}$ is a well-positioned set of G-invariant tensors (of degree 4) for the group G^{der} over $\mathbb{Z}_{(p)}$.

Next we consider the Zariski closure $\mathcal{G}$ of G inside $\mathrm{CSp}_{2g, \mathbb{Z}_{(p)}}$. By assumption, it is reductive. Let $\mathcal{Z} := Z(\mathcal{G})^0$ be the connected center of $\mathcal{G}$, which is a torus over $\mathbb{Z}_{(p)}$ with generic fibre $Z := Z(G)^0$. Also write $\mathcal{C} \subset \mathrm{End}(\mathbb{Z}^{2g}_{(p)})$ for the subalgebra of endomorphisms which commute with the action of $\mathcal{G}$. We claim that the elements of $\mathcal{C}$ form a well-positioned collection of G-invariant tensors (of degree 2) for the group Z over $\mathbb{Z}_{(p)}$. We will not prove this; the essential idea is to reduce to the situation where Z is a split torus. For details we refer to [Va2].

Now we take $\mathcal{T} := \{\pi_{\mathfrak{g}^{\mathrm{der}}}, \beta_{\mathfrak{g}^{\mathrm{der}}}, \beta^*_{\mathfrak{g}^{\mathrm{der}}}\} \cup \mathcal{C}$ as our collection of G-invariant tensors. Notice that the condition $2r_\alpha \leq 2(p-2)$ in (5.6.1) is satisfied, since we are only using tensors of degrees 2 and 4 and since $p \geq 5$. To conclude the proof of the corollary, one considers a faithfully flat $\mathbb{Z}_{(p)}$-algebra R and a free R-module M with an identification $M \otimes_R R[1/p] \cong \mathbb{Q}^{2g} \otimes_\mathbb{Q} R[1/p]$ such that ψ and ψ^*, as well as all tensors in our collection $\mathcal{T}$ are M-integral. Then we know that ψ induces a perfect form ψ_M on M, that the Zariski closure $\mathcal{G}_1$ of $G^{\mathrm{der}} \otimes_\mathbb{Q} R[1/p]$ inside $\mathrm{CSp}(M, \psi_M)$ is semi-simple, and that the Zariski closure $\mathcal{Z}_1$ of $Z \otimes_\mathbb{Q} R[1/p]$ inside $\mathrm{CSp}(M, \psi_M)$ is a torus. We are therefore left with the following question. (In [Va2] it is used implicitly that the answer is affirmative.)

5.9.4 Problem. *Let R be a faithfully flat $\mathbb{Z}_{(p)}$-algebra and let M be a free R-module of finite rank. If $G_{R[1/p]} \subseteq \mathrm{GL}(M[1/p])$ is a reductive subgroup scheme such that the Zariski closures $\mathcal{G}_1$ and $\mathcal{Z}_1$ of respectively its derived subgroup and its connected center are reductive subgroup schemes of $\mathrm{GL}(M)$, does it follow that the Zariski closure of $G_{R[1/p]}$ inside $\mathrm{GL}(M)$ is flat over R and therefore again a reductive subgroup scheme?*

Perhaps the answer to this question is known to experts in this field, in which case we would be interested to hear it. If we assume that the answer is affirmative then Cor. 5.9.3 follows by the arguments given above.

5.9.5 Claim. *Let $(G^{\mathrm{ad}}, X^{\mathrm{ad}})$ be an adjoint Shimura datum of abelian type, and let $p \geq 5$ be a prime number such that $G^{\mathrm{ad}}_{\mathbb{Q}_p}$ is unramified. Then there* *exists a Shimura datum* (G, X) *covering* $(G^{\mathrm{ad}}, X^{\mathrm{ad}})$ *and a closed immersion* $i\colon (G, X) \hookrightarrow (\mathrm{CSp}_{2g,\mathbb{Q}}, \mathfrak{H}_g^{\pm})$ *such that condition (5.6.1) holds for G.*

In [Va2] this statement is claimed as a consequence of a whole chain of constructions, reducing the problem to Cor. 5.9.3.

[?] **5.9.6 Corollary. (Assuming 5.9.2—5.9.5)** *Let* (G, X) *be a Shimura datum of pre-abelian type. Let* $p \geq 5$ *be a prime number such that (notations of 3.21.5)* $p \nmid \delta_G$ *and such that* $G_{\mathbb{Q}_p}$ *is unramified. Let* $K_p \subset G(\mathbb{Q}_p)$ *be a hyperspecial subgroup and let* v *be a prime of* $E(G, X)$ *above* p. *Then there exists an integral canonical model* $\mathcal{M}$ *of* $Sh_{K_p}(G, X)$ *over* $\mathcal{O}_{E,(v)}$. *As a scheme,* $\mathcal{M}$ *is the projective limit of smooth quasi-projective* $\mathcal{O}_{E,(v)}$-*schemes with étale coverings as transition maps.*

§6 Characterizing subvarieties of Hodge type; conjectures of Coleman and Oort

6.1 We now turn to a couple of problems of a somewhat different flavour. Consider a Shimura variety $Sh_K(G, X)$. We have seen in §1 that, depending on the choice of a representation of G, we can view it, loosely speaking, as a "moduli space" for Hodge structures with some given Hodge classes. In this interpretation, the "Shimura subvarieties" would be components of the loci where the Hodge structures have certain additional classes. The type of question that we are interested in here is: "can we give a direct description of these Shimura subvarieties?", and "given an arbitrary subvariety of $Sh_K(G, X)$, can we say something about "how often" it intersects a Shimura subvariety?". More specific questions will be formulated below. First, however, let us make the notion of a Shimura subvariety more precise.

6.2 Definition. Let (G, X) be a Shimura datum. An irreducible algebraic subvariety $S \subseteq Sh_K(G, X)_{\mathbb{C}}$ is called a subvariety of Hodge type if there exist an algebraic subgroup $H \subseteq G$ (defined over $\mathbb{Q}$), an element $\eta \in G(\mathbb{A}_f)$ and a connected component Y_H^+ of the locus

$$Y_H := \{h \in X \mid h\colon \mathbb{S} \to G_{\mathbb{R}} \text{ factors through } H_{\mathbb{R}}\}$$

such that $S(\mathbb{C})$ is the image of $Y_H^+ \times \eta K$ in $Sh_K(G, X)(\mathbb{C}) = G(\mathbb{Q})\backslash X \times G(\mathbb{A}_f)/K$.

If $E(G, X) \subseteq F \subseteq \mathbb{C}$, then an algebraic subvariety $S \subseteq Sh_K(G, X)_F$ is called a subvariety of Hodge type if all components of $S_{\mathbb{C}}$ are of Hodge type. (If S is irreducible then it suffices to check this for one component of $S_{\mathbb{C}}$.)

For example: a point x of $Sh_K(G, X)$, considered as a 0-dimensional subvariety, is of Hodge type if and only if x is a special point. If $Sh_K(G, X) \hookrightarrow A_{g,1,n}$ is a Shimura subvariety of Hodge type then these conditions on x are equivalent to saying that x corresponds to an abelian variety of CM-type (in which case we say that x is a CM-point.)

If $f \colon (G_1, X_1) \hookrightarrow (G_2, X_2)$ is a closed immersion of Shimura varieties and if we have compact open subgroups $K_i \subset G_i(A_f)$ $(i = 1, 2)$ with $f(K_1) \subseteq K_2$, then the connected components of the image of $Sh(f) \colon Sh_{K_1}(G_1, X_1) \to Sh_{K_2}(G_2, X_2)$ are called subvarieties of Shimura type. The subvarieties of Hodge type are precisely the irreducible components of Hecke translates of subvarieties of Shimura type. For further details see [Mo1], Chap. I or [Mo2], section 1.

Now for some of the concrete problems that we are interested in.

6.3 Conjecture. (Coleman, cf. [Co]) *For $g \geq 4$, there are only finitely many smooth projective genus g curves C over $\mathbb{C}$ (taken up to isomorphism) for which $\mathrm{Jac}(C)$ is of CM-type.*

6.4 Conjecture. (Oort, cf. [Oo3]) *Let $Z \hookrightarrow A_{g,1,n} \otimes \mathbb{C}$ be an irreducible algebraic subvariety such that the CM-points on Z are dense for the Zariski topology. Then Z is a subvariety of Hodge type.*

6.5 Let us first make some remarks on the status of these conjectures. Coleman's conjecture, as we phrased it here, is *false* for $g = 4$ and $g = 6$: there exist families of curves $C \to S$ of genus 4 and 6, such that the image of S in $A_{g,1}$ corresponding to the family of Jacobians $\mathrm{Jac}(C/S) \to S$ is (an open part of) a subvariety of Hodge type of dimension > 0. The known examples of this type are given by explicit polynomial equations. For example, let S be the affine line with coordinate λ, and let C_N be the smooth curve over S with affine equation $y^N = x(x - 1)(x - \lambda)$. If $3 \nmid N$ then C is a family of curves of genus $N - 1$ with an automorphism ζ_N of order N given by $(x, y) \mapsto (x, e^{2\pi i/N} \cdot y)$. For $N = 5$ (resp. $N = 7$) we obtain a family of Jacobians $J_N \to S$ with complex multiplication by $\mathbb{Q}[\zeta_5]$ (resp. $\mathbb{Q}[\zeta_7]$), and one computes that the complex embedding given by $\zeta_N \mapsto e^{k \cdot 2\pi i/N}$ has multiplicity 2,1,1,0 for $k = 1, 2, 3, 4$ (resp. multiplicity 2, 2, 1, 1, 0, 0 for $k = 1, \ldots, 6$) on the tangent space. Now the Shimura variety of PEL type parametrizing abelian 4-folds (resp. 6-folds) with complex multiplication by an order of $\mathbb{Q}[\zeta_5]$ (resp. $\mathbb{Q}[\zeta_7]$) and the given multiplicities on the tangent space is 1-dimensional, so the image of S in $A_{4,1}$ (resp. $A_{6,1}$) is an open part of such a subvariety of PEL type. It follows that there are infinitely many values of λ such that $\mathrm{Jac}(\mathcal{C}_\lambda)$ is of CM-type. For further details, and another example of this kind, we refer to [dJN].

For genera $g = 5$ and $g \geq 7$, Coleman's conjecture remains, to our knowledge, completely open. It is plausible that the known counter examples are exceptional, and that examples of such kind only exist for certain "low" genera. Let us point out here that in the above example, we do not find a subvariety of Hodge type if $3 \nmid N$ and $N \geq 8$; this follows from [dJN], Prop. 5.7 and the results of Noot in [No2] (see 6.15 below).

6.6 Oort's conjecture was studied by the author in [Mo1]. The results here are based on a characterization of subvarieties of Hodge type in terms of certain "linearity properties". We will discuss this in more detail below. One of the results in loc. cit., is a proof of Oort's conjecture under an additional assumption. This is a general result, which provides further evidence for the conjecture. Unfortunately, the extra assumption is difficult to verify in practice.

In another direction, one can try to prove the conjecture in concrete cases. The first non-trivial case is to consider subvarieties of a product of two modular curves. After some reduction steps first proved by Chai, André and Edixhoven (see [Ed2]) both found a proof for the conjecture in this case under an additional hypothesis. Both their methods and the hypotheses involved were rather different. Recently, André found an unconditional proof, so that we now have the following result (see [An2]).

6.6.1 Theorem. (André) *Let S_1 and S_2 be modular curves over $\mathbb{C}$, and let $C \subset S_1 \times S_2$ be an irreducible algebraic curve containing infinitely many points (x_1, x_2) such that both $x_1 \in S_1$ and $x_2 \in S_2$ are CM-points (in other words, C contains a Zariski dense set of CM-points). Then C is a subvariety of Hodge type, i.e., either $C = S_1 \times \{x_2\}$, where x_2 is a CM-point of S_2, or $C = \{x_1\} \times S_2$, where x_1 is a CM-point of S_1, or C is a component of a Hecke correspondence.*

6.7 One of the motivations for Oort to formulate his conjecture is its analogy with the Manin-Mumford conjecture, now a theorem of Raynaud (see [Ra2]). We recall the statement:

6.7.1 Theorem. (Raynaud) *Let X be a complex abelian variety, and let $Z \hookrightarrow X$ be an algebraic subvariety which contains a Zariski dense collection of torsion points. Then Z is the translate of an abelian subvariety over a* *torsion point.*

The analogy is obtained by using the following dictionary:

Oort's conjecture	"Manin-Mumford" = Raynaud's thm.
Shimura variety	abelian variety
CM-point (or special point)	torsion point
subvariety of Hodge type	translate of an abelian subvariety over a torsion point

To push the analogy even further, let us mention that one can formulate a conjecture which contains both Oort's conjecture and "Manin-Mumford" as special cases. The idea here is to look at mixed Shimura varieties. Since we have not discussed these in detail, let us mention the following fact: if $S \hookrightarrow A_{g,1,n}$ is a subvariety of Hodge type, and if $X \to S$ is the universal abelian scheme over it, then X can be described as a (component of a) mixed Shimura variety. (See [Pi] and [Mi2] for further examples and details.) The special points on X are the torsion points on fibres X_s of CM-type. However, the axioms of mixed Shimura varieties are too restrictive for our purposes, since, for example, an abelian variety X which is not of CM-type, cannot be described as a mixed Shimura variety. By loosening the axioms somewhat, we are led to what might be called "mixed Kuga varieties" and to the following conjecture, proposed by Y. André in [An1]. (André adds the remark that this is only a tentative statement, which may have to be adjusted.)

6.7.2 Conjecture. *Let G be an algebraic group over $\mathbb{Q}$, let K_∞ be a maximal compact subgroup of $G(\mathbb{R})$, and let Γ be an arithmetic subgroup of $G(\mathbb{Q})$. Suppose that K_∞ is defined over $\mathbb{Q}$, that $G(\mathbb{R})/K_\infty$ has a $G(\mathbb{R})$-invariant complex structure and that the complex analytic space $\Gamma\backslash G(\mathbb{R})/K_\infty$ is algebraizable. Let us call an irreducible algebraic subvariety $S \hookrightarrow \Gamma\backslash G(\mathbb{R})/K_\infty$ a special subvariety if there exists an algebraic subgroup $H \subseteq G$ defined over $\mathbb{Q}$ and an element $g_0 \in G(\mathbb{Q})$ such that $S = \{[g_0 \cdot h] \in \Gamma\backslash G(\mathbb{R})/K_\infty \mid h \in H(\mathbb{R})\}$. Then S is a special subvariety if and only if it contains a Zariski dense collection of special points.*

6.8 Assume Oort's conjecture to be true. Then Coleman's conjecture becomes the question of whether there are positive-dimensional subvarieties of Hodge type $S \hookrightarrow A_{g,1} \otimes \mathbb{C}$ of which an open part is contained in the open Torelli locus $\mathcal{T}_\mathbb{C}^0$ ($:=$ the image of the Torelli morphism $\mathcal{M}_g \otimes \mathbb{C} \to A_{g,1} \otimes \mathbb{C}$).

This seems a difficult question, also if we replace the open Torelli locus by its closure. Hain's paper [H] contains interesting new results about this.[6]

To state Hain's results, let us first consider an algebraic group G over $\mathbb{Q}$ which gives rise to a hermitian symmetric domain X (i.e., an algebraic group of hermitian type), and consider a locally symmetric (or arithmetic) variety $S = \Gamma\backslash X$, where Γ is an arithmetic subgroup of $G(\mathbb{Q})$. If G is $\mathbb{Q}$-simple then we call S a simple arithmetic variety. We say that S is *bad* if it contains a locally symmetric divisor (examples: $G = \mathrm{SO}(n,2)$ or $G = \mathrm{SU}(n,1)$, as well as the case $\dim(S) = 1$); otherwise call S *good*. This is a really a property of G, i.e., it does not depend on Γ and the resulting S. In the next statement we only consider the simple case; this is not a serious restriction since every arithmetic variety has a finite cover which is a product of simple ones.

6.8.1 Theorem. (Hain) *Let S be a simple arithmetic variety which is good in the above sense.*

(i) Suppose $p\colon \mathcal{C} \to S$ is a family of stable curves over S such that the Picard group $\mathrm{Pic}^0(\mathcal{C}_s)$ of every fibre is an abelian variety (i.e., every fibre is a "good" curve: its dual graph is a tree), such that the generic fibre $\mathcal{C}_\eta$ is smooth, and such that the period map $S \to \mathsf{A}_{g,1}$ is a finite map of locally symmetric varieties. Then S is a quotient of the open complex n-ball for some n. (So $G_\mathbb{R}^{\mathrm{ad}} = \mathrm{PSU}(n,1) \times$ (compact factors).)

(ii) Suppose $q\colon Y \to S$ is a family of abelian varieties, such that every fibre Y_s is the Jacobian of a good curve, and such that the period map $S \to \mathsf{A}_{g,1}$ is a finite map of locally symmetric varieties. Write S^{red} (resp. S^{hyp}) for the locus of points such that Y_s is the Jacobian of a reducible (resp. hyperelliptic) curve, and let S^* be the complement of S^{red}, which we assume to be non-empty. Then either S is the quotient of the complex n-ball for some n, or $g \geq 3$, each component of S^{red} has codimension ≥ 2 and $S^* \cap S^{\mathrm{hyp}}$ is a non-empty smooth divisor in S^*.

(We point out that a family $Y \to S$ as in (ii) is not necessarily of the form $\mathrm{Jac}(\mathcal{C}/S) \to S$ for a family $\mathcal{C} \to S$ as in (i), due to the fact that the Torelli morphism is ramified along the hyperelliptic locus. If, in (i), all fibres are smooth then the condition that S is good can be ommitted.)

6.9 The next issue that we want to discuss is the characterization of subvarieties of Hodge type by their property of being "formally linear". Here we

---

[6]We thank R. Hain for sending us a preliminary version of this paper.

owe the reader some explanation. Let us first do the theory over $\mathbb{C}$, which works for arbitrary Shimura varieties.

Consider a Shimura variety $Sh_K = Sh_K(G, X)_{\mathbb{C}}$ over $\mathbb{C}$, and let $S \hookrightarrow Sh_K$ be a subvariety of Hodge type. Then S is a totally geodesic subvariety: if $u \colon X^+ \to Sh_K^0$ is the uniformization of the component $Sh_K^0 \subseteq Sh_K$ containing S, and if $\widetilde{S} \subseteq X^+$ is a component of $u^{-1}(S)$, then $\widetilde{S}$ is a totally geodesic submanifold of the hermitian symmetric domain X^+. This property does not characterize subvarieties of Hodge type; for a trivial example: any point $x \in S$ forms a totally geodesic algebraic subvariety, but $\{x\} \subseteq S$ is a subvariety of Hodge type if and only if x is a special point. Essentially, however, we are dealing with the well-known distinction between "Kuga subvarieties" and subvarieties of Hodge type. In a somewhat less general setting, this distinction was clarified by Mumford in [Mu1]. The same idea works in general, and we have the following characterization (see [Mo1], Thm. II.3.1, or [Mo2]).

6.9.1 Theorem. *Let $S \hookrightarrow Sh_K(G, X)_{\mathbb{C}}$ be an irreducible algebraic subvariety. Then S is a subvariety of Hodge type if and only if (i) S is totally geodesic, and (ii) S contains at least one special point.*

Let us mention that one can also give a description of totally geodesic subvarieties in general (i.e., not necessarily containing a special point). It turns out that they are intimately connected with non-rigidity phenomena. For example, let $Sh_K(G, X)_{\mathbb{C}} \hookrightarrow Sh_{K'}(G', X')_{\mathbb{C}}$ be a closed immersion of Shimura varieties, and suppose that the adjoint group G^{ad} decomposes (over $\mathbb{Q}$) as a product, say $G^{\mathrm{ad}} = G_1 \times G_2$. Correspondingly, there is a decomposition $X^{\mathrm{ad}} = X_1 \times X_2$ of X as a product of (finite unions of) hermitian symmetric domains. Fix a component $X_1^+ \subseteq X_1$, a point $x_2 \in X_2$, and a class $\eta K \in G(\mathbb{A}_f)/K$, and let $S_{\eta K}(X_1^+, x_2)$ denote the image of $X_1^+ \times \{x_2\}$ in $Sh_K(G, X)$ under the map $X \ni x \mapsto [x \times \eta K]$. One can show that $S_{\eta K}(X_1^+, x_2)$ is a totally geodesic algebraic subvariety of $Sh_{K'}(G', X')_{\mathbb{C}}$, and that, conversely, all totally geodesic algebraic subvarieties of $Sh_{K'}(G', X')_{\mathbb{C}}$ are of this form.

After passing to a suitable level (i.e., replacing K by a suitable subgroup of finite index) we can arrange that the component Sh_K^0 of $Sh_K(G, X)_{\mathbb{C}}$ containing $S_{\eta K}(X_1^+, x_2)$ is a product variety $Sh_K^0 = S_1 \times S_2$, with $S_{\eta K}(X_1^+, x_2) = S_1 \times \{s_2\}$ for some point $s_2 \in S_2$. Now assume that G_2 is not trivial, so that $\dim(S_2) > 0$. We see that $S_1 \times \{s_2\}$ is non-rigid: global deformations are obtained by moving the point $s_2 \in S_2$. If (G, X) is of Hodge type, say with $Sh_{K'}(G', X')_{\mathbb{C}} = A_{g,1,n} \otimes \mathbb{C}$ in the above, then we obtain a non-rigid abelian scheme over S_1. (Notice, however, that the non-rigidity may be of a trivial nature, in the sense that all non-rigid factors of the abelian scheme in question are isotrivial.) The gist of the results in [Mo1], §II.4 (see also [Mo2]) is that all non-rigid abelian schemes, and all their deformations, can be described via the above procedure. We refer to loc. cit. for further details.

6.9.2 We can jazz-up the above characterization of subvarieties of Hodge type. This will lead to a formulation very analogous to the results in mixed characteristics, to be discussed next.

The first important remark is that total geodesicness needs to be tested only at one point. More precisely: if $Z \hookrightarrow Sh_K(G, X)_{\mathbb{C}}$ is an irreducible algebraic subvariety, and if $x \in Z$ is a non-singular point of $Sh_K(G, X)_{\mathbb{C}}$, then Z is totally geodesic (globally) if and only if it is totally geodesic at the point x. This is true because $Sh_K(G, X)_{\mathbb{C}}$ has constant curvature.

Next we define a Serre-Tate group structure on the formal completion $\mathfrak{Sh}_x$ of $Sh_K(G, X)_{\mathbb{C}}$ at an arbitrary point x. Here we assume that K is neat, so that $Sh_K(G, X)$ is non-singular. The procedure is the following.

The point x lies in the image Sh^0 of a uniformization map $u\colon X^+ \to Sh_K(G, X)$, which, by our assumption on K, is a topological covering. Choose $\tilde{x} \in X^+$ with $u(\tilde{x}) = x$. We have a Borel embedding

$$X^+ \hookrightarrow \check{X} = G^{\mathrm{ad}}(\mathbb{C})/P_{\tilde{x}}(\mathbb{C}),$$

where $P_{\tilde{x}} \subset G^{\mathrm{ad}}_{\mathbb{C}}$ is the parabolic subgroup stabilizing the point $\tilde{x}$. Using the Hodge decomposition of $\mathfrak{g}_{\mathbb{C}}$ with respect to $\mathrm{Ad} \circ h_{\tilde{x}}$, we obtain a parabolic subgroup $P^-_{\tilde{x}} \subset G^{\mathrm{ad}}_{\mathbb{C}}$ opposite to $P_{\tilde{x}}$. Write $U^-_{\tilde{x}}$ for the unipotent radical of $P^-_{\tilde{x}}$, which is isomorphic to $\widehat{\mathbb{G}}^d_a$ for $d = \dim(X)$. The natural map $U^-_{\tilde{x}}(\mathbb{C}) \to \check{X}$ gives an isomorphism of $U^-_{\tilde{x}}(\mathbb{C})$ onto its image $\mathcal{U} \subset \check{X}$ which is the complement of a divisor $D \subset \check{X}$. On formal completions we obtain an isomorphism

$$\mathfrak{U}_{\tilde{x}} := U^-_{\tilde{x}}{}_{/\{1\}} \xrightarrow{\sim} \mathcal{U}_{/\{\tilde{x}\}} = \check{X}_{/\{\tilde{x}\}} \xrightarrow{u} Sh^0_{/\{x\}} =: \mathfrak{Sh}_x,$$

and in this way $\mathfrak{Sh}_x$ inherits the structure of a formal vector group. This we call the Serre-Tate group structure on $\mathfrak{Sh}_x$. One checks that it is independent of the choice of $\tilde{x}$ above x.

If Z is a subvariety as above, then by taking the formal completion at x, we obtain a formal subscheme $\mathfrak{Z}_x \hookrightarrow \mathfrak{Sh}_x$, and we call Z formally linear at x if $\mathfrak{Z}_x$ is a formal vector subgroup of $\mathfrak{Sh}_x$. Using this terminology we have the following result. (See [Mo2], §5.)

6.9.3 Theorem. Let $Z \hookrightarrow Sh_K(G, X)_{\mathbb{C}}$ be an irreducible algebraic subvariety. If Z is totally geodesic then it is formally linear at all its points. Conversely, if Z is formally linear at some point $x \in Z$, then it is totally geodesic. In particular, Z is a subvariety of Hodge type if and only if (i) Z is formally linear at some point $x \in Z$, and (ii) Z contains at least one special point.

6.10 In mixed characteristics, our notion of formal linearity is based on Serre-Tate deformation theory of ordinary abelian varieties. Almost everything we need is treated in Katz' paper [Ka]; additional references are [DI] and [Me]. Without proofs, we record some statements that are most relevant for our discussion.

Let k be a perfect field of characteristic $p > 0$, and let X_0 be an ordinary abelian variety over k. Set $W = W(k)$, and write $\mathcal{C}_W$ for the category of artinian local W-algebras R with $W/(p) = k \xrightarrow{\sim} R/\mathfrak{m}_R$. The formal deformation functor $\mathcal{D}efo_{X_0} \colon \mathcal{C}_W \to$ Sets is given by

$$\mathcal{D}efo_{X_0}(R) = \{(X, \varphi) \mid X \text{ an abelian scheme over } R;\ \varphi \colon X \otimes k \xrightarrow{\sim} X_0\}/\cong .$$

By the general Serre-Tate theorem, this functor is isomorphic to the formal deformation functor of the p-divisible group $X_0[p^\infty]$. Since X_0 was assumed to be ordinary, the latter is a direct sum $X_0[p^\infty] = G_\mu \oplus G_{\text{ét}}$ of a toroidal and an étale part. For $R \in \mathcal{C}_W$, these two summands both have a unique lifting, say $\widetilde{G}_\mu$ and $\widetilde{G}_{\text{ét}}$ respectively, to a p-divisible group over R. We therefore have

$$\mathcal{D}efo_{X_0}(R) = \{\alpha \in \text{Ext}_R(\widetilde{G}_{\text{ét}}, \widetilde{G}_\mu) \mid \alpha_{|\text{Spec}(k)} \text{ is trivial}\},$$

and in particular we see that $\mathcal{D}efo_{X_0}$ has a natural structure of a group functor.

Fix an algebraic closure $\bar{k}$ of k, write $\overline{W} := W(\bar{k})$, and write $T_p X_0$ for the "physical" Tate module of X_0. The formal deformation functor of $X_0 \otimes \bar{k}$ can be given "canonical coordinates": if $(X, \varphi) \in \mathcal{D}efo_{X_0 \otimes \bar{k}}(R)$ for some $R \in \mathcal{C}_{W(\bar{k})}$, then one associates to X a $\mathbb{Z}_p$-bilinear form

$$q(X/R; -, -) \colon T_p X_0 \times T_p X_0^t \longrightarrow \widehat{\mathbb{G}}_{\text{m}}(R) = 1 + \mathfrak{m}_R$$

and it can be shown that this yields an isomorphism of functors

$$\mathcal{D}efo_{X_0 \otimes \bar{k}} \xrightarrow{\sim} \text{Hom}_{\mathbb{Z}_p}(T_p X_0 \otimes T_p X_0^t, \widehat{\mathbb{G}}_{\text{m}}).$$

If we identify the double dual X^{tt} and X, then we have a symmetry formula $q(X/R; \alpha, \alpha_t) = q(X^t/R; \alpha_t, \alpha)$. Furthermore, if $f_0 \colon X_{0, \bar{k}} \to Y_{0, \bar{k}}$ is a homomorphism of ordinary abelian varieties over $\bar{k}$, then f_0 lifts to a homomorphism $f\colon X \to Y$ over $R \in \mathcal{C}_{\overline{W}}$ if and only if $q\big(X/R; \alpha, f^t(\beta)\big) = q\big(Y/R; f(\alpha), \beta\big)$ for every $\alpha \in T_p X_0$, $\beta \in T_p Y_0$.

Let $\lambda_0\colon X_0 \to X_0^t$ be a principal polarization. Using the induced isomorphism $T_p X_0 \overset{\sim}{\to} T_p X_0^t$ we have $\mathcal{D}efo_{X_{0,\bar{k}}} \cong \mathrm{Hom}(T_p X_0^{\otimes 2}, \widehat{\mathbb{G}}_m)$, and by the previous remarks the formal deformation functor $\mathcal{D}efo_{(X_{0,\bar{k}}, \lambda_0)}$ of the pair $(X_{0,\bar{k}}, \lambda_0)$ is isomorphic to the closed subfunctor $\mathrm{Hom}(\mathrm{Sym}^2(T_p X_0), \widehat{\mathbb{G}}_m)$.

6.11 Let κ be a perfect field of characteristic p with $p \nmid n$, and let $x \in (A_{g,1,n} \otimes \kappa)^{\mathrm{ord}}$ be a closed ordinary moduli point with residue field k. Write $(X_0, \lambda_0, \theta_0)$ for the corresponding p.p.a.v. plus level structure over $\mathrm{Spec}(k)$. The formal completion $\mathfrak{A}_x := \big(A_{g,1,n} \otimes W(\kappa)\big)_{/\{x\}}$ is a formal torus over $\mathrm{Spf}\big(W(k)\big)$; since we consider level n structures with $p \nmid n$, it represents the formal deformation functor $\mathcal{D}efo_{(X_0, \lambda_0)}$. By the above, $\mathfrak{A}_x$ has the structure of a formal torus over $W(k)$, called the Serre-Tate group structure.

Choose a basis $\{\alpha_1, \ldots, \alpha_g\}$ for $T_p X_0$, and set $q_{ij} = q\big(-; \alpha_i, \lambda_0(\alpha_j)\big)$. We have $\mathfrak{A}_x \widehat{\otimes} \overline{W} \cong \mathrm{Spf}(A)$, where $A = \overline{W}[\![q_{ij} - 1]\!]/(q_{ij} - q_{ji})$ with its $\mathfrak{m}$-adic topology, $\mathfrak{m} = (p, q_{ij} - 1)$. If $\mathfrak{X} \to \mathfrak{A} := \mathfrak{A}_x \widehat{\otimes} \overline{W}$ is the universal formal deformation, then there is an explicit description of the Hodge F-crystal $H = H^1_{\mathrm{dR}}(\mathfrak{X}/\mathfrak{A})$: to the chosen basis $\{\alpha_1, \ldots, \alpha_g\}$ one associates an A-basis $\{a_1, \ldots, a_g, b_1, \ldots, b_g\}$ of H such that

(i) the Hodge filtration is given by $\mathrm{Fil}^0 = H \supset \mathrm{Fil}^1 = A \cdot b_1 + \cdots + A \cdot b_g \supset \mathrm{Fil}^2 = (0)$,

(ii) the Gauß-Manin connection is given by $\nabla(a_i) = 0$, $\nabla(b_j) = \sum_i a_i \otimes \mathrm{dlog}(q_{ij})$,

(iii) the Frobenius Φ_H is the φ_A-linear map determined by $\Phi_H(a_i) = a_i$, $\Phi_H(b_j) = p \cdot b_j$.

We will need to work with $\mathfrak{A}_x$ is a slightly more general setting. For this, consider a number field F, a finite prime v of F above p, and write $\mathcal{A}_g = A_{g,1,n} \otimes \mathcal{O}_{(v)}$. Let $x \in \big(\mathcal{A}_g \otimes \kappa(v)\big)^{\mathrm{ord}}$ be a closed ordinary moduli point with residue field k. Set $\mathcal{O}_v = \mathcal{O}_{(v)}^\wedge$, $\Lambda := W(k) \otimes_{W(\kappa(v))} \mathcal{O}_v$, $\overline{\Lambda} := \overline{W} \otimes_{W(\kappa(v))} \mathcal{O}_v$. The formal completion $\mathfrak{A}_x := (\mathcal{A}_g)_{/\{x\}}$ now is a formal torus over Λ. It is simply the pull-back via $\mathrm{Spf}(\Lambda) \to \mathrm{Spf}(W)$ of the formal torus considered above.

6.12 The lifting of X_0 corresponding to the identity element $1 \in \mathfrak{A}_x\big(W(k)\big)$ is called the canonical lifting, and will be denoted X_0^{can}. The liftings over

$W(k)[\zeta_{p^n}]$ corresponding to the torsion points of $\mathfrak{A}_x$ are called the quasi-canonical liftings.

Suppose that k is a finite field, so that X_0 is an abelian variety of CM-type. The canonical lifting X_0^{can} is the unique lifting of X_0 such that all endomorphisms of X_0 lift to X_0^{can}. The quasi-canonical liftings of X_0 are precisely the liftings of X_0 which are of CM-type; they are mutually all isogenous. For proofs see [dJN], section 3, [Me], Appendix, [Mo1], §III.1.

6.13 Definition. Suppose, with the above notations, that $Z \hookrightarrow \mathsf{A}_{g,1,n} \otimes F$ is an algebraic subvariety. Let $\mathcal{Z} \hookrightarrow \mathcal{A}_g$ denote its Zariski closure inside $\mathcal{A}_g$. Suppose that the closed ordinary moduli point x is a point of $\left(\mathcal{Z} \otimes \kappa(v)\right)^{\text{ord}} \hookrightarrow \left(\mathcal{A}_g \otimes \kappa(v)\right)^{\text{ord}}$. Then we say that $\mathcal{Z}$ is formally linear (resp. formally quasi-linear) at x if its formal completion $\mathfrak{Z}_x := \mathcal{Z}_{/\{x\}} \hookrightarrow \mathfrak{A}_x$ is a formal subtorus (resp. if all its (formal) irreducible components are the translate of a formal subtorus over a torsion point).

6.14 Example. Suppose Z is a component of a subvariety of PEL type, parametrizing p.p.a.v. with an action of a given order R in a semi-simple $\mathbb{Q}$-algebra. In particular, we have $\iota_0 \colon R \hookrightarrow \text{End}(X_0)$. Consider the formal subscheme of $\mathfrak{A}_x$ parametrizing liftings X of X_0 such that ι_0 lifts to $\iota \colon R \hookrightarrow \text{End}(X)$. It follows from the facts in 6.10 that this is a union of translates of formal subtori of $\mathfrak{A}_x$ over torsion points. (The reader is encouraged to verify this.) It follows that $\mathcal{Z}$ is formally quasi-linear at x. Moreover, if Z is absolutely irreducible and the order R is maximal at p then $\mathcal{Z}$ is formally linear at x.

The relation between formal linearity and subvarieties of Hodge type is expressed by the following two results, which were obtained by Noot in [No1] (see also [No2]) and the author in [Mo1] (see also [Mo3]), respectively.

6.15 Theorem. (Noot) *Let F be a number field, and let $S \hookrightarrow \mathsf{A}_{g,1,n} \otimes F$ be a subvariety of Hodge type. Let v be a prime of F above p, and write $\mathcal{S}$ for the Zariski closure of S inside $\mathsf{A}_{g,1,n} \otimes \mathcal{O}_{(v)}$. Let x be a closed point in the ordinary locus $\left(\mathcal{S} \otimes \kappa(v)\right)^{\text{ord}}$. Then $\mathcal{S}$ is formally quasi-linear at x. For v outside a finite set of primes of $\mathcal{O}_F$, the formal completion $\mathfrak{S}_x$ of $\mathcal{S}$ at x is a union of formal subtori of $\mathfrak{A}_x$.*

6.16 Theorem. *Let $Z \hookrightarrow \mathsf{A}_{g,1,n} \otimes F$ be an irreducible algebraic subvariety over a number field F. Suppose there is a prime v of $\mathcal{O}_F$ such that the model $\mathcal{Z}$*

of Z *(as above) has formally quasi-linear components at some closed ordinary point* $x \in \left(\mathcal{Z} \otimes \kappa(v)\right)^{\mathrm{ord}}$. *Then* Z *is of Hodge type.*

We refer to [Mo1] and [Mo2] for some applications of 6.16 to Oort's conjecture. Given Z as in 6.4 (which then is defined over a number field), one tries to prove that Z is formally linear at some ordinary point in characteristic p. In general, we do not know how to do this; the main difficulty is that we have little control over the CM-points on Z. With certain additional assumptions, which we will not specify here, one can, however, prove such a statement. See in particular [Mo2], §5.

Notice that 6.16 is a "local" version of Oort's conjecture: an algebraizable irreducible formal subscheme of $\mathfrak{A}_x$ comes from a subvariety of Hodge type if and only if it contains a dense collection of CM-points (= torsion points). (The adjective "algebraizable" is essential.) We think of this local version and of Raynaud's "Manin-Mumford" theorem as "abelian" cases. Morally, the global case of Oort's conjecture is more difficult because it involves non-abelian group structures.

6.17 To finish, let us take one more look at Coleman's conjecture. A naive attempt to disprove it runs as follows: consider the ordinary locus of $\mathcal{M}_{g,\overline{\mathbb{F}}_p}$, and try to lift the corresponding curves to characteristic zero such that the Jacobian remains of CM-type. This does not work so easily, due to the well-known fact that the canonical lifting of a Jacobian in general no longer is a Jacobian. In [DO], Dwork and Ogus give an "abstract" proof of this. ("Abstract" as opposed to the explicit examples demonstrating this fact given by Oort and Sekiguchi in [OS].) They call an ordinary (smooth projective) curve C over a perfect field k of char. p a pre-W_n-canonical curve if, setting $X_0 = \mathrm{Jac}(C)$, the canonical lifting $X_0^{\mathrm{can}} \bmod p^{n+1}$ over $W_n(k)$ is a Jacobian. They then show that the locus Σ_{W_1} of pre-W_1-canonical curves (pre-W_2-canonical in their notations) forms a constructible part of $\mathcal{M}_{g,\overline{\mathbb{F}}_p}^{\mathrm{ord}}$ which is nowhere dense if $g \geq 4$.

For our "naive attempt" this still leaves hope, though. As Dwork and Ogus write, "It would be interesting to study the "deeper" subschemes Σ_{W_n} for higher $n \ldots$". Coleman's conjecture suggests that Σ_{W_∞} should be a finite set of points. Oort's conjecture together with 6.16 lead to another suggestion. Namely, if we write $\tau \colon \mathcal{M}_g \to \mathsf{A}_{g,1}$ for the Torelli morphism, and if $x \in \Sigma_{W_\infty}$ then "locally around $\tau(x)$", the locus $\tau(\Sigma_{W_\infty})$ should be the largest subvariety which is contained in the Torelli locus $\tau(\mathcal{M}_{g,\overline{\mathbb{F}}_p})$ and which is formally linear (purely in characteristic p). It seems that one can prove this by "iterating" the method of [DO]. Unfortunately, our control of the higher-order deformation theory is as yet insufficient to use this to show that Σ_{W_∞} is 0-dimensional.

References

[An1] Y. ANDRÉ, *Distribution des points CM sur les sous-variétés des variétés de modules de variétés abéliennes*, manuscript, April 1997.

[An2] ——, *Finitude des couples d'invariants modulaires singuliers sur une courbe algébrique plane non modulaire*, manuscript, April 1997.

[Aea] A. ASH et al., *Smooth compactification of locally symmetric varieties*, Lie groups: history, frontiers and applications, Vol. IV, Math. Sci Press, Brookline, 1975.

[BB] W.L. BAILY, JR. and A. BOREL, *Compactification of arithmetic quotients of bounded symmetric domains*, Ann. of Math., 84 (1966), pp. 442–528.

[BBM] P. BERTHELOT, L. BREEN and W. MESSING, *Théorie de Dieudonné cristalline II*, Lecture Notes in Mathematics 930, Springer-Verlag, Berlin, 1982.

[BO] P. BERTHELOT and A. OGUS, *F-isocrystals and de Rham cohomology, I*, Inventiones Math., 72 (1983), pp. 159–199.

[BM1] P. BERTHELOT and W. MESSING, *Théorie de Dieudonné cristalline I*, in: Journées de géométrie algébrique de Rennes, Part I, Astérisque 63 (1979), pp. 17-37.

[BM3] ——, *Théorie de Dieudonné cristalline III*, in: The Grothendieck Festschrift, Vol. I, P. Cartier et al., eds., Progress in Math., Vol. 86, Birkhäuser, Boston, 1990, pp. 173–247.

[Bl] D. BLASIUS, *A p-adic property of Hodge classes on abelian varieties*, in: Motives, Part 2, U. Jannsen, S. Kleiman, and J-P. Serre, eds., Proc. of Symp. in Pure Math., Vol. 55, American Mathematical Society, 1994, pp. 293–308.

[Bo1] M.V. BOROVOI, *The Langlands conjecture on the conjugation of Shimura varieties*, Functional Anal. Appl., 16 (1982), pp. 292–294.

[Bo2] ——, *Conjugation of Shimura varieties*, in: Proc. I.C.M. Berkeley, 1986, Part I, pp. 783–790.

[BLR] S. BOSCH, W. LÜTKEBOHMERT, and M. RAYNAUD, *Néron models*, Ergebnisse der Mathematik und ihrer Grenzgebiete, 3. Folge, Band 21, Springer-Verlag, Berlin, 1990.

[Bou]   N. BOURBAKI, *Groupes et algèbres de Lie, Chap. 7 et 8 (Nouveau tirage)*, Éléments de mathématique, Masson, Paris, 1990.

[BT]   F. BRUHAT and J. TITS, *Groupes réductifs sur un corps local, II: Schémas en groupes; Existence d'une donnée radicielle valuée*, Publ. Math. de l'I.H.E.S., N° 60 (1984), pp. 5–84.

[Br]   J.-L. BRYLINSKI, *"1-motifs" et formes automorphes (Théorie arithmétique des domaines de Siegel)*, in: Journées automorphes, Publ. Math. de l'université Paris VII, Vol. 15, Paris, 1983, pp. 43–106.

[Ca]   H. CARAYOL, *Sur la mauvaise réduction des courbes de Shimura*, Compositio Math., 59 (1986), pp. 151–230.

[Ch]   W. CHI, *ℓ-adic and λ-adic representations associated to abelian varieties defined over number fields*, American J. of Math., 114 (1992), pp. 315–353.

[Co]   R. COLEMAN, *Torsion points on curves*, in: Galois representations and arithmetic algebraic geometry, Y. Ihara, ed., Adv. Studies in Pure Math., Vol. 12, North-Holland, Amsterdam, 1987, pp. 235–247.

[dJ]   A.J. DE JONG, *Crystalline Dieudonné module theory via formal and rigid geometry*, Publ. Math. de l'I.H.E.S., N° 82 (1996), pp. 5–96.

[dJN]   A.J. DE JONG and R. NOOT, *Jacobians with complex multiplication*, in: Arithmetic Algebraic Geometry, G. van der Geer, F. Oort and J. Steenbrink, eds., Progress in Math., Vol. 89, Birkhäuser, Boston, 1991, pp. 177–192.

[dJO]   A.J. DE JONG and F. OORT, *On extending families of curves*, J. Alg. Geom., 6 (1997), pp. 545–562.

[De1]   P. DELIGNE, *Travaux de Shimura*, in: Séminaire Bourbaki, Exposé 389, Février 1971, Lecture Notes in Mathematics 244, Springer-Verlag, Berlin, 1971, pp. 123–165.

[De2]   ——, *Théorie de Hodge II*, Publ. Math. de l'I.H.E.S., N° 40 (1972), pp. 5–57.

[De3]   ——, *Variétés de Shimura: interprétation modulaire, et techniques de construction de modèles canoniques*, in: Automorphic Forms, Representations, and *L*-functions, Part 2, A. Borel and W. Casselman, eds., Proc. of Symp. in Pure Math., Vol. XXXIII, American Mathematical Society, 1979, pp. 247–290.

[De4]   ——, *Hodge cycles on abelian varieties (Notes by J.S. Milne)*, in: Hodge cycles, motives, and Shimura varieties, Lecture Notes in Mathematics 900, Springer-Verlag, Berlin, 1982, pp. 9–100.

[De5]   ——, *A quoi servent les motifs?*, in: Motives, Part 1, U. Jannsen, S. Kleiman, and J-P. Serre, eds., Proc. of Symp. in Pure Math., Vol. 55, American Mathematical Society, 1994, pp. 143–161.

[DI]   P. DELIGNE and L. ILLUSIE, *Cristaux ordinaires et coordonnées canoniques (with an appendix by N. Katz)*, Exposé V, in: Surfaces algébriques, J. Giraud, L. Illusie, and M. Raynaud, eds., Lecture Notes in Mathematics 868, Springer-Verlag, Berlin, 1981, pp. 80–137.

[DR]   P. DELIGNE and M. RAPOPORT, *Les schémas de modules de courbes elliptiques*, in: Modular functions of one variable II, Proc. Intern. summer school, Univ. of Antwerp, RUCA, P. Deligne and W. Kuyk, eds., Lecture Notes in Mathematics 349, Springer-Verlag, Berlin, 1973, pp. 143–316.

[DO]   B. DWORK and A. OGUS, *Canonical liftings of Jacobians*, Compositio Math., 58 (1986), pp. 111-131.

[Ed1]   B. EDIXHOVEN, *Néron models and tame ramification*, Compositio Math., 81 (1992), pp. 291–306.

[Ed2]   ——, *Special points on the product of two modular curves*, Univ. Rennes 1, Prépublication 96-26, Octobre 1996.

[Fa1]   G. FALTINGS, *Arithmetic varieties and rigidity*, in: Séminaire de théorie des nombres de Paris, 1982-83, Progress in Math., Vol. 51, Birkhäuser, Boston, 1984, pp. 63–77.

[Fa2]   ——, *Crystalline cohomology and p-adic Galois representations*, in: Algebraic analysis, geometry and number theory, Proc. JAMI inaugural conference, J.-I. Igusa, ed., Johns-Hopkins Univ. Press, Baltimore, 1989, pp. 25–80.

[Fa3]   ——, *Integral crystalline cohomology over very ramified base rings*, preprint, Princeton University (1993?).

[FC]   G. FALTINGS and C-L. CHAI, *Degeneration of abelian varieties*, Ergebnisse der Mathematik und ihrer Grenzgebiete, 3. Folge, Band 22, Springer-Verlag, Berlin, 1990.

[Fo]   J.-M. FONTAINE, *Le corps des périodes p-adiques*, Exposé II, in: Périodes p-adiques, Séminaire de Bures, 1988, Astérisque 223 (1994), pp. 59–101.

[FGA]   A. GROTHENDIECK, *Fondements de la géométrie algébrique*, Extraits du séminaire Bourbaki 1957–1962. Paris: Secrétariat math., 11, R. Pierre Curie, 5.

[EGA]   A. GROTHENDIECK and J. DIEUDONNÉ, *Éléments de géométrie algébrique*, Publ. Math. de l'I.H.E.S., Nᵒˢ 4, 8, 11, 17, 20, 24 and 32 (1960–67).

[SGA1] A. GROTHENDIECK, *SGA 1; Revêtements étales et groupe fondamental*, Lecture Notes in Mathematics 224, Springer-Verlag, Berlin, 1971.

[H] R. HAIN, *Locally symmetric families of curves and Jacobians*, manuscript, June 1997.

[Ha] M. HARRIS, *Arithmetic vector bundles and automorphic forms on Shimura varieties, I*, Inventiones Math., 82 (1985), pp. 151–189.

[Haz] F. HAZAMA, *Algebraic cycles on certain abelian varieties and powers of special surfaces*, J. Fac. Sci. Univ. Tokyo, Sect. IA, Math., 31 (1984), pp. 487–520.

[Il] L. ILLUSIE, *Déformations de groupes de Barsotti-Tate (d'après A. Grothendieck)*, in: Séminaire sur les pinceaux arithmétiques: la conjecture de Mordell, L. Szpiro, ed., Astérisque 127 (1985), pp. 151–198.

[Ka] N.M. KATZ, *Serre-Tate local moduli*, Exposé V^{bis}, in: Surfaces algébriques, J. Giraud, L. Illusie, and M. Raynaud, eds., Lecture Notes in Mathematics 868, Springer-Verlag, Berlin, 1981, pp. 138–202.

[Ko1] R. KOTTWITZ, *Isocrystals with additional structure*, Compositio Math., 56 (1985), pp. 201–220.

[Ko2] ——, *Points on some Shimura varieties over finite fields*, J. of the A.M.S., 5 (1992), pp. 373–444.

[Ko3] ——, *Isocrystals with additional structure. II*, preprint.

[Ku] V. KUMAR MURTY, *Exceptional Hodge classes on certain abelian varieties*, Math. Ann., 268 (1984), pp. 197–206.

[La] R.P. LANGLANDS, *Automorphic representations, Shimura varieties, and motives. Ein Märchen.*, in: Automorphic Forms, Representations, and L-functions, Part 2, A. Borel and W. Casselman, eds., Proc. of Symp. in Pure Math., Vol. XXXIII, American Mathematical Society, 1979, pp. 205–246.

[LR] R.P. LANGLANDS and M. RAPOPORT, *Shimuravarietäten und Gerben*, J. reine angew. Math., 378 (1987), pp. 113–220.

[Me] W. MESSING, *The crystals associated to Barsotti-Tate groups: with applications to abelian schemes*, Lecture Notes in Mathematics 264, Springer-Verlag, Berlin, 1972.

[Mi1] J.S. MILNE, *The action of an automorphism of $\mathbb{C}$ on a Shimura variety and its special points*, in: Arithmetic and geometry, Vol. 1, M. Artin and J. Tate, eds., Progress in Math., Vol. 35, Birkhäuser, Boston, 1983, pp. 239–265.

348 *Models of Shimura varieties*

[Mi2] ——, *Canonical models of (mixed) Shimura varieties and automorphic vector bundles*, in: Automorphic forms, Shimura varieties, and L-functions, L. Clozel and J. S. Milne, eds., Persp. in Math., Vol. 10(I), Academic Press, Inc., 1990, pp. 283–414.

[Mi3] ——, *The points on a Shimura variety modulo a prime of good reduction*, in: The zeta functions of Picard modular surfaces, R. P. Langlands and D. Ramakrishnan, eds., Les Publications CRM, Montréal, 1992, pp. 151–253.

[Mi4] ——, *Shimura varieties and motives*, in: Motives, Part 2, U. Jannsen, S. Kleiman, and J-P. Serre, eds., Proc. of Symp. in Pure Math., Vol. 55, American Mathematical Society, 1994, pp. 447–523.

[MS] J.S. MILNE and K.-Y. SHIH, *Conjugates of Shimura varieties*, in: Hodge cycles, motives, and Shimura varieties, Lecture Notes in Mathematics 900, Springer-Verlag, Berlin, 1982, pp.280–356.

[Mo1] B.J.J. MOONEN, *Special points and linearity properties of Shimura varieties*, Ph.D. thesis, University of Utrecht, 1995.

[Mo2] ——, *Linearity properties of Shimura varieties, I*, to appear in J. Alg. Geom.

[Mo3] ——, *Linearity properties of Shimura varieties, II*, to appear in Compositio Math.

[MZ1] B.J.J. MOONEN and YU.G. ZARHIN, *Hodge classes and Tate classes on simple abelian fourfolds*, Duke Math. Journal, 77 (1995), pp. 553–581.

[MZ2] ——, *Weil classes on abelian varieties*, preprint alg-geom/9612017, to appear in J. reine angew. Math.

[Mor] Y. MORITA, *Reduction mod $\mathfrak{P}$ of Shimura curves*, Hokkaido Math. J., 10 (1981), pp. 209–238.

[Mu1] D. MUMFORD, *A note of Shimura's paper "Discontinuous groups and abelian varieties"*, Math. Ann., 181 (1969), pp. 345–351.

[Mu2] ——, *Abelian varieties*, Tata inst. of fund. res. studies in math., Vol. 5, Oxford University Press, Oxford, 1970.

[Mu3] ——, *Hirzebruch's proportionality in the non-compact case*, Inventiones Math., 42 (1977), pp. 239–272.

[No1] R. NOOT, *Hodge classes, Tate classes, and local moduli of abelian varieties*, Ph.D. thesis, University of Utrecht, 1992.

[No2] ——, *Models of Shimura varieties in mixed characteristic*, J. Alg. Geom., 5 (1996), pp. 187–207.

[Og] A. OGUS, *A p-adic analogue of the Chowla-Selberg formula*, in: *p*-adic analysis, Proceedings, Trento, 1989, F. Baldassari, S. Bosch and B. Dwork, eds., Lecture Notes in Mathematics 1454, Springer-Verlag, Berlin, 1990, pp. 319–341.

[Oo1] F. OORT, *Moduli of abelian varieties and Newton polygons*, C. R. Acad. Sci. Paris, Ser. 1, Math., 312 (1991), pp. 385–389.

[Oo2] ——, *Moduli of abelian varieties in positive characteristic*, in: Barsotti symposium in algebraic geometry, V. Christante and W. Messing, eds., Persp. in Math., Vol. 15, Academic Press, Inc., 1994, pp. 253–276.

[Oo3] ——, *Canonical liftings and dense sets of CM-points*, in: Arithmetic geometry, Proc. Cortona symposium 1994, F. Catanese, ed., Symposia Math., Vol. XXXVII, Cambridge University Press, 1997, pp. 228–234.

[OS] F. OORT and T. SEKIGUCHI, *The canonical lifting of an ordinary Jacobian variety need not be a Jacobian variety*, J. Math. Soc. Japan, 38 (1986), pp. 427–437.

[Pi] R. PINK, *Arithmetical compactification of mixed Shimura varieties*, Ph.D. thesis, Rheinische Friedrich-Wilhelms-Universität Bonn, 1989.

[R1] M. RAPOPORT, *Compactifications de l'espace de modules de Hilbert-Blumenthal*, Compositio Math., 36 (1978), pp. 255–335.

[R2] M. RAPOPORT, *On the bad reduction of Shimura varieties*, in: Automorphic forms, Shimura varieties, and *L*-functions, L. Clozel and J. S. Milne, eds., Persp. in Math., Vol. 10(II), Academic Press, Inc., 1990, pp. 253–321.

[RR] M. RAPOPORT and M. RICHARTZ, *On the classification and specialization of F-crystals with additional structure*, Compositio Math., 103 (1996), pp. 153–181.

[RZ] M. RAPOPORT and TH. ZINK, *Period spaces for p-divisible groups*, Annals of mathematical studies 141, Princeton University Press, Princeton, 1996.

[Ra1] M. RAYNAUD, *Schémas en groupes de type $(p,\ldots,p)$*, Bull. Soc. math. France, 102 (1974), pp. 241–280.

[Ra2] ——, *Sous-variétés d'une variété abélienne et points de torsion*, in: Arithmetic and geometry, Vol. 1, M. Artin and J. Tate, eds., Progress in Math., Vol. 35, Birkhäuser, Boston, 1983, pp. 327–352.

[Ri] K. RIBET, *Hodge classes on certain types of abelian varieties*, Am. J. of Math., 105 (1983), pp. 523–538.

[Se2] J-P. SERRE, *Cohomologie Galoisienne*, Cinquième édition, révisée et complétée, Lecture Notes in Mathematics 5, Springer-Verlag, Berlin, 1973 & 1994.

[Se2]   ——, Letter to John Tate, 2 January, 1985.

[Sh]   G. SHIMURA, *On analytic families of polarized abelian varieties and automorphic functions*, Annals of Math., 78 (1963), pp. 149–192.

[Ta]   S. TANKEEV, *On algebraic cycles on surfaces and abelian varieties*, Math. USSR Izvestia, 18 (1982), pp. 349–380.

[Ti]   J. TITS *Reductive groups over local fields*, in: Automorphic Forms, Representations, and *L*-functions, Part 1, A. Borel and W. Casselman, eds., Proc. of Symp. in Pure Math., Vol. XXXIII, American Mathematical Society, 1979, pp. 29–69.

[Va1]   A. VASIU, *Integral canonical models for Shimura varieties of Hodge type*, Ph.D. thesis, Princeton University, November 1994.

[Va2]   ——, *Integral canonical models for Shimura varieties of preabelian type*, manuscript, E.T.H. Zürich, 31st of October, 1995.

[We]   A. WEIL, *Abelian varieties and the Hodge ring*, Collected papers, Vol. III, [1977c], pp. 421–429.

[Wi]   J-P. WINTENBERGER, *Un scindage de la filtration de Hodge pour certains variétés algébriques sur les corps locaux*, Annals of Math., 119 (1984), pp. 511–548.

WESTFÄLISCHE WILHELMS-UNIVERSITÄT MÜNSTER, MATHEMATISCHES INSTITUT, EINSTEINSTRASSE 62, 48149 MÜNSTER, GERMANY
moonen@math.uni-muenster.de

Euler systems and modular elliptic curves

KARL RUBIN

INTRODUCTION

This paper consists of two parts. In the first we present a general theory of Euler systems. The main results (see §§3 and 4) show that an Euler system for a p-adic representation T gives a bound on the Selmer group associated to the dual module $\mathrm{Hom}(T, \mu_{p^\infty})$. These theorems, which generalize work of Kolyvagin [Ko], have been obtained independently by Kato [Ka1], Perrin-Riou [PR2], and the author [Ru3]. We will not prove these theorems here, or even attempt to state them in the greatest possible generality.

In the second part of the paper we show how to apply the results of Part I and an Euler system recently constructed by Kato [Ka2] (see the article of Scholl [Scho] in this volume) to obtain Kato's theorem in the direction of the Birch and Swinnerton-Dyer conjecture for modular elliptic curves (Theorem 8.1).

Part 1. Generalities

1. SELMER GROUPS ATTACHED TO p-ADIC REPRESENTATIONS

A *p-adic representation* of $G_{\mathbf{Q}} = \mathrm{Gal}(\bar{\mathbf{Q}}/\mathbf{Q})$ is a free $\mathbf{Z}_p$-module T of finite rank with a continuous, $\mathbf{Z}_p$-linear action of $G_{\mathbf{Q}}$. We will assume in addition throughout this paper that T is unramified outside of a finite set of primes. Given a p-adic representation T, we also define

$$V = T \otimes_{\mathbf{Z}_p} \mathbf{Q}_p,$$
$$W = V/T = T \otimes \mathbf{Q}_p/\mathbf{Z}_p.$$

(Note that T determines V and W, and W determines T and V, but in general there may be non-isomorphic $\mathbf{Z}_p$-lattices T giving rise to the same vector space V.)

The following are basic examples of p-adic representations to keep in mind.

Example. If $\rho : G_{\mathbf{Q}} \to \mathbf{Z}_p^\times$ is a continuous character we can take T to be a free, rank-one $\mathbf{Z}_p$-module with $G_{\mathbf{Q}}$ acting via ρ (clearly every one-dimensional

Reprinted from 'Galois Representations in Algebraic Arithmetic Geometry', edited by A. J. Scholl & R. L. Taylor. ©Cambridge University Press 1998

representation arises in this way). For example, when ρ is the cyclotomic character we get
$$T = \mathbf{Z}_p(1) = \varprojlim_n \mu_{p^n}.$$

Example. If A is an abelian variety we can take the p-adic Tate module of A
$$T = T_p(A) = \varprojlim_n A_{p^n}.$$

This is the situation we will concentrate on in this paper, when A is an elliptic curve.

Example. If T is a p-adic representation, then we define the dual representation
$$T^* = \mathrm{Hom}(T, \mathbf{Z}_p(1))$$
and we denote the corresponding vector space and divisible group by
$$V^* = T^* \otimes \mathbf{Q}_p = \mathrm{Hom}(V, \mathbf{Q}_p(1)), \quad W^* = V^*/T^* = \mathrm{Hom}(T, \mu_{p^\infty}).$$

If E is an elliptic curve then the Weil pairing gives an isomorphism $T_p(E)^* \cong T_p(E)$.

Let $\mathbf{Q}_\infty \subset \mathbf{Q}(\mu_{p^\infty})$ denote the cyclotomic $\mathbf{Z}_p$-extension of $\mathbf{Q}$. For every n let $\mathbf{Q}_n \subset \mathbf{Q}(\mu_{p^{n+1}})$ be the extension of $\mathbf{Q}$ of degree p^n in $\mathbf{Q}_\infty$, and let $\mathbf{Q}_{n,p}$ denote the completion of $\mathbf{Q}_n$ at the unique prime above p.

Fix a p-adic representation T as above. We wish to define a Selmer group $\mathcal{S}(\mathbf{Q}_n, W) \subset H^1(\mathbf{Q}_n, W)$ for every n. If v is a place of $\mathbf{Q}_n$ not dividing p, let I_v denote an inertia group of v in $G_{\mathbf{Q}_n}$, let $\mathbf{Q}_{n,v}^{\mathrm{ur}}$ denote the maximal unramified extension of $\mathbf{Q}_{n,v}$, and define
$$H^1_{\mathcal{S}}(\mathbf{Q}_{n,v}, V) = H^1_{\mathrm{ur}}(\mathbf{Q}_{n,v}, V) = \ker(H^1(\mathbf{Q}_{n,v}, V) \to H^1(\mathbf{Q}_{n,v}^{\mathrm{ur}}, V))$$
$$= H^1(\mathbf{Q}_{n,v}^{\mathrm{ur}}/\mathbf{Q}_{n,v}, V^{I_v}).$$

For the unique prime of $\mathbf{Q}_n$ above p, we will ignore all questions about what is the "correct" definition, and we just fix some choice of subspace $H^1_{\mathcal{S}}(\mathbf{Q}_{n,p}, V) \subset H^1(\mathbf{Q}_{n,p}, V)$. (For example, one could choose $H^1_{\mathcal{S}}(\mathbf{Q}_{n,p}, V) = H^1(\mathbf{Q}_{n,p}, V)$ or $H^1_{\mathcal{S}}(\mathbf{Q}_{n,p}, V) = 0$.)

For every place v of $\mathbf{Q}_n$ we now define
$$H^1_{\mathcal{S}}(\mathbf{Q}_{n,v}, W) \subset H^1(\mathbf{Q}_{n,v}, W) \quad \text{and} \quad H^1_{\mathcal{S}}(\mathbf{Q}_{n,v}, T) \subset H^1(\mathbf{Q}_{n,v}, T)$$
to be the image and inverse image, respectively, of $H^1_{\mathcal{S}}(\mathbf{Q}_{n,v}, V)$ under the maps on cohomology induced by the exact sequence
$$0 \to T \to V \to W \to 0.$$

Finally, we define
$$\mathcal{S}(\mathbf{Q}_n, W) = \ker\left(H^1(\mathbf{Q}_n, W) \to \bigoplus_{v \text{ of } \mathbf{Q}_n} H^1(\mathbf{Q}_{n,v}, W)/H^1_{\mathcal{S}}(\mathbf{Q}_{n,v}, W)\right).$$

Of course, this definition depends on the choice we made for $H^1_{\mathcal{S}}(\mathbf{Q}_{n,p}, V)$.

2. EULER SYSTEMS

Fix a p-adic representation T of $G_\mathbf{Q}$ as in §1, and fix a positive integer N divisible by p and by all primes where T is ramified. Define

$$\mathcal{R} = \mathcal{R}(N) = \{\text{squarefree integers } r : (r, N) = 1\}$$

For every prime q which is unramified in T, let Fr_q denote a Frobenius of q in $G_\mathbf{Q}$ and define the characteristic polynomial

$$P_q(x) = \det(1 - \mathrm{Fr}_q x | T) \in \mathbf{Z}_p[x].$$

Since q is unramified in T, P_q is independent of the choice of Frobenius element Fr_q.

Definition. An *Euler system* **c** for T is a collection of cohomology classes

$$c_{\mathbf{Q}_n(\mu_r)} \in H^1(\mathbf{Q}_n(\mu_r), T)$$

for every $r \in \mathcal{R}$ and every $n \geq 0$, such that if $m \geq n$, q is prime, and $rq \in \mathcal{R}$, then

$$\mathrm{Cor}_{\mathbf{Q}_n(\mu_{rq})/\mathbf{Q}_n(\mu_r)} c_{\mathbf{Q}_n(\mu_{rq})} = P_q(q^{-1}\mathrm{Fr}_q^{-1}) c_{\mathbf{Q}_n(\mu_r)},$$

$$\mathrm{Cor}_{\mathbf{Q}_m(\mu_r)/\mathbf{Q}_n(\mu_r)} c_{\mathbf{Q}_m(\mu_r)} = c_{\mathbf{Q}_n(\mu_r)}.$$

Note that this definition depends on N (since $\mathcal{R}$ does), but not in an important way so we will suppress it from the notation.

Remarks. Kolyvagin's original method (see [Ko] or [Ru1]) required the Euler system to satisfy an additional "congruence" condition. The fact that our Euler system "extends in the $\mathbf{Q}_\infty$ direction" (i.e., consists of classes defined over the fields $\mathbf{Q}_n(\mu_r)$ for every n, and not just over $\mathbf{Q}(\mu_r)$) eliminates the need for the congruence condition.

There is some freedom in the exact form of the distribution relation in the definition of an Euler system. It is easy to modify an Euler system satisfying one distribution relation to obtain a new Euler system satisfying a slightly different one.

3. RESULTS OVER Q

We now come to the fundamental applications of an Euler system. For the proofs of Theorems 3.1, 3.2, and 4.1 see [Ka1], [PR2], or [Ru3]. In fact, once the setting is properly generalized the proofs are similar to the original method of Kolyvagin [Ko]; see also [Ru1] and [Ru2].

For this section and the next fix a p-adic representation T as in §1. Fix also a choice of subspaces $H^1_{\mathcal{S}}(\mathbf{Q}_{n,p}, V)$ and $H^1_{\mathcal{S}}(\mathbf{Q}_{n,p}, V^*)$ for every n, so that we have Selmer groups as defined in §1. We assume only that these choices satisfy the following conditions:

- $H^1_{\mathcal{S}}(\mathbf{Q}_{n,p}, V)$ and $H^1_{\mathcal{S}}(\mathbf{Q}_{n,p}, V^*)$ are orthogonal complements under the cup product pairing

$$H^1(\mathbf{Q}_{n,p}, V) \times H^1(\mathbf{Q}_{n,p}, V^*) \to H^2(\mathbf{Q}_{n,p}, \mathbf{Q}_p(1)) = \mathbf{Q}_p,$$

- if $m \geq n$ then

$$\mathrm{Cor}_{\mathbf{Q}_{m,p}/\mathbf{Q}_{n,p}} H^1_{\mathcal{S}}(\mathbf{Q}_{m,p}, V) \subset H^1_{\mathcal{S}}(\mathbf{Q}_{n,p}, V),$$
$$\mathrm{Res}_{\mathbf{Q}_{m,p}} H^1_{\mathcal{S}}(\mathbf{Q}_{n,p}, V) \subset H^1_{\mathcal{S}}(\mathbf{Q}_{m,p}, V)$$

and similarly for V^*.

We will write $H^1_{/\mathcal{S}}(\mathbf{Q}_{n,p}, T) = H^1(\mathbf{Q}_{n,p}, T)/H^1_{\mathcal{S}}(\mathbf{Q}_{n,p}, T)$ (and similarly with T replaced by V or W), and we write $\mathrm{loc}_p^{\mathrm{ram}}$ for the localization map

$$\mathrm{loc}_p^{\mathrm{ram}} : H^1(\mathbf{Q}_n, T) \to H^1_{/\mathcal{S}}(\mathbf{Q}_{n,p}, T).$$

We will make use of two different sets of hypotheses on the Galois representation T. Hypotheses $\mathrm{Hyp}(\mathbf{Q}_\infty, T)$ are stronger than $\mathrm{Hyp}(\mathbf{Q}_\infty, V)$, and will allow us to prove a stronger conclusion.

Hypotheses $\mathrm{Hyp}(\mathbf{Q}_\infty, T)$. (i) There is a $\tau \in G_{\mathbf{Q}_\infty}$ such that
- τ acts trivially on μ_{p^∞},
- $T/(\tau - 1)T$ is free of rank one over $\mathbf{Z}_p$.

(ii) T/pT is an irreducible $\mathbf{F}_p[G_{\mathbf{Q}_\infty}]$-module.

Hypotheses $\mathrm{Hyp}(\mathbf{Q}_\infty, V)$. (i) There is a $\tau \in G_{\mathbf{Q}_\infty}$ such that
- τ acts trivially on μ_{p^∞},
- $\dim_{\mathbf{Q}_p}(V/(\tau - 1)V) = 1$.

(ii) V is an irreducible $\mathbf{Q}_p[G_{\mathbf{Q}_\infty}]$-module.

Theorem 3.1. *Suppose* **c** *is an Euler system for T, and V satisfies* $\mathrm{Hyp}(\mathbf{Q}_\infty, V)$. *If* $\mathbf{c}_{\mathbf{Q}} \notin H^1(\mathbf{Q}, T)_{\mathrm{tors}}$ *and* $[H^1_{/\mathcal{S}}(\mathbf{Q}_p, T) : \mathrm{loc}_p^{\mathrm{ram}}(H^1(\mathbf{Q}, T))]$ *is finite, then $\mathcal{S}(\mathbf{Q}, W^*)$ is finite. In particular if* $\mathrm{loc}_p^{\mathrm{ram}}(\mathbf{c}_{\mathbf{Q}}) \neq 0$ *and* $\mathrm{rank}_{\mathbf{Z}_p} H^1_{/\mathcal{S}}(\mathbf{Q}_p, T) = 1$, *then $\mathcal{S}(\mathbf{Q}, W^*)$ is finite.*

Define $\Omega = \mathbf{Q}(W)\mathbf{Q}(\mu_{p^\infty})$, where $\mathbf{Q}(W)$ denotes the minimal extension of $\mathbf{Q}$ such that $G_{\mathbf{Q}(W)}$ acts trivially on W.

Theorem 3.2. *Suppose* **c** *is an Euler system for T, and T satisfies* $\mathrm{Hyp}(\mathbf{Q}_\infty, T)$. *If $p > 2$ and* $\mathrm{loc}_p^{\mathrm{ram}}(\mathbf{c}_{\mathbf{Q}}) \neq 0$ *then*

$$|\mathcal{S}(\mathbf{Q}, W^*)| \leq |H^1(\Omega/\mathbf{Q}, W)||H^1(\Omega/\mathbf{Q}, W^*)|[H^1_{/\mathcal{S}}(\mathbf{Q}_p, T) : \mathbf{Z}_p \mathrm{loc}_p^{\mathrm{ram}}(\mathbf{c}_{\mathbf{Q}})].$$

Remark. Hypotheses $\mathrm{Hyp}(\mathbf{Q}_\infty, T)$ are satisfied if the image of the Galois representation on T is "sufficiently large." They often hold in practice; see for example the discussion in connection with elliptic curves below. If $\mathrm{rank}_{\mathbf{Z}_p} T = 1$ then $\mathrm{Hyp}(\mathbf{Q}_\infty, T)$ holds with $\tau = 1$.

4. RESULTS OVER $\mathbf{Q}_\infty$

Essentially by proving analogues of Theorem 3.2 for each field $\mathbf{Q}_n$, we can pass to the limit and prove an Iwasawa-theoretic version of Theorem 3.2.

Let Λ denote the Iwasawa algebra

$$\Lambda = \mathbf{Z}_p[[\text{Gal}(\mathbf{Q}_\infty/\mathbf{Q})]] = \varprojlim_n \mathbf{Z}_p[[\text{Gal}(\mathbf{Q}_n/\mathbf{Q})]],$$

so Λ is (noncanonically) isomorphic to a power series ring in one variable over $\mathbf{Z}_p$. If B is a finitely generated torsion Λ-module then there is a pseudo-isomorphism (a Λ-module homomorphism with finite kernel and cokernel)

$$B \to \bigoplus_i \Lambda/f_i\Lambda$$

with nonzero $f_i \in \Lambda$, and we define the characteristic ideal of B

$$\text{char}(B) = \prod_i f_i\Lambda.$$

The characteristic ideal is well-defined, although the individual f_i are not. If B is not a finitely-generated torsion Λ-module we define $\text{char}(B) = 0$.

Define Λ-modules

$$X_\infty = \text{Hom}(\varinjlim_n \mathcal{S}(\mathbf{Q}_n, W^*), \mathbf{Q}_p/\mathbf{Z}_p).$$

and

$$H^1_{\infty,/\mathcal{S}}(\mathbf{Q}_p, T) = \varprojlim_n H^1_{/\mathcal{S}}(\mathbf{Q}_{n,p}, T).$$

(Our requirements on the choices of $H^1_{\mathcal{S}}(\mathbf{Q}_{n,p}, V)$ and $H^1_{\mathcal{S}}(\mathbf{Q}_{n,p}, V^*)$ ensure that if $m \geq n$, the restriction and corestriction maps induce maps

$$\mathcal{S}(\mathbf{Q}_n, W^*) \to \mathcal{S}(\mathbf{Q}_m, W^*) \quad \text{and} \quad H^1_{/\mathcal{S}}(\mathbf{Q}_{m,p}, T) \to H^1_{/\mathcal{S}}(\mathbf{Q}_{n,p}, T)$$

so these limits are well-defined.) If $\mathbf{c}$ is an Euler system let $[\text{loc}_p^{\text{ram}}(\mathbf{c}_{\mathbf{Q}_n})]$ denote the corresponding element of $H^1_{\infty,/\mathcal{S}}(\mathbf{Q}_p, T)$.

Theorem 4.1. *Suppose that V satisfies* $\text{Hyp}(\mathbf{Q}_\infty, V)$, $\mathbf{c}$ *is an Euler system for T,* $[\text{loc}_p^{\text{ram}}(\mathbf{c}_{\mathbf{Q}_n})] \notin H^1_{\infty,/\mathcal{S}}(\mathbf{Q}_p, T)_{\text{tors}}$, *and* $H^1_{\infty,/\mathcal{S}}(\mathbf{Q}_p, T)/\Lambda[\text{loc}_p^{\text{ram}}(\mathbf{c}_{\mathbf{Q}_n})]$ *is a torsion Λ-module. Define*

$$\mathcal{L} = \text{char}(H^1_{\infty,/\mathcal{S}}(\mathbf{Q}_p, T)/\Lambda[\text{loc}_p^{\text{ram}}(\mathbf{c}_{\mathbf{Q}_n})]).$$

Then

(i) *X_∞ is a torsion Λ-module,*
(ii) *there is a nonnegative integer t such that $\text{char}(X_\infty)$ divides $p^t\mathcal{L}$,*
(iii) *if T satisfies $\text{Hyp}(\mathbf{Q}_\infty, T)$ then $\text{char}(X_\infty)$ divides $\mathcal{L}$.*

Part 2. Elliptic curves

The 'Heegner point Euler system' for modular elliptic curves used by Kolyvagin in [Ko] does not fit into the framework of §2, because the cohomology
classes are not defined over abelian extensions of $\mathbf{Q}$. However, Kato [Ka2]
has constructed an Euler system for the Tate module of a modular elliptic
curve, using Beilinson elements in the K-theory of modular curves. We now
describe how, given Kato's Euler system and its essential properties, one can
use the general results above to study the arithmetic of elliptic curves. The
main result is Theorem 8.1 below.

5. LOCAL COHOMOLOGY GROUPS

Suppose E is an elliptic curve defined over $\mathbf{Q}$, and take $T = T_p(E)$, the
p-adic Tate module of E. Then $V = V_p(E) = T_p(E) \otimes \mathbf{Q}_p$ and $W = E_{p^\infty}$.
The Weil pairing gives isomorphisms $V \cong V^*$, $T \cong T^*$, and $W \cong W^*$.

If B is an abelian group, we will abbreviate $B \otimes \mathbf{Z}_p = \varprojlim B/p^n B$, the
p-adic completion of B, and

$$B \otimes \mathbf{Q}_p = (\varprojlim_n B/p^n B) \otimes_{\mathbf{Z}_p} \mathbf{Q}_p, \quad B \otimes \mathbf{Q}_p/\mathbf{Z}_p = (\varprojlim_n B/p^n B) \otimes_{\mathbf{Z}_p} \mathbf{Q}_p/\mathbf{Z}_p.$$

For every n define

$$H^1_{\mathcal{S}}(\mathbf{Q}_{n,p}, V) = \mathrm{image}(E(\mathbf{Q}_{n,p}) \otimes \mathbf{Q}_p \hookrightarrow H^1(\mathbf{Q}_{n,p}, V)),$$

image under the natural Kummer map. Since $V = V^*$, this also fixes a
choice of $H^1_{\mathcal{S}}(\mathbf{Q}_p, V^*)$, and this subgroup is its own orthogonal complement
as required.

Let $\mathrm{III}(E_{/\mathbf{Q}_n})$ denote the Tate-Shafarevich group of E over $\mathbf{Q}_n$.

Proposition 5.1. *With $H^1_{\mathcal{S}}(\mathbf{Q}_{n,p}, V)$ as defined above, $\mathcal{S}(\mathbf{Q}_n, E_{p^\infty})$ is the
classical p-power Selmer group of E over $\mathbf{Q}_n$, so there is an exact sequence*

$$0 \to E(\mathbf{Q}_n) \otimes \mathbf{Q}_p/\mathbf{Z}_p \to \mathcal{S}(\mathbf{Q}_n, E_{p^\infty}) \to \mathrm{III}(E_{/\mathbf{Q}_n})_{p^\infty} \to 0$$

Proof. If $v \nmid p$ then the p-part of $E(\mathbf{Q}_{n,v})$ is finite, and one can check easily
that $H^1_{\mathcal{S}}(\mathbf{Q}_{n,v}, V_p(E)) = 0$. Therefore for every v, $H^1_{\mathcal{S}}(\mathbf{Q}_{n,v}, V_p(E))$ is the
image of $E(\mathbf{Q}_{n,v}) \otimes \mathbf{Q}_p$ under the Kummer map. It follows that for every
v, $H^1_{\mathcal{S}}(\mathbf{Q}_{n,v}, E_{p^\infty})$ is the image of $E(\mathbf{Q}_{n,v}) \otimes \mathbf{Q}_p/\mathbf{Z}_p$ under the corresponding
Kummer map, and so the definition of $\mathcal{S}(\mathbf{Q}_n, E_{p^\infty})$ coincides with the classical
definition of the Selmer group of E. $\square$

For every n let $\tan(E_{/\mathbf{Q}_{n,p}})$ denote the tangent space of $E_{/\mathbf{Q}_{n,p}}$ at the origin
and consider the Lie group exponential map

$$\exp_E : \tan(E_{/\mathbf{Q}_{n,p}}) \xrightarrow{\sim} E(\mathbf{Q}_{n,p}) \otimes \mathbf{Q}_p.$$

Fix a minimal Weierstrass model of E and let ω_E denote the corresponding
holomorphic differential. Then the cotangent space $\mathrm{cotan}(E)$ is $\mathbf{Q}_{n,p}\omega_E$, and

we let ω_E^* be the corresponding dual basis of $\tan(E)$. We have a commutative diagram

$$
\begin{array}{ccc}
\tan(E_{/\mathbf{Q}_{n,p}}) & \xrightarrow{\exp_E} & E(\mathbf{Q}_{n,p}) \otimes \mathbf{Q}_p \\
{\scriptstyle \cdot \omega_E^*} \big\uparrow & & \big\uparrow \\
\lambda_E(\mathfrak{p}_n) & \xrightarrow{\;\sim\;} & \hat{E}(\mathfrak{p}_n) & \xrightarrow{\;\sim\;} & E_1(\mathbf{Q}_{n,p})
\end{array}
$$

where $\hat{E}$ is the formal group of E, λ_E is its logarithm map, $\mathfrak{p}_n$ is the maximal ideal of $\mathbf{Q}_{n,p}$, $E_1(\mathbf{Q}_{n,p})$ is the kernel of reduction in $E(\mathbf{Q}_p)$, and the bottom maps are the formal group exponential followed by the isomorphism of [T] Theorem 4.2. (Note that $\hat{E}(\mathfrak{p}_n)_{\mathrm{tors}} = 0$ because $\mathbf{Q}_p(\hat{E}_p)/\mathbf{Q}_p$ is totally ramified of degree $p - 1$, so λ_E is injective.) Extending λ_E linearly we will view it as a homomorphism defined on all of $E(\mathbf{Q}_{n,p}) \otimes \mathbf{Q}_p$.

Since $V \cong V^*$,

$$\mathrm{Hom}(E(\mathbf{Q}_{n,p}), \mathbf{Q}_p) \cong \mathrm{Hom}(H^1_{\mathcal{S}}(\mathbf{Q}_{n,p}, V), \mathbf{Q}_p) \cong H^1_{/\mathcal{S}}(\mathbf{Q}_{n,p}, V).$$

Thus there is a dual exponential map

$$\exp_E^* : H^1_{/\mathcal{S}}(\mathbf{Q}_{n,p}, V) \xrightarrow{\sim} \mathrm{cotan}(E_{/\mathbf{Q}_{n,p}}) = \mathbf{Q}_{n,p}\omega_E.$$

We write $\exp_{\omega_E}^* : H^1_{/\mathcal{S}}(\mathbf{Q}_{n,p}, V) \xrightarrow{\sim} \mathbf{Q}_{n,p}$ for the composition $\omega_E^* \circ \exp_E^*$ Since $H^1_{/\mathcal{S}}(\mathbf{Q}_{n,p}, T)$ injects into $H^1_{/\mathcal{S}}(\mathbf{Q}_{n,p}, V)$, $\exp_{\omega_E}^*$ is injective on $H^1_{/\mathcal{S}}(\mathbf{Q}_{n,p}, T)$. The local pairing allows us to identify

$$
\begin{array}{ccc}
H^1_{/\mathcal{S}}(\mathbf{Q}_{n,p}, V) & \xrightarrow{\;\sim\;} & \mathrm{Hom}(E(\mathbf{Q}_{n,p}), \mathbf{Q}_p) \\
\big\uparrow & & \big\uparrow \\
H^1_{/\mathcal{S}}(\mathbf{Q}_{n,p}, T) & \xrightarrow{\;\sim\;} & \mathrm{Hom}(E(\mathbf{Q}_{n,p}), \mathbf{Z}_p).
\end{array}
$$

(1)

Explicitly, $z \in H^1_{/\mathcal{S}}(\mathbf{Q}_{n,p}, V)$ corresponds to the map

(2) $$x \mapsto \mathrm{Tr}_{\mathbf{Q}_{n,p}/\mathbf{Q}_p} \lambda_E(x) \exp_{\omega_E}^*(z).$$

Proposition 5.2. $\exp_{\omega_E}^*(H^1_{/\mathcal{S}}(\mathbf{Q}_p, T)) = [E(\mathbf{Q}_p) : E_1(\mathbf{Q}_p) + E(\mathbf{Q}_p)_{\mathrm{tors}}]p^{-1}\mathbf{Z}_p.$

Proof. By (2),

$$\exp_{\omega_E}^*(H^1_{/\mathcal{S}}(\mathbf{Q}_p, T)) = p^a \mathbf{Z}_p$$

where

$$\lambda_E(E(\mathbf{Q}_p)) = p^{-a}\mathbf{Z}_p.$$

We have $\lambda_E(E_1(\mathbf{Q}_p)) = p\mathbf{Z}_p$ and, since $\mathrm{rank}_{\mathbf{Z}_p} E(\mathbf{Q}_p) = 1$,

$$[\lambda_E(E(\mathbf{Q}_p)) : \lambda_E(E_1(\mathbf{Q}_p))] = [E(\mathbf{Q}_p) : E_1(\mathbf{Q}_p) + E(\mathbf{Q}_p)_{\mathrm{tors}}].$$

Thus the proposition follows. $\qquad\square$

6. The p-adic L-function

Let

$$L(E, s) = \sum_{n \geq 1} a_n n^{-s} = \prod_q \ell_q(q^{-s})^{-1}$$

denote the Hasse-Weil L-function of E, where $\ell_q(q^{-s})$ is the usual Euler factor at q. If $N \in \mathbf{Z}^+$ we will also write

$$L_N(E, s) = \sum_{(n,N)=1} a_n n^{-s} = \prod_{q \nmid N} \ell_q(q^{-s})^{-1}$$

for the L-function with the Euler factors dividing N removed. If F is an abelian extension of $\mathbf{Q}$ of conductor f and $\gamma \in \text{Gal}(F/\mathbf{Q})$, define the partial L-function

$$L_N(E, \gamma, F/\mathbf{Q}, s) = \sum_{n \mapsto \gamma} a_n n^{-s}$$

where the sum is over n prime to fN which map to γ under

$$(\mathbf{Z}/f\mathbf{Z})^\times \xrightarrow{\sim} \text{Gal}(\mathbf{Q}(\boldsymbol{\mu}_f)/\mathbf{Q}) \twoheadrightarrow \text{Gal}(F/\mathbf{Q}).$$

If χ is a character of $G_{\mathbf{Q}}$ of conductor f_χ, and $\ker(\chi) = \text{Gal}(\bar{\mathbf{Q}}/F_\chi)$, let

$$L_N(E, \chi, s) = \sum_{(n, f_\chi N)=1} \chi(n) a_n n^{-s} = \prod_{q \nmid f_\chi N} \ell_q(q^{-s}\chi(q))^{-1}$$

$$= \sum_{\gamma \in \text{Gal}(F_\chi/\mathbf{Q})} \chi(\gamma) L_N(E, \gamma, F_\chi/\mathbf{Q}, s).$$

When $N = 1$ we write simply $L(E, \chi, s)$, and then we have

$$(3) \qquad L_N(E, \chi, s) = \prod_{q \mid N} \ell_q(q^{-s}\chi(q)) L(E, \chi, s).$$

If E is modular then these functions all have analytic continuations to $\mathbf{C}$.

Fix a generator $[\zeta_{p^n}]_n$ of $\varprojlim \boldsymbol{\mu}_{p^n}$. Write $G_n = \text{Gal}(\mathbf{Q}_n/\mathbf{Q}) = \text{Gal}(\mathbf{Q}_{n,p}/\mathbf{Q}_p)$. If χ is a character of $\text{Gal}(\mathbf{Q}_\infty/\mathbf{Q})$ of conductor p^n define the Gauss sum

$$\tau(\chi) = \sum_{\gamma \in \text{Gal}(\mathbf{Q}(\boldsymbol{\mu}_{p^n})/\mathbf{Q})} \chi(\gamma) \zeta_{p^n}^\gamma.$$

Fix also an embedding of $\bar{\mathbf{Q}}_p$ into $\mathbf{C}$ so that we can identify complex and p-adic characters of $G_{\mathbf{Q}}$.

The following theorem is proved in [MSD] in the case of good ordinary reduction. See [MTT] for the (even more) general statement.

Theorem 6.1. *Suppose E is modular and E has good ordinary reduction or multiplicative reduction at p. Let $\alpha \in \mathbf{Z}_p^\times$ and $\beta \in p\mathbf{Z}_p$ be the eigenvalues of Frobenius if E has good ordinary reduction at p, and let $(\alpha, \beta) = (1, p)$ (resp. $(-1, -p)$) if E has split (resp. nonsplit) multiplicative reduction. Then there* is a nonzero integer c_E independent of p, and a p-adic L-function $\mathcal{L}_E \in c_E^{-1}\Lambda$ such that for every character χ of $\mathrm{Gal}(\mathbf{Q}_\infty/\mathbf{Q})$ of finite order,

$$\chi(\mathcal{L}_E) = \begin{cases} (1 - \alpha^{-1})^2 L(E,1)/\Omega_E & \text{if } \chi = 1 \text{ and } E \text{ has good reduction at } p \\ (1 - \alpha^{-1}) L(E,1)/\Omega_E & \text{if } \chi = 1 \text{ and } E \text{ is multiplicative at } p \\ \alpha^{-n}\tau(\chi)L(E,\chi^{-1},1)/\Omega_E & \text{if } \chi \text{ has conductor } p^n > 1 \end{cases}$$

where Ω_E is the fundamental real period of E.

If $N \in \mathbf{Z}^+$, define

$$\mathcal{L}_{E,N} = \prod_{q|N, q \neq p} \ell_q(q^{-1}\mathrm{Fr}_q^{-1})\mathcal{L}_E \in \Lambda.$$

Using (3) and Theorem 6.1 one obtains analogous expressions for $\chi(\mathcal{L}_{E,N})$ in terms of $L_N(E,\chi^{-1},1)$.

7. KATO'S EULER SYSTEM

The following theorem of Kato is crucial for everything that follows.

Theorem 7.1 (Kato [Ka1], see also [Scho]). *Suppose that E is modular, and let N be the conductor of E. There are a positive integer r_E independent of p, positive integers $D \not\equiv 1$, $D' \not\equiv 1$ (mod p), and an Euler system $\bar{\mathbf{c}} = \bar{\mathbf{c}}(D,D')$ for $T_p(E)$,*

$$\{\bar{\mathbf{c}}_{\mathbf{Q}_n(\mu_r)} \in H^1(\mathbf{Q}_n(\mu_r), T_p(E)) : r \text{ squarefree}, (r, NpDD') = 1, n \geq 0\}$$

such that for every such $n \geq 0$ and every character $\chi : \mathrm{Gal}(\mathbf{Q}_n/\mathbf{Q}) \to \mathbf{C}^\times$,

$$\sum_{\gamma \in \mathrm{Gal}(\mathbf{Q}_n/\mathbf{Q})} \chi(\gamma) \exp_{\omega_E}^*(\mathrm{loc}_p^{\mathrm{ram}}(\bar{\mathbf{c}}_{\mathbf{Q}_n}^\gamma))$$

$$= r_E DD'(D - \chi^{-1}(D))(D' - \chi^{-1}(D'))L_{Np}(E,\chi,1)/\Omega_E.$$

Proof. See the paper of Scholl [Scho] in this volume (especially Theorem 5.2.7) for the proof of this theorem when $p > 2$ and E has good reduction at p. $\square$

Using the Coleman map $\mathrm{Col}_\infty : H^1_{\infty,/S}(\mathbf{Q}_{n,p}, T) \to \Lambda$ described in the Appendix, we can relate Kato's Euler system to the p-adic L-function.

Corollary 7.2. *With hypotheses and notation as in Theorems 7.1 and 6.1, there is an Euler system $\mathbf{c}$ for $T_p(E)$ such that*

(i) $\exp_{\omega_E}^*(\mathrm{loc}_p^{\mathrm{ram}}(\mathbf{c}_{\mathbf{Q}})) = r_E L_{Np}(E,1)/\Omega_E,$

(ii) $\mathrm{Col}_\infty([\mathrm{loc}_p^{\mathrm{ram}}(\mathbf{c}_{\mathbf{Q}_n})]) = r_E \mathcal{L}_{E,N}.$

Proof. Let $\bar{\mathbf{c}}$ be the Euler system of Theorem 7.1 for some $D, D' \not\equiv 1$ (mod p). Let $\sigma_D \in \mathrm{Gal}(\mathbf{Q}(\mu_{p^\infty})/\mathbf{Q})$ denote the automorphism $\zeta \mapsto \zeta^D$ for $\zeta \in \mu_{p^\infty}$, and similarly for $\sigma_{D'}$.

Since $D, D' \not\equiv 1 \pmod{p}$, $(D - \sigma_D)(D' - \sigma_{D'})$ is invertible in Λ. Let $\rho_{D,D'} \in \mathbf{Z}_p[[G_{\mathbf{Q}}]]$ be any element which restricts to $(D - \sigma_D)^{-1}(D' - \sigma_{D'})^{-1}$ in Λ, and define

$$c_{\mathbf{Q}_n(\mu_r)} = D^{-1} D'^{-1} \rho_{D,D'} \bar{c}_{\mathbf{Q}_n(\mu_r)}.$$

It is clear that $\{c_{\mathbf{Q}_n(\mu_r)}\}$ is still an Euler system, and for every $n \geq 0$ and every character $\chi : \mathrm{Gal}(\mathbf{Q}_n/\mathbf{Q}) \to \mathbf{C}^{\times}$

$$\chi(\rho_{D,D'}) = \chi((D - \sigma_D)(D' - \sigma_{D'}))^{-1} = (D - \chi(D))^{-1}(D' - \chi(D'))^{-1},$$

so by Theorem 7.1

$$\sum_{\gamma \in \mathrm{Gal}(\mathbf{Q}_n/\mathbf{Q})} \chi(\gamma) \exp^*_{\omega_E}(\mathrm{loc}_p^{\mathrm{ram}}(c_{\mathbf{Q}_n}^{\gamma})) = \tau_E L_{Np}(E, \chi, 1)/\Omega_E.$$

When χ is the trivial character this is (i), and as χ varies (ii) follows from Proposition A.2 of the Appendix along with the definition (Theorem 6.1) of $\mathcal{L}_E$ and (3). $\qquad\square$

8. Consequences of Kato's Euler system

Following Kato, we will apply the results of §§3 and 4 to bound the Selmer group of E.

8.1. The main theorem.

Theorem 8.1 (Kato [Ka2]). *Suppose E is modular and E does not have complex multiplication.*

(i) *If $L(E,1) \neq 0$ then $E(\mathbf{Q})$ and $\mathrm{III}(E)$ are finite.*

(ii) *If L is a finite abelian extension of $\mathbf{Q}$, χ is a character of $\mathrm{Gal}(L/\mathbf{Q})$, and $L(E, \chi, 1) \neq 0$ then $E(L)^{\chi}$ and $\mathrm{III}(E_{/L})^{\chi}$ are finite.*

Remarks. We will prove below a more precise version of Theorem 8.1(i).

Kato's construction produces an Euler system for $T_p(E) \otimes \chi$ for every character χ of $G_{\mathbf{Q}}$ of finite order, with properties analogous to those of Theorem 7.1. This more general construction is needed to prove Theorem 8.1(ii). For simplicity we will not treat this more general setting here, so we will only prove Theorem 8.1(i) below. But the method for (ii) is the same.

Theorem 8.1(i) was first proved by Kolyvagin in [Ko], using a system of Heegner points, along with work of Gross and Zagier [GZ], Bump, Friedberg, and Hoffstein [BFH], and Murty and Murty [MM]. The Euler system proof given here, due to Kato, is 'self-contained' in the sense that it replaces all of those other analytic results with the calculation of Theorem 7.1.

Corollary 8.2. *Suppose E is modular and E does not have complex multiplication. Then $E(\mathbf{Q}_{\infty})$ is finitely generated.*

Proof. A theorem of Rohrlich [Ro] shows that $L(E, \chi, 1) \neq 0$ for almost all characters χ of finite order of $\mathrm{Gal}(\mathbf{Q}_\infty/\mathbf{Q})$. Serre's [Se] Théorème 3 shows that $E(\mathbf{Q}_\infty)_{\mathrm{tors}}$ is finite, and the corollary follows without difficulty from Theorem 8.1(ii). □

Remark. When E has complex multiplication, the representation $T_p(E)$ does not satisfy Hypothesis $\mathrm{Hyp}(\mathbf{Q}_\infty, V)$(i) (see Remark 8.5 below), so we cannot apply the results of §§3 and 4 with Kato's Euler system. However, Theorem 8.1 and Corollary 8.2 are known in that case, as Theorem 8.1 for CM curves can be proved using the Euler system of elliptic units. See [CW], [Ru2] §11, and [RW].

8.2. Verification of the hypotheses. Fix a $\mathbf{Z}_p$-basis of T and let

$$\rho_{E,p} : G_{\mathbf{Q}} \to \mathrm{Aut}(T) \xrightarrow{\sim} \mathrm{GL}_2(\mathbf{Z}_p)$$

be the p-adic representation of $G_{\mathbf{Q}}$ attached to E with this basis.

Proposition 8.3. (i) *If E has no complex multiplication, then $T_p(E)$ satisfies hypotheses $\mathrm{Hyp}(\mathbf{Q}_\infty, V)$ and $H^1(\mathbf{Q}(E_{p^\infty})/\mathbf{Q}, E_{p^\infty})$ is finite.*
(ii) *If the p-adic representation $\rho_{E,p}$ is surjective, then $T_p(E)$ satisfies hypotheses $\mathrm{Hyp}(\mathbf{Q}_\infty, T)$ and $H^1(\mathbf{Q}(E_{p^\infty})/\mathbf{Q}, E_{p^\infty}) = 0$.*

Proof. The Weil pairing shows that

$$\rho_{E,p}(G_{\mathbf{Q}(\mu_{p^\infty})}) = \rho_{E,p}(G_{\mathbf{Q}}) \cap \mathrm{SL}_2(\mathbf{Z}_p).$$

If E has no complex multiplication then a theorem of Serre ([Se] Théorème 3) says that the image of $\rho_{E,p}$ is open in $\mathrm{GL}_2(\mathbf{Z}_p)$. It follows that $V_p(E)$ is an irreducible $G_{\mathbf{Q}_\infty}$-representation, and if $\rho_{E,p}$ is surjective then E_p is an irreducible $\mathbf{F}_p[G_{\mathbf{Q}_\infty}]$-representation.

It also follows that we can find $\tau \in G_{\mathbf{Q}(\mu_{p^\infty})}$ such that

$$\rho_{E,p}(\tau) = \begin{pmatrix} 1 & x \\ 0 & 1 \end{pmatrix}$$

with $x \neq 0$, and such a τ satisfies Hypothesis $\mathrm{Hyp}(\mathbf{Q}_\infty, V)$(i). If $\rho_{E,p}$ is surjective we can take $x = 1$, and then τ satisfies Hypothesis $\mathrm{Hyp}(\mathbf{Q}_\infty, T)$(i).
We have

$$H^1(\mathbf{Q}(E_{p^\infty})/\mathbf{Q}, E_{p^\infty}) = H^1(\rho_{E,p}(G_{\mathbf{Q}}), (\mathbf{Q}_p/\mathbf{Z}_p)^2)$$

and the rest of the proposition follows. □

Remark 8.4. Serre's theorem (see [Se] Corollaire 1 of Théorème 3) also shows that if E has no complex multiplication then $\rho_{E,p}$ is surjective for all but finitely many p.

Remark 8.5. The conditions on τ in hypotheses $\mathrm{Hyp}(\mathbf{Q}_\infty, V)$(i) force $\rho_{E,p}(\tau)$ to be nontrivial and unipotent. Thus if E has complex multiplication then there is no τ satisfying $\mathrm{Hyp}(\mathbf{Q}_\infty, V)$(i).

8.3. Bounding $\mathcal{S}(\mathbf{Q}, E_{p^\infty})$.

Theorem 8.6. *Suppose E is modular, E does not have complex multiplication, and $L(E, 1) \neq 0$.*

(i) *$E(\mathbf{Q})$ and $\text{Ш}(E)_{p^\infty}$ are finite.*

(ii) *Suppose in addition that $p \nmid 2r_E$ and $\rho_{E,p}$ is surjective. If E has good reduction at p and $p \nmid |\tilde{E}(\mathbf{F}_p)|$ (where $\tilde{E}$ is the reduction of E modulo p), then*

$$|\text{Ш}(E)_{p^\infty}| \ \text{divides} \ \frac{L_N(E, 1)}{\Omega_E}$$

where N is the conductor of E.

Proof. Recall that $\ell_q(q^{-s})$ is the Euler factor of $L(E, s)$ at q, and that by Proposition 5.1, $\mathcal{S}(\mathbf{Q}, E_{p^\infty})$ is the usual p-power Selmer group of E.

Since $\ell_q(q^{-1})$ is nonzero for every q, Corollary 7.2(i) shows that $\text{loc}_p^{\text{ram}}(\mathbf{c_Q})$ $\neq 0$. By Proposition 8.3(i) and Proposition 5.2 we can apply Theorem 3.1 to conclude that $\mathcal{S}(\mathbf{Q}, E_{p^\infty})$ is finite, which gives (i).

If E has good reduction at p then $p\ell_p(p^{-1}) = |\tilde{E}(\mathbf{F}_p)|$ and

$$[E(\mathbf{Q}_p) : E_1(\mathbf{Q}_p) + E(\mathbf{Q}_p)_{\text{tors}}] \ \text{divides} \ |\tilde{E}(\mathbf{F}_p)|.$$

Therefore if $p \nmid r_E |\tilde{E}(\mathbf{F}_p)|$ then

$$\exp_{\omega_E}^*(H^1_{/\mathcal{S}}(\mathbf{Q}_p, T_p(E))) = p^{-1}\mathbf{Z}_p$$

$$\exp_{\omega_E}^*(\mathbf{Z}_p\text{loc}_p^{\text{ram}}(\mathbf{c_Q})]) = p^{-1}(L_N(E, 1)/\Omega_E)\mathbf{Z}_p$$

by Proposition 5.2 and Corollary 7.2(i) By Proposition 8.3(ii), if $p \neq 2$ we can apply Theorem 3.2, and (ii) follows. $\square$

Remarks. In Corollary 8.9 below, using Iwasawa theory, we will prove that Theorem 8.6(ii) holds for almost all p, even when p divides $|\tilde{E}(\mathbf{F}_p)|$. This is needed to prove Theorem 8.1(i), since $|\tilde{E}(\mathbf{F}_p)|$ could be divisible by p for infinitely many p. However, since $|\tilde{E}(\mathbf{F}_p)| < 2p$ for all primes $p > 5$, we see that if $E(\mathbf{Q})_{\text{tors}} \neq 0$ then $|\tilde{E}(\mathbf{F}_p)|$ is prime to p for almost all p. Thus Theorem 8.1(i) for such a curve follows directly from Theorem 8.6.

The Euler system techniques we are using give an upper bound for the order of the Selmer group, but no lower bound.

8.4. Bounding $\mathcal{S}(\mathbf{Q}_\infty, E_{p^\infty})$. Define

$$\mathcal{S}(\mathbf{Q}_\infty, E_{p^\infty}) = \varinjlim \mathcal{S}(\mathbf{Q}_n, E_{p^\infty}) \subset H^1(\mathbf{Q}_\infty, E_{p^\infty}),$$

and recall that $X_\infty = \text{Hom}(\mathcal{S}(\mathbf{Q}_\infty, E_{p^\infty}), \mathbf{Q}_p/\mathbf{Z}_p)$. Let r_E be the positive integer of Theorem 7.1 and let N be the conductor of E.

Theorem 8.7. *Suppose E is modular, E does not have complex multiplication, and E has good ordinary reduction or nonsplit multiplicative reduction* *at p.* Then X_∞ *is a finitely-generated torsion Λ-module and there is an integer t such that*

$$\text{char}(X_\infty) \quad divides \quad p^t \mathcal{L}_{E,N} \Lambda.$$

If $\rho_{E,p}$ is surjective and $p \nmid r_E \prod_{q|N,q\neq p} \ell_q(q^{-1})$ then $\text{char}(X_\infty)$ divides $\mathcal{L}_E \Lambda$.

If E has split multiplicative reduction at p, the same results hold with $\text{char}(X_\infty)$ replaced by $\mathcal{J}\text{char}(X_\infty)$ where $\mathcal{J}$ is the augmentation ideal of Λ.

Proof. Rohrlich [Ro] proved that $\mathcal{L}_E \neq 0$. Thus the theorem is immediate from Propositions 8.3 and A.2, Corollary 7.2, and Theorem 4.1. □

Corollary 8.8. *Let E be as in Theorem 8.7. There is a nonzero integer M_E such that if p is a prime where E has good ordinary reduction and $p \nmid M_E$, then X_∞ has no nonzero finite submodules.*

Proof. This corollary is due to Greenberg [Gr1], [Gr2]; we sketch a proof here. Let Σ be a finite set of primes containing p, ∞, and all primes where E has bad reduction, and let $\mathbf{Q}_\Sigma$ be the maximal extension of $\mathbf{Q}$ unramified outside of Σ. Then there is an exact sequence

$$(4) \quad 0 \to \mathcal{S}(\mathbf{Q}_\infty, E_{p^\infty}) \to H^1(\mathbf{Q}_\Sigma/\mathbf{Q}_\infty, E_{p^\infty}) \to \oplus_{q\in\Sigma} \oplus_{v|q} H^1_{/S}(\mathbf{Q}_{\infty,v}, E_{p^\infty}).$$

Suppose $q \in \Sigma$, $q \neq p$, and $v \mid q$. If $p \nmid |E(\mathbf{Q}_q)_{\text{tors}}|$ then it is not hard to show that $E(\mathbf{Q}_{\infty,v})$ has no p-torsion, and so by [Gr1] Proposition 2, $H^1(\mathbf{Q}_{\infty,v}, E_{p^\infty}) = 0$. Thus for sufficiently large p the Pontryagin dual of (4) is

$$\varprojlim_n E(\mathbf{Q}_{n,p}) \otimes \mathbf{Z}_p \to \text{Hom}(H^1(\mathbf{Q}_\Sigma/\mathbf{Q}_\infty, E_{p^\infty}), \mathbf{Q}_p/\mathbf{Z}_p) \to X_\infty \to 0.$$

Since $\mathbf{Q}_\infty/\mathbf{Q}$ is totally ramified at p,

$$\varprojlim_n E(\mathbf{Q}_{n,p}) \otimes \mathbf{Z}_p = \varprojlim_n E_1(\mathbf{Q}_{n,p}) = \varprojlim_n \hat{E}(\mathfrak{p}_n)$$

and this is free of rank one over Λ (see for example [PR1] Théorème 3.1 or [Schn] Lemma 6, §A.1). It now follows, using the fact that X_∞ is a torsion Λ-module (Theorem 8.7) and [Gr1] Propositions 3, 4, and 5 that $\text{Hom}(H^1(\mathbf{Q}_\Sigma/\mathbf{Q}_\infty, E_{p^\infty}), \mathbf{Q}_p/\mathbf{Z}_p)$ has no nonzero finite submodules, and by the Lemma on p. 123 of [Gr1] the same is true of X_∞. □

Corollary 8.9. *Suppose E is modular, E does not have complex multiplication, E has good reduction at p, $p \nmid 2r_E M_E \prod_{q|N} \ell_q(q^{-1})$ (where r_E is as in Theorem 7.1 and M_E is as in Corollary 8.8), and $\rho_{E,p}$ is surjective. Then*

$$|\text{III}(E)_{p^\infty}| \quad divides \quad \frac{L(E,1)}{\Omega_E}.$$

Proof. First, if E has supersingular reduction at p then $|\tilde{E}(\mathbf{F}_p)|$ is prime to p, so the corollary follows from Theorem 8.6(ii).

Thus we may assume that E has good ordinary reduction at p. In this case the corollary is a well-known consequence of Theorem 8.7 and Corollary

8.8; see for example [PR1] §6 or [Schn] §2 for details. The idea is that if X_∞ has no nonzero finite submodules and $\text{char}(X_\infty)$ divides $\mathcal{L}_E\Lambda$, then

$$|\mathcal{S}(\mathbf{Q}_\infty, E_{p^\infty})^{\text{Gal}(\mathbf{Q}_\infty/\mathbf{Q})}| \quad \text{divides} \quad \chi_0(\mathcal{L}_{E,N}),$$

where χ_0 denotes the trivial character, and with α as in Theorem 6.1

$$\chi_0(\mathcal{L}_{E,N}) = (1 - \alpha^{-1})^2 \prod_{q|N} \ell_q(q^{-1})(L(E,1)/\Omega_E).$$

On the other hand, one can show that the restriction map

$$\mathcal{S}(\mathbf{Q}, E_{p^\infty}) \to \mathcal{S}(\mathbf{Q}_\infty, E_{p^\infty})^{\text{Gal}(\mathbf{Q}_\infty/\mathbf{Q})}$$

is injective with cokernel of order divisible by $(1 - \alpha^{-1})^2$, and the corollary follows. $\qquad\square$

Proof of Theorem 8.1(i). Suppose E is modular, E does not have complex multiplication, and $L(E,1) \neq 0$. By Theorem 8.6, $E(\mathbf{Q})$ is finite and $\text{III}(E)_{p^\infty}$ is finite for every p. By Corollary 8.9 (and using Serre's theorem, see Remark 8.4) $\text{III}(E)_{p^\infty} = 0$ for almost all p. This proves Theorem 8.1(i). $\qquad\square$

We can also now prove part of Theorem 8.1(ii) in the case where E has good ordinary or multiplicative reduction at p and $L \subset \mathbf{Q}_\infty$. For in that case, by Theorem 8.7, $\chi(\text{char}(\text{Hom}(\mathcal{S}(\mathbf{Q}_\infty, E_{p^\infty}), \mathbf{Q}_p/\mathbf{Z}_p)))$ is a nonzero multiple of $L(E,\chi,1)/\Omega_E$. If $L(E,\chi,1) \neq 0$ it follows that $\mathcal{S}(\mathbf{Q}_\infty, E_{p^\infty})^\chi$ is finite. The kernel of the restriction map $\mathcal{S}(L, E_{p^\infty}) \to \mathcal{S}(\mathbf{Q}_\infty, E_{p^\infty})$ is contained in the finite group $H^1(\mathbf{Q}_\infty/L, E_{p^\infty}^{G_{\mathbf{Q}_\infty}})$, and so we conclude that both $E(L)^\chi$ and $\text{III}(E_{/L})_{p^\infty}^\chi$ are finite.

Appendix. Explicit description of the Coleman map

In this appendix we give an explicit description of the Coleman map from $H^1_{\infty,/\mathcal{S}}(\mathbf{Q}_{n,p}, T)$ to Λ. This map allows us to relate Kato's Euler system with the p-adic L-function $\mathcal{L}_E$.

Suppose for this appendix that E has good ordinary reduction or multiplicative reduction at p. As in Theorem 6.1, let $\alpha \in \mathbf{Z}_p^\times$ and $\beta \in p\mathbf{Z}_p$ be the eigenvalues of Frobenius if E has good ordinary reduction, let $\alpha = 1, \beta = p$ if E has split multiplicative reduction and let $\alpha = -1, \beta = -p$ if E has nonsplit multiplicative reduction.

Recall we fixed before Theorem 6.1 a generator $[\zeta_{p^n}]_n$ of $\varprojlim \mu_{p^n}$. For every $n \geq 0$ define

$$x_n = \alpha^{-n-1}\text{Tr}_{\mathbf{Q}_p(\mu_{p^{n+1}})/\mathbf{Q}_{n,p}} \left(\sum_{k=0}^n \frac{\zeta_{p^{n+1-k}} - 1}{\beta^k} + \frac{\beta}{\beta - 1} \right) \in \mathbf{Q}_{n,p}.$$

Lemma A.1. (i) *If $n \geq m$ then* $\text{Tr}_{\mathbf{Q}_{n,p}/\mathbf{Q}_{m,p}} x_n = x_m$.

(ii) *If χ is a character of G_n then*

$$\chi(\sum_{\gamma\in G_n} x_n^\gamma \gamma) = \begin{cases} \alpha^{-m}\tau(\chi) & \text{if } \chi \text{ has conductor } p^m > 1 \\ (1-\alpha^{-1})(1-\beta^{-1})^{-1} & \text{if } \chi = 1. \end{cases}$$

Proof. Exercise. □

Proposition A.2. (i) *For every $n \geq 0$ there is a $\mathbf{Z}_p[G_n]$-module map*

$$\mathrm{Col}_n : H^1_{/\mathcal{S}}(\mathbf{Q}_{n,p}, T) \to \mathbf{Z}_p[G_n]$$

such that for every $z \in H^1_{/\mathcal{S}}(\mathbf{Q}_{n,p}, T)$ and every nontrivial character χ of G_n of conductor p^m,

$$\chi(\mathrm{Col}_n(z)) = \alpha^{-m}\tau(\chi)\sum_{\gamma\in G_n} \chi^{-1}(\gamma)\exp^*_{\omega_E}(z^\gamma).$$

If χ_0 is the trivial character then

$$\chi_0(\mathrm{Col}_n(z)) = (1-\alpha^{-1})(1-\beta^{-1})^{-1}\sum_{\gamma\in G_n}\exp^*_{\omega_E}(z^\gamma).$$

(ii) *The maps Col_n are compatible as n varies, and in the limit they induce a map of Λ-modules*

$$\mathrm{Col}_\infty : H^1_{\infty,/\mathcal{S}}(\mathbf{Q}_p, T) \to \Lambda.$$

(iii) *The map Col_∞ is injective. If E has split multiplicative reduction at p then the image of Col_∞ is contained in the augmentation ideal of Λ.*

Proof. The proof is based on work of Coleman [Co].

For the curves E which we are considering, $\hat{E}$ is a height-one Lubin-Tate formal group over $\mathbf{Z}_p$ for the uniformizing parameter β. It follows that, writing R for the ring of integers of the completion of the maximal unramified extension of $\mathbf{Q}_p$, $\hat{E}$ is isomorphic over R to the multiplicative group $\mathbf{G}_m$. Fix an isomorphism $\eta : \mathbf{G}_m \xrightarrow{\sim} \hat{E}$, $\eta \in R[[X]]$. We define the p-adic period Ω_p of E

$$\Omega_p = \eta'(0) \in R^\times.$$

This period is unique up to $\mathbf{Z}_p^\times$, and is also characterized by the identity

(5) $$\lambda_E(\eta(X)) = \Omega_p\log(1+X).$$

By [dS] §I.3.2 (4), if ϕ is the Frobenius automorphism of $R/\mathbf{Z}_p$ then

$$\Omega_p^\phi = \alpha^{-1}\Omega_p.$$

By [Co] Theorem 24 (applied to the multiplicative group) there is a power series g in the maximal ideal $(p, X)R[[X]]$ of $R[[X]]$ such that

$$\log(1+g(X)) = \Omega_p^{-1}\left((p-1)\frac{\beta}{\beta-1} + \sum_{k=0}^\infty \sum_{\delta\in\mu_{p-1}\subset\mathbf{Z}_p^\times} \frac{(1+X)^{\delta p^k}-1}{\beta^k}\right).$$

In particular if we set $X = \zeta_{p^{n+1}} - 1$ and use (5),

$$\lambda_E(\eta(g(\zeta_{p^{n+1}} - 1))) = \Omega_p \log(1 + g(\zeta_{p^{n+1}} - 1)) = \alpha^{n+1} x_n.$$

Observe that $\hat{E}(\mathfrak{p}_n R)_{\text{tors}} = 0$ because $\mathbf{Q}_p(\hat{E}_p)/\mathbf{Q}_p$ is totally ramified of degree $p - 1$. Therefore λ_E is injective on $\hat{E}(\mathfrak{p}_n R)$ and so

$$x_n \in \lambda_E(\mathfrak{p}_n R)^{\text{Gal}(\mathbf{Q}_{n,p} R/\mathbf{Q}_{n,p})} = \lambda_E(\mathfrak{p}_n) \subset \lambda_E(E(\mathbf{Q}_{n,p})).$$

Define Col_n on $H^1_{/S}(\mathbf{Q}_{n,p}, T)$ by

$$\text{Col}_n(z) = \left(\sum_{\gamma \in G_n} x_n^\gamma \gamma \right) \left(\sum_{\gamma \in G_n} \exp^*_{\omega_E}(z^\gamma) \gamma^{-1} \right)$$

$$= \sum_{\gamma \in G_n} (\text{Tr}_{\mathbf{Q}_{n,p}/\mathbf{Q}_p} x_n^\gamma \exp^*_{\omega_E}(z)) \gamma$$

Clearly this gives a Galois-equivariant map $H^1_{/S}(\mathbf{Q}_{n,p}, T) \to \mathbf{Q}_p[G_n]$. Since $x_n^\gamma \in \lambda_E(E(\mathbf{Q}_{n,p}))$, (1) and (2) show that $\text{Col}_n(z) \in \mathbf{Z}_p[G_n]$. The equalities of (i) follow from Lemma A.1(ii), and (ii) follows easily.

For (iii), suppose first that E has good ordinary reduction or nonsplit multiplicative reduction at p, so that $\alpha \neq 1$. Then the injectivity of Col_n (and of Col_∞) follows from (i), the nonvanishing of the Gauss sums, and the injectivity of $\exp^*_{\omega_E}$.

If E has split multiplicative reduction at p, then $\alpha = 1$. In this case it follows from (i) that $\ker(\text{Col}_n) = H^1_{/S}(\mathbf{Q}_{n,p}, T)^{G_n}$, which is free of rank one over $\mathbf{Z}_p$, for every n. But one can show using (1) that $H^1_{\infty,/S}(\mathbf{Q}_p, T)$ has no Λ-torsion, so Col_∞ must be injective in this case as well. The assertion about the cokernel is clear from (i), since $\alpha = 1$.   □

REFERENCES

[BFH] Bump, D., Friedberg, S., Hoffstein, J.: Eisenstein series on the metaplectic group and nonvanishing theorems for automorphic L-functions and their derivatives, *Annals of Math.* **131** (1990) 53–127.

[CW]   Coates, J., Wiles, A.: On the conjecture of Birch and Swinnerton-Dyer, *Inventiones math.* **39** (1977) 223–251.

[Co]   Coleman, R.: Division values in local fields. *Invent math.* **53** (1979) 91–116

[dS]   de Shalit, E.: The Iwasawa theory of elliptic curves with complex multiplication, (*Perspec. in Math.* **3**) Orlando: Academic Press (1987)

[Gr1] Greenberg, R.: Iwasawa theory for p-adic representations, *Adv. Stud. in Pure Math.* **17** (1989) 97–137.

[Gr2] Greenberg, R.: Iwasawa theory for p-adic representations II, to appear.

[GZ]   Gross, B., Zagier, D.: Heegner points and derivatives of L-series, *Inventiones math.* **84** (1986) 225–320

[Ka1] Kato, K.: Euler systems, Iwasawa theory, and Selmer groups, to appear

[Ka2] _____: to appear

[Ko]   Kolyvagin, V. A.: Euler systems. In: The Grothendieck Festschrift (Vol. II), P. Cartier, et al., eds., *Prog. in Math* **87**, Boston: Birkhäuser (1990) 435–483.

[MSD] Mazur, B., Swinnerton-Dyer, H.P.F.: Arithmetic of Weil curves, *Inventiones math.* **25** (1974) 1–61

[MTT] Mazur, B., Tate, J., Teitelbaum, J.: On p-adic analogues of the conjectures of Birch and Swinnerton-Dyer, *Invent. math.* **84** (1986) 1–48.

[MM] Murty, K., Murty, R.: Mean values of derivatives of modular *L*-series, *Annals of Math.* **133** (1991) 447–475.

[PR1] Perrin-Riou, B.: Théorie d'Iwasawa p-adique locale et globale, *Invent. math.* **99** (1990) 247–292.

[PR2] _____ : Systèmes d'Euler p-adiques et théorie d'Iwasawa, to appear.

[Ro] Rohrlich, D.: On *L*-functions of elliptic curves and cyclotomic towers, *Invent. math.* **75** (1984) 409–423.

[Ru1] Rubin, K.: The main conjecture. Appendix to: Cyclotomic fields I and II, S. Lang, *Graduate Texts in Math.* **121**, New York: Springer-Verlag (1990) 397–419.

[Ru2] _____ : The "main conjectures" of Iwasawa theory for imaginary quadratic fields, *Invent. math.* **103** (1991) 25–68.

[Ru3] _____ : Euler systems, to appear.

[RW] Rubin, K., Wiles, A.: Mordell-Weil groups of elliptic curves over cyclotomic fields. In: Number Theory related to Fermat's last theorem, *Progress in Math.* **26**, Boston: Birkhauser (1982) 237–254.

[Schn] Schneider, P.: p-adic height pairings, II, *Inventiones math.* **79** (1985) 329–374.

[Scho] Scholl, A.: An introduction to Kato's Euler systems, this volume

[Se] Serre, J-P.: Propriétés Galoisiennes des points d'ordre fini des courbes elliptiques, *Invent. math.* **15** (1972) 259–331.

[T] Tate, J.: Algorithm for determining the type of a singular fiber in an elliptic pencil. In: Modular functions of one variable (IV), *Lecture Notes in Math.* **476**, New York: Springer-Verlag (1975) 33–52.

DEPARTMENT OF MATHEMATICS, OHIO STATE UNIVERSITY, COLUMBUS, OH 43210 USA

E-mail address: rubin@math.ohio-state.edu

Basic notions of rigid analytic geometry

PETER SCHNEIDER

The purpose of my lectures at the conference was to introduce the newcomer to the field of rigid analytic geometry. Precise definitions of the key notions and precise statements of the basic facts were given. But, of course, the limited time did not allow to include any proofs. Instead the emphasis was placed on motivating and explaining the shape of the theory. The positive response from the audience encouraged me to write up the following notes which reproduce my lectures in an essentially unchanged way. I hope that they can serve as a means to quickly grasp the basics of the field. Of course, anybody who is seriously interested has to go on and has to dig into the proper literature.

Rigid or non-archimedean analysis takes place over a field K which is complete with respect to a non-archimedean absolute value $|\ |$. The most important examples are the fields of p-adic numbers $\mathbf{Q}_p$ where p is some prime number. For technical purposes we fix throughout an algebraic closure $\widehat{\overline{K}}$ of K and denote by $\overline{K}$ its completion which again is algebraically closed. The absolute value $|\ |$ extends uniquely to an absolute value $|\ |$ of $\overline{K}$.

Fix a natural number $n \in \mathbf{N}$, and let us consider the "n-dimensional polydisk"

$$\mathbf{B}^n := \{(z_1, \ldots, z_n) \in \widehat{\overline{K}}^n : \max |z_i| \leq 1\}$$

in the n-dimensional vector space over $\widehat{\overline{K}}$. We clearly want this polydisk to be a geometric object (something like a "manifold") in our theory. For this we have to decide which functions on $\mathbf{B}^n$ we will call "analytic". The naive answer would be to follow real or complex analysis and to call a function analytic if it has, locally around each point, a convergent Taylor expansion. But we are dealing here with a non-archimedean metric on $\widehat{\overline{K}}^n$ satisfying the strict triangle inequality. This implies that the topology of $\widehat{\overline{K}}^n$ or $\mathbf{B}^n$ is totally disconnected so that there is a huge supply of functions on $\mathbf{B}^n$ which even are locally constant. For this reason that naive definition cannot lead to satisfying geometric properties. (But it has its use and importance in non-archimedean measure theory. Usually it is qualified as "locally analytic" in contrast to "rigid analytic".)

Reprinted from 'Galois Representations in Arithmetic Algebraic Geometry',
edited by A. J. Scholl & R. L. Taylor. ©Cambridge University Press 1998

Going somehow to the other extreme let

$$f(T_1, \ldots, T_n) = \sum_{\nu_i \geq 0} a_{\nu_1, \ldots, \nu_n} T_1^{\nu_1} \cdot \ldots \cdot T_n^{\nu_n}$$

be an arbitrary formal power series with coefficients in K. The following two properties are quite immediate:

* The power series f converges on all of $\mathbf{B}^n$ if and only if the coefficients tend to zero, i.e., $|a_{\nu_1, \ldots, \nu_n}| \longrightarrow 0$ if $\nu_1 + \ldots + \nu_n \longrightarrow \infty$.

* If f converges on $\mathbf{B}^n$ then we have $f(\mathbf{B}^n(L)) \subseteq L$ for any intermediate field $K \subseteq L \subseteq \widehat{\overline{K}}$ which is finite over K; here $\mathbf{B}^n(L)$ denotes the set of those vectors in $\mathbf{B}^n$ with coordinates in L.

The subalgebra

$$T^n := \{f \in K[[T_1, \ldots, T_n]] \; : \; f \text{ converges on } \mathbf{B}^n\}$$

of the algebra $K[[T_1, \ldots, T_n]]$ of formal power series over K is called a **Tate algebra**. We may and will say that any $f \in T^n$ induces an "analytic function on $\mathbf{B}^n$ defined over K". Why is this a good notion? At first glance it does not seem to be local at all! Certainly we do not want to give up completely the possibility of recognizing the analyticity of a function locally. In order to prepare the way out of this apparent trap we first collect a number of properties of the algebra T^n. The two most basic ones are the following:

1) T^n is a K-Banach algebra w.r.t. the multiplicative norm

$$|f| := \max |a_{\nu_1, \ldots, \nu_n}| \; .$$

2) The Maximum Modulus Principle holds:

$$|f| = \max_{z \in \mathbf{B}^n} |f(z)| \; ;$$

in particular: If $f(z) = 0$ for any $z \in \mathbf{B}^n$ then $f = 0$.

The proof of the Maximum Modulus Principle is actually very easy: By scaling we may assume that $|f| = 1$. We then can reduce f modulo the maximal ideal of K obtaining, because of the convergence criterion, a nonzero polynomial $\bar{f}$ over the residue class field of K. Since the residue field of $\widehat{\overline{K}}$ is infinite we find a point $\bar{z}$ with coordinates in the latter such that $\bar{f}(\bar{z}) \neq 0$. Any lifting $z \in \mathbf{B}^n$ of $\bar{z}$ then satisfies $|f(z)| = 1$.

Next one shows that the Weierstrass theory (preparation theorem, ...) works for T^n. This eventually leads to many ring theoretic properties of T^n:

3) T^n is noetherian and factorial.

4) T^n is Jacobson, i.e., for any ideal $\mathfrak{a} \subseteq T^n$ its radical ideal $\sqrt{\mathfrak{a}}$ is the intersection of all the maximal ideals containing $\mathfrak{a}$.

5) Any ideal in T^n is closed.

6) For any maximal ideal $\mathfrak{m}$ in T^n, the residue field $T^n/\mathfrak{m}$ is a finite extension of K.

This last property is an analogue of Hilbert's Nullstellensatz. It has the interesting consequence that the map

$$\text{Galois orbits in } \mathbf{B}^n(\overline{K}) \xrightarrow{\sim} \text{Max}(T^n)$$
$$z \longmapsto \mathfrak{m}_z := \{f : f(z) = 0\}$$

is a bijection. The inverse map is obtained as follows: For a maximal ideal $\mathfrak{m}$ let φ denote the composite of the projection $T^n \longrightarrow\!\!\!\!\rightarrow T^n/\mathfrak{m}$ and some embedding $T^n/\mathfrak{m} \hookrightarrow \overline{K}$ and put $z_i := \varphi(T_i)$.

In this way the maximal ideal spectrum $\text{Max}(T^n)$ appears as an algebraically defined model for the space $\mathbf{B}^n$. This suggests we should proceed as Grothendieck did in algebraic geometry and define a category of analytic spaces as maximal ideal spectra of certain algebras. (The property 4) is the reason that maximal ideals in contrast to arbitrary prime ideals will suffice.)

Definition:

A K-algebra A is called __affinoid__ if $A \cong T^n/\mathfrak{a}$ for some $n \in \mathbf{N}$ and some ideal $\mathfrak{a}$.

For any affinoid algebra A we have:

* A is noetherian and Jacobson (by 3) and 4)).

* A is a K-Banach algebra with respect to any residue norm (by 5)). Moreover:

    – The topology on A is independent of the chosen residue norm;

    – any homomorphism between affinoid K-algebras is automatically continuous.

We put
$$\text{Max}(A) := \text{set of all maximal ideals of } A \ .$$

By 6), this set depends functorially on A. Also by 6) we may define the so-called supremum or spectral seminorm on A by

$$|f|_{\text{sup}} := \sup_{x \in \text{Max}(A)} |f(x)|$$

where $f(x) := f + \mathfrak{m}_x \in A/\mathfrak{m}_x \hookrightarrow \overline{K}$. It is obviously bounded above by any residue norm. The general **Maximum Modulus Principle** says that

$$|f|_{\mathrm{sup}} = \max_{x \in \mathrm{Max}(A)} |f(x)| \ .$$

If A is reduced then $|\ |_{\mathrm{sup}}$ is a norm which is equivalent to any residue norm. Using the above description of $\mathrm{Max}(T^n)$ as the Galois orbits in $\mathbf{B}^n(\overline{K})$ we obtain (from the metric topology on $\overline{K}$) a "canonical" Hausdorff topology on $\mathrm{Max}(A)$. Of course it is totally disconnected so that our initial problem persists.

In order to emphasize the geometric intuition we write $X := \mathrm{Max}(A)$ from now on. For any functions $g, f_1, \ldots, f_m \in A$ without common zero (i.e., generating the unit ideal $\langle g, f_i \rangle = A$) we introduce the open subset

$$X(\frac{f_\cdot}{g}) := \{x \in X : \max_i |f_i(x)| \le |g(x)|\}$$

of X called a **rational subdomain.** It is not hard to see that the rational subdomains form a basis for the canonical topology on X.

Very important observation:

The K-algebra $A\langle\frac{f_\cdot}{g}\rangle := A\langle T_1, \ldots, T_m\rangle / \langle gT_i - f_i\rangle$ is affinoid and the map

$$\mathrm{Max}(A\langle\frac{f_\cdot}{g}\rangle) \ \longrightarrow \ \mathrm{Max}(A) \ = \ X$$

induced by the obvious algebra homomorphism $A \longrightarrow A\langle\frac{f_\cdot}{g}\rangle$ is a homeomorphism onto $X(\frac{f_\cdot}{g})$.

Comments: $- A\langle T_1, \ldots, T_m\rangle$ is the algebra of all power series in the variables $T_1, \ldots, T_m$ with coefficients in A tending towards 0.

$-$ The affinoid algebra $A\langle\frac{f_\cdot}{g}\rangle$ can be characterized by a universal property which is given solely in terms of the rational subdomain $X(\frac{f_\cdot}{g})$.

$-$ In $A\langle\frac{f_\cdot}{g}\rangle$ we have $|\frac{f_i}{g}| = |$ residue class of $T_i| \le 1$.

$-$ The assertion becomes wrong without the assumption that $\langle g, f_i \rangle$ is the unit ideal (look at $A = K\langle T\rangle$, $m = 1$, $g = T$, and $f_1 = 0$).

This observation allows to define a presheaf $\mathcal{O}_X$ at least on the rational subdomains in X by

$$\mathcal{O}_X(X\langle\frac{f_\cdot}{g}\rangle) := A\langle\frac{f_\cdot}{g}\rangle \ .$$

Main Theorem of Tate:

If $Y, Y_1, \ldots, Y_r \subseteq X$ *are rational subdomains such that* $Y = Y_1 \cup \ldots \cup Y_r$ *then* $\mathcal{O}_X$ *satisfies the sheaf property for that covering, i.e.,*

$$0 \longrightarrow \mathcal{O}_X(Y) \longrightarrow \prod_i \mathcal{O}_X(Y_i) \overset{\longrightarrow}{\underset{\longrightarrow}{}} \prod_{i,j} \mathcal{O}_X(Y_i \cap Y_j)$$

is exact.

This means that the notion of an analytic function on X, i.e., an element of A, is local as far as <u>finite</u> coverings by <u>rational subdomains</u> are concerned! This picture can be "enlarged" by a completely formal construction:

* A subset $U \subseteq X$ is called **admissible open** if there are rational subdomains $U_i \subseteq X$ for $i \in I$ such that

    i. $U = \bigcup_{i \in I} U_i$ (in particular U is open in the canonical top.) and

    ii. for any map $\alpha : Y := \mathrm{Max}(B) \longrightarrow X = \mathrm{Max}(A)$ induced by a homomorphism of affinoid K-algebras $A \longrightarrow B$ with $\mathrm{im}(\alpha) \subseteq U$ the covering $Y = \bigcup_{i \in I} \alpha^{-1}(U_i)$ has a finite subcovering.

* Let V and V_j, for $j \in J$, be admissible open subsets of X such that $V = \bigcup_{j \in J} V_j$; this covering of V is called **admissible** if for any map $\alpha : Y \longrightarrow X$ as above with $\mathrm{im}(\alpha) \subseteq V$ the covering $Y = \bigcup_{j \in J} \alpha^{-1}(V_j)$ can be refined into a finite covering by rational subdomains.

These notions define a Grothendieck topology on X (which is considerably coarser than the canonical topology). The presheaf $\mathcal{O}_X$ extends in a purely formal way (once one knows Tate's theorem) to a sheaf on X with respect to this Grothendieck topology; it is called the structure sheaf of X.

Definition:

The triple $\mathrm{Sp}(A) := (X, \text{ Grothendieck topology}, \mathcal{O}_X)$ *is called an* <u>*affinoid*</u> *variety over* K.

Fact:

Any homomorphism of affinoid K-*algebras* $A \longrightarrow B$ *induces a "morphism of affinoid varieties"* $\mathrm{Sp}(B) \longrightarrow \mathrm{Sp}(A)$.

In order to illustrate these concepts let us go back to the 1-dimensional disk $X = \mathrm{Max}(K\langle T \rangle) = $ "$\{z \in \widehat{\overline{K}} : |z| \leq 1\}$" and look at the simplest example. By construction we have $\mathcal{O}_X(X) = K\langle T \rangle$. Clearly the "unit circle" $V := \{x \in X : |T(x)| = 1\} = X(\frac{1}{T}) = $ "$\{z \in \widehat{\overline{K}} : |z| = 1\}$" is a rational subdomain with

$$\mathcal{O}_X(V) \quad = K\langle T\rangle\langle T'\rangle/\langle TT' - 1\rangle \ = \ K\langle T, T^{-1}\rangle$$

$$:= \{\textstyle\sum_{\nu \in \mathbf{Z}} a_\nu T^\nu \ : \ |a_\nu| \longrightarrow 0 \text{ if } |\nu| \longrightarrow \infty\} \ .$$

We now look at the "open unit disk"

$$U := X \setminus V = \{x \in X : |T(x)| < 1\} \ = \ "\{z \in \widehat{\overline{K}} : |z| < 1\}"$$

and claim that U is admissible open in X. Choose an $\varepsilon = |\pi| \in |K^\times|$ with $0 < \varepsilon < 1$ and put

$$U_n := X(\frac{T^n}{\pi}) \ = \ \{x \in X : |T(x)| \le \varepsilon^{1/n}\} \ = \ "\{z \in \widehat{\overline{K}} : |z| \le \varepsilon^{1/n}\}"$$

for $n \in \mathbf{N}$. These are rational subdomains of X such that $U = \bigcup_{n \in \mathbf{N}} U_n$. Let $\alpha : Y = \mathrm{Max}(B) \longrightarrow X$ be a morphism of affinoid varieties such that $\mathrm{im}(\alpha) \subseteq U$. The Maximum Modulus Principle implies that

$$|\alpha^*(T)|_{\mathrm{sup}} = \max_{y \in Y} |\alpha^*(T)(y)| = \max_{y \in Y} |T(\alpha(y))| < 1 \ .$$

It follows that $\alpha^{-1}(U_n) = Y$ for any sufficiently large n.

This fact does not contradict our geometric intuition that the unit disk is a "connected" space. The point is that $X = U \cup V$ is **not** an admissible covering!

Starting from the affinoid varieties as building blocks one constructs general rigid varieties by the usual gluing procedure.

Definition:

A rigid K-analytic variety is a set X equipped with a Grothendieck topology (consisting of subsets) and a sheaf of K-algebras $\mathcal{O}_X$ such that there is an admissible covering $X = \bigcup U_i$ where each $(U_i, \mathcal{O}_X|U_i)$ is isomorphic to an affinoid variety over K. $^{i \in I}$

A first rather trivial source of examples are admissible open subsets $U \subseteq X = \mathrm{Sp}(A)$; for them $(U, \mathcal{O}_X|U)$ is a rigid K-analytic variety. A much more important source constitute the algebraic varieties over K. There is a natural functor

$$K\text{-schemes locally of finite type} \quad \longrightarrow \quad \text{rigid } K\text{-analytic varieties}$$
$$X \quad \longmapsto \quad X^{\mathrm{an}}$$

together with a natural morphism of locally G-ringed spaces

$$an_X \; : \; X^{\mathrm{an}} \longrightarrow X$$

which has the universal property that any morphism of locally G-ringed spaces $Y \longrightarrow X$ where Y is a rigid K-analytic variety factorizes through X^{an}, i.e., we have a commutative triangle

$$
\begin{array}{ccc}
Y & \dashrightarrow & X^{\mathrm{an}} \\
& \searrow \quad \swarrow an_X & \\
& X &
\end{array}
$$

By a locally G-ringed space we mean a set equipped with a Grothendieck topology consisting of subsets and a sheaf of K-algebras whose stalks are local rings.

In order to demonstrate the existence of X^{an} it suffices, by a gluing argument (made possible by the universal property), to consider the case of an affine K-scheme of finite type $X = \mathrm{Spec}(A)$. As a set we put $X^{\mathrm{an}} := \mathrm{Max}(A)$. To define the analytic structure we fix a representation of A as a quotient

$$A = K[T_1, \ldots, T_d]/\mathfrak{a}$$

of some polynomial algebra and we fix a $c \in K$ with $|c| > 1$. Put

$$U_n := \{x \in X^{\mathrm{an}} \; : \; \max_j |T_j(x)| \le |c|^n\} \quad \text{for } n \ge 0$$

so that

$$X^{\mathrm{an}} = \bigcup_{n \ge 0} U_n \; .$$

We define affinoid K-algebras

$$A_n := K\langle T_1, \ldots, T_d\rangle / \langle P(c^{-n}T_j) \; : \; P \in \mathfrak{a}\rangle \; .$$

From the commutative diagram

$$
\begin{array}{ccccc}
\mathrm{Max}(A_n) & \xrightarrow[c^{-n}T_j \leftarrow T_j]{\sim} & U_n & \subseteq & \text{ball of radius } |c|^n \\
\Big\downarrow{\scriptstyle cT_j} \uparrow {\scriptstyle} \Big\downarrow{\scriptstyle T_j} & & & & \Big\downarrow{\scriptstyle \subseteq} \\
\mathrm{Max}(A_{n+1}) & \xrightarrow[c^{-(n+1)}T_j \leftarrow T_j]{\sim} & U_{n+1} & \subseteq & \text{ball of radius } |c|^{n+1} \\
& & & & \Big\downarrow{\scriptstyle \subseteq} \\
& & X^{\mathrm{an}} & \subseteq & \mathbf{A}^d
\end{array}
$$

it is quite clear that X^{an} has a unique rigid K-analytic structure such that $X^{\mathrm{an}} = \bigcup_n U_n$ is an admissible covering by the affinoid open subsets $U_n \cong \mathrm{Sp}(A_n)$; the morphism $an_X : X^{\mathrm{an}} = \mathrm{Max}(A) \longrightarrow X = \mathrm{Spec}(A)$ is the "inclusion map".

It holds quite generally that X^{an} as a set consists of the closed points of the scheme X.

A quite important feature of a K-affinoid variety $X = \mathrm{Sp}(A)$ is its reduction. In order to describe this construction we need to introduce the valuation ring o in K as well as its residue field k. In A we have the o-subalgebra

$$\overset{\circ}{A} := \{f \in A : |f|_{\mathrm{sup}} \leq 1\} \; ;$$

it contains the ideal

$$\check{A} := \{f \in A : |f|_{\mathrm{sup}} < 1\} \; .$$

The k-algebra

$$\tilde{A} := \overset{\circ}{A}/\check{A}$$

is finitely generated and reduced (the latter since the spectral seminorm is power-multiplicative).

Definition:

The affine k-scheme $\tilde{X} := \mathrm{Spec}(\tilde{A})$ is called the <u>canonical reduction</u> of the affinoid variety X.

This notion obviously is functorial in A. On the level of sets one has the **reduction map**

$$\begin{aligned} red_X : X &\longrightarrow \mathrm{Max}(\tilde{A}) \subseteq \tilde{X} \\ x &\longmapsto \ker(\tilde{A} \to (A/\mathfrak{m}_x)^\sim) = \{f \in \overset{\circ}{A} : |f(x)| < 1\}/\check{A} \; . \end{aligned}$$

It is helpful to realize that $(A/\mathfrak{m}_x)^\sim$ is a finite field extension of k.

Fact: *red_X is surjective.*

We claim that red_X is continuous in the sense that the preimage of any Zariski open subset is admissible open. In order to see this let $f \in \overset{\circ}{A}$ with $|f|_{\mathrm{sup}} = 1$ and let $\tilde{f}$ denote its residue class in $\tilde{A}$. Then $\tilde{f} \notin red_X(x)$ if and only if $|f(x)| \geq 1$. Hence we have

$$red_X^{-1}(\mathrm{Max}(\tilde{A}[\tilde{f}^{-1}])) = X(\frac{1}{f}) \; .$$

Note that also the preimage $red_X^{-1}(V(\tilde{f})) = \{x \in X : |f(x)| < 1\}$ of the Zariski closed zero set $V(\tilde{f})$ of $\tilde{f}$ is admissible open (but i.g. not affinoid).

So far I have described the "classical" approach to rigid analytic geometry which was invented by Tate. More details and full proofs for everything which was said can be found in the book [BGR]. Later on Raynaud saw that rigid geometry can be developed entirely within the framework of formal algebraic geometry. Because of the conceptual as well as technical importance of this approach I want to finish by explaining Raynaud's point of view ([R] or [BL]).

For simplicity we assume that $| \ |$ is a discrete valuation on K. As before o denotes the ring of integers in K. We fix a prime element π in o. Let $o\{\{T_1, \ldots, T_n\}\}$ be the ring of restricted formal power series over o; recall that a formal power series over o is restricted if, for any given $m \geq 1$, almost all its coefficients lie in $\pi^m o$. An o-algebra $\mathcal{A}$ of the form $\mathcal{A} = o\{\{T_1, \ldots, T_n\}\}/\mathfrak{a}$ is called **topologically of finite type**. It is a $\pi\mathcal{A}$-adically complete topological ring and gives rise to the affine formal scheme $\mathrm{Spf}(\mathcal{A})$ over o which is the set of all open prime ideals of $\mathcal{A}$ equipped with the Zariski topology and a certain structure sheaf constructed by localization and completion. Roughly speaking one has

$$\mathrm{Spf}(\mathcal{A}) \ = \ \text{``}\varinjlim_{m}\text{''} \ \mathrm{Spec}(\mathcal{A}/\pi^m \mathcal{A}) \ .$$

The first basic observation is that, for an o-algebra $\mathcal{A}$ topologically of finite type, the tensor product $A := \mathcal{A} \underset{o}{\otimes} K$ is an affinoid K-algebra. The affinoid variety $\mathrm{Sp}(A)$ is called the **general fibre** of the formal scheme $\mathrm{Spf}(\mathcal{A})$. Can we describe the datum $\mathrm{Sp}(A)$ directly in terms of the algebra $\mathcal{A}$?

The set $\mathrm{Max}(A)$:

Consider any prime ideal $\mathfrak{p} \subseteq \mathcal{A}$ such that

1. $\mathfrak{p}$ is not open in $\mathcal{A}$, and

2. $\mathcal{A}/\mathfrak{p}$ is a finitely generated o-module.

We claim that $\mathfrak{p} \otimes K$ is a maximal ideal in A. Condition 1) ensures that the obvious map $o \hookrightarrow \mathcal{A}/\mathfrak{p}$ is injective. Condition 2) implies that $\mathcal{A}/\mathfrak{p}$ is an integral domain finite over o. It follows that $\mathcal{A}/\mathfrak{p}$ is a local ring which is finite and flat over o. Hence $(\mathcal{A}/\mathfrak{p}) \otimes K = A/(\mathfrak{p} \otimes K)$ is a finite field extension of K.

In this way one obtains a bijection

$$\{\mathfrak{p} \subseteq \mathcal{A} \text{ prime ideal with 1), 2)} \ \} \ \overset{\sim}{\longrightarrow} \ \mathrm{Max}(A)$$
$$\mathfrak{p} \ \longmapsto \ \mathfrak{p} \otimes K \ .$$

The rational subdomains (i.e., the Grothendieck topology on $\mathrm{Max}(A)$):

Consider any rational subdomain $X(\frac{f}{g}) \subseteq X := \text{Max}(A)$. There is no loss of generality in assuming that $g, f_1, \ldots, f_r \in \mathcal{A}$. That $g, f_1, \ldots, f_r$ have no common zero means that the ideal $I := \mathcal{A}g + \mathcal{A}f_1 + \ldots + \mathcal{A}f_r$ is open in $\mathcal{A}$. On the other hand, for any open ideal $I \subseteq \mathcal{A}$ and any element $g \in I$, there is a universal construction called **formal blowing-up** of a homomorphism $\mathcal{A} \longrightarrow \mathcal{A}_{I,g}$ of o-algebras topologically of finite type which is universal with respect to making I into a principal ideal generated by g. The morphism of formal schemes $\text{Spf}(\mathcal{A}_{I,g}) \longrightarrow \text{Spf}(\mathcal{A})$ induces in the general fibre the inclusion $X(\frac{f}{g}) \subseteq X$.

The structure sheaf:

As is more or less clear from the above description of the rational subdomains of $X = \text{Sp}(A)$ the structure sheaf $\mathcal{O}_X$ can be reconstructed from the structure sheaves of all the formal blowing-ups of the formal scheme $\text{Spf}(\mathcal{A})$.

Theorem of Raynaud:

The above construction of the "general fibre" induces an equivalence of categories between

   the category of all formal flat o-schemes topologically of finite type in which all formal blowing-ups are inverted

and

   the category of all quasi-compact and quasi-separated rigid K-analytic varieties.

References

[BGR] S. Bosch, U. Güntzer, R. Remmert, Non-Archimedean Analysis, Berlin-Heidelberg-New York 1984
[BL]   S. Bosch, W. Lütkebohmert, Formal and rigid geometry I. Rigid spaces, Math. Ann. 295, 291-317 (1993)
[R]   M. Raynaud, Géométrie analytique rigide d'après Tate, Kiehl, ... Table ronde d'analyse non archimedienne, Bull. Soc. Math. France Mém. 39/40, 319-327 (1974)

MATHEMATISCHES INSTITUT, WESTFÄLISCHE WILHELMS-UNIVERSITÄT, EINSTEINSTR. 62, D-48149 MÜNSTER, GERMANY
pschnei@math.uni-muenster.de
http://www.uni-muenster.de/math/u/schneider

An introduction to Kato's Euler systems

A. J. SCHOLL

to Bryan Birch

Contents

Reprinted from 'Galois Representations in Arithmetic Algebraic Geometry',
edited by A. J. Scholl & R. L. Taylor. ©Cambridge University Press 1998

Introduction

In the conference there was a series of talks devoted to Kato's work on the
Iwasawa theory of Galois representations attached to modular forms. The
present notes are mainly devoted to explaining the key ingredient, which is
the Euler system constructed by Kato, first in the K_2-groups of modular
curves, and then using the Chern class map, in Galois cohomology. This
material is based mainly on the talks given by Kato and the author at the
symposium, as well as a series of lectures by Kato in Cambridge in 1993. In
a companion paper [29] Rubin explains how, given enough information about
an Euler system, one can prove very general finiteness theorems for Selmer
groups whenever the appropriate L-function is non-zero (see §8 of his paper
for precise results for elliptic curves).

Partly because of space, and partly because of the author's lack of understanding, the scope of these notes is limited. There are two particular
restrictions. First, we only prove the key reciprocity law (Theorem 3.2.3 below), which allows one to compute the image of the Euler system under the
dual exponential map, in the case of a prime p of good reduction (actually, for
stupid reasons explained at the end of §2.1, we also must assume p is odd).
Secondly, we say nothing about the case of Galois representations attached
to forms of weight greater than 2. For the most general results, the reader
will need to consult the preprint [17] and Kato's future papers.

Kato's K_2 Euler system has its origins in the work of Beilinson [1] (see
also [30] for a beginner's treatment). Beilinson used cup-products of modular
units to construct elements of K_2 of modular curves. He was able to compute
the regulators of these elements by the Rankin-Selberg method and relate
them to the L-function of the modular curve at $s = 2$, in partial confirmation
of his general conjectures [1; 27] relating regulators and values of L-functions.

Kato discovered that, by using explicit modular units, one obtained norm-
compatible families of elements of K_2. These modular units are the values,
at torsion points, of what are called here *Kato-Siegel functions*. These are
canonical (no indeterminate constant) functions on an elliptic curve (over
any base scheme) with prescribed divisors, which are norm-compatible with
respect to isogenies. Such functions were, over $\mathbb{C}$, first discovered by Siegel
— the associated modular units were studied in depth by Kubert and Lang
[19]. Over $\mathbb{C}$ generalisations of these functions were found by Robert [28]. It

was Kato [16] who first found their elegant algebraic characterisation. In §1 I have given an "arithmetic" modular construction of these functions, which is more complicated than Kato's but at least reveals the key fact behind their existence — namely, the triviality of the 12th power of the sheaf ω on the modular stack. (The Picard group of the modular stack was computed by Mumford [25] many years ago.)

In §2 we turn to K-theory, and give a fairly general construction of the Euler system in K_2 of modular curves, and the norm relations. It is relatively formal to pass from this to an Euler system in Galois cohomology of (say) a modular elliptic curve. The hard part is to show that the cohomology classes one gets are non-trivial if the appropriate L-value is nonzero. In his 1993 Cambridge lectures, Kato explained how this can be regarded as a consequence of a huge generalisation of the explicit reciprocity laws (Artin, Hasse, Iwasawa, Wiles ...) to local fields with imperfect residue field. This is the subject of the preprint [17]. At the Durham conference he sketched a slightly different proof, using the Fontaine-Hyodo-Faltings approach to p-adic Hodge theory. In §3 we give a stripped-down proof of a weak version of one of the reciprocity laws in [17] in the case of good reduction, using a minimal amount of p-adic Hodge theory.

In §4 we explain how Kato uses the Rankin-Selberg integral (very much as Beilinson did) to compute the projection of the the image of the dual exponential into a Hecke eigenspace. Finally in §5 we tie everything together for a modular elliptic curve.

The appendix to §2 (which is the author's only original contribution to this work) is an attempt to extend Kato's methods to other situations. We construct an Euler system in the higher K-groups of (a suitable open part of) Kuga-Sato varieties. This is a precise version of the construction used in [32] (see also §5 of [9] for a summary) to relate archimedean regulators of modular form motives and L-functions. The p-adic applications of these elements remain to be found.

I have many people to thank for their help in the preparation of this paper. Particular mention is due to Jan Nekovář. He encouraged me to think about norm relations in 1994, although in the end that work was overtaken by events, and all that remains of it is the appendix to §2. It is only because of his insistence that §§3–5 exist at all, and his careful reading of much of the paper eliminated many errors (although he is not to be held responsible for those that remain). I am also grateful to Amnon Besser, Spencer Bloch, Kevin Buzzard, John Coates, Ofer Gabber, Henri Gillet, Erasmus Landvogt and Christophe Soulé for useful discussions. Karl Rubin read the original draft of the manuscript and made invaluable suggestions. Above all, it is a great pleasure to thank Kato, the creator of this beautiful and powerful mathematics, for encouraging me to publish this account of his work and for

382 A. J. Scholl

pointing out some blunders in an earlier draft.

This paper was begun while the author was visiting the University of Münster in winter 1996 as a guest of Christopher Deninger, and completed during a stay at the Isaac Newton Institute in 1998. It is a pleasure to thank them for their hospitality.

Notation

If G is a commutative group (or group scheme) and $n \in \mathbb{Z}$ then $[\times n]_G \colon G \to G$ is the endomorphism "multiplication by n", written simply $[\times n]$ if no confusion can occur. We also write $_nG$ and G/n for the kernel and cokernel of $[\times n]$, respectively.

Throughout this paper we use the geometric Frobenius, and normalise the reciprocity laws of class field theory accordingly (see §3.1 below for precise conventions, as well as the remarks following Theorem 5.2.1).

The symbol "=" is used to denote equality or canonical isomorphism. We use the usual notation ":=" to indicate that the right-hand expression is the definition of that on the left (and "=:" for the reflected relation).

1 Kato-Siegel functions and modular units

1.1 Review of modular forms and elliptic curves

We review some well-known facts about the moduli of elliptic curves. See for example [7; 8; 18, Chapter 2]. For any elliptic curve $f \colon E \to S$, with zero-section e, we have the standard invertible sheaf

$$\omega_{E/S} := f_*\Omega^1_{E/S} = e^*\Omega^1_{E/S}.$$

From the second description (as the conormal bundle of the zero-section of E/S) we have the isomorphism $\omega_{E/S} = e^*\mathcal{O}_E(-e)$. Because $\Omega^1_{E/S}$ is free along the fibres of f, in fact $\omega_{E/S} = x^*\Omega^1_{E/S}$ for any section $x \in E(S)$.

The formation of $\omega_{E/S}$ is compatible with basechange — in fancy language, ω is a sheaf on the *modular stack* $\mathcal{M}$ of elliptic curves.

A (meromorphic) modular form of weight k is a rule which assigns to each E/S a section of $\omega_{E/S}^{\otimes k}$, compatible with basechange. By definition this is the same as an element of $\Gamma(\mathcal{M}, \omega^{\otimes k})$. The discriminant $\Delta(E/S)$ is a nowhere-vanishing section of $\omega_{E/S}^{\otimes 12}$ compatible with basechange, and it defines an invertible modular form Δ of weight 12. From this it follows in particular that

- The set of nowhere-vanishing sections of $\omega^{\otimes 12d}$ is $\{\pm\Delta^d\}$, for any integer d.

Let $N \geq 1$ be an integer. The modular stack $\mathcal{M}_{\Gamma_0(N)}$ classifies pairs $(E/S, \alpha)$ where $\alpha \colon E \to E'$ is a cyclic isogeny of degree N of elliptic curves over S. (When N is not invertible on S the definition of cyclic can be found in [18, §3.4].) The functor $(E, \alpha) \mapsto E$ defines a morphism $c \colon \mathcal{M}_{\Gamma_0(N)} \to \mathcal{M}$. A (meromorphic) modular form on $\Gamma_0(N)$ of weight k is a section of $c^* \omega^{\otimes k}$ over $\mathcal{M}_{\Gamma_0(N)}$. Equivalently, it is a rule which associates to each cyclic N-isogeny $\alpha \colon E \to E'$ of elliptic curves over S a section of $\omega_{E/S}^{\otimes k}$, compatible with arbitrary basechange $S' \to S$. As well as Δ, one has the modular form $\Delta^{(N)}$ of weight 12, defined by

$$\Delta^{(N)}(E \xrightarrow{\alpha} E') = \alpha^* \Delta(E').$$

It is invertible exactly where α is étale. In particular, it is invertible on $S \otimes \mathbb{Z}[1/N]$.

Suppose $N = p$ is prime. The reduction of $\mathcal{M}_{\Gamma_0(p)}$ mod p has two irreducible components, one of which parameterises pairs $(E/S, \alpha)$ where α is Frobenius, and the other those pairs where α is Verschiebung. On the first component $\Delta^{(p)}$ vanishes, and on the second it does not.

Let m be the denominator of $(p - 1)/12$. Then $\Delta^{(p)} \cdot \Delta^{-1}$ is the mth power of a modular function $u_p \in \Gamma(\mathcal{M}_{\Gamma_0(p)}, \mathcal{O})$, which is invertible away from characteristic p by the previous remarks. It is a classical fact [26] that

$$\Gamma(\mathcal{M}_{\Gamma_0(p)} \otimes \mathbb{Q}, \mathcal{O}^*) = \langle \mathbb{Q}^*, u_p \rangle.$$

and therefore

$$\Gamma(\mathcal{M}_{\Gamma_0(p)}, \mathcal{O}^*) = \{\pm 1\}.$$

Recall the Kodaira-Spencer map (see e.g. [18, 10.13.10]); if E/S is an elliptic curve and S is smooth over T, one has an $\mathcal{O}_S$-linear map

$$KS = KS_{E/S} \colon \omega_{E/S}^{\otimes 2} \to \Omega_{S/T}^1.$$

If $T = \operatorname{Spec} \mathbb{Q}$ and E/S is the universal elliptic curve over the modular curve $Y(N)$, $N \geq 3$ (the definition is recalled in §2.2 below), then KS is an isomorphism.

If $S \hookrightarrow \overline{S}$, $E \hookrightarrow \overline{E}$ is an extension of E to a curve $\overline{E}/\overline{S}$ of genus 1 (not necessarily smooth), and the identity section $e \in E(S)$ extends to a a a section $e \colon \overline{S} \to \overline{E}$ whose image is contained in the smooth part, then $\omega_{\overline{E}/\overline{S}} := e^* \Omega_{\overline{E}/\overline{S}}$ is an invertible sheaf on $\overline{S}$ extending $\omega_{E/S}$. If $\overline{S}$ is smooth over the base scheme T, and S is the complement in $\overline{S}$ of a divisor S^∞ with relative normal crossings, the Kodaira-Spencer map extends to a homomorphism

$$KS_{\overline{E}/\overline{S}} \colon \omega_{\overline{E}/\overline{S}}^{\otimes 2} \to \Omega_{\overline{S}/T}^1(\log S^\infty). \tag{1.1.1}$$

If $\overline{S} = X(N)_{/\mathbb{Q}}$ for $N \geq 3$ and $\overline{E}$ is the regular minimal model of the universal elliptic curve, then (1.1.1) is an isomorphism.

1.2   Kato-Siegel functions

If $\mathfrak{D}$ is a principal divisor on an elliptic curve over (say) a field, there is in general no 'canonical' function with divisor $\mathfrak{D}$. For certain special divisors, such canonical functions do exist. In their analytic construction they have been used extensively in the theory of elliptic units. Kato observed that they have a completely algebraic characterisation. Here we give a slightly more general, modular, description of such a class of functions.

Theorem 1.2.1. *Let D be an integer with $(6, D) = 1$. There is one and only one rule ϑ_D which associates to each elliptic curve $E \to S$ over an arbitrary base a section $\vartheta_D^{(E/S)} \in \mathcal{O}^*(E - \ker[\times D])$ such that:—*

(i) *as a rational function on E, $\vartheta_D^{(E/S)}$ has divisor $D^2(e) - \ker[\times D]$;*

(ii) *if $S' \to S$ is any morphism, and $g\colon E' = E \times_S S' \to E$ is the basechange, then $g^* \vartheta_D^{(E/S)} = \vartheta_D^{(E'/S')}$;*

(iii) *if $\alpha\colon E \to E'$ is an isogeny of elliptic curves over a connected base S whose degree is prime to D, then*

$$\alpha_* \vartheta_D^{(E/S)} = \vartheta_D^{(E'/S)}$$

(iv) *$\vartheta_{-D} = \vartheta_D$ and $\vartheta_1 = 1$. If $D = MC$ with $M, C \geq 1$ then*

$$[\times M]_* \vartheta_D = \vartheta_C^{M^2} \text{ and } \vartheta_C \circ [\times M] = \vartheta_D / \vartheta_M^{C^2}.$$

In particular, $[\times D]_* \vartheta_D = 1$.

(v) *if $\tau \in \mathbb{C}$ with $\operatorname{Im}(\tau) > 0$ and $E_\tau/\mathbb{C}$ is the elliptic curve whose points are $\mathbb{C}/\mathbb{Z} + \tau \mathbb{Z}$, then $\vartheta_D^{(E_\tau/\mathbb{C})}$ is the function*

$$(-1)^{\frac{D-1}{2}} \Theta(u, \tau)^{D^2} \Theta(Du, \tau)^{-1}$$

where

$$\Theta(u, \tau) = q^{\frac{1}{12}} (t^{\frac{1}{2}} - t^{-\frac{1}{2}}) \prod_{n>0} (1 - q^n t)(1 - q^n t^{-1})$$

and $q = e^{2\pi i \tau}$, $t = e^{2\pi i u}$.

Remarks. (i) We do not require that D be invertible on S.

(ii) Locally for the Zariski topology, any elliptic curve may be obtained by basechange from an elliptic curve over a reduced base. It is therefore enough to restrict to reduced base schemes S.

(iii) Properties (i) and (iii) alone already determine $\vartheta_D^{(E/S)}$ uniquely; any other function with the same divisor is of the form $u\vartheta$, for some $u \in \mathcal{O}^*(S)$, and applying (ii) for the isogenies $[\times 2]$, $[\times 3]$ would give $u^4 = u = u^9$, whence $u = 1$.

(iv) In down-to-earth terms, if $S = \operatorname{Spec} k$ for an algebraically closed field k then for a separable isogeny $\alpha\colon E \to E'$, the property (iii) is just the *distribution relation*

$$\prod_{\substack{x\in E(k) \\ \alpha(x)=y}} \vartheta_D^{(E/k)}(x) = \vartheta_D^{(E'/k)}(y), \quad \text{for any } y \in E'(k).$$

(v) Over $\mathbb{C}$ this theorem was obtained by Robert [28], who proves rather more: he shows that for any elliptic curve $E/\mathbb{C}$ and any finite subgroup $P \subset E$ of order prime to 6, there is a certain canonical function with divisor $\#P(e) - P$ and properties generalising those of ϑ_D. One can prove his more general result in a manner similar to the proof of 1.2.1; in place of the modular form Δ^{D^2-1} one should use $\Delta(E)^{\#P}/\beta^*\Delta(E/P)$, where $\beta\colon E \to E/P$ is the quotient map.

Proof. We begin with the first two conditions. First observe that if S is a spectrum of a field, then the divisor $\ker[\times D] - D^2(e)$ is principal (because D is odd, the sum of $\ker[\times D] - D^2(e)$ in the Jacobian is zero). To give a rule ϑ_D satisfying (i) and (ii) is equivalent to giving, for any elliptic curve E/S, an isomorphism of line bundles on E

$$\mathcal{O}_E(\ker[\times D]) \xrightarrow{\sim} \mathcal{O}_E(D^2 e) \tag{1.2.2}$$

compatible with basechange. We have just observed that the line bundles are isomorphic when restricted to any fibre of E/S. Since we can assume (by remark (ii) above) that S is reduced, the seesaw theorem tells us that to give an isomorphism (1.2.2) is equivalent to giving an isomorphism of the restriction of the bundles to the zero-section. In other words, the existence of ϑ_D is equivalent to finding, for each E/S, a trivialisation of the bundle

$$e^*\mathcal{O}_E(\ker[\times D]) \otimes e^*\mathcal{O}_E(D^2 e)^\vee = e^*[\times D]^*\mathcal{O}_E(e) \otimes e^*\mathcal{O}_E(-D^2 e)$$
$$= e^*\mathcal{O}_E(e)^{\otimes(1-D^2)}$$
$$= \omega_{E/S}^{\otimes(D^2-1)}$$

compatible with base-change. Note that $(6, D) = 1$ implies $D^2 \equiv 1 \pmod{12}$. There are then exactly two non-vanishing sections of $\omega_{E/S}^{\otimes(D^2-1)}$ compatible with arbitrary basechange, namely $\pm\Delta(E/S)^{(D^2-1)/12}$. Choose one of them, and let $\phi^{(E/S)}$ be the corresponding function on $E - \ker[\times D]$. So the rule

$(\phi\colon E/S \mapsto \phi^{(E/S)})$ satisfies properties (i) and (ii). In a moment we will see that exactly one of $\pm\phi$ satisfies (iii). (See also remark 1.2.3 below).

By the basechange compatibility (ii), we are free to make any faithfully flat basechange in order to check (iii). There exists such a basechange over which α factors as a product of isogenies of prime degree. It is therefore enough to verify (iii) when $\deg \alpha = p$ is prime. The quotient

$$g_p(E/S, \alpha) = \alpha_* \phi^{(E/S)} (\phi^{(E'/S)})^{-1} \in \mathcal{O}^*(S)$$

is compatible with basechange. It therefore defines a modular unit $g_p \in \Gamma(\mathcal{M}_{\Gamma_0(p)}, \mathcal{O}^*)$, and so $g_p(E/S, \alpha) \in \{\pm 1\}$ for every $(E/S, \alpha)$. Moreover the sign depends only on p.

To determine the sign, evaluate $g_p(E/\mathbb{F}_p, F_E)$ for an elliptic curve over $\mathbb{F}_p$ and its Frobenius endomorphism. The norm map $F_{E*}\colon \kappa(E)^* \to \kappa(E)^*$ is then the identity map, so $g_p(E, F_E) = 1$, and therefore if p is odd we have $g_p = +1$. Notice that replacing $\phi^{(E/S)}$ by $-\phi^{(E/S)}$ does not change g_p for p odd, but replaces g_2 by $-g_2$. Therefore for exactly one choice $\vartheta_D = \pm\phi$ it will be the case that $g_2 = +1$, so exactly one of these choices satisfies (iii).

Now for property (iv). Evidently ϑ_{-D} also satisfies the characteristic properties (i) and (iii), hence $\vartheta_{-D} = \vartheta_D$. Also $\vartheta_1 = 1$ for the same reason. The function $[\times M]_* \vartheta_D$ has divisor $M^2(C^2(e) - \ker[\times C])$ and is compatible with base change, so we can write $[\times M]_* \vartheta_D = \varepsilon \vartheta_C^{M^2}$ for some $\varepsilon = \pm 1$. Now property (iii) gives

$$\varepsilon \vartheta_C^{M^2} = [\times M]_* \vartheta_D = [\times M]_* [\times 2]_* \vartheta_D =$$
$$= [\times 2]_* [\times M]_* \vartheta_D = [\times 2]_* (\varepsilon \vartheta_C^{M^2}) = \varepsilon^4 \vartheta_C^{M^2} = \vartheta_C^{M^2}$$

and so $\varepsilon = 1$. The same calculation works for $M = D$ by writing $\vartheta_1 = 1$.

If $D = MC$ then the functions $\vartheta_C \circ [\times M]$ and $\vartheta_D / \vartheta_M^{C^2}$ both have divisor

$$C^2 \ker[\times M] - \ker[\times D]$$

hence their ratio is a unit compatible with basechange. The norm compatibility (iii) then shows that this unit equals 1, as in Remark (iii) above. This proves property (iv).

Finally we check (v). Classical formulae (as can for example be found in [39] — the function Θ is essentially the same as the Jacobi theta function ϑ_1) show that

$$F(u, \tau) = \Theta(u, \tau)^{D^2} \Theta(Du, \tau)^{-1}$$

is a function on E_τ with divisor $D^2(e) - \ker[\times D]$, and is $SL_2(\mathbb{Z})$-invariant.[1] Hence $F(u, \tau)$ is a constant multiple (independent of τ) of $\vartheta_D^{(E_\tau/\mathbb{C})}$. As a

---

[1] The $SL_2(\mathbb{Z})$ action is: $\begin{pmatrix} a & b \\ c & d \end{pmatrix} \colon (u, \tau) \mapsto \left(\dfrac{u}{c\tau + d}, \dfrac{a\tau + b}{c\tau + d} \right)$.

formal power series,

$$F = q^{(D^2-1)/12} t^{-D(D-1)/2} \prod_{n \geq 0} \frac{(1 - q^n t)^{D^2}}{1 - q^n t^D} \prod_{n > 0} \frac{(1 - q^n t^{-1})^{D^2}}{1 - q^n t^{-D}}$$

is a unit in the ring of Laurent q-series with coefficients in $\mathbb{Z}[t, 1/t(1 - t^D)]$. So by the q-expansion principle, the constant has to be ± 1. To determine the sign, consider any elliptic curve E_τ defined over $\mathbb{R}$ with 2 real connected components. For such a curve one can assume that $\mathrm{Re}(\tau) = 0$. (To be definite, take E to be the curve

$$Y^2 = X^3 - X$$

for which $\tau = i$.) The real components of E_τ are the images of the line segments $[0, 1]$ and $[\tau/2, 1 + \tau/2]$ in the complex plane. We compute $[\times 2]_* F(u, \tau)$ for such a curve. The explicit formula for $\Theta(u, \tau)$ shows that the first non-vanishing u-derivative of $F(u, \tau)$ at the origin is real and positive. On the interval $[0, 1]$, $F(u, \tau)$ has simple poles at $u = k/D$ $(1 \leq k \leq D-1)$ and so by calculus $(-1)^{(D-1)/2} F(1/2, \tau) > 0$. On the segment $[\tau/2, 1 + \tau/2]$, F is real, finite and non-zero, hence the product $F(\tau/2, \tau) F((1 + \tau)/2, \tau)$ is positive. Therefore at the origin,

$$[\times 2]_* F(u, \tau) = F(\frac{u}{2}, \tau) F(\frac{u+1}{2}, \tau) F(\frac{u+\tau}{2}, \tau) F(\frac{u+1+\tau}{2}, \tau)$$

$$\sim (-1)^{(D-1)/2} \times (\text{positive real}) \times u^{D^2-1}$$

and so $[\times 2]_* F(u, \tau) = (-1)^{(D-1)/2} F(u, \tau)$. $\square$

Remark 1.2.3. As we saw in the proof, ϑ_D corresponds to one of the two nowhere-vanishing modular forms of weight $(D^2 - 1)/12$. Using (v) it is easy to determine which. The form arises by restriction to the zero-section of the composite isomorphism

$$[\times D]^* \mathcal{O}_E(e) \xleftarrow{\sim} \mathcal{O}_E(\ker[\times D]) \xrightarrow[\times \vartheta_D^{(E/S)}]{\sim} \mathcal{O}_E(D^2 e)$$

since $e^* [\times D]^* = e^*$. Therefore the q-expansion is D times the leading coefficient in the expansion of $\vartheta_D^{(E_\tau/\mathbb{C})}$ in powers of t, which from (v) is easily seen to be

$$(-1)^{\frac{D-1}{2}} q^{\frac{D^2-1}{12}} \prod_{n > 0} (1 - q^n)^{2D^2-2} = (-1)^{\frac{D-1}{2}} \Delta(\tau)^{\frac{D^2-1}{12}}$$

Remark. Suppose that E/S is an elliptic curve over an integral base S, and that $P \subset E(S)$ is a finite group of sections. Let

$$\mathfrak{D} = \sum_{x \in P} m_x(x) \in \mathbb{Z}[P]$$

be a divisor with $\sum m_x = 0$ and $\sum m_x x = e$. In the case when S is the spectrum of a field, $\mathfrak{D}$ is principal, but in general this will not be the case. For example, suppose that $P = \{e, x\}$ for a section x of order 2, disjoint from e. Then $\mathfrak{D} = 2(x) - 2(e)$ is principal if and only if $\omega_{E/S}^{\otimes 2} = e^*\mathcal{O}(\mathfrak{D})$ is trivial.

Consider a Dedekind domain R containing $1/2$, and an ideal $A \subset R$ which has order 4 in $\operatorname{Pic} R$. Let $A^4 = (a)$ and let E/R be the elliptic curve given by the affine equation

$$y^2 = x(x^2 - a)$$

over the field of fractions of R. Take an open $U \subset \operatorname{Spec} R$ over which A becomes principal, locally generated by α, say. Then $a = \alpha^4\varepsilon$ for some unit $\varepsilon \in \mathcal{O}(U)^*$, and an equation for E over U is

$$(y/\alpha^3)^2 = (x/\alpha^2)((x/\alpha^2)^2 - \varepsilon).$$

Therefore $\omega_{E/R}$ is locally generated over U by

$$\frac{d(x/\alpha^2)}{y/\alpha^3} = \alpha\frac{dx}{y},$$

i.e. $\omega_{E/R} \simeq A$. So the divisor $2(0,0) - 2(e)$ is not principal on E/R.

1.3   Units and Eisenstein series

Let E be an elliptic curve over an integral base S, let $D > 1$ be an integer prime to 6, and $x \in E(S)$ a section. If x is disjoint from $\ker[\times D]$, then one obtains a unit $\vartheta_D(x) = x^*\vartheta_D \in \mathcal{O}^*(S)$ on the base. In particular, suppose that x is a torsion section of order $N > 1$, with $(N, D) = 1$. Since S is integral, x has order N at the generic point. Under either of the following conditions it is automatic that $x \cap \ker[\times D] = \emptyset$:

- N is invertible on S (then x has order N in every fibre); or

- N is divisible by at least two primes.

In the classical setting one takes S to be a modular curve (over $\mathbb{C}$) and the functions $\vartheta_D(x)$ are the *Siegel units*, studied extensively (see for example [19]). There are at least two ways to form a logarithmic derivative from the pair (ϑ_D, x). The simplest is to form

$$\operatorname{dlog}(\vartheta_D(x)) \in \Gamma(S, \Omega_S^1)$$

which in the classical setting gives weight 2 Eisenstein series. The other way, which leads to weight 1 Eisenstein series, is to first form the "vertical" logarithmic derivative

$$\operatorname{dlog}_v \vartheta_D \in \Gamma(E - \ker[\times D], \Omega_{E/S}^1).$$

Since $\omega_{E/S} = x^* \Omega^1_{E/S}$ (see §1.1) we obtain

$$_D\mathrm{Eis}(x) = \,_D\mathrm{Eis}(E/S, x) := x^* \,\mathrm{dlog}_v \, \vartheta_D \in \Gamma(S, x^*\Omega^1_{E/S}) = \Gamma(S, \omega_{E/S}),$$

a modular form of weight one. Notice that in this construction one can start with *any* function whose divisor is $D^2(e) - \ker[\times D]$, since it will be of the form $g\vartheta_D$ for some $g \in \mathcal{O}^*(S)$, and $\mathrm{dlog}_v \, g = 0$.

From property 1.2.1(iv) we have

$$(\vartheta_D)^{D'^2} \cdot \vartheta_{D'} \circ [\times D] = (\vartheta_{D'})^{D^2} \cdot \vartheta_D \circ [\times D']$$

and therefore

$$D'^2 \, \mathrm{dlog}_v \, \vartheta_D - [\times D']^* \, \mathrm{dlog}_v \, \vartheta_D = D^2 \, \mathrm{dlog}_v \, \vartheta_{D'} - [\times D]^* \, \mathrm{dlog}_v \, \vartheta_{D'}. \qquad (1.3.1)$$

Now $[\times D]^*$ is multiplication by D on global sections of $\Omega^1_{E/S}$. Hence (1.3.1) gives

$$D'^2 \cdot \,_D\mathrm{Eis}(E/S, x) - D' \cdot \,_D\mathrm{Eis}(E/S, D'x)$$
$$= D^2 \cdot \,_{D'}\mathrm{Eis}(E/S, x) - D \cdot \,_{D'}\mathrm{Eis}(E/S, Dx).$$

It follows that for any $D \equiv 1 \pmod{N}$, the section

$$\mathrm{Eis}(E/S, x) := \frac{1}{D^2 - D} \cdot \,_D\mathrm{Eis}(E/S, x) \in \Gamma(S \otimes \mathbf{Z}[\frac{1}{D(D-1)}], \omega) \qquad (1.3.2)$$

is independent of D. Now if $p \nmid 2N$, there exists $D > 1$, $D \equiv 1 \pmod{N}$ with $(D, 6) = 1$ and $p \nmid D(D - 1)$, so one can glue the various $\mathrm{Eis}(E/S, x)$ for the different D to get a section $\mathrm{Eis}(E/S, x) \in \Gamma(S \otimes \mathbf{Z}[1/2N], \omega)$. For *any* D one then has

$$_D\mathrm{Eis}(E/S, x) = D^2 \, \mathrm{Eis}(E/S, x) - D \, \mathrm{Eis}(E/S, Dx).$$

Suppose $E = \mathbf{C}/\Lambda$ is an elliptic curve over $\mathbf{C}$, with $\Lambda = \mathbf{Z}\omega_1 + \mathbf{Z}\omega_2$. Let u be the variable in the complex plane. Using the function $\sigma(z, \Lambda)^{D^2}/\sigma(Dz, \Lambda)$ (Weierstrass σ-function) in place of ϑ_D gives

$$\mathrm{dlog}_v \, \vartheta_D = \left(D^2\zeta(u, \Lambda) - D\zeta(Du, \Lambda)\right) dz$$

and if $x \in E(\mathbf{C}) - \{e\}$ is the torsion point $(a_1\omega_1 + a_2\omega_2)/N \in N^{-1}\Lambda/\Lambda$, with $(N, D) = 1$, then

$$\mathrm{Eis}(E/\mathbf{C}, x) = \sum_{m_i \in \frac{a_i}{N} + \mathbf{Z}} \frac{1}{(m_1\omega_1 + m_2\omega_2)\,|m_1\omega_1 + m_2\omega_2|^s} \Bigg|_{s=0} du. \qquad (1.3.3)$$

On the Tate curve $\mathrm{Tate}(q)$ over $\Lambda_N = \mathbb{Z}[\boldsymbol{\mu}_N]((q^{1/N}))$ there is the canonical differential dt/t, and the level N structure

$$(\mathbb{Z}/N\mathbb{Z})^2 \to \ker[\times N], \quad (a_1, a_2) \bmod N \mapsto \zeta_N^{a_1} q^{a_2/N}.$$

If x is the point $\zeta_N^{a_1} q^{a_2/N}$ then by explicit differentiation of the infinite product in Theorem 1.2.1(v)

$$\mathrm{Eis}(\mathrm{Tate}(q)/\Lambda_N, x) = \left[B_1\!\left(\frac{a_2}{N}\right) - \sum_{n>0}\left(\sum_{\substack{d\in\mathbb{Z},\, d|n \\ \frac{n}{d}\equiv a_2 \bmod N}} \mathrm{sgn}(d)\zeta_N^{a_1 d} \right) q^{n/N} \right] dt/t$$

if $0 \le a_1 < n$, $0 < a_2 < n$. Here $B_1(X) = X - 1/2$ is the Bernoulli polynomial. (In the case $a_2 = 0 \ne a_1$ the constant term is somewhat different.) In particular, $\mathrm{Eis}(\mathrm{Tate}(q)/\Lambda_N, x)$ is holomorphic at infinity.

One can also compute the logarithmic derivative of the unit $\vartheta_D(x) \in \Lambda_N^*$. The result is most interesting if one works with absolute differentials, that is in the module of (q-adically separated) differentials

$$\hat{\Omega}_{\Lambda_N/\mathbb{Z}} = \Lambda_N \cdot d(q^{1/N}) \oplus \Lambda_N/I_N \cdot d\zeta_N$$

where $I_N \subset \Lambda_N$ is the annihilator of $d\zeta_N$ (and equals the ideal generated by the different of $\mathbb{Q}(\boldsymbol{\mu}_N)$). The key point is that the logarithmic derivative of a typical term in the infinite product for $\vartheta_D(x)$ is

$$\mathrm{dlog}\left(1 - \zeta_N^{\pm a_1} q^{(m\pm a_2/N)}\right)$$
$$= \pm \frac{-\zeta_N^{\pm a_1} q^{(m\pm a_2/N)}}{1 - \zeta_N^{\pm a_1} q^{(m\pm a_2/N)}}(a_1 \, \mathrm{dlog}\, \zeta_N + (a_2 \pm mN) \, \mathrm{dlog}\, q^{1/N})$$

whereas the corresponding term for the vertical logarithmic derivative is

$$\mathrm{dlog}(1 - t^{\pm 1} q^m)\Big|_{t=\zeta_N^{a_1} q^{a_2/N}} = \pm \frac{-\zeta_N^{\pm a_1} q^{(m\pm a_2/N)}}{1 - \zeta_N^{\pm a_1} q^{(m\pm a_2/N)}} \, \mathrm{dlog}\, t.$$

Comparing gives the following striking congruence:

Proposition 1.3.4. *If* $x = \zeta_N^{a_1} q^{a_2/N} \in \mathrm{Tate}(q)(\Lambda_N)$, *then*

$$\mathrm{dlog}\, \vartheta_D(x) \equiv \frac{D \mathrm{Eis}(\mathrm{Tate}(q)/\Lambda_N, x)}{\mathrm{dlog}\, t}(a_1 \, \mathrm{dlog}\, \zeta_N + a_2 \, \mathrm{dlog}\, q^{1/N}) \quad \bmod N.$$

2 Norm relations

2.1 Some elements of K-theory

For a regular, separated and noetherian scheme X, the Quillen K-groups $K_i X$, $i \geq 0$, together with the cup-product

$$\cup: K_i X \times K_j X \to K_{i+j} X$$

define a graded ring $K_* X$, which is a contravariant functor in X— for any morphism $f: X' \to X$ of regular schemes there is a graded ring homomorphism $f^*: K_* X \to K_* X'$. If f is *proper*, then there are also pushforward maps $f_*: K_i X' \to K_i X$ (group homomorphisms) which satisfy the projection formula

$$f_*(f^* a \cup b) = a \cup f_* b. \tag{2.1.1}$$

For $i = 1$ there is a canonical monomorphism

$$\mathcal{O}^*(X) \to K_1 X. \tag{2.1.2}$$

For arbitrary f, the restriction of the pullback map f^* to the image of (2.1.2) is pullback on functions; if f is finite and flat, then the pushforward map f_* restricts to the norm map on functions.

In this section we are concerned with K_2. The cup-product in this case is the *universal symbol map*

$$\begin{aligned} \mathcal{O}^*(X) \otimes \mathcal{O}^*(X) &\to K_2 X \\ u \otimes v &\mapsto \{u, v\} \end{aligned}$$

which is alternating and satisfies the Steinberg relation: $\{u, 1 - u\} = 0$ if u, $1 - u \in \mathcal{O}^*(X)$. If $X = \operatorname{Spec} F$ for a field F, then the symbol map induces an *isomorphism*

$$K_2 X = K_2 F \xrightarrow{\sim} \Lambda^2 F^* / (\text{Steinberg relation})$$

by Matsumoto's theorem.

Returning for a moment to the general situation, let Y be a smooth (not necessarily proper) variety over a number field F. Write $\overline{Y} = Y \otimes_F \overline{\mathbb{Q}}$, $G_F = \operatorname{Gal}(\overline{\mathbb{Q}}/F)$, and let p be prime. Then if $H^{j+1}(\overline{Y}, \mathbb{Q}_p(n))$ has no G_F-invariants, there is an *Abel-Jacobi homomorphism*

$$K_{2n-j-1} Y \to H^1(G_F, H^j(\overline{Y}, \mathbb{Q}_p)(n)).$$

The condition that $H^0(G_F, H^{j+1}(\overline{Y}, \mathbb{Q}_p(n))) = 0$ can often be checked just by considering weights; if for example Y is proper, then by considering the action of an unramified Frobenius and using Deligne's theorem (Weil conjectures) one sees that it holds if $j + 1 \neq 2n$.

In the case of interest here, Y is a curve and $j = 1$, and $n = 2$. Then the Abel-Jacobi map is even defined integrally:

$$AJ_2 \colon K_2 Y \to H^1(G_F, H^1(\overline{Y}, \mathbf{Z}_p)(2)). \qquad (2.1.3)$$

It is constructed as follows. There is a theory of Chern classes from higher K-theory to étale cohomology: these are functorial homomorphisms, for each $q \geq 0$ and $n \in \mathbf{Z}$:

$$c_{q,n} \colon K_q Y \to H^{2n-q}(Y, \mathbf{Z}_p(n)).$$

Here the cohomology on the right-hand side is continuous étale cohomology. These maps are not multiplicative, but can be made into a multiplicative map by the Chern character construction.. All we need to know here is that if α, $\alpha' \in K_1 Y$ then

$$c_{1,1}(\alpha) \cup c_{1,1}(\alpha') = -c_{2,2}(\alpha \cup \alpha') \qquad (2.1.4)$$

(see for example [33, p.28]). One writes $\mathrm{ch} = -c_{2,2}$.

The étale cohomology of Y is related to that of $\overline{Y}$ by the Hochschild-Serre spectral sequence:

$$E_2^{i,j} = H^i(G_F, H^j(\overline{Y}, \mathbf{Z}_p)(n)) \Rightarrow H^{i+j}(Y, \mathbf{Z}_p(n)).$$

Let $Y \hookrightarrow X$ be the smooth compactification of Y, so that $Y = X - Z$ for a finite $Z \subset X$. The $\mathbf{Z}_p$-module $H^0(\overline{Y}, \mathbf{Z}_p) = H^0(\overline{X}, \mathbf{Z}_p)$ is free of rank equal to the number of components of $\overline{X}$, and $H^2(\overline{X}, \mathbf{Z}_p) = H^0(\overline{X}, \mathbf{Z}_p)(-1)$. The module $H^1(\overline{X}, \mathbf{Z}_p)$ is the Tate module of the Jacobian of $\overline{X}$, hence is free. There is an exact sequence

$$0 \to H^1(\overline{X}, \mathbf{Z}_p) \to H^1(\overline{Y}, \mathbf{Z}_p) \to H^0(\overline{Z}, \mathbf{Z}_p)(-1)$$
$$\xrightarrow{\gamma} H^2(\overline{X}, \mathbf{Z}_p) \to H^2(\overline{Y}, \mathbf{Z}_p) \to 0.$$

The map γ is the Gysin homomorphism, mapping the class of a point $z \in \overline{Z}$ to the class of the component of $\overline{X}$ to which it belongs. Therefore all the modules $H^j(\overline{Y}, \mathbf{Z}_p)$ are free[2]. Moreover if α is an eigenvalue of a geometric Frobenius acting on $H^j(\overline{Y}, \mathbf{Q}_p)$ at a prime $v \nmid p$ of good reduction, then α is an algebraic integer satisfying

$$|\alpha| = \begin{cases} 1 & \text{if } j = 0, \\ N(v)^{1/2} \text{ or } N(v) & \text{if } j = 1, \text{ and} \\ N(v) & \text{if } j = 2. \end{cases}$$

---

[2]In the case to be considered later, Y is actually *affine*, in which case one even has $H^2(\overline{Y}, \mathbf{Z}_p) = 0$.

Therefore when $n = 2$ the first column $\{E_2^{0,j}\}$ of the spectral sequence vanishes. The exact sequence of lowest degree terms then becomes:

$$0 \to H^2(G_F, H^0(\overline{Y}, \mathbb{Z}_p)(2)) \to H^2(Y, \mathbb{Z}_p(2))$$
$$\xrightarrow{e_2} H^1(G_F, H^1(\overline{Y}, \mathbb{Z}_p)(2)) \to H^3(G_F, H^0(\overline{Y}, \mathbb{Z}_p)(2)).$$

Composing the "edge homomorphism" e_2 with ch $= -c_{2,2}$ defines the Abel-Jacobi homomorphism (2.1.3) — the minus sign is chosen because of (2.1.4). Notice also that the last group $H^3(G_F, H^0(\overline{Y}, \mathbb{Z}_p)(2))$ is zero if p is odd, and killed by 2 in general (see for example [24]).

We also need the Chern character into de Rham cohomology. For a Noetherian affine scheme $X = \operatorname{Spec} R$ there are homomorphisms for each $q \geq 0$

$$\operatorname{dlog} = \operatorname{dlog}_R \colon K_q R \to \Omega^q_{R/\mathbb{Z}}$$

satisfying:

(i) $\operatorname{dlog}(a \cup b) = \operatorname{dlog} a \wedge \operatorname{dlog} b$;

(ii) If $b \in R^* \subset K_1 R$ then $\operatorname{dlog} b = b^{-1} db \in \Omega^1_{R/\mathbb{Z}}$;

(iii) On $K_0 R$, dlog is the degree map.

(iv) If R'/R is a finite flat extension of regular rings, then $\operatorname{tr}^\Omega_{R'/R} \circ \operatorname{dlog}_{R'} = \operatorname{dlog}_R \circ \operatorname{tr}^K_{R'/R}$.

In (iv), $\operatorname{tr}^K_{R'/R} \colon K_q R' \to K_q R$ is the proper push-forward for $\operatorname{Spec} R' \to \operatorname{Spec} R$ (also called the transfer), and $\operatorname{tr}^\Omega_{R'/R} \colon \Omega^q_{R'/\mathbb{Z}} \to \Omega^q_{R/\mathbb{Z}}$ is the trace map for differentials. Since this compatibility does not seem to be documented in the literature we make some remarks about it. What follows was suggested in conversation with Gillet and Soulé.

To check the compatibility we can work locally on $\operatorname{Spec} R$, and thus assume that R is local. Therefore R' is a free R-module of rank d say. Choosing a basis gives a matrix representation $\mu \colon R' \hookrightarrow M_d(R)$. We get for every $n \geq 1$ corresponding inclusions $GL_n(R') \hookrightarrow GL_{nd}(R)$, which in the limit give an inclusion $GL(R') \hookrightarrow GL(R)$. This induces by functoriality the transfer on $K_q(-) = \pi_q(BGL(-)^+)$.

One way to define the map dlog is to use Hochschild homology (see for example [22, 1.3.11ff.]). There is a simplicial R-module $C_\bullet(R)$ with $C_q(R) = R^{\otimes q+1}$ (tensor product over $\mathbb{Z}$), whose homology is Hochschild homology $HH_*(R)$. There is also a pair of R-linear maps $\Omega^q_{R/\mathbb{Z}} \xrightarrow{\epsilon_q} HH_q(R) \xrightarrow{\pi_q} \Omega^q_{R/\mathbb{Z}}$, whose composite is multiplication by $q!$. The map π_q is given by $r_0 \otimes r_1 \otimes \cdots \otimes r_q \mapsto r_0 dr_1 \wedge \cdots \wedge r_q$.

There is a map Dtr: $H_*(GL(R), \mathbb{Z}) \to HH_*(R)$, the *Dennis trace* (see [22, 8.4.3], which maps $r \in R^* \subset H_1(GL(R), \mathbb{Z})$ to the homology class of $r^{-1} \otimes r \in C_1(R)$. Assume that $q!$ is invertible in R. Then composing on one side with the Hurewicz map $K_q(R) \to H_q(GL(R), \mathbb{Z})$, and on the other with $(q!)^{-1}\pi_q$, defines the map dlog: $K_q(R) \to \Omega^q_{R/\mathbb{Z}}$, for any $q > 0$. It is not too hard to check directly (an exercise from [22, Ch.8]) that if $a, b \in R^*$ then dlog$\{a, b\} = (ab)^{-1} da \wedge db$, which is the only part of (i) needed in what follows.

There is a trace map tr^{HH} on Hochschild homology: the representation $R' \hookrightarrow M_d(R)$ induces by functoriality a map $HH_*(R') \to HH_*(M_d(R))$, and by Morita invariance [22, 1.2.4] we have an isomorphism $HH_*(M_d(R) \xrightarrow{\sim} HH_*(R)$.

Still under the hypothesis that $q!$ is invertible, the maps ε_q and $(q!)^{-1}\pi_q$ make $\Omega^q_{R/\mathbb{Z}}$ a direct factor of $HH_q(R)$. One can then *define* the trace map $\text{tr}^\Omega_{R'/R}$ as the composite $(q!)^{-1}\pi_q \circ \text{tr}^{HH}_{R'/R} \circ \varepsilon_q$. (This approach to trace maps is due to Lipman [21] — see also Hübl's thesis [13].) It now is a simple exercise to check the compatibility (iv), the essential point being the transitivity [22, E1.2.2] of the generalised trace.

We need all of this only for $q \leq 2$. This means that in the reciprocity law 3.2.3 and all its consequences we need to assume that p is odd.

2.2 Level structures

Let E/S be an elliptic curve. Then for every positive integer N which is invertible on S, there exists[3] a moduli scheme $S(N)$, which is finite and étale over S, and which represents the functor on S-schemes T

$$S(N)(T) = \left\{ \begin{array}{c} \text{level } N \text{ structures on } E \times_S T \\ \alpha \colon (\mathbb{Z}/N)^2_{/T} \xrightarrow{\sim} \ker[\times N]_{/T} \end{array} \right\}$$

More generally, for pairs (M, N) of positive integers invertible on S there is a scheme $S(M, N)$ which represents the functor

$$S(M, N)(T) = \left\{ \begin{array}{c} \text{monomorphisms of } S\text{-group schemes} \\ \alpha \colon (\mathbb{Z}/M \times \mathbb{Z}/N)_{/T} \hookrightarrow E_{/T} \end{array} \right\}$$

The group $GL_2(\mathbb{Z}/N)$ acts freely on $S(N)$ on the right, with quotient S. One has $S(N, N) = S(N)$; in general $S(M, N)$ is a quotient of $S(N')$ where $N' = \text{lcm}(M, N)$. One usually writes $S_1(N)$ for $S(N, 1)$, and we will also write $S_1'(N)$ for $S(1, N)$. Of course $S_1(N)$ and $S_1'(N)$ are isomorphic, but they are different as quotients of $S(N)$. We have a lattice of subgroups of $GL_2(\mathbb{Z}/N\mathbb{Z})$, and a corresponding diagram of quotients of $S(N)$:

---

[3] To avoid overloading the notation we do not include the dependence on E in the notation.

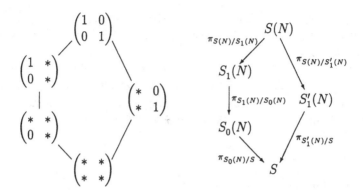

Over $S(N)$ there is a canonical level N structure $\alpha_N \colon (\mathbb{Z}/N)^2 \xrightarrow{\sim} \ker[\times N] \subset E$, and we let y_N, $y'_N \in E(S(N))$ be the images of the generators $(1,0)$, $(0,1)$. Then y_N already belongs to $E(S_1(N))$, and y'_N to $E(S'_1(N))$.

$S_1(N)$ is canonically isomorphic to the open subscheme of $\ker[\times N]$ consisting of points in the kernel whose order is exactly N; and

$$\ker[\times N] = \coprod_{M \mid N} S_1(M). \tag{2.2.1}$$

The scheme $S_0(N)$ parameterises $\Gamma_0(N)$-structures on E/S; in other words, $S_0(N)(T)$ is functorially the set of cyclic subgroup schemes of rank N of $E \times_S T$. In the case $T = S_1(N)$, the morphism $\pi_{S_1(N)/S_0(N)} \colon S_1(N) \to S_0(N)$ classifies the cyclic subgroup generated by y_N.

If $M \mid M'$ and $N \mid N'$ then there is a canonical level-changing map

$$\pi_{S(M',N')/S(M,N)} \colon S(M', N') \to S(M, N)$$

induced by the inclusion $\mathbb{Z}/M \times \mathbb{Z}/N \hookrightarrow \mathbb{Z}/M' \times \mathbb{Z}/N'$. One has $y_M = (M'/M)y_{M'}$ and likewise for y'.

We also recall that all the above moduli schemes can be defined for integers M, N which are not invertible on S, using Drinfeld level structures, see [18, *passim*]. They are finite and flat over S.

Recall finally that for a positive integer N which is the product of two coprime integers ≥ 3, there is a universal elliptic curve with level N structure over the modular curve $Y(N)/\mathbb{Z}$. We shall use the standard notations $Y_?(N)$, $Y(M, N)$ without comment.

2.3 Norm relations for $\Gamma(\ell)$-structure

Now fix an integer $D > 1$ which is prime to 6. On each basechange $E \times_S T$ (where T is one of the above moduli schemes) there is the canonical function

$\vartheta_D^{(E_T/T)}$, which by Theorem 1.2.1(ii) is simply the pullback of $\vartheta_D^{(E/S)}$. Since E and D will be fixed in the discussion that follows we shall write all these functions simply as ϑ.

Consider the case of prime level ℓ. Write $y = y_\ell$, $y' = y'_\ell$, and abbreviate $S_? = S_?(\ell)$ ($? = 1$ or 0). Fix $x \in E(S)$ such that $D\ell x$ does not meet the zero section of E. Let $\lambda \colon E \times_S S_0 \to \widetilde{E}$ be the quotient by the canonical subgroup scheme of rank ℓ, generated by y. Let $\tilde{x} \in \widetilde{E}(S_0)$ be the composite

$$S_0 \xrightarrow{\ x\ } E \times_S S_0 \xrightarrow{\ \lambda\ } \widetilde{E}.$$

Write $\tilde{\vartheta} = \vartheta_D^{(\widetilde{E}/S_0)} \in \Gamma(\widetilde{E} - \ker[\times D], \mathcal{O}^*)$.

Lemma 2.3.1.

$$N_{S_1/S}(\vartheta(x+y)) = \vartheta(\ell x)\vartheta(x)^{-1}. \tag{N1}$$

$$N_{S(\ell)/S_1}(\vartheta(x+y')) = \vartheta(\ell x) \prod_{a \in \mathbb{Z}/\ell} \vartheta(x+ay)^{-1}. \tag{N2}$$

$$N_{S_1/S_0}(\vartheta(x+y)) = \tilde{\vartheta}(\tilde{x})\vartheta(x)^{-1} \tag{N3}$$

$$N_{S(\ell)/S_1}(\vartheta(x+y')) = \vartheta(\ell x)\tilde{\vartheta}(\tilde{x})^{-1}. \tag{N4}$$

$$N_{S_0/S}(\tilde{\vartheta}(\tilde{x})) = \vartheta(x)^\ell \vartheta(\ell x). \tag{N5}$$

Proof. (N1) By (2.2.1) there is a Cartesian square

$$
\begin{array}{ccc}
S_1 \amalg S & \xrightarrow{(x+y,x)} & E \\
\downarrow & & \downarrow{\scriptstyle[\times\ell]} \\
S & \xrightarrow{\ \ell x\ } & E
\end{array}
$$

Hence

$$
\begin{aligned}
N_{S_1/S}(\vartheta(x+y))\vartheta(x) &= N_{S_1 \amalg S/S}((x+y,x)^*\vartheta) \\
&= (\ell x)^*[\times\ell]_*\vartheta \quad \text{since the square is Cartesian} \\
&= (\ell x)^*\vartheta \quad \text{by 1.2.1(iii)} \\
&= \vartheta(\ell x).
\end{aligned}
$$

(N2) The same argument, applied to the Cartesian square

$$
\begin{array}{ccc}
S(\ell) \amalg \coprod_{a \in \mathbb{Z}/\ell} S_1 & \xrightarrow{(x+y',x+ay)} & E \times_S S_1 \\
\downarrow & & \downarrow{\scriptstyle[\times\ell]} \\
S_1 & \xrightarrow{\ \ell x\ } & E \times_S S_1
\end{array}
$$

(N3) This comes from the Cartesian square:

$$S_1 \amalg S_0 \xrightarrow{(x+y,x)} E \times_S S_0$$
$$\downarrow \qquad\qquad \downarrow\lambda$$
$$S_0 \xrightarrow{\tilde{x}} \widetilde{E}$$

The remaining relations (N4) and (N5) are obtained by combining (N1)–(N3) and using

$$N_{S_1/S_0}(\vartheta(x+y)) = \prod_{a\in(\mathbb{Z}/\ell)^*} \vartheta(x+ay). \qquad \square$$

Lemma 2.3.2. *The norm relations* (N1)–(N5) *hold without the hypothesis that ℓ is invertible on S.*

Proof. Choose an auxiliary integer $r > 2$ prime to ℓ. Then after replacing S by an étale basechange there exists a level r structure $\beta_r: (\mathbb{Z}/r)^2 \to \ker[\times r]$ on E, with r invertible on S. Let $\mathcal{E}^{\mathrm{univ}} \to Y(r)$ be the universal elliptic curve with level r structure over $\mathbb{Z}[1/r]$. Then there is a unique morphism $\xi: S \to \mathcal{E}^{\mathrm{univ}} - \ker[\times \ell D]$ which classifies the triple $(E/S, \beta_r, x)$: there is a Cartesian square

$$E \longrightarrow \mathcal{E}^{\mathrm{univ}} \times_{Y(r)} (\mathcal{E}^{\mathrm{univ}} - \ker[\times\ell D])$$
$$\downarrow \qquad\qquad\qquad \downarrow^{pr_2}$$
$$S \xrightarrow{\xi} \mathcal{E}^{\mathrm{univ}} - \ker[\times\ell D]$$

such that β_r is the inverse image of the canonical level r structure on $\mathcal{E}^{\mathrm{univ}}$, and x is the pullback of the diagonal section $\mathcal{E}^{\mathrm{univ}} \to \mathcal{E}^{\mathrm{univ}} \times \mathcal{E}^{\mathrm{univ}}$.

By the basechange compatibility of ϑ_D, it is enough to verify the norm relations in this universal setting; but the inclusion $Y(r) \otimes \mathbb{Z}[1/\ell] \hookrightarrow Y(r)$ induces an injection on $\mathcal{O}^*$. Thus we reduce to the case in which ℓ is invertible on S. $\qquad \square$

Now consider two auxiliary integers D, D' with $(6\ell, DD') = 1$, and write $\vartheta = \vartheta_D^{(E/S)}$ and $\vartheta' = \vartheta_{D'}^{(E/S)}$. Following tradition we write $N_{?/?}$ for the pushforward maps $\pi_{?/?*}$ on K_2, but the group operation in K_2 will be written additively (for consistency with the higher K-theory case to be considered below).

Proposition 2.3.3. *In $K_2 S$ the following identity holds:*

$$N_{S(\ell)/S}\{\vartheta(x+y), \vartheta'(x'+y')\}$$
$$= \{\vartheta(\ell x), \vartheta'(\ell x')\} + \ell\{\vartheta(x), \vartheta'(x)\} - N_{S_0/S}\{\tilde{\vartheta}(\tilde{x}), \tilde{\vartheta}'(\tilde{x}')\}$$

Proof. Compute using the projection formula (2.1.1) and the norm relations
2.3.1:

$$N_{S(\ell)/S}\{\vartheta(x+y), \vartheta'(x'+y')\}$$

$$= N_{S_1/S}\{\vartheta(x+y), N_{S(\ell)/S_1}\vartheta'(x'+y')\}$$

$$= N_{S_1/S}\{\vartheta(x+y), \vartheta'(\ell x')\tilde{\vartheta}'(\tilde{x}')^{-1}\} \qquad\qquad \text{by (N4)}$$

$$= N_{S_0/S}\{\tilde{\vartheta}(\tilde{x})\vartheta(x)^{-1}, \vartheta'(\ell x')\tilde{\vartheta}'(\tilde{x}')^{-1}\} \qquad\qquad \text{by (N3)}$$

$$= -N_{S_0/S}\{\tilde{\vartheta}(\tilde{x}), \tilde{\vartheta}'(\tilde{x}')\} - (\ell+1)\{\vartheta(x), \vartheta'(\ell x')\}$$
$$\qquad + \{\vartheta(x), N_{S_0/S}\tilde{\vartheta}'(\tilde{x}')\} + \{N_{S_0/S}\tilde{\vartheta}(\tilde{x}), \vartheta'(\ell x')\}$$

$$= -N_{S_0/S}\{\tilde{\vartheta}(\tilde{x}), \tilde{\vartheta}'(\tilde{x}')\} - (\ell+1)\{\vartheta(x), \vartheta'(\ell x')\}$$
$$\qquad + \{\vartheta(x), \vartheta'(x')^{\ell}\vartheta'(\ell x')\} + \{\vartheta(x)^{\ell}\vartheta(\ell x), \vartheta'(\ell x')\} \quad \text{by (N5)}$$

$$= -N_{S_0/S}\{\tilde{\vartheta}(\tilde{x}), \tilde{\vartheta}'(\tilde{x}')\} + \ell\{\vartheta(x), \vartheta'(x')\} + \{\vartheta(\ell x), \vartheta'(\ell x')\} \qquad \square$$

Now suppose that S is a modular curve of level prime to ℓ, and E is the universal elliptic curve. Therefore $S = Y_H := Y(N)/H$ for some subgroup $H \subset GL_2(\mathbb{Z}/N)$, $E = \mathcal{E}^{\mathrm{univ}} \xrightarrow{f} Y_H$, and $S_0(\ell) = Y_{H,\ell} := Y_0(\ell, N)/H$. It is then possible to rewrite the above norm relation using the Hecke and the diamond operators, whose definitions we briefly recall.

The centre $(\mathbb{Z}/N)^* \subset GL_2(\mathbb{Z}/n)$ acts on Y_H and $\mathcal{E}^{\mathrm{univ}}$ on the right, defining the diamond operators $\langle a \rangle \in \mathrm{Aut}\, Y_H$, $\langle a \rangle_{\mathcal{E}} \in \mathrm{Aut}\,\mathcal{E}$ for $a \in (\mathbb{Z}/N)^*$. In modular language, the B-valued points of Y_H are pairs $(X/B, [\alpha_n]_H)$, where X/B is an elliptic curve and $[\alpha_N]_H$ is an H-equivalence class of level N structures $(\mathbb{Z}/N)^2 \to X$. Then $\langle a \rangle\colon (X/B, [\alpha_N]_H) \mapsto (X/B, [a\alpha_N]_H)$ is an automorphism of Y_H. The B-valued points of $\mathcal{E}^{\mathrm{univ}}$ are triples $(X/B, [\alpha_N]_H, z)$ with $z \in X(B)$, and the automorphism $\langle a \rangle_{\mathcal{E}}$ of $\mathcal{E}^{\mathrm{univ}}$ is given by $(X/B, [\alpha_N]_H, z) \mapsto (X/B, [a\alpha_N]_H, z)$.

Recall also [18, (9.4.1)] that the e_N pairing defines a morphism

$$e_N\colon Y_H \to \mathrm{Spec}\, \mathbb{Z}[\boldsymbol{\mu}_N]^{\det H} \tag{2.3.4}$$

$$(E/S, \alpha_N) \longmapsto e_N(\alpha_N\begin{pmatrix} 1/N \\ 0 \end{pmatrix}, \alpha_N\begin{pmatrix} 0 \\ 1/N \end{pmatrix}) \tag{2.3.5}$$

and the restriction of $\langle a \rangle^*$ to $\mathbb{Z}[\boldsymbol{\mu}_N]^{\det H}$ is then the map $\zeta \mapsto \zeta^{a^2}$ (since $\det\langle a \rangle = a^2$).

The is a commutative diagram [5, (3.17)]

$$
\begin{array}{ccccc}
\mathcal{E}^{\mathrm{univ}} & \xrightarrow{\;[\times \ell]\;} & \mathcal{E}^{\mathrm{univ}} & \xrightarrow[\sim]{\;\langle \ell \rangle_{\mathcal{E}}\;} & \mathcal{E}^{\mathrm{univ}} \\
& {\scriptstyle f}\searrow \quad {\scriptstyle f}\downarrow & & & {\scriptstyle f}\downarrow \\
& & Y_H & \xrightarrow[\langle \ell \rangle]{\sim} & Y_H
\end{array}
$$

An introduction to Kato's Euler systems 399

and if $x\colon Y_H \to \mathcal{E}^{\mathrm{univ}}$ is any N-torsion section, $\langle\ell\rangle_{\mathcal{E}} \circ \ell x = x \circ \langle\ell\rangle$. Therefore by the basechange property 1.2.1(ii),

$$\tilde{\vartheta}_D(\ell x) = (\ell x)^* \vartheta_D = (\ell x)^* \langle\ell\rangle_E^* \vartheta_D = \langle\ell\rangle^* \vartheta_D(x) = \langle\ell^{-1}\rangle_* \vartheta_D(x).$$

The scheme S_0 is the quotient $Y_{H,\ell} := Y_0(N\ell, N)/H$, and $Y_0(N\ell, N)$ classifies triples (X, α_N, C) with $C \subset X$ a subgroup of rank ℓ. One then has the standard diagram [5, (3.16)]

$$
\begin{array}{ccccccccc}
\mathcal{E}^{\mathrm{univ}} & \xleftarrow{c_{\mathcal{E}}:=pr_1} & \mathcal{E}^{\mathrm{univ}} \times_{Y_H} Y_{H,\ell} & \xrightarrow{\lambda} & \widetilde{\mathcal{E}^{\mathrm{univ}}} & \xrightarrow[\sim]{v} & \mathcal{E}^{\mathrm{univ}} \times_{Y_H} Y_{H,\ell} & \xrightarrow{c_{\mathcal{E}}} & \mathcal{E}^{\mathrm{univ}} \\
\downarrow & & \downarrow & & \downarrow & & \downarrow & & \downarrow \\
Y_H & \xleftarrow{c} & Y_{H,\ell} & = = & Y_{H,\ell} & \xrightarrow{\sim}{w} & Y_{H,\ell} & \xrightarrow{c} & Y_H \\
& & & & \downarrow & & \downarrow & & \\
& & & & \mathrm{Spec}\,\mathbb{Z}[\mu_N]^H & \xrightarrow{\zeta \mapsto \zeta^\ell} & \mathrm{Spec}\,\mathbb{Z}[\mu_N]^H & &
\end{array}
$$

with the brace labeled $w_{\mathcal{E}}$ over the top row.

in which the first, third and fourth squares in the top row are Cartesian. The Hecke operator T_ℓ is by definition the correspondence $c_*(cw)^*$ on Y_H, and the correspondence $c_{\mathcal{E}*}(c_{\mathcal{E}}w_{\mathcal{E}})^*$ on $\mathcal{E}^{\mathrm{univ}}$. All the horizontal arrows are compatible with the level N structure on $\mathcal{E}^{\mathrm{univ}}$ and the quotient level N structure on $\widetilde{\mathcal{E}^{\mathrm{univ}}}$, hence $c_{\mathcal{E}} \circ w_{\mathcal{E}} \circ x = x \circ c \circ w$, and therefore

$$\tilde{\vartheta}(\tilde{x}) = (\lambda \circ x)^* \tilde{\vartheta} = (\lambda \circ x)^*(c_{\mathcal{E}} \circ v)^* \vartheta = (c \circ w)^* \vartheta(x)$$

using as always the basechange property 1.2.1(ii). Therefore

$$T_\ell\{\vartheta(x), \vartheta'(x')\} = c_*(c \circ w)^*\{\vartheta(x), \vartheta'(x')\} = N_{S_0(\ell)/S}\{\tilde{\vartheta}(\tilde{x}), \tilde{\vartheta}'(\tilde{x}')\}.$$

Finally write $z = x + y_\ell$, $z' = x' + y'_\ell$, so that $\ell x = \ell z$ and $x = (\ell z) \circ \langle\ell^{-1}\rangle$. Observe that the norm relation is invariant under the action of $GL_2(\mathbb{Z}/\ell)$, so that y_ℓ, y'_ℓ can be replaced by any basis for the ℓ-torsion of E. This yields the following reformulation of 2.3.3:

Proposition 2.3.6. *If $S = Y_H$ is a modular curve of level prime to ℓ and z, z' are torsion sections of $E/S(\ell)$ whose projections onto $\ker[\times\ell]$ are linearly independent, then*

$$N_{S(\ell)/S}\{\vartheta(z), \vartheta'(z')\} = (1 - T_\ell \circ \langle\ell\rangle_* + \ell\langle\ell\rangle_*)\{\vartheta(\ell z), \vartheta'(\ell z')\}.$$

2.4 Norm relations for $\Gamma(\ell^n)$-structure

We now consider norm relations in the tower $\{S(\ell^m, \ell^n)\}$.

Lemma 2.4.1. *If $m > 1$, $n \geq 0$ and $x \in E(S(\ell^m, \ell^n))$ is any section, then*

$$N_{S(\ell^m,\ell^n)/S(\ell^{m-1},\ell^n)}: \vartheta(x + y_{\ell^m}) \mapsto \vartheta(\ell x + y_{\ell^{m-1}}). \qquad (N6)$$

Proof. If ℓ is invertible on S this follows from the Cartesian square

$$
\begin{array}{ccc}
S(\ell^m, \ell^n) & \xrightarrow{\;x+y_{\ell^m}\;} & E \\[4pt]
\downarrow & & \downarrow {\scriptstyle [\times \ell]} \\[4pt]
S(\ell^{m-1}, \ell^n) & \xrightarrow{\;\ell x + y_{\ell^{m-1}}\;} & E
\end{array}
$$

In the general case one reduces to the universal situation exactly as in 2.3.2.
$\square$

Proposition 2.4.2. *If $(6\ell, DD') = 1$ and m, $n > 1$ then for all x, $x' \in E(S(\ell^{m-1}, \ell^{n-1}))$,*

$$N_{S(\ell^m,\ell^n)/S(\ell^{m-1},\ell^{n-1})}: \{\vartheta(x + y_{\ell^m}), \vartheta'(x' + y_{\ell^n}')\}$$
$$\mapsto \{\vartheta(\ell x + y_{\ell^{m-1}}), \vartheta'(\ell x' + y_{\ell^{n-1}}')\}.$$

Proof. This follows from the lemma since

$$N_{S(\ell^m,\ell^n)/S(\ell^{m-1},\ell^{n-1})}\{\vartheta(x + y_{\ell^m}), \vartheta'(x' + y_{\ell^n}')\}$$
$$= N_{S(\ell^m,\ell^{n-1})/S}\{\vartheta(x + y_{\ell^m}), N_{S(\ell^m,\ell^n)/S(\ell^m,\ell^{n-1})}\vartheta'(x' + y_{\ell^n}')\}$$
$$= N_{S(\ell^m,\ell^{n-1})/S}\{\vartheta(x + y_{\ell^m}), \vartheta'(\ell x' + y_{\ell^{n-1}}')\}$$
$$= \{\vartheta(\ell x + y_{\ell^{m-1}}), \vartheta'(\ell x' + y_{\ell^{n-1}}')\} \qquad\qquad \square$$

If E/S is a modular family over $S = Y_H$ of level prime to ℓ, then $E(Y_H)$ is finite of order prime to ℓ. Therefore

$$E(Y_H(\ell^m, \ell^n))_{\text{torsion}} = (\mathbf{Z}/\ell^m \times \mathbf{Z}/\ell^n) \times (\text{prime to } \ell),$$

so there is a well-defined projection onto $(\mathbf{Z}/\ell)^2 = \ker[\times \ell]$. Computing as in 2.3.6 we get:

Proposition 2.4.3. *Suppose that $S = Y_H$ is a modular curve of level prime to ℓ and that z, z' are torsion sections of E over $Y_H(\ell^m, \ell^n)$, with m, $n > 1$. If the projections of $\{z, z'\}$ into $\ker[\times \ell]$ are linearly independent, then*

$$N_{Y_H(\ell^m,\ell^n)/Y_H(\ell^{m-1},\ell^{n-1})}\{\vartheta(z), \vartheta'(z')\} = \{\vartheta(\ell z), \vartheta'(\ell z')\}. \qquad \square$$

2.5 Norm relations for products of Eisenstein series

We shall repeat the construction of the last paragraph for products of the form

$$_D\mathrm{Eis}(E/S, x) \cdot {}_{D'}\mathrm{Eis}(E/S, x') \in \Gamma(S, \omega_{E/S}^2).$$

If $g: S' \to S$ is a finite and flat morphism of smooth T-schemes and $E' = E \times_S S'$ then there are trace maps

$$\mathrm{tr}_g = \mathrm{tr}_{S'/S} \colon g_*\mathcal{O}_{S'} \to \mathcal{O}_S, \quad g_*\Omega^1_{S'/T} \to \Omega^1_{S/T}$$

as well as a trace map on modular forms, defined to be the composite

$$\mathrm{tr}_g = \mathrm{tr}_{S'/S} \colon \Gamma(S', \omega_{E'/S'}^{\otimes k}) = \Gamma(S', g^*\omega_{E/S}^{\otimes k})$$

$$= \Gamma(S, g_*\mathcal{O}_{S'} \otimes_{\mathcal{O}_S} \omega_{E/S}^{\otimes k}) \xrightarrow{\mathrm{tr}_g} \Gamma(S, \omega_{E/S}^{\otimes k}).$$

The Kodaira-Spencer map (§1.1) and the trace are not compatible.

Proposition 2.5.1. *The diagrams below commute:*

$$
\begin{array}{ccc}
\omega_{E'/S'}^{\otimes 2} & \xrightarrow{KS_{E'/S'}} & \Omega^1_{S'/T} \\
\uparrow{\scriptstyle \imath} & & \uparrow{\scriptstyle g^*} \\
g^*\omega_{E/S}^{\otimes 2} & \xrightarrow{g^*KS_{E/S}} & g^*\Omega^1_{S/T}
\end{array}
\qquad
\begin{array}{ccc}
\omega_{E'/S'}^{\otimes 2} & \xrightarrow{KS_{E'/S'}} & \Omega^1_{S'/T} \\
\uparrow{\scriptstyle \imath} & & \downarrow{\scriptstyle \mathrm{tr}_g} \\
g^*\omega_{E/S}^{\otimes 2} & \xrightarrow{\deg(g)\cdot g^*KS_{E/S}} & g^*\Omega^1_{S/T}
\end{array}
$$

Proof. The first commutes because of the functoriality of the Kodaira-Spencer map. Then applying $\mathrm{tr}_g \circ g^* = \deg g$ gives the second. $\square$

Lemma 2.5.2. *The notation as in Lemma 2.4.1,*

$$\mathrm{tr}_{S(\ell^m, \ell^n)/S(\ell^{m-1}, \ell^n)} \colon {}_D\mathrm{Eis}(x + y_{\ell^m}) \mapsto \ell^{-1} {}_D\mathrm{Eis}(\ell x + y_{\ell^{m-1}}).$$

Proof. Since $[\times \ell]_* \vartheta_D = \vartheta_D$, we have $\mathrm{tr}_{[\times \ell]} \colon \mathrm{dlog}\, \vartheta_D \mapsto \mathrm{dlog}\, \vartheta_D$. But on global sections of $\Omega^1_{E/S}$, $\mathrm{tr}_{[\times \ell]}$ is multiplication by ℓ. Therefore the diagram

$$
\begin{array}{ccc}
\Gamma(E - \ker[\times D], \Omega^1) & \xrightarrow{(x+y_{\ell^n})^*} & \Gamma(S(\ell^n, \ell^{n-1}), \omega) \\
{\scriptstyle \mathrm{tr}_{[\times \ell]}}\downarrow & & \downarrow{\scriptstyle \ell\,\mathrm{tr}_{S(\ell^n, \ell^{n-1})/S(\ell^{n-1})}} \\
\Gamma(E - \ker[\times D], \Omega^1) & \xrightarrow{(\ell x + y_{\ell^{n-1}})^*} & \Gamma(S(\ell^{n-1}), \omega)
\end{array}
$$

commutes, which gives the result. $\square$

Corollary 2.5.3. *The notations being as in 2.4.3, let g be the projection $g: Y_H(\ell^m, \ell^n) \to Y_H(\ell^{m-1}, \ell^{n-1})$. Then*

$$\mathrm{tr}_g \colon {}_D\mathrm{Eis}(z) \cdot {}_{D'}\mathrm{Eis}(z') \qquad \mapsto \ell^{-2} {}_D\mathrm{Eis}(\ell z) \cdot {}_{D'}\mathrm{Eis}(\ell z') \qquad (2.5.4)$$

$$\mathrm{tr}_g \colon KS\big({}_D\mathrm{Eis}(z) \cdot {}_{D'}\mathrm{Eis}(z')\big) \mapsto \ell^2 KS\big({}_D\mathrm{Eis}(\ell z) \cdot {}_{D'}\mathrm{Eis}(\ell z')\big) \qquad (2.5.5)$$

Proof. Follows from the preceding two lemmas, since $\deg(g) = \ell^4$. $\square$

A   Appendix: Higher K-theory of modular varieties

A.1   Eisenstein symbols

Let $f\colon E \to S$ be an elliptic curve, and assume from now on that S is a regular scheme. For any integer $k > 0$, write E^k for the fibre product $E \times_S \cdots \times_S E$ — it is an abelian scheme of dimension k over S.

In [2], Beilinson discovered a family of canonical elements of $K_{k+1}(E^k)$. More precisely, he defined a canonical map

$$\mathbb{Q}[E_{\mathrm{tors}}]^{\mathrm{degree}=0} \to K_{k+1}(E^k) \otimes \mathbb{Q}$$

which he called the Eisenstein symbol. Here we make a modified construction which gives a norm-compatible system.

Let Γ_k be the semidirect product of the symmetric group $\mathfrak{S}_k$ and μ_2^k, which acts on E^k as follows:

- $\mathfrak{S}_k$ acts by permuting the copies of E;

- the ith copy of μ_2 acts as multiplication by ± 1 in the ith factor of the product.

There is a character $\varepsilon_k\colon \Gamma_k \to \mu_2$ which is the identity on each factor μ_2 and the sign character on the symmetric group.

Γ_k has a natural realisation in $GL_k(\mathbb{Z})$ as the set of all permutation matrices with entries ± 1. Geometrically it is the group of orthogonal symmetries of a cube in n-space. In terms of this representation, ε_k is just determinant. For any $\mathbb{Z}[\Gamma_k]$-module M, write $M(\varepsilon_k)$ for the ε_k-isotypical component of $M \otimes \mathbb{Z}[1/2 \cdot k!]$.

For $x \in E(S)$ we shall consider the inclusion

$$i_x\colon E^k \to E^{k+1}$$
$$(u_1,\ldots,u_k) \mapsto (x - u_1, u_1 - u_2, \ldots, u_{k-1} - u_k, u_k).$$

whose image is the subscheme

$$\left\{ (v_1,\ldots,v_{k+1}) \;\middle|\; \sum_1^{k+1} v_i = x \right\} \subset E^{k+1}.$$

For any integer $D \neq 0$ such that x is disjoint from $\ker[\times D]$, we define the following open subschemes of E^k:

$$U_{D,x}^{k\prime} = i_x^{-1}(E - \ker[\times D])^{k+1})$$
$$U_{D,x}^k = \bigcap_{\gamma \in \Gamma_k} \gamma(U_{D,x}^{k\prime}).$$

Observe that $U_{D,x}^{k'}$ and $U_{D,x}^{k}$ are stable under translation by $\ker[\times D]^k$, and there is an étale covering

$$[\times D]\colon U_{D,x}^{k} \to U_{1,Dx}^{k}.$$

We prove below the following lemma.

Lemma A.1.1. *If $z \in E(S)$ is any section disjoint from e, the inclusion $U_{1,z}^{k} \hookrightarrow (E - \{\pm z\})^k$ induces an isomorphism*

$$K_*(E - \{\pm z\})^k(\varepsilon_k) \xrightarrow{\sim} K_*U_{1,z}^{k}(\varepsilon_k).$$

Using this lemma we define K-theory elements, whenever $(6, D) = 1$:

$$\vartheta_D^{[k]} = pr_1^*(\vartheta_D) \cup \cdots \cup pr_{k+1}^*(\vartheta_D) \quad \in K_{k+1}(E - \ker[\times D])^{k+1}$$

$$^{(1)}\vartheta_D^{[k]}(x) = i_x^*(\vartheta_D^{[k]}) \quad \in K_{k+1}U_{D,x}^{k'}$$

$$^{(2)}\vartheta_D^{[k]}(x) = \frac{1}{\#\Gamma_k}\sum_{\gamma\in\Gamma_k}\varepsilon_k(\gamma)\gamma^*(^{(1)}\vartheta_D^{[k]}(x)) \in K_{k+1}(U_{D,x}^{k})(\varepsilon_k)$$

$$^{(3)}\vartheta_D^{[k]}(x) = [\times D]_*{}^{(2)}\vartheta_D^{[k]}(x) \quad \in K_{k+1}(E - \{\pm Dx\})^k(\varepsilon_k)$$

We call $^{(i)}\vartheta_D^{[k]}(x)$ *Eisenstein symbols*. For $k = 0$ we simply define i_x to be the section $x \in E(S)$, and the Eisenstein symbol then becomes a Siegel unit:

$$^{(i)}\vartheta_D^{[0]}(x) = \vartheta_D(x) \in \mathcal{O}^*(S).$$

If $\alpha_1, \ldots, \alpha_{k+1}\colon \widetilde{E} \to E$ are isogenies of degree prime to D, then by repeated application of 1.2.1(iii) one get the norm-compatibility

$$(\alpha_1, \ldots, \alpha_{k+1})_*(\vartheta_D^{[k]}) = \vartheta_D^{[k]} \tag{A.1.2}$$

Actually this is only of interest when all the α_i are equal.

Proof of Lemma A.1.1. For any T/S

$$U_{1,z}^{k}(T) = \left\{(u_1, \ldots, u_k) \in E(T)^k \; \middle| \; \begin{array}{l} \text{for all } i,\, u_i \neq e, \pm z; \\ \text{for all } i \neq j,\, u_i \pm u_j \neq 0. \end{array}\right\}$$

The complementary divisor $E^k - U_{1,z}^{k}$ is the union of the two divisors

$$V^k = \{(u_i) \mid \text{for some } i,\, u_i = e\} \cup \{(u_i) \mid \text{for some } i \neq j,\, u_i \pm u_j = e\}$$

and

$$W_z^k = \{(u_i) \mid \text{for some } i,\, u_i = \pm z\}$$

As S is regular, the K-groups in the lemma can be computed in K'-theory. From the localisation sequence, it is then enough to show that $K'_*(V^k - W_z^k)(\varepsilon_k)$ vanishes. This is a special case of the following:

Lemma A.1.3. *Let $V' \subset V^k$ be any Γ_k-invariant open subscheme. Then $K'_*(V')(\varepsilon_k)$ is trivial.*

Proof. Define a sequence of reduced closed subschemes

$$V^k = V^k_{[1]} \supset V^k_{[2]} \supset \cdots \supset V^k_{[k+1]} = \emptyset$$

inductively, by writing $V^k_{[r+1]}$ for the smallest closed subset of $V^k_{[r]}$ such that $V^k_{[r]} - V^k_{[r+1]}$ is smooth over S. Write $V'_{[r]} = V^k_{[r]} \cap V'$. Then from the definition of V^k it is easy to see that:

(i) $V^k_{[r]}$ is a union of closed subsets each given by the vanishing of a certain collection of expressions u_i, $u_i \pm u_j$, which are permuted by Γ_k;

(ii) This gives a decomposition of $V'_{[r]} - V'_{[r+1]}$ as a disjoint union $\coprod V'_{[r]\mu}$, of open and closed pieces, permuted by Γ_k, in such a way that for each μ there is some $\gamma_\mu \in \Gamma_k$ which acts trivially on $V'_{[r]\mu}$ and for which $\varepsilon_k(\gamma_\mu) = -1$.

This forces $K'_*(V'_{[r]} - V'_{[r+1]})(\varepsilon_k) = 0$ for each $r \geq 1$. In fact, if

$$c = \sum c_\mu \in K'_*(V'_{[r]} - V'_{[r+1]}) \otimes \mathbb{Z}[1/2 \cdot k!]$$

$$= \bigoplus_\mu K'_*(V'_{[r]\mu}) \otimes \mathbb{Z}[1/2 \cdot k!]$$

then $\gamma^*_\mu(c) = \varepsilon_k(\gamma_\mu)c = -c$, whereas the μ-component of $\gamma^*_\mu(c)$ is evidently $+c_\mu$ by (ii). Now using the long exact sequences

$$K'_*(V'_{[r+1]}) \to K'_*(V'_{[r]}) \to K'_*(V'_{[r]} - V'_{[r+1]}) \to \cdots$$

inductively (beginning with $r = k - 1$) we deduce that $K'_*(V')(\varepsilon_k) = 0$. $\square$

A.2   Norm relations in higher K-groups

Here we find norm relations for the Eisenstein symbols and for cup-products, analogous to those in sections 2.3 and 2.4.

For the $\Gamma(\ell)$-structure norm relations, we use the same notation as in 2.3. In addition, write $\hat{\lambda}$: $\widetilde{E} \to E \times_S S_0$ for the isogeny dual to λ, and λ^k, $\hat{\lambda}^k$ for the isogenies on E^k, $\widetilde{E}^k$. Consider the push-forward for the morphisms:

$$[\times \ell] \times \pi_{S_1/S} \colon E^k \times_S S_1 \to E^k$$

$$\lambda^k \times \pi_{S_1/S_0} \colon E^k \times_S S_1 \to \widetilde{E}^k$$

$$\hat{\lambda}^k \times \pi_{S_0/S} \colon \widetilde{E}^k \to E^k$$

Fix $i \in \{1, 2, 3\}$ and abbreviate $\vartheta^{[k]}_D(x) = {}^{(i)}\vartheta^{[k]}_D(x)$. The symbol $\tilde{\vartheta}^{[k]}_D$ will denote the analogue on $\widetilde{E}^k$ of $\vartheta^{[k]}_D$.

Lemma A.2.1. *The following relations hold in K_{k+1}:*

$$([\times \ell] \times \pi_{S_1/S})_* \vartheta_D^{[k]}(x + y) = \vartheta_D^{[k]}(\ell x) - [\times \ell]_* \vartheta_D^{[k]}(x) \qquad \text{(EN1)}$$

$$(\lambda^k \times \pi_{S_1/S_0})_* \vartheta_D^{[k]}(x + y) = \tilde{\vartheta}_D^{[k]}(\tilde{x}) - \lambda_*^k \vartheta_D^{[k]}(x) \qquad \text{(EN3)}$$

$$(\hat{\lambda}^k \times \pi_{S_0/S})_* \tilde{\vartheta}_D^{[k]}(\tilde{x}) = \ell [\times \ell]_* \vartheta_D^{[k]}(x) + \vartheta_D^{[k]}(\ell x). \qquad \text{(EN5)}$$

Proof. Here (EN1) and (EN3) are to be understood on $U_{D,x}^k$, and (EN5) on $(\tilde{\lambda}^k)^{-1}(U_{D,x}^k)$. The relations (EN1) and (EN3) are proved just as (N1) and (N3), by considering the Cartesian diagrams:

$$
\begin{array}{ccc}
E^k \times_S S_1 \amalg E^k & \xrightarrow{(i_{x+y}, i_x)} & E^{k+1} \\
{\scriptstyle ([\times\ell]\times\pi_{S_1/S}, [\times\ell])} \downarrow & & \downarrow {\scriptstyle [\times\ell]} \\
E^k & \xrightarrow{i_{\ell x}} & E^{k+1}
\end{array}
$$

and

$$
\begin{array}{ccc}
E^k \times_S S_1 \amalg E^k \times_S S_0 & \xrightarrow{(i_{x+y}, i_x)} & E^{k+1} \times_S S_0 \\
{\scriptstyle (\lambda^k \times \pi_{S_1/S_0}, \lambda^k)} \downarrow & & \downarrow {\scriptstyle \lambda^{k+1}} \\
\tilde{E}^k & \xrightarrow{i_{\tilde{x}}} & \tilde{E}^{k+1}
\end{array}
$$

and using the norm-compatibility (2.5.4). Applying $\hat{\lambda}^k \times \pi_{S_0/S}$ to (EN3) gives (EN5). $\qquad\square$

We now consider cup-products of the form $\vartheta_D^{[k]} \cup \vartheta_{D'}$ in K_{k+2}. Consider the factorisation of multiplication by ℓ:

$$
E^k \times_S S(\ell) \xrightarrow{\text{id}\times\pi_{S(\ell)/S_1}} E^k \times_S S_1 \xrightarrow{\lambda^k \times \pi_{S_1/S_0}} \tilde{E}^k \xrightarrow{\hat{\lambda}^k \times \pi_{S_0/S}} E^k
$$

$$\underbrace{\phantom{E^k \times_S S(\ell) \xrightarrow{\text{id}} E^k \times_S S_1 \xrightarrow{\lambda} \tilde{E}^k \xrightarrow{\hat{\lambda}} E^k}}_{[\times\ell]\times\pi_{S(\ell)/S}}$$

We compute:

$$
\begin{aligned}
(\lambda^k \times \pi_{S(\ell)/S_0})_* [\vartheta_D^{[k]}(x + y) \cup \vartheta_{D'}(x' + y')] & \\
= (\lambda^k \times \pi_{S_1/S_0})_* [\vartheta_D^{[k]}(x + y) \cup (\vartheta_{D'}(\ell x') - \tilde{\vartheta}_{D'}(\tilde{x}'))] & \quad \text{by (N4)} \\
= (\tilde{\vartheta}_D^{[k]}(\tilde{x}) - \lambda_*^k \vartheta_D^{[k]}(x)) \cup (\vartheta_{D'}(\ell x') - \tilde{\vartheta}_{D'}(\tilde{x}')) & \quad \text{by (EN3)}
\end{aligned}
$$

We need to compute the image of this cup-product under $(\hat{\lambda}^k \times \pi_{S_0/S})_*$. Taking the terms in turn:

$(\hat{\lambda}^k \times \pi_{S_0/S})_*$:

$\tilde{\vartheta}_D^{[k]}(\tilde{x}) \cup \vartheta_{D'}(\ell x') \mapsto ([\times \ell]_* \vartheta_D^{[k]}(x)^\ell + \vartheta_D^{[k]}(\ell x)) \cup \vartheta_{D'}(\ell x')$ by (EN5)

$\lambda_*^k \vartheta_D^{[k]}(x) \cup \tilde{\vartheta}_{D'}(\tilde{x}') \mapsto [\times \ell]_* \vartheta_D^{[k]}(x) \cup (\ell \vartheta_{D'}(x') + \vartheta_{D'}(\ell x'))$ by (N5)

$\lambda_*^k \vartheta_D^{[k]}(x) \cup \vartheta_{D'}(\ell x') \mapsto (\ell + 1)([\times \ell]_* \vartheta_D^{[k]}(x) \cup \vartheta_{D'}(\ell x'))$ as $\deg \pi_{S_0/S} = \ell + 1$

Combining these gives the required generalisation of 2.3.3:

Proposition A.2.2.

$$([\times \ell] \times \pi_{S(\ell)/S})_* (\vartheta_D^{[k]}(x + y) \cup \vartheta_{D'}(x' + y')) = \vartheta_D^{[k]}(\ell x) \cup \vartheta_{D'}(\ell x')$$
$$- (\hat{\lambda}^k \times \pi_{S_0/S})_* (\tilde{\vartheta}_D^{[k]}(\tilde{x}) \cup \tilde{\vartheta}_{D'}(\tilde{x}')) + \ell([\times \ell]_* \vartheta_D^{[k]}(x) \cup \vartheta_{D'}(x')). \qquad \square$$

Having got this far the analogue of 2.4.2 presents no further difficulty:

Proposition A.2.3. *If* $n > 1$ *and* $x, x' \in E(S(\ell^{n-1}))$ *then*

$$([\times \ell] \times \pi_{S(\ell^n)/S(\ell^{n-1})})_* (\vartheta_D^{[k]}(x + y_{\ell^n}) \cup \vartheta_{D'}(x' + y_{\ell^n}'))$$
$$= \vartheta_D^{[k]}(\ell x + y_{\ell^{n-1}}) \cup \vartheta_{D'}(\ell x' + y_{\ell^{n-1}}'). \qquad \square$$

3   The dual exponential map

3.1   Notations

In this section K will denote a finite extension of $\mathbb{Q}_p$ with ring of integers $\mathfrak{o}$. We fix an algebraic closure $\bar{K}$ of K. Write $\bar{\mathfrak{o}}$ for the integral closure of $\mathfrak{o}$ in $\bar{K}$, and G_K for the Galois group of $\bar{K}$ over K. We normalise all p-adic valuations such that $v(p) = 1$. Let $\widehat{\bar{K}}$ be the completion of $\bar{K}$, and write $\widehat{\bar{\mathfrak{o}}}$ for its valuation ring. Fix a uniformiser π_K of $\mathfrak{o}$.

We fix for each $n > 0$ a primitive p^n-th root of unity ζ_{p^n} in $\bar{K}$ such that $\zeta_{p^{n+1}}^p = \zeta_{p^n}$. Write $K_n = K(\zeta_{p^n})$ and denote by $\mathfrak{o}_n$ the valuation ring of K_n. Put $\mathfrak{d}_n =$ the relative different of K_n/K.

For a topological G_K-module M write $H^i(K, M)$ for the *continuous* Galois cohomology groups [38].

The cyclotomic character $\chi_{\text{cycl}} \colon G_K \to \mathbb{Z}_p^*$ is defined by $g(\zeta_{p^n}) = \zeta_{p^n}^{\chi_{\text{cycl}}(g)}$, for every $g \in G_K$ and $n > 0$. Its logarithm is a homomorphism from G_K to $\mathbb{Z}_p$, often viewed as an element of $H^1(K, \mathbb{Z}_p)$.

We normalise the reciprocity law of local class field theory in such a way that if L/K is unramified, then the norm residue symbol $(\pi_K, L/K)$ equals the geometric Frobenius (inverse of the Frobenius substitution $x \mapsto x^q$). This implies that for any $u \in \mathfrak{o}^*$ we have $\chi_{\text{cycl}}(u, K^{\text{ab}}/K) = N_{K/\mathbb{Q}_p}(u)$.

3.2 The dual exponential map for H^1 and an explicit reciprocity law

Let V be a continuous finite-dimensional representation of G_K over $\mathbb{Q}_p$. Suppose that V is *de Rham* (for generalities about p-adic representations, see for example [12]). Let $DR(V) = (B_{\mathrm{dR}} \otimes_{\mathbb{Q}_p} V)^{G_K}$ be the associated filtered K-vector space, with the decreasing filtration $DR^i(V)$ (induced from the filtration on B_{dR}). Then Kato has defined a *dual exponential map* [16, §II.1.2]

$$\exp^* : H^1(K, V) \to DR^0(V)$$

which is the composite:

$$H^1(K, V) \to H^1(K, B_{\mathrm{dR}}^0 \otimes_{\mathbb{Q}_p} V) = H^1(K, \mathrm{Fil}^0(B_{\mathrm{dR}} \otimes_K DR(V))) \simeq DR^0(V).$$

The last isomorphism comes from Tate's computation [37] of the groups $H^i(K, \widehat{\overline{K}}(j))$:

$$H^i(K, \widehat{\overline{K}}(j)) = 0 \quad \text{unless } j = 0 \text{ and } i = 0 \text{ or } 1; \text{ and}$$

$$K = H^0(K, \widehat{\overline{K}}) \xrightarrow[\sim]{\cup \log \chi_{\mathrm{cycl}}} H^1(K, \widehat{\overline{K}}) \tag{3.2.1}$$

together with the isomorphisms $B_{\mathrm{dR}}^j / B_{\mathrm{dR}}^{j+1} \simeq \widehat{\overline{K}}(j)$.

The group $H^1(K, V)$ classifies extensions $0 \to V \to V' \to \mathbb{Q}_p(0) \to 0$ of p-adic Galois representations, and the extension V' is de Rham if and only if its class lies in $\ker(\exp^*)$. (This follows from [3], remark before 3.8 and Lemma 3.8.1.) In particular, the kernel of $\exp^*$ is the Bloch-Kato subgroup $H_g^1(K, V) \subset H^1(K, V)$.

In some cases one can define and study the dual exponential map without reference to B_{dR}. For example, if $V = H^1(A, \mathbb{Q}_p(1))$ for an abelian variety A/K, it can be defined just using the exponential map for the analytic group $A(K)$. More generally, if the filtration on $DR(V)$ satisfies $DR^1(V) = 0$, then one only needs to use the Hodge-Tate decomposition

$$\widehat{\overline{K}} \otimes_{\mathbb{Q}_p} V \xrightarrow{\sim} \bigoplus_{i \in \mathbb{Z}} \widehat{\overline{K}}(-i) \otimes_K \mathrm{gr}^i DR(V) \tag{3.2.2}$$

since then by (3.2.1) $\exp^*$ is the natural map from $H^1(K, V)$ to

$$H^1(K, \widehat{\overline{K}} \otimes_{\mathbb{Q}_p} V) \xrightarrow[(3.2.2)]{\sim} \bigoplus_i H^1(K, \widehat{\overline{K}}(-i) \otimes_K \mathrm{gr}^i DR(V))$$

$$= H^1(K, \widehat{\overline{K}}) \otimes_K DR^0(V) \xleftarrow[(3.2.1)]{\sim} DR^0(V)$$

In what follows we shall be concerned with the case $V = H^1(Y_{\bar{K}}, \mathbb{Q}_p)(1)$ for a smooth $\mathfrak{o}$-scheme Y, which is the complement in a smooth proper $\mathfrak{o}$-scheme X of a divisor Z with relatively normal crossings. Write

$$H^i_{dR}(Y/\mathfrak{o}) = H^i(X, \Omega^{\bullet}_{X/\mathfrak{o}}(Z))$$

(the hypercohomology of the de Rham complex of differentials with logarithmic singularities along Z). Then $DR(V)$ is just de Rham cohomology with a shift of filtration:

$$DR^{-1}(V) = DR(V) = H^1_{dR}(Y/\mathfrak{o}) \otimes_{\mathfrak{o}} K$$
$$DR^0(V) = H^0(X, \Omega^1_{X/\mathfrak{o}}(\log Z)) \otimes_{\mathfrak{o}} K = \mathrm{Fil}^1 H^1_{dR}(Y/\mathfrak{o}) \otimes_{\mathfrak{o}} K$$
$$DR^1(V) = 0$$

Moreover the Hodge-Tate decomposition has an explicit description, essentially thanks to the work of Fontaine [11] and Coleman [4]. To compute $\exp^*$ one just needs to know the projection

$$\pi_1 \colon \widehat{\bar{K}} \otimes_{\mathbb{Q}_p} V \xrightarrow[(3.2.2)]{\sim} \widehat{\bar{K}} \otimes_{\mathfrak{o}} H^0(X, \Omega^1_{X/\mathfrak{o}}(\log Z)) \oplus \widehat{\bar{K}}(1) \otimes_{\mathfrak{o}} H^1(X, \mathcal{O}_X)$$

$$\longrightarrow \widehat{\bar{K}} \otimes_{\mathfrak{o}} H^0(X, \Omega^1_{X/\mathfrak{o}}(\log Z)).$$

It is the limit of the maps given by the diagram

$$H^1(Y_{\bar{K}}, \mu_{p^n}) \xrightarrow[\sim]{(1)} H^0_{\mathrm{Zar}}(Y_{\bar{K}}, \mathcal{O}^*/p^n) \xleftarrow[\sim]{(2)} H^0_{\mathrm{Zar}}(Y_{\bar{\mathfrak{o}}}, \mathcal{O}^*/p^n)$$

with $\pi_1 \pmod{p^n}$ going to, and dlog going down to,

$$H^0(X, \Omega^1_{X/\mathfrak{o}}(\log Z)) \otimes \bar{\mathfrak{o}}/p^n$$

Remarks. (i) The isomorphism labelled **(1)** comes about as follows. Generally, let S be a scheme on which m is invertible, with $\mu_m \subset \mathcal{O}_S$. An element of $H^1(S, \mu_m)$ is an isomorphism class of finite étale coverings $S' \to S$, Galois with group μ_m. Given such an S'/S, there is an open (Zariski) covering $\{U_i\}$ of S and units $f_i \in \mathcal{O}^*(U_i)$ such that $S' \times U_i = U_i[\sqrt[m]{f_i}]$. It is easy to see that $\{f_i\}$ is a well-defined element of $H^0(Y_{\bar{K}}, \mathcal{O}^*_S/m)$, and moreover that the map thus obtained fits into an exact sequence

$$H^1_{\mathrm{Zar}}(S, \mu_m) \to H^1_{\text{ét}}(S, \mu_m) \xrightarrow{(1)} H^0_{\mathrm{Zar}}(S, \mathcal{O}^*_S/m) \to H^2_{\mathrm{Zar}}(S, \mu_m).$$

If S is irreducible (as is the case here) then every non-empty Zariski open subset is connected, so μ_m is flasque for the Zariski topology, and the map **(1)** is an isomorphism.

(ii) The inclusion $j\colon Y_{\bar{K}} \hookrightarrow Y_{\bar{\sigma}}$ induces an isomorphism $H^0(Y_{\bar{\sigma}}, \mathcal{O}^*/p^n) \xrightarrow{\sim} H^0(Y_{\bar{K}}, \mathcal{O}^*/p^n)$ denoted **(2)** in the diagram. To see this, consider the effect of multiplication by p^n on the exact sequence

$$0 \to \mathcal{O}^*_{Y_{\bar{\sigma}}} \to j_* \mathcal{O}^*_{Y_{\bar{K}}} \xrightarrow{v} \mathbb{Q}_{Y_{\bar{k}}} \to 0$$

(the last map is the p-adic valuation along the special fibre, taking values in the constant sheaf $\mathbb{Q}$). This shows that $\mathcal{O}^*_{Y_{\bar{\sigma}}}/p^n \xrightarrow{\sim} (j_* \mathcal{O}^*_{Y_{\bar{K}}})/p^n$. It is therefore enough to show that $(j_* \mathcal{O}^*_{Y_{\bar{K}}})/p^n \xrightarrow{\sim} j_*(\mathcal{O}^*_{Y_{\bar{K}}}/p^n)$, because then

$$H^0(Y_{\bar{\sigma}}, \mathcal{O}^*/p^n) \xrightarrow{\sim} H^0(Y_{\bar{\sigma}}, j_*(\mathcal{O}^*_{Y_{\bar{K}}}/p^n)) = H^0(Y_{\bar{K}}, \mathcal{O}^*/p^n).$$

By passing to the direct limit, we can replace $\bar{K}$ by a finite extension of K. Now consider more generally an open immersion $U \hookrightarrow S$, where S is a separated noetherian scheme which is integral and regular in codimension 1. Suppose that m is invertible on U, and that $\mu_m \subset \mathcal{O}_U$. Then $R^i j_* \mu_m = 0$ for $i > 0$ as μ_m is Zariski flasque. The exact sequences

$$0 \to \mu_m \to \mathcal{O}^* \to (\mathcal{O}^*)^m \to 0$$
$$0 \to (\mathcal{O}^*)^m \to \mathcal{O}^* \to \mathcal{O}^*/m \to 0.$$

give a short exact sequence

$$0 \to (j_* \mathcal{O}^*_U)/m \to j_*(\mathcal{O}^*_U/m) \to {}_m R^1 j_* \mathcal{O}^*_U \to 0.$$

But because S is regular in codimension one, the divisor sequence

$$1 \to \mathcal{O}^*_S \to \mathcal{K}^*_S \xrightarrow{\mathrm{div}} \coprod_{\mathrm{codim}(x)=1} i_{x*} \mathbb{Z} \to 0$$

is exact, and therefore $R^1 j_* \mathcal{O}^*_U = 0$.

(iii) The simplest case (which is, however, not enough for our purposes) is when $Y = X$ is proper, when this recipe reduces to that given by Coleman: an element of $H^1(X_{\bar{K}}, \mu_{p^n}) = \mathrm{Pic}\, X_{\bar{K}}[p^n]$ is the class $[D]$ of a divisor D on $X_{\bar{K}}$ such that $p^n D = \mathrm{div}(g)$ is principal. One can assume that the divisor of g on $X_{\bar{\sigma}}$ is precisely the closure D^c of D. Put

$$\omega = \mathrm{dlog}\, g \in H^0(X_{\bar{\sigma}}, \Omega^1_{X_{\bar{\sigma}}/\bar{\sigma}}(\mathrm{supp}\, D^c)).$$

Then because the residues of ω at $\mathrm{supp}\, D$ are $\equiv 0 \pmod{p^n}$, one has

$$\omega \pmod{p^n} \in H^0(X, \Omega^1_{X/\mathfrak{o}}) \otimes \bar{\mathfrak{o}}/p^n$$

and this defines $\pi_1([D]) \pmod{p^n} = \omega \pmod{p^n}$. (Coleman even defines such a map in the case of bad reduction.) Unfortunately we know of no reference for this description of the Hodge-Tate decomposition in the non-proper case.

Now assume that X is a smooth and proper curve over $\mathfrak{o}$, and that Y is affine. Then Z is a finite étale $\mathfrak{o}$-scheme. Recall (see also the following section) that the different $\mathfrak{d}_n = \mathfrak{d}_{K_n/K}$ is the annihilator in $\mathfrak{o}_n$ of $\Omega_{\mathfrak{o}_n/\mathfrak{o}}$. If $K/\mathbb{Q}_p$ is unramified, then $\mathfrak{o}_n = \mathfrak{o}[\zeta_{p^n}]$ and therefore $\Omega_{\mathfrak{o}_n/\mathfrak{o}}$ is generated by $\mathrm{dlog}\,\zeta_{p^n}$, and moreover $\mathfrak{d}_n = p^n(\zeta_p - 1)^{-1}\mathfrak{o}_n$.

Theorem 3.2.3. *Suppose that $K/\mathbb{Q}_p$ is unramified (and that $p > 2$). There exists an integer c such that for every $n > 0$ the following diagram commutes up to p^c-torsion:*

$$
\begin{array}{ccc}
K_2(Y \otimes \mathfrak{o}_n) \otimes \mu_{p^n}^{\otimes -1} & \xrightarrow{\;\;ch\;\;} & H^2(Y \otimes K_n, \mu_{p^n}) \\[2mm]
\Big\downarrow{\scriptstyle \mathrm{dlog}} & & \Big\downarrow{\scriptstyle \text{Hochschild-Serre}} \\[2mm]
H^0(X \otimes \mathfrak{o}_n, \Omega^2_{X \otimes \mathfrak{o}_n/\mathfrak{o}}(\log Z))(-1) & & H^1(K_n, H^1(Y \otimes \bar{K}, \mu_{p^n})) \\[2mm]
\Big\| & & \Big\downarrow{\scriptstyle \pi_1 \;(\mathrm{mod}\; p^{n-1})} \\[2mm]
\Omega^1_{\mathfrak{o}_n/\mathfrak{o}}(-1) \otimes \mathrm{Fil}^1 H^1_{\mathrm{dR}}(Y/\mathfrak{o}) & & H^1(K_n, \bar{\mathfrak{o}}/p^{n-1}) \otimes_{\mathfrak{o}} \mathrm{Fil}^1 H^1_{\mathrm{dR}}(Y/\mathfrak{o}) \\[2mm]
{\scriptstyle \mathrm{dlog}\,\zeta_{p^n} \otimes [\zeta_{p^n}]^{-1} \mapsto 1} \Big\downarrow{\wr} & & \Big\uparrow{\scriptstyle \cup (1/p^n)\log \chi_{\mathrm{cycl}}} \\[2mm]
\mathfrak{o}_n/\mathfrak{d}_n \otimes_{\mathfrak{o}} \mathrm{Fil}^1 H^1_{\mathrm{dR}}(Y/\mathfrak{o}) & \xrightarrow{\hspace{2cm}} & \mathfrak{o}_n/p^{n-1} \otimes_{\mathfrak{o}} \mathrm{Fil}^1 H^1_{\mathrm{dR}}(Y/\mathfrak{o})
\end{array}
$$

Corollary 3.2.4. *(The explicit reciprocity law) The following diagram commutes:*

$$
\begin{array}{ccc}
\varprojlim_n \left(K_2(Y \otimes \mathfrak{o}_n) \otimes \mu_{p^n}^{\otimes -1} \right) & \xrightarrow{\;\;HS \circ ch\;\;} & \varprojlim_n H^1(K_n, H^1(Y \otimes \bar{K}, \mu_{p^n})) \\[2mm]
\Big\downarrow{\scriptstyle \mathrm{dlog}} & & \Big\downarrow \\[2mm]
\varprojlim_n H^0(X \otimes \mathfrak{o}_n, \Omega^2_{X \otimes \mathfrak{o}_n/\mathfrak{o}}(\log Z))(-1) & & \varprojlim_{n \geq m} H^1(K_m, H^1(Y \otimes \bar{K}, \mu_{p^n})) \\[2mm]
\Big\| & & \Big\downarrow{\wr} \\[2mm]
\varprojlim_n \Omega^1_{\mathfrak{o}_n/\mathfrak{o}}(-1) \otimes \mathrm{Fil}^1 H^1_{\mathrm{dR}}(Y/\mathfrak{o}) & & H^1(K_m, H^1(Y \otimes \bar{K}, \mathbb{Z}_p)(1)) \\[2mm]
\Big\downarrow & & \Big\downarrow{\scriptstyle \exp^*} \\[2mm]
\varprojlim_n \mathfrak{o}_n/\mathfrak{d}_n \otimes_{\mathfrak{o}} \mathrm{Fil}^1 H^1_{\mathrm{dR}}(Y/\mathfrak{o}) & \xrightarrow{(\frac{1}{p^n}\,\mathrm{tr}_{K_n/K_m})_{n \geq m}} & K_m \otimes_{\mathfrak{o}} \mathrm{Fil}^1 H^1_{\mathrm{dR}}(Y/\mathfrak{o})
\end{array}
$$

Remarks. (i) The assumption that $K/\mathbb{Q}_p$ is unramified is not essential for the proof, and is only included to simplify the statement. In general the situation is completely analogous to 3.3.15 below. The case $p = 2$ is excluded only because we do not know a reference for the compatibility of the trace maps in this case, cf. §2.1.

(ii) The maps "Hochschild-Serre" comes from the Hochschild-Serre spectral sequence with finite coefficients (cf. §2.1); since Y is affine, $H^2(Y \otimes \bar{K}, \mu_{p^n}) = 0$.

(iv) For a discussion of the map dlog, see §2.1. *A priori* its target is the group $H^0(Y \otimes o_n, \Omega^2_{X \otimes o_n/o})(-1)$. We just explain why its image is contained in the submodule of differentials with logarithmic singularities along Z. By making an unramified basechange, one is reduced to the case when Z is a union of sections. Let A be the local ring of $X \otimes o_n$ at a closed point of Z, and t a local equation for Z. Then by the localisation sequence, one sees that $K_2 A[t^{-1}]$ is generated by $K_2 A$ and symbols $\{u, t\}$ with $u \in o_n^*$, and $\mathrm{dlog}\{u,t\} = u^{-1}du \wedge \mathrm{dlog}\, t$.

Proof. First we explain precisely what are the transition maps in the various inverse systems in the diagram. In the Galois cohomology groups they are given by corestriction and reduction mod p^n. The finite flat morphisms $Y \otimes o_{n+1} \to Y \otimes o_n$ induce compatible trace maps (cf. §2.1)

$$K_2(Y \otimes o_{n+1}) \to K_2(Y \otimes o_n) \quad \text{and} \quad \Omega^2_{Y \otimes o_{n+1}/o} \to \Omega^2_{Y \otimes o_n/o}$$

which are the maps in the first and second inverse systems in the left-hand side of the diagram. In the system $(\Omega^1_{o_n/o})_n$ the transition maps are trace, and in the remaining system $(o_n/\partial_n)_n$ the maps are $\frac{1}{p}\mathrm{tr}_{K_{n+1}/K_n}$. (For the compatibility of these various maps, see 3.3.12 below.)

From the discussion above, the diagram below commutes:

$$
\begin{array}{ccc}
H^1(K_m, H^1(Y \otimes \bar{K}, \mathbb{Q}_p)(1)) & \xrightarrow{\pi_1} & H^1(K_m, \widehat{\bar{K}} \otimes_o \mathrm{Fil}^1 H^1_{\mathrm{dR}}(Y/o)) \\
{\scriptstyle \exp^*}\downarrow & & \| \\
K_m \otimes_o \mathrm{Fil}^1 H^1_{\mathrm{dR}}(Y/o) & \xrightarrow[\cup \log \chi_{\mathrm{cycl}}]{\sim} & H^1(K_m, \widehat{\bar{K}}) \otimes_o \mathrm{Fil}^1 H^1_{\mathrm{dR}}(Y/o)
\end{array}
$$

To deduce the corollary from the theorem it is thus only necessary to take inverse limits and use the commutativity (cf. Proposition 3.3.10 below) of the following diagram

$$
\begin{array}{ccc}
K_n & \xrightarrow{\frac{1}{p^n}\log \chi_{\mathrm{cycl}}} & H^1(K_n, \widehat{\bar{K}}) \\
{\scriptstyle \frac{1}{p^{n-m}}\mathrm{tr}_{K_n/K_m}}\downarrow & & \downarrow{\scriptstyle \mathrm{cor}} \\
K_m & \xrightarrow{\frac{1}{p^m}\log \chi_{\mathrm{cycl}}} & H^1(K_m, \widehat{\bar{K}}).
\end{array}
\qquad (3.2.5)
$$

$\square$

Remarks. (i) Consider the special case $Y = \mathbb{A}^1 - \{0\} = \mathrm{Spec}\, o[t, t^{-1}]$. Let $(u_n) \in \varprojlim o_n^*$ be a universal norm. By applying the corollary to the norm-compatible symbols $\{u_n, t\} \in K_2(Y \otimes o_n)$ one recovers a form of Iwasawa's cyclotomic explicit reciprocity law, which will be proved more directly in 3.3.15 below.

(ii) Theorem 3.2.3 is proved in section 3.4 below. It is much easier than the general cases considered by Kato in [17], first because one is not working with coefficients in a general formal group, and secondly because the assumption that $X/\mathfrak{o}$ is smooth makes for considerable simplifications. In the non-smooth case there is an analogous statement which is needed to compute the image of Kato's Euler system when p divides the conductor.

3.3   Fontaine's theory

We shall review here some of the theory of differentials for local fields developed by Fontaine [11], and as a warm-up for the next section, show how it gives a version of Iwasawa's explicit reciprocity law.

Recall (see for example [34, §III.6–7]) that if K'/K is a finite extension then its valuations ring $\mathfrak{o}'$ equals $\mathfrak{o}[x]$ for some $x \in \mathfrak{o}'$. This implies that the module of Kähler differentials $\Omega_{\mathfrak{o}'/\mathfrak{o}}$ is a cyclic $\mathfrak{o}'$-module, generated by dx, and that its annihilator is the relative different $\mathfrak{d}_{K'/K}$.

The module $\Omega_{\bar{\mathfrak{o}}/\mathfrak{o}}$ equals the direct limit of $\Omega_{\mathfrak{o}'/\mathfrak{o}}$ taken over all finite extensions of K in $\bar{K}$. In particular, it is torsion.

Theorem 3.3.1. [11] *There is a short exact sequence of $\bar{\mathfrak{o}}$-modules*

$$0 \to \mathfrak{a}(1) \to \bar{K}(1) \overset{\alpha}{\to} \Omega_{\bar{\mathfrak{o}}/\mathfrak{o}} \to 0$$

where $\mathfrak{a} = \mathfrak{a}_{\bar{\mathfrak{o}}/\mathfrak{o}}$ is the fractional ideal

$$\mathfrak{a}_{\bar{\mathfrak{o}}/\mathfrak{o}} = (\zeta_p - 1)^{-1} \mathfrak{d}_{K/\mathbb{Q}_p}^{-1} \bar{\mathfrak{o}} \subset \bar{K}$$

and where $\alpha \colon \bar{K}(1) := \mathbb{Z}_p(1) \otimes \bar{K} \to \Omega_{\bar{\mathfrak{o}}/\mathfrak{o}}$ is the unique $\bar{\mathfrak{o}}$-linear map satisfying

$$\alpha([\zeta_{p^m}]_m \otimes p^{-n}) = \text{dlog}\, \zeta_{p^n} = \frac{d\zeta_{p^n}}{\zeta_{p^n}}$$

for any $n \geq 0$.   □

Remark 3.3.2. In particular, for any $n \geq 0$ the annihilator of $\text{dlog}\, \zeta_{p^n} \in \Omega_{\bar{\mathfrak{o}}/\mathfrak{o}}$ is $p^n \mathfrak{a} \cap \bar{\mathfrak{o}}$.

From 3.3.1 we get the fundamental canonical isomorphism

$$\widehat{\mathfrak{a}}_{\bar{\mathfrak{o}}/\mathfrak{o}}(1) \overset{\sim}{\to} T_p \Omega_{\bar{\mathfrak{o}}/\mathfrak{o}} \tag{3.3.3}$$

which is $\bar{\mathfrak{o}}$-linear, and maps $(\zeta_{p^n})_n \in \mathbb{Z}_p(1) \subset \widehat{\mathfrak{a}}(1)$ to $(\text{dlog}\, \zeta_{p^n})_n$.

Suppose that $K''/K'/K$ are finite extensions. Then there is an exact sequence of differentials

$$\Omega_{\mathfrak{o}'/\mathfrak{o}} \otimes_{\mathfrak{o}'} \mathfrak{o}'' \to \Omega_{\mathfrak{o}''/\mathfrak{o}} \to \Omega_{\mathfrak{o}''/\mathfrak{o}'} \to 0$$

(the "first exact sequence", [23, 26.H]), which is exact on the left as well by the multiplicativity of the different (or alternatively by the argument in the footnote on page 420). Passing to the direct limit over K'' gives a short exact sequence

$$0 \to \Omega_{\mathfrak{o}'/\mathfrak{o}} \otimes_{\mathfrak{o}'} \bar{\mathfrak{o}} \to \Omega_{\bar{\mathfrak{o}}/\mathfrak{o}} \to \Omega_{\bar{\mathfrak{o}}/\mathfrak{o}'} \to 0 \qquad (3.3.4)$$

At this point, recall that for any short exact sequence $0 \to X \to Y \to Z \to 0$ of abelian groups, there is an inverse system of long exact sequences

$$0 \to {}_{p^n}X \to {}_{p^n}Y \to {}_{p^n}Z \to X/p^n \to Y/p^n \to Z/p^n \to 0. \qquad (3.3.5)$$

If the inverse systems ${}_{p^n}M$ (for $M = X$, Y, Z) satisfy the Mittag-Leffler condition (ML) then the inverse limit sequence

$$0 \to T_pX \to T_pY \to T_pZ \to \varprojlim X/p^n \to \varprojlim Y/p^n \to \varprojlim Z/p^n \to 0$$

is also exact (a special case of EGA 0, 13.2.3). Note that $({}_{p^n}M)$ satisfies (ML) in two particular cases:

- the torsion subgroup of M is p-divisible (then ${}_{p^n}M \to {}_{p^{n-1}}M$ is surjective);

- the p-primary torsion subgroup of M has finite exponent (then $({}_{p^n}M)$ is ML-zero).

Applying these considerations to (3.3.4), since $\Omega_{\bar{\mathfrak{o}}/\mathfrak{o}}$ and $\Omega_{\bar{\mathfrak{o}}/\mathfrak{o}'}$ are divisible and $\Omega_{\mathfrak{o}'/\mathfrak{o}}$ is killed by a power of p, we get an exact sequence

$$0 \to T_p\Omega_{\bar{\mathfrak{o}}/\mathfrak{o}} \to T_p\Omega_{\bar{\mathfrak{o}}/\mathfrak{o}'} \to \Omega_{\mathfrak{o}'/\mathfrak{o}} \otimes_{\mathfrak{o}'} \bar{\mathfrak{o}} \to 0. \qquad (3.3.6)$$

Now pass to continuous Galois cohomology. This gives a long exact sequence since the surjection in (3.3.6) has a continuous set-theoretic section (this is obvious here as $\Omega_{\mathfrak{o}'/\mathfrak{o}} \otimes_{\mathfrak{o}'}\bar{\mathfrak{o}}$ is discrete). We are only interested in the connecting map, and define δ to be the composite homomorphism:

$$\delta = \delta_{K'} : \Omega_{\mathfrak{o}'/\mathfrak{o}} \hookrightarrow H^0(K', \Omega_{\mathfrak{o}'/\mathfrak{o}} \otimes_{\mathfrak{o}'} \bar{\mathfrak{o}}) \xrightarrow{\text{connecting}} H^1(K', T_p\Omega_{\bar{\mathfrak{o}}/\mathfrak{o}}).$$

The map "reduction mod p^n" : $T_p\Omega_{\bar{\mathfrak{o}}/\mathfrak{o}} \to {}_{p^n}\Omega_{\bar{\mathfrak{o}}/\mathfrak{o}}$ induces a map on cohomology, which when composed with $\delta_{K'}$ gives

$$\delta_{K'} \;(\text{mod } p^n) : \Omega_{\mathfrak{o}'/\mathfrak{o}} \to H^1(K', {}_{p^n}\Omega_{\bar{\mathfrak{o}}/\mathfrak{o}}).$$

Lemma 3.3.7. *(i) The following diagram commutes:*

$$
\begin{array}{ccc}
\mathfrak{o}'^* & \xrightarrow{\text{Kummer}} & H^1(K', \mu_{p^n}) \\
\text{dlog} \downarrow & & \downarrow \text{dlog} \\
\Omega_{\mathfrak{o}'/\mathfrak{o}} & \xrightarrow{\delta_{K'} \bmod p^n} & H^1(K', {}_{p^n}\Omega_{\bar{\mathfrak{o}}/\mathfrak{o}})
\end{array}
$$

(ii) For any nonzero $x \in \mathfrak{o}'$

$$
\delta_{K'}(dx)\ (\bmod\ p^n) = x\,\text{dlog}(\text{Kummer}(x))
$$

Proof. (i) Simply compute: if $u \in \mathfrak{o}'^*$ then fix a sequence (u_m) in $\bar{\mathfrak{o}}^*$ with $u_0 = u$, $u_{m+1}^p = u_m$. The composite $\text{dlog} \circ \text{"Kummer"}$ maps u to the class of the cocycle

$$
g \mapsto \text{dlog}(u_n^{g-1}) \in {}_{p^n}\Omega_{\bar{\mathfrak{o}}/\mathfrak{o}}.
$$

Now compute the effect of $\delta_{K'}$ on $\text{dlog}\,u$: first lift $\text{dlog}\,u$ in the exact sequence (3.3.6) to the element $(\text{dlog}\,u_m)_m \in T_p(\Omega_{\bar{\mathfrak{o}}/\mathfrak{o}'})$, then act by $g-1$ to get the desired cocycle. So the commutativity is trivial.

(ii) If x is a unit this is equivalent to (i). For the general case one simply calculates as in (i).   □

Lemma 3.3.8. *Let $n \geq 1$ and assume that $\mu_{p^n} \subset K'$. If $p \neq 2$, then the diagram*

$$
\begin{array}{ccc}
\mathfrak{o}'/p^n(1) & \xrightarrow{\cup \frac{1}{p^n} \log \chi_{\text{cycl}}} & H^1(K', \mathfrak{a}_{\bar{\mathfrak{o}}/\mathfrak{o}}/p^n)(1) \\
{}_{1 \otimes [\zeta_{p^n}] \mapsto \text{dlog}\,\zeta_{p^n}} \downarrow & & \wr \downarrow {}_{(3.3.3)} \\
\Omega^1_{\mathfrak{o}'/\mathfrak{o}} & \xrightarrow{\delta_{K'} \bmod p^n} & H^1(K', {}_{p^n}\Omega^1_{\bar{\mathfrak{o}}/\mathfrak{o}})
\end{array}
$$

commutes. For $p = 2$ it commutes mod 2^{n-1}.

Proof. All the maps are $\mathfrak{o}'$-linear, so it is enough to compute the image of $1 \otimes [\zeta_{p^n}]$. We have $\chi_{\text{cycl}}(g) \equiv 1\ (\bmod\ p^n)$ for all $g \in G_K$, hence $\log \chi_{\text{cycl}}(g) \equiv 0\ (\bmod\ p^n)$ and so if $p \neq 2$ then

$$
\frac{1}{p^n} \log \chi_{\text{cycl}}(g) \equiv \frac{1}{p^n}(\chi_{\text{cycl}}(g) - 1) \quad \bmod\ p^n.
$$

In the proof of 3.3.7 one can take $u_m = \zeta_{p^{m+n}}$ for all $m \geq 0$, and then $\delta_{K'}(\text{dlog}\,\zeta_{p^n}) \in H^1(K', T_p(\Omega_{\bar{\mathfrak{o}}/\mathfrak{o}}))$ is represented by the cocycle

$$
g \mapsto (\text{dlog}\,\zeta_{p^{m+n}}^{g-1})_m \in T_p(\Omega_{\bar{\mathfrak{o}}/\mathfrak{o}})
$$

and $\zeta_{p^{m+n}}^{g-1} = \zeta_{p^{m+n}}^{\chi_{\text{cycl}}(g)-1} = \zeta_{p^m}^{(\chi_{\text{cycl}}(g)-1)/p^n}$. Applying the inverse of (3.3.3) maps this to the class of the cocycle

$$g \mapsto \frac{1}{p^n}(\chi_{\text{cycl}}(g) - 1) \otimes (\zeta_{p^m})_m \qquad \in \mathfrak{a}_{\bar{\mathfrak{o}}/\mathfrak{o}}(1)$$

$$\equiv \frac{1}{p^n} \log \chi_{\text{cycl}}(g) \mod p^n.$$

The reader will make the necessary modifications when $p = 2$. $\qquad\square$

We now need some elementary facts about cyclotomic extensions of local fields. Our chosen normalisation of the reciprocity law of local class field theory identifies the homomorphisms

$$\log \chi_{\text{cycl}} \in H^1(K, \mathbb{Z}_p) = \text{Hom}_{\text{cts}}(\text{Gal}(\bar{K}/K), \mathbb{Z}_p)$$

and

$$\log \circ N_{K/\mathbb{Q}_p} : K^* \longrightarrow \mathbb{Z}_p.$$

As observed in the proof of the previous lemma, if $\mu_{p^m} \subset K$ then $\log \chi_{\text{cycl}} \equiv 0 \pmod{p^m}$.

Lemma 3.3.9. *Suppose that* $\mu_{p^m} \subset K$. *Then for any finite extension* K'/K *the diagram*

$$\begin{array}{ccc}
\mathfrak{o}' & \xrightarrow{\cup \frac{1}{p^m} \log \chi_{\text{cycl}}} & H^1(K', \widehat{\mathfrak{o}}) \\
{\scriptstyle \text{tr}_{K'/K}} \downarrow & & \downarrow {\scriptstyle \text{cor}} \\
\mathfrak{o} & \xrightarrow{\cup \frac{1}{p^m} \log \chi_{\text{cycl}}} & H^1(K, \widehat{\mathfrak{o}})
\end{array}$$

commutes.

Proof. The statement follows from the projection formula for cup-product in group cohomology, since on H^0 the corestriction

$$\text{cor} : H^0(K', \widehat{\mathfrak{o}}) = \mathfrak{o}' \longrightarrow H^0(K, \widehat{\mathfrak{o}}) = \mathfrak{o}$$

equals $\text{tr}_{K'/K}$. $\qquad\square$

Recall that $K_n := K(\zeta_{p^n})$. Let ℓ be the largest integer such that $\mu_{p^\ell} \subset K$. Then if $n > m \geq \ell$, direct calculation gives

$$\text{tr}_{K_n/K_m}\left(\mathfrak{o}[\zeta_{p^n}]\right) = \begin{cases} p^{n-m}\mathfrak{o}[\zeta_{p^m}] & \text{if } m > 0 \\ p^{n-1}\mathfrak{o} & \text{if } m = 0. \end{cases}$$

Define, for any $n > m \geq 0$

$$t_{n,m} := \frac{1}{p^{n-m}} \text{tr}_{K_n/K_m} : K_n \to K_m.$$

Proposition 3.3.10. *If* $n > m \geq \max(\ell, 1)$ *the diagram*

$$
\begin{array}{ccc}
\mathfrak{o}[\zeta_{p^n}] & \xrightarrow{\cup \frac{1}{p^n-1} \log \chi_{cycl}} & H^1(K_n, \widehat{\mathfrak{o}}) \\
t_{n,m} \downarrow & & \downarrow \text{cor} \\
\mathfrak{o}[\zeta_{p^m}] & \xrightarrow{\cup \frac{1}{p^m-1} \log \chi_{cycl}} & H^1(K_m, \widehat{\mathfrak{o}})
\end{array}
$$

is commutative. If $n > \ell = 0$, *the diagram*

$$
\begin{array}{ccc}
\mathfrak{o}[\zeta_{p^n}] & \xrightarrow{\cup \frac{1}{p^n-1} \log \chi_{cycl}} & H^1(K_n, \widehat{\mathfrak{o}}) \\
pt_{n,0} \downarrow & & \downarrow \text{cor} \\
\mathfrak{o} & \xrightarrow{\cup \log \chi_{cycl}} & H^1(K, \widehat{\mathfrak{o}})
\end{array}
$$

is commutative.

Proof. For $n = 1$, the second diagram commutes by 3.3.9 with $K' = K_1$. By transitivity of trace and corestriction, the lemma will be proved if we verify the commutativity of the first diagram for $n = m + 1 > 1$. Take the diagram of 3.3.9 for K_{m+1}/K_m and factorise:

$$
\begin{array}{ccc}
\mathfrak{o}[\zeta_{p^{m+1}}] & \xrightarrow{\cup p^{-m} \log \chi_{cycl}} & H^1(K_{m+1}, \widehat{\mathfrak{o}}) \\
\downarrow \text{tr}_{K_{m+1}/K_m} & & \downarrow \text{cor} \\
p\mathfrak{o}[\zeta_{p^m}] \xrightarrow{\cup p^{-m} \log \chi_{cycl}} & H^1(K_m, \mathfrak{o}_m) \longrightarrow & H^1(K_m, \widehat{\mathfrak{o}}) \\
\uparrow \times p & \cup p^{1-m} \log \chi_{cycl} & \\
\mathfrak{o}[\zeta_{p^m}] & &
\end{array}
$$

The bottom triangle commutes since $H^1(K_m, \mathfrak{o}_m) = \text{Hom}_{cts}(\text{Gal}(\overline{K}/K_m), \mathfrak{o}_m)$ is torsion-free. Hence the entire diagram is commutative, and going round the outside gives what we need. $\square$

Now consider $\partial_n = \partial_{K_n/K}$. From the definition of ∂_n^{-1} as the largest fractional ideal of K_n whose trace is contained in $\mathfrak{o}$, it is an easy exercise to check

$$
\partial_n^{-1} \subset p^{-n} \mathfrak{o}[\zeta_{p^n}].
$$

By [37, Propn. 5] the difference $v_p(\partial_n) - n$ is bounded, so for some c independent of n,

$$
p^c \mathfrak{o}_n \subset \mathfrak{o}[\zeta_{p^n}] \subset \mathfrak{o}_n.
$$

Since $\Omega_{o_n/o}$ is cyclic with annihilator $\mathfrak{d}_n$, the homomorphism

$$o[\zeta_{p^n}]/p^n \xrightarrow{\; x \mapsto x\, \mathrm{dlog}\, \zeta_{p^n}\;} \Omega_{o_n/o} \qquad (3.3.11)$$

is well-defined, and its kernel and cokernel are killed by a bounded power of p, by remark 3.3.2.

Proposition 3.3.12. *Let $n > m \geq \max(\ell, 1)$. Then the diagram*

$$
\begin{array}{ccc}
o[\zeta_{p^n}] & \xrightarrow{\;\times\, \mathrm{dlog}\, \zeta_{p^n}\;} & \Omega_{o_n/o} \\[4pt]
{\scriptstyle t_{n,m}}\downarrow & & \downarrow{\scriptstyle \mathrm{tr}} \\[4pt]
o[\zeta_{p^m}] & \xrightarrow{\;\times\, \mathrm{dlog}\, \zeta_{p^m}\;} & \Omega_{o_m/o}
\end{array}
$$

commutes.

Proof. It is enough to compute what happens when $m = n - 1$. Taking $1, \zeta_{p^n}, \ldots, \zeta_{p^n}^{p-1}$ as basis for $o[\zeta_{p^n}]$ over $o[\zeta_{p^{n-1}}]$, for $1 \leq j < p$

$$\mathrm{tr}(\zeta_{p^n}^j \, \mathrm{dlog}\, \zeta_{p^n}) = \mathrm{tr}(j^{-1} d\zeta_{p^n}^j) = j^{-1} d(\mathrm{tr}\, \zeta_{p^n}^j) = 0$$

and for $j = 0$

$$\mathrm{tr}(\mathrm{dlog}\, \zeta_{p^n}) = \mathrm{dlog}(N_{K_n/K_{n-1}} \zeta_{p^n}) = \mathrm{dlog}\, \zeta_{p^{n-1}}. \qquad \square$$

Therefore passing to the inverse limit gives a homomorphism

$$\varprojlim_{t_{-,-}} o[\zeta_{p^n}] = \varprojlim_{t_{-,-}} o[\zeta_{p^n}]/p^n \longrightarrow \varprojlim_{\mathrm{tr}} \Omega_{o_n/o} \qquad (3.3.13)$$

which becomes an isomorphism when tensored with $\mathbb{Q}$. (If $K/\mathbb{Q}_p$ is unramified, then (3.3.13) is itself an isomorphism.) By [38, Proposition 2.2], the canonical map $H^1(K_m, \mathbb{Z}_p) \to \varprojlim_n H^1(K_m, \mathbb{Z}/p^n)$ is an isomorphism. Inverting both of these arrows yields a diagram

$$
\begin{array}{ccc}
\mathbb{Q} \otimes \varprojlim_{\mathrm{norm}} o_n^* & \xrightarrow{\;\text{Kummer}\;} & \mathbb{Q} \otimes \varprojlim_n H^1(K_n, \mu_n) \\[6pt]
{\scriptstyle \mathrm{dlog}}\downarrow & & \downarrow{\scriptstyle \zeta_{p^n} \mapsto 1} \\[6pt]
\mathbb{Q} \otimes \varprojlim_{\mathrm{trace}} \Omega_{o_n/o} & & \mathbb{Q} \otimes \varprojlim_n H^1(K_n, \mathbb{Z}/p^n) \\[6pt]
{\scriptstyle \mathrm{dlog}\, \zeta_{p^n} \mapsto 1}\downarrow{\scriptstyle \wr} & & \downarrow{\scriptstyle \mathrm{cor}} \qquad (3.3.14) \\[6pt]
\mathbb{Q} \otimes \varprojlim_{t_{-,-}} o[\zeta_{p^n}]/p^n & & \mathbb{Q} \otimes \varprojlim_n H^1(K_m, \mathbb{Z}/p^n) \\[6pt]
{\scriptstyle (t_{n,m})_n}\downarrow & & \downarrow \\[6pt]
K_m & \xrightarrow[\; \cup \frac{1}{p^m} \log \chi_{\mathrm{cycl}} \;]{\;\sim\;} & H^1(K_m, \widehat{\widehat{K}})
\end{array}
$$

where down the right-hand side all the inverse limits are with respect to the corestriction maps and reduction mod p^n. We then have the following version of the classical explicit reciprocity law of Artin-Hasse and Iwasawa. Without loss of generality we can assume m chosen so that $\mu_{p^{m+1}} \not\subset K$.

Theorem 3.3.15. *The diagram* (3.3.14) *is commutative.*

Proof. At finite level, replacing $\mathbb{Z}/p^n$ with $\bar{\mathfrak{o}}/p^n$, one has the diagram:

$$
\begin{array}{ccc}
\mathfrak{o}_n^* & \xrightarrow{\text{Kummer}} & H^1(K_n, \mu_{p^n}) \\
\downarrow{\scriptstyle \text{dlog}} & & \downarrow{\scriptstyle \text{dlog}} \\
\Omega^1_{\mathfrak{o}_n/\mathfrak{o}} & \xrightarrow{\delta_{K_n} \bmod p^n} & H^1(K_n, p^n\Omega_{\bar{\mathfrak{o}}/\mathfrak{o}}) \\
\uparrow{\scriptstyle 1 \mapsto \text{dlog}\,\zeta_{p^n}} & & \uparrow{\scriptstyle 1 \mapsto \text{dlog}\,\zeta_{p^n}} \\
\mathfrak{o}[\zeta_{p^n}]/p^n & \xrightarrow{\cup \frac{1}{p^n}\log\chi_{\text{cycl}}} & H^1(K_n, \bar{\mathfrak{o}}/p^n) \\
\downarrow{\scriptstyle t_{n,m}} & & \downarrow{\scriptstyle \text{cor}} \\
\mathfrak{o}[\zeta_{p^m}]/p^n & \xrightarrow{\cup \frac{1}{p^m}\log\chi_{\text{cycl}}} & H^1(K_m, \bar{\mathfrak{o}}/p^n).
\end{array}
$$

with $\zeta_{p^n} \mapsto 1$ bracketing the right-hand maps.

This is for $m > 0$; for $m = 0$ the bottom arrow should read $p^{-1}\mathfrak{o}/p^n \to H^1(K, p^{-1}\bar{\mathfrak{o}}/p^n)$. The top two squares commute by 3.3.7, 3.3.8 respectively. The bottom square commutes up to p-torsion by (3.2.5). All maps are compatible with passing to the inverse limit. As remarked after equation (3.3.11), the left-hand map labelled "$1 \mapsto \text{dlog}\,\zeta_{p^n}$" has cokernel and kernel killed by a bounded power of p, and by (3.3.3) the same is true for the one on the right. Therefore passing to the limit and tensoring with $\mathbb{Q}$ one obtains the theorem. $\square$

Remark. One can use 3.3.7(ii) to describe the image of an arbitrary element of K_n^* under the Kummer map in a similar way.

Here is the relation with the usual form of the explicit reciprocity law. Let $u = (u_n)_n \in \varprojlim \mathfrak{o}_n^*$ be a universal norm. Its image down the left hand side of the diagram (3.3.14) equals (with an obvious abuse of notation)

$$
\Phi(u) := \lim_{n\to\infty} \frac{1}{p^{n-m}} \operatorname{tr}_{K_n/K_m} \left(\frac{\text{dlog}\,u_n}{\text{dlog}\,\zeta_{p^n}} \right) \in K_m.
$$

Going round the other way, use the expression of the Kummer map in terms of the Hilbert symbol, which we write as a bilinear map $[-,-]_n : K_n^* \times K_n^* \to \mathbb{Z}/p^n$ given by

$$
\left(\sqrt[p^n]{x} \right)^{(a, K_n^{\text{ab}}/K_n)-1} = \zeta_{p^n}^{[x,a]_n}
$$

Thus u_n is mapped to the cocycle in $H^1(K_n, \mathbb{Z}/p^n)$ which takes the norm residue symbol $(a, K_n^{\mathrm{ab}}/K_n)$ to $[u_n, a]_n$. By the compatibility of the norm residue symbol with norm and corestriction, one gets that the image of the family u in $H^1(K_m, \mathbb{Z}_p)$ is represented by the cocycle (i.e. homomorphism)

$$(a, K_m^{\mathrm{ab}}/K_m) \longmapsto \lim_{n \to \infty} [u_n, a]_n \in \mathbb{Z}_p$$

Therefore the reciprocity law says that this homomorphism, and the homomorphism

$$g \longmapsto p^{-m} \Phi(u) \log \chi_{\mathrm{cycl}}(g)$$

represent the same cohomology class in $H^1(K_m, \widehat{\overline{K}})$.

Proposition 3.3.16. [35, III.A7 ex. 2] *Let* $c_K \colon \mathrm{Hom}_{\mathbb{Q}_p}(K, K) \to K$ *be the unique map such that for all* $T \in \mathrm{Hom}_{\mathbb{Q}_p}(K, K)$ *and all* $x \in K$,

$$\mathrm{tr}([\times x] \circ T) = \mathrm{tr}_{K/\mathbb{Q}_p}(x\, c_K(T)).$$

Then the diagram

$$
\begin{array}{ccc}
\mathrm{Hom}_{\mathbb{Q}_p}(K, K) & \xrightarrow{c_K} & K \\
{\scriptstyle \circ \log}\downarrow & & \downarrow{\scriptstyle \wr\, \log \chi_{\mathrm{cycl}}} \\
\mathrm{Hom}_{\mathrm{cts}}(\mathfrak{o}^*, K) & & \\
{\scriptstyle \text{local CFT}}\uparrow{\scriptstyle \wr} & & \\
\mathrm{Hom}_{\mathrm{cts}}(G_K^{\mathrm{ab}}, K) & \longrightarrow & H^1(G_K, \widehat{\overline{K}})
\end{array}
$$

is commutative. □

Remark. Because of the normalisation of the reciprocity law of local class field theory used here (see §3.1), this differs from the statement in [35] by a sign.

Now the composite

$$\mathrm{Hom}_{\mathbb{Q}_p}(K, \mathbb{Q}_p) \hookrightarrow \mathrm{Hom}_{\mathbb{Q}_p}(K, K) \xrightarrow{c_K} K$$

is the inverse of the isomorphism $K \xrightarrow{\sim} \mathrm{Hom}_{\mathbb{Q}_p}(K, \mathbb{Q}_p)$ given by the trace form. Therefore, for every $a \in \mathfrak{o}_m^*$,

$$\lim_{n \to \infty} [u_n, a]_n = p^{-m}\, \mathrm{tr}_{K_m/\mathbb{Q}_p}(\Phi(u) \log a)$$

which is the "limit form" of the classical explicit reciprocity law [20, Ch. 9, Thm. 1.2].

3.4   Big local fields

This section reviews the generalisation by Hyodo [14, esp. §4] and Faltings [10, §2] of Fontaine's theory to local fields with imperfect residue field. We consider fields $L \supset \mathbb{Q}_p$ such that:

> L is complete with respect to a discrete valuation, and its residue field ℓ satisfies $[\ell : \ell^p] = p^r < \infty$. (3.4.1)

Fix such a field L, and write A for its ring of integers. If $R \subset A$ is any subring, define

$$\widehat{\Omega}_{A/R} := \varprojlim \Omega_{A/R}/p^n \Omega_{A/R}.$$

Fix also an algebraic closure $\bar{L}$ of L, and let $\bar{A}$ be the integral closure of A in $\bar{L}$. For any B with $A \subset B \subset \bar{A}$ and any subring $R \subset B$ set

$$\widehat{\Omega}_{B/R} = \varinjlim_{A'} \widehat{\Omega}_{A'/A' \cap R}$$

the limit running over all finite extensions A'/A contained in B.

Let $K \subset L$ be a finite extension of $\mathbb{Q}_p$, with ring of integers $\mathfrak{o}$ and uniformiser π_K. Then π_K is prime in A if and only if $A/\mathfrak{o}$ is formally smooth (by [23], (28.G) and Theorems 62, 82).

Let L'/L be a finite extension with valuation ring A'. Then A' is finite over A (being the normalisation of a complete DVR in a finite extension), and is a relative complete intersection (by EGA IV 19.3.2). Therefore the first exact sequence of differentials is exact on the left as well[4]

$$0 \to A' \otimes_A \Omega_{A/\mathfrak{o}} \to \Omega_{A'/\mathfrak{o}} \to \Omega_{A'/A} \to 0.$$

------

[4]More generally, if A'/A is a relative complete intersection of integral domains which is generically smooth, then for any $R \subset A$ the first exact sequence is exact on the left. For an elementary proof, write A' as the quotient B/I of a polynomial algebra B over A by an ideal I generated by a regular sequence. Then one has a split exact sequence

$$0 \to \Omega_{A/R} \otimes B \to \Omega_{B/R} \to \Omega_{B/A} \to 0 \tag{1}$$

as well as exact sequences, for $? = A$ or R,

$$I/I^2 \to \Omega_{B/?} \otimes_B A' \to \Omega_{A'/?} \to 0. \tag{2}$$

Applying the tensor product $\otimes_B A'$ to (1), and using (2) and the snake lemma, gives the exact sequence

$$N_{A'/A} \to A' \otimes_A \Omega_{A/R} \to \Omega_{A'/R} \to \Omega_{A'/A} \to 0.$$

where $N_{A'/A} = \ker(I/I^2 \to \Omega_{B/A} \otimes_B A')$. Since A'/A is generically smooth the map $I/I^2 \to \Omega_{B/A}$ is generically an injection, hence $N_{A'/A}$ is torsion. Now I/I^2 is projective since I is a regular ideal; therefore $N_{A'/A} = 0$.

As in (3.3.5), we get an exact sequence of inverse systems

$$_{p^n}\Omega_{A'/A} \to A' \otimes_A \Omega_{A/o}/p^n \to \Omega_{A'/o}/p^n \to \Omega_{A'/A}/p^n \to 0.$$

Since $\Omega_{A'/A}$ is a finite A'-module, the inverse system $(_{p^n}\Omega_{A'/A})$ is ML-zero, and so passing to the inverse limit gives an exact sequence:

$$0 \to A' \otimes_A \widehat{\Omega}_{A/o} \to \widehat{\Omega}_{A'/o} \to \Omega_{A'/A} \to 0. \tag{3.4.2}$$

Proposition 3.4.3. *(i)* $\widehat{\Omega}_{A/o}$ *is a finite A-module, generated by elements of the form dy, $y \in A^*$.*

(ii) If $T_1, \ldots T_r \in A$ are elements whose whose images in ℓ form a p-basis, then $\{\mathrm{dlog}\, T_i\}$ is a basis for the vector space $\widehat{\Omega}_{A/o} \otimes_A L$.

(iii) If π_K is prime in A, then $\widehat{\Omega}_{A/o}$ is free over A.

Proof. By [23] pp. 211–212, A is a finite extension of a complete DVR B in which p is prime. Then $A_0 = Bo$ is a complete DVR with uniformiser π_K, and A/A_0 is finite and totally ramified. Let $k = o/\pi_K o$. One knows (*loc. cit.*, Thm. 86) that the image of $\{dT_i\}$ is an ℓ-basis for $\Omega_{\ell/k}$, and therefore (by Nakayama's lemma) $\widehat{\Omega}_{A_0/o} = \bigoplus A_0 \cdot dT_i = \bigoplus A_0 \cdot \mathrm{dlog}\, T_i$, proving (iii). To deduce (i) and (ii), it is enough to apply the exact sequence (3.4.2) to $A/A_0/o$. □

Taking the direct limit of (3.4.2) over all finite extensions L'/L, one gets an exact sequence

$$0 \to \bar{A} \otimes_A \widehat{\Omega}_{A/o} \to \widehat{\Omega}_{\bar{A}/o} \to \Omega_{\bar{A}/A} \to 0$$

of $\bar{A}$-modules. Now apply (3.3.5) again. Since $x\, dy = pz^{p-1}x\, dz$ if $y = z^p$, one sees (using 3.4.3(i)) that $\widehat{\Omega}_{\bar{A}/o}$ and $\Omega_{\bar{A}/A}$ are divisible. Therefore, since $\widehat{\Omega}_{A/o}$ is finitely generated, one can pass to the limit to get an exact sequence of $\widehat{\bar{A}}$-modules

$$0 \to T_p(\widehat{\Omega}_{\bar{A}/o}) \to T_p(\Omega_{\bar{A}/A}) \xrightarrow{\pi} \widehat{\bar{A}} \otimes_A \widehat{\Omega}_{A/o} \to 0. \tag{3.4.4}$$

Because $\widehat{\Omega}_{A/o}$ is a finite A-module, the map π has a continuous set-theoretic section (write $\widehat{\Omega}_{A/o} = P \oplus N$ with P free and N torsion; over $\widehat{\bar{A}} \otimes P$ one has a continuous linear section of π by freeness, and $\widehat{\bar{A}} \otimes N$ is discrete, so over it one can take any section).

One then has Hyodo's generalisation [14, (4-2-2)] of 3.3.1 (see also [10, §2b)]):

Proposition 3.4.5. *Let* $\mathfrak{a}_{\bar{o}/o}$ *be as in 3.3.1 above, and put* $\mathfrak{a}_{\bar{A}/o} = \mathfrak{a}_{\bar{o}/o}\bar{A} \subset \bar{L}$. *Then there is an exact sequence of* $\bar{A}$*-modules and Galois-equivariant maps*

$$0 \to \mathfrak{a}_{\bar{A}/o}(1) \overset{\subseteq}{\to} \bar{L}(1) \overset{\alpha}{\to} \widehat{\Omega}_{\bar{A}/o} \overset{\beta}{\to} \bar{L}^r \to 0$$

where α *is given by the same formula as in 3.3.1, and where the map* β *is a split surjection, with right inverse*

$$\bar{L}^r \to \widehat{\Omega}_{\bar{A}/o}$$

$$(a_1/p^n, \ldots, a_r/p^n) \mapsto \sum a_i \, \mathrm{dlog}(T_i^{p^{-n}}) \quad (a_i \in \bar{A}). \qquad \square$$

Remark. Hyodo states this only in the case $K = \mathbb{Q}_p$, but his proof works in general. The key point (which underlies Faltings' approach to p-adic Hodge theory) is that the extension $\bar{L}/L(\mu_{p^\infty}, T_i^{p^{-\infty}})$ is almost unramified (cf. the proof of Proposition 3.4.12 below), which shows that $\widehat{\Omega}_{\bar{A}/o}$ is generated as an $\widehat{\bar{A}}$-module by the forms $\mathrm{dlog}\,\zeta_{p^n}$, $\mathrm{dlog}\,T_i^{p^{-n}}$.

Corollary 3.4.6. *There is a unique isomorphism*

$$\widehat{\mathfrak{a}}_{\bar{A}/o}(1) \overset{\sim}{\longrightarrow} T_p(\widehat{\Omega}_{\bar{A}/o}) \qquad (3.4.7)$$

which maps $(\zeta_{p^n})_n \in \mathbb{Z}_p(1)$ *to* $(\mathrm{dlog}\,\zeta_{p^n})_n$. $\qquad \square$

Remark. Comparing (3.3.3) and (3.4.7) we have in particular

$$_{p^n}\widehat{\Omega}_{\bar{A}/o} = {}_{p^n}\Omega_{\bar{o}/o} \otimes_{\bar{o}} \bar{A}. \qquad (3.4.8)$$

Now consider as before the connecting homomorphism attached to the Galois cohomology of (3.4.4), for a (not necessarily finite) extension L'/L contained in $\bar{L}$:

$$\delta_{L'/L} \colon \widehat{\Omega}_{A/o} \otimes_A \widehat{A}' \to H^1(L', T_p \widehat{\Omega}_{\bar{A}/o}) \qquad (3.4.9)$$

For $L' = L$ we write δ_L for $\delta_{L'/L}$. If L'/L is finite the maps $\delta_{L'}$, $\delta_{L'/L}$ are related by a commutative diagram

$$(3.4.10)$$

(because the exact sequence (3.4.4) is functorial in A). If L'/L is infinite, we define $\delta_{L'} \colon \widehat{\Omega}_{A'/o} \to H^1(L', T_p \widehat{\Omega}_{\bar{A}/o})$ as the direct limit of the maps $\delta_{L''}$, for finite subextensions $L \subset L'' \subset L'$; the analogue of (3.4.10) still holds.

The following lemma is proved just the same way as 3.3.7.

Lemma 3.4.11. *For any algebraic extension L'/L, the following diagram commutes:*

$$
\begin{array}{ccc}
A'^* & \xrightarrow{\ \text{Kummer}\ } & H^1(L', \mu_{p^n}) \\[4pt]
\text{dlog}\Big\downarrow & & \Big\downarrow\text{dlog} \\[4pt]
\widehat{\Omega}_{A'/\mathfrak{o}} & \xrightarrow{\ \delta_{L'}\ \bmod p^n\ } & H^1(L', {}_{p^n}\widehat{\Omega}_{\bar{A}/\mathfrak{o}})
\end{array}
$$
$\hspace{2cm}\square$

Proposition 3.4.12. *Let L_∞/L be an algebraic extension which contains all p-th power roots of unity, with valuation ring A_∞, and whose residue field extension is separable. Suppose that $r = 1$, so that $[l : l^p] = p$. Then for $j \geq 2$, $H^j(L_\infty, T_p\widehat{\Omega}_{\bar{A}/\mathfrak{o}})$ is killed by the maximal ideal $\mathfrak{m}_\infty \subset A_\infty$, and the kernel and cokernel of*

$$
\delta_{L_\infty/L} : \widehat{\Omega}_{A/\mathfrak{o}} \otimes_A \widehat{A_\infty} \to H^1(L_\infty, T_p\widehat{\Omega}_{\bar{A}/\mathfrak{o}})
$$

are killed by a power of p.

Proof. Initially there is no need to make any assumption on r. Choose units $T_1, \ldots, T_r \in A^*$ whose images in l form a p-basis. Consider the extensions $M = L(T_i^{p^{-\infty}}, \ldots T_i^{p^{-\infty}})$ and $M_\infty = ML_\infty$. Let B, B_∞ be the valuation rings of M, M_∞. Then the residue field of M is perfect, so Tate's theory [37] applies; in particular, the groups $H^i(M_\infty, \widehat{\bar{A}})$ are $\mathfrak{m}_\infty$-torsion for $i > 0$. Therefore, using the Hochschild-Serre spectral sequence and the fact that $\mathfrak{m}_\infty^2 = \mathfrak{m}_\infty$, the inflation map

$$
H^j(M_\infty/L_\infty, \widehat{B_\infty}) = H^j(M_\infty/L_\infty, H^0(M_\infty, \widehat{\bar{A}})) \to H^j(L_\infty, \widehat{\bar{A}}) \qquad (3.4.13)
$$

is an isomorphism up to $\mathfrak{m}_\infty$-torsion. Now by Kummer theory and the hypothesis on the residue fields, $\mathrm{Gal}(M_\infty/L_\infty) \simeq \mathbb{Z}_p(1)^r$ (the isomorphism being determined by the choice of $\{T_i\}$). Therefore if $r = 1$

$$
H^j(M_\infty/L_\infty, \widehat{B_\infty}(1)) = 0 \quad \text{for all } j > 1, \text{ and} \qquad (3.4.14)
$$
$$
H^1(M_\infty/L_\infty, \widehat{B_\infty}(1)) \simeq (\widehat{B_\infty})_{\mathbb{Z}_p(1)}. \qquad (3.4.15)
$$

Now by 3.4.6 there exists a (non-canonical!) isomorphism of $\mathrm{Gal}(\bar{L}/L_\infty)$-modules $T_p\widehat{\Omega}_{\bar{A}/\mathfrak{o}} \simeq \widehat{A}$. Combining this and equations (3.4.14) and (3.4.13), one sees that that $H^j(L_\infty, T_p\widehat{\Omega}_{\bar{A}/\mathfrak{o}})$ is killed by $\mathfrak{m}_\infty$ for all $j > 1$.

For the second part, we compute the coinvariants in (3.4.15). First observe that the ring $A' = A[T^{p^{-n}}]$ is finite over A, and that $\pi_A A'$ is a maximal ideal in it. Therefore A' is a discrete valuation ring, hence is the valuation ring of $L(T^{p^{-n}})$. It follows that any element of $\widehat{B_\infty}$ has the form

$$b = \sum_{a \in \mathbb{Q}_p/\mathbb{Z}_p} b_a T^a$$

where $T = T_1$ and $b_a \in \widehat{A_\infty}$, with $b_a \to 0$ as $|a|_p \to \infty$. Let $\gamma \in \operatorname{Gal}(M_\infty/L_\infty)$ be the topological generator for which $\gamma(T^{1/p^r}) = \zeta_{p^r} T^{1/p^r}$, for each $r \geq 1$. If b is divisible by $(1 - \zeta_p)$, then $b = b_0 + (1 - \gamma)b'$, where

$$b' = \sum_{0 \neq x/p^r \in \mathbb{Q}_p/\mathbb{Z}_p} (1 - \zeta_{p^r}^x)^{-1} b_{x/p^r} T^{x/p^r} \in \widehat{B_\infty}.$$

From this one sees that the inclusion $\widehat{A_\infty} \subset \widehat{B_\infty}$ induces an injection

$$\widehat{A_\infty} \hookrightarrow H^1(M_\infty/L_\infty, \widehat{B_\infty}(1)) \tag{3.4.16}$$

whose cokernel is killed by $(1 - \zeta_p)$. Now there is a diagram

$$
\begin{array}{ccc}
\widehat{A_\infty} & \xrightarrow{(3.4.16)} & H^1(M_\infty/L_\infty, \widehat{B_\infty}(1)) \\
\Big\downarrow{\scriptstyle 1 \mapsto \mathrm{dlog}\, T} & & \Big\downarrow{\scriptstyle \mathrm{infl}} \\
& & H^1(L_\infty, \widehat{A}(1)) \\
& & \Big\downarrow{\scriptstyle (3.4.7)} \\
\widehat{\Omega}_{A/o} \otimes \widehat{A_\infty} & \xrightarrow{\delta_{L_\infty/L}} & H^1(L_\infty, T_p \widehat{\Omega}_{\bar{A}/o})
\end{array}
$$

in which the vertical arrows have kernel and cokernel killed by a power of p (by 3.4.3, 3.4.6 and (3.4.13)). It remains to check that it is commutative, which having got this far is an easy exercise. $\square$

   A similar computation can be carried out for all $r > 1$, using the isomorphism $\operatorname{Gal}(M_\infty/L_\infty) \simeq \mathbb{Z}_p(1)^r$ and the Koszul complex. In this way Hyodo computed the cohomology of $\widehat{L}(j)$ over L, generalising Tate's result. His final result (not needed here) is:

Theorem 3.4.17. [14, Theorem 1] *There are canonical isomorphisms*

$$H^q(L, \widehat{L}(j)) \xrightarrow{\sim} \begin{cases} \widehat{\Omega}_{A/o}^q \otimes \mathbb{Q} & \text{if } j = q \\ \widehat{\Omega}_{A/o}^{q-1} \otimes \mathbb{Q} & \text{if } j = q - 1 \\ 0 & \text{otherwise} \end{cases}$$

compatible with cup-product. For $j = q - 1 = 0$ it is given by cup-product with $\log \chi_{\mathrm{cycl}}$ and for $q = j = 1$ by (3.4.7) and (3.4.9). $\square$

3.5 Proof of Theorem 3.2.3

Theorem 3.2.3 is proved by reducing to the setting of the previous section. Recall that X is a smooth and proper curve over the ring of integers $\mathfrak{o}$ of a finite unramified extension $K/\mathbb{Q}_p$. Assume that X is connected and that $\Gamma(X, \mathcal{O}_X) = \mathfrak{o}$ (otherwise first replace K by an unramified extension). Let $\eta \in X$ be the generic point of the special fibre. Write also:

$$A = \widehat{\mathcal{O}_{X,\eta}}; \quad L = \text{field of fractions of } A;$$

$$L_n = L(\mu_{p^n}); \quad A_n = \text{integral closure of } A \text{ in } L_n;$$

The fields L, L_n satisfy the hypothesis (3.4.1), with $r = 1$. There is an obvious localisation map $\phi \colon \operatorname{Spec} A \to Y$. Note that since $A/\mathfrak{o}$ is formally smooth we actually have $A_n = A \otimes \mathfrak{o}_n$; and by 3.4.3, $\widehat{\Omega}_{A/\mathfrak{o}}$ is a free A-module of rank 1. Now use the fact that the map

$$\phi^* \colon \operatorname{Fil}^1 H^1_{\mathrm{dR}}(Y/\mathfrak{o}) = H^0(X, \Omega^1_{X/\mathfrak{o}}(\log Z)) \to \widehat{\Omega}_{A/\mathfrak{o}}$$

is injective and its cokernel is torsion-free (this holds because the fibres of $X/\mathfrak{o}$ are connected). This means that the diagram in Theorem 3.2.3 can be localised to $\operatorname{Spec} A$ without losing information. We shall write down the localised diagram and then explain why it implies 3.2.3.

Proposition 3.5.1. *There exists an integer c such that for every $n > 0$ the following diagram commutes up to p^c-torsion:*

$$
\begin{array}{ccc}
(K_2(A_n) \otimes \mu_{p^n}^{\otimes -1})^0 & \xrightarrow{\text{ch}} & H^2(L_n, \mu_{p^n})^0 \\
\Big\downarrow{\text{dlog}} & & \Big\downarrow{\text{Hochschild-Serre}} \\
\Omega^2_{A_n/\mathfrak{o}}(-1) & & H^1(K_n, H^1(L\bar{K}, \mu_{p^n})) \\
\| & & \| \\
\Omega^1_{\mathfrak{o}_n/\mathfrak{o}}(-1) \otimes_{\mathfrak{o}} \widehat{\Omega}_{A/\mathfrak{o}} & & H^1(K_n, (A\bar{\mathfrak{o}})^*/p^n) \\
\Big\downarrow{\text{dlog}\,\zeta_{p^n} \otimes [\zeta_{p^n}]^{-1} \mapsto 1}\wr & & \Big\downarrow{\text{dlog}} \\
\mathfrak{o}_n/\partial_n \otimes_{\mathfrak{o}} \widehat{\Omega}_{A/\mathfrak{o}} & \xrightarrow{\cup (1/p^n)\log\chi_{\text{cycl}}} & H^1(K_n, \widehat{\Omega}_{A/\mathfrak{o}} \otimes \bar{\mathfrak{o}}/p^{n-1})
\end{array}
$$

Remarks. (i) Since $A/\mathfrak{o}$ is formally smooth, the valuation ring of $L\bar{K}$ is simply $A\bar{\mathfrak{o}}$.

(ii) We have written

$$H^2(L_n, \mu_{p^n})^0 = \ker\big(\text{res}\colon H^2(L_n, \mu_{p^n}) \to H^2(L\bar{K}, \mu_{p^n})\big)$$
$$(K_2(A_n) \otimes \mu_{p^n}^{\otimes -1})^0 = \ker\big(\text{ch}\colon K_2(A_n) \otimes \mu_{p^n}^{\otimes -1} \to H^2(L\bar{K}, \mu_{p^n})\big).$$

The map marked "Hochschild-Serre" is then the first edge-homomorphism from the Hochschild-Serre spectral sequence.

(iii) Concerning the bottom right-hand corner: the natural map is

$$\text{dlog}: (A\bar{\mathfrak{o}})^*/p^n \to \widehat{\Omega}_{A\bar{\mathfrak{o}}/\mathfrak{o}}/p^n$$

but as $A/\mathfrak{o}$ is formally smooth

$$\widehat{\Omega}_{A\bar{\mathfrak{o}}/\mathfrak{o}} = (\widehat{\Omega}_{A/\mathfrak{o}} \otimes \bar{\mathfrak{o}}) \oplus (\Omega_{\bar{\mathfrak{o}}/\mathfrak{o}} \otimes_{\mathfrak{o}} A)$$

and the second summand is divisible.

(iv) To deduce Theorem 3.2.3 from the proposition, it is enough, by what has already been said, to show that there is a map from the diagram in 3.2.3 to the diagram above. Since the composite $K_2(Y \otimes \mathfrak{o}_n) \to K_2(A_n) \to H^2(L\bar{K}, \mu_{p^n}^{\otimes 2})$ factors through $H^2(Y \otimes \bar{K}, \mu_{p^n}^{\otimes 2}) = 0$, one obtains the map $K_2(Y \otimes \mathfrak{o}_n) \to K_2(A_n)^0$. The only remaining thing to check is that the diagram

$$
\begin{array}{ccc}
H^1(K_n, H^1(Y \otimes \bar{K}, \mu_{p^n})) & \longrightarrow & H^1(K_n, H^1(L\bar{K}, \mu_{p^n})) \\
\Big\downarrow {\scriptstyle \pi_1 \ (\text{mod } p^n)} & & \Big\| \\
& & H^1(K_n, (A\bar{\mathfrak{o}})^*/p^n) \\
& & \Big\downarrow {\scriptstyle \text{dlog}} \\
H^1(K_n, \bar{\mathfrak{o}}(1)/p^n) \otimes_{\mathfrak{o}} \text{Fil}^1 H^1_{\text{dR}}(Y/\mathfrak{o}) & \longrightarrow & H^1(K_n, \widehat{\Omega}_{A/\mathfrak{o}} \otimes \bar{\mathfrak{o}}/p^n)
\end{array}
$$

commutes, but this follows from the description of π_1 given in §3.2.

Proof of 3.5.1. We reduce the diagram to the (smaller) diagrams in the following three lemmas. By (3.4.7), $_{p^n}\widehat{\Omega}_{\bar{A}/\mathfrak{o}}$ is free over $\bar{A}/p^n$ of rank one, and by $_{p^n}\widehat{\Omega}_{\bar{A}/\mathfrak{o}}^{\otimes 2}$ we mean its tensor square as $\bar{A}/p^n$-module.

Lemma 3.5.2. *For any m, n the diagram below commutes:*

$$
\begin{array}{ccc}
K_2(A_m) & \xrightarrow{\text{ch}} & H^2(L_m, \mu_{p^n}^{\otimes 2}) \\
\Big\downarrow {\scriptstyle \text{dlog}} & & \Big\downarrow \\
\widehat{\Omega}^2_{A_m/\mathfrak{o}} & \xrightarrow{\Lambda^2 \delta_{L_m}} & H^2(L_m, {_{p^n}}\widehat{\Omega}_{\bar{A}/\mathfrak{o}}^{\otimes 2})
\end{array}
$$

in which the unlabelled arrow is induced by $\text{dlog}: \mu_{p^n} \to {_{p^n}}\widehat{\Omega}_{\bar{A}/\mathfrak{o}}$.

Proof. Since A_m is local the symbol $A_m^* \otimes A_m^* \to K_2(A_m)$ is surjective. Since the Chern character is compatible with cup-product, the compatibility follows by Lemma 3.4.11.   □

This reduces the computation to Galois cohomology. Write

$$H^2(L_{n,\,p^n}\widehat{\Omega}^{\otimes 2}_{\bar{A}/o})^0 = \ker\left[H^2(L_{n,\,p^n}\widehat{\Omega}^{\otimes 2}_{\bar{A}/o}) \to H^2(L\bar{K},\,_{p^n}\widehat{\Omega}^{\otimes 2}_{\bar{A}/o})\right].$$

Lemma 3.5.3. *(i) The composite map*

$$\widehat{\Omega}^2_{\bar{A}_n/o} \xrightarrow{\wedge^2 \delta_{L_n}} H^2(L_{n,\,p^n}\widehat{\Omega}^{\otimes 2}_{\bar{A}/o}) \longrightarrow H^2(L\bar{K},\,_{p^n}\widehat{\Omega}^{\otimes 2}_{\bar{A}/o})$$

equals zero.
(ii) The following diagram is commutative:

$$\Omega_{o_n/o} \otimes \widehat{\Omega}_{\bar{A}/o} = \widehat{\Omega}^2_{\bar{A}_n/o} \xrightarrow{\wedge^2 \delta_{L_n}} H^2(L_{n,\,p^n}\widehat{\Omega}^{\otimes 2}_{\bar{A}/o})^0$$

$$\delta_{K_n/K}\otimes\mathrm{id}\Big\downarrow \qquad\qquad\qquad\qquad\qquad\qquad\qquad \Big\uparrow$$

$$H^1(K_{n,\,p^n}\widehat{\Omega}_{\bar{o}/o} \otimes_o \widehat{\Omega}_{\bar{A}/o}) \qquad\qquad \text{Hochschild-Serre}\Big|$$

$$H^1(\mathrm{id}\otimes\delta_{L\bar{K}/L})\Big\downarrow$$

$$H^1(K_{n,\,p^n}\widehat{\Omega}_{\bar{o}/o} \otimes_{\bar{o}} H^1(L\bar{K},\,_{p^n}\widehat{\Omega}_{\bar{A}/o})) \;=\!=\; H^1(K_n, H^1(L\bar{K},\,_{p^n}\widehat{\Omega}^{\otimes 2}_{\bar{A}/o}))$$

(iii) The map $H^1(\mathrm{id}\otimes\delta_{L\bar{K}/L})$ has kernel and cokernel killed by a bounded power of p.

Proof. (i) The cup-product $\wedge^2\delta_{L_n}$ factorises as

$$\Omega_{o_n/o} \otimes \widehat{\Omega}_{\bar{A}/o} \xrightarrow{\delta_{K_n}\otimes\delta_{L_n/L}} H^1(K_{n,\,p^n}\Omega_{\bar{o}/o}) \otimes H^1(L_{n,\,p^n}\widehat{\Omega}_{\bar{A}/o}) \to H^2(L_{n,\,p^n}\widehat{\Omega}^{\otimes 2}_{\bar{A}/o})$$

and so its composition with the restriction to $L\bar{K}$ is zero (it factors through $H^1(\bar{K},\,_{p^n}\Omega_{\bar{o}/o}) \otimes H^1(L\bar{K},\,_{p^n}\widehat{\Omega}_{\bar{A}/o}) = 0$).
(ii) The bottom equality comes from (3.4.8). The commutativity is a general fact. We have groups

$$\Gamma = \mathrm{Gal}(\bar{L}/L_n) \supset \Delta = \mathrm{Gal}(\bar{L}/L\bar{K}), \quad \Gamma/\Delta = \mathrm{Gal}(L\bar{K}/L_n) = \mathrm{Gal}(\bar{K}/K_n)$$

and two exact sequences of Γ-modules

$$0 \to A \to B \to C \to 0$$
$$0 \to A' \to B' \to C' \to 0$$

given by (3.3.6) and (3.4.4) respectively. On the first Δ acts trivially. So we have the following diagram

$$H^0(\Gamma/\Delta, C) \otimes H^0(\Gamma, C') \xrightarrow{\delta \otimes \delta'} H^1(\Gamma/\Delta, A) \otimes H^1(\Gamma, A')$$

$$\downarrow{\delta \otimes \mathrm{id}} \qquad\qquad\qquad \downarrow{\cup}$$

$$\ker\left[\mathrm{res}\colon H^2(\Gamma, A \otimes A') \to H^2(\Delta, A \otimes A')\right]$$

$$\downarrow{\mathrm{HS}}$$

$$H^1(\Gamma/\Delta, A) \otimes H^0(\Gamma, C') \qquad\qquad H^1(\Gamma/\Delta, H^1(\Delta, A \otimes A'))$$

$$\cup\downarrow \qquad\qquad\qquad\qquad \uparrow$$

$$H^1(\Gamma/\Delta, A \otimes H^0(\Delta, C')) \xrightarrow{H^1(\mathrm{id} \otimes \delta')} H^1(\Gamma/\Delta, A \otimes H^1(\Delta, A'))$$

and it is a simple, if tedious, exercise to check this commutes.

(iii) Follows from Proposition 3.4.12 applied to $L_\infty = L\bar{K}$. □

Lemma 3.5.4. *The following diagram commutes:*

$$H^1(L\bar{K}, \mu_{p^n}^{\otimes 2}) \xrightarrow{\mathrm{dlog} \otimes \mathrm{dlog}} H^1(L\bar{K}, {}_{p^n}\widehat{\Omega}_{\bar{A}/o}^{\otimes 2})$$

$$\uparrow{\mathrm{Kummer}}{\wr} \qquad\qquad \|$$

$$(A\bar{o})^*/p^n(1) \qquad\qquad {}_{p^n}\widehat{\Omega}_{\bar{o}/o} \otimes_{\bar{o}} H^1(L\bar{K}, {}_{p^n}\widehat{\Omega}_{\bar{A}/o})$$

$$\downarrow{\mathrm{dlog}} \qquad\qquad \uparrow{\mathrm{id} \otimes \delta_{L\bar{K}/L}}$$

$$\bar{o}/p^n(1) \otimes \widehat{\Omega}_{A/o} \xrightarrow{x \otimes \zeta_{p^n} \otimes \omega \mapsto x\,\mathrm{dlog}\,\zeta_{p^n} \otimes \omega} {}_{p^n}\Omega_{\bar{o}/o} \otimes_o \widehat{\Omega}_{A/o}$$

Proof. This follows from Lemma 3.4.11. □

As $K/\mathbb{Q}_p$ is unramified, we have $\mathfrak{a} = \mathfrak{a}_{\bar{o}/o} = (\zeta_p - 1)^{-1}\bar{o}$ by 3.3.1, so $\mathfrak{d}_n\bar{o} = p^n\mathfrak{a}$ and $\bar{o}/\mathfrak{d}_n\bar{o} \hookrightarrow \mathfrak{a}/p^n$.

We now can make a big diagram:

$$(K_2 A_n \otimes \mathbb{Z}/p^n)^0 \longrightarrow H^2(L_n, \mu_{p^n}^{\otimes 2})^0 \longrightarrow H^1(K_n, H^1(L\bar{K}, \mu_{p^n}^{\otimes 2})) \longleftarrow$$

$$\downarrow \qquad\qquad \downarrow \qquad\qquad \downarrow$$

$$\widehat{\Omega}_{A_n/o}^2 \longrightarrow H^2(L_n, {}_{p^n}\widehat{\Omega}_{\bar{A}/o}^{\otimes 2})^0 \longrightarrow H^1(K_n, H^1(L\bar{K}, {}_{p^n}\widehat{\Omega}_{\bar{A}/o}^{\otimes 2}))$$

$$\| \qquad\qquad\qquad\qquad \|$$

$$\Omega_{o_n/o} \otimes \widehat{\Omega}_{A/o} \longrightarrow H^1(K_n, {}_{p^n}\Omega_{\bar{o}/o} \otimes_o \widehat{\Omega}_{A/o}) \xrightarrow{(*)} H^1(K_n, {}_{p^n}\Omega_{\bar{o}/o} \otimes_{\bar{o}} H^1(L\bar{K}, {}_{p^n}\widehat{\Omega}_{\bar{A}/o}))$$

$$\wr\downarrow{\mathrm{dlog}\,\zeta_{p^n} \mapsto 1 \otimes \zeta_{p^n}} \qquad \uparrow$$

$$o_n/\mathfrak{d}_n(1) \otimes \widehat{\Omega}_{A/o} \xrightarrow{(\dagger)} H^1(K_n, \mathfrak{a}/p^n(1) \otimes_o \widehat{\Omega}_{A/o}) \longleftarrow H^1(K_n, (A\bar{o})^*/p^n)(1)$$

To save space we have not labelled most of the arrows: they can be found in the corresponding places in the subdiagrams 3.5.2–3.5.4, apart from the arrow labelled (†), which is $\cup(1/p^n)\log\chi_{\text{cycl}}$. The top left square commutes by 3.5.2, and the top right square by functoriality of the Hochschild-Serre spectral sequence. The rectangle in the middle commutes by 3.5.3, and the bottom left square by 3.3.8. The remaining part of the diagram (the right-hand hexagon) commutes by 3.5.4

Going round the outside of the diagram in both directions gives two maps

$$(K_2 A_n \otimes \mathbb{Z}/p^n)^0 \longrightarrow H^1(K_n, \mathfrak{a}/p^n(1) \otimes_o \widehat{\Omega}_{A/o})$$

and it is enough to show that their difference is killed by a bounded power of p. This follows from the commutativity of the diagram, since the kernel of the arrow marked (∗) is killed by a bounded power of p, by 3.5.3(iii).   □

4   The Rankin-Selberg method

In this section we calculate the projection of the product of two weight one Eisenstein series onto a cuspidal Hecke eigenspace, using the Rankin-Selberg integral. In order to separate the Euler factors more easily, we work semi-adelically, regarding modular forms as functions on $(\mathbb{C} - \mathbb{R}) \times GL_2(\mathbf{A}_f)$. The passage from classical to adelic modular forms is well-known, but we review the correspondence briefly in §4.2 since there is more than one possible normalisation. The same applies to the discussion of Eisenstein series in section §4.3.

4.1   Notations

G denotes the algebraic group GL_2, with the standard subgroups

$$P = \begin{pmatrix} * & * \\ 0 & * \end{pmatrix}, \quad U = \begin{pmatrix} 1 & * \\ 0 & 1 \end{pmatrix}, \quad Z = \left\{ \begin{pmatrix} a & 0 \\ 0 & a \end{pmatrix} \right\}$$

If R is a ring and H is G or any of the above subgroups, write H_R for the group of R-valued points of H. If $R \subset \mathbb{R}$ then H_R^+ denotes $\{h \in H_R \mid \det(h) > 0\}$.

The ring of finite adeles of $\mathbb{Q}$ is $\mathbf{A}_f = \widehat{\mathbb{Z}} \otimes_\mathbb{Z} \mathbb{Q}$. If $\chi = \prod \chi_p : \mathbf{A}_f^* / \mathbb{Q}_{>0}^* \to \mathbb{C}^*$ is a character (continuous homomorphism) and M is a multiple of the conductor of χ, then $\chi_{\text{mod } M} : (\mathbb{Z}/M\mathbb{Z})^* \to \mathbb{C}^*$ denotes the associated (not necessarily primitive) Dirichlet character: for $a \in \widehat{\mathbb{Z}}^*$, $\chi(a) = \chi_{\text{mod } M}(a \bmod M)$. Of course this means that if $(p, M) = 1$ then $\chi_{\text{mod } M}(p \bmod M) = \chi_p(p)^{-1}$.

Write finite idelic and p-adic modulus as $|-|_f$, $|-|_p$, and archimedean absolute value as $|-|_\infty$. If there can be no confusion we drop the subscripts.

Write also H_f, H_p in place of $H_{\mathbf{A}_f}$, $H_{\mathbf{Q}_p}$, and define the standard congruence subgroups

$$G_p \supset K_p = G_{\mathbf{Z}_p} \supset K_0(p^\nu) = \left\{ h \in K_p \mid h \equiv \begin{pmatrix} * & * \\ 0 & * \end{pmatrix} \bmod p^\nu \right\}$$

$$\supset K_1(p^\nu) = \left\{ h \in K_p \mid h \equiv \begin{pmatrix} 1 & * \\ 0 & * \end{pmatrix} \bmod p^\nu \right\}.$$

Haar measure on all the groups encountered is to be normalised in the usual way: on $\mathbf{Q}_p$ the additive measure dx gives $\mathbf{Z}_p$ measure 1, and on $\mathbf{Q}_p^*$ the multiplicative measure d^*x gives $\mathbf{Z}_p^*$ measure 1. On G_f, G_p the subgroups $G_{\hat{\mathbf{Z}}}$, K_p have measure 1.

Fix additive characters $\psi_p \colon \mathbf{Q}_p \to \mathbf{C}^*$, $\psi_f = \prod \psi_p \colon \mathbf{A}_f \to \mathbf{C}^*$ by requiring $\psi_p(x/p^n) = e^{2\pi i x/p^n}$ for every $x \in \mathbf{Z}$.

$\mathfrak{H}$ denotes the upper-half plane, and $\mathfrak{H}^\pm = \mathbf{C} - \mathbf{R}$. The group $G_{\mathbf{R}}$ acts on $\mathfrak{H}^\pm$ by linear fractional transformations. Put

$$j(\gamma, \tau) = \det \gamma \cdot (c\tau + d)^{-1} \quad \text{if } \tau \in \mathfrak{H}^\pm, \ \gamma = \begin{pmatrix} a & b \\ c & d \end{pmatrix} \in G_{\mathbf{R}}$$

so that $j(\gamma, \tau)(1, -\tau)\gamma^{-1} = (1, -\gamma(\tau))$.

Write $S(\mathbf{A}_f^2)$ for the space of locally constant functions $\mathbf{A}_f^2 \to \mathbf{C}$ of compact support. The group G_f acts on $S(\mathbf{A}_f^2)$ by the rule

$$(g\phi)(\underline{x}) = \phi(g^{-1}\underline{x}), \quad \phi \in S(\mathbf{A}_f^2), \ \underline{x} \in \mathbf{A}_f^2. \tag{4.1.1}$$

If $\delta \in \mathbf{A}_f^*$ and $\phi \in S(\mathbf{A}_f^2)$, write $[\delta]\phi$ for the function

$$[\delta]\phi \colon \underline{x} \mapsto \phi(\delta^{-1}\underline{x}). \tag{4.1.2}$$

So in particular, if ϕ is the characteristic function of an open compact subset $X \subset \mathbf{A}_f^2$, then $[\delta]\phi$ is the characteristic function of δX.

4.2   Adelic modular forms

In the adelic setting, a holomorphic modular form of weight k is a function

$$F \colon \mathfrak{H}^\pm \times G_f \to \mathbf{C}$$

which is holomorphic in the first variable, and satisfies:

(i) For every $\gamma \in G_{\mathbf{Q}}$, $F(\gamma(\tau), \gamma g) = j(\gamma, \tau)^{-k} F(\tau, g)$;

(ii) There exists an open compact subgroup $K \subset G_f$ such that $F(\tau, gh) = F(\tau, g)$ for all $h \in K$;

(iii) F is holomorphic at the cusps.

Any modular form F has a Fourier expansion

$$F(\tau, g) = \sum_{m \in \mathbb{Q}} a_m(g) e^{2\pi i m \tau}, \quad \tau \in \mathfrak{H}, \ g \in G_f$$

where $a_m(g) = 0$ when $m < 0$ (this is the meaning of condition (iii)). Put $A(g) = a_1(g)$, the *Whittaker function* attached to F. Then A is a locally constant function on G_f which satisfies

$$A\left(\begin{pmatrix} 1 & b \\ 0 & 1 \end{pmatrix} g\right) = \psi_f(-b) A(g) \quad \text{for all } b \in \mathbb{A}_f. \tag{4.2.1}$$

One can recover the remaining Fourier coefficients (apart from the constant term) from $A(g)$ by

$$a_m(g) = m^k A\left(\begin{pmatrix} m & 0 \\ 0 & 1 \end{pmatrix} g\right) \quad \text{if } 0 < m \in \mathbb{Q}.$$

It is convenient to introduce the normalised Whittaker function

$$A^*(g) = A(g) |\det g|_f^{-k/2}. \tag{4.2.2}$$

The group G_f acts on adelic modular forms by right translation, and the translates of F by G_f generate an admissible representation, call it π. From the definition π is isomorphic to the representation generated by $A(g)$; the representation generated by $A^*(g)$ is isomorphic to the twist $\pi \otimes |\det|_f^{-k/2}$. If π is irreducible it has a factorisation $\pi = \otimes' \pi_p$, and the centre of G_f acts on the space of π via a character. With the normalisation used here, there is a (finite order) character $\varepsilon : \mathbb{A}_f^* / \mathbb{Q}_{>0}^* \to \mathbb{C}^*$ with $\varepsilon(-1) = (-1)^k$ such that

$$\pi \begin{pmatrix} a & 0 \\ 0 & a \end{pmatrix} = \varepsilon(a) |a|_f^k \quad \text{for all } a \in \mathbb{A}_f^*.$$

This means that

$$A^*\left(\begin{pmatrix} a & 0 \\ 0 & a \end{pmatrix} g\right) = \varepsilon(a) A^*(g) \quad \text{for all } a \in \mathbb{A}_f^* \text{ and } g \in G_f.$$

If F comes from a newform on $\Gamma_1(N)$ then $A(g)$ is factorisable: there are functions $A_p : G_p \to \mathbb{C}$, satisfying $A_p(1) = 1$, $\prod A_p(g_p) = A(g)$ for all $g = (g_p) \in G_f$ and such that

$$A_p\left(\begin{pmatrix} a & 0 \\ 0 & a \end{pmatrix} \begin{pmatrix} 1 & b \\ 0 & 1 \end{pmatrix} g\right) = \varepsilon_p(a) |a|_p^k \psi_p(-b) A_p(g) \quad \text{for all } a \in \mathbb{Q}_p^* \text{ and } b \in \mathbb{Q}_p. \tag{4.2.3}$$

Suppose A_p is K_p-invariant. Then π_p is unramified, and its local L-function is given by

$$L(\pi_p, s) = \sum_{r \geq 0} A_p \begin{pmatrix} p^r & 0 \\ 0 & 1 \end{pmatrix} p^{r(k-s)} = \sum_{r \geq 0} A_p^* \begin{pmatrix} p^r & 0 \\ 0 & 1 \end{pmatrix} p^{r(k/2-s)}$$

$$= \frac{1}{(1 - \alpha_p p^{-s})(1 - \alpha_p' p^{-s})}$$

where

$$\alpha_p + \alpha_p' = p^{k/2} A_p^* \begin{pmatrix} p & 0 \\ 0 & 1 \end{pmatrix} \quad \text{and} \quad \alpha_p \alpha_p' = \varepsilon_p(p) p^{k-1}$$

(this is the normalisation of L-functions which gives the functional equation for $s \leftrightarrow k - 1 - s$; it differs from that of Jacquet-Langlands by a shift).

Complex conjugation of Fourier coefficients defines an involution of the space of modular forms. In representation theoretic terms, this becomes the isomorphism

$$\bar{\pi} \simeq \pi \otimes \varepsilon^{-1}. \tag{4.2.4}$$

If $\lambda: \mathbf{A}_f^* / \mathbb{Q}_{>0}^* \to \mathbb{C}^*$ is any character of finite order and F is an adelic modular form of weight k, so is

$$F \otimes \lambda: (\tau, g) \mapsto \lambda(\det g) F(\tau, g). \tag{4.2.5}$$

To go from adelic to classical modular forms, let $K(n)$ be the standard level n subgroup of $G_{\hat{\mathbb{Z}}}$. Then

$$G_{\mathbb{Q}} \backslash \mathfrak{H}^{\pm} \times G_f / K(n) = G_{\mathbb{Z}} \backslash \mathfrak{H}^{\pm} \times G_{\hat{\mathbb{Z}}} / K(n) \simeq Y(n)(\mathbb{C}).$$

where the last isomorphism is normalised in such a way that the point $(\tau, h) \in \mathfrak{H}^{\pm} \times GL_2(\mathbb{Z}/n\mathbb{Z})$ corresponds to the elliptic curve $E_\tau = \mathbb{C}/(\mathbb{Z} + \tau\mathbb{Z})$ with level structure

$$\alpha_{\tau,h}: \underline{v} \mapsto (1/n, -\tau/n) \cdot h \cdot \underline{v}$$

$$(\mathbb{Z}/n\mathbb{Z})^2 \to (\frac{1}{n}\mathbb{Z} + \frac{\tau}{n}\mathbb{Z})/(\mathbb{Z} + \tau\mathbb{Z}) = \ker[\times n]_{E_\tau}.$$

Write z for the coordinate on E_τ, and let F be an adelic modular form which is invariant under $K(n)$. It corresponds to the classical modular form over $\mathbb{C}$

$$(E_\tau, \alpha_{\tau,h}) \longmapsto F(\tau, \tilde{h}) du^{\otimes k} \in H^0(E_\tau, \omega^{\otimes k})$$

where $h \in G_{\mathbb{Z}/n\mathbb{Z}}$ and $\tilde{h} \in G_{\hat{\mathbb{Z}}}$ is any lifting of h.

The map (2.3.4) $e_N: Y(N) \to \mathrm{Spec}\,\mathbb{Q}(\mu_N)$ is then given on complex points by

$$e_N: (\tau, g) \mapsto \psi_f(\pm \det g/N) \quad \text{if } g \in G_{\hat{\mathbb{Z}}}$$

(the sign depends on the normalisation of the e_N-pairing).

4.3 Eisenstein series

Here we establish notations for Eisenstein series in the framework of the previous section. The results quoted can be obtained easily from those found in classical references (probably [31, Chapter VII] is closest to what is found here).

Let $\phi \in \mathcal{S}(\mathbf{A}_f^2)$, with the action (4.1.1) of G_f. The series

$$E_{k,s}(\phi)(\tau, g) = \sum_{0 \neq \underline{m} \in \mathbb{Q}^2} (g\phi)(\underline{m})(m_1 - m_2\tau)^{-k}|m_1 - m_2\tau|^{-2s}$$

is absolutely convergent for $k + 2\operatorname{Re}(s) > 2$, with a meromorphic continuation, and satisfies

$$E_{k,s}(\phi)(\gamma\tau, \gamma g) = j(\gamma, \tau)^{-k}|j(\gamma, \tau)|^{-2s}E_{k,s}(\phi)(\tau, g) \quad \text{for all } \gamma \in G_{\mathbb{Q}}.$$

The functions $E_k(\phi) := E_{k,0}(\phi)$ are holomorphic (and therefore modular of weight k) if $k \geq 3$ or $k = 1$; if $k = 2$ they are holomorphic provided that $\int_{\mathbf{A}_f^2} \phi = 0$.

The map $\phi \mapsto E_{k,s}(\phi)$ is G_f-equivariant. In particular, if $\delta = d \in \mathbb{Q}^*$, then in the notation of (4.1.2)

$$E_{k,s}([d]\phi) = d^{-k}|d|_\infty^{-2s}E_{k,s}(\phi). \tag{4.3.1}$$

One can rewrite the Eisenstein series as a sum over the group. If $f: G_f \to \mathbb{C}$ is a locally constant function satisfying

$$f\left(\begin{pmatrix} a & b \\ 0 & d \end{pmatrix} g\right) = a^{-k}|a|_\infty^{-2s}f(g) \quad \text{for all } a, d \in \mathbb{Q}^*, b \in \mathbf{A}_f \tag{4.3.2}$$

then define

$$E_{k,s,f}(\tau, g) = \sum_{\gamma \in P_{\mathbb{Q}}^+ \backslash G_{\mathbb{Q}}^+} f(\gamma g)j(\gamma, \tau)^k|j(\gamma, \tau)|^{2s}. \tag{4.3.3}$$

The relation between the two definitions is that $E_{k,s}(\phi) = E_{k,s,f}$ with

$$f(g) = \sum_{x \in \mathbb{Q}^*} (g\phi)\begin{pmatrix} x \\ 0 \end{pmatrix} x^{-k}|x|_\infty^{-2s}.$$

Moreover every $E_{k,s,f}$ is an $E_{k,s}(\phi)$ for some ϕ.

In the normalisation used here, the Whittaker function of $E_k(\phi)$ is

$$B(g) = \frac{(2\pi i)^k}{(k-1)!} \lim_{s \to 0} \sum_{y \in \mathbb{Q}^*} y^{-k}|y|_\infty^{1-2s} \int_{\mathbf{A}_f} \psi_f(-x/y)(g\phi)\begin{pmatrix} x \\ y \end{pmatrix} dx$$

which can be obtained without too much difficulty from the classical formulae
— see e.g. [31, pp.156–7 & 164ff.].

One can decompose the Eisenstein series under the action of the centre of
G_f. It is more convenient to replace f and B by the normalised functions
(cf. (4.2.2) above)

$$f^*(g) = f(g)|\det g|_f^{-(k+2s)/2}, \quad B^*(g) = B(g)|\det g|_f^{-k/2}$$

and then to write

$$f^*(g) = \sum_\chi f_\chi^*(g), \quad B^*(g) = \sum_\chi B_\chi^*(g)$$

where the functions $f_\chi^*(g)$, $B_\chi^*(g)$ are zero unless $\chi(-1) = (-1)^k$, in which
case

$$f_\chi^*(g) = \int_{\mathbb{A}_f^*/\mathbb{Q}_{>0}^*} \chi(a) f^* \left(\begin{pmatrix} a^{-1} & 0 \\ 0 & a^{-1} \end{pmatrix} g \right) d^*a$$

$$= 2|\det g|_f^{-(k+2s)/2} \int_{\mathbb{A}_f^*} \chi(x)|x|_f^{k+2s}(g\phi) \begin{pmatrix} x \\ 0 \end{pmatrix} d^*x$$

and

$$B_\chi^*(g) = \int_{\mathbb{A}_f^*/\mathbb{Q}_{>0}^*} \chi(a) B^* \left(\begin{pmatrix} a^{-1} & 0 \\ 0 & a^{-1} \end{pmatrix} g \right) d^*a$$

$$= 2\frac{(2\pi i)^k}{(k-1)!}|\det g|_f^{-k/2} \int_{\mathbb{A}_f \times \mathbb{A}_f^*} \chi(y)|y|_f^{k+2s-1}\psi_f(-x/y)(g\phi) \begin{pmatrix} x \\ y \end{pmatrix} dx\,d^*y \Bigg|_{s=0}$$

If $\phi = \prod \phi_p$ is factorisable, with ϕ_p equal to the characteristic function of $\mathbb{Z}_p^2$
for almost all p, then the expressions above factorise and one has

$$f_\chi^*(g) = 2 \prod_p f_{\chi_p}^*(g_p), \quad B_\chi^*(g) = 2\frac{(2\pi i)^k}{(k-1)!} \prod_p B_{\chi_p}^*(g_p) \quad \text{for } g = (g_p) \in G_f$$

where the functions $f_{\chi_p}^*$, $B_{\chi_p}^*$ are given by local integrals

$$f_{\chi_p}^*(g_p) = |\det g_p|_p^{-(k+2s)/2} \int_{\mathbb{Q}_p} \chi_p(x)|x|_p^{k+2s}(g_p\phi_p) \begin{pmatrix} x \\ 0 \end{pmatrix} d^*x \qquad (4.3.4)$$

$$B^*_{\chi_p}(g_p) = |\det g_p|_p^{-k/2} \int_{\mathbb{Q}_p \times \mathbb{Q}_p^*} \chi_p(y)|y|_p^{k+2s-1}\psi_p(-x/y)(g_p\phi_p)\begin{pmatrix} x \\ y \end{pmatrix} dx\, d^*y \Bigg|_{s=0}$$

$$= |\det g_p|_p^{-k/2} \int_{\mathbb{Q}_p \times \mathbb{Q}_p^*} \chi_p(y)|y|_p^{k+2s}\psi_p(-x)(g_p\phi_p)\begin{pmatrix} xy \\ y \end{pmatrix} dx\, d^*y \Bigg|_{s=0}$$

(4.3.5)

In fact the integral in (4.3.5) is a finite sum, because the x-integral is a finite linear combination of integrals of the form

$$\int_{t+p^\nu\mathbb{Z}_p} \psi_p(-x/y)\, dx$$

which vanish if y is sufficiently close to 0. So one can omit s from the formula.

Because of (4.3.2) and (4.2.3) the functions $f^*_{\chi_p}$ and $B^*_{\chi_p}$ are determined by their restrictions to the subgroup

$$\begin{pmatrix} \mathbb{Q}_p^* & 0 \\ 0 & 1 \end{pmatrix} K_p \subset G_p$$

and these are given by

$$f^*_{\chi_p}\left(\begin{pmatrix} m & 0 \\ 0 & 1 \end{pmatrix} h\right) = \chi_p(m)|m|^{(k+2s)/2} f^*_{\chi_p}(h)$$

(4.3.6)

$$B^*_{\chi_p}\left(\begin{pmatrix} m & 0 \\ 0 & 1 \end{pmatrix} h\right) = |m|_p^{-k/2+1} \int_{\mathbb{Q}_p \times \mathbb{Q}_p^*} \chi_p(y)|y|_p^{k-1}\psi_p(-mx/y)(h\phi_p)\begin{pmatrix} x \\ y \end{pmatrix} dx\, d^*y$$

(4.3.7)

$$= |m|_p^{-k/2+1} \int_{\mathbb{Q}_p \times \mathbb{Q}_p^*} \chi_p(y)|y|_p^{k}\psi_p(-mx)(h\phi_p)\begin{pmatrix} xy \\ y \end{pmatrix} dx\, d^*y$$

(4.3.8)

4.4 The Rankin-Selberg integral

Let F, G be adelic modular forms of weights $k+l$, k respectively, at least one of which is a cusp form, and let $E_{l,s,f}$ be the Eisenstein series (4.3.3). The product

$$\Omega = E_{l,s,f}G\overline{F}y^{k+l+s-2}|\det g|^{-k-l-s}d\tau \wedge d\overline{\tau}$$

is a left $G_{\mathbb{Q}}^+$-invariant form on $\mathfrak{H} \times G_f$, and the aim of this and the following sections is to compute the inner product

$$\langle E_{l,s,f}G, F \rangle := \int_{G_{\mathbb{Q}}^+\backslash \mathfrak{H} \times G_f} \Omega\, dg$$

which is a Rankin-Selberg integral.

Proposition 4.4.1. *Let $A(g)$, $B(g)$ be the Whittaker functions of F and G. Then*

$$\langle E_{l,s,f}G, F\rangle = -\frac{i\Gamma(k+l+s-1)}{(4\pi)^{k+l+s-1}}$$

$$\times \int\limits_{\mathbf{A}_f^* \times G_{\hat{\mathbf{z}}}} f\left(\begin{pmatrix} m & 0 \\ 0 & 1 \end{pmatrix} h\right) B\left(\begin{pmatrix} m & 0 \\ 0 & 1 \end{pmatrix} h\right) \overline{A\left(\begin{pmatrix} m & 0 \\ 0 & 1 \end{pmatrix} h\right)} |m|^{-k-l-s-1} d^*m\, dh.$$

Proof. A very similar calculation is done in [30, §5]; here we simply write the equations with little comment:

$$\langle E_{l,s,f}G, F\rangle = \int\limits_{G_{\mathbf{Q}}^+\backslash\mathfrak{H}\times G_f} \sum_{\gamma \in P_{\mathbf{Q}}^+\backslash G_{\mathbf{Q}}^+} f(\gamma g)j(\gamma, \tau)^l |j(\gamma, \tau)|^{2s}$$

$$\times G(\tau, g)\overline{F(\tau, g)}y^{k+l+s-2} |\det g|_f^{-k-l-s}d\tau \wedge d\bar{\tau}\, dg$$

$$= -2i \int\limits_{P_{\mathbf{Q}}^+\backslash\mathfrak{H}\times G_f} f(g)G(\tau, g)\overline{F(\tau, g)}y^{k+l+s-2} |\det g|_f^{-k-l-s}dx\, dy\, dg$$

$$= -2i \int\limits_{P_{\mathbf{Q}}^+\backslash\mathfrak{H}\times G_f} f(g) \sum_{m\in\mathbf{Q}_{>0}^*} B\overline{A}\left(\begin{pmatrix} m & 0 \\ 0 & 1 \end{pmatrix} g\right)$$

$$\times e^{-4\pi my}y^{k+l+s-2} |\det g|_f^{-k-l-s}dx\, dy\, dg$$

$$= -2i \int\limits_{Z_{\mathbf{Q}}U_{\mathbf{Q}}\backslash\mathfrak{H}\times G_f} f(g)B(g)\overline{A(g)}e^{-4\pi y}y^{k+l+s-2} |\det g|_f^{-k-l-s}dx\, dy\, dg$$

$$= -\frac{2i\Gamma(k+l+s-1)}{(4\pi)^{k+l+s-1}} \int\limits_{Z_{\mathbf{Q}}N_{\mathbf{Q}}\backslash\mathbb{R}\times G_f} f(g)B(g)\overline{A(g)}|\det g|_f^{-k-l-s}dx\, dg$$

To get the final result, use the parameterisation

$$\pi\colon \mathbf{A}_f \times \mathbf{A}_f^* \times G_{\hat{\mathbf{z}}}/\{\pm 1\} \longrightarrow G_f/Z_{\mathbf{Q}}$$

$$(b, m, h) \longmapsto \begin{pmatrix} 1 & b \\ 0 & 1 \end{pmatrix}\begin{pmatrix} m & 0 \\ 0 & 1 \end{pmatrix} h$$

in terms of which integration is given by

$$\int\limits_{G_f/Z_{\mathbf{Q}}} \Phi(g)\, dg = \int\limits_{\mathbf{A}_f \times \mathbf{A}_f^* \times G_{\hat{\mathbf{z}}}/\{\pm 1\}} (\pi^*\Phi)(b, m, h)|y|^{-1}db\, d^*m\, dh$$

Since $f(g)B(g)\overline{A(g)}$ is invariant by $g \mapsto \begin{pmatrix} 1 & b \\ 0 & 1 \end{pmatrix} g$, the integral in the last line above splits as a product

$$\int\limits_{\mathbb{Q}\backslash \mathbb{R} \times \mathbb{A}_f} dx \, db \int\limits_{\mathbb{A}_f^* \times G_{\widehat{\mathbb{Z}}}/\{\pm 1\}} f B \overline{A} \left(\begin{pmatrix} m & 0 \\ 0 & 1 \end{pmatrix} h \right) d^* m \, dh$$

and the first factor equals 1 by choice of Haar measure. □

Now suppose:

- F is a cusp form, belonging to an irreducible $\pi = \otimes' \pi_p$, with central character ε, whose Whittaker function $A(g) = \prod A_p(g_p)$ is factorisable;

- $G = E_k(\phi')$ is an Eisenstein series and $f = f_\phi$ for factorisable functions $\phi = \prod \phi_p$, $\phi' = \prod \phi'_p \in \mathcal{S}(\mathbb{A}_f^2)$.

Then the integral in the previous proposition can be decomposed under the action of the centre and then factorised, giving:

Proposition 4.4.2. *Under the above hypotheses:*

$$\langle E_{l,s}(\phi) E_k(\phi'), F \rangle = C \sum_{\chi,\chi'} \prod_p I_p(\chi_p, \chi'_p)$$

where $C = \dfrac{i^{k-1} \Gamma(k+l+s-1)}{2^{k+2l+2s-4}(k-1)!}$ *and*

$$I_p(\chi_p, \chi'_p) = \int\limits_{\mathbb{Q}_p^* \times K_p} f^*_{\chi_p} B^*_{\chi'_p} \overline{A^*_p} \left(\begin{pmatrix} m & 0 \\ 0 & 1 \end{pmatrix} h \right) |m|^{-1} d^* m \, dh$$

and $f^*_{\chi_p}$, $B^*_{\chi'_p}$ *are as in* (4.3.4), (4.3.5) *above. The sum is over all pairs of characters* χ, χ': $\mathbb{A}_f^* / \mathbb{Q}_{>0}^* \to \mathbb{C}^*$ *such that* $\chi\chi' = \varepsilon$ *and* $\chi(-1) = (-1)^l$.

Remark 4.4.3. It will become clear in the computation that follows that the sum over characters is actually a finite sum.

4.5 Local integrals

Write char$\begin{bmatrix} X \\ Y \end{bmatrix}$ for the characteristic function of a subset $X \times Y \subset \mathbb{A}_f^2$. The next proposition will compute the local integral $I_p(\chi_p, \chi'_p)$ for almost all primes.

Proposition 4.5.1. *Suppose that*

$$\phi_p = \phi'_p = \phi^0_p := \mathrm{char}\begin{bmatrix} \mathbb{Z}_p \\ \mathbb{Z}_p \end{bmatrix}$$

and that A_p is K_p-invariant, with $A_p(1) = 1$. Then

$$I_p(\chi_p, \chi'_p) = \begin{cases} L(\pi_p, k+l+s-1)L(\pi_p \otimes \chi'^{-1}_p, l+s) & \text{if } \chi_p \text{ is unramified} \\ 0 & \text{otherwise.} \end{cases}$$

Remark 4.5.2. Since π_p has a K_p-invariant vector, ε is unramified at p. Therefore since $\chi\chi' = \varepsilon$, either both or neither of χ_p, χ'_p are unramified.

Proof. (See [15, §15.9].) If $\phi_p = \phi'_p = \phi^0_p$ then by (4.3.6) for $h \in K_p$, $m \in \mathbb{Q}^*_p$

$$f^*_{\chi_p}\left(\begin{pmatrix} m & 0 \\ 0 & 1 \end{pmatrix} h\right) = \chi_p(m)|m|^{(l+2s)/2} \int\limits_{\mathbb{Z}_p - \{0\}} \chi_p(x)|x|^{l+2s} d^* x$$

$$= \begin{cases} L(\chi_p, l+2s) & \text{if } \chi_p \text{ is unramified} \\ 0 & \text{otherwise.} \end{cases}$$

Moreover by (4.3.7)

$$B^*_{\chi'_p}\left(\begin{pmatrix} m & 0 \\ 0 & 1 \end{pmatrix} h\right) = |m|^{1-k/2} \int\limits_{\mathbb{Z}_p \times \mathbb{Z}_p - \{0\}} \chi'_p(y)|y|^{k-1}\psi_p(-mx/y)dx\,dy$$

where the x-integral equals 1 is $m/y \in \mathbb{Z}_p$ and vanishes otherwise. The y-integral then vanishes if $\chi'_p|_{\mathbb{Z}^*_p} \neq 1$, giving

$$B^*_{\chi'_p}\left(\begin{pmatrix} m & 0 \\ 0 & 1 \end{pmatrix} h\right) = \begin{cases} p^{-rk/2} \sum\limits_{0 \leq j \leq r} \chi'_p(p)^j p^{(r-j)(k-1)} \\ \qquad \text{if } r = v_p(m) \geq 0 \text{ and } \chi'_p \text{ is unramified} \\ 0 \qquad \text{otherwise.} \end{cases}$$

$$(4.5.3)$$

Thus $I_p(\chi_p, \chi'_p) = 0$ unless both χ, χ' are unramified at p, which we now assume. Then $f^*_{\chi_p}$, $B^*_{\chi'_p}$ are K_p-invariant and

$$I_p(\chi_p, \chi'_p) = \sum_{r \in \mathbb{Z}} p^r f^*_{\chi_p} B^*_{\chi'_p} \overline{A^*_p}\begin{pmatrix} p^r & 0 \\ 0 & 1 \end{pmatrix}$$

$$= L(\chi_p, l+2s) \sum_{r \geq 0} \chi(p)^r p^{r(1-l/2-s)} B^*_{\chi'_p} \overline{A^*_p}\begin{pmatrix} p^r & 0 \\ 0 & 1 \end{pmatrix}.$$

Now from (4.5.3)

$$\sum_{r \geq 0} B^*_{\chi_p} \begin{pmatrix} p^r & 0 \\ 0 & 1 \end{pmatrix} T^r = \frac{1}{(1 - \chi'_p(p)p^{-k/2}T)(1 - p^{k/2-1}T)}$$

and therefore by [15, Lemma 15.9.4] one gets

$$I_p(\chi_p, \chi'_p) = L(\chi_p, l + 2s) \frac{L(\bar{\pi}_p \otimes \chi_p \chi'_p, k + l + s - 1)L(\bar{\pi}_p \otimes \chi_p, l + s)}{L(\varepsilon_p^{-1}\chi'_p\chi_p^2, l + 2s)}.$$

Since $\varepsilon = \chi\chi'$ and $\bar{\pi} \simeq \pi \otimes \varepsilon^{-1}$ (4.2.4) the result follows. $\qquad \square$

Corollary 4.5.4. *Under the hypotheses of Proposition 4.4.2, let S be a finite set of primes such that, for every $p \notin S$, $\phi_p = \phi'_p = \phi^0_p$, A_p is K_p-invariant and $A_p(1) = 1$. Then*

$$\langle E_{l,s}(\phi)E_k(\phi'), F \rangle =$$
$$C \cdot L_S(\pi, k + l + s - 1) \sum_{\chi, \chi'} L_S(\pi \otimes \chi'^{-1}, l + s) \prod_{p \in S} I_p(\chi_p, \chi'_p)$$

where the sum is over characters χ, χ' unramified outside S, with $\chi\chi' = \varepsilon$ and $\chi(-1) = (-1)^l$.

Here L_S denotes the L-function with Euler factors at all $p \in S$ removed. At other primes we use the following choice for ϕ_p:

Proposition 4.5.5. *Let $t \in \mathbb{Q}_p$ with $v_p(t) = -\nu < 0$. Suppose that*

$$\phi_p = \phi_p^{1,t} := \mathrm{char}\left[\frac{t + \mathbb{Z}_p}{\mathbb{Z}_p}\right].$$

Let $m \in \mathbb{Q}_p^*$, $h = \begin{pmatrix} a & b \\ c & d \end{pmatrix} \in K_p$. Then

$$f^*_{\chi_p}\left(\begin{pmatrix} m & 0 \\ 0 & 1 \end{pmatrix} h\right) = \begin{cases} \left(1 - \frac{1}{p}\right)^{-1} p^{\nu(l+2s-1)}\chi_p(amt)|m|^{l/2+s} \\ \qquad \qquad \text{if } \mathrm{cond}\,\chi_p \leq \nu \text{ and } h \in K_0(p^\nu) \\ 0 \qquad \qquad \text{otherwise.} \end{cases}$$

Proof. Straightforward calculation from (4.3.4). $\qquad \square$

At the bad primes for F we are going to choose ϕ'_p in such a way as to make the local factor be simply a constant.

The standard way to achieve this is to use a suitable Atkin-Lehner operator to replace the coefficient of q^n in the q-expansion by zero whenever $p|n$. In representation-theoretic language, this means to use the vector in the Kirillov model which is the characteristic function of $\mathbb{Z}_p^*$. (See [6, Thm. 2.5.6], and also compare [30, 4.5.4].) For Eisenstein series it is easy to write down a parameter ϕ_p' which does the trick, although possibly this does not give the best constant in 4.6.3 below.

Proposition 4.5.6. *Suppose*

$$\phi_p' = \phi_p^{2,t'} := \operatorname{char}\begin{bmatrix} t' + \mathbb{Z}_p \\ \mathbb{Z}_p^* \end{bmatrix} - \frac{1}{p}\operatorname{char}\begin{bmatrix} t' + p^{-1}\mathbb{Z}_p \\ \mathbb{Z}_p^* \end{bmatrix}$$

where $t' \in \mathbb{Q}_p^*$, $v_p(t') = -\mu \le 0$. Then for all $h = \begin{pmatrix} a & b \\ c & d \end{pmatrix} \in K_0(p^{\mu+1}) \cap K_0(p^2)$

$$B_{\chi_p}^*\left(\begin{pmatrix} m & 0 \\ 0 & 1 \end{pmatrix}h\right) = \begin{cases} \chi_p'(-amt') \displaystyle\int_{p^{-\mu}\mathbb{Z}_p^*} \chi_p'(y)^{-1}\psi_p(y)\,d^*y \\ \qquad\qquad \text{if } \operatorname{cond}\chi_p' \le \mu \text{ and } m \in \mathbb{Z}_p^* \\ 0 \qquad\qquad\quad \text{otherwise.} \end{cases}$$

Remark. In the special case $\mu = 0$, this becomes

$$\phi_p' = \phi_p^{2,1} = \operatorname{char}\begin{bmatrix} \mathbb{Z}_p \\ \mathbb{Z}_p^* \end{bmatrix} - \frac{1}{p}\operatorname{char}\begin{bmatrix} p^{-1}\mathbb{Z}_p \\ \mathbb{Z}_p^* \end{bmatrix}$$

and for all $h \in K_0(p^2)$ and all $m \in \mathbb{Q}_p^*$

$$B_{\chi_p}^*\left(\begin{pmatrix} m & 0 \\ 0 & 1 \end{pmatrix}h\right) = \begin{cases} 1 & \text{if } \chi_p' \text{ is unramified and } |m| = 1 \\ 0 & \text{otherwise.} \end{cases}$$

Proof. First consider what happens when $\phi_p' = \operatorname{char}\begin{bmatrix} t' + \mathbb{Z}_p \\ \mathbb{Z}_p^* \end{bmatrix}$. For every $h \in K_0(p^{\mu+1})$ one has

$$h\phi_p' = \operatorname{char}\begin{bmatrix} at' + \mathbb{Z}_p \\ \mathbb{Z}_p^* \end{bmatrix}$$

which gives

$$B_{\chi_p}^*\left(\begin{pmatrix} m & 0 \\ 0 & 1 \end{pmatrix}h\right) = |m|^{1-k/2}\int_{(at'+\mathbb{Z}_p)\times\mathbb{Z}_p^*} \chi_p'(y)\psi_p(-mx/y)\,dx\,d^*y \qquad (4.5.7)$$

and the x-integral vanishes for $m \notin \mathbb{Z}_p$, and equals $\psi_p(-amt'/y)$ otherwise. Therefore for $m \in \mathbb{Z}_p$ (4.5.7) becomes

$$|m|^{1-k/2} \int_{\mathbb{Z}_p^*} \chi_p'(y)\psi_p(-amt'/y)d^*y$$

$$= |m|^{1-k/2}\chi_p'(-amt') \int_{mt'\mathbb{Z}_p^*} \chi_p'(y)^{-1}\psi_p(y)\,d^*y$$

Now if $B: G_p \to \mathbb{C}$ is *any* Whittaker function (i.e. satisfies (4.2.1) and is locally constant) which is invariant under $K_1(p^\nu)$, some $\nu \geq 1$, then the function

$$\widetilde{B} = B - \frac{1}{p} \sum_{x \bmod p} \begin{pmatrix} 1 & p^{-1}x \\ 0 & 1 \end{pmatrix} B$$

satisfies, for every $h \in K_0(p^\nu) \cap K_0(p^2)$,

$$\widetilde{B}\left(\begin{pmatrix} m & 0 \\ 0 & 1 \end{pmatrix} h \right) = \text{char}_{\mathbb{Z}_p^*}(m) B\left(\begin{pmatrix} m & 0 \\ 0 & 1 \end{pmatrix} h \right).$$

Since

$$(t' + p^{-1}\mathbb{Z}_p) \times \mathbb{Z}_p^* = \coprod_{x \bmod p} \begin{pmatrix} 1 & p^{-1}x \\ 0 & 1 \end{pmatrix} [(t' + \mathbb{Z}_p) \times \mathbb{Z}_p^*]$$

the result follows. $\qquad\square$

From Propositions 4.5.5 and 4.5.6 one obtains:

Proposition 4.5.8. *Suppose that A_p is invariant under $K_1(p^\nu)$ and that $A_p(1) = 1$. Let t, $t' \in \mathbb{Q}_p^*$ with $v_p(t) = -\nu$, $v_p(t') = -\mu$ and $\nu > \mu \geq 0$, $\nu \geq 2$. Then if $\phi_p = \phi_p^{1,t}$ and $\phi_p' = \phi_p^{2,t'}$,*

$$I_p(\chi_p, \chi_p') = \begin{cases} \left(1 - \dfrac{1}{p^2}\right)^{-1} p^{\nu(l+2s-2)}\chi_p(t)\chi_p'(-t') \displaystyle\int_{p^{-\mu}\mathbb{Z}_p^*} \chi_p'(y)^{-1}\psi_p(y)\,d^*y & \text{if } \operatorname{cond}\chi_p' \leq \mu \\ 0 & \text{otherwise.} \end{cases}$$

Remark. Since $\varepsilon = \chi\chi'$ one has $\operatorname{cond}\chi_p' \leq \mu \implies \operatorname{cond}\chi_p \leq \nu$.

Proof. If $\operatorname{cond}\chi_p' > \mu$ then $I_p(\chi_p, \chi_p') = 0$. Otherwise, if $h \in K_0(p^\nu)$ and A_p is $K_1(p^\nu)$-invariant, one has $A_p^*(h) = \varepsilon_p(a)A_p^*(1) = \chi_p\chi_p'(a)A_p^*(1)$, so that

$I_p(\chi_p, \chi_p')$ equals

$$\left(1 - \frac{1}{p}\right)^{-1} p^{\nu(l+2s-1)} \int_{p^{-\mu}\mathbb{Z}_p^*} \chi_p'(y)^{-1}\psi_p(y)\,d^*y \int_{K_0(p^\nu)} \chi_p(at)\chi_p'(-at')\overline{A_p^*(h)}\,dh$$

$$= \mathrm{vol}\,K_0(p^\nu) \left(1 - \frac{1}{p}\right)^{-1} p^{\nu(l+2s-1)} \int_{p^{-\mu}\mathbb{Z}_p^*} \chi_p'(y)^{-1}\psi_p(y)\,d^*y\chi_p(t)\chi_p'(-t')\overline{A_p^*(1)}$$

and $\mathrm{vol}\,K_0(p^\nu) = [K_p : K_0(p^\nu)]^{-1} = (1 + 1/p)^{-1}p^{-\nu}$.   □

There is just one more case to consider, in order to compute the image of the Euler system in the cyclotomic tower.

Proposition 4.5.9. *Suppose that*

$$\phi_p' = (\phi_p^{1,t'})^{\mathrm{tr}} = \mathrm{char}\left[\begin{matrix} \mathbb{Z}_p \\ t' + \mathbb{Z}_p \end{matrix}\right], \quad v_p(t') = -\nu < 0.$$

Then for all $m \in \mathbb{Q}_p^*$ *and* $h = \begin{pmatrix} a & b \\ c & d \end{pmatrix} \in K_0(p^\nu)$,

$$B_{\chi_p'}^* \left(\begin{pmatrix} m & 0 \\ 0 & 1 \end{pmatrix} h\right) = \begin{cases} \left(1 - \dfrac{1}{p}\right)^{-1} p^{\nu(k-2)}|m|^{(1-k/2)}\chi_p'(dt')\psi_p(-mb/d) \\ \qquad \qquad if\ \mathrm{cond}\,\chi_p' \le \nu\ and\ v_p(m) \ge -\nu \\ 0 \qquad \qquad otherwise. \end{cases}$$

Proof. One has $h\phi_p' = \mathrm{char}\left[\begin{matrix} bt' + \mathbb{Z}_p \\ dt' + \mathbb{Z}_p \end{matrix}\right]$ and $dt' + \mathbb{Z}_p = dt'(1 + p^\nu\mathbb{Z}_p)$, therefore by (4.3.7)

$$B_{\chi_p'}^* = |m|^{-k/2+1} \int_{dt'(1+p^\nu\mathbb{Z}_p)} \chi_p'(y)|y|^{k-1}\left(\int_{bt'+\mathbb{Z}_p} \psi_p(-mx/y)\,dx\right) d^*y$$

$$= \begin{cases} 0 \qquad \qquad if\ v_p(m) < -\nu \\ |m|^{-k/2+1}\chi_p'(dt')p^{\nu(k-1)} \displaystyle\int_{1+p^\nu\mathbb{Z}_p} \chi_p'(y)\psi_p(-mb/dy)\,d^*y \quad if\ v_p(m) \ge -\nu. \end{cases}$$

If $v_p(m) \ge -\nu$ and $y \in 1 + p^\nu\mathbb{Z}_p$ then $\psi_p(-mb/dy) = \psi_p(-mb/d)$ and the result follows.   □

Corollary 4.5.10. *Suppose there exists a character* $\lambda_p \colon \mathbb{Q}_p^* \to \mathbb{C}^*$ *such that* $\pi_p \otimes \lambda_p^{-1}$ *is unramified, and that* A_p^* *is the twist of the spherical vector; that is,* $A_p^* \otimes \lambda_p^{-1} \colon g \mapsto \lambda_p(\det g)^{-1}A_p(g)$ *is* K_p-*invariant and* $A_p(1) = 1$. *Put*

$$\phi_p = \phi_p^{1,t}, \quad \phi_p' = (\phi_p^{1,t'})^{\mathrm{tr}} \quad with\ v_p(t) = v_p(t') = -\nu < 0.$$

Then if $\chi_p' \lambda_p^{-1}$ is unramified and cond $\lambda_p \leq \nu$,

$$I_p(\chi_p, \chi_p') = \left(1 - \frac{1}{p}\right)^{-1}\left(1 - \frac{1}{p^2}\right)^{-1} p^{\nu(k+l+2s-4)}\chi_p(t)\chi_p'(t')L(\pi_p \otimes \chi_p'^{-1}, l+s)$$

and otherwise $I_p(\chi_p, \chi_p') = 0$.

Proof. The central character $\varepsilon_p = \chi_p \chi_p'$ of π_p is λ_p^2 times an unramified character, so one of $\chi_p \lambda_p^{-1}$, $\chi_p' \lambda_p^{-1}$ is unramified if and only if both are. By Propositions 4.5.5 and 4.5.9, $I_p(\chi_p, \chi_p') = 0$ whenever cond χ_p or cond $\chi_p' > \nu$. Otherwise

$$I_p(\chi_p, \chi_p') = \left(1 - \frac{1}{p}\right)^{-2} p^{\nu(k+l+2s-3)}\chi_p(t)\chi_p'(t')$$

$$\times \int_{\mathbb{Q}_p^* \times K_0(p^\nu)} \chi_p(am)\chi_p'(d)|m|^{s+(l-k)/2}\overline{A_p^*\left(\begin{pmatrix} m & 0 \\ 0 & 1 \end{pmatrix} h\right)}\, d^*m\, dh$$

$$= \left(1 - \frac{1}{p}\right)^{-2} p^{\nu(k+l+2s-3)}\chi_p(t)\chi_p'(t')$$

$$\times \sum_{r\geq 0} \chi_p(p)^r p^{-r(l-k+2s)/2}\lambda_p(p)^{-r}\overline{A_p^* \otimes \lambda_p^{-1}\begin{pmatrix} p^r & 0 \\ 0 & 1 \end{pmatrix}}$$

$$\times \int_{K_0(p^\nu)} \chi_p(a)\chi_p'(d)\lambda(\det h)\, dh.$$

The integral in the last expression vanishes unless $\chi_p \lambda_p^{-1}$ and $\chi_p' \lambda_p^{-1}$ are unramified, in which case $I_p(\chi_p, \chi_p')$ becomes

$$\left(1 - \frac{1}{p}\right)^{-1}\left(1 - \frac{1}{p^2}\right)^{-1} p^{\nu(k+l+2s-4)}\chi_p(t)\chi_p'(t')$$

$$\times \sum_{r\geq 0}(\chi_p\lambda_p^{-1})(p)^r \overline{A_p^* \otimes \lambda_p^{-1}\begin{pmatrix} p^r & 0 \\ 0 & 1 \end{pmatrix}} p^{-r(l-k+2s)/2} \qquad (4.5.11)$$

$$= \left(1 - \frac{1}{p}\right)^{-1}\left(1 - \frac{1}{p^2}\right)^{-1} p^{\nu(k+l+2s-4)}\chi_p(t)\chi_p'(t')L(\bar{\pi}_p \otimes \chi_p, l+s)$$

giving the desired expression from (4.2.4).   □

Remark 4.5.12. One can do exactly the same computation merely assuming that $A_p^* \otimes \lambda_p^{-1}$ is $K_1(p^\nu)$-invariant; $I_p(\chi_p, \chi_p')$ vanishes unless $\chi_p' \lambda_p^{-1}$ is unramified, in which case one gets the formula (4.5.11).

4.6 Putting it all together

We change notation slightly from the previous sections. Begin with a cusp form F of weight $k+l$, generating an irreducible $\pi = \otimes' \pi_p$, and whose Whittaker function $A(g) = \prod A_p(g_p)$ factorises, with $A_p(1) = 1$ for all p. Let ε be the character of the centre of G_f on A^*. Assume that the following data is given:

(i) Disjoint finite sets of primes S and T, such that if $p \notin S$ then A_p is K_p-invariant. (In particular this means that ε is unramified outside S.)

(ii) A character $\lambda \colon \mathbf{A}_f^* / \mathbb{Q}_{>0}^* \to \mathbb{C}^*$, unramified outside T.

(iii) For each $p \in S$, elements $t_p, t_p' \in \mathbb{Q}_p^*$ with $v_p(t_p) = -\nu_p$, $v_p(t_p') = -\mu_p$, such that $\nu_p > \mu_p \geq 0$, $\nu_p \geq 2$, and A_p is $K_1(p^{\nu_p})$-invariant.

(iv) For each $p \in T$, elements $t_p, t_p' \in \mathbb{Q}_p^*$ with $v_p(t_p) = v_p(t_p') = -\nu_p$ where $\nu_p \geq \max(\operatorname{cond} \lambda_p, 1)$.

Put

$$N = \prod_{p \in S} p^{\nu_p}, \quad M = \prod_{p \in S} p^{\mu_p}, \quad R = \prod_{p \in T} p^{\nu_p}.$$

Denote by t, t' the finite ideles whose components at primes $p \in S \cup T$ are t_p, t_p', and which are 1 elsewhere. Pick $y \in \mathbb{Z}$ such that

$$y \equiv -\frac{p^{\mu_p} t_p'}{N t_p} \quad \bmod p^{\mu_p} \quad \text{for all } p \in S \tag{4.6.1}$$

(note that the right-hand side belongs to $\mathbb{Z}_p^*$) — thus y is well-defined mod M. The integral to compute is

$$\langle E_{l,s}(\phi) E_k(\phi') \otimes \lambda^{-1}, F \rangle = \langle E_{l,s}(\phi) E_k(\phi'), F \otimes \lambda \rangle$$

where

$$F \otimes \lambda(\tau, g) = \lambda(\det g) F(\tau, g)$$

is the twist of F by λ, and ϕ, ϕ' are given as follows:

- For $p \in S$, $\phi_p = \phi_p^{1, t_p}$ and $\phi_p' = \phi_p^{2, t_p'}$.

- For $p \in T$, $\phi_p = \phi_p^{1, t_p}$ and $\phi_p' = (\phi_p^{1, t_p'})^{\mathrm{tr}}$.

- For $p \notin S \cup T$, $\phi_p = \phi_p' = \phi_p^0$.

We can then assemble the previous calculations. Put $\chi = \varepsilon\lambda\theta$ and $\chi' = \lambda\theta^{-1}$ for a variable character θ — thus $\chi\chi' = \varepsilon\lambda^2$, the central character of $A^* \otimes \lambda$. Then only those θ satisfying the following conditions contribute to the sum:

- $\theta(-1) = (-1)^k\lambda(-1)$;

- If $p \notin S$ then θ_p is unramified;

- If $p \in S$ then $\operatorname{cond}\theta_p \le \mu_p$.

So $\operatorname{cond}\theta|M$, $\operatorname{cond}\varepsilon|N$ and $\operatorname{cond}\lambda|R$. This gives:

$$\langle E_{l,s}(\phi)E_k(\phi'), F \otimes \lambda\rangle = C \cdot L_{S\cup T}(\pi \otimes \lambda, l + k + s - 1)$$

$$\times \sum_{\substack{\theta(-1)=(-1)^k\lambda(-1) \\ \operatorname{cond}\theta|M}} L_{S\cup T}(\pi \otimes \theta, l+s) \prod_{p\in S\cup T} I_p(\chi_p, \chi'_p)$$

by 4.5.4, 4.5.8 and 4.5.9, where

$$\prod_{p\in S} I_p(\chi_p, \chi'_p) = N^{l+2s-2} \prod_{p\in S}\left(1 - \frac{1}{p^2}\right)^{-1} \chi_p(t_p)\chi'_p(-t'_p) \int_{p^{-\mu_p}\mathbf{Z}_p^*} \chi'_p(y)^{-1}\psi_p(y)\, d^*y$$

with

$$\prod_{p\in S}\chi_p(t_p)\chi'_p(-t'_p) = \prod_{p\in S}\lambda_p(p)^{-\mu_p-\nu_p}\varepsilon_p(t_p)\theta_p(-t_p/t'_p)$$

$$= \theta_{\operatorname{mod} M}(y)\prod_{p\in S}\varepsilon_p(t_p)\theta_p(p)^{\mu_p}\lambda_p(MN)^{-1},$$

and

$$\prod_{p\in T} I_p(\chi_p, \chi'_p) =$$

$$R^{k+l+2s+4}\prod_{p\in T}\left(1 - \frac{1}{p}\right)^{-1}\left(1 - \frac{1}{p^2}\right)^{-1}\chi_p(t_p)\chi'_p(t'_p)L(\pi_p \otimes \theta_p, l+s)$$

and for each $p \in T$, $\chi_p(t_p)\chi'_p(t'_p) = \varepsilon_p(t_p)\lambda_p(t_pt'_p)$. This gives

$$\langle E_{l,s}(\phi)E_k(\phi'), F \otimes \lambda\rangle =$$

$$C \cdot N^{l+2s-2}R^{k+l+2s-4}\prod_{p\in S\cup T}\left(1 - \frac{1}{p^2}\right)^{-1}\prod_{p\in T}\left(1 - \frac{1}{p}\right)^{-1}$$

$$\times \varepsilon(t)\prod_{p\in T}\lambda_p(MNt_pt'_p) \cdot L_{S\cup T}(\pi \otimes \lambda, l+k+s-1)$$

$$\times \sum_{\substack{\theta(-1)=(-1)^k\lambda(-1) \\ \operatorname{cond}\theta|M}}\left(\prod_{p|M}\theta_p(p)^{\mu_p}\int_{p^{-\mu_p}\mathbf{Z}_p^*}\chi'_p(y)^{-1}\psi_p(y)\, d^*y\right.$$

$$\left.\times \theta_{\operatorname{mod} M}(y) \cdot L_S(\pi \otimes \theta, l+s)\right). \qquad (4.6.2)$$

In the third line of this expression, the product over $p|M$ can be rewritten in terms of a classical Gauss sum as

$$\varphi(M)^{-1} \sum_{x \in (\mathbb{Z}/M\mathbb{Z})^*} \theta_{\text{mod } M}(x) e^{2\pi i x/M}.$$

(Here φ is Euler's totient function.) The sum over characters θ in (4.6.2) then becomes (combining odd and even characters)

$$\frac{1}{2}\varphi(M)^{-1} \sum_{\text{cond}\,\theta|M} \sum_{x \in (\mathbb{Z}/M\mathbb{Z})^*} \sum_{\substack{m \geq 1 \\ (m,N)=1}} \left[1 + (-1)^k \theta \lambda(-1)\right]$$

$$\times \theta_{\text{mod } M}(xy^{-1}m^{-1}) e^{2\pi i x/M} a_m m^{-l-s}$$

$$= \sum_{\substack{m \geq 1 \\ (m,N)=1}} \frac{1}{2}\left[\psi_f(my/M) + (-1)^k \lambda(-1)\psi_f(-my/M)\right] a_m m^{-l-s}$$

by the character orthogonality relations.

For $\alpha \in \mathbb{A}_f$ write

$$L_S(\pi, s; \alpha) = \sum_{\substack{m \geq 1 \\ (m,N)=1}} \psi_f(m\alpha) a_m m^{-s}$$

for the twisted Dirichlet series.

Theorem 4.6.3. *Under the above hypotheses*

$$\langle E_{l,s}(\phi) E_k(\phi') \otimes \overline{\lambda}, F \rangle = C N^{l+2s-2} R^{k+l+2s} \#GL_2(\mathbb{Z}/R\mathbb{Z})^{-1}$$

$$\times \varepsilon(t) \prod_{p \in T} \lambda_p(MNt_p t_p') \prod_{p \in S} \left(1 - \frac{1}{p^2}\right)^{-1} L_{S \cup T}(\pi \otimes \lambda, l + k + s - 1)$$

$$\times \left(L_S(\pi, l + s; y/M) + (-1)^k \lambda(-1) L_S(\pi, l + s; -y/M)\right)$$

with C as in Proposition 4.4.2.   □

5 The Euler systems

5.1 Modular curves

We can at last give Kato's construction of an Euler system in the Galois cohomology of the modular curve $Y(N)$ over a family of abelian extensions of $\mathbb{Q}$. We assume throughout that p is a prime not dividing N.

Pick auxiliary integers D, $D' > 1$ which are prime to $6Np$, and put

$$\mathcal{R}'_p = \{\text{squarefree positive integers prime to } NpDD'\}$$
$$\mathcal{R}_p = \{r = r_0 p^m \mid r_0 \in \mathcal{R}'_p, \ m \geq 1\}$$

We suppose that, for each $r \in \mathcal{R}_p$, we are given points z_r, $z'_r \in \mathcal{E}^{\mathrm{univ}}(Y(Nr)) \simeq (\mathbb{Z}/Nr)^2$, such that:

- If r and $rs \in \mathcal{R}_p$, then $sz^{(\prime)}_{rs} = z^{(\prime)}_r$ — i.e., one has elements of the inverse limit

$$(z_r)_r, \ (z'_r)_r \in \varprojlim_{r \in \mathcal{R}_p} \ker[\times Nr] \simeq \mathbb{Z}_p^2 \times \prod_{\ell \nmid NpDD'} \mathbb{Z}/\ell$$

of torsion points on the universal elliptic curve.

- For every $r \in \mathcal{R}_p$, the points Nz_r and Nz'_r generate $\ker[\times r]$ (in particular, the orders of z_r, z'_r are multiples of r).

- If $r = p^m$ then the orders of z_r, z'_r are divisible by a prime other than p.

Remarks. (i) The first condition implies that there exists $e \in \mathbb{Z}_p^*$ such that for every $r = r_0 p^m \in \mathcal{R}_p$, the Weil pairing of z_r and z'_r is

$$e_{Nr}(z_r, z'_r) = \zeta_{p^m}^{er_0^{-1}} \times (\text{prime-to-}p \text{ root of 1}). \tag{5.1.1}$$

(ii) The third condition is really only added for convenience. It ensures that for every r the points $z^{(\prime)}_r$ are not of prime power order, which means that they do not meet the zero section of $\mathcal{E}^{\mathrm{univ}}/Y(Nr)$ in any characteristic.

It follows from (ii) that the modular units $\vartheta_D(z_r)$, $\vartheta_{D'}(z'_r)$ actually belong to $\mathcal{O}^*(Y(Nr)_{/\mathbb{Z}})$, for any $r \in \mathcal{R}_p$. Define

$$\widetilde{\sigma}_r = \{\vartheta_D(z_r), \vartheta_{D'}(z'_r)\} \in K_2\big(Y(Nr)\big)$$

and also

$$\sigma_r = N_{Y(Nr)/Y(N) \otimes \mathbb{Q}(\mu_r)} \widetilde{\sigma}_r \in K_2\big(Y(N) \otimes \mathbb{Q}(\mu_r)\big);$$

by what was just said, these belong to the images of K_2 of the models over Spec $\mathbb{Z}$.

Let $T_\ell = T_{\ell, Y(N)}$, $\langle a \rangle = \langle a \rangle_{Y(N)}$ denote the Hecke correspondence and diamond operators as in §2.3 above. If $(\ell, r) = 1$ write $\mathrm{Frob}_\ell \in \mathrm{Gal}(\mathbb{Q}(\mu_r)/\mathbb{Q})$ for the *geometric* Frobenius automorphism, so that $\mathrm{Frob}_\ell = \varphi_\ell^{-1}$ where $\varphi_\ell \colon \zeta_r \mapsto \zeta_r^\ell$ is the arithmetic Frobenius substitution. For every finite field extension

L'/L write simply $N_{L'/L}$ for the norm map $K_2(Y(N) \otimes L') \to K_2(Y(N) \otimes L)$. Notice that if $\ell \nmid Nr$ then

$$N_{Y(Nr)/Y(N) \otimes \mathbb{Q}(\mu_r)} \circ T_{\ell, Y(Nr)} = (T_{\ell, Y(N)} \otimes \varphi_\ell) \circ N_{Y(Nr)/Y(N) \otimes \mathbb{Q}(\mu_r)}$$

$$N_{Y(Nr)/Y(N) \otimes \mathbb{Q}(\mu_r)} \langle \ell \rangle_{Y(Nr)} = (\langle \ell \rangle_{Y(N)} \otimes \varphi_\ell^2) \circ N_{Y(Nr)/Y(N) \otimes \mathbb{Q}(\mu_r)}$$

since T_ℓ acts as φ_ℓ on the constant field and $\langle \ell \rangle$ acts as φ_ℓ^2, by §2.3. Therefore from 2.3.6 and 2.4.3 one obtains:

Theorem 5.1.2. *Let $r \in \mathcal{R}_p$. Then:*

(i) $N_{\mathbb{Q}(\mu_{rp})/\mathbb{Q}(\mu_r)} \sigma_{rp} = \sigma_r$.

(ii) *If ℓ is prime and $(\ell, NDD'r) = 1$ then*

$$N_{\mathbb{Q}(\mu_{\ell r})/\mathbb{Q}(\mu_r)} \sigma_{\ell r} = (1 - T_\ell \langle \ell \rangle_* \otimes \mathrm{Frob}_\ell + \ell \langle \ell \rangle_* \otimes \mathrm{Frob}_\ell^2) \sigma_r.$$

Now write

$$\mathbb{T}_{p,N} = H^1(Y(N) \otimes_\mathbb{Q} \overline{\mathbb{Q}}, \mathbb{Z}_p(1))$$

and consider, for $r = r_0 p^m$, the homomorphisms

$$\varprojlim_{n \geq m} K_2(Y(N) \otimes \mathbb{Q}(\mu_{r_0 p^n})) \otimes \mu_{p^n}^{\otimes -1}$$

$$\downarrow \text{AJ}$$

$$\varprojlim_{n \geq m} H^1(\mathbb{Q}(\mu_{r_0 p^n}), H^1(Y(N) \otimes_\mathbb{Q} \overline{\mathbb{Q}}, \mu_{p^n}))$$

$$\downarrow \text{cor}$$

$$\varprojlim_{n \geq m} H^1(\mathbb{Q}(\mu_{r_0 p^m}), H^1(Y(N) \otimes_\mathbb{Q} \overline{\mathbb{Q}}, \mu_{p^n}))$$

$$\|$$

$$H^1(\mathbb{Q}(\mu_r), \mathbb{T}_{p,N})$$

By 5.1.2(i), the family $\{\sigma_{r_0 p^n} \otimes [\zeta_{p^n}]^{-1}, \; n \geq m\}$ is an element of the first group. (This twisting of elements of K_2 was used first by Soulé.) Let

$$\xi_r = \xi_r(Y(N)) \in H^1(\mathbb{Q}(\mu_r), \mathbb{T}_{p,N})$$

be its image. On the one hand, $\mathrm{Gal}(\mathbb{Q}(\mu_r)/\mathbb{Q})$ acts on $H^1(\mathbb{Q}(\mu_r), \mathbb{T}_{p,N})$, since $\mathbb{T}_{p,N}$ is a $\mathrm{Gal}(\overline{\mathbb{Q}}/\mathbb{Q})$-module; on the other, the level N Hecke operators T_ℓ, $\langle \ell \rangle$ act by functoriality. Also Frob_ℓ acts as ℓ^{-1} on μ_{p^n}. By Theorem 5.1.2 the classes ξ_r therefore satisfy Euler system-like identities:

Corollary 5.1.3. *(i) For all* $r \in \mathcal{R}_p$, $\mathrm{cor}_{\mathbb{Q}(\mu_{r_p})/\mathbb{Q}(\mu_r)} \xi_{rp} = \xi_r$.
(ii) If ℓ is prime and r, $\ell r \in \mathcal{R}_p$ then

$$\mathrm{cor}_{\mathbb{Q}(\mu_{\ell r})/\mathbb{Q}(\mu_r)} \xi_{\ell r} = \left(1 - \ell^{-1} T_\ell \langle \ell \rangle_* \mathrm{Frob}_\ell + \ell^{-1} \langle \ell \rangle_* \mathrm{Frob}_\ell^2\right) \xi_r.$$

In the next section we will pass to an elliptic curve and get an Euler system in the sense of §2 of [29].

Recall from §1.3 the definition of the weight 1 Eisenstein series (for any N and any $D > 1$ which is prime to $6N$)

$$_D\mathrm{Eis}(z) = z^* \, \mathrm{dlog}_v \, \vartheta_D \in H^0(X(N), \omega).$$

defined for any $0 \neq z \in \mathcal{E}^{\mathrm{univ}}(Y(N)) = (\mathbb{Z}/N\mathbb{Z})^2$. The form $_D\mathrm{Eis}(z)$ extends to $X(N)_{\mathbb{Z}}$ provided the order of z is divisible by at least 2 primes.

Recall from §1.1 the Kodaira-Spencer isomorphism

$$KS_N := KS_{Y(N)} \colon H^0(X(N), \omega^{\otimes 2}) \xrightarrow{\sim} H^0(X(N), \Omega^1_{X(N)/\mathbb{Q}}(\log \text{ cusps}))$$

identifying holomorphic modular forms of weight 2 and differentials with at worst simple poles at cusps. Let $Y(N)^{\mathrm{ord}}$ be the complement in $Y(N)_{/\mathbb{Z}}$ of the (finite) set of supersingular points in characteristic dividing N. The scheme $Y(N)^{\mathrm{ord}}$ is smooth over $\mathbb{Z}[\mu_N]$ by [18, Cor. 10.9.2].

Proposition 5.1.4. *The Kodaira-Spencer map divided by N extends to a homomorphism of sheaves on $Y(N)^{\mathrm{ord}}$*

$$\frac{1}{N} KS_N \colon \omega^{\otimes 2}_{\mathcal{E}^{\mathrm{univ}}/Y(N)^{\mathrm{ord}}} \to \Omega^1_{Y(N)^{\mathrm{ord}}/\mathbb{Z}[\mu_N]}$$

with logarithmic singularities at the cusps.

Proof. The Kodaira-Spencer map takes the modular form $f(q^{1/N})\,(dt/t)^{\otimes 2}$ to the differential $f(q^{1/N})\,dq/q = Nf(q^{1/n})\,\mathrm{dlog}(q^{1/N})$. So on q-expansions it is divisible by N. The result follows by the q-expansion principle. $\square$

Remark. One knows that (always assuming that N is the product of two coprime integers, each ≥ 3) the scheme $X(N)$ is regular. Therefore the morphism $e_N \colon X(N) \to \mathrm{Spec}\,\mathbb{Z}[\mu_N]$ is a local complete intersection (being a flat morphism of finite type between regular schemes, EGA IV 19.3.2). Therefore the sheaf of relative differentials extends to an invertible sheaf on $X(N)_{/\mathbb{Z}}$, namely the relative dualising sheaf (sheaf of regular differentials), and one can then show that $(1/N)KS_N$ extends to an *isomorphism* of invertible sheaves on all of $X(N)_{/\mathbb{Z}}$

$$\frac{1}{N} KS_N \colon \omega^2 \to \Omega^{\mathrm{reg}}_{X(N)/\mathbb{Z}[\mu_N]}(\log \text{ cusps}).$$

This is not needed in what follows.

Because $Y(N)^{\text{ord}}$ is smooth over $\mathbb{Z}[\mu_N]$, one has

$$\Omega^2_{Y(N)^{\text{ord}}/\mathbb{Z}} = \Omega^1_{Y(N)^{\text{ord}}/\mathbb{Z}[\mu_N]} \otimes \Omega^1_{\mathbb{Z}[\mu_N]/\mathbb{Z}}$$

and $\Omega^1_{\mathbb{Z}[\mu_N]/\mathbb{Z}}$ is killed by N and generated by $\text{dlog}(\zeta_N)$.

Proposition 5.1.5. *Let z, $z' \in \mathcal{E}^{\text{univ}}(Y(N)_{/\mathbb{Z}})$ be disjoint from $\ker[\times DD']$. In $\Omega^2_{Y(N)^{\text{ord}}/\mathbb{Z}}$ the identity*

$$\text{dlog}\{\vartheta_D(z), \vartheta_{D'}(z')\} = \frac{1}{N} KS_N\big({}_D\text{Eis}(z) \cdot {}_{D'}\text{Eis}(z')\big) \otimes \text{dlog}\, e_N(z, z')$$

holds.

Proof. This can be checked on q-expansions. Suppose that on the completion of $Y(N)$ along a cusp we have fixed an isomorphism of $\mathcal{E}^{\text{univ}}$ with the Tate curve $\text{Tate}(q)$ over $\mathbb{Z}[\mu_N]((q^{1/N}))$, and that z, z' are the points $z = \zeta_N^{a_1} q^{a_2/N}$, $z' = \zeta_N^{b_1} q^{b_2/N}$. Applying the congruence 1.3.4 and the fact that $e_N(z, z') = \zeta_N^{a_2 b_1 - a_1 b_2}$ one get the desired result. (We have normalised the e_N-pairing as in [18, (2.8.5.3)].)   □

We can now give Kato's description of the image of the Euler system $\{\xi_r\}$ under the dual exponential map (see §3.2 above)

$$\exp^*_p \colon H^1(\mathbb{Q}(\mu_r), \mathbb{T}_{p,N}) \to \mathbb{Q}_p \otimes_{\mathbb{Q}} \mathbb{Q}(\mu_r) \otimes_{\mathbb{Q}} \text{Fil}^1 H^1_{\text{dR}}(Y(N)/\mathbb{Q})$$

recalling that $\text{Fil}^1 H^1_{\text{dR}}(Y(N)/\mathbb{Q}) = H^0(X(N), \Omega^1_{X(N)/\mathbb{Q}}(\log \text{cusps}))$.

Define the following differentials on the modular curve in terms of the weight 1 Eisenstein series:

$$\widetilde{\omega}_r = \frac{1}{Nr} KS_{Nr}\big({}_D\text{Eis}(z_r) \cdot {}_{D'}\text{Eis}(z'_r)\big) \in H^0(X(Nr), \Omega^1(\log \text{cusps})).$$

$$\omega_r = \text{tr}_{X(Nr)/X(N) \otimes \mathbb{Q}(\mu_r)} \widetilde{\omega}_r \in H^0(X(N) \otimes \mathbb{Q}(\mu_r), \Omega^1(\log \text{cusps})) \quad (5.1.6)$$

Theorem 5.1.7. *For every $r \in \mathcal{R}_p$,*

$$\exp^*_p \xi_r = \frac{e}{r}\, \omega_r$$

where $e \in \mathbb{Z}_p^*$ is as in (5.1.1).

Proof. By 5.1.5 we have in $H^0(X(Nr)^{\text{ord}}, \Omega^2_{X(Nr)/\mathbb{Z}}(\log \text{cusps}))$ the identity

$$\text{dlog}\, \widetilde{\sigma}_r = \widetilde{\omega}_r \otimes \text{dlog}\, e_{Nr}(z_r, z'_r).$$

Now take $r = r_0 p^m$ and tensor with $\mathbb{Z}_p$. Then by (5.1.1)

$$\text{dlog}\, \widetilde{\sigma}_r = r_0^{-1} e\, \widetilde{\omega}_r \otimes \text{dlog}\, \zeta_{p^m} \in H^0(X(Nr)^{\text{ord}} \otimes \mathbb{Z}_p, \Omega^2(\log \text{cusps})).$$

Taking the trace to $X(N) \otimes \mathbb{Q}(\mu_r)$ gives, using the compatibility (§2.1) of trace and transfer

$$\mathrm{dlog}\, \sigma_r = r_0^{-1} e\, \omega_r \otimes \mathrm{dlog}\, \zeta_{p^m}.$$

Let $\mathfrak{o}_n$ be the ring of integers of $\mathbb{Q}_p(\mu_{p^n})$. By the explicit reciprocity law 3.2.3,

$$\exp_p^* \xi_r = \lim_{n \to \infty} \frac{r_0^{-1} e}{p^n}\, \mathrm{tr}_{Y(N) \otimes \mathfrak{o}_n / Y(N) \otimes \mathfrak{o}_m}\, \omega_{r_0 p^n}.$$

But since $\mathrm{tr}_{Y(Nr_0 p^n)/Y(Nr)}\, \widetilde{\omega}_{r_0 p^n} = p^{n-m}\widetilde{\omega}_r$ by 2.5.3, this gives the desired formula. $\qquad \square$

5.2 Elliptic curves

Suppose that $E/\mathbb{Q}$ is a modular elliptic curve of conductor N_E, with a Weil parameterisation

$$\varphi_E \colon X_0(N_E) \to E.$$

Choose a prime p not dividing $2N_E$, and write $T_p(E) = H^1(\overline{E}, \mathbb{Z}_p)(1)$ — of course, this is the same as the Tate module of E, but it is better to think in terms of cohomology, especially if we were to work more generally with any weight 2 eigenform (with character). Let the L-series of E be

$$L(E,s) = \sum_{n \geq 1} a_n n^{-s}$$

(again, this is best thought of here as the L-series attached to the motive $h^1(E)$). Let N be any positive multiple of N_E with $(N, p) = 1$. (The actual choice of N is to be made later.) Consider the composite morphism

$$\varphi_{E,N} \colon X(N) \to X_0(N_E) \xrightarrow{\varphi_E} E.$$

There are Galois-equivariant maps of restriction and direct image

$$H^1(X(N) \otimes_\mathbb{Q} \overline{\mathbb{Q}}, \mathbb{Z}_p(1)) \xrightarrow{\text{restriction}} H^1(Y(N) \otimes_\mathbb{Q} \overline{\mathbb{Q}}, \mathbb{Z}_p(1)) = \mathbb{T}_{p,N}$$

$$\Big\downarrow \varphi_{E,N*}$$

$$H^1(E \otimes_\mathbb{Q} \overline{\mathbb{Q}}, \mathbb{Z}_p(1)) = T_p(E)$$

Now the Manin-Drinfeld theorem (or rather its proof) implies that there is an idempotent Π_N^{cusp} in the Hecke algebra (with rational coefficients) which induces for every p a left inverse to the map labelled "restriction". So for some positive integer h_E (independent of p) the composite map

$$h_E \, \varphi_{E,N*} \circ \Pi_N^{\text{cusp}} : \mathbb{T}_{p,N} \to T_p(E)$$

is well-defined. Choose D, D' prime to $6Np$, and systems (z_r), (z_r') as in the previous section.

Theorem 5.2.1. *Define for* $r \in \mathcal{R}_p$

$$\xi_r(E) := (h_E \, \varphi_{E,N*} \circ \Pi_N^{\text{cusp}}) \xi_r(Y(N)) \in H^1(\mathbb{Q}(\boldsymbol{\mu}_r), T_p(E)).$$

Then the family $\{\xi_r(E)\}$ *is an Euler system for* $T_p(E)$*; that is,*

- *For every* $r \in \mathcal{R}_p$, $\text{cor}_{\mathbb{Q}(\boldsymbol{\mu}_{rp})/\mathbb{Q}(\boldsymbol{\mu}_r)} \xi_{rp}(E) = \xi_r(E)$;

- *If* ℓ *is prime and* $(\ell, NDD'r) = 1$ *then*

$$\text{cor}_{\mathbb{Q}(\boldsymbol{\mu}_{\ell r})/\mathbb{Q}(\boldsymbol{\mu}_r)} \xi_{\ell r}(E) = (1 - \ell^{-1} a_\ell \text{Frob}_\ell + \ell^{-1} \text{Frob}_\ell^2) \, \xi_r(E).$$

where $\text{Frob}_\ell \in \text{Gal}(\mathbb{Q}(\boldsymbol{\mu}_r)/\mathbb{Q})$ *is the geometric Frobenius.*

Remark. Actually, Rubin considers cohomology classes not over $\mathbb{Q}(\boldsymbol{\mu}_r)$ but rather over the subfield $\mathbb{Q}_{m-1}(\boldsymbol{\mu}_{r_0})$, where $r = r_0 p^m$ and $\mathbb{Q}_{m-1}/\mathbb{Q}$ is the unique extension of degree p^{m-1} contained in the cyclotomic $\mathbb{Z}_p$-extension of $\mathbb{Q}$. To get an Euler system in the precise sense of [29, §2], one should therefore take the corestriction of $\xi_r(E)$ to $\mathbb{Q}_{m-1}(\boldsymbol{\mu}_{r_0})$. Note that his formula for the norm relation differs from that here, as we are using geometric Frobenius: as Nekovář has explained to us, the relation (ii) can be rewritten more conceptually as $\text{cor}(\xi_{\ell r}) = Q_\ell(\text{Frob}_\ell) \xi_r$, where $Q_\ell(x) = \det(1 - \text{Frob}_\ell x \mid T_p(E)^*(1))$. Writing $P_\ell(x) = Q_\ell(\ell^{-1} x)$ one gets the same formula as in *loc. cit.*

Proof. The first statement follows directly from the corresponding statement 5.1.3(i) for $\xi_{rp}(Y(N))$. The second follows from 5.1.3(ii) together with the fact that $\langle \ell \rangle = 1$ and $T_\ell = a_\ell$ on $T_p(E)$.   □

On differentials, the projector Π_N^{cusp} is the identity on cusp forms and annihilates Eisenstein series. Put

$$\omega_r^{\text{cusp}} = \Pi_N^{\text{cusp}}(\omega_r) \in H^0(X(N) \otimes \mathbb{Q}(\boldsymbol{\mu}_r), \Omega^1).$$

Then Theorem 5.1.7 gives:

$$\exp_p^* \xi_r(E) = \frac{e h_E}{r} \varphi_{E,N*}(\omega_r^{\text{cusp}}). \tag{5.2.2}$$

To compute this in terms of the L-function, use the Rankin-Selberg integral from §4. Fix a differential ω_E on $E/\mathbb{Q}$ such that $\varphi_E^* \omega_E$ is a newform on $X_0(N_E)$, which we write as $2\pi i F(\tau, g) d\tau$ for a weight 2 cusp form F whose Whittaker function satisfies:

- A_q is K_q-invariant if $q \nmid N_E$, and is $K_0(q^\nu)$-invariant if $\mathrm{ord}_q(N_E) = \nu > 0$;

- $A_q(1) = 1$ for all q.

This means that $A_q(q) = q^{-2}a_q$ for every $q \nmid N_E$, and that $L(E, s) = L(\pi, s)$ where π is the representation of G_f generated by F.

At this point it would be wise to recall that we have in §3.1 normalised the reciprocity law of local class field theory to take uniformisers to geometric Frobenius. This gives the classical isomorphism:

$$\hat{\mathbb{Z}}^* \longhookrightarrow \mathbb{A}_{\mathbb{Q}}^* \xrightarrow{\text{global CFT}} \mathrm{Gal}(\mathbb{Q}^{\mathrm{ab}}/\mathbb{Q})$$
$$a \longmapsto (\zeta_n \mapsto \zeta_n^a)$$

If λ is any idele class character of conductor M, with associated Dirichlet character $\lambda_{\mathrm{mod}\,M} \colon (\mathbb{Z}/M\mathbb{Z})^* \to \mathbb{C}^*$, we then have

$$L(\pi \otimes \lambda^{-1}, s) = L(E, \lambda_{\mathrm{mod}\,M}, s) := \sum_{(m,M)=1} a_m \lambda_{\mathrm{mod}\,M}(m) m^{-s}$$

We also define the incomplete and twisted L-series

$$L_N(E, \lambda_{\mathrm{mod}\,M}, s) := \sum_{(m,MN)=1} a_m \lambda_{\mathrm{mod}\,M}(m) m^{-s}$$
$$L_N(E, s; \alpha) := \sum_{(m,N)=1} e^{2\pi i m \alpha} a_m m^{-s}$$

as in §4 above. Now put

- δ_E = number of connected components of $E(\mathbb{R})$;

- Ω_E^+ = fundamental real period of ω_E;

- Ω_E^- = $\delta_E \times$ fundamental imaginary period of ω_E

so that

$$\int_{E(\mathbb{C})} \omega_E \wedge \bar{\omega}_E = \Omega_E^+ \Omega_E^- \in i\mathbb{R}$$

The set of complex points $\mathrm{Spec}\,\mathbb{Q}(\mu_r)(\mathbb{C})$ is the set of primitive r^{th} roots of unity $\{e^{2\pi i x/r}\}$ in $\mathbb{C}$, which we identify with $(\mathbb{Z}/r\mathbb{Z})^*$. Write $\iota_x \colon \mathbb{Q}(\mu_r) \longhookrightarrow \mathbb{C}$ for the corresponding embedding $\zeta_r \mapsto e^{2\pi i x/r}$. Suppose $\lambda \colon \mathbb{A}_f^*/\mathbb{Q}_{>0}^* \to \mathbb{C}^*$ is a character of conductor dividing r. Then we can compute

$$\sum_{x \in (\mathbb{Z}/r\mathbb{Z})^*} \int_{E(\mathbb{C})} \lambda_{\mathrm{mod}\,r}(x) \cdot \iota_x \varphi_{E,N*}(\omega_r^{\mathrm{cusp}}) \wedge \bar{\omega}_E$$

$$= \int_{Y(N)(\mathbb{C}) \times (\mathbb{Z}/r\mathbb{Z})^*} \lambda_{\mathrm{mod}\,r}(x) \cdot \iota_x \omega_r^{\mathrm{cusp}} \wedge \varphi_{E,N}^* \bar{\omega}_E$$

$$= \int_{Y(N)(\mathbb{C}) \times (\mathbb{Z}/r\mathbb{Z})^*} \lambda_{\mathrm{mod}\,r}(x) \cdot \iota_x \omega_r \wedge \varphi_{E,N}^* \bar{\omega}_E$$

(since cusp forms and Eisenstein series are orthogonal)

$$= \int_{Y(Nr)(\mathbb{C})} (\lambda_{\mathrm{mod}\,Nr} \circ e_{Nr}) \cdot \tilde{\omega}_r \wedge \varphi_{E,Nr}^* \bar{\omega}_E \qquad (5.2.3)$$

where the map $e_{Nr} \colon Y_{Nr}(\mathbb{C}) \to \mathrm{Spec}\,\mathbb{Q}(\mu_{Nr})(\mathbb{C}) = (\mathbb{Z}/Nr\mathbb{Z})^*$ is that defined in (2.3.4).

At this point we need to choose the parameters z_r, z_r' of the Euler system ξ_r in such a way that the expression (5.2.3) can be computed using Theorem 4.6.3. In fact it will be necessary to replace $\xi_r(E)$ by a certain linear combination of Euler systems. The choices to be made are best broken down into a number of steps:

Step 1: Fix a prime p with $p \nmid N_E$, and $\varepsilon \in \{\pm 1\}$. We will restrict to characters λ with $\lambda(-1) = \varepsilon$.

Step 2: If $\alpha = y/M \in \mathbb{Q}$, the value of the twisted Dirichlet series at $s = 1$ is a period integral

$$L_{MN_E}(E, 1; \alpha) = - \int_\alpha^{i\infty} \sum_{(n, MN_E)=1} a_n q^n \, 2\pi i \, d\tau$$

and one knows that this is a rational multiple of a period along a closed path in $X(N)(\mathbb{C})$, for suitable N. Moreover the cusp form $\sum_{(n,N)=1} a_n q^n \, dq/q$ is obtained from the eigenform $\varphi^* \omega_E$ by applying a suitable Hecke operator. It follows that for any $\alpha \in \mathbb{Q}$,

$$L_{MN_E}(E, 1; \alpha) - \varepsilon L_{MN_E}(E, 1; -\alpha) \qquad (5.2.4)$$

is a rational multiple of Ω_E^ε. Moreover, one can find α with denominator prime to any chosen integer for which (5.2.4) is nonzero, by [36].

We choose an $\alpha = y/M$ with $M > 0$ and $(M, y) = (M, p) = 1$, and for which (5.2.4) is non-zero. By what has been just said, there will be a finite collection of such y/M which will cover all possible choices of p. We then take

$$N = \prod_{q|MN_E} q^{\nu_q}, \quad \nu_q = \max(2, \operatorname{ord}_q(N_E), \operatorname{ord}_q(M) + 1).$$

Step 3: Fix auxiliary integers D, $D' > 1$ with $(DD', 6pN_E) = 1$ and $D \equiv D' \equiv 1 \pmod M$. Let $r = r_0 p^m \in \mathcal{R}_p$; thus $m \geq 1$ and $r_0 > 0$ is squarefree and coprime to $pDD'N$. In the notation of §4.6 we put $R = r$, $T = \{q|r\}$, $S = \{q|N\}$ and choose the ideles t, $t' \in \mathbf{A}_f^*$ to have local components

$$t_q = \begin{cases} 1 & \text{if } q \nmid Nr \\ (Nr)^{-1} & \text{if } q|Nr \end{cases}; \quad t'_q = \begin{cases} 1 & \text{if } q \nmid Mr \\ -r^{-1}y|M|_q & \text{if } q|M \\ (Mr)^{-1} & \text{if } q|r \end{cases}$$

Then (4.6.1) holds, and $t \in (Nr)^{-1}\hat{\mathbf{Z}}^*$, $t' \in (Mr)^{-1}\hat{\mathbf{Z}}^*$. In §4.6 this data then determines functions ϕ, $\phi' \in \mathcal{S}(\mathbf{A}_f^2)$. Let $\delta \in \hat{\mathbf{Z}}^*$ be the finite unit idele

$$\delta_q = \begin{cases} D & \text{if } q|Nr; \\ 1 & \text{otherwise} \end{cases}$$

and set, by analogy with (1.3.2),

$$_D\phi = D^2\phi - D[\delta]\phi$$

in the notation of (4.1.2). Likewise define δ' and $_{D'}\phi'$ in the obvious way. Since $(Nr, D) = 1$, if $\operatorname{cond}(\lambda)|r$ we have

$$\lambda(\delta) = \prod_{q|Nr} \lambda_q(D) = \lambda_{\operatorname{mod} r}(D). \tag{5.2.5}$$

Step 4: We have $\phi = \operatorname{char}[(t + \hat{\mathbf{Z}}) \times \hat{\mathbf{Z}}] = \operatorname{char}[(Nr)^{-1} + \hat{\mathbf{Z}} \times \hat{\mathbf{Z}}]$. Choose $z_r \in \mathcal{E}^{\operatorname{univ}}(Y(Nr))$ to be the point which in complex coordinates is

$$\frac{1}{Nr} \in (\frac{1}{Nr}\mathbf{Z} + \frac{\tau}{Nr}\mathbf{Z})/(\mathbf{Z} + \tau\mathbf{Z}) \simeq (\mathbf{Z}/Nr\mathbf{Z})^2.$$

For different r the points z_r are compatible: $\ell z_{\ell r} = z_r$. We then can use (1.3.3) to write the Eisenstein series in terms of the complex parameterisation as

$$_D\operatorname{Eis}(z_r) = E_1(_D\phi)\, du. \tag{5.2.6}$$

Step 5: The function ϕ' has local components

$$\phi_q' = \begin{cases} \text{char}[\mathbb{Z}_q \times \mathbb{Z}_q] & \text{if } q \nmid Nr; \\ \text{char}[\mathbb{Z}_q \times (1/Mr + \mathbb{Z}_q)] & \text{if } q|r; \\ \text{char}\begin{bmatrix} t_q' + \mathbb{Z}_q \\ \mathbb{Z}_q^* \end{bmatrix} - q^{-1}\text{char}\begin{bmatrix} t_q' + q^{-1}\mathbb{Z}_q \\ \mathbb{Z}_q^* \end{bmatrix} & \text{if } q|N. \end{cases}$$

The last expression can be rewritten as

$$\text{char}\begin{bmatrix} t_q' + \mathbb{Z}_q \\ \mathbb{Z}_q \end{bmatrix} - [q]\text{char}\begin{bmatrix} q^{-1}t_q' + q^{-1}\mathbb{Z}_q \\ \mathbb{Z}_q \end{bmatrix}$$

$$- q^{-1}\text{char}\begin{bmatrix} t_q' + q^{-1}\mathbb{Z}_q \\ \mathbb{Z}_q \end{bmatrix} + q^{-1}[q]\text{char}\begin{bmatrix} q^{-1}t_q' + q^{-2}\mathbb{Z}_q \\ \mathbb{Z}_q \end{bmatrix}.$$

Now by (4.3.1) there exist a finite set of points $z_{r,j}' \in \mathcal{E}^{\text{univ}}(Y(Nr))$ and constants $b_j \in N^{-1}\mathbb{Z}$ which are independent of r, such that

$$\sum_j b_j \cdot {}_{D'}\text{Eis}(z_{r,j}') = E_1({}_{D'}\phi')\,du$$

and $\ell z_{\ell r,j}' = z_{r,j}'$. Moreover the differences $z_{r,j}' - z_{r,i}'$ will be N-torsion, and in complex coordinates $Nz_{r,j}'$ will be the point

$$(-Nt' \bmod \hat{\mathbb{Z}})\tau \in (\frac{1}{r}\mathbb{Z} + \frac{\tau}{r}\mathbb{Z})/(\mathbb{Z} + \tau\mathbb{Z}) \simeq (\mathbb{Z}/r\mathbb{Z})^2.$$

It follows that

$$e_{Nr}(z_r, z_{r,j}') = \zeta_{p^m}^{-(Mr_0)^{-1}} \times (\text{prime-to } p \text{ root of } 1)$$

and thus that the constant e of (5.1.1) equals $(-M^{-1}) \in \mathbb{Z}_p^*$.

Step 6: Put $\tilde{\sigma}_{r,j} = \{\vartheta_D(z_r), \vartheta_{D'}(z_{r,j}')\}$, and let $\xi_{r,j}(E)$ be the associated Euler system for $T_p(E)$. The required Euler system is then

$$c_r = \sum_j b_j\xi_{r,j}(E) \in H^1(\mathbb{Q}(\mu_r), T_p(E)).$$

We can now compute the dual exponential of c_r. Put $\tilde{\omega}_{r,j}$ for the differential on $Y(Nr)$ constructed from $(z_r, z_{r,j}')$. The Kodaira-Spencer map takes $(dt/t)^{\otimes 2}$ to dq/q, and therefore $du^{\otimes 2}$ to $(2\pi i)^{-1}d\tau$. Therefore

$$\sum_j b_j \cdot \tilde{\omega}_{r,j} = \frac{(2\pi i)^{-1}}{Nr}E_1({}_D\phi)E_1({}_{D'}\phi')\,d\tau.$$

We then get

$$\sum_{\gamma\in\mathrm{Gal}(\mathbb{Q}(\mu_r)/\mathbb{Q})=(\mathbb{Z}/r\mathbb{Z})^*} \lambda_{\mathrm{mod}\,r}(\gamma)\,\exp_p^*c_r^\gamma$$

$$= \left(\frac{-M^{-1}h_E}{r\Omega_E^+\Omega_E^-}\sum_j b_j \int_{Y(Nr)(\mathbb{C})} (\lambda_{\mathrm{mod}\,Nr}\circ e_{Nr})\cdot\widetilde\omega_{r,j}\wedge\varphi_{E,Nr}^*\bar\omega_E\right)\omega_E$$

$$= -\frac{h_E\#GL_2(\mathbb{Z}/Nr\mathbb{Z})}{MNr^2\Omega_E^+\Omega_E^-}\langle E_1(_D\phi)E_1(_{D'}\phi')\otimes\lambda,F\rangle\omega_E.$$

By Theorem 4.6.3, taking $k=l=1$ and $s=0$,

$$\langle E_1(\phi)E_1(\phi')\otimes\lambda,F\rangle = C'r^{-2}\#GL_2(\mathbb{Z}/r\mathbb{Z})^{-1}\prod_{q|r}\lambda_q(MNt_qt_q')^{-1}$$

$$\times L_{Nr}(E,\lambda_{\mathrm{mod}\,r},1)\big(L_N(E,1;y/M)-\lambda(-1)L_N(E,1;-y/M)\big)$$

for some $C'\in\mathbb{Q}^*$, depending only on E, M and N. Moreover, using (5.2.5) and the hypothesis that $D\equiv D'\equiv 1\pmod M$,

$$\langle E_1(_D\phi)\,E_1(_{D'}\phi')\otimes\lambda,F\rangle = C'r^2\#GL_2(\mathbb{Z}/r\mathbb{Z})^{-1}$$

$$\times \prod_{q|r}\lambda_q(MNt_qt_q')^{-1}DD'(D-\lambda_{\mathrm{mod}\,r}(D)^{-1})(D'-\lambda_{\mathrm{mod}\,r}(D')^{-1})$$

$$\times L_{Nr}(E,\lambda_{\mathrm{mod}\,r},1)\big(L_N(E,1;y/M)-\lambda(-1)L_N(E,1;-y/M)\big).$$

Now for $q|r$ we have $t_qt_q'=(MNr^2)^{-1}$, so $\prod_{q|r}\lambda_q(MNt_qt_q')=1$ since $\mathrm{cond}(\lambda)|r$. Combining everything one gets the final result:

Theorem 5.2.7. *Let $E/\mathbb{Q}$ be a modular elliptic curve of conductor N_E. Fix a non-zero 1-form $\omega_E\in\Omega^1(E/\mathbb{Q})$, with real and imaginary periods Ω_E^+, Ω_E^-. Let p be a prime not dividing N_E. Then there is an integer M prime to p, and for every pair of integers D, $D'>1$ with $(DD',6pN_E)=1$ and $D\equiv D'\equiv 1$ (mod M) an Euler system:*

$$c_r=c_r(E,p,D,D')\in H^1(\mathbb{Q}(\mu_r),T_p(E)),\quad r=r_0p^m,\ r_0\ squarefree\ and$$
$$prime\ to\ pMN_E,\ m\ge 1$$

such that for each r and each character λ: $\mathrm{Gal}(\mathbb{Q}(\mu_r)/\mathbb{Q})\simeq(\mathbb{Z}/r\mathbb{Z})^*\to\mathbb{C}^*$ with $\lambda(-1)=\pm1$

$$\sum_{\gamma\in\mathrm{Gal}(\mathbb{Q}(\mu_r)/\mathbb{Q})}\lambda(\gamma)\exp_p^*c_r^\gamma =$$
$$C_E^\pm DD'(D-\lambda(D)^{-1})(D'-\lambda(D')^{-1})\frac{L_{rMN_E}(E,\lambda,1)}{\Omega_E^\pm}\omega_E$$

for some constant $C_E^\pm$, depending only on E.

In the special case $r=p^m$ this is (with minor modifications of notation) Theorem 7.1 of [29].

458   A. J. Scholl

References

[1] A. A. Beilinson: *Higher regulators and values of L-functions* J. Soviet Math. **30** (1985), 2036–2070

[2] — : *Higher regulators of modular curves*. Applications of algebraic K-theory to algebraic geometry and number theory (Contemporary Mathematics **5** (1986)), 1–34

[3] S. Bloch, K. Kato: *Tamagawa numbers of motives*. In: The Grothendieck Festschrift, Vol. I. Progress in Mathematics **86**, 333–400 (Birkhäuser, 1990)

[4] R. Coleman: *Hodge-Tate periods and Abelian integrals*. Inventiones math. **78** (1984) 351–379

[5] P. Deligne: *Formes modulaires et représentations ℓ-adiques* Sém. Bourbaki, éxposé 355. Lect. notes in mathematics **179**, 139–172 (Springer, 1969)

[6] — : *Formes modulaires et représentations de GL₂*. In: Modular functions of one variable II. Lect. notes in mathematics **349**, 55–105 (Springer, 1973)

[7] — : *Courbes elliptiques: formulaire*. In: Modular functions of one variable IV. Lect. notes in mathematics **476**, 79–88 (Springer, 1975)

[8] —, M. Rapoport: *Les schémas de modules des courbes elliptiques*. Modular functions of one variable II. Lect. notes in mathematics **349**, 143–316 (Springer, 1973)

[9] C. Deninger, A. J. Scholl: *The Beilinson Conjectures*. In: L-functions and Arithmetic (ed. J. H. Coates, M. J. Taylor), Cambridge University Press (1991), 173–209

[10] G. Faltings: *Hodge-Tate structures and modular forms*. Math.Ann. **278** (1987) 133–149

[11] J-M. Fontaine: *Formes différentielles et modules de Tate des variétés abeliennes sur les corps locaux*. Inventiones math. **65** (1982), 379–409

[12] — : *Représentations p-adiques semi-stables*. In: Périodes p-adiques. Astérisque **223** (1994)

[13] R. Hübl: *Traces of differential forms and Hochschild homology*. Lect. notes in mathematics **1368**, 1989

[14] O. Hyodo: *On the Hodge-Tate decomposition in the imperfect residue field case.* Crelle **365** (1986), 97–113

[15] H. Jacquet *Automorphic forms on GL(2) II.* Lect. notes in mathematics **278** (Springer, 1972)

[16] K. Kato: *Lectures on the approach to Iwasawa theory for Hasse-Weil L-functions via* B_{dR}. In: Arithmetic Algebraic Geometry. Lect. notes in mathematics **1553**, 50–163 (Springer, 1993)

[17] — : *Generalized explicit reciprocity laws.* Preprint, 1994

[18] N. Katz, B. Mazur: *Arithmetic moduli of elliptic curves.* Ann. of Math. Studies **108** (1985)

[19] D. Kubert, S. Lang: *Units in the modular function field II.* Math. Ann. **218** (1975), 175–189

[20] S. Lang: *Cyclotomic Fields I and II.* 2nd edition, Springer-Verlag, 1990

[21] J. Lipman: *Traces and residues of differential forms via Hochschild homology.* Contemporary Mathematics **61**. American Math. Society, 1987

[22] J-L. Loday: *Cyclic homology.* Springer-Verlag, 1992

[23] H. Matsumura: *Commutative Algebra.* 2nd edition, Benjamin/Cummings, 1980

[24] J. S. Milne: *Arithmetic duality theorems.* Academic Press, 1986

[25] D. Mumford: *Picard groups of moduli problems.* In: Arithmetic Algebraic Geometry (O. F. G. Schilling, ed.). Harper and Row (1965)

[26] A. Ogg: *Rational points on certain elliptic modular curves.* Proc. Symp. Pure Math. **24**, 221–231 (AMS, 1973)

[27] M. Rapoport, N. Schappacher, P. Schneider (eds.): *Beilinson's conjectures on special values of L-functions.* Academic Press, 1988

[28] G. Robert: *Concernant la relation de distribution satisfaite par la fonction φ associée à un réseau complexe.* Inventiones math. **100** (1990), 231–257

[29] K. Rubin: These proceedings, 351–366

[30] N. Schappacher, A. J. Scholl: *Beilinson's theorem on modular curves.* In [27], 273–304

[31] B. Schoeneberg: *Modular forms.* Springer, 1974

[32] A. J. Scholl: *Higher regulators and L-functions of modular forms.* Book in preparation

[33] P. Schneider: *Introduction to the Beilinson conjectures.* In [27], 1–35

[34] J-P. Serre: *Corps locaux.* Hermann, 1968

[35] — : *Elliptic curves and abelian ℓ-adic representations.* Benjamin/Cummings, 1968

[36] G. Shimura: *On the periods of modular forms.* Math. Annalen **229** (1977), 211–221

[37] J. Tate: *p-divisible groups.* Proceedings of a conference on local fields, Driebergen (T. A. Springer, ed.), 158–183, Springer-Verlag (1967)

[38] — : *Relations between K_2 and Galois cohomology.* Inventiones math. **36** (1976), 257–274

[39] E. Whittaker, G. L. Watson: *A course in modern analysis.* 4th edition, Cambridge, 1958

DEPARTMENT OF MATHEMATICAL SCIENCES, UNIVERSITY OF DURHAM, SOUTH ROAD, DURHAM DH1 3LE, ENGLAND
a.j.scholl@durham.ac.uk

La distribution d'Euler-Poincaré d'un groupe profini

JEAN-PIERRE SERRE

à John Tate

Lorsqu'on désire calculer des groupes de cohomologie, il y a intérêt à disposer de résultats généraux simples, du genre "dualité" ou "formule d'Euler-Poincaré". Un exemple typique (dû à Tate, [14], §2) est celui de la cohomologie du groupe profini $G = \mathrm{Gal}(\overline{K}/K)$, où K est une extension finie de $\mathbf{Q}_p$. Un autre exemple (dû à Lazard, [7], p.11) est celui où G est un groupe de Lie p-adique compact sans torsion.

Pour calculer des caractéristiques d'Euler-Poincaré dans d'autres cas (par exemple celui d'un groupe p-adique compact pouvant avoir de la torsion), il est commode de définir une certaine distribution $\mu_{G,p}$ sur le groupe G considéré (p désignant un nombre premier fixé). Cette distribution est la *distribution d'Euler-Poincaré* de G. Tout revient ensuite à déterminer $\mu_{G,p}$, par exemple à montrer que c'est 0 pour certains couples (G,p). C'est là l'objet du présent travail. Les principaux résultats sont résumés au §1 ci-après.

Table des matières

Reprinted from 'Galois Representations in Arithmetic Algebraic Geometry', edited by A. J. Scholl & R. L. Taylor. ©Cambridge University Press 1998

§1. Enoncé des résultats

1.1. Notations

La lettre p désigne un nombre premier, fixé dans tout ce qui suit.

On note G un groupe profini (cf. [12]), et $\mathbf{U}_G$ l'ensemble des sous-groupes ouverts normaux de G. Le groupe G est limite projective des groupes finis G/U, pour $U \in \mathbf{U}_G$.

Un élément s de G est dit *régulier* (ou "p-régulier") si ses images dans les G/U ($U \in \mathbf{U}_G$) sont d'ordre premier à p. Cela revient à dire que l'ordre de s dans G (au sens profini, cf. [12], I.1.3) est premier à p.

L'ensemble des éléments réguliers de G est noté G_{reg}. C'est une partie compacte de G. On a $G_{\text{reg}} = \varprojlim (G/U)_{\text{reg}}$.

1.2. G-modules et caractères de Brauer

On note $C_G(p)$, ou simplement C_G, la catégorie des G-modules discrets qui sont des $\mathbf{F}_p$-espaces vectoriels de dimension finie. C'est la limite inductive des catégories $C_{G/U}(p)$, pour $U \in \mathbf{U}_G$.

Si A est un objet de C_G (ce que nous écrirons "$A \in C_G$"), on note

$$\varphi_A \colon G_{\text{reg}} \to \mathbf{Z}_p$$

le *caractère de Brauer* de A (n°3.3). C'est une fonction localement constante sur G_{reg}.

1.3. Cohomologie

Si $A \in C_G$, les groupes de cohomologie $H^i(G, A)$ sont des $\mathbf{F}_p$-espaces vectoriels, nuls pour $i < 0$. Nous ferons dans tout ce § les deux hypothèses suivantes sur le couple (G, p) :

(1.3.1) *On a* $\dim H^i(G, A) < \infty$ *pour tout* $i \in \mathbf{Z}$ *et tout* $A \in C_G$.
(Par "dim" on entend la dimension sur le corps $\mathbf{F}_p$.)

(1.3.2) *On a* $\operatorname{cd}_p(G) < \infty$, autrement dit il existe un entier d tel que $H^i(G, A) = 0$ pour tout $i > d$ et tout $A \in C_G$, cf. [12], I.3.1.

Ces deux hypothèses permettent de définir la *caractéristique d'Euler-Poincaré* de A :

$$e(G, A) = \sum (-1)^i \dim H^i(G, A).$$

C'est un entier, qui dépend de façon additive de A.

1.4. Distribution d'Euler-Poincaré

Si X est un espace compact totalement discontinu, une distribution μ sur X, à valeurs dans $\mathbf{Q}_p$, est une forme linéaire

$$f \mapsto\ <f, \mu> = \int f(x)\mu(x)$$

sur l'espace vectoriel des fonctions localement constantes sur X, à valeurs dans $\mathbf{Q}_p$ (n°3.1).

Théorème A (n°3.4)—*Il existe une distribution μ_G et une seule sur l'espace G_{reg}, à valeurs dans $\mathbf{Q}_p$, qui a les deux propriétés suivantes :*
(1.4.1) *Pour tout $A \in C_G$, la caractéristique d'Euler-Poincaré de A est donnée par :*

$$e(G, A) = <\varphi_A, \mu_G>\ .$$

(1.4.2) *μ_G est invariante par les automorphismes intérieurs $s \mapsto gsg^{-1}$ $(g \in G)$, ainsi que par $s \mapsto s^p$.*

La distribution μ_G sera appelée la *distribution d'Euler-Poincaré* de G. On la note $\mu_{G,p}$ lorsqu'on veut préciser p. D'après (1.4.1), la caractéristique d'Euler-Poincaré d'un G-module A s'obtient en "intégrant" le produit du caractère de Brauer φ_A de A par la distribution μ_G.

1.5. Exemples

(1.5.1) Dans le cas, dû à Tate, où $G = \text{Gal}(\overline{K}/K)$, K étant une extension finie de $\mathbf{Q}_p$, on a $e(G, A) = -d \cdot \dim A$, avec $d = [K : \mathbf{Q}_p]$, cf. [14]. Comme $\dim A = \varphi_A(1)$, cela signifie que $\mu_G = -d \cdot \delta_1$, où δ_1 est la distribution de Dirac en l'élément 1 de G, cf. n°6.1.
(1.5.2) Si l'ordre du centre de G est divisible par p, on a $\mu_G = 0$ (cf. n°5.2), autrement dit $e(G, A) = 0$ pour tout $A \in C_G$.

1.6. Détermination de μ_G à partir des $H^i_c(U, \mathbf{Q}_p)$

Si $U \in \mathbf{U}_G$, on définit (au moyen de cochaînes continues) les groupes de cohomologie $H^i_c(U, \mathbf{Q}_p)$. Ce sont des $\mathbf{Q}_p$-espaces vectoriels de dimension finie, nuls si $i > \text{cd}_p(G)$. Le groupe fini G/U opère de façon naturelle sur $H^i_c(U, \mathbf{Q}_p)$. Soit $h^i_U : G/U \to \mathbf{Q}_p$ le caractère de la représentation ainsi obtenue. On pose

$$h_U = \sum (-1)^i h^i_U.$$

Le caractère virtuel h_U est nul en dehors de G_{reg}. De plus, la distribution μ_G est égale à la limite des h_U, au sens suivant (n°4.4, th.4.4.3) :

Théorème B—*Soit* $f: G_{\mathrm{reg}} \to \mathbf{Q}_p$ *une fonction constante* (mod U). *On a :*

$$<f, \mu_G> = (G : U)^{-1} \sum f(s) h_U(s),$$

où la somme porte sur les éléments s de $(G/U)_{\mathrm{reg}}$.

(Une fonction f sur G_{reg}, ou sur G, est dite "constante (mod U)" si $f(s)$ ne dépend que de l'image de s dans G/U.)

1.7. Le cas analytique

Supposons maintenant que G soit un *groupe de Lie p-adique compact* sans élément d'ordre p. D'après Lazard [7], le groupe G possède les propriétés (1.3.1) et (1.3.2), avec $\mathrm{cd}_p(G) = \dim G$. La distribution μ_G peut alors s'expliciter de la manière suivante :

Soit Lie G l'algèbre de Lie de G. Si $g \in G$, notons $\mathrm{Ad}(g)$ l'automorphisme de Lie G défini par l'automorphisme intérieur $x \mapsto gxg^{-1}$, et posons :

$$F(g) = \det(1 - \mathrm{Ad}(g^{-1})).$$

La fonction F ainsi définie sur G est localement constante, et nulle en dehors de G_{reg}. De plus, elle est égale à h_U pour tout $U \in \mathbf{U}_G$ assez petit (n°7.2). Vu le th.B, on en déduit (cf. n°7.3) :

Théorème C—*La distribution μ_G est égale au produit de F par la distribution de Haar dg de G* (normalisée pour que sa masse soit égale à 1).

En d'autres termes, on a la formule :

$$e(G, A) = \int \varphi_A(g) \det(1 - \mathrm{Ad}(g^{-1})) dg \quad \text{pour tout } A \in C_G.$$

Corollaire—*On a $\mu_G = 0$ si et seulement si le centralisateur de tout élément de G est de dimension > 0.*

C'est le cas lorsque G est un sous-groupe ouvert de $\underline{G}(\mathbf{Q}_p)$, où $\underline{G}$ est un $\mathbf{Q}_p$-groupe algébrique connexe de dimension > 0, cf. n°7.4.

1.8. Une application

Soit G un sous-groupe ouvert compact de $\mathbf{GL}_n(\mathbf{Z}_p)$, et soit I le G-module discret $(\mathbf{Q}_p/\mathbf{Z}_p)^n$. Supposons $1 < n < p - 1$. On démontre (cf. n°8.3) :

Théorème D—(a) *Les groupes de cohomologie $H^i(G, I)$ sont des p-groupes finis, nuls pour $i \geq n^2$.*

(b) *Si l'on pose $h^i(G, I) = \log_p |H^i(G, I)|$, on a $\sum(-1)^i h^i(G, I) = 0$.*
(Autrement dit, le produit alterné des ordres des $H^i(G, I)$ est égal à 1.)

Lorsque $n = 2$ et $p \geq 5$, ce résultat m'avait été commandé par J. Coates, qui en avait besoin pour des calculs de caractéristiques d'Euler-Poincaré de groupes de Selmer, cf. [4].

§2. Caractères de Brauer : rappels

2.1. Représentants multiplicatifs et trace de Brauer

Soit k un corps parfait de caractéristique p, et soit $W(k)$ l'anneau des vecteurs de Witt de k. On note K le corps des fractions de $W(k)$. Le cas le plus important pour la suite est celui où $k = \mathbf{F}_p$, $W(k) = \mathbf{Z}_p$ et $K = \mathbf{Q}_p$. Si $x \in k$, on note $\overline{x}$ son *représentant multiplicatif* dans K, autrement dit l'élément $(x, 0, 0, \dots)$ de $W(k)$. Lorsque x est une racine de l'unité, on peut caractériser $\overline{x}$ comme l'unique racine de l'unité de $W(k)$, de même ordre que x, et dont l'image dans k par l'isomorphisme $W(k)/pW(k) = k$ est égale à x.

L'automorphisme de Frobenius $x \mapsto x^p$ de k définit un automorphisme de $W(k)$ (et aussi de K) que nous noterons F. Pour tout $x \in k$, le représentant multiplicatif de x^p est $F(\overline{x}) = (\overline{x})^p$. Un élément w de $W(k)$ est fixé par F si et seulement si l'on a $w \in \mathbf{Z}_p$.

Soit n un entier > 0, et soit $f(T) = T^n + \dots$ un polynôme unitaire de degré n à coefficients dans k. Ecrivons f sous la forme

$$f(T) = \prod(T - x_i),$$

où $x_1, \dots, x_n$ appartiennent à une extension galoisienne finie k' de k. Définissons un polynôme $\overline{f}(T)$ par $\overline{f}(T) = \prod(T - \overline{x}_i)$. Les coefficients de $\overline{f}(T)$ sont des éléments de $W(k')$ invariants par $\mathrm{Gal}(k'/k)$; ils appartiennent donc à $W(k)$. On a $\overline{f} \equiv f \pmod{p}$.

Soit A un espace vectoriel de dimension finie n sur k, et soit u un élément de $\mathrm{End}(V)$. Soit $f(T) = \det(T - u)$ le polynôme caractéristique de u et soit $\overline{f}(T) = T^n - a_1 T^{n-1} + \dots$ le polynôme correspondant ; on a $\overline{f} \in W(k)[T]$. Le coefficient a_1 sera appelé la *trace de Brauer* de u, et noté $\mathrm{Tr}_{\mathrm{Br}}(u)$; c'est la somme des représentants multiplicatifs des valeurs propres de u (dans une extension convenable de k). On a par construction :

$$\mathrm{Tr}_{\mathrm{Br}}(u) \equiv \mathrm{Tr}(u) \pmod{p} \qquad \text{et} \qquad \mathrm{Tr}_{\mathrm{Br}}(u^p) = F(\mathrm{Tr}_{\mathrm{Br}}(u)).$$

Remarque. Lorsque $k = \mathbf{F}_p$, on a $\mathrm{Tr}_{\mathrm{Br}}(u) \in \mathbf{Z}_p$. Si l'on représente u par une matrice (u_{ij}) et si $U = (U_{ij})$ est une matrice à coefficients dans $\mathbf{Z}_p$ telle que $U_{ij} \equiv u_{ij} \pmod{p}$ pour tout i,j, on vérifie facilement la formule :

$$\mathrm{Tr}_{\mathrm{Br}}(u) = \lim \mathrm{Tr}(U^{p^m}) \quad \text{pour } m \to \infty,$$

la limite étant prise pour la topologie naturelle de $\mathbf{Z}_p$. Cela fournit une définition de $\mathrm{Tr}_{\mathrm{Br}}(u)$ "sans sortir de $\mathbf{Z}_p$".

2.2. Caractères de Brauer des groupes finis

Supposons G fini, et écrivons son ordre sous la forme $p^a m$, avec $(p, m) = 1$. Le corps k est dit "assez gros pour G" (cf. [11], Chap.14) s'il contient toutes les racines m-ièmes de l'unité.

Notons $C_{G,k}$ la catégorie des $k[G]$-modules de type fini; lorsque $k = \mathbf{F}_p$, c'est la catégorie notée C_G au n°1.2. Si A est un objet de $C_{G,k}$, et si $s \in G$, on note s_A l'automorphisme correspondant du k-espace vectoriel A. On pose :

$$\varphi_A(s) = \mathrm{Tr}_{\mathrm{Br}}(s_A).$$

Bien que cette définition ait un sens pour tout $s \in G$, on se borne à $s \in G_{\mathrm{reg}}$, i.e. s d'ordre premier à p (les autres éléments de G ne fournissent pas d'information supplémentaire : si $s \in G_{\mathrm{reg}}$ est la p'-composante d'un élément g de G, on a $\varphi_A(g) = \varphi_A(s)$, comme on le vérifie facilement). La fonction

$$\varphi_A \colon G_{\mathrm{reg}} \to W(k)$$

est appelée le *caractère de Brauer* de A. Les propriétés suivantes sont bien connues (cf. e.g. [5], Chap.IV ou [11], Chap.18) :

(2.2.1) φ_A *est une fonction centrale* (i.e. invariante par automorphismes intérieurs).

(2.2.2) *On a* $\varphi_A(s^p) = F(\varphi_A(s))$ *pour tout* $s \in G_{\mathrm{reg}}$.

(2.2.3) φ_A *ne dépend que des quotients de Jordan-Hölder de* A (i.e. le semi-simplifié de A a même caractère de Brauer que A).

(2.2.4) *Si* A *et* A' *sont semi-simples et ont même caractère de Brauer, ils sont isomorphes.*

(2.2.5) *Les caractères de Brauer des différents* $k[G]$-*modules simples sont linéairement indépendants sur* K. *Si* k *est assez gros, ils forment une base de l'espace vectoriel des fonctions centrales sur* G_{reg}.

2.3. Le cas où $k = \mathbf{F}_p$

Supposons $k = \mathbf{F}_p$, auquel cas le caractère de Brauer φ_A d'un objet A de C_G est à valeurs dans $\mathbf{Z}_p$. D'après (2.2.1), φ_A est une fonction centrale sur G_{reg}, et d'après (2.2.2), on a :

(2.3.1) $\varphi_A(s^p) = \varphi_A(s)$ pour tout $s \in G_{\mathrm{reg}}$.

Choisissons un ensemble de représentants Σ_G des classes d'objets *simples* de C_G.

Proposition 2.3.2—*Les caractères* φ_S, $S \in \Sigma_G$, *forment une* $\mathbf{Q}_p$-*base de l'espace des fonctions centrales sur* G_{reg}, *à valeurs dans* $\mathbf{Q}_p$, *invariantes par* $s \mapsto s^p$.

Soit F l'espace des fonctions en question. Les φ_S appartiennent à F et sont linéairement indépendants d'après (2.2.5). Il reste à voir que tout élément f de F est combinaison linéaire des φ_S. Choisissons un corps fini k contenant $\mathbf{F}_p$, et assez gros pour G (cf. n°2.2). D'après (2.2.5), on peut écrire f sous la forme

$$f = \sum a_T \varphi_T,$$

où les modules T sont des $k[G]$-modules, et les coefficients a_T appartiennent au corps des fractions K de $W(k)$. Chacun des modules T défini par restriction des scalaires à $\mathbf{F}_p$ un $\mathbf{F}_p[G]$-module T^0. Si $n = [k : \mathbf{F}_p]$, on a :

$$(2.3.3) \qquad \varphi_{T^0}(s) = \sum_{i=0}^{n-1} \varphi_T(s^{p^i}).$$

On déduit de là :

$$(2.3.4) \qquad n.f = \sum a_T \varphi_{T^0}.$$

En décomposant les φ_{T^0} en combinaisons linéaires des φ_S, cela donne :

$$(2.3.5) \qquad f = \sum b_S \varphi_S, \text{ avec } b_S \in K.$$

Mais f et les φ_S sont des fonctions à valeurs dans $\mathbf{Q}_p$, et les φ_S sont linéairement indépendants ; on a donc $b_S \in \mathbf{Q}_p$ pour tout S, ce qui achève la démonstration.

Remarque. Soit $R(C_G)$ le groupe de Grothendieck de la catégorie C_G. La prop.2.3.2 revient à dire que l'application $A \mapsto \varphi_A$ définit un isomorphisme de $\mathbf{Q}_p \otimes R(C_G)$ sur l'algèbre des fonctions centrales sur G_{reg}, à valeurs dans $\mathbf{Q}_p$ et invariantes par $s \mapsto s^p$.

2.4. Le cas où $k = \mathbf{F}_p$ (suite)

Proposition 2.4.1—*Soit $c: \Sigma_G \to \mathbf{Q}_p$ une application. Il existe une fonction centrale θ^c et une seule sur G, à valeurs dans $\mathbf{Q}_p$, qui ait les propriétés suivantes :*
(2.4.2) θ^c *est nulle en dehors de* G_{reg}.
(2.4.3) θ^c *est invariante par* $s \mapsto s^p$.
(2.4.4) *Pour tout* $S \in \Sigma_G$, *on a*

$$c(S) = \frac{1}{|G|} \sum \theta^c(s) \varphi_S(s).$$

(Dans (2.4.4), on considère le produit $\theta^c \varphi_S$ comme une fonction sur G nulle en dehors de G_{reg}, et la somme porte sur tous les éléments s de G.)

Soit $R(s, s')$ la relation d'équivalence sur G_{reg} : "il existe $i \in \mathbf{Z}$ tel que s' et s^{p^i} soient conjugués". Soit $s_1, \ldots, s_r$ un système de représentants de G_{reg} (mod R) ; notons C_i la classe d'équivalence de s_i ; l'ensemble G_{reg} est réunion disjointe des C_i. Soit f_i la fonction égale à $1/|C_i|$ sur C_i et à 0 ailleurs. D'après la prop.2.3.2 on peut écrire f_i sous la forme

$$f_i = \sum a_{i,S}\, \varphi_S, \quad \text{avec } a_{i,S} \in \mathbf{Q}_p.$$

468   Jean-Pierre Serre

On définit alors θ^c en donnant sa valeur en s_i pour tout i :

(2.4.5) $\theta^c(s_i) = |G| \sum_S a_{i,S}\, c(S).$

On vérifie par un simple calcul que la fonction θ^c ainsi définie a les propriétés voulues ; son unicité résulte du fait que les φ_S forment une base de
l'espace des fonctions sur G_{reg} (mod R).

Remarque. On peut écrire θ_c en termes de *caractères de modules projectifs*
de la manière suivante :

Si $S \in \Sigma_G$, notons P_S son enveloppe projective (cf. [11], Chap.14) ; c'est un
$\mathbf{F}_p[G]$-module projectif dont le plus grand quotient semi-simple est isomorphe
à S ; il est unique, à isomorphisme près. Ce module est la réduction modulo
p d'un $\mathbf{Z}_p[G]$-module projectif $\tilde{P}_S$, de type fini, dont le caractère (au sens
usuel) est noté Φ_S. On sait ([11], Chap.18) que Φ_S est nul en dehors de G_{reg}
et que sa restriction à G_{reg} est le caractère de Brauer de P_S. Notons Φ_S^* la
fonction $s \mapsto \Phi_S(s^{-1})$; c'est le caractère du dual de $\tilde{P}_S$. Soit D_S l'algèbre
des endomorphismes de S ; c'est un corps, extension finie de $\mathbf{F}_p$; posons
$d_S = [D_S : \mathbf{F}_p]$. La fonction θ^c de la prop.2.4.1 peut s'écrire de façon simple
comme combinaison linéaire des Φ_S^* :

Proposition 2.4.6—*On a :*

$$\theta^c = \sum c(S) d_S^{-1} \Phi_S^*,$$

où la somme porte sur les éléments S de Σ_G.

Cela se voit en remarquant que, pour tout $T \in \Sigma_G$, on a (cf. [11], Chap.18) :

$$\frac{1}{|G|} \sum_S \Phi_S^*(s)\varphi_T(s) = \dim \text{Hom}^G(S,T) = d_S^{-1}\delta_{ST},$$

où δ_{ST} est le symbole de Kronecker (1 si $S = T$ et 0 sinon).

Corollaire 2.4.7—*Si $c(S)d_S^{-1}$ appartient à $\mathbf{Z}$ pour tout S, la fonction θ_c
est le caractère d'un $\mathbf{Z}_p[G]$-module projectif "virtuel"* (i.e., c'est le caractère
d'un élément du groupe de Grothendieck $P(G)$ de la catégorie des $\mathbf{Z}_p[G]$-
modules projectifs de type fini).

C'est clair.

Nous aurons besoin dans la suite d'une propriété de "passage au quotient"
pour θ^c :

Soit N un sous-groupe normal de G. L'ensemble $\Sigma_{G/N}$ s'identifie à une
partie de Σ_G. L'application $c \colon \Sigma_G \to \mathbf{Q}_p$ définit donc des "fonctions θ^c" à la
fois pour G et pour G/N ; notons ces fonctions θ_G^c et $\theta_{G/N}^c$.

Proposition 2.4.8—*Pour tout $x \in G/N$, on a :*

$$\theta^c_{G/N}(x) = \frac{1}{|N|} \sum \theta^c_G(s),$$

où la somme porte sur les $s \in G$ d'image x dans G/N.

En effet, si l'on note θ' la fonction sur G/N définie par le membre de droite, il est clair que θ' a les propriétés réclamées dans la prop.2.4.1, relativement à G/N ; on a donc bien $\theta' = \theta^c_{G/N}$.

§3. La distribution μ_G

3.1. Distributions sur un espace compact totalement discontinu

Soit X un espace compact totalement discontinu. On sait (cf. e.g. [2], TG.II.32, cor.à la prop.6) que les ouverts fermés de X ("clopen subsets") forment une base de la topologie de X, de sorte que X est limite projective d'ensembles finis.

Soit R un anneau commutatif. On note $C(X;R)$ la R-algèbre des fonctions localement constantes sur X, à valeurs dans R. On a :

$$C(X;R) = R \otimes C(X;\mathbf{Z}).$$

Une *distribution* μ sur X, à valeurs dans R, est une R-forme linéaire

$$\mu\colon C(X;R) \to R.$$

Si f est un élément de $C(X;R)$, $\mu(f)$ est également noté $<f, \mu>$, ou aussi $\int f(x)\mu(x)$.

Remarque. Si l'on préfère "mesure" à "intégration", on peut voir une distribution comme une fonction $U \mapsto \mu(U)$, définie sur les ouverts fermés de X, à valeurs dans R, et additive :

$$\mu(U \cup U') = \mu(U) + \mu(U') \quad \text{si } U \cap U' = \emptyset.$$

Le *support* d'une distribution μ se définit à la façon habituelle : c'est la plus petite partie fermée Y de X telle que μ soit nulle sur $X - Y$ (i.e. $<f, \mu> = 0$ pour toute f nulle sur Y). Le support de μ se note $\mathrm{Supp}(\mu)$.

Si X' est une partie fermée de X, les distributions sur X' peuvent être identifiées (par prolongement par 0) aux distributions sur X à support contenu dans X'. On fera souvent cette identification par la suite.

3.2. Exemples de distributions

3.2.1. Si x est un point de X, la *distribution de Dirac* δ_x en x est la forme linéaire $f \mapsto f(x)$.

3.2.2. Soit G un groupe profini. Supposons que R soit une **Q**-algèbre. Si f est une fonction localement constante sur G, à valeurs dans R, choisissons $U \in \mathbf{U}_G$ tel que f soit constante mod U, et posons

$$\mu(f) = \frac{1}{(G:U)} \sum_{x \in G/U} f(x).$$

Il est clair que $\mu(f)$ ne dépend pas du choix de U. On obtient ainsi une distribution sur G. Si Z est une partie ouverte et fermée de G, $\mu(Z)$ n'est autre que la mesure de Z relativement à la mesure de Haar de G (normalisée pour que sa masse totale soit 1). Pour cette raison, nous appellerons μ la *distribution de Haar* de G.

3.2.3. Si μ est une distribution sur X, et F une fonction localement constante, on définit le *produit* $F.\mu$ de F et de μ par la formule :

$$<f, F.\mu> = <f.F, \mu> .$$

3.2.4. Si $h: X \to X'$ est une application continue de X dans un espace compact totalement discontinu X', et si μ est une distribution sur X, *l'image* $h\mu$ de μ par h est définie par la formule

$$<f', h\mu> = <f' \circ h, \mu> \quad \text{pour tout } f' \in C(X', R).$$

3.3. Distribution associée à une fonction additive de modules

On revient maintenant aux notations du §1, et l'on note G un groupe profini. Si A est un objet de C_G, on note φ_A son *caractère de Brauer*, défini par :

(3.3.1) $\varphi_A(s) = \text{Tr}_{\text{Br}}(s_A)$ pour $s \in G_{\text{reg}}$.
(Pour la définition de Tr_{Br}, voir n°2.1.)

L'action de G sur A se factorise par un quotient fini G/U, avec $U \in \mathbf{U}_G$. Il en résulte que φ_A est constant mod U. C'est donc une *fonction localement constante sur G_{reg}, à valeurs dans $\mathbf{Z}_p$*. De plus, φ_A dépend additivement de A : si

$$0 \to A \to B \to C \to 0$$

est une suite exacte dans C_G, on a $\varphi_B = \varphi_A + \varphi_C$.

Notons Σ_G un ensemble de représentants des objets simples de C_G, et soit $c: \Sigma_G \to \mathbf{Q}_p$ une application. Si $A \in C_G$ a une suite de Jordan-Hölder dont les quotients successifs sont $S_1, \ldots, S_m \in \Sigma_G$, on pose

$$c(A) = c(S_1) + \cdots + c(S_m).$$

La fonction c ainsi définie sur C_G est additive au sens ci-dessus. On va voir qu'on peut l'exprimer en termes des φ_A :

Théorème 3.3.2—*Il existe une distribution μ^c sur G_{reg}, à valeurs dans $\mathbf{Q}_p$, et une seule, telle que :*

(3.3.3) $c(A) = {<}\varphi_A, \mu^c{>}$ *pour tout $A \in C_G$.*

(3.3.4) μ^c *est invariante par l'action des automorphismes intérieurs, ainsi que par $s \mapsto s^p$.*

(Noter que l'application $s \mapsto s^p$ est un homéomorphisme de l'espace compact G_{reg} sur lui-même, ce qui donne un sens à (3.3.4).)

Lorsque G est fini, ce résultat a déjà été démontré (prop.2.4.1). On va se ramener à ce cas :

Soit $U \in \mathbf{U}_G$. D'après la prop.2.4.1 appliquée à G/U, il existe une fonction θ_U sur G/U, et une seule, qui soit à valeurs dans $\mathbf{Q}_p$ et possède les propriétés suivantes :

a) θ_U est une fonction centrale, invariante par $s \mapsto s^p$.

b) Pour tout $A \in C_{G/U}$, on a

$$c(A) = \frac{1}{(G:U)} \sum_{s \in G/U} \theta_U(s)\varphi_A(s).$$

Si $f \in C(G_{\text{reg}}; \mathbf{Q}_p)$ est constante (mod U), on définit ${<}f, \mu^c{>}$ par la formule

(3.3.5) $\qquad\qquad {<}f, \mu^c{>} = \dfrac{1}{(G:U)} \sum \theta_U(s)f(s),$

où la somme porte sur les éléments s de $(G/U)_{\text{reg}}$. Il résulte de la prop.2.4.8 que ${<}f, \mu^c{>}$ ne dépend pas du choix de U. La fonction

$$f \mapsto {<}f, \mu^c{>}$$

ainsi définie répond évidemment aux conditions imposées. Son unicité se démontre par un argument analogue : on se ramène au cas où G est fini, déjà traité dans la prop.2.4.1.

3.4. Démonstration du théorème A du n°1.4

Supposons que G satisfasse aux hypothèses du n°1.3, à savoir $\mathrm{cd}_p(G) < \infty$ et $\dim H^i(G, A) < \infty$ pour tout $i \in \mathbf{Z}$ et tout $A \in C_G$. La caractéristique d'Euler-Poincaré

$$e(G, A) = \sum (-1)^i \dim H^i(G, A)$$

est alors définie pour tout $A \in C_G$, et c'est une fonction additive de A, à valeurs dans $\mathbf{Z}$ (donc aussi à valeurs dans $\mathbf{Q}_p$). Le th.3.3.2, appliqué à

$c: A \mapsto e(G, A)$, fournit alors une distribution μ^c sur G_{reg} ; c'est la *distribution d'Euler-Poincaré* μ_G cherchée. D'après (3.3.3), on a :

$$(3.4.1) \qquad e(G, A) = \int \varphi_A(s)\mu_G(s) \quad \text{pour tout } A \in C_G.$$

Exemples.

Je me borne à deux cas élémentaires ; on en verra d'autres plus loin.

(3.4.2) Supposons G *fini*. Son ordre est premier à p (sinon, $\text{cd}_p(G)$ serait infini). On a alors $e(G, A) = \dim H^0(G, A)$, ce qui peut aussi s'écrire :

$$e(G, A) = \frac{1}{m} \sum_s \varphi_A(s), \quad \text{avec } m = |G|, \text{ cf. [11], 18.1.ix.}$$

En comparant avec (3.4.1), on voit que $\mu_G = \frac{1}{m} \sum_s \delta_s$, où δ_s est la distribution de Dirac en s. En d'autres termes, μ_G *est la distribution de Haar de* G.

(3.4.3) Supposons que G soit un *pro-p-groupe*. Le seul élément régulier de G est l'élément neutre. Il en résulte que μ_G est un multiple $E.\delta_1$ de la distribution de Dirac en ce point. Comme $\varphi_A(1) = \dim A$, la formule (3.4.1) s'écrit :

$$e(G, A) = E. \dim A.$$

En prenant $A = \mathbf{F}_p$ (avec action triviale de G—d'ailleurs aucune autre n'est possible), on voit que $E = e(G, \mathbf{F}_p)$. L'entier E est appelé la *caractéristique d'Euler-Poincaré* du pro-p-groupe G, cf. [12], I.4.1.exerc.

3.5. Interprétation de μ_G en termes de modules projectifs

Revenons à la situation du th.3.3.2, dans le cas où c est la fonction $A \mapsto e(G, A)$. Si $U \in \mathbf{U}_G$, on a associé à U (et c) une certaine fonction centrale θ_U sur G/U, qui détermine μ_G sur les fonctions f qui sont constantes (mod U) :

$$(3.5.1) \qquad <f, \mu_G> = \frac{1}{(G:U)} \sum \theta_U(s) f(s).$$

Proposition 3.5.2—*La fonction θ_U est le caractère d'un $\mathbf{Z}_p[G/U]$-module projectif virtuel.*

D'après le cor.2.4.7, il suffit de montrer que, pour tout objet simple S de $C_{G/U}$, le nombre

$$<\varphi_S, \mu_G> = e(G, S)$$

est un entier divisible par $d_S = \dim \text{End}^G(S)$. Or c'est clair, car chacun des $\mathbf{F}_p$-espaces vectoriels $H^i(G, S)$ a une structure naturelle d'espace vectoriel sur le corps $\text{End}^G(S)$, et sa dimension sur $\mathbf{F}_p$ est donc divisible par d_S.

Remarque. Soit $P(G/U)$ le groupe de Grothendieck de la catégorie des $\mathbf{Z}_p[G/U]$-modules projectifs de type fini, et soit P_U l'élément de $P(G/U)$ correspondant à θ_U d'après la prop.3.5.2. On peut se demander s'il existe une définition *purement homologique* de P_U. C'est bien le cas. On peut en effet montrer que le foncteur cohomologique $A \mapsto (H^i(G, A))$, défini sur la catégorie des G/U-modules qui sont des p-groupes abéliens finis, est de la forme

$$A \mapsto (H^i(\mathrm{Hom}^{G/U}(K_U, A))),$$

où $K_U = (K_{i,U})$ est un complexe de $\mathbf{Z}_p[G/U]$-modules projectifs de type fini. (L'existence d'un tel K_U non nécessairement de type fini est standard. Le fait qu'on puisse le choisir de type fini résulte par exemple de [6], 0.11.9.1 et 0.11.9.2.) Si $K_{i,U}^*$ désigne le dual de $K_{i,U}$, on montre que l'on a :

$$(3.5.3) \qquad P_U = \sum (-1)^i K_{i,U}^* \quad \text{dans } P(G/U).$$

C'est l'interprétation homologique cherchée.

De ce point de vue, la distribution μ_G apparaît comme un élément canonique de la limite projective des groupes de Grothendieck $P(G/U)$.

3.6. La $\mathbf{Z}_p$-mesure de $\mathrm{Cl}\, G$ définie par μ_G

Soit $\mathrm{Cl}\, G$ l'espace (compact) des classes de conjugaison de G, et soit $\pi \colon G \to \mathrm{Cl}\, G$ la projection canonique. Soit $\mu_G^0 = \pi \mu_G$ l'image par π de μ_G (cf. 3.2.4). C'est une distribution sur $\mathrm{Cl}\, G$, à valeurs dans $\mathbf{Q}_p$; son support est contenu dans la partie régulière $\mathrm{Cl}\, G_{\mathrm{reg}}$ de $\mathrm{Cl}\, G$.

Proposition 3.6.1—*La distribution μ_G^0 est à valeurs dans $\mathbf{Z}_p$.*

Il faut montrer que, si f est une fonction centrale localement constante sur G, à valeurs dans $\mathbf{Z}_p$, on a $<f, \mu_G> \in \mathbf{Z}_p$. Soit $U \in \mathbf{U}_G$ tel que f soit constante (mod U). On a

$$<f, \mu_G> = \frac{1}{(G:U)} \sum \theta_U(s) f(s),$$

et l'on vient de voir que θ_U est le caractère d'un élément de $P(G/U)$. On est donc ramené à prouver le résultat suivant :

Lemme 3.6.2—*Soient Γ un groupe fini et χ le caractère d'un $\mathbf{Z}_p[\Gamma]$-module projectif de type fini. Si f est une fonction centrale sur Γ, à valeurs dans $\mathbf{Z}_p$, on a $\frac{1}{|\Gamma|} \sum \chi(s) f(s) \in \mathbf{Z}_p$.*

Il suffit de considérer le cas où f est la fonction caractéristique d'une classe de conjugaison C de Γ. Si x est un élément de C, et $Z(x)$ son centralisateur, on a

$$\frac{1}{|\Gamma|} \sum \chi(s) f(s) = \frac{|C|}{|\Gamma|} \chi(x) = \frac{1}{|Z(x)|} \chi(x).$$

474   Jean-Pierre Serre

Or on sait ([5], Chap.IV,cor.2.5) que ce nombre a une valuation p-adique ≥ 0.
D'où le résultat cherché.

Remarque. Une distribution à valeurs dans $\mathbf{Z}_p$ est parfois appelée une
mesure p-adique (cf. e.g. [9], n°1.3). On peut l'utiliser pour intégrer des fonctions plus générales que les fonctions localement constantes, par exemple des
fonctions continues p-adiques. Dans le cas de Cl G, une telle mesure peut
s'interpréter comme un élément du *groupe de Hattori-Stallings* $T(\mathbf{Z}_p[[G]])$,
quotient de $\mathbf{Z}_p[[G]]$ par l'adhérence du sous-groupe additif engendré par les
$xy - yx$, cf. [9], *loc.cit.* De ce point de vue, le fait que le support de μ_G^0 soit
contenu dans l'ensemble des classes régulières (et aussi que μ_G^0 soit invariante
par $s \mapsto s^p$) est à rapprocher des résultats de Zaleskii [16] et Bass [1] dans le
cas discret.

§4. Représentations $\mathbf{Z}_p$-linéaires et $\mathbf{Q}_p$-linéaires

Les G-modules considérés jusqu'ici étaient des groupes abéliens finis de type
$(p, \ldots, p)$. Nous allons maintenant nous occuper de cas plus généraux, par
exemple de $\mathbf{Q}_p$-espaces vectoriels de dimension finie. Cela permettra d'ob-
tenir une caractérisation simple de la distribution μ_G (n°4.4). C'est cette
caractérisation qui sera utilisée par la suite.

4.1. Cohomologie continue : le cas des $\mathbf{Z}_p$-modules de type fini

On suppose que G satisfait à la condition de finitude (1.3.1) :

$$\dim H^i(G, A) < \infty \quad \text{pour tout } i \in \mathbf{Z} \text{ et tout } A \in C_G.$$

Soit L un $\mathbf{Z}_p$-module de type fini, sur lequel G opère continûment (pour la
topologie p-adique de L). Pour tout $n \geq 0$, $L/p^n L$ est un G-module discret,
au sens usuel. Les *groupes de cohomologie continue* $H_c^i(G, L)$ sont définis de
l'une des deux façons (équivalentes) suivantes :

$$(4.1.1) \qquad H_c^i(G, L) = \varprojlim H^i(G, L/p^n L),$$

la limite projective étant prise pour $n \to \infty$.

(4.1.2) Si $C_c(G, L)$ désigne le *complexe des cochaînes continues* sur G à
valeurs dans L, on définit $H_c^i(G, L)$ comme $H^i(C_c(G, L))$.

L'équivalence de ces deux définitions se voit en remarquant que $C_c(G, L)$
est limite projective des complexes de cochaînes $C(G, L/p^n L)$. Comme les
groupes de cohomologie de ces complexes sont finis (grâce à l'hypothèse faite
sur G), les homomorphismes naturels

$$H^i(C_c(G, L)) \to \varprojlim H^i(C(G, L/p^n L))$$

sont des isomorphismes (cf. par exemple [6], 13.1.2 et 13.2.3). D'où le résultat cherché.

Par construction, les $H^i(G, L)$ sont des pro-p-groupes commutatifs, donc des $\mathbf{Z}_p$-modules topologiques compacts.

Si $0 \to L' \to L \to L'' \to 0$ est une suite exacte, on a une suite exacte de cohomologie correspondante :

$$\cdots \to H_c^i(G, L') \to H_c^i(G, L) \to H_c^i(G, L'') \to H_c^{i+1}(G, L') \to \cdots;$$

c'est clair si l'on utilise la définition (4.1.2).

Proposition 4.1.3—*Les $H_c^i(G, L)$ sont des $\mathbf{Z}_p$-modules de type fini.*

En utilisant la suite exacte ci-dessus, on se ramène au cas où L est sans torsion. Si l'on pose $H^i = H_c^i(G, L)$, on a alors une suite exacte de cohomologie :

$$\cdots \to H^i \xrightarrow{p} H^i \to H^i(G, L/pL) \to \cdots$$

Comme $H^i(G, L/pL)$ est fini, il en résulte que H^i/pH^i est fini. Comme H^i est un pro-p-groupe commutatif, cela entraîne que H^i est topologiquement de type fini, d'où la proposition.

4.2. Caractéristique d'Euler-Poincaré des $\mathbf{Z}_p$-modules de type fini

On conserve les hypothèses précédentes, et l'on suppose en outre que $\operatorname{cd}_p(G) < \infty$, de sorte que les caractéristiques d'Euler-Poincaré $e(G, A)$ sont définies, ainsi que la distribution μ_G.

Soit L un $\mathbf{Z}_p$-module de type fini, sur lequel G opère continûment. Les $H_c^i(G, L)$ sont des $\mathbf{Z}_p$-modules de type fini, nuls pour $i > \operatorname{cd}_p(G)$. Notons $\operatorname{rg} H_c^i(G, L)$ le *rang* de $H_c^i(G, L)$ comme $\mathbf{Z}_p$-module, autrement dit la dimension du $\mathbf{Q}_p$-espace vectoriel $\mathbf{Q}_p \otimes H_c^i(G, L)$.

Proposition 4.2.1—*Supposons L sans torsion. On a alors*

$$\sum(-1)^i \operatorname{rg} H_c^i(G, L) = e(G, L/pL) = \sum(-1)^i \dim H^i(G, L/pL).$$

Posons, comme au n°4.1, $H^i = H_c^i(G, L)$ et utilisons la suite exacte de cohomologie

$$\cdots \to H^i \xrightarrow{p} H^i \to H^i(G, L/pL) \to \cdots$$

On obtient ainsi des suites exactes

$$0 \to H^i/pH^i \to H^i(G, L/pL) \to H_p^{i+1} \to 0,$$

où X_p désigne le noyau de la multiplication par p dans le groupe abélien X. On en déduit

$$e(G, L/pL) = \sum(-1)^i(\dim H^i/pH^i - \dim H_p^i).$$

Or, si M est un $\mathbf{Z}_p$-module de type fini, on a rg $M = \dim M/pM - \dim M_p$. La formule ci-dessus peut donc s'écrire

$$e(G, L/pL) = \sum(-1)^i \mathrm{rg}\, H^i,$$

ce qui démontre la proposition.

Remarque. Les arguments donnés ci-dessus sont standard ; ils s'appliquent chaque fois qu'un foncteur cohomologique possède des propriétés de finitude (par exemple en cohomologie ℓ-adique).

Avec les hypothèses ci-dessus, posons :

$$(4.2.2) \qquad e(G, L) = \sum(-1)^i \mathrm{rg}\, H_c^i(G, L),$$

de sorte que la prop.4.2.1 revient à dire que $e(G, L) = e(G, L/pL)$.

Notons χ_L le caractère de la représentation de G dans L (ou dans $\mathbf{Q}_p \otimes L$, cela revient au même). C'est une fonction centrale sur G, continue (mais pas nécessairement localement constante), et à valeurs dans $\mathbf{Z}_p$.

Proposition 4.2.3—*La restriction de χ_L à G_{reg} coïncide avec le caractère de Brauer $\varphi_{L/pL}$ de L/pL.*

C'est là une propriété bien connue des caractères de Brauer. Rappelons la démonstration :

Si $m = \mathrm{rg}\, L$, on peut identifier L à $\mathbf{Z}_p \times \cdots \times \mathbf{Z}_p$ (m facteurs), et l'action de G sur L est donnée par un homomorphisme continu

$$\rho\colon G \to \mathbf{GL}_m(\mathbf{Z}_p).$$

Si s est un élément de G_{reg}, $\rho(s)$ est régulier dans $\mathbf{GL}_m(\mathbf{Z}_p)$, donc d'ordre fini r premier à p ; en effet, la composante première à p de l'ordre de $\mathbf{GL}_m(\mathbf{Z}_p)$ est finie. Les valeurs propres de $\rho(s)$ dans une extension convenable de $\mathbf{Q}_p$ sont des racines de l'unité d'ordre divisant r ; ce sont les représentants multiplicatifs de leurs réductions mod p. La trace de $\rho(s)$ est donc égale à la trace de Brauer de sa réduction (mod p), ce qui démontre la formule cherchée.

Corollaire 4.2.4—*Le caractère χ_L est localement constant sur G_{reg}, et l'on a :*

$$e(G, L) = \int \chi_L(s)\mu_G(s).$$

Cela résulte des prop.4.2.1 et 4.2.3, compte tenu du fait que

$$e(G, L/pL) = \int \varphi_{L/pL}(s)\mu_G(s).$$

4.3. Le cas des $\mathbf{Q}_p$-espaces vectoriels

Soit V un $\mathbf{Q}_p$-espace vectoriel de dimension finie, sur lequel G opère continûment. Comme G est compact, il laisse stable un *réseau* L de V, autrement dit un $\mathbf{Z}_p$-sous-module de type fini de V qui engendre V. Si l'on note $C_c(G, V)$ le complexe des cochaînes continues de G à valeurs dans V, on a

$$(4.3.1) \qquad C_c(G,V) = \bigcup C_c(G, p^{-n}L).$$

Les groupes de cohomologie de ce complexe sont notés $H_c^i(G, V)$. Il résulte de (4.3.1) que l'on a :

$$(4.3.2) \qquad H_c^i(G, V) = \mathbf{Q}_p \otimes H_c^i(G, L).$$

En particulier, les $H_c^i(G, V)$ sont des $\mathbf{Q}_p$-espaces vectoriels de dimension finie, nuls pour $i > \mathrm{cd}_p(G)$. On pose

$$(4.3.3) \qquad e(G, V) = \sum (-1)^i \dim H_c^i(G, V),$$

où dim désigne la dimension sur $\mathbf{Q}_p$. Notons χ_V le caractère de la représentation de G dans V. On a $\chi_V = \chi_L$ et les formules ci-dessus, combinées au cor.4.2.4, donnent :

Proposition 4.3.4—*Le caractère χ_V est localement constant sur G_{reg}, et l'on a*

$$e(G, V) = \int \chi_V(s)\mu_G(s).$$

4.4. Détermination de μ_G à partir des $H_c^i(U, \mathbf{Q}_p)$

Soit $U \in \mathbf{U}_G$. Le groupe U satisfait aux mêmes hypothèses de finitude que G. Les groupes de cohomologie continue $H_c^i(U, \mathbf{Q}_p)$ sont donc définis (l'action de U sur $\mathbf{Q}_p$ étant triviale). Posons :

$$(4.4.1) \qquad H_U^i = H_c^i(U, \mathbf{Q}_p).$$

Les H_U^i sont des $\mathbf{Q}_p$-espaces vectoriels de dimension finie; par exemple, $H_U^0 = \mathbf{Q}_p$, et $H_U^1 = \mathrm{Hom}_c(U, \mathbf{Q}_p)$, groupe des homomorphismes continus de G dans $\mathbf{Q}_p$. Le groupe G/U opère de façon naturelle sur chaque H_U^i (à cause de l'action de G sur U par automorphismes intérieurs). On obtient ainsi des représentations linéaires de dimension finie de G/U. Notons h_U^i les caractères de ces représentations, et posons :

$$(4.4.2) \qquad h_U = \sum (-1)^i h_U^i.$$

La fonction h_U est un caractère virtuel de G/U, à valeurs dans $\mathbf{Q}_p$ (et même dans $\mathbf{Z}_p$).

Théorème 4.4.3—*Le caractère h_U est égal à la fonction θ_U introduite au §3, n°3.5.*

Cela revient à dire que, si f est une fonction sur G constante (mod U), on a :

$$(4.4.4) \qquad <f, \mu_G> = \frac{1}{(G:U)} \sum_{s \in G/U} h_U(s) f(s).$$

Notons tout de suite une conséquence de th.4.4.3 :

Corollaire 4.4.5—*Le caractère h_U est nul en dehors de $(G/U)_{\text{reg}}$.*

(Bien entendu, ceci ne s'étend pas aux h_U^i ; par exemple, pour $i = 0$, on a $h_U^i = 1$.)

Démonstration du théorème 4.4.3.

On a tout d'abord :

Lemme 4.4.5—*La formule (4.4.4) est vraie lorsque $f = \chi_V$, où V est une représentation $\mathbf{Q}_p$-linéaire de dimension finie de G/U.*

Le membre de gauche de (4.4.4) est :

$$(4.4.6) \quad <\chi_V, \mu_G> = e(G, V) = \sum (-1)^i \dim H_c^i(G, V), \quad \text{cf. 4.3.4.}$$

D'autre part, la suite spectrale des extensions de groupes (ou un argument de corestriction) montre que :

$$(4.4.7) \qquad H_c^i(G, V) = H^0(G/U, H_c^i(U, V)).$$

Comme U opère trivialement sur V, on a :

$$(4.4.8) \qquad H_c^i(U, V) = H_c^i(U, \mathbf{Q}_p) \otimes V = H_U^i \otimes V.$$

Le caractère de la représentation de G/U sur cet espace est $h_U^i \cdot \chi_V$. La dimension du sous-espace fixé par G/U est donc $\frac{1}{(G:U)} \sum h_U^i(s) \chi_V(s)$. Vu (4.4.7), cela donne :

$$(4.4.9) \qquad \dim H_c^i(G, V) = \frac{1}{(G:U)} \sum h_U^i(s) \chi_V(s).$$

En combinant ceci avec (4.4.6), on obtient bien la formule cherchée :

$$<\chi_V, \mu_G> = \frac{1}{(G:U)} \sum h_U(s) \chi_V(s).$$

Posons maintenant $\psi = h_U - \theta_U$. D'après ce que l'on vient de voir, on a $\sum \psi(s) \chi(s) = 0$ pour tout caractère χ d'une représentation de G/U sur $\mathbf{Q}_p$. Or h_U et θ_U sont des caractères de représentations virtuelles de G/U sur

$\mathbf{Q}_p$: c'est clair pour h_U, et pour θ_U, cela résulte de la prop.3.5.2. Le lemme élémentaire suivant montre alors que $\psi = 0$ (ce qui démontre 4.4.3) :

Lemme 4.4.10—*Soient Γ un groupe fini, K un corps de caractéristique zéro, et ψ le caractère d'une représentation virtuelle de Γ sur K. Supposons que $\sum \psi(s)\chi(s) = 0$ pour tout caractère χ de Γ sur K. Alors $\psi = 0$.*

Soit $\psi = \sum n_S \chi_S$ la décomposition de ψ en combinaison linéaire de caractères de $K[\Gamma]$-modules simples. Si l'on prend pour V le dual de la représentation simple S, on a

$$\sum \psi(s)\chi_V(s) = \sum \psi(s)\chi_S(s^{-1}) = |\Gamma|.d_S.n_S,$$

où $d_S = [\text{End}(S) : K]$. L'hypothèse faite sur ψ entraîne donc $n_S = 0$, d'où $\psi = 0$.

4.5. Interprétation de μ_G en termes de représentations admissibles

Soit K un corps de caractéristique 0. Une *représentation admissible* de G sur K est un K-espace vectoriel E sur lequel G agit de façon K-linéaire, en satisfaisant aux deux conditions suivantes :

(a) L'action de G est continue, autrement dit le fixateur d'un point de E est un sous-groupe ouvert de G.

(b) Pour tout $U \in \mathbf{U}_G$, le sous-espace E^U de E fixé par U est de dimension finie.

(Noter que E est réunion des E^U, d'après (a).)

A une telle représentation de G est associée une *distribution-trace* μ_E, à valeurs dans K, caractérisée par la formule

$$(4.5.1) \qquad \mu_E(f) = \text{Tr} f_E,$$

pour toute fonction localement constante f sur G, à valeurs dans K, f_E désignant l'endomorphisme de E défini par f. (Si $x \in E$, on choisit U fixant x tel que f soit constante (mod U), et l'on définit $f_E(x)$ comme la moyenne sur G/U des $f(s)sx$; l'opérateur f_E est de rang fini, ce qui donne un sens à (4.5.1).)

Pour tout $i \geq 0$, posons $E_i = \varinjlim H_c^i(U, \mathbf{Q}_p)$, où la limite inductive est prise par rapport aux homomorphismes de restriction

$$\text{res}: H_c^i(U, \mathbf{Q}_p) \to H_c^i(U', \mathbf{Q}_p) \quad (\text{pour } U' \subset U),$$

qui sont des inclusions. L'espace E_i est une représentation admissible de G ; on a $E_i^U = H_c^i(U, \mathbf{Q}_p)$ pour tout U. Soit μ_i la distribution-trace correspondante. L'énoncé suivant n'est qu'une reformulation du th.4.4.3 :

Théorème 4.5.2—*On a* $\mu_G = \sum (-1)^i \mu_i$.

Exemples. $(i = 0, 1)$

Pour $i = 0$, on a $E_0 = \mathbf{Q}_p$, avec action triviale de G. La distribution-trace associée μ_0 est la *distribution de Haar* de G, cf. 3.2.2.

Pour $i = 1$, on a $E_1^U = \mathrm{Hom}_c(U, \mathbf{Q}_p)$; ainsi, E_1 est l'espace vectoriel des *germes d'homomorphismes continus* de G dans $\mathbf{Q}_p$, avec l'action naturelle de G sur cet espace (provenant des automorphismes intérieurs).

*Généralisation.*A la place de la représentation triviale de G sur $\mathbf{Q}_p$, on peut prendre une représentation $\mathbf{Q}_p$-linéaire continue V de dimension finie, et définir $E_i(V)$ comme la limite inductive des $H^i_c(U, V)$. La somme alternée des distributions-traces des $E_i(V)$ est une distribution sur G, qui est égale à $\chi_V \cdot \mu_G$, où χ_V est le caractère de V ; cela se vérifie par un calcul analogue à celui du n°4.4.

§5. Quelques propriétés de μ_G

Dans ce §, on suppose que G possède les propriétés de finitude (1.3.1) et (1.3.2) permettant de définir μ_G.

5.1. Restriction à un sous-groupe ouvert

Proposition 5.1.1—*Si G' est un sous-groupe ouvert de G, on a*

$$\mu_{G'} = (G : G') . \mu_G | G',$$

où $\mu_G | G'$ désigne la restriction de μ_G à G'.

Il faut prouver que, si f est une fonction localement constante sur G_{reg}, nulle en dehors de G'_{reg}, on a :

$$(5.1.2) \qquad <f, \mu_G> = (G : G'). <f, \mu_G> .$$

Choisissons $U \in \mathbf{U}_G$ contenu dans G', et tel que f soit constante (mod U). Si l'on note h_U (resp. h'_U) le caractère virtuel de G/U (resp. de G'/U) défini au n°4.4, on a d'après (4.4.4) :

$$<f, \mu_{G'}> = (G' : U)^{-1} \sum_{s \in G'/U} f(s) h'_U(s)$$

et

$$<f, \mu_G> = (G : U)^{-1} \sum_{s \in G/U} f(s) h_U(s).$$

Mais il est clair que h'_U est la restriction de h_U à G'/U. D'où la formule (5.1.2), puisque $(G : G').(G' : U) = (G : U)$.

5.2. Invariance de μ_G par translation par le centre de G

Soit $Z(G)$ le centre de G.

Proposition 5.2.1—*La distribution μ_G est invariante par les translations $s \mapsto sz$, pour $z \in Z(G)$.*

Vu la caractérisation de μ_G au moyen des h_U (n°4.4), il suffit de voir que $h_U(s) = h_U(sz)$ pour tout $s \in G/U$. Or l'on a

$$h_U = \sum (-1)^i h_U^i,$$

où h_U^i est le caractère de la représentation naturelle de G/U dans $H_c^i(U, \mathbf{Q}_p)$. Comme l'action de G/U sur $H_c^i(U, \mathbf{Q}_p)$ provient de celle de G sur U par automorphismes intérieurs, les éléments s et sz agissent de la même façon (puisque z appartient à $Z(G)$), d'où $h_U(s) = h_U(sz)$, ce qui démontre la proposition.

Corollaire 5.2.2—*Si le support de μ_G est égal à un point, on a $Z(G) = 1$.*

En effet, la prop.5.2.1 entraîne que ce support est stable par multiplication par $Z(G)$.

Corollaire 5.2.3—*Si $\mu_G \neq 0$, l'ordre de $Z(G)$ est premier à p.*

Supposons $\mu_G \neq 0$, et soit s un élément de $\mathrm{Supp}(\mu_G)$. Si l'ordre de $Z(G)$ est divisible par p, choisissons un élément $z \neq 1$ du p-sous-groupe de Sylow de $Z(G)$. Comme s est d'ordre premier à p, la p-composante de sz est z. Il en résulte que sz n'appartient pas à G_{reg}, donc n'appartient pas à $\mathrm{Supp}(\mu_G)$, ce qui contredit la prop.5.2.1.

Remarque. Les corollaires ci-dessus sont analogues au *théorème de Gottlieb* pour les groupes discrets, cf. Stallings [13]. Dans le cas profini, des résultats voisins avaient déjà été obtenus par Nakamura (cf. [8], ainsi que [9], th.1.3.2).

5.3. Groupes à dualité

Posons $n = \mathrm{cd}_p(G)$. Supposons que G possède un *module dualisant I* ayant les propriétés suivantes :

(5.3.1) *I est isomorphe à $\mathbf{Q}_p/\mathbf{Z}_p$ (comme groupe abélien).*

(5.3.2) $H^n(G, I) = \mathbf{Q}_p/\mathbf{Z}_p$.

(5.3.3) *Si A est un G-module fini p-primaire, et si $B = \mathrm{Hom}(A, I)$, le cup-produit*

$$H^i(G, A) \times H^{n-i}(G, B) \to H^n(G, I) = \mathbf{Q}_p/\mathbf{Z}_p \quad (i = 0, 1, \dots, n)$$

met en dualité les groupes finis $H^i(G, A)$ et $H^{n-i}(G, B)$.

(Lorsque G est un pro-p-groupe, ces propriétés signifient que G est un "groupe de Poincaré de dimension n", cf. [12], I.4.5.)

L'action de G sur I se fait par un homomorphisme $\epsilon\colon G \to \mathbf{Z}_p^* = \mathrm{Aut}(\mathbf{Q}_p/\mathbf{Z}_p)$, appelé le *caractère dualisant* de G.

Si $A \in C_G$, le G-module $B = \mathrm{Hom}(A, I)$ appartient à C_G, et son caractère de Brauer est donné par la formule :

$$(5.3.4) \qquad\qquad \varphi_B(s) = \epsilon(s)\varphi_A(s^{-1}).$$

Vu (5.3.3), les caractéristiques d'Euler-Poincaré de A et de B sont liées par la relation :

$$(5.3.5) \qquad\qquad e(G, A) = (-1)^n e(G, B).$$

Proposition 5.3.6—*On a $\epsilon.\mu_G = (-1)^n \mu_G^*$, où μ_G^* désigne l'image de la distribution μ_G par $s \mapsto s^{-1}$.*

(Le produit $\epsilon.\mu_G$ a un sens, car la restriction de ϵ à G_{reg} prend ses valeurs dans le sous-groupe d'ordre $p - 1$ de $\mathbf{Z}_p^*$, donc est localement constante.)

Posons $\mu' = (-1)^n \epsilon^{-1} \mu_G^*$. Il est clair que μ' est invariante par conjugaison, ainsi que par $s \mapsto s^p$. D'autre part, pour tout $A \in C_G$, on a, d'après (5.3.4) et (5.3.5) :

$$<\varphi_A, \mu'> = (-1)^n <\varphi_B, \mu_G> = (-1)^n e(G, B) = e(G, A) = <\varphi_A, \mu_G> \ .$$

Vu l'unicité de μ_G (cf. th.A), cela entraîne $\mu' = \mu_G$, d'où (5.3.6).

Voici une autre conséquence de la dualité :

Proposition 5.3.7—*Soit $U \in \mathbf{U}_G$, et soit $H_U^n = H_c^n(U, \mathbf{Q}_p)$, cf. n°4.4. On a :*

$$\dim H_U^n = \begin{cases} 1 & \text{si } \epsilon(U) = 1 \\ 0 & \text{sinon.} \end{cases}$$

En termes des μ_i du n°4.5, ceci entraîne :

Corollaire 5.3.8—*On a $\mu_n = 0$ si et seulement si $\mathrm{Im}(\epsilon)$ est infini* (c'est-à-dire, si $\mathrm{scd}_p(G) = n$, cf. [12], Chap.I, prop.19).

Démonstration de (5.3.7)

On sait que U possède les mêmes propriétés que G, avec le même module dualisant I (cf. [12], *loc.cit.*, ainsi que les Appendices de Tate et de Verdier). On peut donc supposer que $U = G$. On applique alors la dualité (5.3.3) au module $\mathbf{Z}/p^m\mathbf{Z}$, avec opérateurs triviaux. Si $\epsilon = 1$ on en déduit que $H^n(G, \mathbf{Z}/p^m\mathbf{Z})$ est dual de $H^0(G, \mathbf{Z}/p^m\mathbf{Z}) = \mathbf{Z}/p^m\mathbf{Z}$, donc est isomorphe à $\mathbf{Z}/p^m\mathbf{Z}$. On obtient ainsi $H_c^n(G, \mathbf{Z}_p) = \mathbf{Z}_p$, d'où $H_G^n = \mathbf{Q}_p$. Si $\epsilon \neq 1$, il existe $s \in G$ tel que $\epsilon(s) \not\equiv 1 \pmod{p^r}$ pour un $r > 0$. Le même argument que ci-dessus montre que $p^r.H_c^n(G, \mathbf{Z}_p) = 0$, d'où $H_G^n = 0$.

§6. Exemples galoisiens

6.1. Le cas local : énoncé

On note K une extension de $\mathbf{Q}_p$ de degré fini d, et l'on pose

$$(6.1.1) \qquad G = \mathrm{Gal}(\overline{K}/K),$$

où $\overline{K}$ est une clôture algébrique de K. D'après Tate ([14]—voir aussi [12], II.5.7), le groupe G possède les propriétés de finitude (1.3.1) et (1.3.2), avec $\mathrm{cd}_p(G) = 2$. De plus, Tate démontre que

$$(6.1.2) \qquad e(G, A) = -d.\dim A \text{ pour tout } A \in C_G.$$

En termes de μ_G, cela donne :

Théorème 6.1.3—*On a $\mu_G = -d.\delta_1$, où δ_1 est la distribution de Dirac en l'élément neutre de G.*

6.2. Démonstration du théorème (6.1.3)

Vu la prop.5.1.1, on peut supposer que $K = \mathbf{Q}_p$, i.e. $d = 1$. Si l'on introduit les caractères h_U^i et h_U du n°4.4, on a

$$h_U = h_U^0 - h_U^1 + h_U^2,$$

et tout revient à montrer que $h_U = -r_{G/U}$, où $r_{G/U}$ est le caractère de la représentation régulière de G/U. Cela résulte de l'énoncé plus précis suivant :

Proposition 6.2.1—*Pour tout $U \in \mathbf{U}_G$, on a $h_U^0 = 1$, $h_U^1 = 1 + r_{G/U}$ et $h_U^2 = 0$.*

L'assertion relative à h_U^0 est triviale. Celle relative à h_U^2 provient du fait (également dû à Tate) que G est un groupe à dualité dont le caractère dualisant est le caractère cyclotomique, dont l'image est infinie, ce qui permet d'appliquer (5.3.7).

Reste le cas de h_U^1, qui est le caractère de la représentation naturelle de G/U sur $H_c^1(U, \mathbf{Q}_p) = \mathrm{Hom}_c(U, \mathbf{Q}_p)$. Notons K_U l'extension galoisienne de $\mathbf{Q}_p$ correspondant à U ; on a $G/U = \mathrm{Gal}(K_U/\mathbf{Q}_p)$. La théorie du corps de classes permet d'identifier $\mathrm{Hom}_c(U, \mathbf{Q}_p)$ à $\mathrm{Hom}_c(K_U^*, \mathbf{Q}_p)$. Or, on a une suite exacte :

$$(6.2.2) \qquad 0 \to \mathrm{Unit}(K_U) \to K_U^* \to \mathbf{Z} \to 0,$$

où $\mathrm{Unit}(K_U)$ désigne le groupe des unités du corps local K_U.

On déduit de là une suite exacte de G/U-modules :

$$(6.2.3) \qquad 0 \to \mathbf{Q}_p \to H_U^1 \to \mathrm{Hom}_c(\mathrm{Unit}(K_U), \mathbf{Q}_p) \to 0;$$

cette suite est scindée car G/U est fini. De plus, le logarithme p-adique

$$\log: \mathrm{Unit}(K_U) \twoheadrightarrow K_U$$

est un isomorphisme local, et l'on voit facilement qu'il donne un isomorphisme de $\mathrm{Hom}_c(\mathrm{Unit}(K_U), \mathbf{Q}_p)$ sur $\mathrm{Hom}_c(K_U, \mathbf{Q}_p)$, lequel s'identifie au dual K'_U de K_U (comme $\mathbf{Q}_p$-espace vectoriel). Via la forme bilinéaire $\mathrm{Tr}(xy)$, on peut identifier K'_U à K_U. Finalement, on obtient un isomorphisme de $\mathbf{Q}_p[G/U]$-modules :

(6.2.4) $$H^1_U = \mathbf{Q}_p \oplus K_U.$$

D'après le théorème de la base normale, la représentation de G/U dans K_U est isomorphe à la représentation régulière. D'ou $h^1_U = 1 + r_{G/U}$, ce qui achève la démonstration.

Remarques.

1) En termes des *représentations admissibles* E_i du n°4.5, la proposition ci-dessus signifie que l'on a

$$E_0 = \mathbf{Q}_p, \qquad E_1 = \mathbf{Q}_p \oplus \overline{K}, \qquad E_2 = 0,$$

l'action de G sur $\overline{K}$ (resp. sur $\mathbf{Q}_p$) étant l'action naturelle (resp. l'action triviale).

2) Si l'on suppose que K est, non un corps p-adique, mais un corps p'-adique, avec $p' \neq p$, les mêmes arguments que ci-dessus montrent que $h^0_U = h^1_U = 1$ et $h^2_U = 0$, d'où $h_U = 0$ pour tout $U \in \mathbf{U}_G$, et $\mu_G = 0$. On retrouve ainsi un autre résultat de Tate, cf. [12], II.5.4.

6.3. Le cas global : énoncé

Soit K un corps de nombres algébriques de degré d, et soit S un ensemble fini de places de K, contenant l'ensemble S_∞ des places archimédiennes, ainsi que les places de caractéristique résiduelle p. Soit K_S l'extension galoisienne maximale de K non ramifiée en dehors de S, et soit $G = \mathrm{Gal}(K_S/K)$. D'après Tate ([14]), le groupe G possède la propriété de finitude (1.3.1). Supposons $p \neq 2$, ou K totalement imaginaire. On a alors $\mathrm{cd}_p(G) = 2$, de sorte que (1.3.2) est satisfaite, et que les caractéristiques d'Euler-Poincaré $e(G, A)$ sont définies pour tout $A \in C_G$. Le calcul de $e(G, A)$ a été fait par Tate (cf. [15]). Le résultat s'énonce de la manière suivante :

Si $v \in S_\infty$ est une place réelle, notons c_v le "Frobenius réel" correspondant ; c'est un élément d'ordre 2 de G, défini à conjugaison près ; il appartient à G_{reg}. Pour tout G-module A, on note A_v le sous-groupe de A fixé par c_v.

Soit $A \in C_G$. Posons :

$$e_v(A) = \begin{cases} -\dim A & \text{si } v \text{ est une place complexe,} \\ \dim A_v - \dim A & \text{si } v \text{ est une place réelle.} \end{cases}$$

La formule de Tate est alors (cf. [15]) :

Théorème 6.3.1—$e(G, A) = \displaystyle\sum_{v \in S_\infty} e_v(A)$ *pour tout* $A \in C_G$.

Si φ_A est le caractère de Brauer de A, on a :

(6.3.2) $\dim A = \varphi_A(1)$ et $\dim A_v = (\varphi_A(1) + \varphi_A(c_v))/2$.

Cela permet de récrire (6.3.1) sous la forme :

(6.3.3) $$e(G, A) = -\frac{d}{2}\varphi_A(1) + \frac{1}{2}\sum \varphi_A(c_v),$$

où la somme porte sur les places réelles de K.

Passons maintenant à μ_G. Si v est une place réelle, notons μ_v l'image de la distribution de Haar de G par l'application $g \mapsto gc_v g^{-1}$; c'est l'unique distribution invariante par conjugaison, de masse totale 1 et de support la classe de conjugaison de c_v. Il est clair que (6.3.3) est équivalente à l'énoncé suivant :

Théorème 6.3.4—*On a* $\mu_G = -\frac{d}{2}\delta_1 + \frac{1}{2}\sum \mu_v$.

Remarques.

1) Comme dans le cas local, cette formule peut se démontrer en explicitant les H^i_U ($i = 0, 1, 2$). Nous laissons les détails au lecteur.

2) Si $p = 2$, et si K a au moins une place réelle, on a $\mathrm{cd}_p(G) = \infty$, de sorte que $e(G, A)$ et μ_G ne sont pas définis. Toutefois, Tate a montré comment on peut modifier $e(G, A)$ de façon à avoir encore une formule simple. La recette qu'il utilise revient à considérer la *cohomologie relative* de G modulo la famille des groupes de décomposition locaux (G_v) associés aux places archimédiennes (i.e. $G_v = \{1\}$ si v est complexe et $G_v = \{1, c_v\}$ si v est réelle). On trouve que la distribution associée est égale à $-d \cdot \delta_1$, tout comme dans le cas local.

§7. Groupes de Lie p-adiques

Dans ce §, G est un groupe de Lie p-adique compact (cf. [3]) ; on suppose que G n'a pas d'élément d'ordre p. On note Lie G l'algèbre de Lie de G ; c'est une $\mathbf{Q}_p$-algèbre de Lie de dimension finie.

D'après Lazard ([7], complété par [10]), G satisfait aux conditions de finitude nécessaires pour que μ_G soit définie. On a en particulier

$$\mathrm{cd}_p(G) = \dim G = \dim \mathrm{Lie}\, G.$$

7.1. La fonction F

Soit $g \in G$. On note $\mathrm{Ad}(g)$ l'automorphisme de Lie G défini par l'automorphisme intérieur $x \mapsto gxg^{-1}$, cf. [3], III.152. On pose :

(7.1.1) $\qquad\qquad F(g) = \det(1 - \mathrm{Ad}(g^{-1}))$.

Proposition 7.1.2—(i) *On a $F(g) = 0$ si et seulement si le centralisateur* $Z(g)$ *de g est de dimension > 0* (i.e. *si $Z(g)$ est infini*).
(ii) *La fonction F est une fonction centrale, nulle en dehors de G_{reg}.*

Soit $z(g)$ le sous-espace de Lie G fixé par $\mathrm{Ad}(g)$. On sait ([3], III.234) que $z(g)$ est l'algèbre de Lie du groupe $Z(g)$. On a donc $\dim Z(g) > 0$ si et seulement si $z(g) \neq 0$, c'est-à-dire si 1 est valeur propre de $\mathrm{Ad}(g)$, autrement dit si $F(g) = 0$. D'où (i).

Le fait que F soit une fonction centrale est clair. D'autre part, si g n'appartient pas à G_{reg}, on peut écrire g sous la forme $g = us$, avec $us = su$, $s \in G_{\mathrm{reg}}$, u d'ordre une puissance de p (finie ou infinie) et $u \neq 1$. Comme G est sans p-torsion, u est d'ordre p^∞, autrement dit l'adhérence C_u du groupe cyclique engendré par u est isomorphe à $\mathbf{Z}_p$. Comme C_u est contenu dans $Z(g)$, cela montre que $\dim Z(g) > 0$, d'où $F(g) = 0$ d'après (i), ce qui démontre (ii).

Remarque. On verra plus loin que F est *localement constante*. Cela pourrait aussi se prouver directement.

7.2. Détermination du caractère h_U

Si $U \in \mathbf{U}_G$, posons $H_U^i = H_c^i(U, \mathbf{Q}_p)$, cf. n°4.4, et soit h_U^i le caractère de la représentation naturelle de G/U sur cet espace.

D'après Lazard ([7], V.2.4.10), H_U^i s'identifie à un sous-espace de $H^i(\mathrm{Lie}\, G, \mathbf{Q}_p)$, et l'on a même :

(7.2.1) $\qquad\qquad H_U^i = H^i(\mathrm{Lie}\, G, \mathbf{Q}_p)$

si U est assez petit. (Précisons qu'il s'agit ici de la cohomologie de l'algèbre de Lie Lie G, à coefficients dans $\mathbf{Q}_p$, l'action de Lie G sur $\mathbf{Q}_p$ étant triviale.)

De plus, cette identification est canonique, donc compatible avec l'action de G par automorphismes intérieurs. On déduit de là :

Lemme 7.2.2—*Supposons U assez petit. Alors, pour tout $g \in G$, $h_U^i(g)$ est la trace de l'automorphisme de $H^i(\mathrm{Lie}\, G, \mathbf{Q}_p)$ défini par $\mathrm{Ad}(g)$.*

Le lemme suivant permet de calculer la somme alternée de ces traces :

Lemme 7.2.3—*Soit L une algèbre de Lie de dimension finie sur un corps k, et soit s un automorphisme de L. Pour $i \geq 0$, soit s_i l'automorphisme de $H^i(L, k)$ défini par s (par transport de structure). On a alors :*

$$\sum (-1)^i \mathrm{Tr}(s_i) = \det(1 - s^{-1}).$$

Notons C le complexe des cochaînes alternées de L à valeurs dans k, et soit C^i sa composante homogène de degré i. L'invariance des caractéristiques d'Euler-Poincaré par passage à la cohomologie donne :

$$(7.2.4) \qquad \sum(-1)^i \mathrm{Tr}(s_i) = \sum(-1)^i \mathrm{Tr}(s|C^i),$$

où $s|C^i$ désigne l'automorphisme de C^i défini par s. Or $C^i = \wedge^i L^*$, où L^* est le dual de L. Cette identification transforme $s|C^i$ en $\wedge^i s^*$, où $s^* = {}^t s^{-1}$ est le contragrédient de s. En appliquant à s^* la formule bien connue :

$$\sum(-1)^i \mathrm{Tr}(\wedge^i x) = \det(1 - x)$$

on obtient :

$$(7.2.5) \qquad \sum(-1)^i \mathrm{Tr}(s|C^i) = \det(1 - s^*) = \det(1 - s^{-1}).$$

D'où le lemme.

Posons maintenant $h_U = \sum(-1)^i h_U^i$, cf. n°4.4. Les lemmes 7.2.2 et 7.2.3 entraînent :

Proposition 7.2.6—*Si $U \in \mathbf{U}_G$ est assez petit, on a :*

$$h_U(g) = \det(1 - \mathrm{Ad}(g^{-1})) = F(g)$$

pour tout $g \in G$.

Ceci montre que F est constante (mod U), donc *localement constante*. Vu 4.4.5, cela démontre aussi que F est nulle en dehors de G_{reg}, ce que l'on savait déjà.

7.3. Détermination de μ_G

Choisissons $U \in \mathbf{U}_G$ assez petit pour que $h_U = F$, cf. 7.2.6. D'après le théorème 4.4.3, combiné à la formule (3.5.1), on a donc

$$(7.3.1) \qquad <f, \mu_G> = \frac{1}{(G:U)} \sum_{s \in G/U} F(s) f(s),$$

pour toute fonction f sur G qui est constante (mod U). Or le membre de droite est égal à $<f, \mu'>$, où μ' est le produit de F et de la distribution de Haar de G, cf. 3.2.2. On a ainsi

$$<f, \mu_G - \mu'> = 0$$

pour toute f constante (mod U). Comme ceci est vrai pour tout U assez petit, on en déduit que $\mu_G - \mu' = 0$. D'où :

Théorème 7.3.2—*La distribution d'Euler-Poincaré* μ_G *est égale au produit par* F *de la distribution de Haar de* G.

Si l'on note "ds" la distribution de Haar, on a donc :

$$e(G, A) = \int \varphi_A(s)F(s)ds \qquad \text{pour tout } A \in C_G.$$

(Bien que φ_A ne soit défini que sur G_{reg}, cette intégrale a un sens puisque F est nulle en dehors de G_{reg}).

Corollaire 7.3.3—*Le support de* μ_G *est l'ensemble des éléments* s *de* G *dont le centralisateur* $Z(s)$ *est fini.*

Cela résulte de la prop.7.1.2 (i).

Corollaire 7.3.4—*On a* $\mu_G = 0$ *si et seulement si le centralisateur de tout élément de* G *est infini.*

C'est clair.

Remarques.

1) Le cor.7.3.3 équivaut à dire que $\text{Supp}(\mu_G)$ est égal à la réunion des classes de conjugaison de G qui sont *ouvertes* dans G. Ces classes sont en nombre fini (elles forment en effet une partition de l'espace compact G_{reg} en ouverts fermés). La $\mathbf{Z}_p$-mesure sur $\text{Cl}\, G$ définie par μ_G (n°3.6) a donc un *support fini*.

2) D'après Lazard [7], V.2.5.8, G est un "groupe à dualité" (au sens du n°5.3) dont le caractère dualisant ϵ est donné par :

$$(7.3.5) \qquad \epsilon(g) = \det \text{Ad}(g) \qquad \text{pour tout } g \in G.$$

(Lazard se borne au cas où G est un pro-p-groupe, mais sa démonstration s'applique au cas général.)

Vu (5.3.6), ceci entraîne l'identité $\epsilon.\mu_G = (-1)^n \mu_G^*$, i.e. :

$$(7.3.6) \qquad \epsilon(g)F(g) = (-1)^n F(g^{-1}) \qquad \text{pour tout } g \in G,$$

ce qui se déduit aussi directement de (7.1.1) et (7.3.5).

Exemples.

1) Si G est un pro-p-groupe de dimension $n > 0$, on a $\mu_G = 0$ puisque $G_{\text{reg}} = \{1\}$ et que le centralisateur de 1 est infini; on retrouve un résultat connu, cf. [12], I.4.1, exerc.(e).

2) Si $n = 0$, G est fini d'ordre premier à p, on a Lie $G = 0$, $F = 1$, et μ_G est la distribution de Haar de G, cf. 3.4.2.

3) Supposons $p \neq 2$, et soit G le groupe diédral p-adique, produit semi-direct d'un groupe $\{1, c\}$ d'ordre 2 par un groupe U isomorphe à $\mathbf{Z}_p$, l'action de c sur U étant $u \mapsto u^{-1}$. On a :

$$n = \dim \text{Lie}\, G = 1.$$

L'action de U sur Lie G est triviale alors que celle de c est $x \mapsto -x$. D'où $F = 0$ sur U, et $F = 2$ sur $cU = G - U$. On en déduit que μ_G a pour support $G - U$ (qui est la classe de conjugaison de c), et que sa masse totale est 1. Autrement dit, si $A \in C_G$, on a :

$$(7.3.7) \qquad e(G, A) = \varphi_A(c) = 2.\dim A^c - \dim A,$$

où A^c est le sous-espace de A fixé par c.

7.4. Un cas où $\mu_G = 0$

Proposition 7.4.1—*Soit $\underline{G}$ un groupe algébrique connexe sur $\mathbf{Q}_p$, de dimension > 0. Soit G un sous-groupe ouvert compact de $\underline{G}(\mathbf{Q}_p)$ ne contenant pas d'élément d'ordre p. Alors $\mu_G = 0$.*

Il faut montrer que la fonction $F: G \to \mathbf{Z}_p$ est égale à 0. Tout d'abord on a $F(1) = \det(1-1) = 0$ puisque $\dim G > 0$. Comme F est localement constante, cela montre que $F = 0$ dans un voisinage U de l'élément neutre. Mais F est la restriction à G de la fonction "morphique" $\underline{F}: g \mapsto \det(1 - \mathrm{Ad}(g^{-1}))$, définie sur la variété $\underline{G}$. Comme $\underline{G}$ est lisse et connexe, U est dense dans $\underline{G}$ pour la topologie de Zariski. Le fait que F soit 0 sur U entraîne donc $\underline{F} = 0$, d'où le résultat cherché.

Variante. On peut aussi démontrer directement que, pour tout point g de $\underline{G}$, la dimension du centralisateur de g est > 0 (puisqu'il en est ainsi pour l'élément générique).

Remarques.

1) Un argument analogue montre que $\mu_G = 0$ si $\dim G > 0$, et si l'image de

$$\mathrm{Ad}: G \to \mathbf{GL}(\mathrm{Lie}\, G)$$

est connexe pour la topologie de Zariski de $\mathbf{GL}(\mathrm{Lie}\, G)$.

2) Supposons que 7.4.1 s'applique, donc que $\mu_G = 0$. Par définition de μ_G, cela signifie que $e(G, A) = 0$ pour tout $A \in C_G$. Plus généralement, soit A un G-module discret qui soit un p-groupe fini, et soit $\chi(G, A)$ sa caractéristique d'Euler-Poincaré au sens multiplicatif, autrement dit le produit alterné des ordres $|H^i(G, A)|$. On a alors :

$$(7.4.2) \qquad \chi(G, A) = 1.$$

En effet, comme $\chi(G, A)$ est une fonction multiplicative de A, on peut se ramener par dévissage au cas où $pA = 0$, de sorte que A appartient à C_G, et que

$$(7.4.3) \qquad \chi(G, A) = p^{e(G,A)} = 1,$$

puisque $e(G, A) = 0$.

§8. Une application

Dans les §§ précédents, il ne s'agissait que de G-modules *de type fini* (sur $\mathbf{F}_p$, $\mathbf{Z}_p$ ou $\mathbf{Q}_p$, suivant les cas). Or le problème du calcul de la caractéristique d'Euler-Poincaré se pose pour tout G-module A dont les groupes de cohomologie sont de type fini (sur l'anneau de base considéré), ce qui n'entraîne nullement que A soit lui-même de type fini. C'est d'un cas de ce genre que nous allons nous occuper.

8.1. Les données

Comme au n°7.4, on part d'un groupe algébrique $\underline{G}$ sur $\mathbf{Q}_p$, et d'un sous-groupe ouvert compact G de $\underline{G}(\mathbf{Q}_p)$.

On se donne une représentation linéaire $\rho:\underline{G} \to \mathbf{GL}_V$, où V est un $\mathbf{Q}_p$-espace vectoriel de dimension finie. On note L un $\mathbf{Z}_p$-réseau de V stable par G (il en existe, puisque G est compact). Le G-module auquel on va s'intéresser est $I = V/L$; si $m = \dim V$, I est isomorphe comme groupe abélien à la somme directe de m copies de $\mathbf{Q}_p/\mathbf{Z}_p$.

On se donne également un plongement h du groupe multiplicatif $\mathbf{G}_m$ dans le centre de $\underline{G}$, et l'on suppose que V est homogène de poids $\neq 0$ pour l'action de $\mathbf{G}_m$: autrement dit, il existe un entier $r \neq 0$ tel que $\rho(h(t)).v = t^r v$ pour tout point t de $\mathbf{G}_m$ et tout point v de V.

(*Remarque.* Les données ci-dessus sont celles que l'on rencontre (ou que l'on espère rencontrer) dans le théorie des motifs sur un corps de nombres K : $\underline{G}$ est le groupe de Galois motivique, h est l'homomorphisme de $\mathbf{G}_m$ dans $\underline{G}$ associé au poids, V est la réalisation p-adique du motif, L est une $\mathbf{Z}_p$-forme de V, et G est l'image de $\mathrm{Gal}(\overline{K}/K)$ dans $\underline{G}(\mathbf{Q}_p)$.)

Dans ce qui suit, on identifie $\mathbf{G}_m$ à un sous-groupe de $\underline{G}$ au moyen de h, et l'on note $\underline{PG}$ le groupe quotient $\underline{G}/\mathbf{G}_m$. Soit PG l'image de G dans $\underline{PG}(\mathbf{Q}_p)$. On fait les hypothèses suivantes :

(8.1.1) $\underline{G}$ *est connexe, de dimension > 1.*

(8.1.2) *Le groupe PG n'a pas de p-torsion.*

(8.1.3) *Si $p = 2$, G ne contient pas $h(-1)$.*

On a alors :

Théorème 8.1.4—*Sous les hypothèses ci-dessus, les groupes de cohomologie $H^i(G, I)$ sont des p-groupes finis, nuls si $i \geq \dim G$, et le produit alterné de leurs ordres :*

$$\chi(G, I) = \prod |H^i(G, I)|^{(-1)^i}$$

est égal à 1.
(Rappelons que $I = V/L$.)

8.2. Démonstration du théorème 8.1.4

Soit N le noyau de la projection $G \to PG$. Comme N est un sous-groupe ouvert compact de $\mathbf{G}_m(\mathbf{Q}_p) = \mathbf{Q}_p^*$, c'est un sous-groupe ouvert de $\mathbf{Z}_p^*$; de plus, si $p = 2$, N ne contient pas -1 d'après (8.1.3). On peut donc écrire N sous la forme

$$(8.2.1) \qquad N = C \times U,$$

où C est cyclique d'ordre divisant $p - 1$, et U est isomorphe à $\mathbf{Z}_p$. On va utiliser la suite spectrale relative à l'extension de groupes :

$$1 \to N \to G \to PG \to 1.$$

Comme $\mathrm{cd}_p(N) = 1$, nous n'avons à nous occuper de $H^i(N, I)$ que pour $i = 0$ et $i = 1$. On a :

Lemme 8.2.2—(i) $H^0(N, I)$ *est un p-groupe fini.*
(ii) $H^1(N, I) = 0$.

Comme $H^i(N, I)$ se plonge dans $H^i(U, I)$, il suffit de prouver (i) et (ii) avec N remplacé par U. Or, si u est un générateur topologique de U, on a

$$(8.2.3) \qquad H^0(U, I) = \mathrm{Ker}(u - 1 \colon I \to I)$$

et

$$(8.2.4) \qquad H^1(U, I) = \mathrm{Coker}(u - 1 \colon I \to I).$$

Mais, si l'on identifie u à un élément de $\mathbf{Z}_p^*$, on sait que u agit sur I par u^r, où r est un entier $\neq 0$. Puisque u est d'ordre infini, on a $u^r - 1 \neq 0$, et comme I est un groupe divisible, $u - 1 \colon I \to I$ est surjectif, d'où (ii). De plus, le noyau de $u - 1 \colon I \to I$ est fini, ce qui démontre (i). (De façon plus précise, si v est la valuation p-adique de $u^r - 1$, $H^0(U, I)$ est isomorphe à $L/p^v L$.)

Vu (8.2.2), la suite spectrale $H^i(PG, H^j(N, I)) \Rightarrow H^{i+j}(G, I)$ dégénère en un isomorphisme :

$$(8.2.5) \qquad H^i(G, I) = H^i(PG, A),$$

où $A = H^0(N, I)$ est un PG-module fini d'ordre une puissance de p.

Or PG est un sous-groupe ouvert compact de $\underline{PG}(\mathbf{Q}_p)$, sans p-torsion d'après (8.1.2). On a donc $\mathrm{cd}_p(PG) = \dim \underline{PG} = \dim G - 1$, ce qui montre que les $H^i(PG, A)$ sont finis pour tout i, et nuls si $i > \dim G - 1$. Enfin, l'hypothèse (8.1.1) équivaut à dire que $\underline{PG}$ est connexe de dimension > 0. En appliquant à $\underline{PG}$ la prop.7.4.1, on voit que $\mu_{PG} = 0$, d'où $\chi(PG, A) = 1$ d'après (7.4.2). Comme $\chi(G, I) = \chi(PG, A)$ d'après (8.2.5), cela achève la démonstration.

8.3. Démonstration du théorème D du n°1.8

Le théorème D est le cas particulier du th.8.1.4 où $\underline{G} = \mathbf{GL}_n$ et $V = (\mathbf{Q}_p)^n$, $L = (\mathbf{Z}_p)^n$, $I = (\mathbf{Q}_p/\mathbf{Z}_p)^n$. Le plongement $h: \mathbf{G}_m \to \underline{G}$ est le plongement évident (homothéties). On a $\underline{PG} = \mathbf{PGL}_n$. La condition (8.1.1) est satisfaite puisque $\dim G = n^2 > 1$. La condition (8.1.3) est satisfaite puisque $p > n+1 > 2$. Le fait que (8.1.2) soit satisfaite résulte du lemme élémentaire suivant :

Lemme 8.3.1—*Si $p > n+1$, les groupes $\mathbf{GL}_n(\mathbf{Q}_p)$ et $\mathbf{PGL}_n(\mathbf{Q}_p)$ ne contiennent pas d'élément d'ordre p.*

Si $s \in \mathbf{GL}_n(\mathbf{Q}_p)$ est d'ordre p, l'une de ses valeurs propres, z, est une racine primitive p-ième de l'unité. Mais on sait que $z, z^2, \ldots, z^{p-1}$ sont conjuguées par $\mathrm{Gal}(\overline{\mathbf{Q}}_p/\mathbf{Q}_p)$. Donc $z, z^2, \ldots, z^{p-1}$ sont des valeurs propres de s, ce qui entraîne $n \geq p - 1$.

Si $s \in \mathbf{PGL}_n(\mathbf{Q}_p)$ est d'ordre p, soit x un représentant de s dans $\mathbf{GL}_n(\mathbf{Q}_p)$, et soit $d = \det(x)$. Notons t l'homothétie x^p. On a

$$d^p = \det(x^p) = \det(t) = t^n.$$

Mais n et p sont premiers entre eux, puisque $p > n+1$. Il en résulte que t est de la forme $t = \theta^p$, avec $\theta \in \mathbf{Q}_p^*$. L'élément $y = \theta^{-1}x$ est alors un élément d'ordre p de $\mathbf{GL}_n(\mathbf{Q}_p)$, ce qui est impossible comme on vient de le voir.

Cela achève la démonstration du théorème D.

Remarque. On laisse au lecteur le soin d'étendre le lemme 8.3.1 à d'autres groupes réductifs que $\mathbf{GL}_n$ et $\mathbf{PGL}_n$. Par exemple, si $\underline{G}$ est un groupe de type E_8 sur $\mathbf{Q}_p$, $\underline{G}(\mathbf{Q}_p)$ n'a pas de p-torsion si $p \neq 2,3,5,7,11,13,19,31$.

Bibliographie

[1] H. Bass—*Euler characteristics and characters of discrete groups*, Invent. math. **35** (1976), 155–196.

[2] N. Bourbaki—*Topologie Générale*, Chap. 1 à 4, Hermann, Paris, 1971.

[3] N. Bourbaki—*Groupes et Algèbres de Lie*, Chap. 3, *Groupes de Lie*, Hermann, Paris, 1971.

[4] J. Coates et S. Howson—*Euler characteristics and elliptic curves*, Proc. Nat. Acad. USA **94** (Oct. 1997), 1115–11117.

[5] W. Feit—*The Representation Theory of Finite Groups*, North-Holland, New York, 1982.

[6] A. Grothendieck—*Eléments de Géométrie Algébrique*, (rédigés avec la collaboration de J. Dieudonné), Chap. 0, Publ. Math. IHES **11** (1961), 349–423.

[7] M. Lazard—*Groupes analytiques p-adiques*, Publ. Math. IHES **26** (1965), 389–603.

[8] H. Nakamura—*On the pro-p Gottlieb theorem*, Proc. Japan Acad. **68** (1992), 279–292.

[9] H. Nakamura—*Galois rigidity of pure sphere braid groups and profinite calculus*, J. Math. Sci. Univ. Tokyo **1** (1994), 71–136.

[10] J-P. Serre—*Sur la dimension cohomologique des groupes profinis*, Topology **3** (1965), 413–420 (=*Oe.* 66).

[11] J-P. Serre—*Représentations Linéaires des Groupes Finis*, 5e édition corrigée, Hermann, Paris, 1998.

[12] J-P. Serre—*Cohomologie Galoisienne*, 5e édition révisée et complétée, Lect. Notes in Math. **5**, Springer-Verlag, 1994.

[13] J. Stallings—*Centerless groups—An algebraic formulation of Gottlieb's theorem*, Topology **4** (1965), 129–134.

[14] J. Tate—*Duality theorems in Galois cohomology over number fields*, Proc. Int. Congress Stockholm (1962), 288–295.

[15] J. Tate—*On the conjectures of Birch and Swinnerton-Dyer and a geometric analog*, Sém. Bourbaki 1965/1966, n°**306**, Benjamin Publ., New York, 1966.

[16] A.E. Zaleskii—*Sur un problème de Kaplansky* (en russe), Dokl. Akad. Nauk SSSR **203** (1972), 749–751 (trad. anglaise : Soviet Math. Dokl. **13** (1972), 449–452).

COLLÈGE DE FRANCE, 3, RUE D'ULM, F-75005 PARIS, FRANCE
serre@dmi.ens.fr

Printed in the United States
By Bookmasters